AF595476

Remote Sensing of Volcanic Processes and Risk

Remote Sensing of Volcanic Processes and Risk

Editors

Francesca Cigna
Deodato Tapete
Zhong Lu
Susanna K. Ebmeier

MDPI • Basel • Beijing • Wuhan • Barcelona • Belgrade • Manchester • Tokyo • Cluj • Tianjin

Editors

Francesca Cigna
Italian Space Agency (ASI)
Italy

Deodato Tapete
Italian Space Agency (ASI)
Italy

Zhong Lu
Southern Methodist University
USA

Susanna K. Ebmeier
University of Leeds
UK

Editorial Office
MDPI
St. Alban-Anlage 66
4052 Basel, Switzerland

This is a reprint of articles from the Special Issue published online in the open access journal *Remote Sensing* (ISSN 2072-4292) (available at: https://www.mdpi.com/journal/remotesensing/special_issues/rs_vpr).

For citation purposes, cite each article independently as indicated on the article page online and as indicated below:

LastName, A.A.; LastName, B.B.; LastName, C.C. Article Title. *Journal Name* **Year**, *Volume Number*, Page Range.

ISBN 978-3-0365-0126-0 (Hbk)
ISBN 978-3-0365-0127-7 (PDF)

Cover image courtesy of Taro Taylor.

Contents

About the Editors

Francesca Cigna (Ph.D.), Researcher in Earth observation and data analytics at the Italian Space Agency (ASI), with a Ph.D. in Earth sciences, and an M.Sc. and B.Sc. in environment and territory engineering. Her key areas of interest and specialty are satellite radar, InSAR, natural and man-made hazards and risks, shallow geological processes, ground instability and land changes, landscape archaeology and cultural heritage. She leads research on Earth observation from space-borne platforms, image processing, and time series analysis, mainly focused on radar through past, current, and future SAR missions such as the COSMO-SkyMed constellation, ESA's ERS-1/2 and ENVISAT, and Copernicus Sentinel-1. She collaborates with the Committee for Earth Observation Satellites (CEOS) Working Group Disasters. She serves as an Associate Editor of the *International Journal of Applied Earth Observation and Geoinformation* (ISSN 0303-2434), and *Science of Remote Sensing* (ISSN 2666-0172).

Deodato Tapete (Ph.D.), Researcher in Earth observation and data analytics at the Italian Space Agency (ASI). With more than 10 years of research experience in Earth sciences, natural hazards and remote sensing, he is specialized in synthetic aperture radar imaging and interferometry for deformation monitoring, hazard assessment, and archaeological prospection, as well as in assessment of anthropogenic impacts on the environment, natural resources and cultural heritage. He is involved in CEOS Working Group Disasters and serves as an Associate Editor of 'Environmental Remote Sensing' of *Remote Sensing* (ISSN 2072-4292) journal, and as Editor in Chief of the 'Natural Hazards' Section of *Geosciences* (ISSN 2076-3263) journal.

Zhong Lu (Ph.D.), Shuler-Foscue Professor at Roy M. Huffington Department of Earth Sciences, Southern Methodist University, Dallas, Texas, USA (www.smu.edu/dedman/lu). His research interests include technique developments of InSAR processing and their applications to the study of volcano, landslide, and human-induced geohazards. He has published more than 200 peer-reviewed journal articles and book chapters focused on InSAR techniques and applications, and a book on "InSAR Imaging of Aleutian Volcanoes: Monitoring a Volcanic Arc from Space" with co-author Dan Dzurisin. He is a member of NASA-ISRO satellite radar (NISAR) Science Team (2012–present), Editor-in-Chief of the journal *GeoHazards*, Senior Associate Editor of the journal *Remote Sensing*, Associate Editor of journal *Frontier in Earth Sciences*, and a member of the editorial boards of *International Journal of Image and Data Fusion*, and journal *Geomatics, Natural Hazards and Risk*.

Susanna K. Ebmeier (Ph.D.), NERC Independent Research Fellow at the University of Leeds (UK), with PhD in Earth sciences, and BSc in geophysics. Her areas of expertise include volcanology, satellite remote sensing, InSAR, volcano deformation, volcanic emissions, and source separation. She investigates different aspects of volcanic activity using satellite remote sensing. She is interested in the physical processes that control the character of volcanic activity, including magma storage, edifice growth and collapse, and volcanic interactions. An important aspect of her research using satellite imagery is exploring how it can be made useful for volcano monitoring and assessing volcanic hazards. She also has an interest in the ways in which the emission of gases affects the Earth's environment. She co-leads the Volcano Demonstrator within the CEOS Working Group Disasters.

Preface to "Remote Sensing of Volcanic Processes and Risk"

Remote sensing data and methods are increasingly implemented in assessments of volcanic processes and risk. This happens thanks to their ability to provide a spectrum of observation and measurement opportunities to accurately sense the dynamics, magnitude, frequency, and impacts of volcanic activity in the ultraviolet (UV), visible (VIS), infrared (IR), and microwave domains.

This book includes research papers published in the Special Issue "Remote Sensing of Volcanic Processes and Risk" of the journal Remote Sensing.

Launched in mid-2018, the Special Issue comprises 1 editorial and 19 research papers on the use of satellite, aerial, and ground-based remote sensing to detect thermal features and anomalies, investigate lava and pyroclastic flows, predict the flow path of lahars, measure gas emissions and plumes, and estimate ground deformation.

The strong multi-disciplinary character of the approaches employed for volcano monitoring and the combination of a variety of sensor types, platforms, and methods that emerge from the papers testify the current scientific and technology trends toward multi-data and multi-sensor monitoring solutions.

The research advances presented in the published papers are achieved thanks to a wealth of data including but not limited to the following: thermal IR from satellite missions (e.g., MODIS, VIIRS, AVHRR, Landsat-8, Sentinel-2, ASTER, TET-1) and ground-based stations (e.g., FLIR cameras); digital elevation/surface models from airborne sensors (e.g., light detection and ranging (LiDAR), or 3D laser scans) and satellite imagery (e.g., tri-stereo Pléiades, SPOT-6/7, PlanetScope); airborne hyperspectral surveys; geophysics (e.g., ground-penetrating radar, electromagnetic induction, magnetic survey); ground-based acoustic infrasound; ground-based scanning UV spectrometers; ground-based and satellite synthetic aperture radar (SAR) imaging (e.g., TerraSAR-X, Sentinel-1, Radarsat-2).

Data processing approaches and methods include change detection, offset tracking, interferometric SAR (InSAR), photogrammetry, hotspots and anomalies detection, neural networks, numerical modeling, inversion modeling, wavelet transforms, and image segmentation. Some authors also share codes for automated data analysis and demonstrate methods for post-processing standard products that are made available for end-users, and are expected to stimulate the research community to exploit them in other volcanological application contexts.

The geographic scope is global, with case studies in Chile, Peru, Ecuador, Guatemala, Mexico, Hawai'i, Alaska, Kamchatka, Japan, Indonesia, Vanuatu, Réunion Island, Ethiopia, Canary Islands, Greece, Italy, and Iceland.

The added value of the published research lies in the demonstration that remote sensing technologies can improve our knowledge of volcanoes that pose a threat to local communities; back-analysis and critical revision of recent volcanic eruptions and unrest periods; and improvement of modeling and prediction methods. Therefore, this Special Issue provides not only a collection of forefront research in remote sensing applied to volcanology, but also a selection of case studies demonstrating the societal impact that this scientific discipline can potentially have in volcanic hazard and risk management.

Francesca Cigna, Deodato Tapete , Zhong Lu, Susanna K. Ebmeier

Editors

Editorial

Remote Sensing of Volcanic Processes and Risk

Francesca Cigna [1,*], Deodato Tapete [1] and Zhong Lu [2]

[1] Italian Space Agency, Via del Politecnico snc, 00133 Rome, Italy; deodato.tapete@asi.it
[2] Roy M. Huffington Department of Earth Sciences, Southern Methodist University, Dallas, TX 75275-0395, USA; zhonglu@smu.edu
* Correspondence: francesca.cigna@asi.it

Received: 29 July 2020; Accepted: 3 August 2020; Published: 10 August 2020

Abstract: Remote sensing data and methods are increasingly being embedded into assessments of volcanic processes and risk. This happens thanks to their capability to provide a spectrum of observation and measurement opportunities to accurately sense the dynamics, magnitude, frequency, and impacts of volcanic activity in the ultraviolet (UV), visible (VIS), infrared (IR), and microwave domains. Launched in mid-2018, the Special Issue "Remote Sensing of Volcanic Processes and Risk" of *Remote Sensing* gathers 19 research papers on the use of satellite, aerial, and ground-based remote sensing to detect thermal features and anomalies, investigate lava and pyroclastic flows, predict the flow path of lahars, measure gas emissions and plumes, and estimate ground deformation. The strong multi-disciplinary character of the approaches employed for volcano monitoring and the combination of a variety of sensor types, platforms, and methods that come out from the papers testify the current scientific and technology trends toward multi-data and multi-sensor monitoring solutions. The research advances presented in the published papers are achieved thanks to a wealth of data including but not limited to the following: thermal IR from satellite missions (e.g., MODIS, VIIRS, AVHRR, Landsat-8, Sentinel-2, ASTER, TET-1) and ground-based stations (e.g., FLIR cameras); digital elevation/surface models from airborne sensors (e.g., Light Detection And Ranging (LiDAR), or 3D laser scans) and satellite imagery (e.g., tri-stereo Pléiades, SPOT-6/7, PlanetScope); airborne hyperspectral surveys; geophysics (e.g., ground-penetrating radar, electromagnetic induction, magnetic survey); ground-based acoustic infrasound; ground-based scanning UV spectrometers; and ground-based and satellite Synthetic Aperture Radar (SAR) imaging (e.g., TerraSAR-X, Sentinel-1, Radarsat-2). Data processing approaches and methods include change detection, offset tracking, Interferometric SAR (InSAR), photogrammetry, hotspots and anomalies detection, neural networks, numerical modeling, inversion modeling, wavelet transforms, and image segmentation. Some authors also share codes for automated data analysis and demonstrate methods for post-processing standard products that are made available for end users, and which are expected to stimulate the research community to exploit them in other volcanological application contexts. The geographic breath is global, with case studies in Chile, Peru, Ecuador, Guatemala, Mexico, Hawai'i, Alaska, Kamchatka, Japan, Indonesia, Vanuatu, Réunion Island, Ethiopia, Canary Islands, Greece, Italy, and Iceland. The added value of the published research lies on the demonstration of the benefits that these remote sensing technologies have brought to knowledge of volcanoes that pose risk to local communities; back-analysis and critical revision of recent volcanic eruptions and unrest periods; and improvement of modeling and prediction methods. Therefore, this Special Issue provides not only a collection of forefront research in remote sensing applied to volcanology, but also a selection of case studies proving the societal impact that this scientific discipline can potentially generate on volcanic hazard and risk management.

Keywords: volcano monitoring; gas emissions; magma accumulation; edifice growth and collapse; volcanic unrest; lava flows; pyroclastic flows; ash plumes; thermal anomalies; volcano deformation

1. Aims and Goals

Understanding volcanic processes and hazards, assessing the associated risk for exposed communities, critical infrastructure, and business, and enhancing risk awareness are activities of vital importance toward risk mitigation (e.g., [1]). Remote sensing observations are increasingly being embedded into assessments of volcanic processes and risk, thanks to their capability to provide a spectrum of opportunities to accurately sense the dynamics, magnitude, frequency, and impacts of volcanic activity in the ultraviolet (UV), visible (VIS), infrared (IR), and microwave domains (e.g., [2–4]). Crucial is their potential to monitor volcanoes where no ground sensor networks exist, as well as otherwise inaccessible locations.

Launched in mid-2018, the Special Issue "Remote Sensing of Volcanic Processes and Risk" [5] of *Remote Sensing* aimed to gather original research articles, reviews, technical notes, and letters on the use of satellite, aerial, and ground-based remote sensing data and methods to sense volcanic processes (e.g., deformation, lava and pyroclastic flows, gas emissions and plumes) and assess the associated hazard and risk. One of the key goals of the Special Issue was to collect research studies combining two or more remote sensing methods or types of data, integrating remote sensing with in situ observations (e.g., GPS benchmark surveying, topographic leveling, seismic and geochemical data) or embedding remotely sensed information into volcanic processes, hazard and risk assessment models, near-real-time monitoring, early warning, and decision-making. Submissions of articles and/or review papers on global or continental volcano databases, monitoring, and models were equally encouraged.

Developed as part of a successful series of thematic volumes promoted by the "Remote Sensing in Geology, Geomorphology and Hydrology" section of *Remote Sensing*, this project follows on from the Special Issue "Volcano Remote Sensing" published in 2015–2016 [6], and it was collaboratively led by an international team of four Guest Editors based in Europe and the USA: Dr Francesca Cigna and Dr Deodato Tapete from the Italian Space Agency in Italy, Prof Zhong Lu from the Southern Methodist University in Texas, and Dr Susanna K. Ebmeier from the University of Leeds in the UK.

This editorial provides an overview of the research papers composing the Special Issue, an outline of the data and methods used by the contributing authors (see Section 2), and some statistics on the editorial and peer-review process, as well as on the initial scientific impact made during the first months following the closure of the Call for Papers (see Section 3). Conclusions and an outlook to the future are also provided (see Section 4), together with directions to other thematic issues on volcano remote sensing and opportunities in MDPI journals for further reading and specialist contributions on this topic.

2. Overview, Data and Methods

The Special Issue comprises 19 research papers, namely 17 articles, 1 technical note, and 1 letter. Figure 1 shows a pictorial word cloud of the thematic keywords used by the 19 papers, where the strong multi-disciplinary character of the approaches employed for volcano monitoring, and the combination of a variety of sensor types and platforms are apparent. A summary of the remote sensing data and methods used, and the areas of interest investigated, is provided in Table 1.

Table 1. Summary of remote sensing data, methods, and areas of interest discussed in the 19 research papers composing the Special Issue (sorted in ascending order, according to the publication date). Notation: DEM, Digital Elevation Model; IR, InfraRed; TIR, Thermal IR; SAR, Synthetic Aperture Radar; InSAR, Interferometric SAR; Ground-Based InSAR, GBInSAR; NDVI, Normalized Difference Vegetation Index; NN, Neural Networks; UV, UltraViolet; VIS, Visible; VHR, Very High Resolution.

Paper Reference	Data and Methods	Areas of Interest
Plank et al. 2018 [7] [1]	satellite TIR data (TET-1, MODIS, VIIRS); ash coverage, change detection (TET-1, Landsat-8)	Villarrica (Chile)
Bredemeyer et al. 2018 [8] [1]	ground-based scanning UV spectrometer (Mini-DOAS; SO_2); multi-component Gas Analyzer System (Multi-GAS) survey (molar H_2O/SO_2 ratio); satellite SAR data, differential InSAR (TerraSAR-X)	Láscar (Chile)
Marchese et al. 2018 [9] [1]	satellite TIR data (MODIS, AVHRR, Sentinel-2, Landsat-8 OLI); volcanic hotspot detection (RST_{VOLC} algorithm)	Etna (Italy)
Di Traglia et al. 2018 [10] [1]	TIR and VIS cameras; GBInSAR; ground-based seismic signals and NN-based analysis to detect incipient landslides; high-resolution DEM (LiDAR, tri-stereo Pléiades imagery), topographic change detection	Stromboli (Italy)
Papageorgiou et al. 2019 [11] [1]	satellite SAR data, Multi-Temporal InSAR, MT-InSAR (Sentinel-1, Radarsat-2, TerraSAR-X, ERS-1/2, ENVISAT); inversion modeling (Volcano and Seismic source Model, VSM)	Santorini (Greece)
Aufaristama et al. 2019 [12] [1]	airborne hyperspectral data (AisaFENIX sensor, NERC Airborne Research Facility); VHR airborne photographs	Holuhraun (Iceland)
Sansivero and Vilardo 2019 [13] [2]	ground-based TIR data (FLIR cameras), thermal anomalies detection (Matlab© code ASIRA, Automated System of IR Analysis)	Phlegraean Fields and Vesuvius (Italy)
Rogic et al. 2019 [14] [1]	satellite TIR data (Landsat-5/7, ASTER Global Emissivity Database); numerical modeling (MAGFLOW); sample chemical composition (X-Ray Fluorescence), reflectance spectra (Fourier Transform IR, FTIR spectroscopy)	Etna (Italy)
Lombardo et al. 2019 [15] [1]	geostationary satellite IR data (Meteosat Second Gen. SEVIRI); IR wavelet transform; radiance time series analysis	Etna (Italy)
Gomez-Ortiz et al. 2019 [16] [1]	ground-penetrating radar; electromagnetic induction; magnetic survey; detection of geothermal anomalies	Timanfaya (Canary Islands, Spain)
Cando-Jácome and Martínez-Graña 2019 [17] [1]	satellite SAR, differential InSAR (Sentinel-1); satellite optical data (Sentinel-2); morphometric indices (SAGA GIS), mass movement analysis (SHALSTAB); lahars flow path prediction	Fuego (Guatemala)
de Michele et al. 2019 [18] [3]	satellite optical data, volcanic cloud-top height (VCTH) estimation (Landsat-8, Meteosat Second Gen. SEVIRI, MODIS)	Etna (Italy)
Dávila et al. 2019 [19] [1]	satellite optical data, image segmentation for lava flow mapping (SPOT-6/7, EO-1 ALI); digital surface model (SPOT-6)	Colima (Mexico)
Laiolo et al. 2019 [20] [1]	satellite TIR data (MODIS MIROVA, Sentinel-2); infrasound arrays; seismic tremor data	Etna (Italy)
Delle Donne et al. 2019 [21] [1]	UV camera data (SO_2 flux monitoring); satellite (MODIS MIROVA) and ground-based (monitoring cameras) TIR data; seismic tremor data	Etna (Italy)
Mania et al. 2019 [22] [1]	satellite SAR, pixel offset (TerraSAR-X); webcam imagery; ground-based seismic data	Bezymianny (Kamchatka, Russia)
De Angelis et al. 2019 [23] [1]	ground-based acoustic infrasound; ash plume modeling	Etna (Italy), Santiaguito and Fuego (Guatemala), Tunghurahua (Ecuador), Sakurajima (Japan), Sabancaya (Peru), Augustine and Redoubt (Alaska, USA)
Valade et al. 2019 [24] [1]	satellite optical and SAR data, tropospheric monitoring (Sentinels); ground-based seismic data; artificial intelligence (Convolutional NN)	Erta Ale (Ethiopia), Fuego (Guatemala), Kilauea (Hawai'i, USA), Anak Krakatau (Indonesia), Ambrym (Vanuatu), Piton de la Fournaise (Réunion Island, France)
Aldeghi et al. 2019 [25] [1]	satellite optical data (CubeSats, i.e., PlanetScope); NDVI; visual analysis	Fuego (Guatemala)

[1] Article. [2] Technical note. [3] Letter.

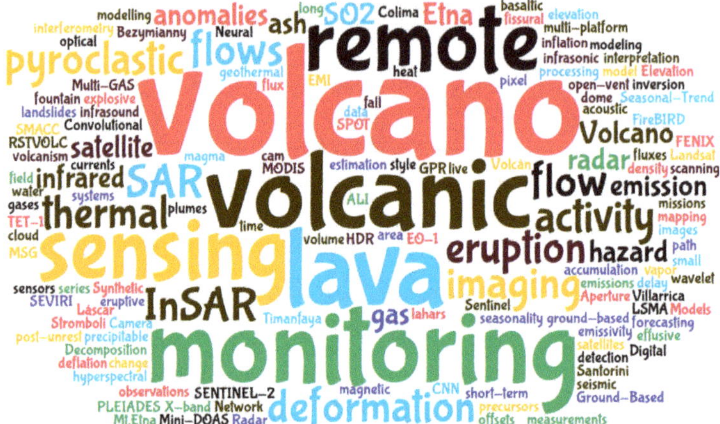

Figure 1. Thematic keywords of the 19 research papers composing the Special Issue "Remote Sensing of Volcanic Processes and Risk" [5] of *Remote Sensing* (created with wordclouds.com).

Short accounts for each paper are also provided in the next sections, according to the following subdivision:

- Remote sensing of thermal features and anomalies
- Investigation, mapping, and prediction of lava flows and lahars
- Monitoring gas emissions and volcanic plumes
- Ground deformation analysis based on SAR and InSAR
- Multi-data and multi-sensor monitoring of volcanoes

This subdivision is an attempt to organize the papers by the dominant scope, main character of the research, and/or peculiar type of remote sensing approach exploited. Therefore, such subdivision may appear a simplification for those papers encompassing more than one volcanological aspect or focusing on more than one volcanic parameter or observable. Consequently, the last group collects those papers that encompass multi-disciplinary approaches and encourage the integration of multi-sensor data.

It is worth noting that a common ground across the papers is the promotion of novel, integrated, or improved monitoring solutions to increase the existing observation capabilities based on either satellite, aerial, or terrestrial devices. Remote sensing, in this regard, represents an opportunity, in the context that less than 10% of the approximately 1500 active subaerial volcanoes around the world are monitored regularly on the ground, as recalled in [14].

2.1. Remote Sensing of Thermal Features and Anomalies

Satellite sensors such as MODerate resolution Imaging Spectroradiometer (MODIS), Meteosat Second Generation Spinning Enhanced Visible and InfraRed Imager (SEVIRI), Advanced Very-High-Resolution Radiometer (AVHRR), and Landsat-8 Thermal InfraRed Sensor (TIRS), currently provide effective means for the thermal remote sensing of volcanoes at the global scale. Several papers published in the Special Issue demonstrate how these satellite data are nowadays well established across the volcanological community for research and the operational monitoring of volcanic thermal features (e.g., vents, geysers, hot springs, lava flows, lava domes).

SEVIRI is among the most used satellite sensors, owing to its high temporal resolution ranging from 15 min (Earth's full disk) to 5 min (rapid scan mode over Europe and Northern Africa). Exploiting the time series collected in real time from a Meteosat-8 ground station antenna operating since 2010 at the National Institute of Geophysics and Volcanology (INGV) in Rome, Lombardo et al. [15] tested a new detection method based on the wavelet transform of SEVIRI InfraRed (IR) data, to investigate eruptive processes and discriminate different styles of volcanic activity. In particular, a statistical analysis was performed on wavelet smoothed data derived from SEVIRI Mid-IR (MIR) radiance collected over Mount Etna in Italy from 2011 to 2017, when the volcano changed its eruptive style from predominantly effusive to more explosive (2011–2015), and a vigorous Strombolian activity started from the south-east crater producing a small lava flow (February 2017). The results (validated through ground-based information and literature references) suggested a relationship between the rate of increment in radiance (and thus temperature) and the nature of the volcanic process causing that increment. Statistical analysis of SEVIRI MIR radiance trends highlighted the involvement of at least two different heating mechanisms for eruptions at Mount Etna. Since the methodology applies to data acquired at the onset of the eruption, prediction of what will be its dominant eruptive style at later stages is feasible, with a degree of trustworthiness in the first eight hours from the beginning of the eruption.

Instead, Marchese et al. [9] integrated AVHRR and MODIS observations in the Robust Satellite Techniques–volcanoes (RST_{VOLC}) algorithm within the satellite-based system developed at the Institute of Methodologies for Environmental Analysis (IMAA) of the National Research Council (CNR) of Italy to monitor Italian volcanoes in near-real time. The authors investigated the eruptive events occurring in May 2016 at Mount Etna, and the fumarolic emissions recorded before a small degassing vent (of approximately 20–30 m in diameter) opened within the Voragine crater (VOR). AVHRR and MODIS observations were integrated in RST_{VOLC} to generate hotspot products (i.e., JPG, Google Earth KML, and ASCII files) a few minutes after the sensing time. The detection of volcanic thermal anomalies is based on two local variation indices. The first identifies anomalous signal variations in the MIR band of the AVHRR (channel 3: 3.55–3.93 μm) and MODIS (channels 21/22: 3.929–3.989 μm) sensors, where hot magmatic surfaces reach the peak of thermal emissions. The second index allows the minimization of spurious effects associated with non-volcanological signal fluctuations. The results indicated that the Strombolian eruption of 21 May 2016 lasted longer than reported by field observations, or that a short-lived event occurred in the late afternoon of the same day. Furthermore, the intensity of fumarolic emissions changed before 7 August 2016, as a possible preparatory phase of the hot degassing activity occurring at VOR. These outcomes matched with the evidence found in Sentinel-2 MSI (Multispectral Instrument) and Landsat-8 OLI (Operational Land Imager) data.

While the above space missions remain reference data sources for the community, the paper by Plank et al. [7] is an excellent example of the advances in space-borne sensor development, to achieve higher spatial resolution and sensitivity with regard to thermal anomalies. The authors demonstrated the applicability of the Technology Experiment Carrier-1 (TET-1), i.e., the first of two small experimental satellites of the German Aerospace Center (DLR)'s FireBIRD mission. TET-1 was launched in July 2012, followed by BIROS (Berlin IR Optical System) in June 2016. Both satellites are flying sun-synchronously in a low-Earth orbit at approximately 500 km altitude, with a repetition rate of approximately five days for one satellite, depending on its geographic location, and up to less than 3 days with ± 30° across-track acquisitions. The sensors operate two IR cameras, one in the Mid Wave IR (MWIR) and one in the Long Wave IR (LWIR), as well as a three-channel camera in the visible (VIS: RED and GREEN) and Near IR (NIR). The MWIR, LWIR, and RED channel are installed in the nadir position, while the GREEN and NIR channel are oriented off-nadir. TET-1 operates a push broom sensor system with a ground sampling distance of 178 m for the thermal channels, corresponding to a pixel resolution of 356 m due to staggering. The MWIR channel of TET-1 is more suitable for the detection of high-temperature events than LWIR (higher temperature events have a higher radiant power at shorter wavelengths according to Wien's displacement law). In particular, the authors analyzed a time series of nine TET-1 thermal images acquired before and during the March/April 2015 eruption of Villarrica volcano (Chile), which is one of the most active volcanoes in the South Andes Volcanic Zone. The temperature, area coverage, and radiant power of the detected thermal hotspots were derived at the subpixel level and compared with observations derived from MODIS and Visible IR Imaging Radiometer Suite (VIIRS) data. TET-1 images allowed thermal anomalies to be detected nine days before the eruption. After the decrease of the radiant power following the 3 March 2015 eruption, a stronger increase of the radiant power was observed on 25 April 2015. Since the eruption caused ash coverage of the glacier at the eastern flank, surface changes were also investigated. The comparison with higher resolution multispectral data from Landsat-8 highlighted that the event was well captured in TET-1 imagery.

Alongside satellite observations, Thermal IR (TIR) ground-based observations are another important source of information to investigate thermal features, as well as volcanic plumes and gases, lava flows, lava lakes, and fumarole fields. Sansivero and Vilardo [13] rightly pointed out that, in the past, such observations were mostly collected during a limited time span (e.g., eruption phases, field campaigns). The evidence that thermal precursors can be successfully detected before eruptions based on more continuous TIR observations has recently stimulated the increasing installation of permanent ground TIR stations at active volcanoes across the globe. However, such augmented availability of data has not yet been followed by an equal development of software packages allowing the processing of continuous TIR time series and near real-time automated analysis of large datasets. In this context, Sansivero and Vilardo [13] aimed to fill the gap by presenting an operational processing chain developed in the Matlab© environment (Automated System of IR Analysis, ASIRA) that allows the detection and quantification of possible changes in time and space of the ground-surface thermal features. ASIRA allows effective removal of the seasonal component of IR temperature time series, and an estimation of radiative heat fluxes of thermal anomaly areas. The ASIRA code was applied to process TIR frames acquired at night by the stations of the TIRNet surveillance network operated by INGV's Osservatorio Vesuviano (INGV-OV) at Phlegraean Fields volcanic area in Italy. The results show the effectiveness of this method. The Matlab© code of ASIRA and the Operative Manual are included in the Supplementary Materials of the paper, thus offering a tool for scholars to implement the analysis elsewhere with similar data.

Finally, it is important to not forget the impact that geophysical techniques can have on the detection of thermal anomalies associated to volcanic geothermal systems. Active volcanic areas are characterized by high-enthalpy geothermal systems that exhibit high-temperature zones (>150–200 °C) at ground level. While convective hydrothermal systems are most common, Hot Dry Rock (HDR) geothermal systems consisting of subsurface zones with very low fluid content are rarer. Gomez-Ortiz et al. [16] focused their attention on HDR, thus providing an original contribution to the still very few studies on such geothermal systems based on magnetotellurics, transient electromagnetics, electrical resistivity, magnetics, self-potential, and seismic. The reason for this paucity of specialist literature is that the presence of vapor as the dominant phase (instead of hot water) makes this kind of system more difficult to interpret using electromagnetic methods. The authors demonstrate how near-surface geophysical modeling can be implemented to image thermal anomalies in HDR geothermal systems through a case study in the Timanfaya volcanic area in Lanzarote, Canary Islands. There, thermal anomalies are still present as a consequence of the historical eruptive activity that occurred between 1730 and 1736. In particular, the authors combined ground-penetrating radar (GPR), electromagnetic induction (EMI), and magnetic prospecting to characterize the geophysical signature of the high ground temperature areas. GPR revealed that when the material is homogeneous, the signature of the reflections is more intense in the areas with high temperature values. Similarly, a variation in the subsurface distribution of the thermal anomaly was detected for the first time through the analysis of GPR at different periods. The resistivity models obtained from the inversion of EMI data demonstrated that high-resistive areas are associated with high-temperature zones. Magnetic data showed that the zones with high-temperature values are associated with magnetic lows due to the demagnetization of the volcanic materials when they are heated to temperatures close to or higher than the Curie point of the involved magnetic minerals. Therefore, the authors conclude that the combined use of GPR, EMI, and magnetic prospecting methods is effective to locate and study both the geometry at depth and seasonal variability of geothermal areas associated with HDR systems.

2.2. *Investigation, Mapping, and Prediction of Lava Flows and Lahars*

Aufaristama et al. [12] recalled that in high eruption frequency areas, lava flows often overlap each other and may exhibit similar spectral signatures, which makes their discrimination difficult. Spectral range and resolution (i.e., number of spectral bands), as well as the spatial resolution of satellite images, can further constrain the success of separation between different spectral signatures. Hyperspectral satellite data still provide limited spatial resolution (e.g., Earth Observing-1 or EO-1 Hyperion with a ground resolution of 30 m), so airborne sensors may help, but dedicated acquisition surveys are needed. A demonstration of such capabilities is discussed by the authors based on the analysis of hyperspectral data that they collected in 622 channels with spectral range from approximately 400 to 2500 nm by using the AisaFENIX sensor onboard the Natural Environment Research Council (NERC) Airborne Research Facility, five months after the Holuhraun 2014–2015 lava flow, in NE Iceland. This event lasted about six months (31 August 2014 to 27 February 2015) and produced a bulk volume of approximately 1.44 km^3 of basaltic lava, i.e., a diverse surface environment to investigate and characterize lava deposits. The objective was to retrieve the main lava surface type contributing to the signal recorded by airborne hyperspectral data at the very top surface of Holuhraun, on the area around the eruptive fissures vent, for which very high-resolution aerial photographs of the lava field (0.5 m spatial resolution) were available for comparison, and validation of the unmixing results. The data were atmospherically corrected using the QUick Atmospheric Correction (QUAC) algorithm, and the Sequential Maximum Angle Convex Cone (SMACC) method was used to find spectral endmembers and their abundances throughout the airborne hyperspectral image. In total, 15 endmembers (i.e., representing pure surface materials in a hyperspectral image) were estimated and categorized into six groups based on the shape of the endmembers: (1) basalt; (2) hot material; (3) oxidized surface; (4) sulfate mineral; (5) water; and (6) noise. This was required, since the amplitude varied due to illumination conditions, spectral variability, and topography. The respective abundances from each endmember group were retrieved using fully constrained Linear Spectral Mixture Analysis (LSMA). The authors conclude that the combination of SMACC and LSMA methods offers an optimum and a fast selection for volcanic products segregation. However, ground-truthing spectra are recommended for further analysis.

Instead, Dávila et al. [19] relied on satellite images to revise and improve the chronological reconstruction of the different eruptive phases that occurred from September 2014 to September 2016 at Volcán de Colima in Mexico, including the eruption on 10–11 July 2015, which was the most violent since the 1913 Plinian eruption. Their analysis mainly relied on satellite products. SPOT-6 dual-stereoscopic and tri-stereopair images were used to generate the Digital Surface Models (DSMs) and therefore estimate the volumes of lava flows and the main pyroclastic flow deposits. SPOT-6/7 and EO-1 ALI (Advanced Land Imager) data were in parallel combined to better define the spatial distribution of the lava flows prior to and after the volcanic activity of July 2015. Then, pre- and post-eruption DSMs were used as topographic inputs to calibrate lava flow simulation software (i.e., the Etna Lava Flow Model, ELFM) to simulate different paths that lava might follow when emitted from a volcanic vent. Through this methodology, the authors were able to estimate that the total volume of the magma that erupted during the 2014–2016 event was approximately 40×10^7 m^3, i.e., one order of magnitude lower than that of the 1913 Plinian eruption. A larger magma volume stored in the magma chamber would have been necessary, and, as observed in the precursory activity of the 1913 eruption, dome destruction would have been accompanied by explosive events for the 2015 event to be similar to the 1913 one.

In the context of recent advances in space-borne sensors, the research by Aldeghi et al. [25] is relevant in its attempt to demonstrate the novel Earth Observation (EO) technology of PlanetScope images with 3 m pixel size in an operational scenario of "rapid response" mapping of eruption deposits, which could be eroded or removed by rainfall soon after emplacement, and thus trigger secondary volcanic hazards such as lahars. Constellations of small satellites ('CubeSats') such as PlanetScope are bringing a new paradigm in the EO arena, given that they achieve very high spatial resolution at high cadence (up to less than 1 h) by means of numerous cheap satellites allowing for multiple scene acquisitions within a few minutes in the overlapping region. Aldeghi et al. [25] tested PlanetScope 16-bit calibrated orthorectified surface reflectance data with a positional accuracy of better than 10 m, which was collected during the 31 January–2 February 2018 eruption of the Fuego volcano in Guatemala, to map lava flows, Pyroclastic Density Currents (PDCs), and tephra falls through visual analysis, the Normalized Difference Vegetation Index (NDVI) difference method, and trial-and-error approach using single bands. The authors found that high-resolution visible images can be a good alternative for lava flow mapping, provided that enough contrast with the background is achieved, and they allow for the detailed mapping of structural and morphological changes associated with the volcanic activity. In particular, the scar at the head of Barranca Honda was immediately identified after the eruption and could be interpreted as either a collapse or erosion feature associated with the generation or transit of PDCs through that area. As expected, the NDVI difference approach was not suitable for detecting changes in areas that were originally non-vegetated, and thus, no good results could be achieved for PDCs confined to the channels on the flanks of the volcanic edifice. On the contrary, for tephra fall mapping analysis, the method provided a much better alternative than visual mapping, given that the gradational boundaries of the tephra fall deposits (i.e., the transition from areas with heavy tephra fall to areas with no tephra fall) are smooth and may be difficult to define visually.

The Fuego volcano is also the subject of the paper by Cando-Jácome and Martínez-Graña [17], who discussed the integrated results of satellite remote sensing and geospatial analysis to complement hazard maps generated in the aftermath of an eruption. In particular, the authors focused on the strong eruption occurred on 3 June 2018, when a dense cloud of 10 km-high volcanic ash and destructive pyroclastic flows caused approximately 200 deaths and huge economic losses in the nearby region. After the eruption, two scenarios of lahars for medium and heavy rains based on the numerical models were produced using the LAHARZ software, which was developed at the United States Geological Survey. To improve lahar mapping, Sentinel-1 Synthetic Aperture Radar (SAR) data were processed by means of Differential Interferometric SAR (DInSAR) to locate areas of ground deformation on the volcano flanks, where lahars could have formed and been triggered. To determine the trajectory of the lahars, parameters and morphological indices—accumulation of flow, topographic wetness index, length–magnitude factor of the slope—were analyzed with the software System for Automated Geoscientific Analysis (SAGA), and a slope stability analysis was performed using the SHAllow Landslide STABility software (SHALSTAB) based on the Mohr–Coulomb theory and its parameters. The application of this complementary methodology provided a more accurate response of the areas destroyed by primary and secondary lahars in the vicinity of the volcano.

For crisis management of effusive volcanic events, lava flow 'distance-to-run' is a key parameter to predict. However, lava flow is a complex surface feature to observe using remote sensing, given that temperature, texture, vesicularity, and thickness vary across the moving material. Rogic et al. [14] focused their attention on the emissivity—the efficiency with which a surface radiates its thermal energy at various wavelengths—that is in close relationship with land surface temperatures and radiant fluxes and, as such, impacts directly on the prediction of lava flow behavior. Since emissivity is seldom measured and mostly assumed, the authors attempted a multi-stage experiment, combining laboratory-based Fourier Transform IR (FTIR) analyses, remote sensing data, and numerical modeling. In particular, they tested the capacity for reproducing emissivity using the ASTER Global Emissivity Database (GED) built by NASA's Jet Propulsion Laboratory, while assessing the spatial heterogeneity of emissivity. To this purpose, the chemical composition of 10 rock samples was analyzed through X-Ray Fluorescence (XRF), and emissivity was retrieved from both reflectance and radiance data at ambient/low and high temperatures using FTIR spectroscopy. The laboratory–satellite emissivity values were used to establish a realistic land surface temperature from Landsat-7 ETM+ to obtain an instant temperature–radiant flux and eruption rate results for the 2001 Mount Etna eruption, which gave rise to an outstanding pattern of seven different fast-developing lava flows. Forward-modeling tests were conducted on the 2001 'aa' lava flow by means of the MAGFLOW Cellular Automata code. Good correlation was found between laboratory (FTIR) and space-borne (ASTER GED) data for the same target area, and at specific TIR wavelengths, by exhibiting an emissivity range/error of ≤0.03. However, the authors concluded that this emissivity information is 'static', relates to the solidified (cooled) product, and does not reflect the range of temperatures involved at an active lava flow or the emissivity/temperature trend seen in the high-temperature FTIR results. Furthermore, the theoretical empirical approaches and modeling indicated that a 0.2 variation in emissivity may result in significant changes to the prediction of lava flow 'distance-to-run' estimates. Indeed, the tests conducted by the authors provided differences of up to approximately 600 m in the simulated lava flow 'distance-to-run' for a range of emissivity values. Therefore, this study highlighted the need to assess the role and significance of emissivity, not only as a 'static' and uniform value across all wavelengths and temperatures, but also to take its response to thermal gradient into account.

2.3. Monitoring Gas Emissions and Volcanic Plumes

Dissolved gases in magmas are the main drivers of most volcanic eruptions. Changes in their composition and fluxes can be proxies of subtle changes in the rate of magma ascent and degassing within shallow volcano plumbing systems. For example, this is the case of volcanic sulfur dioxide (SO_2) emissions in plumbing systems located at less than 3 km depth.

In this regard, the paper by Delle Donne et al. [21] provided an innovative contribution through a robust experimental demonstration to constrain the degassing regimes and eruptive behavior of the Mount Etna volcano in 2016. The technological focus of the paper is the development and testing of a novel algorithm for the real-time automatic processing of UltraViolet (UV) camera data and visualization of SO_2 flux time-series. Automation is meant to solve a known limitation of permanent UV camera systems that are extensively used to monitor volcanic SO_2 emissions, but produce streams of data whose processing is still time-consuming and labor-intensive. To obtain SO_2 emissions associated with diverse volcanic processes and dynamics—including quiescent (passive) degassing, explosive eruptions (Strombolian activity/lava fountaining), and effusive eruptions, and therefore capture switches between these different phases—the authors exploited the UV camera system installed at the Montagnola site, at about a 3 km distance from the active summit vents, and streaming real-time SO_2 flux results through a Wi-Fi data link. Measurements were also carried out during an ongoing lava fountaining event, which is not so common to find in the specialist literature. The results were validated through MODIS satellite-based thermal data obtained from the MIROVA (Middle IR Observation of Volcanic Activity) system, ground-based thermal data streamed by monitoring cameras of INGV's Osservatorio Etneo (INGV-OE), and seismic tremor data. All these independent datasets showed coherent temporal variations that validated the use of UV cameras for detecting subtle changes in volcanic and degassing activity. Pre-paroxysm SO_2 fluxes were found to have consistent values (of approximately 2000 t/d) during the three episodes. Similarly, the highest SO_2 fluxes (from 3000 up to 5200 t/d on a daily average basis) were identified during the three eruptive sequences, while post-eruptive fluxes were systematically characterized by reduced degassing (<1000 t/d). If confirmed by future observations, these results may bring implications for identifying switches in volcanic activity regime. Therefore, this paper is novel because it has demonstrated a recent advance in instrumental volcanic gas monitoring as well as for the quantitative information published therein.

Since (at least) the 2010 eruption of Eyjafjallajökull in Iceland, even non-experts know that airborne volcanic ash represents a direct threat to aviation that can cause disruption to flight operations and damage infrastructure. During eruptions, warnings can be issued based on the outputs of models of atmospheric ash transport which, among other parameters, exploit the rate at which the material is ejected from volcanic vents.

Large-scale eruptions involving the injection of hot gas-laden pyroclasts into the atmosphere generate infrasound acoustic waves with frequencies typically <20 Hz, which can travel distances of up to several thousands of kilometers. Such low-frequency waves can be detected with ground-based acoustic infrasound instrumentation. Given the increasing implementation of this instrumentation by researchers and practitioners, the review by De Angelis et al. [23] is timely in that it provides an insightful assessment of the developments and lessons learnt in this field. Focusing on acoustic infrasound at local distances (i.e., within 10–15 km from eruptive vents), the authors review near-field (<10 km from the vent) linear acoustic wave theory, evaluate recent advances in volcano infrasound modeling and inversion, and comment on the advantages and current limitations of these methods. Among the highlights, it is worth mentioning the recent introduction of numerical modeling to approximate the atmosphere's impulse response in the presence of realistic topography, and how this approach has been integrated within inversion workflows. The authors also stress that the temporal resolution offered by acoustic infrasound in retrieving eruption source parameters in order to inform ash plume rise and transport models remains unmatched, and how this capability can be exploited for the rapid assessment of airborne eruption hazards.

Volcanic Cloud-Top Height (VCTH), as a Plume Elevation Model (PEM), is one of the most critical parameters to retrieve, because it affects the quantitative estimation of volcanic cloud ash and gases parameters, the mass eruption rate needed for the transport and deposition models, and the definition of the most dangerous zone for air traffic.

In a logic sequence with their previous publication focused on Landsat-8 OLI Level 0 raw data [26], de Michele et al. [18] presented a method to extract VCTH from orthorectified Level 1 data, i.e., the standard product available free of charge for end users. The concept behind this retrieval method is that the physical distance between the panchromatic sensor (PAN) and the multispectral sensors (MS), both onboard Landsat-like satellites, yields a baseline and a time lag between the PAN and MS image acquisitions during a single passage of the satellite. This information can be used to extract a spatially detailed map of VCTH from virtually any multispectral push broom system, namely PEM. While adapting such PEM methodology to the standard Landsat-8 products, the authors aimed to simplify the procedure for routine monitoring, offering an opportunity to produce PEM maps. They implemented this approach on the episodes that occurred at Mount Etna on 26 October 2013 and compared the results with independent VCTH measures from the geostationary SEVIRI and the polar MODIS. The analysis highlighted a good agreement with the Landsat-8 VCTH product, thus corroborating the accuracy and reliability of the proposed method.

2.4. Ground Deformation Analysis Based on SAR and InSAR

Owing to their capability to collect data in all weather conditions, SAR sensors, either space-borne or ground-based, are undoubtedly advantageous for monitoring volcanic activity. Countless examples can be found in the specialist literature on the use of SAR images and their derived products by means of Interferometric SAR (InSAR) processing. According to a recent estimate, over 500 volcanoes worldwide have now been the subject of InSAR measurements [2].

In this Special Issue, SAR and InSAR are well represented to monitor the growth of lama domes [22], examine a post-unrest period [11], and to investigate the radar path delays due to the water vapor contained in volcanic gas plumes [8].

Monitoring the growth of lava domes is crucial, given that explosive eruptions can make outer flanks unstable until they collapse and cause pyroclastic flows, which may move very quickly down the slopes and impact on regions several kilometers away and/or pose threat for aviation. An interesting contribution toward an effective monitoring solution is presented by Mania et al. [22]. The authors combined seismic data, camera monitoring, and Mimatsu diagrams with change detection maps and pixel offset tracking based on TerraSAR-X SpotLight SAR images to understand the dome growth mechanisms acting during the January 2016–June 2017 eruption sequence at Bezymianny, an andesitic dome-building volcano in Kamchatka, Russia. In particular, camera monitoring allowed the approximate identification of topographic changes at Bezymianny's flank. This assessment was refined using ground motion estimated from the pixel offset tracking algorithm, owing to the pixel spacing of 0.9×1.25 m in the slant-range and azimuth directions provided by TerraSAR-X, alongside the selected descending viewing geometry overcoming visibility issues due to the foreshortening and shadowing of the flanks. The results revealed clear morphometric changes preceding eruptions that were associated with intrusions and extrusions.

In particular, seven to nine months of precursory ground motion were captured and were interpreted as a rigid body extruded at the summit prior to the first documented effusive December 2016–February 2017 eruption. Besides exogenous growth, the SAR amplitude images also unveiled distinct, recurrent endogenous growth stages as Bezymianny's dome bulged northwards multiple times. Based on this evidence, the authors developed a conceptual model of volcanic growth at Bezymianny, thus proving how the integration of satellite observations with other remote sensing data can generate an improved understanding of lava dome building processes.

Papageorgiou et al. [11] exploited multi-sensor satellite SAR datasets (i.e., Sentinel-1, Radarsat-2, and TerraSAR-X, and previously published ERS-1/2 and ENVISAT) processed with Multi-Temporal InSAR (MT-InSAR), as well as inversion modeling based on Volcano and Seismic source Model (VSM), to examine the post-unrest period of the Santorini volcano in Greece in 2012–2017. In the last century, volcanic activity up to the most recent eruption in 1950 was intertwined with the building of the intra-caldera islets of Palea and Nea Kameni. The latest volcano reactivation was followed by the restless period of 2011, but this did not produce an eruption. The geodetic analysis of the MT-InSAR data confirmed the new volcano state after the unrest period. The post-unrest response to the 2011–2012 inflation episode is well explained by a shallow sill-like source at 2 km depth. This is located just above the approximately 4 km-deep inflation source responsible for the 2011–2012 uplift. The authors also used ERS-1/2 and ENVISAT data from 1992 to 2010 in order to interpret the similarity between the pre- and post-unrest volcano deformation. The presence of a steady subsidence source at Nea Kameni, in accordance with the pre-unrest period, led to the re-evaluation of the 2011–2012 unrest. The interpretation model suggested the co-existence of the Kameni source during the unrest, although having a lower impact compared to the larger deformation induced by the inflation source.

If the above paper is a further demonstration of what can be achieved with InSAR to understand volcanoes, it is to be acknowledged that this technique is largely affected by changes in atmospheric refractivity, in particular changes in distribution of water vapor (H_2O) in the atmospheric column. Atmospheric contributions to Differential InSAR (DInSAR) data often have similar magnitudes and wavelengths as the actual ground deformation signal. To remove such interference from interferograms, scholars usually either implement time-space-based filtering or model atmospheric contribution, with the latter being based on prediction of the atmospheric phase delay along the satellite line-of-sight and its compensation by means of high-resolution numerical weather models. However, Bredemeyer et al. [8] rightly pointed out that weather models are typically not able to capture atmospheric disturbances due to continuously degassing volcanoes. Consequently, the large and variable amounts of water vapor in volcanic plumes may cause differential phase errors in InSAR measurements due to the reduction of radar propagation velocity within the plume above and downwind of the volcano, which are notably well captured by short-wavelength X-band SAR systems (e.g., TerraSAR-X). In turn, this may lead to the misinterpretation of ground motions from an interferogram. Inversely, the estimation of the Precipitable Water Vapor (PWV) content in the plume at the time of SAR acquisitions is the key to overcome this limitation.

To investigate the radar path delays due to water vapor contained in the volcanic gas plume, Bredemeyer et al. [8] selected Láscar volcano, in the dry Atacama Desert of Northern Chile. This choice proved to be very effective, given that Láscar is among the most active volcanoes of the central Andes, the second largest emission source of volcanic gases in Northern Chile, and is located in one of the driest areas on the Earth, where background atmospheric PWV is very low most of the year with generally less than 1 mm total water column. The authors estimated water vapor contents based on SO_2 emission measurements from a scanning UV spectrometer (Mini-DOAS) station installed at Láscar volcano, which were scaled by H_2O/SO_2 molar mixing ratios obtained during a Multi-component Gas Analyzer System (Multi-GAS) survey on the volcano crater rim. This methodological approach was justified in light of challenging direct measurements of volcanic water vapor emissions by means of optical remote sensing. Based on these estimates, the authors obtained daily average PWV contents inside the volcanic gas plume of a 0.2–2.5 mm equivalent water column, which translates to a Slant

Wet Delay (SWD) in DInSAR data of 1.6–20 mm. By combining these estimates with high-resolution TerraSAR-X DInSAR observations at Láscar volcano, the authors demonstrated that gas plume-related refractivity changes are significant and detectable in DInSAR measurements.

2.5. Multi-Data and Multi-Sensor Monitoring of Volcanoes

In the last decade, the scientific literature on the remote sensing of volcanic hazard and risk is increasingly exploiting the integration of different observation capabilities, instrumentation and devices, and data (e.g., [27,28]). This trend is also observed in this Special Issue, particularly with the papers by Valade et al. [24], Laiolo et al. [20], and Di Traglia et al. [10].

From a satellite data point of view, in the context of the EO revolution that the European Commission's Copernicus Programme has opened with free accessibility to an increasingly large volume of data and observations from different satellite platforms, the volcano monitoring platform MOUNTS (Monitoring Unrest from Space) presented by Valade et al. [24] is among the best examples of current operational infrastructure enabling users to understand the temporal evolution of volcanic activity and eruptive products based on the integration of multi-sensor satellite imagery with in situ and other remote sensing data. MOUNTS monitors 17 volcanoes, and its results are published in the form of both geocoded images and time series of relevant parameters through an open-access website, as they are generated from processing Sentinel-2 VIS and Short-Wave IR (SWIR) and Sentinel-5P TROPOspheric Monitoring Instrument (TROPOMI) data for thermal and SO_2 monitoring purposes, respectively. Additionally, one of the most interesting features is the pre-trained Convolutional Neural Network (CNN) that is incorporated into the processing pipeline to detect large deformation in InSAR interferograms and generate wrapped interferograms, coherence maps, unwrapped interferograms, SAR intensity image, deformation, and decorrelation time series in less than 24 h. The CNN approach is compared with methods published in the literature, and its performance is tested on the same volcanoes (i.e., Erta Ale in Ethiopia and Mount Etna in Italy). The paper also provides a portfolio of recent eruptions (Erta Ale 2017, Fuego 2018, Kilauea 2018, Anak Krakatau 2018, Ambrym 2018, and Piton de la Fournaise 2018–2019) to demonstrate the MOUNTS products and the utility of its interdisciplinary approach.

At Mount Etna, Laiolo et al. [20] tested the hypothesis that the combination of multiple datasets can help for the detection of short- and long-term precursors preceding these events. Open-vent basaltic volcanoes can indeed alternate continuous emissions of magmatic-related products into the atmosphere with sporadic more energetic phenomena, such as paroxysmal explosions or flank eruptions. The authors chose the main effusive event that occurred on 24 December 2018 and the successive resumption of the summit explosive activity, and they combined heat flux data derived by the MODIS MIROVA sensor to calculate and track the evolution of time-averaged lava discharge rates and erupted volumes. Instead, they used the high spatial resolution of Copernicus Sentinel-2 multispectral images to locate the thermal activity at the multiple active summit vents. Infrasonic arrays and tremor amplitude measures were used to track the intensity, the frequency, and the source of the explosive events occurring at summit craters. Based on such data integration, the authors could record the shifting from open-vent conditions, represented by sustained summit Strombolian activity, to the 24–26 December flank effusion promoted by a 2 km-long feeder dyke intrusion. The dyke propagation lasted for almost 3 h, during which magma migrated from the central conduit system to the lateral vent, at a mean speed of 0.15–0.20 m/s. An accurate estimate of the lava volume from the summit outflows and lateral effusive episode was achieved.

Remaining in southern Italy, Di Traglia et al. [10] conceived a well-structured multi-sensor study combining in situ and remote sensing measurements to characterize the run-up phase and the phenomena that occurred during the August–November 2014 flank eruption at Stromboli volcano in Italy. In particular, the authors relied on TIR and VIS cameras from the Camera Monitoring Network of INGV-OE; ground displacement recorded by the permanent-sited Ku-band Ground-Based InSAR (GBInSAR) device; seismic signals (band 0.02–10 Hz) from INGV-OV network, including amplitude of volcanic tremor, amplitude of explosion quakes, inclination of the seismic polarization in the Very-Long-Period (VLP) band (0.05–0.5 Hz), and a neural network-based analysis of seismic signals to detect signals related to landslides occurring along the Sciara del Fuoco slope; and finally high-resolution Digital Elevation Models (DEMs) reconstructed based on pre-eruptive 2012 LiDAR data and post-eruptive 2017 tri-stereo Pléiades-1 imagery, and related topographic change detection. With such a wealth of data, the authors found that the explosive activity peaked between 5 and 6 August 2014, whereas the GBInSAR device recorded a drastic increase in the displacement rate since the morning of 6 August, which was consistent with a strong inflation of the crater terrace. Ground displacement started to show evidence of sliding in the crater terrace after the 6 August 2014 evening, and this was also corroborated by seismic signals. The breaching of the summit cone with emplacement of a landslide along the Sciara del Fuoco was anticipated by the GBInSAR measurements, as observed by the live camera and recorded by the seismic data. Based on topographic change detection, a total volume of $3.07 \pm 0.37 \times 10^6$ m^3 of lava flow field emplaced on the steep Sciara del Fuoco slope was estimated. This volume was below the limit of $6.5 \pm 1 \times 10^6$ m^3 expected for triggering a paroxysmal explosion.

3. Statistics, Altmetrics, and Impact

3.1. Editorial and Peer-Review Process

The four Guest Editors handled a total of 25 manuscript submissions over the 10 months when the Call for Papers was disseminated and the system was open for submissions, namely from 22 June 2018 to 30 April 2019 [5]. One more manuscript was handled by another Editorial Board Member of *Remote Sensing* and later added to the Special Issue given its very good fit with the thematic goals of the Special Issue.

In total, more than 100 authors contributed to the submitted manuscripts, and a few of them co-authored more than one submission.

A team of 48 anonymous experts in the field of volcano remote sensing helped the Guest Editors to ensure a rigorous peer-reviewing process during the course of the 15 month-long Special Issue project (i.e., June 2018–September 2019 [29,30]), for both the 19 manuscripts that were finally published and those that were not. At least 3 reviewers provided feedback on each manuscript on average, and some of them were called upon to assess more than one manuscript in their specialist field of expertise. These numbers provide a quantitative metric of the enormous effort behind this Special Issue and the active engagement of the scientific community who voluntarily contributed to review the research papers.

The average time from submission to acceptance was 46 days, while the average time from acceptance to online publication was 8 days. The first paper was published on 30 August 2018 [7], while the last was published on 16 September 2019 [25].

3.2. Altmetrics and Impact

To gather an understanding of the impact of the 19 published papers as of mid July 2020, i.e., 10 months after the publication of the last paper of the Special Issue, MDPI's article metrics powered by TrendMD were exploited. TrendMD uses technologies such as Google Analytics by Google Inc. to track visitors' use of and interaction with webpages, and it therefore allows the monitoring of views and downloads of each paper.

The analysis of metrics for the 19 papers showed that since the publication of the first article in August 2018, the Special Issue received more than 25,500 views in total and was reached by 37 readers per day on average over the 22 month-long time span between August 2018 and July 2020. These numbers provide a sense of the visibility that this Special Issue has gathered across the journal readership. Detailed metrics for each research paper are shown in Figure 2.

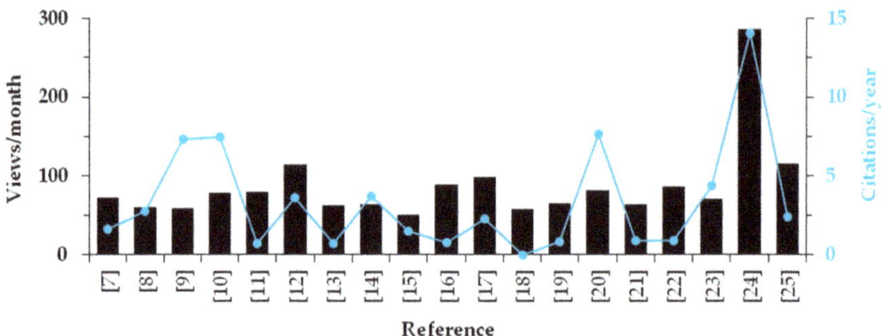

Figure 2. Views and citations attracted by the 19 research papers composing the Special Issue "Remote Sensing of Volcanic Processes and Risk" [5] of *Remote Sensing*.

While it is reasonable to imagine that the majority of the readers are researchers and scientists and not necessarily staff from organizations with statutory responsibilities that include monitoring volcanic hazards (the so-called "volcano observatories" according to [2]), it is also true that the open access policy with which this Special Issue is published at least removes one of the common barriers to the accessibility of scientific papers. Therefore, it is hoped that this would facilitate stakeholders to come across these publications and take some benefit from the knowledge about state-of-the-art technologies and up-to-date scientific insights into some of the most studied volcanoes in the world.

The immediate impact of the research published in the Special Issue, at least across the scientific community, can be inferred from the overall 84 citations in the indexed literature received as of mid July 2020, in the first few months after publication. Many of the citations of the 19 papers were made by articles published in *Remote Sensing*, while others were made by articles in different scientific journals in the fields of natural hazards, applied Earth sciences, remote sensing, Earth observation, environmental and Earth sciences. While generally most papers received 1 to 5 citations, four apparent positive outliers are the research articles by Laiolo et al. [20], Marchese et al. [9], Di Traglia et al. [10], and Valade et al. [24], with 9, 12, 12, and 15 citations received so far, respectively.

In particular, metrics for the article by Valade et al. [24] show a boosted performance in terms of total views, with more than 3600 reached since its publication in June 2019 and as of mid July 2020, i.e., nearly 300 views/month (Figure 2). The article was also mentioned among the 'highly cited papers' of *Remote Sensing* at the beginning of March 2020. Moreover, this research has attracted attention in news media and, among others, it has been featured in a dedicated article by National Geographic [31].

4. Conclusions and an Outlook to the Future

The present Special Issue provides a collection of papers, the scientific quality and reliability of which has been assessed by a multidisciplinary network of expert and authoritative scientists in different fields including, but not limited to, volcanology, risk assessment, geophysics, and remote sensing applied to volcanic hazard and risk.

While the variety of the methods and case studies discussed in the papers cannot be exhaustive and representative of the whole spectrum of scientific research on this topic, this Special Issue definitely provides an assortment of the most recent achievements in monitoring techniques and scientific knowledge of volcanoes that, for different reasons, are not only scientifically interesting to study, but more importantly are of potential concern for the safety of the local communities that could be impacted.

Volcanic hazard and risk has been for long a topical theme for MDPI journals. Therefore, this Special Issue should be considered as a contribution (from the remote sensing point of view) to a wider editorial series. In this regard, a number of opportunities for scholars interested in volcano remote sensing are currently available, and they can be considered both to access further articles and to contribute to the scientific literature on this specialist topic. These include the following Special Issues of MDPI journals *Remote Sensing*, *Sensors*, and *Applied Sciences* that are currently open for submissions:

- "Data Processing and Modeling on Volcanic and Seismic Areas" in *Applied Sciences* [32]
- "Applications of Remote Sensing in Earthquakes, Volcanic and Tsunami Events" in *Remote Sensing* [33]
- "Quantitative Volcanic Hazard Assessment and Uncertainty Analysis in Satellite Remote Sensing and Modeling" in *Remote Sensing* [34]
- "Volcano Monitoring: From the Magma Reservoir to Eruptive Processes" in *Applied Sciences* [35]
- "Ground-Based Imaging of Active Volcanic Phenomena" in *Remote Sensing* [36]
- "Satellite Remote Sensing for Volcanic Applications" in *Sensors* [37]
- "Remote Sensing for Volcano Systems Monitoring" in *Remote Sensing* [38]
- "Volcanic Processes Monitoring and Hazard Assessment Using Integration of Remote Sensing and Ground-Based Techniques" in *Remote Sensing* [39]
- "Volcanic Impacts on the Environment and Health Hazards" in *Remote Sensing* [40]

The articles already published and soon to be published in the above thematic volumes will definitely contribute, together with the 19 papers published in the present Special Issue, to the exceptionally lively and stimulating discussion on the use of EO and remote sensing data and technology to monitor volcanic processes and risks, and to the consolidation of a topical theme that is increasingly being investigated across MDPI publications at the level that it has become a cross-journal topic.

Author Contributions: Conceptualization, F.C., D.T. and Z.L.; formal analysis, F.C. and D.T.; data curation, F.C. and D.T.; visualization, F.C., D.T. and Z.L.; writing—original draft preparation, F.C. and D.T.; writing—review and editing, Z.L. All authors have read and agreed to the published version of the manuscript.

Funding: This research received no external funding.

Acknowledgments: This Special Issue was developed by Francesca Cigna, Deodato Tapete, Zhong Lu, and Susanna K. Ebmeier following on from the invitation sent to Francesca Cigna by Richard Gloaguen, *Remote Sensing* "Remote Sensing in Geology, Geomorphology and Hydrology" Section Editor-in-Chief. The Guest Editors would like to acknowledge the over 100 authors who contributed to this Special Issue with their papers and express their sincere gratitude to the 48 anonymous reviewers for the dedication, time, and expertise to provide their feedback on the submitted manuscripts and help the authors enhance the scientific quality of their papers. Susanna K. Ebmeier is sincerely acknowledged for her work and commitment in handling manuscripts submitted to the Special Issue. The whole *Remote Sensing* Editorial Office and the team of Assistance Editors, and in particular Sofia Zhao and Nelson Peng at MDPI Branch Office in Beijing, are greatly acknowledged for the help and assistance to setup, kick-off, manage, and publish this Special Issue.

Conflicts of Interest: The authors declare no conflict of interest.

References

1. Loughlin, S.C.; Sparks, S.; Brown, S.K.; Jenkins, S.F.; Vye-Brown, C. *Global Volcanic Hazards and Risk*; Cambridge University Press: Cambridge, UK, 2015; pp. 1–80.
2. Ebmeier, S.K.; Andrews, B.J.; Araya, M.C.; Arnold, D.W.D.; Biggs, J.; Cooper, C.; Cottrell, E.; Furtney, M.; Hickey, J.; Jay, J.; et al. Synthesis of global satellite observations of magmatic and volcanic deformation: Implications for volcano monitoring & the lateral extent of magmatic domains. *J. Appl. Volcanol.* **2018**, *7*, 1–26. [CrossRef]
3. Pyle, D.M.; Mather, T.A.; Biggs, J. Remote sensing of volcanoes and volcanic processes: Integrating observation and modelling-introduction. *Geol. Soc. Spec. Publ.* **2013**, *380*, 1–13. [CrossRef]
4. Francis, P.W. Remote sensing of volcanoes. *Adv. Space Res.* **1989**, *9*, 89–92. [CrossRef]
5. Cigna, F.; Tapete, D.; Lu, Z.; Ebmeier, S.K. MDPI Remote Sensing: Special Issue "Remote Sensing of Volcanic Processes and Risk". Available online: https://www.mdpi.com/journal/remotesensing/special_issues/rs_vpr (accessed on 24 July 2020).
6. Lu, Z.; Webley, P. MDPI Remote Sensing: Special Issue "Volcano Remote Sensing". Available online: https://www.mdpi.com/journal/remotesensing/special_issues/volcano (accessed on 24 July 2020).

7. Plank, S.; Nolde, M.; Richter, R.; Fischer, C.; Martinis, S.; Riedlinger, T.; Schoepfer, E.; Klein, D. Monitoring of the 2015 Villarrica Volcano Eruption by Means of DLR's Experimental TET-1 Satellite. *Remote Sens.* **2018**, *10*, 1379. [CrossRef]
8. Bredemeyer, S.; Ulmer, F.G.; Hansteen, T.H.; Walter, T.R. Radar path delay effects in volcanic gas plumes: The case of Láscar Volcano, Northern Chile. *Remote Sens.* **2018**, *10*, 1514. [CrossRef]
9. Marchese, F.; Neri, M.; Falconieri, A.; Lacava, T.; Mazzeo, G.; Pergola, N.; Tramutoli, V. The contribution of multi-sensor infrared satellite observations to monitor Mt. Etna (Italy) Activity during May to August 2016. *Remote Sens.* **2018**, *10*, 1948. [CrossRef]
10. Di Traglia, F.; Calvari, S.; D'Auria, L.; Nolesini, T.; Bonaccorso, A.; Fornaciai, A.; Esposito, A.; Cristaldi, A.; Favalli, M.; Casagli, N. The 2014 effusive eruption at stromboli: New insights from in situ and remote-sensing measurements. *Remote Sens.* **2018**, *10*, 2035. [CrossRef]
11. Papageorgiou, E.; Foumelis, M.; Trasatti, E.; Ventura, G.; Raucoules, D.; Mouratidis, A. Multi-sensor SAR geodetic imaging and modelling of santorini volcano post-unrest response. *Remote Sens.* **2019**, *11*, 259. [CrossRef]
12. Aufaristama, M.; Hoskuldsson, A.; Ulfarsson, M.O.; Jonsdottir, I.; Thordarson, T. The 2014-2015 lava flow field at Holuhraun, Iceland: Using airborne hyperspectral remote sensing for discriminating the lava surface. *Remote Sens.* **2019**, *11*, 476. [CrossRef]
13. Sansivero, F.; Vilardo, G. Processing thermal infrared imagery time-series from volcano permanent ground-based monitoring network. Latest methodological improvements to characterize surface temperatures behavior of thermal anomaly areas. *Remote Sens.* **2019**, *11*, 553. [CrossRef]
14. Rogic, N.; Cappello, A.; Ferrucci, F. Role of emissivity in lava flow "Distance-to-Run" estimates from satellite-based volcano monitoring. *Remote Sens.* **2019**, *11*, 662. [CrossRef]
15. Lombardo, V.; Corradini, S.; Musacchio, M.; Silvestri, M.; Taddeucci, J. Eruptive Styles Recognition Using High Temporal Resolution Geostationary Infrared Satellite Data. *Remote Sens.* **2019**, *11*, 669. [CrossRef]
16. Gomez-Ortiz, D.; Blanco-Montenegro, I.; Arnoso, J.; Martin-Crespo, T.; Solla, M.; Montesinos, F.G.; Vélez, E.; Sánchez, N. Imaging thermal anomalies in hot dry rock geothermal systems from near-surface geophysical modelling. *Remote Sens.* **2019**, *11*, 675. [CrossRef]
17. Cando-Jácome, M.; Martínez-Graña, A. Determination of primary and secondary lahar flow paths of the Fuego Volcano (Guatemala) using morphometric parameters. *Remote Sens.* **2019**, *11*, 727. [CrossRef]
18. De Michele, M.; Raucoules, D.; Corradini, S.; Merucci, L.; Salerno, G.; Sellitto, P.; Carboni, E. Volcanic cloud top height estimation using the plume elevation model procedure applied to orthorectified Landsat 8 data. test case: 26 October 2013 Mt. Etna eruption. *Remote Sens.* **2019**, *11*, 785. [CrossRef]
19. Dávila, N.; Capra, L.; Ferrés, D.; Gavilanes-Ruiz, J.C.; Flores, P. Chronology of the 2014–2016 eruptive phase of Volcán De Colima and volume estimation of associated lava flows and pyroclastic flows based on optical multi-sensors. *Remote Sens.* **2019**, *11*, 1167. [CrossRef]
20. Laiolo, M.; Ripepe, M.; Cigolini, C.; Coppola, D.; Della Schiava, M.; Genco, R.; Innocenti, L.; Lacanna, G.; Marchetti, E.; Massimetti, F.; et al. Space-and ground-based geophysical data tracking of magma migration in shallow feeding system of mount etna volcano. *Remote Sens.* **2019**, *11*, 1182. [CrossRef]
21. Delle Donne, D.; Aiuppa, A.; Bitetto, M.; D'Aleo, R.; Coltelli, M.; Coppola, D.; Pecora, E.; Ripepe, M.; Tamburello, G. Changes in SO2 Flux Regime at Mt. Etna Captured by Automatically Processed Ultraviolet Camera Data. *Remote Sens.* **2019**, *11*, 1201. [CrossRef]
22. Mania, R.; Walter, T.R.; Belousova, M.; Belousov, A.; Senyukov, S.L. Deformations and morphology changes associated with the 2016-2017 eruption sequence at Bezymianny volcano, Kamchatka. *Remote Sens.* **2019**, *11*, 1278. [CrossRef]
23. De Angelis, S.; Diaz-Moreno, A.; Zuccarello, L. Recent developments and applications of acoustic infrasound to monitor volcanic emissions. *Remote Sens.* **2019**, *11*, 1302. [CrossRef]
24. Valade, S.; Ley, A.; Massimetti, F.; D'Hondt, O.; Laiolo, M.; Coppola, D.; Loibl, D.; Hellwich, O.; Walter, T.R. Towards global volcano monitoring using multisensor sentinel missions and artificial intelligence: The MOUNTS monitoring system. *Remote Sens.* **2019**, *11*, 1528. [CrossRef]
25. Aldeghi, A.; Carn, S.; Escobar-Wolf, R.; Groppelli, G. Volcano monitoring from space using high-cadence planet CubeSat images applied to Fuego volcano, Guatemala. *Remote Sens.* **2019**, *11*, 2151. [CrossRef]
26. De Michele, M.; Raucoules, D.; Arason, T. Volcanic Plume Elevation Model and its velocity derived from Landsat 8. *Remote Sens. Environ.* **2016**, *176*, 219–224. [CrossRef]

27. Duda, K.A.; Ramsey, M.; Wessels, R.; Deh, J. Optical Satellite Volcano Monitoring: A Multi-Sensor Rapid Response System. In *Geoscience and Remote Sensing*; InTech: London, UK, 2009; pp. 473–496.
28. Bignami, C.; Chini, M.; Amici, S.; Trasatti, E. Synergic Use of Multi-Sensor Satellite Data for Volcanic Hazards Monitoring: The Fogo (Cape Verde) 2014–2015 Effusive Eruption. *Front. Earth Sci.* **2020**, *8*, 22. [CrossRef]
29. Remote Sensing Editorial Office. Acknowledgement to Reviewers of Remote Sensing in 2018. *Remote Sens.* **2019**, *11*, 127. [CrossRef]
30. Remote Sensing Editorial Office. Acknowledgement to Reviewers of Remote Sensing in 2019. *Remote Sens.* **2020**, *12*, 327. [CrossRef]
31. Snow, J. Volcano forecasts could soon be a reality as AI reads satellite photos. *Natl. Geogr. Mag.* **2019**, *11*, 1528.
32. Bonforte, A.; Cannavò, F. MDPI Applied Sciences: Special Issue "Data Processing and Modeling on Volcanic and Seismic Areas". Available online: https://www.mdpi.com/journal/applsci/special_issues/volcanic_seismic (accessed on 24 July 2020).
33. Adriano, B.; Gokon, H.; Liu, W.; Wienland, M.; Koch, M. MDPI Remote Sensing: Special Issue "Applications of Remote Sensing in Earthquakes, Volcanic and Tsunami Events". Available online: https://www.mdpi.com/journal/remotesensing/special_issues/earthquakes_volcanic_tsunami_events (accessed on 24 July 2020).
34. Del Negro, C.; Ramsey, M.; Hérault, A.; Ganci, G. MDPI Remote Sensing: Special Issue "Quantitative Volcanic Hazard Assessment and Uncertainty Analysis in Satellite Remote Sensing and Modeling". Available online: https://www.mdpi.com/journal/remotesensing/special_issues/volcano_rs (accessed on 24 July 2020).
35. Beauducel, F.; Rizzo, A.L. MDPI Applied Sciences: Special Issue "Volcano Monitoring: From the Magma Reservoir to Eruptive Processes". Available online: https://www.mdpi.com/journal/applsci/special_issues/Volcano_Monitoring (accessed on 24 July 2020).
36. Bani, P.; Tamburello, G.; Pering, T. MDPI Remote Sensing: Special Issue "Ground Based Imaging of Active Volcanic Phenomena". Available online: https://www.mdpi.com/journal/remotesensing/special_issues/GBI_VP (accessed on 24 July 2020).
37. Pergola, N.; Plank, S.; Marchese, F.; Ramsey, M. MDPI Sensors: Special Issue "Satellite Remote Sensing for Volcanic Applications". Available online: https://www.mdpi.com/journal/sensors/special_issues/RS_VA (accessed on 24 July 2020).
38. Tizzani, P.; Solaro, G.; Castaldo, R. MDPI Remote Sensing: Special Issue "Remote Sensing for Volcano Systems Monitoring". Available online: https://www.mdpi.com/journal/remotesensing/special_issues/monitoring_volcano_systems (accessed on 24 July 2020).
39. Calvari, S.; Bonaccorso, A.; Cappello, A.; Giudicepietro, F.; Sansosti, E. MDPI Remote Sensing: Special Issue "Volcanic Processes Monitoring and Hazard Assessment Using Integration of Remote Sensing and Ground-Based Techniques". Available online: https://www.mdpi.com/journal/remotesensing/special_issues/Volcanic_Monitoring_Hazard_Assessment (accessed on 24 July 2020).
40. Laiolo, M. MDPI Remote Sensing: Special Issue "Volcanic Impacts on the Environment and Health Hazards". Available online: https://www.mdpi.com/journal/remotesensing/special_issues/volcanic_hazards (accessed on 24 July 2020).

Article

Monitoring of the 2015 Villarrica Volcano Eruption by Means of DLR's Experimental TET-1 Satellite

Simon Plank [1,*], Michael Nolde [1], Rudolf Richter [2], Christian Fischer [3], Sandro Martinis [1], Torsten Riedlinger [1], Elisabeth Schoepfer [1] and Doris Klein [1]

1 German Remote Sensing Data Center, German Aerospace Center (DLR), 82234 Oberpfaffenhofen, Germany; Michael.Nolde@dlr.de (M.N.); Sandro.Martinis@dlr.de (S.M.); Torsten.Riedlinger@dlr.de (T.R.); Elisabeth.Schoepfer@dlr.de (E.S.); Doris.Klein@dlr.de (D.K.)
2 Remote Sensing Technology Institute, German Aerospace Center DLR, 82234 Oberpfaffenhofen, Germany; Rudolf.Richter@dlr.de
3 Institute of Optical Sensor Systems, German Aerospace Center DLR, 12489 Berlin, Germany; C.Fischer@dlr.de
* Correspondence: simon.plank@dlr.de; Tel.: +49-8153-28-3460

Received: 17 July 2018; Accepted: 28 August 2018; Published: 30 August 2018

Abstract: Villarrica Volcano is one of the most active volcanoes in the South Andes Volcanic Zone. This article presents the results of a monitoring of the time before and after the 3 March 2015 eruption by analyzing nine satellite images acquired by the Technology Experiment Carrier-1 (TET-1), a small experimental German Aerospace Center (DLR) satellite. An atmospheric correction of the TET-1 data is presented, based on the Advanced Spaceborne Thermal Emission and Reflection Radiometer (ASTER) Global Emissivity Database (GDEM) and Moderate Resolution Imaging Spectroradiometer (MODIS) water vapor data with the shortest temporal baseline to the TET-1 acquisitions. Next, the temperature, area coverage, and radiant power of the detected thermal hotspots were derived at subpixel level and compared with observations derived from MODIS and Visible Infrared Imaging Radiometer Suite (VIIRS) data. Thermal anomalies were detected nine days before the eruption. After the decrease of the radiant power following the 3 March 2015 eruption, a stronger increase of the radiant power was observed on 25 April 2015. In addition, we show that the eruption-related ash coverage of the glacier at Villarrica Volcano could clearly be detected in TET-1 imagery. Landsat-8 imagery was analyzed for comparison. The information extracted from the TET-1 thermal data is thought be used in future to support and complement ground-based observations of active volcanoes.

Keywords: volcanic thermal anomalies; change detection; Villarrica Volcano; small satellites; FireBIRD; TET-1

---

1. Introduction

1.1. Villarrica Volcano

Villarrica Volcano (39°25′12″S, 71°55′48″W) is one of the most active volcanoes of the South Andes Volcanic Zone. This volcano belongs to the currently eight volcanoes on Earth with confirmed active lava lakes [1]. The other seven are Kilauea (Halema'uma'u crater, Hawaii [2]), Ambrym (Vanuatu [3]), Masaya (Nicaragua [4]), Nyiragongo (Democratic Republic of Congo [5]), Erta 'Ale (Ethiopia [6]), Erebus (Antarctica [7]), and recently observed since 2014 Nyamuragira (Democratic Republic of Congo [8,9]).

Villarrica Volcano is a basaltic-andesitic stratovolcano with a height of 2847 m above sea level (a.s.l.). Its summit is covered by approximately 30 km^2 of large glaciers, according to measurements taken in 2007 [10]. Based on infrasound measurements performed by Ripepe et al. [11] and gas composition investigations performed by Shinohara and Witter [12] both aforementioned groups suggest that at Villarrica Volcano degassing occurs very close to equilibrium with the magma and not by bursting of small gas bubbles at the surface of the magma column as occurs at other open-system volcanoes e.g., Stromboli. Over 50 historical eruptions are reported at Villarrica Volcano since the 16th century. These historic eruptions have ranged from effusive lava-producing eruptions to explosive eruptions up to the Volcanic Eruption Index (VEI) '3' [13]. According to Palma et al. [14], spattering and associated Strombolian eruptions as well as fire fountains have been observed at Villarrica Volcano.

Since the last eruptive episode in 1984, an actively degassing lava lake of width approximately 20 m to 30 m, located at depths of 50 m to over 150 m within the funnel shaped summit crater [15], has filled the crater [1,9,10]. Although the summit crater of Villarrica Volcano has a diameter of approximately 150 m, measurements based on inclinometer and laser range finder performed in 2000 and 2001 showed that the vent in the crater floor that leads to the surface of the lava lake has a diameter of only approximately 2 m to 30 m [9]. Measurements by Goto and Johnson [16], performed in January 2010, showed that the lava lake was located within a cylindrical cavity approximately 24 m below a 65 m diameter spatter roof overhang, which grew from the repeated agglutination and accumulation of ejected spatter. The diameter of the vent in the roof was approximately 10 m [16].

In this article we analyzed the active period before and after the 3 March 2015 eruption. During the eruption, which began at 03:08 on 3 March 2015 local Chile time (07:08 Coordinated Universal Time UTC) and lasted 55 min, a volume of approximately 4.7 ± 1.0 million m^3 erupted [17]. This is classified as a VEI '2' eruption, and produced intense tephra fallout, scoria flows, and a 20 km long lahar [18].

1.2. Satellite-Based Thermal Remote Sensing of Volcanoes

Satellite-based thermal remote sensing of volcanoes aims at (1) the early detection of volcanic activity to support decision makers and civil security authorities with respect to early warning activities [19,20] and (2) the monitoring of the spatiotemporal evolution of the volcanic eruptions to enhance our understanding of volcanic processes (see the reviews [21–24]).

To perform the first aim on a global scale, Earth observation satellite missions with a high temporal repetition rate and a large spatial coverage, such as the Moderate Resolution Imaging Spectroradiometer (MODIS) or Sentinel-3, are required. For instance, Wright et al. [25] described an automated MODIS data-based hotspots detection processor for near real-time thermal monitoring of volcanoes (MODVOLC). Kervyn et al. [26] proposed an updated version (MODLEN) that is also able to detect cooler lava, which was not possible with MODVOLC. For example, MODLEN enables the monitoring of the Tanzanian volcano Oldoinyo Lengai which erupts natro-carbonite lava at temperatures of ~585 °C. Coppola et al. [27,28] developed the Middle InfraRed Observation of Volcanic Activity (MIROVA) system, an enhancement of the aforementioned MODVOLC and MODLEN approaches.

Spampinato et al. [29] compared, at the Nyiragongo lava lake, ground measurements of FLIR infrared cameras and Spinning Enhanced Visible and Infrared Imager (SEVIRI) satellite data. Both observations showed similar values of the measured radiant power.

Blackett [30] showed the capabilities of thermal Landsat-8 imagery for analysis of volcanic activity, based on a case study of the Paluweh Volcano, Indonesia in April 2013.

The Technology Experiment Carrier-1 (TET-1), a small experimental German Aerospace Center (DLR) satellite, does not provide a continuous global coverage. However, due to its higher spatial resolution and higher sensitivity with regards to thermal anomalies, a more detailed analysis of volcanic activity is possible than with data from the aforementioned satellite missions. Section 2.1 provides more details on TET-1. Fischer et al. [31] demonstrated the high temperature event detection capabilities of TET-1 by showing the detection of the volcanic hotspot at the 22 February 2015 acquisition as one example. Zakšek et al. [32] investigated the August–November 2014 Stromboli Volcano eruption by combining TET-1 imagery and thermal ground-based observations.

In this study we analyzed a time series of nine TET-1 thermal images that were acquired before and after the 3 March 2015 Villarrica Volcano eruption (cf. Table 1). The temperature, area coverage, and radiant power of the detected thermal hotspots were derived at subpixel level and compared with observations derived from MODIS and Visible Infrared Imaging Radiometer Suite (VIIRS) data. In addition, to show a further application of TET-1 imagery in the field of volcano monitoring, the eruption-related ash coverage of the glacier at Villarrica Volcano was investigated by means of TET-1 imagery and the results were compared with higher resolution multispectral data from Landsat-8.

Table 1. Satellite imagery analyzed; MS = Multispectral.

Acquisition Date & Time	Local Time, Chile	Satellite	Sensor Type	MODIS Water Vapor Acquisition Date & Time
22 February 2015, 14:29 UTC	10:29	Landsat-8	MS, thermal	-
22 February 2015, 17:23 UTC	13:23	TET-1	Thermal	22 February 2015, 18:10 UTC
9 March 2015, 04:32 UTC	00:32	TET-1	Thermal	9 March 2015, 03:45 UTC
9 March 2015, 17:31 UTC	13:31	TET-1	Thermal	9 March 2015, 19:05 UTC
10 March 2015, 14:29 UTC	10:29	Landsat-8	MS, thermal	-
12 March 2015, 04:35 UTC	00:35	TET-1	Thermal	12 March 2015, 04:15 UTC
12 March 2015, 17:32 UTC	13:32	TET-1	Thermal	12 March 2015, 15:20 UTC
21 March 2015, 04:36 UTC [1]	00:36	TET-1	Thermal	21 March 2015, 04:10 UTC
26 March 2015, 14:29 UTC	10:29	Landsat-8	MS, thermal	-
27 March 2015, 17:33 UTC	14:33	TET-1	Thermal	27 March 2015, 18:50 UTC
25 April 2015, 04:35 UTC	00:35	TET-1	Thermal	25 April 2015, 03:05 UTC
25 April 2015, 17:49 UTC [1]	13:49	TET-1	Thermal	25 April 2015, 18:20 UTC

[1] TET-1 scenes used for calibration (cf. Section 2.2.2).

2. Materials and Methods

2.1. Satellite Data

A time series of nine TET-1 thermal images acquired before and during the March/April 2015 eruption of Villarrica Volcano were analyzed (Table 1). In addition, three Landsat-8 scenes were used for validation purposes. Moreover, the thermal hotspots derived by the VIIRS and by MODIS were used as reference for the comparison described in Section 3.2.

TET-1 is the first of two satellites of DLR's FireBIRD mission. TET-1 was launched in July 2012, followed by BIROS (Berlin InfraRed Optical System) in June 2016. Both satellites are flying sun-synchronously in a low-Earth orbit at approximately 500 km altitude. The repetition rate of one satellite was approximately five days, depending on its geographic location. Nowadays, with both satellites in orbit, the potential repetition rate is less than three days with $\pm 30^\circ$ across track acquisitions. The sensors operate two infrared cameras, one in the mid wave infrared (MWIR) and one in the long wave infrared (LWIR), as well as a three-channel camera in the visible (VIS: RED and GREEN) and near infrared (NIR). The MWIR, LWIR, and RED channel are installed in nadir position, while the GREEN and NIR channel are oriented off-nadir (Table 2).

Table 2. Characteristics of the TET-1 cameras.

Channel	Wavelength (μm)	Looking Angle (°)
GREEN	0.460–0.560	6° forward
RED	0.565–0.725	Nadir
NIR	0.790–0.930	6° backward
MWIR	3.400–4.200	Nadir
LWIR	8.500–9.300	Nadir

TET-1 operates a push broom sensor system with a ground sampling distance of 178 m for the thermal channels. This corresponds to a pixel resolution of 356 m due to staggering. The MWIR channel of TET-1 is more suitable for the detection of high temperature events than LWIR, since higher temperature events have a higher radiant power at the shorter wavelengths according to the Wien's displacement law.

Landsat-8 is the latest satellite of the Landsat series launched on 11 February 2013 by the National Aeronautics and Space Administration (NASA) and operated by the United States Geological Survey (USGS). Landsat-8 acquires multispectral (VIS, NIR, and short wave infrared (SWIR)) imagery with a 30 m spatial resolution, panchromatic imagery with 15 m resolution, and LWIR imagery with 100 m resolution.

2.2. Methods

Figure 1 shows the processing workflow which is described in the Sections 2.2.1–2.2.3 in more detail.

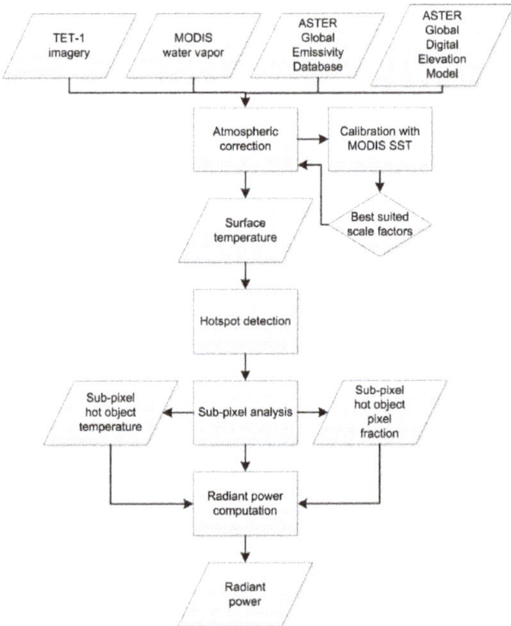

Figure 1. Workflow of the TET-1 image processing.

2.2.1. Atmospheric Correction of TET-1 Thermal Imagery

Radiometrically calibrated top-of-atmosphere (TOA) radiance data of TET-1 are used as input for the processing. The imagery is a stack of the coregistered nadir looking channels, i.e., the LWIR band and also the RED band in case of day time scenes, using the MWIR band as a master image.

To achieve surface radiances and temperatures in the MWIR and LWIR, the TET-1 data has to be atmospherically corrected. The atmospheric correction is based on look-up tables derived from the MODTRAN-5 radiative transfer code [33]. Moreover, the following additional external information is used for the atmospheric correction. (1) Atmospheric water vapor strongly influences the signal. Due to the available channels of TET-1 (Table 2), water vapor cannot be derived directly from TET-1 data. Therefore, an external source was used to get information about the water vapor content during the time of the TET-1 acquisition. The MODIS water vapor product of the MODIS acquisition with the shortest temporal baseline to TET-1 acquisition is used for this purpose (Table 1). (2) Moreover, the Advanced Spaceborne Thermal Emission and Reflection Radiometer (ASTER) Global Digital Elevation

Model (GDEM) is used during the atmospheric correction, as the water vapor also depends on the topographical elevation of the ground pixel.

(3) Another important factor in order to derive the TET-1 surface radiance is the emissivity of the surface. This data can be derived from the ASTER Global Emissivity Database (GED). Since ASTER has no MWIR channel, the emissivity can only be derived from data in the LWIR. For the applied processing, the emissivity of the ASTER bands 8.6 μm and 9.1 μm is used, because these bands match best with the LWIR of TET-1. Since a high spatial resolution global MWIR emissivity database is not available, the assumption of emissivity $\epsilon(MWIR) = 1$ is often made. However, Salisbury and D'Aria [34], as well as Giglio et al. [35], indicate that the approximation $\epsilon(MWIR) = \epsilon(LWIR)$ is more realistic than $\epsilon(MWIR) = 1$. Therefore, we employ this approximation, i.e., for the final processing of the surface radiance in LWIR and MWIR the same emissivity values are used. The atmospheric correction of the thermal TET-1 imagery is described in more detail in a past paper [36]. Next, according to the Planck's equation the surface radiance in LWIR and in MWIR is converted into the corresponding surface temperatures.

2.2.2. Calibration by means of MODIS Sea Surface Temperature

To control the radiometric quality of the TET-1 thermal imagery, the atmospherically corrected TET-1 surface temperature was tested against the MODIS sea surface temperature (SST); as the temperature over sea surfaces is more homogenous than over land surfaces. Two of the nine TET-1 scenes analyzed in this study cover cloud-free areas over the sea. For these two TET-1 scenes, one was derived during day time and one during night time (bold in Table 1); the optimal scale factor was derived by minimizing the difference between the TET-1 and the MODIS SST and consecutively applied to the radiometry. This was performed by an empirical test of different scale factors (cf. Section 3.1). These optimal scale factors for day and night time TET-1 scenes (cf. Section 3.1) were applied to all other day or night time TET-1 scenes, respectively (Table 1).

2.2.3. Hotspot Detection and Subpixel Analysis

Next, by using the background temperature T_b and the bi-spectral approach from Dozier [37] one can estimate (I1) the subpixel temperature T_{sub} of the hot areas and (2) the pixel fraction p which is covered by a hot object. Thus, despite the relatively coarse resolution of TET-1 (around 356 m), it is possible to derive the temperature of hot objects much smaller than a pixel (subpixel).

In addition, the radiant power Φ in watts [W] is derived as the difference of the power due to the subpixel temperature T_{sub} and the background temperature T_b using Wien's law (Equation (1)).

$$\Phi = \sigma\, \epsilon \left(\mathrm{T}_{\mathrm{sub}}^{4} - \mathrm{T}_{\mathrm{b}}^{4}\right) A_{sub}, \tag{1}$$

with σ being the Stefan–Boltzmann constant $\left[\frac{\mathrm{W}}{\mathrm{m}^2\mathrm{K}^4}\right]$, ϵ being the LWIR emissivity and A_{sub} (m^2) being the subpixel area of the hot object, which is obtained from the pixel size and the fractional subpixel area p.

We compared the results of the proposed method with the results of the Zhukov approach [38] for different case studies and found a good agreement of the results of both approaches, e.g., Halle et al. [39].

2.2.4. Detection of Surface Changes

Besides the detection and analysis of high temperature events, the imagery of the TET-1 mission is also suited for detection of larger changes on the Earth's surface due to volcanic eruptions. The March 2015 Villarrica eruption caused ash coverage of the glacier at the eastern flank. To detect the changes within the area covered by the glacier, the difference of the RED bands and the difference of the MWIR surface temperatures of the latest pre-event (22 February 2015, 17:23 UTC) and the next post-event, acquired at the same daytime (9 March 2015, 17:31 UTC), were computed.

3. Results

3.1. Optimal Scale Factors

The Tables 3 and 4 show the difference of the TET-1 SST and the MODIS SST for a day time and a night time TET-1 scene regarding different scale factors. The best suited scale factors, i.e., the ones with the lowest difference of TET-1 and MODIS SST, are marked in bold. These optimal scale factors were then applied to all the other day or night time TET-1 scenes, respectively. Prior to this study, we compared SST from TET-1 and MODIS for a series of other study sites.

Table 3. Difference of the TET-1 and MODIS SST for day time TET-1 scene 25 April 2015 17:49 UTC, using a MODIS SST dataset acquired 23 min later.

Scale Factor	MWIR Δ MODIS SST to TET-1 SST (K)	LWIR Δ MODIS SST to TET-1 SST (K)
1.00	+2.84	+2.33
1.05	+1.04	**−0.62**
1.10	**+0.02**	−3.48

Table 4. Difference of the TET-1 and MODIS SST for night time TET-1 scene 21 March 2015 04:36 UTC, using a MODIS SST dataset acquired 26 min earlier.

Scale Factor	MWIR Δ MODIS SST to TET-1 SST (K)	LWIR Δ MODIS SST to TET-1 SST (K)
1.00	+4.07	+4.98
1.05	+2.85	+2.17
1.10	+1.54	**−0.73**
1.15	**+0.28**	−3.52

Figure 2a shows the pre-eruption (22 February 2015) surface temperature derived from the MWIR channel after applying the atmospheric correction described in Section 2.2.1 in a 3D representation, using the optimal scale factors (cf. Tables 3 and 4). A TanDEM-X digital elevation model (DEM) is used as a source for the topographic information. At this daytime TET-1 scene, the cool glacier can very well be distinguished from the much warmer bare rock areas which are located at a lower elevation. The even lower elevated areas are covered by vegetation. For comparison, also see the false color Landsat-8 image acquired on the same date (Figure 2b).

Snow and ice are characterized by a very high reflectivity of solar irradiation. In contrast to this, bare rocks have a much lower reflectivity and therefore a much higher absorption of solar irradiation. The absorbed solar energy shifted to longer wavelengths and was emitted as thermal energy. The reflectivity of vegetation lies in between the reflectivity of snow/ice and bare rocks. Therefore, bare rocks show a higher temperature.

Figure 3 shows the derived surface temperature for all nine TET-1 acquisitions (cf. Table 1). The corresponding optimal scale factors of the Tables 3 and 4 were applied. One clearly sees a difference between the daytime and the night time acquisitions (cf. also Table 1). In contrast to the daytime image, the vegetation covered area shows slightly higher temperatures at night time than the bare rock areas, which cool down during the night. It is important to note that for all TET-1 MWIR surface temperature images the same minimum-maximum stretch was used, which makes the separation of bare rocks and vegetated areas in night time images more complicated, but enables a better comparison of the different acquisitions.

The change of the glacier coverage, visible when comparing Figure 3a,c is analyzed in Section 3.3.

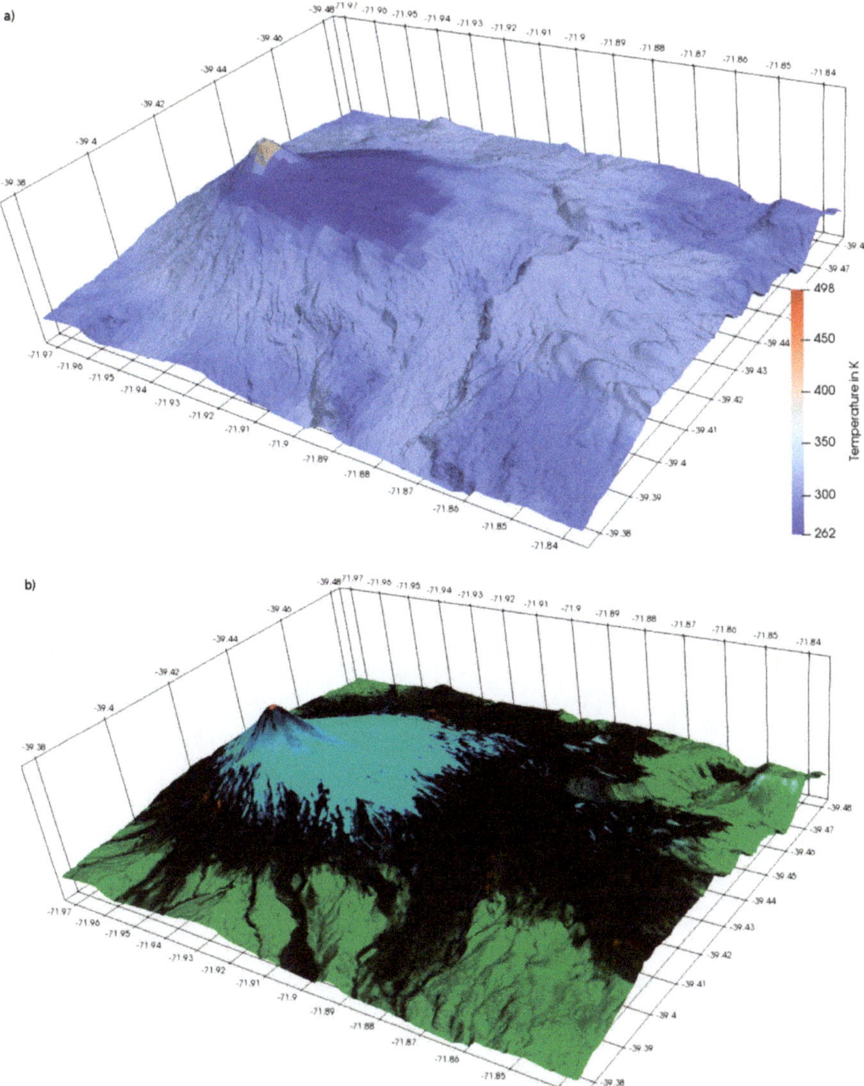

Figure 2. Villarrica Volcano observed on 22 February 2015: (**a**) The surface temperature derived from the atmospherically corrected TET-1 MWIR channel, overlaid on a TanDEM-X DEM. A minimum-maximum stretch over the observed values was applied. (**b**) Landsat-8 false color (SWIR2/NIR/GREEN) image, overlaid on a TanDEM-X DEM. The thermal hotspot at the Volcano summit is shown by red colors. Snow and ice covered areas appear in blue, bare rocks in black, and vegetated areas in green.

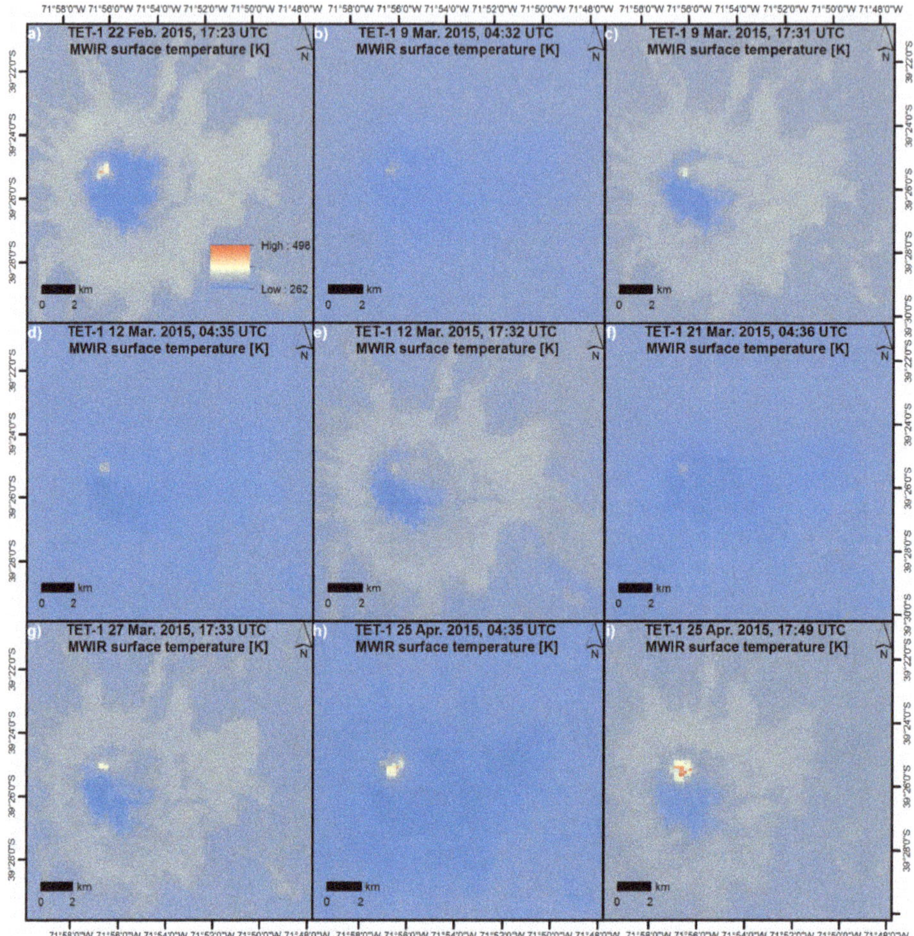

Figure 3. Surface temperature derived from the atmospherically corrected TET-1 MWIR channel. (**a**,**c**,**e**,**g**,**i**) daytime acquisitions. (**b**,**d**,**f**,**h**) Night time acquisitions. The same minimum-maximum stretching of the surface temperature was applied to all images. The legend of figure (**a**) is valid for all TET-1 acquisitions.

3.2. Hotspot Detection and Subpixel Analysis

The MWIR surface temperature images (Figure 3) show the highest temperatures at the summit crater, with a strong temporal variation. We can see a strong signal in the 22 February 2015 image which was acquired nine days before the 3 March 2015 Villarrica Volcano eruption. After this first eruption, the summit crater still shows higher temperatures than the remaining part of the area of interest during all nine acquisitions dates. The 27 March 2015 scene and the two acquisitions on 25 April 2015, especially the one on 17:49 UTC, show a temperature increase at the summit crater. Figure 3 shows the surface temperature averaged within the area of each pixel.

The subpixel analysis, described in Section 2.2.3, provides more details on the temperature of the hot parts, such as liquid lava at the summit crater lava lake. Figure 4 shows the temperatures at subpixel level for the detected hot objects. Three hotspots with temperatures up to 572 K were detected at the pre-eruption TET-1 acquisition (22 February 2015). The hotspot temperatures measured from 9 to 12 March 2015 vary from 486 K to 592 K. For the 21 March 2015 TET-1 acquisition, also showing the lowest MWIR surface temperatures (Figure 3f), no hotspots were detected. The acquisition of 27 March 2015 and the two acquisitions on 25 April 2015 show higher subpixel temperatures with up to 715 K, 828 K, and 956 K, respectively.

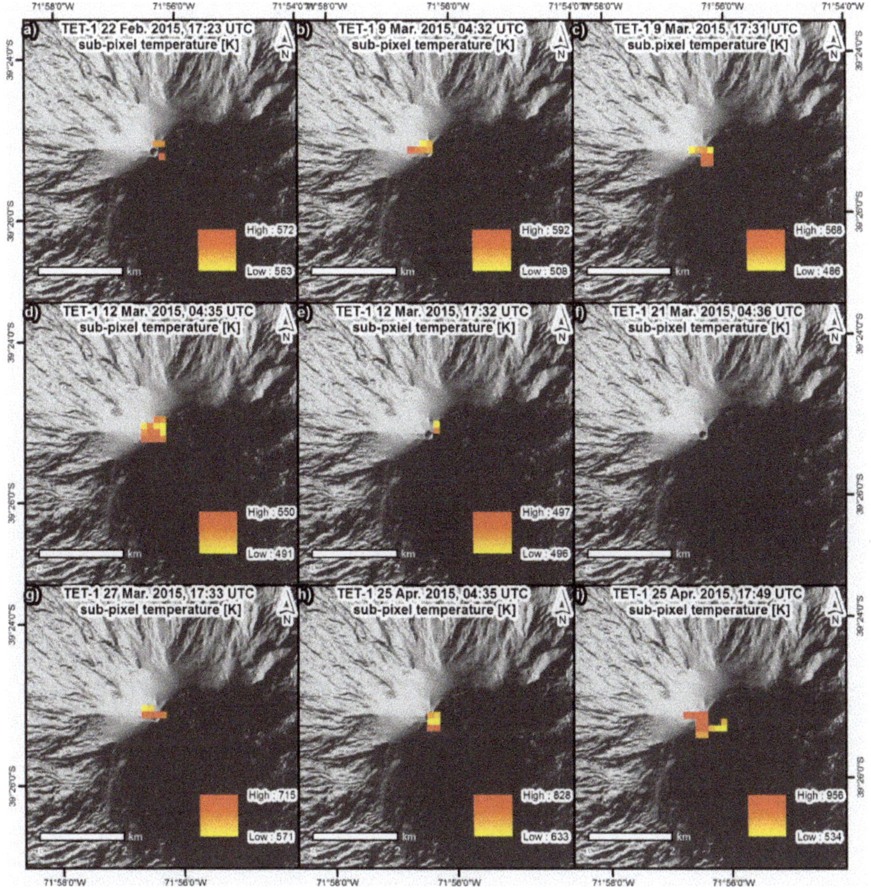

Figure 4. Subpixel temperatures of the hot areas at the summit crater derived from the nine TET-1 acquisitions (cf. Table 1). Background TanDEM-X DEM © DLR.

Figure 5 shows the corresponding subpixel area for the aforementioned hotspots, i.e., the fraction of a pixel which is covered by the hotspot. Finally, the radiant power, derived from the subpixel temperature and the pixel fraction (cf. Section 2.2.3), is shown in Figure 6. The highest radiant power of a single pixel, amounting to 97 MW, was detected at the 25 April 2015 17:49 UTC TET-1 acquisition. Table 5 summarizes the radiant power integrated over all hotspot pixels detected within each TET-1 acquisition, restricted to the area of Villarrica Volcano. The integrated radiant power for the 25 April 2015, 17:49 UTC TET-1 acquisition is higher by the order of one magnitude compared to the other acquisitions.

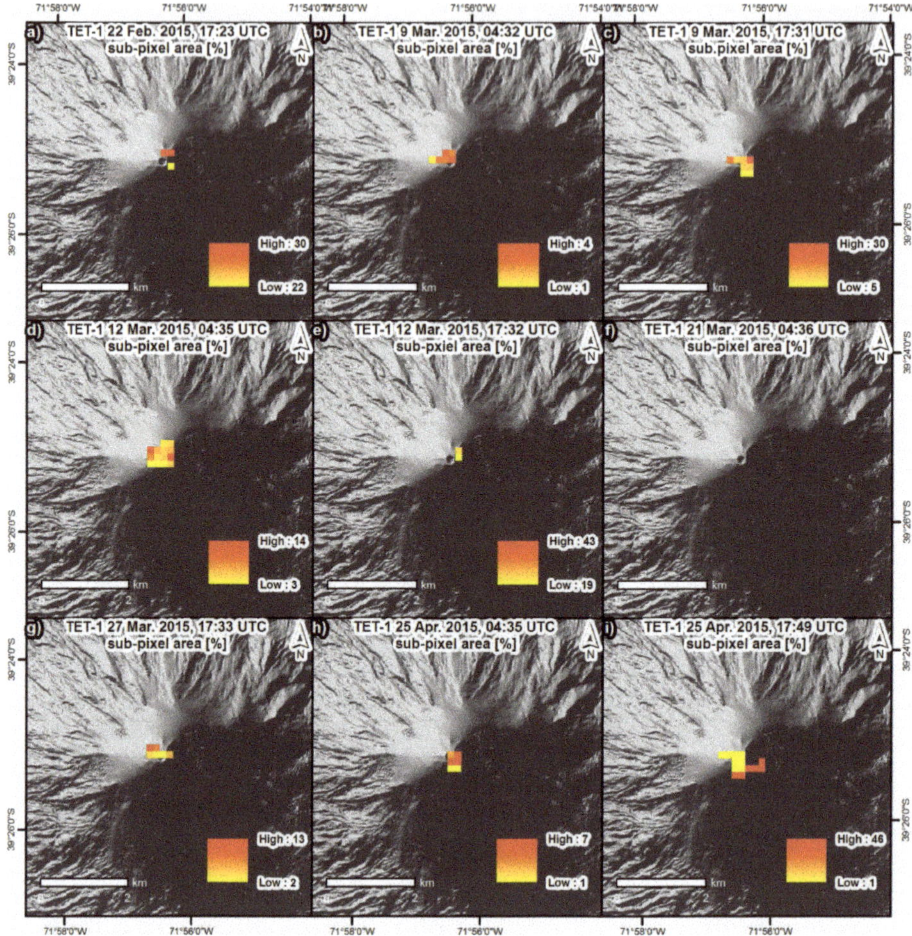

Figure 5. Subpixel area (percentage fraction of a pixel) of the hot areas at the summit crater derived from the nine TET-1 acquisitions (cf. Table 1). Background TanDEM-X DEM © DLR.

Table 5. Radiant power integrated over all detected hotspots. NO = No hotspots detected on that date by the corresponding sensors.

TET-1 Acquisition Date & Time [UTC]; Off-Nadir Angle	TET-1 Integrated Radiant Power [MW]	MODIS Acquisition: Time Difference to TET-1	MODIS Integrated Radiant Power [MW]	VIIRS Acquisition: Time Difference to TET-1	VIIRS Integrated Radiant Power [MW]
22 February 2015, 17:23; +19.4°	96.7	−11 h 12 min	24.3	−11 h 45 min	32.7
9 March 2015, 04:32; −21.9°	15.7	-	*NO*	+1 h 25 min	4.1
9 March 2015, 17:31; +6.0°	76.5	-	*NO*	-	*NO*
12 March 2015, 04:35; −20.5°	63.9	-	*NO*	+26 min	1.9
12 March 2015, 17:32; +5.0°	24.5	-	*NO*	-	*NO*
21 March 2015, 04:36; −28.5°	*NO*	-	*NO*	-	*NO*
27 March 2015, 17:33; +4.2°	56.7	−11 h 47 min	18.0	-	*NO*
25 April 2015, 04:35; +8.4°	62.4	-	*NO*	−4 min +1 h 40 min	1.9 110.5
25 April 2015, 17:49; −18.0°	599.0	+23 min	211.6	-	*NO*

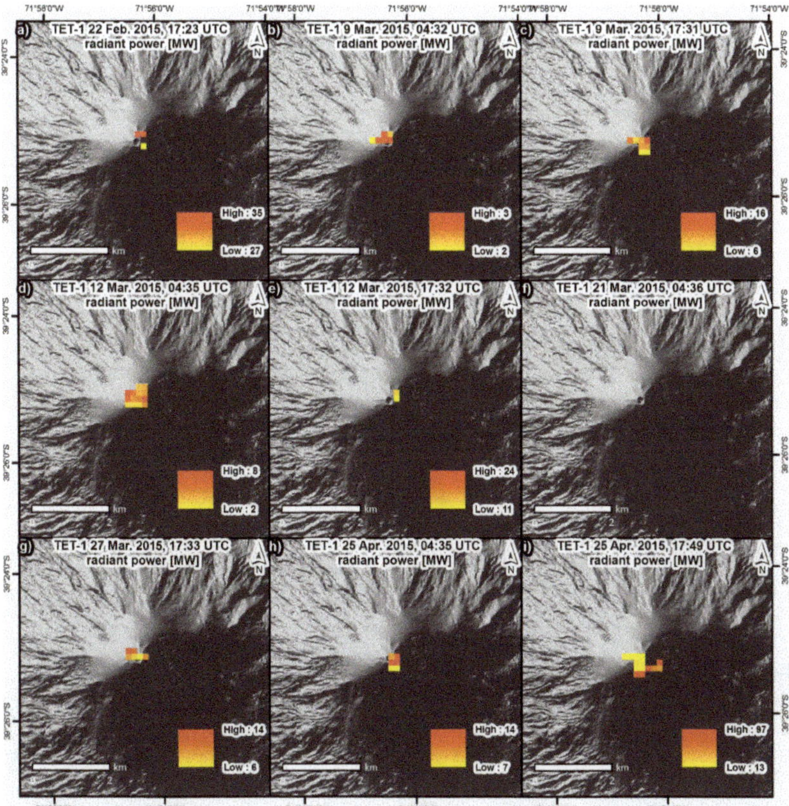

Figure 6. Radiant power of the hot areas at the summit crater derived from the nine TET-1 acquisitions (cf. Table 1). Background TanDEM-X DEM © DLR.

In addition, Table 5 and Figure 7 show the integrated radiant power of the MODIS [40] and VIIRS [41] acquisitions (derived from [42]) of the corresponding TET-1 acquisition dates. This table is discussed in detail in Section 4.

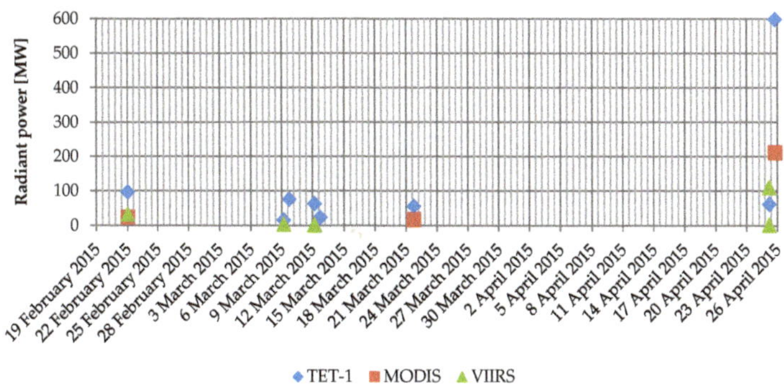

Figure 7. Temporal evolution of the radiant power integrated over all detected hotspots (cf. Table 5).

3.3. Detection of Surface Changes

Figure 8 shows the RED band and the MWIR surface temperature of the latest pre-event (22 February 2015, 17:23 UTC) and the next post-event TET-1 scene acquired at the same daytime (9 March 2015, 17:31 UTC). Snow and ice coverage is generally characterized by high reflectivity in the RED band and lower thermal emission in the MWIR band. In the post-event images (Figure 8, second row), one clearly sees the decrease in the snow and ice covered area at the eastern flank of Villarrica Volcano. Figure 8g,h highlight this area in red, which was derived from the difference of the pre- and post-event MWIR surface temperature by setting a empirically derived threshold of $\Delta \geq 30$ K. An additional criterion was the decrease of the reflectivity of the RED TET-1 channel.

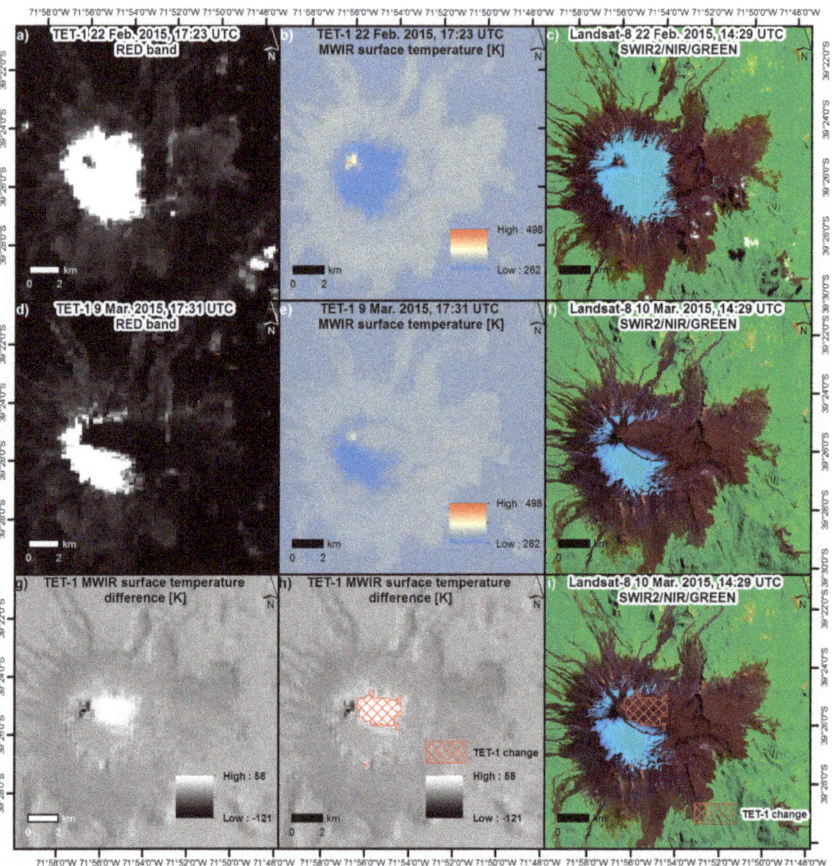

Figure 8. (**a**,**d**) RED band of the TET-1 pre-event (22 February 2015) and post-event (9 March 2015) acquisitions. (**b**,**e**) MWIR surface temperature of the TET-1 pre-event (22 February 2015) and post-event (9 March 2015) acquisitions. (**c**,**f**) False color composite (SWIR2/NIR/GREEN) of the Landsat-8 pre-event (22 February 2015) and post-event (10 March 2015) acquisitions. (**g**) Difference of the surface temperature of the pre- and post-event TET-1 MWIR bands. Detected change overlaid on the difference of the surface temperature of the pre- and post-event TET-1 MWIR bands (**h**) and on the post-event Landsat-8 acquisition (**i**).

The third column shows a false color composite of a pre-event (22 February 2015) and a post-event (10 March 2015) Landsat-8 acquisition as a reference. An overlay of the changed area derived from TET-1 MWIR data onto the Landsat-8 post-event scene acquired one day later shows a very good agreement between the detected change area and the area where the glacier was covered by ash due to the volcanic eruption.

4. Discussion

4.1. Calibration by Means of MODIS Sea Surface Temperature

To control the radiometric quality of the TET-1 data, a scale factor was applied as described in Section 2.2.2. By empirically testing different scale factors, the optimal scale factor was derived by minimizing the difference between the atmospherically corrected TET-1 and the MODIS SST. We assume stable temperatures during the short time differences of less than 26 min between the TET-1 and the corresponding MODIS acquisitions.

In the ideal case individual optimal scale factors should have been obtained from and applied to all single TET-1 acquisitions. However, for only two of the nine TET-1 acquisitions an area over the sea was free of clouds. Thus, optimal scale factors could be obtained only for these two TET-1 acquisitions. The optimal scale factor of the 25 April 2015, 17:49 UTC day time scene was applied to all other day time acquisitions. The optimal scale factor of the 21 March 2015, 04:36 UTC night time image was applied to all other night time scenes. Nevertheless, the Tables 3 and 4 show a very small derivation of the temperature differences between the two optimal scale factors for the day and night time scene. In addition, we see that even in the worst case of scale factor 1.00, meaning no 'correction' of the radiometry, the maximum difference between the TET-1 and the MODIS SST was 5K. Therefore, we assume that reasonable scale factors were applied to all nine TET-1 acquisitions analyzed in this study.

4.2. Comparison with MODIS and VIIRS Hotspots

Table 5 showed the integrated radiant power for each TET-1 acquisition and also the one of the MODIS and VIIRS scenes acquired on the same dates for comparison. There are different reasons for a "missed" hotspot by MODIS or VIIRS, while thermal activity could be detected by TET-1. First, the volcano might be covered by clouds or a volcanic ash plume during an overfly of MODIS or VIIRS. Second, TET-1 is more sensitive for detection of thermal activity than the two other sensors, especially MODIS. Third, a different volcanic thermal activity can be assumed at the different acquisition times of the three sensors.

The following discusses the single acquisitions in more detail. The radiant power of the 22 February 2015, 17:23 UTC and the 27 March 2015 TET-1 acquisitions were not comparable with the corresponding radiant power derived from MODIS or VIIRS data, since the TET-1 acquisitions were taken more than 11 h later. However, the MODIS and VIIRS hotspots confirm the activity of Villarrica Volcano at these two dates.

The radiant power derived from the 9 March 2015, 04:32 UTC TET-1 acquisition is in the same order of magnitude as the one derived from the corresponding VIIRS acquisition (05:57 UTC). In contrast to this, for the 12 March 2015, 04:35 UTC TET-1 scene there is a stronger difference with the corresponding VIIRS scene which was acquired 26 min later. However, both sensors, TET-1 and VIIRS, confirm the thermal activity of Villarrica Volcano for that date.

On 21 March 2015, no hotspot was detected by TET-1. This matches the observations by VIIRS and MODIS. The TET-1 25 April 2015, 04:35 UTC shows a higher radiant power than the one derived from the VIIRS scene acquired 4 min earlier (04:31). However, a second VIIRS acquisition taken 1.5 h later shows a strong increase of the radiant power compared to the first VIIRS acquisition of that date. The radiant power measured by TET-1 is in the middle between these two VIIRS acquisitions. Finally, the high thermal activity detected at the 25 April 2015, 17:49 UTC TET-1 acquisition was confirmed by the MODIS scene acquired 23 min later. However, the radiant power detected by TET-1 is more than twice as high as the one measured by MODIS.

Overall, these aspects show the major differences of the obtained results, due to the different resolution (and sensitivity), and thus underlines the potential of high resolution thermal infrared sensor systems for this kind of investigation.

4.3. Comparison with Independent Observations

Moussallam et al. [1] reported that the lava lake at the Villarrica summit crater briefly disappeared on 25 February 2015 and reappeared at the surface on 28 March 2015. The analysis of the TET-1 imagery showed lower radiant power values after the 3 March 2015 eruption than at the TET-1 scene 22 February 2015, three days before the disappearance of the crater lava lake at the surface (cf. Section 3.2). Higher radiant power was again observed on the 25 April 2015, 17:49 UTC acquisition. The subpixel temperatures already showed for the 27 March 2015 TET-1 scene, i.e., one day before the reappearance of the crater lava lake at the surface, and the two following TET-1 acquisitions on 25 April 2015, higher values than before.

4.4. Influence of the Off-nadir Angle and the Depth/Width of the Crater

As mentioned in Section 1.1, the lava lake at the funnel shaped summit crater of Villarrica Volcano is approximately 20 m to 30 m wide and located at depths of 50 m to over 150 m [15]. As the crater lava lake disappeared on 25 February 2015 (cf. Section 4.3) [1], these conditions are valid for the time before the 3 March 2015 eruption, i.e. for the 22 February 2015 TET-1 acquisition. The off-nadir angle of the center line of this TET-1 acquisition, where Villarrica Volcano is located, is 19.4°. The deeper the location of the crater lava lake, the smaller is the percentage of the crater lava lake which is visible for the TET-1 sensor. When assuming a crater lava lake width of 30 m, the maximum depth of the crater lava lake (after which the full crater lava lake is in the shadow and not directly visible anymore for the TET-1 sensor under the aforementioned off-nadir angle) is approximately 90 m (maximum 60 m depth, for a width of 20 m). Consequently, we can assume that a part of the crater lava lake was covered by shadow and not visible for the TET-1 sensor. Therefore, there is a high probability that the values of the radiant power presented in Section 3.2 were underestimated. Except for the aforementioned disappearance of the crater lava lake on 25 February 2015 and reappearance at the surface on 28 March 2015 [1], no further information about the depth of the crater lava lake was available for the time after the 3 March 2015 eruption. Nevertheless, we can also assume an influence of the depth of the crater lava lake on the measured radiant power at the other TET-1 acquisitions. Besides the depths of the crater lava lake, the type of volcanic activity (spanning from the lava lake to Strombolian) also influences the detectability from spaceborne sensors [43].

4.5. Detection of Surface Changes

Regarding the spatial resolution of the thermal channels and the repetition rate, the TET-1 satellite is in between the class of the low spatial resolution/high repetition rate sensors, such as MODIS and Sentinel-3, and the class of high spatial resolution/lower repetition rate sensors such as Landsat-8. Its high flexibility and the aforementioned characteristics make the TET-1 satellite well-suited for rapid detection of larger changes at the Earth's surface. Section 3.3 showed the application of TET-1 data for the detection of the ash coverage of the Villarrica glacier caused by the 3 March 2015 eruption. This flexibility allowed TET-1 to acquire the first image after the 3 March 2015 eruption earlier than the first available post-eruption Landsat-8 acquisition.

We found that the difference of the pre- and post-event MWIR surface temperature is much better suited for the detection of changes in the glacier area than the difference of the reflectivity of the pre- and post-event RED bands. The best results could be obtained by a combined analysis of both channels.

5. Conclusions

Villarrica Volcano is, with over 50 eruptions reported since the 16th century, one of the most active volcanoes of the South Andes Volcanic Zone. A time series of nine thermal images of the first satellite Technology Experiment Carrier-1 (TET-1) of the German Aerospace Center's (DLR) FireBIRD mission was analyzed to study the time before and after the 3 March 2015 eruption. We presented an atmospheric correction of the TET-1 data. The emissivity information was derived from the Advanced Spaceborne Thermal Emission and Reflection Radiometer (ASTER) Global Emissivity Database and the corresponding water vapor data from the Moderate Resolution Imaging Spectroradiometer (MODIS) acquisition with the shortest temporal baseline to the TET-1 acquisitions.

The detected thermal anomalies were investigated at subpixel level by deriving the subpixel temperature, the percentage area coverage of a pixel, and the radiant power. These observations were compared with hotspot information derived from MODIS and Visible Infrared Imaging Radiometer Suite (VIIRS) data. Analysis of TET-1 data showed thermal activity of Villarrica Volcano nine days before the 3 March 2015 eruption. The measured radiant power showed a decrease after this first eruption. An increase of the volcanic activity was again observed on 25 April 2015.

In addition to the analysis of the thermal hotspots at subpixel level, also the eruption-related ash coverage of the glacier at Villarrica Volcano was investigated by means of TET-1 imagery. The changes detected using TET-1 imagery matched well with the reference information derived from higher spatial resolution Landsat-8 imagery.

In summary, the information extracted from the thermal data of the flexible FireBIRD mission could be used in future to support and complement ground-based observations of active volcanoes.

Author Contributions: S.P. designed the experiments, performed the data analysis, and wrote the paper. R.R. supported the atmospheric correction of the TET-1 imagery. C.F. supported the TET-1 data analysis. M.N., S.M., T.R., E.S., and D.K. supported the study and provided suggestions for its improvement.

Funding: This research was funded, in part, by the German Federal Ministry of Education and Research (BMBF) under grant no. 03G0876 (project RIESGOS).

Acknowledgments: The TanDEM-X DEM data was kindly provided by DLR (Proposal ID: DEM_GEOL1424). The authors would like to thank the three anonymous reviewers for their very constructive remarks.

Conflicts of Interest: The authors declare no conflicts of interest.

References

1. Moussallam, Y.; Bani, P.; Curtis, A.; Banie, T.; Moussellam, M.; Peters, N.; Schipper, C.I.; Aiuppa, A.; Giudice, G.; Amigo, A.; et al. Sustaining Persistent Lava Lakes: Observations from High-Resolution Gas Measurements at Villarrica Volcano, Chile. *Earth Planet. Sci. Lett.* **2016**, *454*, 237–247. [CrossRef]
2. Patrick, M.R.; Orr, T.; Sutton, A.J.; Lev, E.; Thelen, W.; Fee, D. Shallowly driven fluctuations in lava lake outgassing (gas pistoning), Kilauea Volcano. *Earth Planet. Sci Lett.* **2016**, *433*, 326–338. [CrossRef]
3. Bani, P.; Oppenheimer, C.; Tsanev, V.; Carn, S.; Cronin, S.; Crimp, R.; Calkins, J.; Charley, D.; Lardy, M.; Roberts, T. Surge in Sulphur and halogen degassing from Ambrym volcano, Vanuatu. *Bull. Volcanol.* **2009**, *71*, 115–1168. [CrossRef]
4. Rymer, H.; de Vries, B.V.; Stix, J.; Williams-Jones, G. Pit crater structure and processes governing persistent activity at Masaya Volcano, Nicaragua. *Bull. Volcanol.* **1998**, *59*, 345–355. [CrossRef]
5. Harris, A.J.L.; Flynn, L.P.; Rothery, D.A.; Oppenheimer, C.; Sherman, S.B. Mass flux Measurements at Active Lava Lakes: Implications for Magma Recycling. *J. Geophys. Res. Solid Earth* **1999**, *104*, 7117–7136. [CrossRef]
6. Sawyer, G.M.; Oppenheimer, C.; Tsanev, V.; Yirgu, G. Magmatic degassing at Erta 'Ale volcano, Ethiopia. *J. Volcanol. Geotherm. Res.* **2008**, *178*, 837–846. [CrossRef]
7. Calkins, J.; Oppenheimer, C.; Kyle, P.R. Ground-based thermal imaging of lava lakes at Erebus volcano, Antarctica. *J. Volcanol. Geotherm. Res.* **2008**, *177*, 695–704. [CrossRef]
8. Campion, R. New laval lake at Nyamuragira volcano revealed by combined ASTER and OMI SO2 measurements. *Geophys. Res. Lett.* **2014**, *41*, 7485–7492. [CrossRef]

9. Coppola, D.; Campion, R.; Laiolo, M.; Cuoco, E.; Balagizi, C.; Ripepe, M.; Cigolini, C.; Tedesco, D. Birth of a lava lake: Nyamulagira volcano 2011–2015. *Bull. Volcanol.* **2016**, *78*. [CrossRef]
10. Rivera, A.; Bown, F. Recent glacier variations on active ice capped volcanoes in the Southern Volcanic Zone (37°–46°S), Chilean Andes. *J. South Am. Earth Sci.* **2013**, *45*, 345–356. [CrossRef]
11. Ripepe, M.; Marchetti, E.; Bonadonna, C.; Harris, A.J.L.; Pioli, L.; Ulivieri, G. Monochromatic infrasonic tremor driven by persistent degassing and convection at Villarrica Volcano, Chile. *Geophys. Res. Lett.* **2010**, *37*. [CrossRef]
12. Shinohara, H.; Witter, J.B. Volcanic gases emitted during mild Strombolian activity of Villarrica volcano, Chile. *Geophys. Res. Lett.* **2005**, *32*. [CrossRef]
13. Witter, J.B.; Kress, V.C.; Delmelle, P.; Stix, J. Volatile degassing, petrology, and magma dynamics of the Villarrica Lava Lake, Southern Chile. *J. Volcanol. Geotherm. Res.* **2004**, *134*, 303–337. [CrossRef]
14. Palma, J.L.; Calder, E.S.; Basualto, D.; Blake, S.; Rothery, D.A. Correlations between SO2 flux, seismicity, and outgassing activity at the open vent of Villarrica volcano, Chile. *J. Geophys. Res. Solid Earth* **2008**, *113*, B10. [CrossRef]
15. Aiuppa, A.; Bitetto, M.; Francofonte, V.; Velasquez, G.; Parra, C.B.; Giudice, G.; Liuzzo, M.; Moretti, R.; Moussallam, Y.; Peters, N.; et al. CO_2-gas precursor to the March 2015 Villarrica volcano eruption. *Geochem. Geophys. Geosyst.* **2017**, *18*, 2120–2132. [CrossRef]
16. Goto, A.; Johnson, J.B. Monotonic infrasound and Helmholtz resonance at Volcan Villarrica (Chile). *Geophys. Res. Lett* **2011**, *38*, L06301. [CrossRef]
17. Bertin, D.; Amigo, A.; Bertin, L. Erupción del volcán Villarrica 2015: Productos emitidos y volumen involucrado. In Proceedings of the XIV Congreso Geológico Chileno, La Serena, Chile, 4–8 October 2015; Volume III, pp. 249–252.
18. Johnson, J.B.; Palma, J.L. Lahar infrasound associated with Volcán Villarrica's 3 March 2015 eruption. *Geophys. Res. Lett.* **2015**, *42*, 6324–6331. [CrossRef]
19. Harris, A.J.L.; De Groeve, T.; Carn, S.A.; Garel, F. Risk evaluation, Detection and Simulation during Effusive Eruption Disasters. In *Detecting, Modelling and Responding to Effusive Eruptions*; Harris, A.J.L., De Groeve, T., Garel, F., Carn, S.A., Eds.; Geological Society: London, UK, 2016; Volume 426, pp. 1–22. [CrossRef]
20. Harris, A.J.L. Effusive crisis at Piton de la Fournaise 2014–2015: A Review of a Multi-National Response Model. *J. App. Volcanol.* **2017**, *6*. [CrossRef]
21. Harris, A.J.L. *Thermal Remote Sensing of Active Volcanoes. A User's Manual*; Cambridge University Press: Cambridge, UK, 2013; p. 736. [CrossRef]
22. Blackett, M. An Overview on Infrared Remote Sensing of Volcanic Activity. *J. Imaging* **2017**, *3*, 13. [CrossRef]
23. Harris, A.J.L.; Baloga, S.M. lava discharge rates from satellite-measured heat flux. *Geophys. Res. Lett.* **2009**, *36*, L19302. [CrossRef]
24. Harris, A.J.L.; Dehn, J.; Calvari, S. Lava effusion rate definition and measurement: A Review. *Bull. Volcanol.* **2007**, *70*, 1–22. [CrossRef]
25. Wright, R.; Flynn, L.; Garbeil, H.; Harris, A.; Piler, E. Automated volcanic eruption detection using MODIS. *Remote Sens. Environ.* **2002**, *82*, 135–155. [CrossRef]
26. Kervyn, M.; Ernst, G.G.J.; Harris, A.J.L.; Belton, F.; Mbede, E.; Jacobs, P. Thermal remote sensing of the low-intensity carbonatite volcanism of Oldoinyo Lengai, Tanzania. *Int. J. Remote Sens.* **2008**, *29*, 6467–6499. [CrossRef]
27. Coppola, D.; Laiolo, M.; Cigolini, C.; Delle Donne, D.; Ripepe, M. Enhanced volcanic hot-spot detection using MODIS IR data: Results from the MIROVA system. In *Detecting, Modelling and Responding to Effusive Eruptions*; Harris, A.J.L., De Groeve, T., Garel, F., Carn, S.A., Eds.; Geological Society: London, UK, 2016; Volume 426. [CrossRef]
28. Coppola, D.; Laiolo, M.; Cigolini, C.; Orozco, G. The 2008 "Silent" Eruption of Nevados de Chillán (Chile) detected from space: Effusive rates and trends from the MIROVA system. *J. Volcanol. Geotherm. Res.* **2016**, *327*, 322–329. [CrossRef]
29. Spampinato, L.; Ganci, G.; Hernández, P.A.; Calvo, D.; Pérez, N.M.; Calvari, S.; Del Negro, C.; Yalire, M.M. Thermal insights into the dynamics of Nyiragongo lava lake from gound and satellite measurements. *J. Geophys. Res. Solid Earth* **2013**, *118*, 5771–5784. [CrossRef]
30. Blackett, M. Early Analysis of Landsat-8 Thermal Infrared Sensor Imagery of Volcanic Activity. *Remote Sens.* **2014**, *6*, 2282–2295. [CrossRef]

31. Fischer, C.; Klein, D.; Kerr, G.; Stein, E.; Lorenz, E.; Frauenberger, O.; Borg, E. Data Validation and Case Studies using the TET-1 Thermal Infrared Satellite System. *Int. Arch. Photogramm. Remote Sens. Spat. Inf. Sci.* **2015**, *XL-7/W3*, 1177–1182. [CrossRef]
32. Zakšek, K.; Hort, M.; Lorenz, E. Satellite and Ground Based Thermal Observation of the 2014 Effusive Eruption at Stromboli Volcano. *Remote Sens.* **2015**, *7*, 17190–171211. [CrossRef]
33. Berk, A.; Hawes, F.; van den Bosch, J.; Anderson, G.P. *MODTRAN5.4.0 User's Manual*; Spectral Sciences Inc.: Burlington, MA, USA, 2016.
34. Salisbury, J.W.; D'Aria, D.M. Emissivity of terrestrial materials in the 3–5 μm atmospheric window. *Remote Sens. Environ.* **1994**, *47*, 345–361. [CrossRef]
35. Giglio, L.; Kendall, J.D.; Justice, C.O. Evaluation of global fire detection algorithms using simulated AVHRR infrared data. *Int. J. Remote Sens.* **1999**, *20*, 1947–1985. [CrossRef]
36. Klein, D.; Richter, R.; Strobl, C.; Schläpfer, D. Solar Influence on Fire Radiative Power. *IEEE Trans. Geosci. Remote Sens.* under review.
37. Dozier, J. A method for satellite identification of surface temperature fields of subpixel resolution. *Remote Sens. Environ.* **1981**, *11*, 221–229. [CrossRef]
38. Zhukov, B.; Lorenz, E.; Oertel, D.; Wooster, M.; Roberts, G. Space-borne detection and characterization of fires during the bi-spectral infrared detection (BIRD) experimental small satellite mission (2001–2004). *Remote Sens. Environ.* **2006**, *100*, 29–51. [CrossRef]
39. Halle, W.; Asam, S.; Borg, E.; Fischer, C.; Frauenberger, O.; Lorenz, E.; Klein, D.; Nolde, M.; Paproth, C.; Plank, S.; et al. FireBIRD–small satellite or wild fire assessment. In Proceedings of the International Geoscience and Remote Sensing Symposium IGARSS, Valencia, Spain, 22–27 July 2018.
40. Justice, C.; Giglio, L.; Boschetti, L.; Roy, D.; Csiszar, I.; Morisette, J.; Kaufman, Y. *Modis Fire Products*; Algorithm Technical Background Document; NASA Goddard Space Flight Center: Greenbelt, MD, USA, 2006.
41. Visible Infrared Imaging Radiometer Suite (VIIRS) 375 m Active Fire Detection and Characterization Algorithm Theoretical Basis Document 1.0. Available online: https://viirsland.gsfc.nasa.gov/PDF/VIIRS_activefire_375m_ATBD.pdf (accessed on 19 June 2018).
42. NASA. Fire Information for Resource Management System (FIRMS). Available online: https://firms.modaps.eosdis.nasa.gov (accessed on 19 June 2018).
43. Aiuppa, A.; Maarten de Moor, J.; Arellano, S.; Coppola, D.; Francofonte, V.; Galle, B.; Giudice, G.; Liuzzo, M.; Mendoza, E.; Saballos, A.; et al. Tracking Formation of a Lava Lake from Ground and Space: Masaya Volcano (Nicaragua), 2014–2017. *Geochem. Geophys. Geosys.* **2018**, *19*, 496–515. [CrossRef]

Article

Radar Path Delay Effects in Volcanic Gas Plumes: The Case of Láscar Volcano, Northern Chile

Stefan Bredemeyer [1,*], Franz-Georg Ulmer [2], Thor H. Hansteen [1] and Thomas R. Walter [3]

1 GEOMAR Helmholtz Centre for Ocean Research Kiel, Research Division 4: Dynamics of the Ocean Floor, Wischhofstr. 1–3, 24148 Kiel, Germany; thansteen@geomar.de
2 German Aerospace Center (DLR), Remote Sensing Technology Institute, Oberpaffenhofen, 82234 Weßling, Germany; info@informathik.de
3 GFZ German Research Centre for Geosciences, Section 2.1: Physics of Volcanoes and Earthquakes, Telegrafenberg, 14473 Potsdam, Germany; twalter@gfz-potsdam.de
* Correspondence: sbredemeyer@geomar.de; Tel.: +49-431-600-2135

Received: 28 August 2018; Accepted: 18 September 2018; Published: 21 September 2018

Abstract: Modern volcano monitoring commonly involves Interferometric Synthetic Aperture Radar (InSAR) measurements to identify ground motions caused by volcanic activity. However, InSAR is largely affected by changes in atmospheric refractivity, in particular by changes which can be attributed to the distribution of water (H_2O) vapor in the atmospheric column. Gas emissions from continuously degassing volcanoes contain abundant water vapor and thus produce variations in the atmospheric water vapor content above and downwind of the volcano, which are notably well captured by short-wavelength X-band SAR systems. These variations may in turn cause differential phase errors in volcano deformation estimates due to excess radar path delay effects within the volcanic gas plume. Inversely, if these radar path delay effects are better understood, they may be even used for monitoring degassing activity, by means of the precipitable water vapor (PWV) content in the plume at the time of SAR acquisitions, which may provide essential information on gas plume dispersion and the state of volcanic and hydrothermal activity. In this work we investigate the radar path delays that were generated by water vapor contained in the volcanic gas plume of the persistently degassing Láscar volcano, which is located in the dry Atacama Desert of Northern Chile. We estimate water vapor contents based on sulfur dioxide (SO_2) emission measurements from a scanning UV spectrometer (Mini-DOAS) station installed at Láscar volcano, which were scaled by H_2O/SO_2 molar mixing ratios obtained during a multi-component Gas Analyzer System (Multi-GAS) survey on the crater rim of the volcano. To calculate the water vapor content in the downwind portion of the plume, where an increase of water vapor is expected, we further applied a correction involving estimation of potential evaporation rates of water droplets governed by turbulent mixing of the condensed volcanic plume with the dry atmosphere. Based on these estimates we obtain daily average PWV contents inside the volcanic gas plume of 0.2–2.5 mm equivalent water column, which translates to a slant wet delay (SWD) in DInSAR data of 1.6–20 mm. We used these estimates in combination with our high resolution TerraSAR-X DInSAR observations at Láscar volcano, in order to demonstrate the occurrence of repeated atmospheric delay patterns that were generated by volcanic gas emissions. We show that gas plume related refractivity changes are significant and detectable in DInSAR measurements. Implications are two-fold: X-band satellite radar observations also contain information on the degassing state of a volcano, while deformation signals need to be interpreted with care, which has relevance for volcano observations at Láscar and for other sites worldwide.

Keywords: gas emission monitoring; X-band InSAR; scanning Mini-DOAS; Multi-GAS; volcanic gases; precipitable water vapor; radar path delay; Láscar volcano

1. Introduction

Volcano monitoring is effectively based on multi-parametric datasets, often characterizing surface deformation, degassing, seismicity and temperature changes in time and space. However, even though comparison of independent datasets is common, they are only rarely integrated into a common evaluation scheme to enhance the relative strengths of the individual applied methods. Seismic and deformation data for instance are increasingly being integrated to improve spatio-temporal localization of magma movement beneath a volcano, which enables to better constrain the geometries of magma pathways and reservoirs, and to derive more precise physical models of the associated processes [1,2]. Gas emission and deformation data, however, have not yet been deeply considered as integrated information, though it has been shown that variations in degassing and deformation are often intimately coupled [3]. A combined analysis of time series measurements of volcanic degassing and deformation rates proved to be a powerful tool for the evaluation of volumetric changes in magma reservoirs due to the accumulation or discharge of volcanic gases (e.g., [4,5]). Dynamics of degassing are furthermore critical for understanding pressure fluctuations in a magmatic and hydrothermal system, which additionally may manifest as temporary deformation of the volcanic edifice [6–8]. However, these datasets are commonly considered separately, and the fate of gas emissions downwind of the emission source is typically not taken into account, which is why the contribution of gas emissions on the radar signal used for deformation measurements has not been comprehensively demonstrated. Differential interferometric synthetic aperture radar (DInSAR) measurements allow for the detection of mm-scale line of sight (LOS) displacements of the observed surface [9]. It has been well demonstrated that the accuracy of satellite radar ground deformation measurements is affected by changes in atmospheric refractivity, in particular by changes which can be attributed to the highly variable distribution of water vapor in the observed atmospheric column [10,11].

Atmospheric contributions to DInSAR data often have similar magnitudes and wavelengths as the actual ground deformation signal, and thus they need to be removed from interferograms, when deformation measurements are the purpose of monitoring. There have been two main strategies for this, (i) time-space-based filtering, and (ii) modeling of atmospheric contribution [12]. In particular, the latter is highly successful, as interfering contributions from a moist atmosphere can coarsely be predicted and compensated by means of high-resolution numerical weather models (e.g., MM5, and WRF), which provide the information required to predict the atmospheric phase delay in the LOS of an interferometric measurement [13–16].

Degassing volcanoes; however, produce their own atmospheric disturbances, which are typically not well captured by weather models [17,18]. The large and variable amounts of water vapor in volcanic plumes may cause differential phase errors in interferometric measurements due to reduction of radar propagation velocity within the plume above and downwind of the volcano [19–22].

The leeward sector of a degassing volcano is hence prone to be affected by pronounced differential phase signatures, which can be misinterpreted as a deforming ground surface, and may thus obscure the real ground deformation information of an interferogram. Rosen et al. [19] described the possible consequences that such propagation delays would have on DInSAR measurements using the examples of phase signatures that occurred on Kilauea volcano, Hawaii. Similarly, Wadge et al. [20] observed enhanced tropospheric delays in radar data of Soufrière Hills. Instead of using weather models for the correction of their DInSAR data, these authors exploited data from Global Positioning System (GPS) measurements on the leeward side of the volcano. Based on local SO_2 flux and gas compositional measurements, and assuming a homogeneous distribution of water vapor inside the volcanic gas plume, they however obtained an estimate of the gas plume related radar delay of only about 0.05 mm, which is far below, and thus negligible with respect to InSAR resolution. Gas plume related disturbances are nevertheless also identified elsewhere in interferograms of degassing volcanoes (e.g., at Pico do Fogo as described in González et al. [21]), but remained to be quantitatively demonstrated and validated for a selected showcase.

To reach this aim, the precipitable water vapor (PWV) content in the plume at the time of SAR acquisitions has to be determined. PWV is the amount of water vapor (in kg·m^{-2} or mm) vertically integrated in an atmospheric column. Space borne radar interferometric delay measurements in principle can be used to map the spatial distribution of water vapor in the atmosphere [23–25], if the contributions from topographic phase and ground deformation are known, or when independent PWV estimates are used as a reference. Accordingly, those measurements also enable to map the spatial extent of a volcanic gas plume when the strength of the degassing source is known, and independent PWV estimates are integrated into analysis.

Even though water vapor by far is the largest and most abundant component in the gas emissions of most volcanoes, volcanic water vapor emissions to date only rarely have been measured and quantified by means of remote sensing (and preferentially by means of indirect methods as e.g., in [26]), due to the missing contrast relative to the high water vapor concentrations in the ambient atmosphere, which commonly preclude its quantification in the volcanic cloud. During the present decade, however, some progress has been made in this respect, and volcanologists have adopted and adjusted several technologies known from meteorological applications to remotely map and quantify volcanic water vapor emissions despite a variable background atmosphere. Fiorani et al. [27] for instance, successfully measured water vapor fluxes of Stromboli volcano in Italy using a ground-based LiDAR system, and Bryan et al. [28] introduced a combined millimeter-wave radar/radiometer system (also ground-based), which enables 3-dimensional mapping of the spatial distribution of ash and water vapor in eruptive clouds. These methods have in common that they use active sensors, which transmit a signal that is able to penetrate the volcanic cloud and thus allow measuring the reflection, refraction, or scattering of the signal occurring inside the cloud. Such methods thus are largely superior to methods deploying passive optical imaging sensors (such as UV- and IR-spectroscopical methods), which typically are strongly affected by the presence of liquid water and aerosols [29], and therefore fail to work in very dense clouds. Ground-based active sensors are, however, rather exotic, and are cost- and labor-intensive, and thus have not yet routinely been used for permanent gas emission monitoring purposes.

We thus used a combination of more common sensors, which enable unsupervised continuous monitoring. Because direct measurements of volcanic water vapor emissions by means of optical remote sensing are challenging [30], we measured SO_2 column density profiles in conjunction with molar H_2O/SO_2 ratios in this work, in order to estimate the apparent PWV contents in the gas plume above a volcano. These estimates were integrated as an a-priori knowledge into the analysis of DInSAR observations in order to investigate the propagation delay that was generated by volcanic gas emissions. We demonstrate the repeated contribution of the volcanic gas plume to the radar delay in the interferometric measurements and thus determine the plume-induced delay variations which can be falsely interpreted as a deformation signal.

2. Study Area

The study was conducted at Láscar volcano, one of the most active volcanoes of the central Andes [31,32], which has been identified as the second largest emission source of volcanic gases in Northern Chile [33]. The volcano is located east of the Salar de Atacama basin (23.37°S, 67.73°W), on the western margin of the Puna plateau (Figure 1a), which is one of the driest areas on Earth. Much of the plateau around the 5592 m high Láscar volcano has an altitude of more than 4 km above sea level. Background atmospheric PWV over the region is very low most of the year, and generally less than 1 mm total water column [34]. Cloudless skies, an extremely low atmospheric water content, and the high altitude lead to an exceptionally high atmospheric transparency [35], which in general is conducive for remote sensing applications and hence for the purpose of our study.

The volcanic edifice of Láscar comprises two truncated intersecting composite cones, termed Western and Eastern edifice hereafter, and hosts 5 nested craters, which are aligned along an ENE–WSW trending lineament [31,36,37]. During the past two decades Láscar exhibited several periods of cyclic dome growth and collapses, which were punctuated by occasional explosive vulcanian to plinian eruptions [38]. Recent activity is characterized by gradual subsidence [39,40] and persistently strong fumarolic degassing from the remnants of a dacitic dome [41] located in the westernmost crater of the Eastern edifice (Figure 1b,c). The dome remnants cover an area of about 40,000 m^2 (about 230 m in diameter) in the central depression of the 800 m wide and 400 m deep crater. Numerous high temperature fumaroles (about 300 °C) are distributed on top of the dome and the surrounding inner crater walls [41–43]. Several low temperature fumaroles and diffuse vent sites are located in the eastern craters of the Eastern edifice and on the southern crater rim, but the high temperature gases ascending from the dome and the fumaroles in the active crater are the main emission source [41]. These hot gas emissions produce a persistent thermal anomaly in the middle of the crater, which has been subject of numerous remote sensing studies using infrared satellite imagery (e.g., [44–48]). Gas emissions from Láscar volcano are very moist in contrast to the dry atmospheric background. Water vapor emissions from Láscar account for more than 90% of its total gas emissions, approaching rates of several metric kilotons per day [33]. The persistently degassing, and largely unvegetated stratovolcano (Figure 1d) thus provides ideal conditions to examine radar propagation delays caused by refractivity variations in volcanic gas.

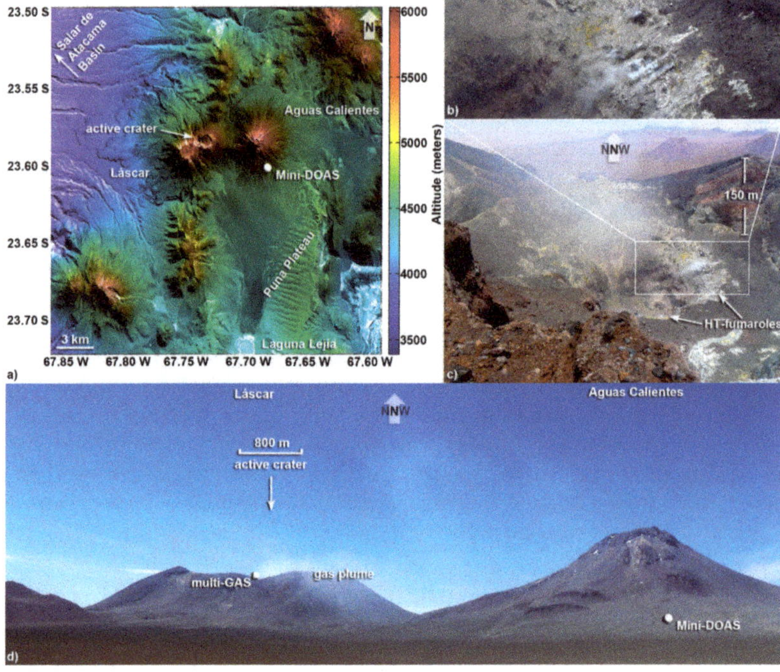

Figure 1. Láscar and adjacent Aguas Calientes volcanoes. The location of the scanning Mini-DOAS station is indicated by the *white dot* on the southern flank of Aguas Calientes volcano. (**a**) Aster visible range composite of 29 April 2013 draped onto SRTM-1 grid; (**b**) Close-up of high temperature fumaroles in the active crater; (**c**) View from the southern crater rim towards NNW into the active crater of Láscar showing fumarolic activity during the Multi-GAS survey on the Southern crater rim on 2 December 2012. The *white bounding box* framing the area shown in (**b**) roughly spans 200 m in width and 100 m in height; (**d**) Panoramic view from South towards NNW showing a dispersed gas plume emanating from Láscars' active crater on 5 December 2012, and drifting towards SE, which corresponds to the main transport direction during daylight time. Distance between Mini-DOAS in the foreground and active crater in the background of the image is about 6 km.

3. Data and Methods

In this section, we will introduce a method that enables us to isolate and map gas plume related phase delay contributions in interferometric radar measurements by means of correlating a-priori information on the temporal variations of PWV contents in the volcanic gas plume with the signal strength variations at each range azimuth position of a DInSAR time series. Application of the method further requires involving several different prior constraints encompassing the temporal variations of all other processes that potentially contribute to total DInSAR phase in order to derive an estimate of the gas plume related interferometric signal in the DInSAR time series. In the following we will specify which data were used for our case study spanning the period October 2013 to February 2014 (later often simply referred to as the considered period), present the techniques used to derive the necessary a-priori knowledge, and describe how the estimation technique works.

3.1. SO_2 Column Density Retrieval

SO_2 path-length concentrations were measured in the plume of Láscar volcano using a NOVAC-type stationary scanning Mini-DOAS instrument (Figure 1; [49]; NOVAC: Network for the Observation of Volcanic and Atmospheric Change) that was permanently installed in April 2013 by the Chilean Observatorio Volcanológico De los Andes del Sur (OVDAS). The DOAS station (Lejía DOAS) was deployed east of Láscar (at Lat $-23.387°$, Long $-67.678°$), according to prevailing westerly wind directions. The DOAS-instrument scans across the sky from horizon to horizon along a semi-conical surface (Figure 2a) and measures the spectra of the incoming scattered sunlight at 51 angular steps of 3.6° [49]. Each complete scan takes about 5–10 min and yields a crosswind SO_2 total column profile, i.e., perpendicular to the transport direction of the volcanic gas plume. Measurements are conducted during daylight, i.e., when UV intensities are sufficiently high to achieve an acceptable signal-to-noise ratio. At these latitudes the required conditions are met during a time interval approximately ranging from about 11:00 a.m. to 10:00 p.m. (UTC) in austral summer, and 12:00 a.m. to 09:00 p.m. (UTC) in austral winter.

SO_2 differential slant column densities (SCDs) were retrieved from the sunlight spectra by means of the DOAS method [50], as implemented in an automated evaluation routine of the NOVAC-software [49]. Evaluation was performed in the wavelength range 310–325 nm using absorption cross sections of SO_2 [51], and O_3 [52], which were convolved to the spectral resolution of the instrument. In addition, a Ring spectrum was included in the DOAS-fit, in order to avoid the filling in of the Fraunhofer lines of the solar spectrum as a result of rotational Raman scattering [53]. Shift correction was applied to the measured spectra using the known features of a pixel-wavelength calibrated solar spectrum, which was derived from the Solar flux atlas of Kurucz et al. [54].

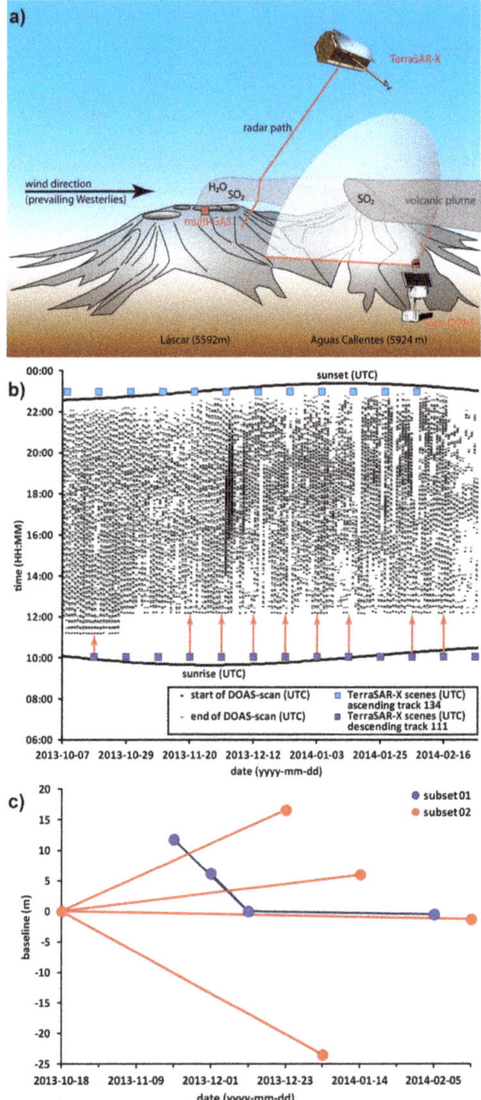

Figure 2. Spatio-temporal relationships between the methods used for the gas plume estimate. (**a**) Sketch showing the measurement geometry and location of the Multi-GAS and scanning Mini-DOAS instruments at Láscar volcano during an overpass of TerraSAR-X. Refractive delay of the radar occurs inside the volcanic plume, which is heading towards East, due to predominantly westerly winds. See text for discussion; (**b**) Date versus time plot depicting acquisition times of scanning DOAS measurements and SAR images. Measurement times are indicated using Coordinated Universal Time (UTC), which is offset by +3 h with respect to Chile Summer Time (CLST). *Red arrows* indicate the temporal offset between SAR images and scanning DOAS data chosen for analysis; (**c**) Spatial and temporal baselines of SAR images used in DInSAR time series subsets 01 (*blue*) and 02 (*red*). Master scenes of *subset 01* and *subset 02* are from 18 October 2013 and 12 December 2013, respectively, and each computed interferogram is represented by a line between the corresponding two images.

3.2. Estimation of Water Vapor Contents in the Láscar Plume

Water vapor contents in the Láscar gas plume were obtained by scaling the SO_2 SCDs with the molar H_2O/SO_2 ratio determined from gas concentration measurements, which were in turn conducted using a multi-component Gas Analyzer System [55] during a field survey on 2 December 2012 (see [33] for details), that is roughly one year in advance of the period that we consider here for the analysis of our SAR and Mini-DOAS observations (Figure 2b). The Multi-GAS instrument basically consists of an air pump feeding ambient air into a suite of electrochemical sensors for the measurement of SO_2, H_2S and H_2, and a non-dispersive closed-path IR spectrometer for measuring CO_2 and H_2O concentrations. Gas concentrations were measured for several hours in a dilute and partly condensed volcanic plume crossing the southern rim of the active crater (roughly at Lat −23.366°, Long −67.734°; see Figure 2a), and the signals of all sensors were recorded at 0.5 Hz by a data-logger, yielding a large number of measurements. The raw data was processed according to [56,57], in order to identify the characteristic molar gas ratios [33].

Multi-GAS instruments, however, exclusively measure gaseous water vapor and do not take into account the liquid water content (LWC) in the volcanic cloud. Due to the obvious condensed nature of the plume at the crater rim, our PWV contents were thus estimated on the basis of the maximum molar H_2O/SO_2 ratio obtained for the measurement period, which was used to scale the SO_2 SCDs yielding a first PWV estimate. This first estimate, we however consider to be representative solely for the proximal portion of the volcanic plume (see Appendix A for details on the conversion of SO_2 SCDs to PWV contents).

To estimate PWV contents also from the distal portion of the volcanic plume, we additionally consider condensation and evaporation processes, which change the liquid-to-vapor ratio during transport of the volcanic plume. LWCs are typically elevated close to the emission source and decrease with increasing downwind distance, due to gradual entrainment of dry ambient air, causing evaporation of cloud droplets (e.g., [58]). This in turn results in an increase of water vapor in the downwind portion of the volcanic cloud, which may further be promoted by the presence of aerosols [59], and which is expected to be reflected as an enhancement of the radar delay in interferometric measurements.

Potential evaporation rates were thus determined for the measurement periods of the scanning DOAS and then used to upscale our fixed molar gas ratio, which was thereby adjusted to increase proportionally to evaporation rates, which on the other hand strongly depend on wind speeds over arid areas (see Appendix B for details on evaporation calculation). The resulting variable ratio was used to upscale the SO_2 SCDs at each scan angle of the plume cross sections as described above (and in Appendix A), yielding a second PWV estimate, which we deem to be representative for the distal part of the gas plume.

3.3. SAR Data and InSAR Methods

The aforementioned water vapor maps into the used DInSAR data, which are now briefly described. TerraSAR-X was tasked to acquire high resolution spotlight SAR scenes covering a 10×10 km large area around Láscar volcano on both ascending and descending passes of the orbit. TerraSAR-X flies along the day-night boundary of the Earth in a near-polar sun-synchronous dusk-dawn orbit with an 11-day repeat cycle. SAR observations of a given point on Earth's surface are thus conducted within fixed time windows, which at Láscar correspond to day times of about 10:00 a.m. (UTC) on descending nodes and 11:00 p.m. (UTC) on ascending nodes of the orbit. The DInSAR time series considered here covers a 5 months period from 18 October 2013 to 16 February 2014 (Figure 2b). DInSAR preparation and analysis was done using the dedicated modular InSAR software system GENESIS from DLR [60]. A set of 9 SAR images from descending track 111 was chosen according to availability of complementary scanning DOAS data (Figure 2b), and these images were used to create two temporally overlapping disjoint interferogram time series, termed *subset 01* and *subset 02* hereafter. The subsets consist of 3 and 4 temporally interconnected interferograms, respectively, i.e., the interferograms of each subset are based on a common master scene (Figure 2c; Table 1). SAR images were combined to form interferograms with very small spatial baselines, which accordingly resulted in different temporal baselines for each interferometric pair as depicted in Figure 2c and Table 1. The very small spatial baselines chosen here minimize phase contributions from topography and prevent from unwrapping failures due to a coarse digital elevation model (DEM).

Table 1. Combinations of SAR acquisitions used in DInSAR time series subsets 01 & 02, and their respective spatial and temporal baselines.

Master Scene Date (yyyy-mm-dd)	Slave Scene Date (yyyy-mm-dd)	Spatial Baseline (m)	Temporal Baseline (Days)
	Subset 01		
2013-12-12	2013-11-20	11.8	−22
2013-12-12	2013-12-01	6.2	11
2013-12-12	2014-02-05	−0.5	55
	Subset 02		
2013-10-18	2013-12-23	16.6	66
2013-10-18	2014-01-03	−23.5	77
2013-10-18	2014-01-14	6	88
2013-10-18	2014-02-16	−1.3	121

3.4. Determination of Gas Plume Related Phase Delays

PWV contents determined for the gas plumes of each SAR acquisition were used to compute the corresponding differential phase delays encountered in each interferogram. For this purpose we first determined the slant wet delay (SWD) that such PWV contents theoretically would produce in the LOS of a SAR image. Determination of the theoretical SWD is straightforward, since it can be inferred from the zenith path delay (ZPD), which is linearly related to the vertically integrated water vapor (IWV) contents, where 1 mm in PWV contents roughly results in a 6.5 mm ZPD of microwave signals [61,62], and the ZPD accordingly is

$$\delta^{ZP} = \Pi_{T_S}^{-1}\, IWV, \tag{1}$$

where δ^{ZP} is the ZPD in (mm), IWV is the columnar integrated water vapor in (kg·m^{-2}), which is approximately equal to PWV contents in (mm), and $\Pi_{T_S}^{-1}$ (≈ 6.5) is a dimensionless conversion factor that is approximated using surface temperature T_S in (K). Calculation of the SWD thus merely requires further conversion of the ZPD, taking into account the incidence angle of acquisitions, in order to obtain the corresponding delay along the LOS. The conversion factor used for the calculation of the theoretical SWD is

$$\delta^{SW} = \delta^{ZP}\frac{1}{cos(\theta)}, \tag{2}$$

where δ^{SW} corresponds to the path delay along the LOS, and θ to the incidence angle of SAR observations. The incidence angle of the SAR acquisitions we used in this work (TerraSAR-X descending track 111, see Section 3.3 for details) was inclined 37° with respect to zenith path resulting in a conversion factor of $1/cos(37) = 1.25$.

Finally, the differences between PWV contents estimated for the acquisition dates of the master image and the slave images were calculated ($PWV_{DIFF} = PWV_{master} - PWV_{slave}$), and used to determine the differential SWD (dSWD) that such differential PWV contents theoretically would produce in the gas plumes of each interferogram.

3.5. DInSAR Preparation

Estimation of the gas plume related radar path delays in interferometric SAR measurements requires several pre-processing steps, involving the removal of large-scale phase contributions related to topography and refractivity changes in the troposphere, the latter of which are commonly referred to as the atmospheric phase screen (APS). First, the raw interferograms were corrected using the baseline information of the interferograms in conjunction with a DEM from the Shuttle Radar Topography Mission (SRTM-1), which provides elevation data at a resolution of one arc second grid spacing (i.e., a relative horizontal precision of ± 15 m) and a relative vertical precision of ± 6 m. The results are terrain corrected differential interferograms (DInSARs), which however still contain some small-scale topographical errors, due to the coarse resolution of the DEM.

The terrain corrected DInSARs were then unwrapped and further corrected by means of path delay difference maps obtained from numerical weather hindcasts of the WRF, using a horizontal grid spacing of 900 m and a vertical division of 51 eta levels for the simulations. The WRF simulations were initialized with the ERA interim dataset from ECMWF at 06:00 a.m. (UTC) of each SAR acquisition date, and the 900 m fine-resolution domain covering the area of interest further was embedded into two larger coarse-resolution domains using 2700 and 13,500 m grids, respectively, which were used to update the boundary conditions of the inner high resolution domain. This correction was done in order to reduce atmospheric phase contributions, and in particular to avoid altitude correlated distortions known as stratification effect, which is the most prominent atmospheric disturbance in the presence of strong topography [15]. After removal of the linearly stratified atmosphere, the resulting DInSARs, however, still contain phase contributions that were caused by small-scale lateral refractivity variations of a turbulent atmosphere and the volcanic steam plume, which we decomposed as described in the following section.

3.6. DInSAR Decomposition

To extract the gas plume related phase contribution from our interferograms, we used a wavelet-based DInSAR decomposition (WBDD) technique, which enabled us to integrate our estimates of the PWV contents in the volcanic gas plume into DInSAR analysis and thus allowed for the separate evaluation of gas plume related phase delays and their comparison with respect to all other phase contributions. The WBDD technique [63] was developed to deal with temporally unconnected DInSAR time series subsets and small stack sizes, and does not use the isotropy assumption to mitigate the atmospheric effect. The algorithm is based on Dual-tree complex wavelet transform ($DT\text{-}\mathbb{C}WT$), which is being used in a wide range of applications in the field of signal and image processing [64]. The $DT\text{-}\mathbb{C}WT$ calculates the complex transform of a signal using two separate discrete wavelet transform decompositions (tree a and tree b), which may be regarded as equivalent to filtering the input signal with two separate banks of bandpass filters, that separately produce the real and imaginary coefficients of the complex wavelet. The $DT\text{-}\mathbb{C}WT$ allows to capture both frequency and location information of a multidimensional signal, i.e., the times and locations at which these frequencies occur.

The WBDD technique requires at least two temporally interconnected interferograms as an input and the workflow is as follows (Figure 3). First, the $DT\text{-}\mathbb{C}WT$ representations of all interferograms in the DInSAR time series are derived, i.e., complex wavelet coefficients are determined for each range azimuth position, where the phase of the coefficient determines the exact position of the wavelet in the interferogram and the modulus of the coefficient corresponds to the strength of the wavelet at given position. Second, each of these complex wavelet time series is analyzed using so-called *priors*, respectively *prior time series* containing information on the temporal evolution of any of the processes that possibly contributed to the total phase of the interferograms. That is, we are including prior information on the signal strength variations related to (i) inaccuracies of the DEM, (ii) linear deformation of the ground surface, and (iii) changing refractive properties of the ambient atmosphere and (iv) the volcanic gas plume (details on the determination and compilation of these *priors* are given in the following section). This prior information allows assigning each complex wavelet to their likely causes, depending on their best temporal correlation with the prior knowledge, and therefore enables to decompose the interferograms into signals related to the *priors*. The decomposition thus yields one separate radar propagation path delay estimate for each *prior* that was incorporated in the decomposition analysis of the DInSAR time series.

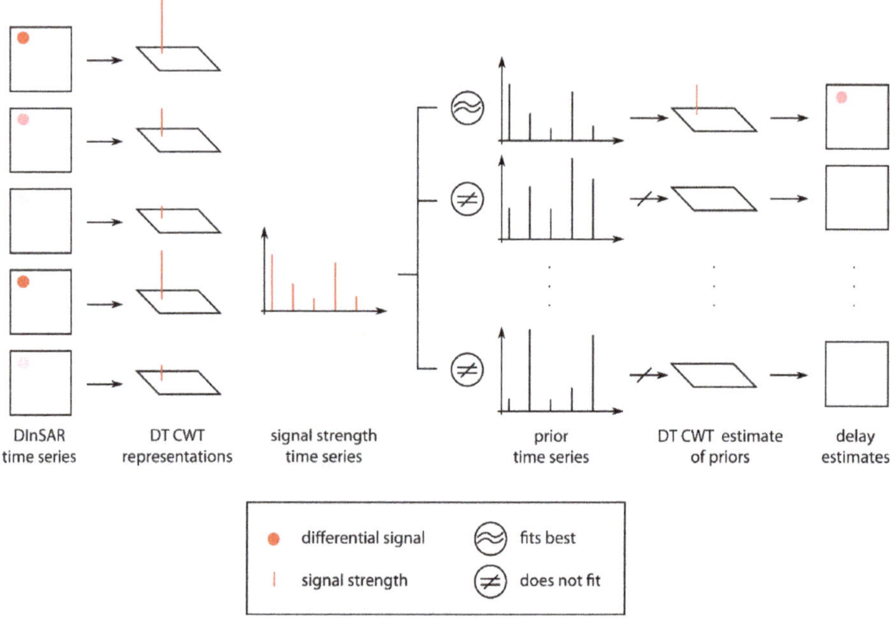

Figure 3. Technical diagram depicting the processing steps of the WBDD algorithm. In a DInSAR time series each DInSAR pixel (respectively range azimuth position) is characterized by a certain temporal evolution of its differential signal strength. Computation of the *DT-ℂWT* representations of each DInSAR enables to capture the temporal evolution of the signal strength of all SAR pixels of the DInSAR time series by means of a small number of complex wavelet coefficients. Similarly each process which causes changes in SAR signal strength can be described by a time series, which reflects the temporal variations of the process (e.g., variations of water vapor contents in the volcanic gas plume, variations in spatial baseline, relative humidity, ground temperature, and pressure). The algorithm uses the time series of these processes as an a priori knowledge of a related possible change in SAR signal strength and assigns the SAR signals to their likely causes by comparison of the *prior time series* with the temporal evolution of SAR signal strength at each range azimuth position, which decomposes the interferograms into different phase screens.

3.7. Priors Used for DInSAR Decomposition

To model the signal strength variations associated with different sources/processes affecting the interferometric measurements we used the following *prior time series* reflecting the temporal behavior of the corresponding processes. The dSWD time series, which were determined following the descriptions given in Section 3.4, were used as prior information for our algorithm, aiming to determine exclusively those interferometric phase variations, which are characterized by a similar temporal evolution as that of the PWV contents in the gas plume. As a result, we obtained the intersecting set of all gas plume related phase contributions contained in the DInSAR time series. Thus, our gas plume related phase delay estimates exclusively cover the area in which gas plumes of the different interferograms overlap.

The spatial and temporal baseline histories of the interferograms were included as prior information in order to capture small-scale topography related phase contributions, which were caused by errors in the DEM, and respectively by linear ground surface deformation. This works, because the effect of the DEM error typically evolves proportionally to the spatial baseline history of a DInSAR time series [65], and linear ground surface deformation evolves proportionally to the temporal baseline history. The values of the baseline histories were taken as they are and did not require any further computation, since these already represent differential values.

Furthermore, we estimated the refractivity related phase contributions of the ambient atmosphere in order to prevent them from leaking into the gas plume estimate, which involved incorporation of several *priors* in the analysis. In our approach we separately consider non-repeating phase contributions, which are caused by turbulent mixing of the atmosphere, and repeating phase contributions, which are rather related to the vertical stratification of the atmosphere and orographic effects, and thus typically correlate with the underlying topographic relief. *Dirac-function priors* were used to model APS spatial variation at each SAR acquisition, encompassing all non-recurrent phase contributions, which e.g., resulted from the turbulent transport of atmospheric and volcanic water vapor in the lower part of the troposphere, and thus were exclusively present at the times of single SAR observations. For this purpose one *Dirac-function prior* was included for each of the SAR scenes (using positive *Dirac-functions* for master scenes, and negative *Dirac-functions* for slave scenes), providing the benefit that no external information is required for this type of estimate, which in the following will be referred to as so-called *single event APS estimate*. In addition to *single event APS estimation*, we incorporated *prior time series* containing information on the differential temporal variations of relative humidity, surface temperature and surface pressure, which were derived from the detailed WRF simulation (described in Section 3.5) over a single spot on the summit of the volcano. This was done in order to also cover repeatedly occurring small-scale disturbances, which are related to stratification and to recurring formation of localized orographic clouds, and thus had not yet been captured by the *single event APS estimates*. Moreover, this approach decomposes the refractivity related propagation delay into different contributions of the single physical processes, which contribute to the variations of the tropospheric delay (compare with Equation (1) in Smith and Weintraub [66]), and thus enables us to separately examine the small-scale phase delay effects of the different physical processes in the troposphere.

4. Results

4.1. PWV Contents in the Volcanic Gas Plume of Láscar

During the days of the Multi-GAS survey in December 2012 and in the investigated period, the activity of Láscar volcano was characterized by quiescent degassing from the active crater (Figure 1). Webcam and visual on-site observations confirmed the frequent wind-blown dispersal of a white gas plume in south-easterly direction over the Puna plateau (Figure 1d). This gas plume therefore very often is also fully captured by the well-located scanning DOAS station. SO_2 SCDs measured within the gas plume on the days with TerraSAR-X acquisitions ranged from 2.5×10^{16} to 2.8×10^{18} molecules·cm^{-2} and averaged at 7.2×10^{17} molecules·cm^{-2}. SO_2 SCDs exhibited diurnal variations, which were on average slightly enhanced during the early morning and evening hours.

The Multi-GAS measurements conducted during the field survey in December 2012 showed considerable fluctuations of the water contents at nearly constant CO_2/S molar ratios, suggesting variable water loss due to condensation of water inside the cooling plume prior to the sampling by the instrument [33]. This in turn resulted in a particularly low average molar H_2O/SO_2 ratio of 9.37:1, as Multi-GAS instruments exclusively measure gaseous water vapor and do not take into account the LWC in the volcanic cloud. Using the values of Tamburello et al. [33] we thus determined a maximum molar H_2O/SO_2 ratio of 34:1, by means of which we finally arrived at a water content constituting roughly 90 mol.% of total gas emissions, which is in the range of typical values (85–99 mol.%) for high temperature gases from arc volcanoes elsewhere (e.g., Table 5 in Oppenheimer et al. [67]).

Our PWV estimates were thus obtained by conversion of SO_2 SCDs at each scan angle of the plume cross sections using the fixed molar H_2O/SO_2 ratio of 34:1 (Figure 4a; see Appendix A for details). The amount of water vapor in the centerline of the plume was determined by finding the maximum PWV contents of each DOAS scan, and these were used to constrain the daily average PWV contents for days with TerraSAR-X acquisitions. Estimated daily average PWV contents obtained from the centerline of the plume ranged from 0.007 to 0.015 mm at the times of SAR observations (Figure 4b), translating to SWDs of 0.07 to 0.1 mm (Table 2, columns 3 and 8). Additionally, the average amount of water vapor in each DOAS scan was determined and used to calculate the daily average PWV contents of the bulk plume, i.e., by taking into account the whole plume cross sections at this time. Daily average bulk plume PWV contents ranged from 0.002 to 0.0035 mm (Figure 4b), corresponding to SWDs of 0.02 to 0.03 mm. Differential SWDs from the plume center were used as a *prior time series* (Table 3, column 3) for the estimation of the gas plume related phase contribution in the proximal portion of the gas plume. These values reflect H_2O vapor contents at the crater rim, and are thus not representative for PWV contents arriving at the scanning plane of the scanning DOAS (Figure 2a), since they do not take into account downwind evaporation.

Potential evaporation rates were thus determined for the measurement periods of the scanning DOAS (Figure 4c; see Appendix B for details on evaporation calculation), and used to upscale the H_2O/SO_2 ratio, which was thereby adjusted to increase proportionally to evaporation rates, which in turn strongly depend on wind speeds (Figure 4d) over arid areas. The resulting variable ratio was then used to upscale the SO_2 SCDs at each scan angle of the plume cross sections as described above (and in Appendix A), at this time, however, taking into account the effect of downwind evaporation (Figure 4e).

We assume that such adjusted estimates provide a more reasonable approximation of the temporal variations in PWV encountered above various points of the plume cross sections. Estimated daily average PWV contents obtained from the centerline of the plume ranged from 0.8 to 9.6 mm at the times of SAR observations (Figure 4f), translating to SWDs of 6.4 to 77.0 mm (Table 2, columns 4 and 9). Daily average bulk plume PWV contents ranged from 0.2 to 2.5 mm (Figure 4f), which correspond to SWDs of 1.6 to 20 mm (Table 2, columns 5 and 10). The daily average bulk plume PWV contents were used for the determination of the second dSWD *prior time series* (Table 3, column 4), which in turn was used for the estimation of the gas plume related phase contribution in the distal portion of the volcanic gas plume.

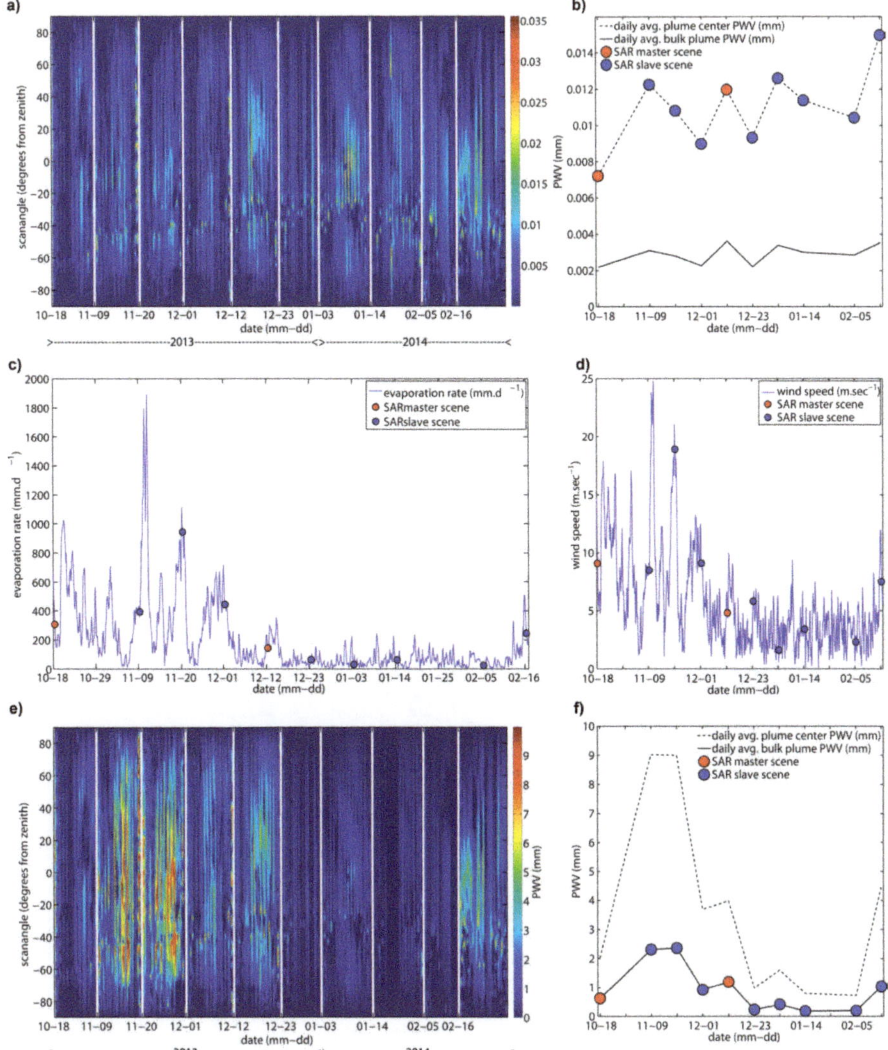

Figure 4. (**a**) Time series of PWV contents determined for plume cross-sections recorded by scanning DOAS on days with TerraSAR-X acquisitions (18 December 2013 to 16 February 2014). The time scale is not linear. Result without consideration of evaporation; (**b**) Daily average PWV contents in the centerline of the plume (*stippled line*) and the bulk plume (*solid line*), which are representative for the moisture distribution above the crater rim on days with available SAR acquisitions; (**c**) Time series of potential evaporation rates at plume height above the summit of Láscar volcano. Evaporation rates at the times of SAR observations are indicated by *red* (master scenes) and *blue* (slave scenes) *dots*; (**d**) Time series of wind speeds at summit altitude used for calculation of potential evaporation rates; (**e**) Time series of PWV contents in plume cross-sections considering downwind evaporation; (**f**) Daily average PWV content in the centerline of the plume (*stippled line*) and the bulk plume (*solid line*) downwind of the volcano on days with available SAR acquisitions.

Table 2. Combinations of SAR acquisitions used in DInSAR time series subsets 01 and 02, and corresponding estimates of PWV contents inside the plume, and the ambient atmosphere are listed along with conversion factors $\Pi_{T_s}^{-1}$ used for calculation of theoretical SWDs.

Scene	Date (yyyy-mm-dd)	PWV (mm)				$\Pi_{T_s}^{-1}$ (mm)	SWD (mm)			
		Crater Rim Plume Center	Downwind Plume Center	Downwind Bulk Plume	Average Background Atmosphere		Crater rim Plume Center	Downwind Plume Center	Downwind Bulk Plume	Average Background Atmosphere
					Subset 01					
Master	2013-12-12	0.012	4.3	1.3	2.1	6.42	0.096	34.59	10.38	17.25
Slave	2013-11-20	0.011	9.4	2.4	1.0	6.58	0.089	77.01	20.16	8.46
Slave	2013-12-01	0.009	4.3	1.1	0.7	6.39	0.072	34.44	8.61	5.97
Slave	2014-02-05	0.010	0.8	0.2	4.3	6.41	0.084	6.43	1.76	34.85
totals		**0.042**	**18.8**	**5.0**	**8.3**		**0.341**	**152.46**	**40.91**	**66.52**
					Subset 02					
Master	2013-10-18	0.007	2.2	0.7	0.7	6.51	0.059	17.98	5.51	5.31
Slave	2013-12-23	0.009	1.1	0.3	5.2	6.43	0.075	8.91	2.10	41.48
Slave	2014-01-03	0.013	1.8	0.5	3.0	6.38	0.101	14.15	3.64	23.96
Slave	2014-01-14	0.011	0.9	0.2	3.6	6.39	0.091	7.27	1.67	29.06
Slave	2014-02-16	0.015	4.9	1.1	1.7	6.44	0.121	39.30	9.11	13.78
totals		**0.056**	**10.9**	**2.7**	**14.1**		**0.447**	**87.60**	**22.03**	**113.58**

Table 3. *Priors* used in the WBDD analysis include two gas plume related differential slant wet delay (dSWD) time series for the proximal and distal portions of the plume, differential time series of relative humidity, surface temperature and surface pressure, as well as spatial and temporal baselines. *Dirac-function priors* for the *single event APS* estimation are not displayed.

Master Scene Date (yyyy-mm-dd)	Slave Scene Date (yyyy-mm-dd)	Gas plume Related Priors (dSWD)		Troposphere Related Priors			Topography Related Priors	
		Crater Rim Plume Center (mm)	Downwind Bulk Plume (mm)	Relative Humidity (%)	Surface Temperature (K)	Surface Pressure (hPa)	Spatial Baseline (m)	Temporal Baseline (Days)
				Subset 01				
2013-12-12	2013-11-20	0.007	−9.78	−0.05	4.60	0.96	11.8	−22
2013-12-12	2013-12-01	0.024	1.77	18.09	−3.32	−1.49	6.2	−11
2013-12-12	2014-02-05	0.013	8.62	−37.95	0.05	0.02	−0.5	55
				Subset 02				
2013-10-18	2013-12-23	−0.016	3.40	−48.55	−4.57	−0.19	16.6	66
2013-10-18	2014-01-03	−0.042	1.87	−17.55	−3.05	1.85	−23.5	77
2013-10-18	2014-01-14	−0.032	3.84	−29.09	−4.35	−1.09	6	88
2013-10-18	2014-02-16	−0.062	−3.60	−9.42	−2.73	−0.45	−1.3	121

4.2. Gas Plume Related Phase Delays

PWV contents obtained for the distal portion of the volcanic gas plume (Figure 5a) were used to exemplify how the DInSAR decomposition analysis can be used to map the corresponding refractivity anomalies in the DInSAR time series. To this end, we used the entire DInSAR data set described in Section 3.3, and applied all *priors* listed in Section 3.7. The differential interferograms, which were coarsely corrected according to the steps described in Section 3.5, show a relatively large amount of coherent pixels (Figure 5b). Pronounced phase changes are visible around the summit and on the flanks of the volcano, representing the sum of phase contributions from individual sources including slow and linear ground surface deformation [40], residual topography due to errors in the DEM, and refractivity fluctuations in atmosphere and volcanic gas plume.

The coarsely corrected DInSARs were decomposed (Section 3.6) in order to obtain the intersecting set of repeating gas plume related phase contributions, which are present in all interferograms of the analyzed DInSAR time series. Owing to the different temporal development of PWV contents which were determined for the two different sections of the plume (Figure 4b,f), we accordingly retrieved two contrasting phase delay results for the ascending part of the plume above the crater (Figure 6a) and the downwind portion of the gas plume (Figure 6b). The phase delay map obtained for the *prior time series* of the gas plume related SWDs above the crater rim (Table 3, column 3) displays a weak altitude correlated shortening of the delay over an area which is largely confined to the summit region of the volcano (Figure 6a). The phase delay map obtained for the downwind portion of the gas plume (Figure 6b) in contrast shows a lenticular pattern that indicates lengthening of the radar propagation path over an area that agrees well with the most common eastward plume transport directions observed during spring and late summer of that period (see Appendix A, Figure A1c,d).

DInSAR decomposition further yielded phase delay maps for each of the other *priors* that were included in the analysis. A detailed description of these maps (which are displayed in Figures 6 and A4) would however go beyond the scope of this paper and thus can be found in the Appendixs E and F.

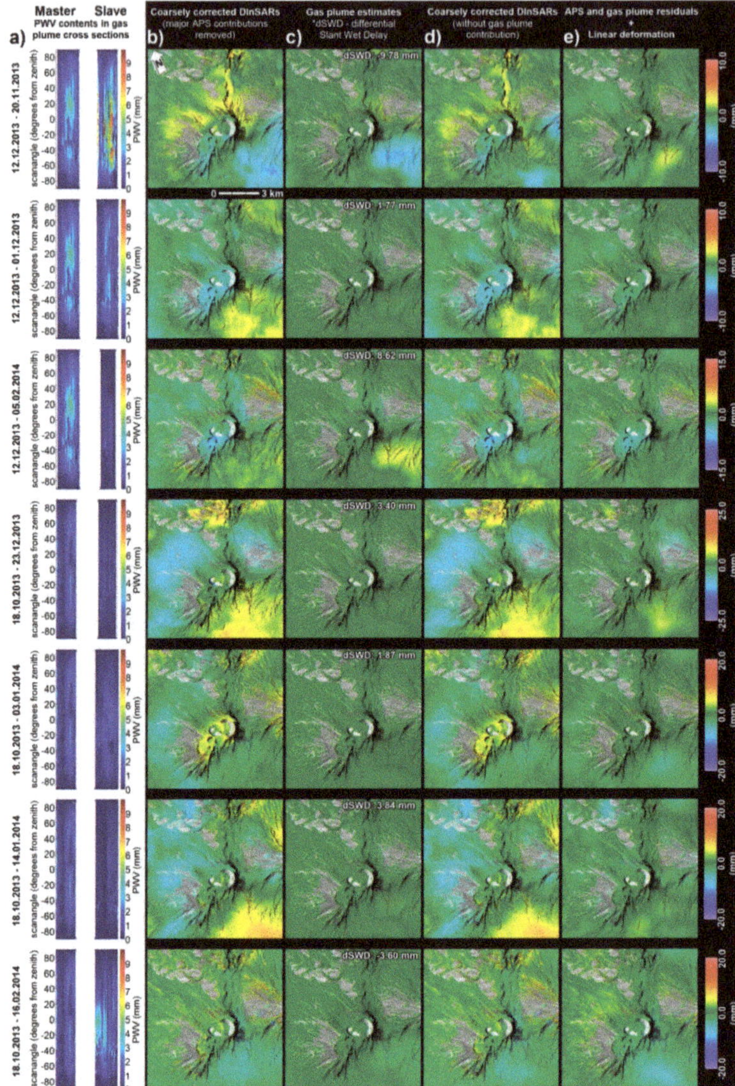

Figure 5. (**a**) PWV contents in plume cross-sections obtained for the downwind portion of the volcanic gas plume on acquisition dates of SAR master and slave scenes, and (**b–e**) corresponding DInSAR maps (10 × 10 km in azimuth and range direction). Scales of the DInSAR maps indicate range change in units of mm, and are unique to each image using the same scale bounds for the sake of comparability with other phase contributions. Incoherent areas are masked out; (**b**) Coarsely corrected DInSAR maps used for decomposition analysis and retrieval of the gas plume estimate. DEM and major APS contributions derived from WRF were removed; (**c**) Phase difference maps of the gas plume estimates obtained for the downwind portion of the volcanic gas plume. Corresponding theoretical dSWDs are indicated in the upper right corner of each map; (**d**) DInSAR maps from 5b, where the phase contributions of the gas plume were removed; (**e**) Linear deformation estimates obtained from the refined correction using the delay estimates obtained from WBDD analysis.

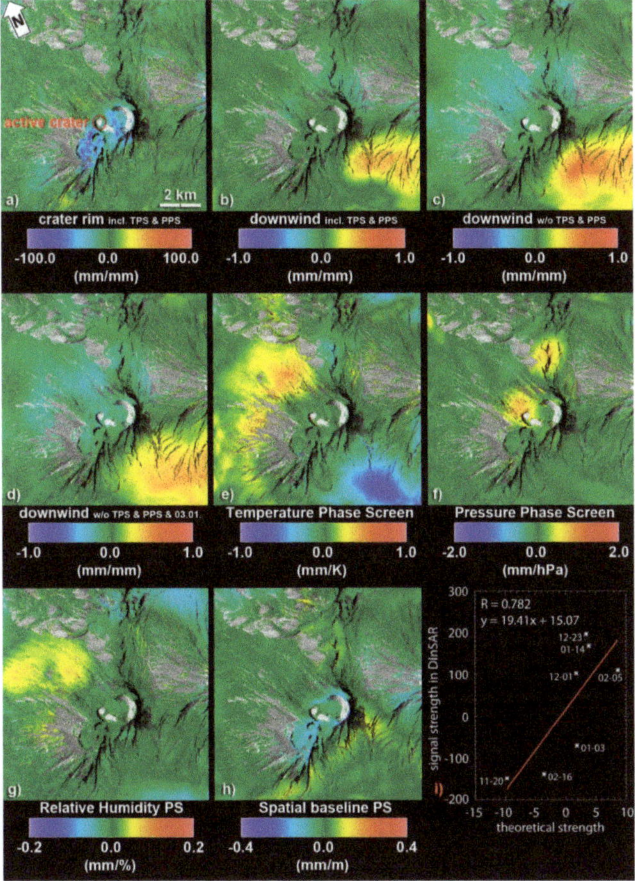

Figure 6. (**a–h**) Delay correlation maps depicting the estimated repeating patterns of SAR delays obtained from *priors* included in the WBDD analysis. Location of active crater is indicated by the *red circle*; (**a**) Delay estimate for the gas plume aloft the crater (**b–d**) Delay estimates for the downwind portion of the gas plume (**b**) with and (**c**) without additional removal of pressure and temperature dependent phase screens by including/excluding respective *priors* in the WBDD analysis; (**d**) Gas plume estimate as in c), however omitting the SAR observation of 3 January 2014. Scales correspond to estimated delay (mm) per theoretical delay (mm). Scale bar limits of the gas plume estimate obtained for the proximal part of the plume above the crater indicate that theoretical delays are by a factor 100 smaller than the delay estimate. Upper and lower limits of the scale bars of the estimates obtained for the downwind portion of the gas plume are equal to unity, indicating concurrence of theoretical and estimated delay. The red to yellow colored signature in the lower right corner of the image indicates a lengthening of the delay, where the effect of H_2O emissions is large. The affected area corresponds to the most common plume transport directions (see Appendix D, Figure A3a,b) and locations where the volcanic plume regularly touches the ground (fumigation); (**e**) Estimates of surface temperature and (**f**) surface pressure related delays. Scales indicate mm estimated delay per Kelvin, respectively hPa of the input *prior*; (**g**) Delay estimate obtained for the relative humidity *prior* (**h**) Delay estimate for the spatial baseline *prior*; (**i**) Scatter plot of theoretical delay (mm) versus determined signal strength of the gas plume contribution in the DInSARs used for the gas plume estimate in (**b**). Each asterisk symbol represents one of the DInSARs and corresponding slave dates (mm-dd) are indicated.

4.3. Isolation of the Gas Plume Related Signal in DInSARs

The phase delay map obtained for the downwind portion of the gas plume (Figure 6b) was used to determine the gas plume related phase contributions in each DInSAR image of the time series. For this purpose the theoretical dSWDs, which were used as a *prior* for this estimate (Table 3, column 4), were multiplied by the respective phase delay estimate, which resulted in simple phase difference maps depicting the repeating gas plume related phase contribution present in each of the DInSAR maps (Figure 5c). The phase difference maps use the same scale bounds as those used in Figure 5b in order to display the relative effect of the gas plume related phase delay on the interferometric measurements with respect to other phase contributions. Each of these maps reveals a lenticular phase difference pattern, which equals the pattern that was already observed in the corresponding phase delay map (Figure 6b), and which persists throughout the entire DInSAR time series. As a consequence of the mathematical operation used to retrieve these maps, the amplitude of the pattern varies in proportion to the a-priori constraints on the differences in gas plume PWV content and is most pronounced in those DInSARs, in which the differential PWV contents determined for the epochs of master and slave SAR acquisitions were the highest (compare Figure 5a,c). Moreover, it is interesting to note that the contrast between plume related and atmospheric signals was more pronounced in the DInSAR images of *subset 01*.

As a next step, the gas plume related phase difference maps depicted in Figure 5c were subtracted from the coarsely corrected DInSAR maps depicted in Figure 5b in order to mitigate the gas plume related phase contribution. The resulting DInSAR maps (Figure 5d) differ clearly from the coarsely corrected DInSAR maps, which still contain the gas plume related signal. As expected, the removal of the gas plume related signal is most obvious in the DInSARs of *subset 01*, where the mitigation in two cases (DInSARs 1 and 3) even resulted in a change to the opposite delay direction over the area which was affected by the gas plume (Figure 5d, rows 1 and 3).

4.4. Residuals of the Refractivity Related Phase Delay and Ground Deformation

To verify the efficiency of our decomposition analysis, we now examine the residual phase, which persists when all estimated refractivity and DEM error related phase contributions have been removed from the DInSAR maps. To this end, the complete set of refractivity and DEM error related phase delay estimates obtained from WBDD analysis as shown in Figures 6 and A4 (and further detailed in Sections 4.2 and 5.3 and the Appendix) were multiplied with the values of their respective input *prior time series* to obtain phase difference maps corresponding to each of the DInSARs. Similarly the *single event APS estimates* of individual SAR acquisitions (Figure A4) were combined to form phase difference maps for each interferogram by subtraction of the slave APS from the master APS. The resulting phase difference maps were then subtracted from the coarsely corrected DInSARs (Figure 5b) obtaining interferograms, which mainly contain the residual phase and topographic contributions attributed to a deforming ground surface (Figure 5e). This refined correction more or less efficiently removed small-scale phase delay patterns, which were caused by the volcanic gas plume and a turbulent atmosphere, as well as small-scale DEM errors related to acquisition geometry, respectively spatial baseline, leaving only some faint and blurry signatures of low amplitude.

In between the blurry residuals of non-deformation related phase contributions a putative small-scale ground deformation signal can be observed in the central of the three north-eastern summit craters of Láscar. The signature occurs in each of the DInSARs, and indicates a progressively subsiding crater floor confined to an arcuate fault, where Pavez et al. [39] detected co-eruptive subsidence of the crater floor during the 1995 eruption, and where Richter et al. [40] identified linear subsidence during the period 2012–2016.

4.5. Validation of the Gas Plume Signal Estimate

The phase delay estimates which were obtained by decomposition analysis can be validated by means of relating the position and strength of the estimated DInSAR signal to the input prior information. We thus compared the signal strength of the lens-shaped gas plume pattern in each of the DInSAR maps (Figure 5b) to the strength of the theoretical dSWDs (Table 3, column 4), which served as a *prior* for the corresponding gas plume estimate (Figure 6b). For this purpose the signal strength S of the gas plume related phase contribution was determined for each interferogram using the amplitude of the gas plume pattern in the phase delay estimate obtained for the downwind portion of the plume (orange color in Figure 6b) by means of following Equation (3).

$$S = sgn(\vec{X} \cdot \vec{Y}) \sqrt{|\vec{X} \cdot \vec{Y}|} \tag{3}$$

here sgn denotes the sign function and $\vec{X} \cdot \vec{Y} = \sum_i X_i Y_i$ is the dot product of vectors $\vec{X}$ and $\vec{Y}$, where $\vec{X}$ corresponds to the amplitude of the gas plume estimate obtained from WBDD and $\vec{Y}$ is the associated measured signal in each of the DInSARs, which were used in the analysis. The comparison revealed a moderate positive linear relationship between signal strength S and theoretical strength (Table 3, column 4), which is fairly well represented by the regression line depicted in Figure 6i, since 61% ($R^2 = 0.61$) of the total variation in measured signal strength can be explained by the linear relationship between theoretical strength and signal strength. Thus, in principle, this regression line can be used to determine the gas plume related phase delays even of those DInSARs for which no corresponding ground-based gas emission measurements are available.

5. Discussion

5.1. Why the Distinction between Periods of Volcano Deformation and Enhanced Degassing is Important

This work constitutes the first serious attempt to quantify gas plume related phase delays in differential interferometric radar measurements of a persistently degassing and deforming volcano. We propose a method that enables us to attain gas plume detection by means of satellite-based radar interferometry and demonstrate that gas plume related phase delays in DInSARs can be isolated and mapped, if a-priori constraints on the gas plume PWV contents obtained from independent gas measurements are included in the analysis of DInSAR data.

The distinction between periods of local volcano deformation and enhanced degassing can be of vital importance both for early warning purposes and for a better understanding of volcano-tectonic processes. Degassing and deforming volcanoes often display a variety of deformation sources, which reflect the complex interplay between (i) magma emplacement/retreat, (ii) accumulation/discharge of volcanic gases and (iii) changes in hydrothermal activity, as e.g., was observed at Campi Flegrei in Italy [68,69], or at Lastarria in Northern Chile [70]. Retention and loss of magmatic and hydrothermal volatiles play a central role in virtually any of the processes that lead to volcano deformation. Gas emission measurements on deforming volcanoes thus provide essential auxiliary information on the ultimate cause of individual deformation observations [71,72], and deformation measurements likewise may deliver explanations for observed changes in the amount and composition of gas emissions [73].

5.2. How Methodological Limitations Can be Turned into Benefit

Area-wide precise ground deformation measurements by means of InSAR require adequate mitigation of atmospheric phase contributions, which may be challenging in small SAR datasets. In particular, repeatedly occurring spatially correlated errors, which e.g., may be caused by orographic clouds and volcanic gas plumes, are difficult to remove [74,75]. Insufficient resolution of meteorological and DEMs used for correction of DInSAR data adds further uncertainties to retrievals of ground deformation. Moreover, most modern SAR satellites revolve Earth on a sun-synchronous dusk-dawn orbit, and record their data along the day-night boundary, i.e., at a time, when emission rates of continuously degassing volcanoes typically are the highest [76–78], which in turn increases the risk of having disturbances due to volcanic water vapor emissions in InSAR measurements. Such noise may, however, be transformed into data, as soon as phase contributions of different sources can be distinguished, which can be accomplished by integration of a-priori constraints.

In our approach we derive prior constraints of volcanic H_2O emissions using measurement techniques, which are commonly used in volcano monitoring, hence providing the benefit of data availability for a comparatively large number of volcanoes. Volcano monitoring generally demands rapid response, in order to assess the volcanic hazard in a timely manner, and the specific information which would be relevant to identify more rapid ground deformations often may be confined to a limited number of available SAR observations. Therefore it seems highly desirable to develop a method that allows removing disturbances of any kind even from small sets of interferograms, where conventional methods as persistent scatterer interferometry and the small baseline subset algorithm would fail. The presented method addresses the problems of the different measurement methods mentioned above, and exploits the fact that DInSAR analysis allows for mapping of water vapor distribution.

In the following subsections, we will discuss the phase delay effects that can be observed in the phase delay maps which were obtained for the proximal and distal sections of the plume, and relate our observations to variations in gas emission rates. Furthermore, we will examine the relative and absolute effects of these phase delays with respect to (seasonal) variations in water vapor contents of the ambient atmosphere. Finally, we will assess the benefits and drawbacks associated with our approach.

5.3. Phase Delay Effects above the Crater and Downwind of Láscar

We now elucidate the contrasting phase delay effects that are observed in the phase delay estimates obtained for the proximal and distal portions of the gas plume (Figure 6a,b).

The phase delay map determined for the downwind portion of the gas plume (Figure 6b) shows a lenticular pattern that indicates lengthening of the radar propagation path due to enhanced moisture over an area that agrees well with the most common eastward plume transport directions observed during spring and late summer of that period (see Appendix D, Figure A1c,d). Westward drifting plumes, which generally may have occurred during early summer (see Appendix D, Figure A1c,d and Figure A2), did obviously not occur at times of the SAR acquisitions we used to obtain our gas plume related phase delay estimates, as these would have prohibited detection of a repeatedly occurring phase delay pattern. Moreover, such westward drifting plumes would less likely leave any repeating traces in the SAR signal, due to a less pronounced humidity/refractivity contrast between plume and surrounding atmosphere in early summer. This is further supported by the fact that easterly winds over the Atacama region typically have a lower velocity (see Appendix D, Figure A2), and tend to be less stable with respect to their direction (see Appendix D, Figure A1c,d), resulting in unsteady plume transport directions, which additionally impedes formation of pronounced repeating patterns. Wind speeds were generally larger during periods with eastward plume transport (spring and late summer; see Appendix D, Figures A1a and A2), causing plume transport to be more turbulent, and thus the volcanic plume was very likely less condensed, and instead contained more vapor during that period.

Additional phase delay estimates of the distal portion of the gas plume were processed without (Figure 6c,d) additionally including prior information on the temporal behavior of surface temperature and pressure in order to show the effect of APS mitigation by the corresponding temperature and pressure phase screens (Figure 6e,f). For the computation of the estimate displayed in Figure 6d we furthermore reduced the number of input DInSARs omitting the interferogram that was formed from SAR acquisitions of 18 October 2013 and 03 January 2014 (Figure 5, row 5). Prior information on relative humidity variations were included in all of the mentioned cases, yielding a relative humidity related phase delay estimate (Figure 6g), which was mitigated from all other phase delay estimates.

Comparison of the resulting gas plume phase delay estimates clearly shows that the repeating patterns of the gas plume signal (Figure 6c,d) can readily be reduced (Figure 6b), when a surface temperature and air pressure correction (Figure 6e,f) is included in the WBDD analysis. The gas plume related phase delay estimate depicted in Figure 6b thus clearly represents the result in which the separation of gas plume related phase delays from other phase contributions was most successful.

The phase delay map obtained for the prior time series of the gas plume related SWDs above the crater rim (Table 3, column 3) shows a weak altitude correlated shortening of the delay over an area which is largely confined to the summit region of the volcano (Figure 6a). The location of this phase delay signature is thus consistent with the assumption that the dSWDs, which were used as a *prior* for this estimate (Table 3, column 3), are representative for the temporal evolution of SAR signal strength variations measured above the crater. The observed shortening of the delay, however, contradicts the hypothesis that this signature can be ascribed to enhanced moisture over the respective area, since a water vapor field would produce a lengthening of the delay (compare Equation (1) in Smith and Weintraub [66]). Furthermore, the phase delay patterns in this estimate strongly resemble the patterns observed in the DEM error related phase delay estimate (Figure 6h), which was obtained using the temporal history of the spatial baseline (Table 3, column 8), suggesting that the phase delays, that the algorithm attributed to the gas plume above the crater, are in fact rather related to errors in the DEM (detailed in Appendix F). This suspicion is further substantiated, if one compares the shape of the two corresponding *priors*, which strongly resemble each other in appearance. As a consequence of this similarity, the signal obviously was assigned to both *priors* by the WBDD algorithm, since this signal does not perfectly match to any of the *priors*. This phase delay estimate thus is a good example for a case in which the algorithm failed to isolate possible gas plume related phase delays from other phase contributions. Thus, care must be taken to avoid misinterpretation of model results.

5.4. Gas Emission Rates from Láscar Volcano

Gas emission rates from Láscar volcano were determined and compared to our PWV estimates (Figure 4b,f), in order to illustrate how much water vapor has been emitted from the volcano to produce the corresponding PWV contents in the gas plume, and to relate our observations to variations in volcanic degassing activity. SO_2 emission rates were calculated from scanning DOAS measurements using wind field information obtained from GDAS1 soundings provided in the web-based archive of the National Oceanic and Atmospheric Administration's (NOAA) Air Resources Laboratory [79]. These were then upscaled by the H_2O/SO_2 molar ratio of 34:1 obtained from Multi-GAS, in order to retrieve the corresponding H_2O-fluxes (Figure 7a).

Assuming a constant H_2O/SO_2 ratio and a constantly strong condensation of the emitted water vapor, the H_2O (SO_2) emission rates ranged from 5 to 100 $\text{kt}\cdot\text{day}^{-1}$ (150 to 2900 $\text{t}\cdot\text{day}^{-1}$) and averaged at 12 $\text{kt}\cdot\text{day}^{-1}$ (350 $\text{t}\cdot\text{day}^{-1}$) during the considered period. Based on these assumptions we estimate that the total errors of our upscaled H_2O emission rates likely were around -15 to +45% and thus even more skewed towards underestimation than the error envelopes of corresponding SO_2-fluxes, which were constrained to about -10 to +30% (see e.g., [80] for error determination). Gas emission rates of October and November 2013 showed more pronounced variations and were on average slightly enhanced (about 19 kilotons of H_2O and respectively 550 tons of SO_2 per day) with respect to the rest of the period (10 kilotons of H_2O and respectively 300 tons of SO_2 per day). The enhanced degassing activity of October and November further was accompanied by elevated heat dissipation from the main active crater as was suggested by intermittent night-time webcam observations of illuminated gas plumes (obtained on several days during the first half of October and lastly on 20 November 2013; see Figure 7a), which indicated the presence of incandescent material inside the active crater [81]. Such observations support the repeated influx of fresh magma from depth feeding the high-temperature gas exhalations at the surface of the crater bottom, thus suggesting a concurrent increase of magmatic gas emissions and progressive drying of the ambient hydrothermal system (e.g., [41]), which may have altered gas ratios temporarily. Contemporaneous measurements of SO_2

emission rates confirmed this suspicion and appropriately showed sharp increases over the course of each of these days (Figure 7a).

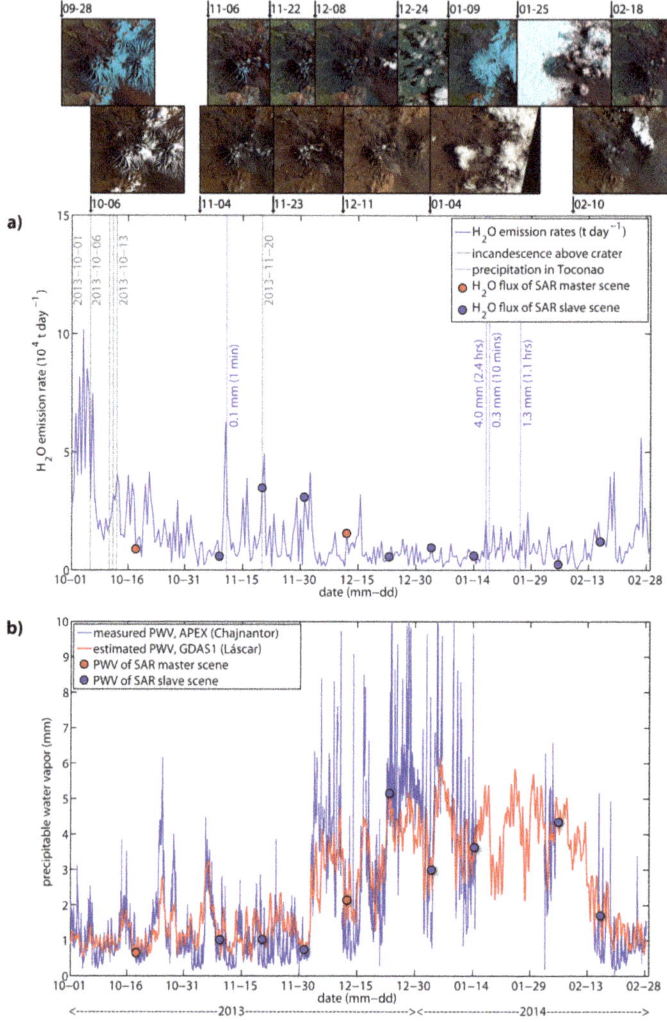

Figure 7. (**a**) H_2O emission rates from Láscar volcano along with a sequence of Aster, Landsat-8 and EO-1 ALI scenes depicting snow and cloud coverage (upper row: false-color composites of SWIR, NIR and visible red spectral bands; lower row: true-color images using a combination of visible red, green and blue spectral bands). Days with incandescence observations are indicated by *stippled grey vertical lines*. Precipitation events that were recorded in Toconao (on 10 November 2013, 17 January 2014, and 26 January 2014) are indicated by *vertical blue lines*, including information on amount and duration of the events; (**b**) Background atmospheric PWV contents estimated from GDAS1 soundings above Láscar volcano (*red curve*), compared to radiometer measurements conducted at APEX (*blue curve*). Good agreement between the two curves reflects similar weather conditions caused by a similar morphologic exposition of both sites, and may additionally be attributed to the coarse spatial resolution of the GDAS1 data. PWV estimates for the times of SAR observations are indicated by *red* (master scene) and *blue* (slave scene) *dots*, respectively.

Bearing these observations in mind, we compared the gas emission rates with the estimated PWV contents of the gas plume (Figure 4b,f) revealing that gas emission rates evolved differently than the PWV contents obtained for the gas plume above the crater (Figure 4b), but displayed a very similar temporal evolution as that of the PWV contents, which were obtained for the downwind portion of the gas plume (Figure 4f). The latter thus likewise were enhanced during the period of increased volcanic activity.

5.5. Estimation of Background Atmospheric PWV Contents

Atmospheric disturbances due to volcanic gas plumes are intuitively expected to be more pronounced during periods which are characterized by strong variations of gas emission rates and low ambient atmospheric humidity. Determination of PWV contents in the local atmosphere was thus done to be able to approach the relative and absolute effects of the plume on the specific SAR acquisitions. In the Atacama region the driest conditions are typically found during the period late March to early December, and between local midnight and 11:00 a.m. [82]. Background PWV contents in the atmosphere above the volcano were retrieved from vertical atmospheric GDAS1 profiles provided by NOAA (see Appendix C for equations). These estimates were also cross-checked with PWV estimates obtained from radiometer measurements (Figure 7b) provided by the Atacama Pathfinder EXperiment (APEX), which is located about 30 km north of Láscar at a very similar altitude of 5100 m above sea level (m.a.s.l.), thus reflecting similar weather conditions.

Because PWV is a measure of how much water is available for potential rainfall, it dictates precipitation intensity, which thus reversely can be used to infer accompanying PWV. To give an impression of when to expect enhanced PWV contents and to further complete the picture of the hydrological cycle of the study area, we will first briefly describe the characteristic seasonal patterns according to which precipitation typically occurs at Láscar, and narrow down the periods during which precipitation actually was encountered in the course of the observation interval of our case-study.

The considered period covers parts of the "dry season", which is characterized by dry and cold prevailing westerly synoptic winds, and is spanning most of the year (here October, November and second half of February are included), as well as the "wet season" ranging from December to February, which is characterized by less stable atmospheric conditions, resulting in more humidity being transported to the Andes by frequently recurrent warm easterly winds [83]. As a consequence of this easterly moisture source, precipitation on the leeward slopes of the Puna plateau exhibits a marked seasonality in particular at elevations below 3500 m.a.s.l., where about 90% of the annual precipitation ($<$20 mm$\cdot$year^{-1}) falls during the (austral) summer months, that is from December to March [84]. The associated rainfall events typically are very short, but can become quite intense. During the considered period, a weather station (operated by the Instituto de Investigaciónes Agropecuarias) located at 2500 m.a.s.l. in Toconao, a small village roughly 30 km NW of Láscar, recorded 4 events of precipitation which amounted to a total of 5.7 mm (equivalent to liters per square meter) within only 3.6 h (vertical blue lines in Figure 7a). At elevations above 3500 m.a.s.l. the situation is slightly different and precipitation ($<$200 mm$\cdot$year^{-1}) occurs even during the dry winter period, mostly in the form of snow with the greatest snowfall frequency happening to occur near the Tropic of Capricorn [85], where also Láscar coincidentally is located. Láscar volcano thus commonly is almost completely (except for the hot bottom of the currently active crater) covered with several meters of snow during the "dry season", which however melt and sublimate due to the warm winds of the "wet season". Such snowmelt events hence constitute an important source for groundwater recharge, which potentially affect the hydrothermal activity of Láscar. The temporal development of the snow and cloud cover encountered at Láscar during the period of our case-study is illustrated by a sequence of satellite images (Aster, Landsat-8, and EO-1 ALI) in the top panel of Figure 7a in order to give a visual impression of the situation associated with different epochs of the temporal development of atmospheric moisture.

Measured and estimated atmospheric PWV contents were at average below 1 mm during October and November 2013, sharply increased to 5 mm in December 2013, stayed enhanced through the first half of February 2014, and finally decreased again to 1 mm in the second half of February 2014 (see Figure 7b and also Table 2, column 6). The comparatively low PWV values determined for the times of SAR observations used in this work reflect the pronounced atmospheric transparency around sunrise, and the resulting SWDs (Table 2, column 11) are in good agreement (R^2 = 0.96; see Appendix G for detailed statistics) with the amplitudes of associated phase delays observed in the corresponding DInSARs (Figures 5b and A4).

It is interesting to note that the apparent SO_2 (H_2O) emission rates generally were particularly low whenever the ambient atmospheric water vapor levels were elevated, which is clearly visible especially during the "wet season" (compare Figure 7a,b). This observation suggests the presence of low hanging clouds beneath the volcanic gas plume, which may have caused attenuation of the absorption signal due to contributions of scattered light that did not pass through the plume [86]. Moreover, a fraction of the degassing SO_2 may have been scrubbed by dissolution into the aqueous phase e.g., during possible events of fresh water influx into the hydrothermal system [87], though this would most likely only have a marginal effect on the emissions of the hot degassing dome of Láscar, which is less probably being deeply infiltrated by permeating waters that interact with ascending magmatic fluids, than the peripheral low-temperature vent sites. Another mechanism, which may be considered to explain the observed diminution of SO_2-fluxes is the dissolution of emitted SO_2 in the water droplets of low hanging clouds [88], which, however, usually is not particularly effective over a short distance, and the associated scavenging rate of SO_2 emissions is generally estimated being on the order of merely a few percent of total emitted SO_2 per hour. This in turn leads to relatively long atmospheric lifetimes of SO_2 emissions, which were shown to range between several hours to days according to weather conditions [89].

Estimated PWV contents inside the volcanic gas plume (Figure 4b,f) and modeled atmospheric PWV contents outside the plume (Figure 7b) were compared in order to assess the humidity/refractivity contrast between ambient atmosphere and volcanic plume. The comparison revealed that PWV contents at the downwind plume centerline (0.8 to 9.6 mm) were generally larger than the PWV contents of the ambient air (<1 mm) during periods with dry atmospheric conditions, i.e., most of the year. During early summer the situation was different, and the downwind plume PWV content (<1 mm) was less contrasting with respect to the PWV content of the atmospheric background (2–5 mm). PWV contents of the downwind bulk volcanic plume (0.2 to 2.5 mm) were accordingly much less contrasting most of the year when compared to background atmosphere, and were slightly higher than ambient atmospheric humidity only during the first three SAR observations of the time series (on 18 October 2013, 20 November 2013, and 01 December 2013). Future studies exploiting a larger InSAR dataset of Láscar might therefore well capture these seasonalities in gas plume and atmospheric PWV contents, the latter of which were already identified at Láscar [39] but also elsewhere, e.g., at Campi Flegrei and Vesuvius [90]. PWV contents determined for the proximal part of the plume above the crater rim were in general negligible with respect to PWV contents of the ambient atmosphere (0.007 to 0.015 mm on the centerline of the plume, and 0.002 to 0.0035 mm averaged over the bulk plume).

The corresponding SWDs obtained for the proximal part of the plume above the crater rim ranged from 0.07 to 0.1 mm in the centerline of the plume (Table 2, column 8) and from 0.02 to 0.03 mm in the bulk plume and thus were negligible compared to the InSAR accuracy. The SWDs in the more distal part of the plume downwind of the crater, however, were on the order of background atmospheric SWDs (5.3 to 41.5 mm; compare with Table 2, column 11), and ranged from 1.6 to 20 mm in the bulk volcanic plume and from 6.4 to 77.0 mm along the plume centerline (Table 2, columns 9 and 10), which is more than large enough to produce a detectable effect in interferometric measurements, as can be seen in the corresponding phase delay maps (Figure 5c).

Our results thus collectively suggest that refractivity changes in volcanic gas plumes may, particularly under dry climatic conditions, have a significant effect on DInSAR measurements, hence requiring extreme caution, when alleged deformation signals are detected above or downwind of visibly degassing volcanoes, especially if these are located at high elevations.

5.6. Limitations of the Proposed Method

Our approach was hampered by several practical limitations, which were mainly related to spatial data coverage and temporal coincidence of the different measurement techniques that we combined.

Simultaneous measurements of Multi-GAS and DOAS instruments were unfortunately not available for the period considered for DInSAR analysis, and thus our PWV estimates had to rely on the gas concentration measurements obtained during a short Multi-GAS survey conducted about one year prior to the period considered here. This is of particular relevance, since Láscar exhibited elevated SO_2 emission rates and incandescence during the first two months of this period, suggesting a slightly enhanced magmatic activity, which likely was accompanied by temporarily altered gas ratios. At the same time, the snow cover on the summit of the volcano was observed to be reducing strongly (satellite images in Figure 7), which may be an indication for concurrent fresh water influx into the peripheral hydrothermal system. Thus, there is evidence for both, elevated emissions of magmatic SO_2 and increased hydrothermal H_2O contributions of meteoric origin. The assumption that molar H_2O/SO_2 ratios stayed more or less constant throughout the entire period of our case study therefore may not be perfectly appropriate. Availability of data from a stationary Multi-GAS instrument certainly would have helped to better characterize the temporal variations of molar H_2O/SO_2 ratios, and thus to better quantify the water vapor contents at the instance of gas sampling by the Multi-GAS.

Moreover, volcanic gas plumes typically contain variable amounts of condensed liquid water in addition to gaseous water, which are not taken into account by the measurements of Multi-GAS instruments, and thus unequivocally lead to an underestimation of the water vapor contents, which are expected to increase in the downwind portion of the volcanic gas plume due to evaporation. Such an underestimation could be avoided, if water vapor contents were mapped in situ across the whole extent of the gas plume, e.g., by means of airborne gas composition measurements, though this approach would not be very practicable but rather labor and cost-intensive in the view of satellite revisit times of 11 days.

Additionally, the DOAS measurements are generally restricted to daylight conditions and thus deviate from SAR acquisition times by up to 2 h (Figure 2b). Due to this deviation in acquisition times and because the proposed DInSAR decomposition analysis requires only one representative PWV value for each SAR image (see Sections 3.6 and 3.7), we thus used daily average gas plume PWV contents, in order to determine the SWD that such PWV contents theoretically would produce in the LOS of a SAR image. In the view of strong diurnal degassing variations, which are characterized by pronounced gas emission maxima around sunrise and sunset [77,78], such time averaged values however inevitably lead to systematic underestimation of the water content in the volcanic plume at the times of the SAR observations.

Furthermore, our analyses have shown that different locations within the gas plume may exhibit a different temporal evolution of the PWV contents, which may appropriately be captured by inclusion of corresponding *priors*. However, we point out that modeling the evolution of water vapor contents over the entire extent of the gas plume even by means of two *prior time series* still is a simplification. Prudent sampling of the gas emission data in time and space hence constitutes an important prerequisite in order to obtain a-priori information, which yields feasible results from DInSAR decomposition analysis.

This becomes particularly clear, if one notes that the position of the plume center varied during the measurement period, because the gas plume of Láscar somewhat meanders within the scan range of the scanning DOAS. Transport of the plume changed between north-easterly and south-easterly directions approximately every 10 min, which roughly corresponds to the duration of 1 complete DOAS scan. This resulted in different plume positions in consecutive scans, which is why the sequences of consecutive scans in Figure 4a,e have a corrugated appearance. We thus assume that the resulting daily average plume center statistic may not be representative for the effect that the bulk plume has on SAR measurements further downwind of the volcano, since the gas distribution along plume cross sections is highly irregular when the plume meanders. For this reason, we chose to use the PWV contents of the bulk gas plume instead of those from the plume center in order to estimate related phase delay in the downwind portion of the gas plume.

Further limitations of our approach lie within the applicability of the WBDD technique. The method makes extensive use of ancillary data, which have to be made available to obtain reasonable results. To successfully isolate the plume-induced signal from total phase, it is necessary to additionally include the *priors* of all potentially interfering signals in the analysis, i.e., the temporal and spatial baseline histories, which were used as *priors* for the estimation of phase delays related to DEM error and linear deformation, as well as several *priors* defining the atmospheric phase delay of each acquisition. We consider that, even when all required *priors* are being used, the estimated gas plume related signal still may additionally contain phase information related to any other interferometric signal that coincidently correlates with the gas plume related signal (as was shown in Section 5.3 for the phase delay estimates depicted in Figure 6a,h). In the presence of a non-linearly deforming ground surface it may additionally be necessary to include prior information on the temporal behavior of the associated deformation signal, which e.g., could be derived from deformation measurements of independent ground-based methods, such as GPS, or tiltmeters. Last, but not least, the DInSAR decomposition analysis exclusively yields reasonable results for the gas plume estimate when there is significant overlap between gas plumes of different SAR acquisitions, since "solitary" plumes would be captured by the *single event APS estimates* causing a gap in the temporal evolution of the repeating pattern. To capture the whole gas plume at the times of SAR acquisitions instead of only the intersecting area of multiple different gas plumes definitely would have been an asset.

5.7. Advantages of the Proposed Method

The WBDD technique provides a framework for data fusion, enabling to integrate a variety of external information, such as weather data, results from gas emission monitoring or even independent deformation measurements into the analysis of DInSAR time series. These time series may be very short and it is shown that the combination of only three SAR acquisitions already enables us to generate high-quality DInSARs including estimates of their corresponding phase screens (as demonstrated in Appendix H). The algorithm does not require any information on the location of the plume, in order to find the related "disturbance" in the interferometric signal. This is of particular advantage to the application described here, since we lack precise information on the heading of the plume at the times of SAR observations, apart from a rough approximation that can be derived from webcam images and wind directions offered by the weather models. Also, no information about the interferometric signal strength, rather merely the shape of the prior time series is used, because prior information about the relations between signal strength and amplitudes of the prior time series is allotted by chance, which is why the prior information can be scaled arbitrarily without influencing the location of the associated filtered phase delay patterns. Correspondingly, the radar path delay estimates can be validated by the position and strength of the estimated signal in relation to the corresponding prior information (Section 4.5). The gas plume is transported by advection, therefore has a direction, and is correspondingly anisotropic. For this reason, the WBDD technique arguably is the best choice for the detection of gas plumes in interferometric measurements. Additionally, the ability to use temporally unconnected time series subsets enables to use only those DInSARs which have a very small spatial baseline. This helps avoid unwrapping errors due to erroneous DEMs.

6. Summary and Conclusions

Here, we presented a novel time-space-based filtering method that allows for the isolation and mitigation of virtually any conceivable non-intrinsic error source that affects interferometric ground deformation measurements, by means of including prior constraints on their temporal behavior in the

DInSAR analysis. Including *prior* information on gas plume related refractivity variations furthermore enabled us to identify and map spatially correlated repeating interferometric phase delay patterns caused by refractivity changes within volcanic gas plumes that were drifted in a similar direction during several subsequent SAR acquisitions. Apart from potentially being a new tool to detect volcanic gas plumes in DInSAR data, the WBDD algorithm thus provides the possibility to mitigate the plume-induced phase delay, where deformation measurements are the main purpose of monitoring.

Similarly to the small baseline subset algorithm our technique operates on a pixel-by-pixel basis on areas that show sufficient coherence throughout the entire DInSAR time series. The algorithm requires only a relatively short time series comprising at least 2 temporally interconnected DInSARs, respectively 3 observations in order to capture the plume related phase delay, and when more than 2 interferograms are available, these do not necessarily have to be temporally connected. Moreover, the method allows for iterative determination of PWV contents in the volcanic gas plume by matching the estimated phase delays with theoretical phase delays derived from ground-based gas emission measurements, which resulted in reasonably realistic values for the gas plume of Láscar volcano.

To our knowledge this is the first time that phase delay effects in volcanic water vapor emissions were quantitatively investigated by means of radar interferometry. Examination of PWV contents and associated phase delay effects in the volcanic gas plume of Láscar yielded daily average bulk plume PWV contents of 0.2 to 2.5 mm water column, which would generate plume wide excess path delays in the range of 1.6 to 20 mm. The corresponding H_2O emission rates ranged from 5 to 100 kt·day^{-1} and remarkably displayed a similar temporal behavior as the PWV contents that were obtained for the downwind portion of the gas plume. Our observations consider that the locations of the interferometric patterns in the phase delay estimates are consistent with determined locations of the volcanic plume, and variations in phase delay amplitudes are in good agreement with variations in the strength of the emission source and humidity of the ambient atmosphere. Therefore there is ample evidence, that the DInSAR phase variations detected by the WBDD algorithm are indeed related to refractivity variations in the volcanic plume. Additional independent data on plume location and chemistry at the time of SAR acquisitions would help to further corroborate our results as a proof of concept.

We have illustrated that integrating a-priori information from gas emission measurements and meteorological data into the analysis of DInSAR time series provides previously unknown possibilities to efficiently decompose the radar signal into different phase contributions. The implications of this new technique are wide, ranging from improved deformation measurements to simultaneous quantification of volcanic outgassing activity, DEM error estimation, and meteorological applications, such as mapping and quantification of turbulently transported atmospheric PWV in each SAR acquisition. Our results furthermore encourage the use of data from ground-based gas emission monitoring networks, such as NOVAC for calibration and validation of satellite-based measurements on active volcanoes.

Author Contributions: Conceptualization, S.B. and F.-G.U.; Methodology, S.B. and F.-G.U.; Software, F.-G.U.; Validation, F.-G.U. and S.B.; Formal Analysis, F.-G.U. and S.B.; Investigation, S.B. and F.-G.U.; Data Curation, S.B. and F.-G.U.; Writing-Original Draft Preparation, S.B., F.-G.U., T.H.H. and T.R.W.; Writing-Review & Editing, S.B., T.H.H., T.R.W., and F.-G.U.; Visualization, S.B. and F.-G.U.; Supervision, T.R.W. and T.H.H.; Project Administration, T.R.W. and T.H.H.; Funding Acquisition, T.R.W. and T.H.H.

Funding: This project was funded by the Helmholtz Association through the "Remote Sensing and Earth System Dynamics Alliance", which was realized in the framework of the Initiative and Networking Fund, and by the GEOMAR Helmholtz-Centre for Ocean Research Kiel (project-id: HA-310/IV010). This is a contribution to VOLCAPSE, a research project funded by the European Research Council under the European Union's H2020 Programme/ERC consolidator grant no. ERC-CoG 646858.

Acknowledgments: The authors would like to thank Claudia Bucarey from Observatorio Volcanológico De los Andes del Sur (OVDAS) for installation and maintenance of the permanent scanning DOAS instrument at Láscar volcano, providing an invaluable gas emission data set. TerraSAR-X High Resolution Spotlight scenes were provided by DLR and were tasked and acquired by the proposal WA1642. Further acknowledged is the Atacama Pathfinder EXperiment for providing archived weather and radiometer data, and the Instituto de Investigaciónes Agropecuarias for providing the precipitation data of their weather station in Toconao on the server of the Dirección Meteorológica de Chile (Meteochile). The Aster, Landsat-8 and EO-1 ALI images used in this study are made available by the U.S. Geological Survey through the USGS EarthExplorer platform (https://earthexplorer.usgs.gov/). Michael Eineder is thanked for fruitful discussions and encouragements regarding the work on this paper. Martin Zimmer, Christian Kujawa, Elske de Zeeuw-van Dalfsen, Nicole Richter, Mehdi Nikkho, Jacky Salzer, Giancarlo Tamburello, and Ayleen Gaete for joint field work activities and discussions. The editors and anonymous reviewers are thanked for their comments that helped improving the manuscript.

Conflicts of Interest: The authors declare no conflict of interest.

Appendix A. Estimation of PWV Contents in the Volcanic Cloud

Water vapor contents in the volcanic plume were estimated by scaling the SO_2 column density profiles measured by our scanning DOAS with the molar H_2O/SO_2 ratio derived from gas compositional measurements of the Multi-GAS instrument. For this purpose we used the 'raw' differential SO_2 SCD output of the NOVAC-software [91], which in our case provides SO_2 path length concentrations in units of ppm·m, that is SO_2 concentration distributed along the effective light paths captured at each observing angle of the scans. Several conversions of the 'raw' data were necessary, in order to estimate the precipitable water vapor content of the volcanic cloud in a vertical atmospheric column right above each point of the plume cross-sections.

The light paths associated with the conical viewing geometry of the scanning DOAS instrument are generally inclined with respect to the vertical, entailing amplification of the absorption signal in relation to the signal that would be obtained along the more direct zenith light path due to extension of the distance that the incoming light has to travel through the absorbing air mass. Extension of the effective light path along inclined viewing directions is typically taken into account by means of the so-called air mass factor (AMF), which relates the measured SO_2 SCDs to their respective vertical column densities (VCDs). The SO_2 VCDs thus simply correspond to the ratio of SO_2 SCDs and air mass factor as expressed in Equation (A1) (e.g., [92]).

$$SO_2\ VCD\ (ppm{\cdot}m) = \frac{SO_2\ SCD}{AMF_G} \tag{A1}$$

Here, we used a geometrical approximation of the air mass factor AMF_G, that can be derived for each viewing direction of a scan by means of Equation (A2) [91]

$$AMF_G = \sqrt{\frac{(cos(\beta)\ cos(\delta)\ + sin(\beta)\ cos(\theta)\ sin(\delta))^2 + (sin(\beta)\ sin(\theta))^2}{(cos(\beta)\ sin(\delta) +\ sin(\beta)\ cos(\theta)\ cos(\delta))^2} + 1}, \tag{A2}$$

where θ is the scan angle (or zenith angle, which corresponds to the angle between observing elevation and zenith position), $\delta = 0$ is the tilt of the scanner of our DOAS station at Láscar, and $\beta = 60$ is half the opening angle of the conical scanning surface. More sophisticated radiative transfer corrections were not performed additionally. The resulting SO_2 VCDs are given as path length concentrations in units of ppm·m, and were used to approximate the corresponding PWV contents in units of mm water column. This required several steps of conversion, which were combined in Equation (A3).

$$PWV\ (mm) = 10\ mm \times \frac{SO_2\ VCD(ppm{\cdot}m) \times 2.5 \times 10^{15} SO_2\ molecules{\cdot}cm^{-2} \times H_2O/SO_2}{3.34 \times 10^{22} H_2O\ molecules{\cdot}cm^{-2}} \tag{A3}$$

Single steps of this conversion equation were derived from the following relationships. A path length concentration of 1 $ppm{\cdot}m\ SO_2$ roughly corresponds to a column density of $2.5 \times 10^{15} SO_2\ molecules{\cdot}cm^{-2}$ at standard conditions for temperature and pressure, that is 298.15 K and 1 atmosphere air pressure (e.g., [91]). The corresponding number of vertically integrated water vapor molecules per square centimeter thus is readily obtained by scaling the SO_2 VCDs in units of molecules·cm^{-2} with the H_2O/SO_2 molar ratio determined from the Multi-GAS measurements. H_2O VCDs in units of molecules·cm^{-2} further can be converted to the total amount of water vapor present in a vertical atmospheric column, also known as column integrated water vapor (IWV), which is commonly stated as the vertically integrated mass of water vapor per unit area (e.g., kg·m^{-2}). At conditions of 1 atmosphere air pressure and 273.15 K one can find that an IWV of $1\ g\ H_2O\ cm^{-2}$ $(=10\,kg\ H_2O\ m^{-2})$ roughly corresponds to a column density of 6.02214×10^{23} $molecules{\cdot}mol^{-1}/18.015\ g{\cdot}mol^{-1} = 3.34 \times 10^{22}\ H_2O\ molecules{\cdot}cm^{-2}$, and respectively to 10 mm of an equivalent column of liquid water (A4) (pp.3 and 4 of the Appendix in McClatchey et al. [93]; or p.3 in Wagner et al. [94]).

$$3.34 \times 10^{22} H_2O\ molecules{\cdot}cm^{-2} = 1\ g{\cdot}cm^{-2} = 10\ kg{\cdot}m^{-2} \approx 10\ mm\ water\ column \quad \text{(A4)}$$

Conversion of IWV in units of kg·m^{-2} to PWV in units of mm water column is obtained using Equation (A5) (e.g., [61])

$$PWV(mm) = \frac{IWV}{\rho_{LW}}, \quad \text{(A5)}$$

where the density of liquid water ρ_{LW} typically roughly is equal to $1\ g{\cdot}cm^{-3}$, and thus a water vapor column density of $1\ kg{\cdot}m^{-2}$ roughly corresponds to 1 mm height of an equivalent column of liquid water.

Appendix B. Compensation of Downwind Evaporation

LWCs in the plume above the crater rim were unknown, thus it was not possible to determine the actual evaporation rate from a predefined amount of liquid water. Considering increasing water vapor concentrations, respectively increasing H_2O/SO_2 ratios in the downwind portion of the volcanic plume, we thus calculated the drying power of the ambient air, respectively potential evaporation rates at plume height, based on climatic variables obtained from GDAS1 soundings, and terrain surface roughness. Potential evaporation rates were then used to scale our fixed H_2O/SO_2 ratio in order to compensate for downwind evaporation.

Potential evaporation rates correspond to the amount of liquid water that can be transformed to vapor through evaporation, if sufficient water is available (e.g., [95]). Assuming an unlimited availability of water, evaporation can in such a situation continue until the air above the evaporating surface approaches saturation humidity, and therefore actual evaporation cannot exceed potential evaporation. In the present case, availability of liquid water does not seem to be a limiting factor inside the plume, since we made the observation that the plume typically still is partly condensed when it arrives at the scanning DOAS. We thus assume that scaling by the potential evaporation rate yields a feasible approximation of the increase in water vapor at the expense of liquid water in the downwind portion of the volcanic cloud. We further assume that downwind dilution of water vapor due to dispersion of the plume is largely covered by the variations of the SO_2 column densities measured by the DOAS, since both water vapor and SO_2 are equally affected by diffusion and turbulent mixing processes (e.g., [96]). The amount of evaporation occurring inside the plume is largely controlled by the humidity of the entrained ambient air and the degree of turbulent mixing with air parcels of the plume, which is mainly governed by wind speed and atmospheric stability. Evaporation is thus commonly more pronounced during periods with dry atmospheric conditions and high wind velocities, which cause the transport of the plume to be more turbulent, and it is less significant, when the atmosphere is more humid and less turbulent, due to small transport velocities.

Evaporation rates were calculated by an aerodynamic (respectively mass-transfer) method (e.g., [97]), using temperatures, dew point temperatures, and wind speeds obtained from GDAS1 soundings provided by the National Oceanic and Atmospheric Administration (NOAA) [79]. The aerodynamic approach can be expressed by a Dalton-type Equation (A6) [98,99],

$$E_P\left(cm{\cdot}sec^{-1}\right) = f(u)(e_s - e), \quad \text{(A6)}$$

in which E_P is the rate of potential evaporation (cm·sec^{-1}), $f(u)$ is a wind speed function (cm·sec^{-1}·hPa^{-1}), respectively the vapor transfer coefficient, which is based on a logarithmic vertical wind velocity profile, and the second term $(e_s - e)$ is the humidity term corresponding to the vapor pressure deficit of air (hPa). The product of both is typically referred to as the drying power of air. The wind function of the Dalton-type equation can be written as (A7) (Equation (4) in [100])

$$f(u)\left(cm{\cdot}sec^{-1}{\cdot}hPa^{-1}\right) = \frac{\rho_{AIR}\,\epsilon k^2}{\rho_{LW}\,P}\,\frac{u_1}{ln(z_1/z_0)^2}, \tag{A7}$$

where ρ_{AIR} and ρ_{LW} correspond to the density of moist air and liquid water (g·cm^{-3}), respectively. $\epsilon = 0.622$ is the ratio of the molar mass of water vapor (18.015 $g{\cdot}mol^{-1}$) to the average molar mass of dry air (28.9647 $g{\cdot}mol^{-1}$), and $k = 0.4$ is the dimensionless von Kármán's constant describing the logarithmic velocity profile of a turbulent air flow near a rough boundary with a no-slip condition [101], i.e., it is assumed that wind speed is zero directly above the surface. u_1 refers to the wind speed (cm·sec^{-1}) at measurement height z_1 (cm above ground surface), and P is barometric pressure (mbar or hPa) at measurement height z_1. z_0 is the surface roughness height (cm above surface), which corresponds to the average height of obstacles in the trajectory of the wind [102]. The density of moist air ρ_{AIR} ($g{\cdot}cm^{-3}$) was calculated using the ideal gas law (A8)

$$\rho_{AIR}\left(g{\cdot}cm^{-3}\right) = \frac{0.1P}{R_d(1+0.608qv)(T+273.15)}, \tag{A8}$$

where $R_d\left(J{\cdot}kg^{-1}{\cdot}K^{-1}\right) = 287.04$ is the gas constant for dry air, P is barometric pressure (hPa), T air temperature (°C), and qv specific humidity (g·g^{-1}), which in turn was approximated by Equation (A9).

$$qv\left(g{\cdot}g^{-1}\right) = \epsilon\frac{e}{P} \tag{A9}$$

Here, $\epsilon = 0.622$ again is the ratio of the molar mass of water vapor to the average molar mass of dry air, e is the partial pressure of water vapor in hPa, and P is barometric pressure in hPa.

Saturation vapor pressure e_s (hPa) was calculated using the August–Roche–Magnus formula (A10) [103] with coefficients determined by Sonntag [104] (Table 1 in [105]).

$$e_s\,(hPa) = 6.112exp\,\frac{17.62T}{243.12+T}, \tag{A10}$$

where T is temperature in Celsius degree. Water vapor partial pressure e (hPa) was calculated as a function of relative humidity and saturation vapor pressure using Equation (A11)

$$e\,(hPa) = \frac{RH}{100}e_s, \tag{A11}$$

where $RH/100$ is the fractional relative humidity, and e_s saturation vapor pressure at the ground level of the GDAS1 sounding. Relative humidity (%) was obtained by comparison of actual water vapor partial pressure e and saturation vapor pressure e_s, using Equation (A12) with coefficients determined by Alduchov and Eskridge [105].

$$RH\,(\%) = 100\frac{e}{e_s} = 100\frac{exp(17.625T_D)/(243.04+T_D)}{exp(17.625T)/(243.04+T)}, \tag{A12}$$

where T and T_D are air temperatures, and respectively dew point temperatures in Celsius degree. Density of liquid water ρ_{LW} (kg·m^{-3}) was calculated from temperatures T (°C) at plume height using Equation (A13) that was empirically determined by Jones and Harris [106].

$$\begin{aligned}\rho_{LW}\left(kg{\cdot}m^{-3}\right) = {} & 999.85308 + 6.32693\times10^{-2}\,T - 8.523829\times10^{-3}\,T^2 \\ & -6.943248\times10^{-5}T^3 - 3.821216\times10^{-7}T^4\end{aligned} \tag{A13}$$

The roughness height z_0 was approximated by Equation (A14), which was experimentally determined by Plate and Quraishi [107] in wind tunnel experiments, and which yields fair results in the absence of more precise information [108].

$$z_0\ (cm) = 0.15h, \tag{A14}$$

where h is the average height of roughness elements (cm). As in this work the volcano is the main obstacle, and since the evaporating surface, respectively the volcanic plume typically is located roughly at summit altitude of the volcano, the average height of roughness elements relevant for the turbulent mixing of the plume with the atmosphere, was assumed to be equal to the height of the volcano above the surrounding plateau (about 750 m).

Appendix C. Estimation of PWV Contents in the Atmosphere

Background PWV contents in the atmosphere above the volcano were approximated by means of an empirically determined regression Equation (A15), which is based on the experimental results of Garrison and Adler [109] and proved to be applicable over a broad range of climatic conditions (e.g., [110]). The calculation requires climatic variables including pressure, dew point temperature and temperature, which were obtained from vertical atmospheric GDAS1 profiles provided by NOAA.

$$PWV(mm) = \frac{4.1173\ RH\ P}{1013.25(273.15+T)} e_s + 0.2, \tag{A15}$$

where RH is relative humidity (%), which was calculated using Equation (A12), P is barometric pressure in hPa, T is temperature in Celsius degree, and e_s saturation vapor pressure at the ground level of the GDAS1 sounding was calculated using Equation (A10). Note that we reduced the intercept of the equation from +2 to +0.2 mm water column, in order to adapt the formula to hyper-arid conditions that are characterized by average PWV contents of less than 1 mm.

Appendix D. Wind Field during SAR Acquisitions

Wind, in particular over arid regions, plays a major role for the transport of humidity in the atmosphere, which is also true for the humidity emitted from volcanoes. In order to understand humidity variations in the atmosphere above a volcano, it is thus necessary to know about the local wind field. The temporal variations of the wind field above Láscar volcano were examined using atmospheric soundings of the Global Data Assimilation System (GDAS1) with one-degree grid spacing provided by NOAA. The soundings were concatenated to form a time series of wind speed and direction encountered at Láscar volcano during the period of interest (Figure A1a,b). Wind speeds at plume altitude ranged from 0.1 to 18 m·sec^{-1} during the period of interest, and generally decreased throughout the whole atmospheric profile during austral summer (December to February) (Figure A1a). High altitude winds over the region usually are prevailing westerly throughout the year (orange colors in the upper part of Figure A1b), and are interrupted by short periods, which are dominated by easterly wind anomalies during austral summer. Surface winds typically are oriented according to topography, and exhibit pronounced diurnal variations (alternating blue and orange colors in the lower part of Figure A1b). These diurnal variations are characterized by valley winds (anabatic upslope breezes depicted by orange colors in the lower part of Figure A1b) that typically occur during daytime due to insolation of mountain flanks, and increasing wind speeds towards the evening (light blue colors in the lower part of Figure A1a), whereas rather calm mountain winds (katabatic downslope breezes and over-hill flows) prevail during the night (depicted by blue colors in the lower part of Figure A1b and dark blue colors in Figure A1a). Diurnal variations in surface wind directions are more prominent during summer, frequently resulting in a characteristic diurnal pattern of plume transport directions. We observed that in the morning hours of a summer day the plume typically drifts towards west or northwest, turning over south towards east before noon, where it stays more or less steady until sunset. Downslope winds frequently force the plume to follow the steep morphology of the volcano.

Time series of wind directions in the range of typical plume heights, namely several 100 m above and below the summit of Láscar volcano were extracted from the 500 and 550 mbar pressure levels of the GDAS1 soundings, respectively, in order to narrow down the range of possible plume transport directions at the times of SAR acquisitions (Figure A1c,d). Wind directions at the time of SAR acquisitions of track 111, which were conducted around sunrise at about 10:00 a.m. UTC (Figure 2b), were predominantly westerly during spring 2013 and late summer 2014, whereas easterly directions prevailed during early summer 2013. Wind directions were generally very similar over the broad altitude range of both pressure levels, indicating relatively stable wind conditions. Pronounced differences in wind directions above and below summit were mainly observed during early summer, when winds tended to be less stable with respect to the direction, and were characterized by low velocities.

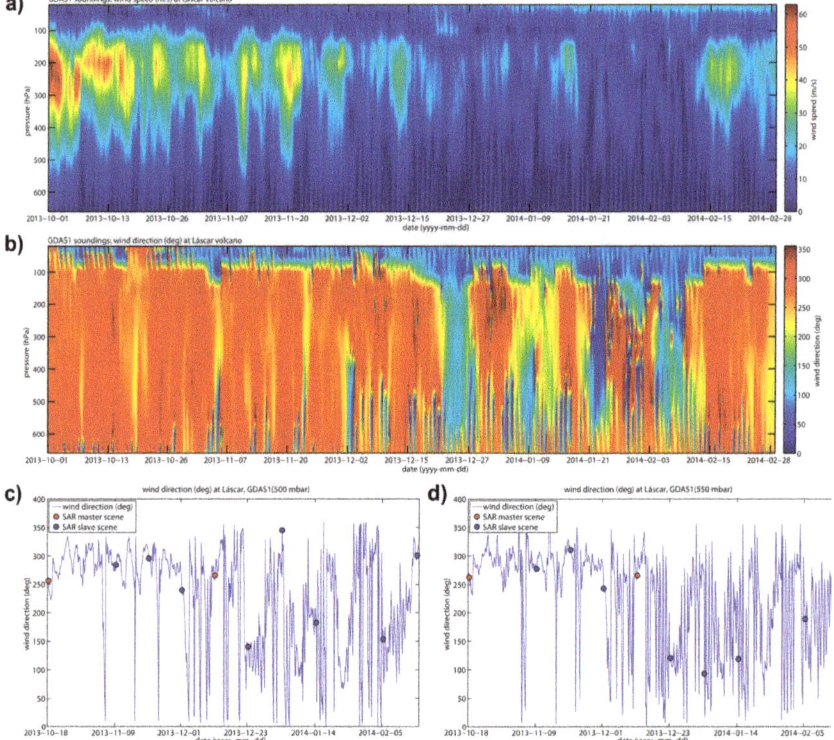

Figure A1. (**a**,**b**) Time series of vertical atmospheric profiles depicting variations in (**a**) wind speed and (**b**) wind direction. Note the weak winds accompanied by strong variations of wind directions during austral summer (ranging from December 2013 to February 2014). (**c**,**d**) Time series of wind directions some 100 m (**c**) above and (**d**) below the summit of Láscar (500 and 550 mbar pressure levels of the GDAS1 soundings, respectively). Wind directions at the time of SAR observations are indicated by *red* (master scene) and *blue dots* (slave scene). Wind directions during austral summer (ranging from December 2013 to February 2014) were predominantly easterly at the time of SAR observations.

We further complemented our observations by means of an additional high resolution wind field analysis, which is based on 3-dimensional gridded hindcasts of the Weather Research and Forecasting model (WRF). This we did, in order to resolve how the wind field evolved from turbulent flow close to the ground surface to laminar flow at higher altitude during the times of SAR acquisitions, because the gas plume typically is being emitted at the interface between surface flows and higher altitude flows. To this end, the WRF simulation was run using a 900 m horizontal grid spacing and a vertical division of 51 (terrain-following) eta levels.

Surface wind fields obtained from the lowest eta level of the WRF simulation, reveal that katabatic (downslope) winds prevailed during most SAR acquisition times (Figure A2).

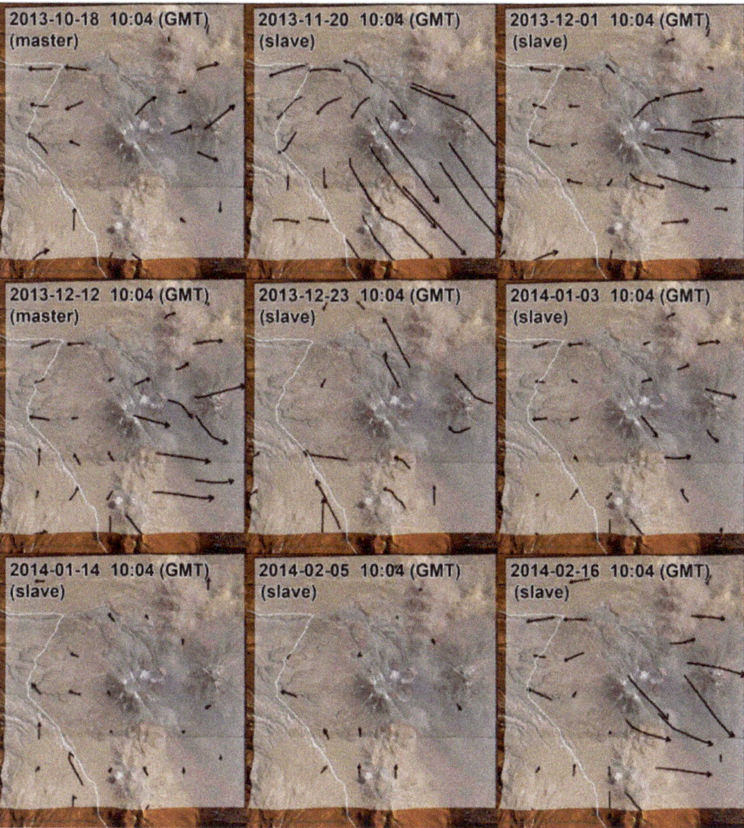

Figure A2. Time series of surface wind fields obtained from the lowest eta level of WRF, determined for acquisition times of each SAR image. Katabatic (downslope) mountain winds prevail at the time of SAR acquisitions, which were recorded during the early morning hours at about 10:04 a.m. (UTC), respectively 07:04 a.m. (CLST).

The wind fields of all acquisition times were averaged, in order to match the cumulative SAR delay estimates produced by WBDD. This was done for the first eta level, which represents the wind conditions closest to the surface (Figure A3a). Changing conditions with increasing altitude were captured by means of gradually increasing the number of eta levels, which were included into the averaged wind field. In other words eta levels 1–5, 1–10, and 1–15 were combined to yield averaged wind fields (Figure A3b–d). Average wind direction of the terrain following winds close to the surface was predominantly towards southeast aloft the plateau, and heading downslope, which is mainly oriented towards west, over the western flank of the plateau (Figure A3a). Gradual integration of more eta levels into the average wind field nicely illustrates the turning direction of the winds with increasing altitude (Figure A3a–d), finally arriving at upper tropospheric westerly trade wind directions, which are typical for this region (Figure A3d).

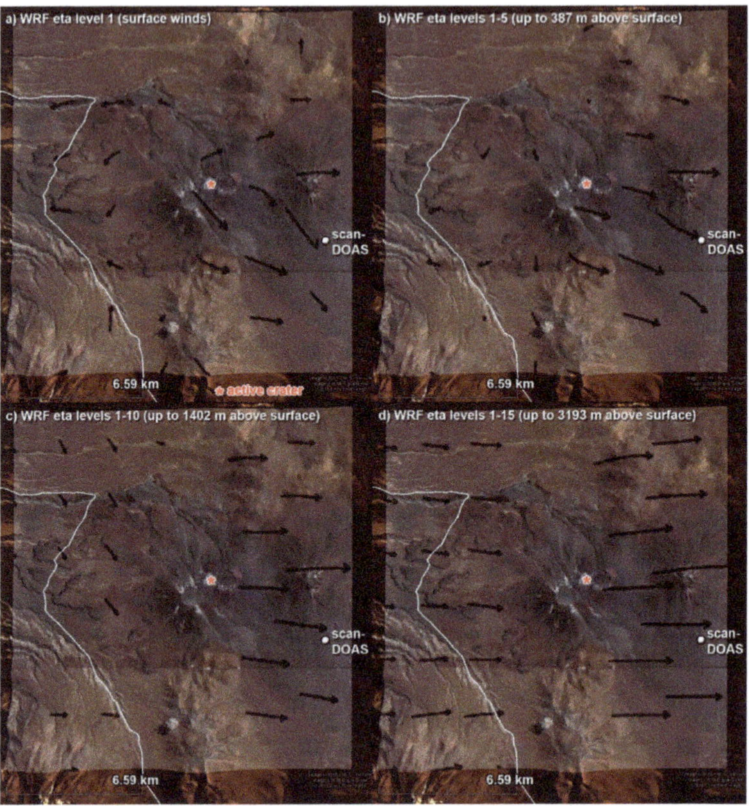

Figure A3. Averaged wind fields combining wind fields of all SAR observation times. (**a**) Average surface winds from eta level 1 (**b**) Average winds up to 387 m above surface (**c**) Average winds up to 1402 m above surface (**d**) Average winds up to 3193 m above surface.

Appendix E. The Decomposed APS: Non-Repeating and Repeating Atmospheric Phase Delays

E.1. Single Event APS Estimates

Delay maps of *single event APSs* were generated to mitigate non-repeating atmospheric disturbances from the gas plume estimate, and were used to infer the meteorological conditions during each SAR observation (Figure A4). These APS estimates are presented in the form of simple phase delay maps, where the scale indicates lengthening or shortening of the radar delay in units of millimeters. Such phase delay maps are snapshots of the meteorological situation, reflecting the spatial distribution of water vapor fields at the times of SAR acquisitions. Atmospheric disturbances aloft volcanoes are generally more pronounced and show more complex flow patterns on the lee side of the volcanic edifice [18]. This anisotropic distribution of turbulent atmospheric patterns can be attributed to the presence of volcanic gas plumes on the one hand, but also to orographic effects that govern the transport of moist air over mountainous terrain [17,20]. Orographic effects comprise diabatic heating of air masses over insolated mountain flanks and orographic lifting of air masses that are pushed by the wind, causing upslope advection of moist air that is forced to rise following the steep topography, and to cool adiabatically causing an increase of the relative humidity.

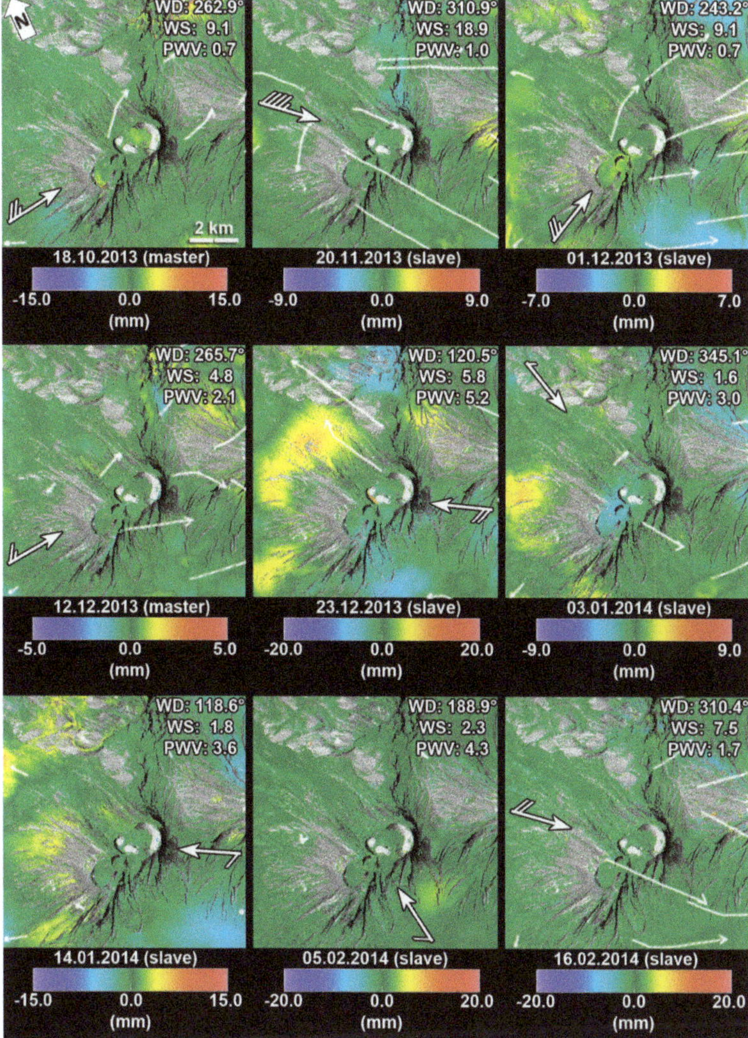

Figure A4. *Single event APS estimates* superimposed by surface wind fields obtained from the lowest eta level of WRF (*thin white arrows*). Scales indicate range change in millimeters, and are unique to each image, in order to enhance contrast by depicting the full range of each image. Katabatic (downslope) mountain winds prevail at the time of SAR acquisitions recorded during the early morning hours (10:04 a.m. GMT, local time is offset −3 h). *Wind barbs* indicate wind directions and wind speeds above the summit of Láscar volcano, which were obtained from GFS hindcasts at the times of SAR acquisitions. The *barbs* are displaced upstream in order not to cover the delay signatures of the summit area. The individual lines of the *barbs* represent the wind speeds in units of knots (half strokes correspond to 5 knots and full strokes correspond to 10 knots). Wind directions (clockwise degrees from North), wind speed (m·sec^{-1}) and estimated average PWV contents (mm) are additionally indicated in the upper right corner of each image.

APS delay patterns were compared to wind directions and atmospheric PWV contents estimated for the times of SAR acquisitions to assure that the modeled and measured wind directions and PWV estimates are consistent with the location and strength of associated phase delay patterns. Wind directions determined for the summit region of Láscar were prevailing westerly during the early morning hours of the "dry season", whereas wind directions of the "wet season" were predominantly easterly (see Figure A1c,d, Figures A2 and A3). This is reflected by the spatial distribution of cloud shaped signatures, which are confined to the plateau east of the volcano during the "dry season", while pronounced cloud shaped signatures are mainly confined to the western flank of the volcano during early summer, indicating that moist air has been transported at low-altitude by easterly winds towards the edge of the plateau during the "wet season" (period comprising SAR acquisitions of 23 December 2013 to 14 January 2014). Wind directions are thus consistent with the spatial distribution of observed phase delay patterns and PWV estimates agree with the strength of these patterns. Enhanced humidity variations encountered during SAR acquisitions of early summer can further be ascribed to the SAR acquisition strategy. Space based SARs typically repeat their observations at the same local time, causing SAR observations of early summer to be more affected by atmospheric disturbances, since they are recorded later with respect to sunrise, due to variations in the length of the day (Figure 2b). At Láscar volcano this effect is additionally enforced by more humid conditions that generally occur during that period.

E.2. Repeating Atmospheric Phase Delays

The phase delay estimates obtained for air temperature, air pressure and relative humidity *priors* comprise refractivity related phase contributions, which repeatedly occurred in all DInSARs of the time series and thus were not captured by the *single event APS estimates*. Such phase contributions therefore may contain residues of the stratification, which have not yet been removed through the coarse atmospheric correction that was performed prior to WBDD analysis utilizing the phase delay simulations obtained from the WRF, as well as repeating orographic effects, which may occur in multiple interferograms due to similar weather conditions.

Phase delay patterns in the air temperature dependent phase screen (Figure 6e) are asymmetrically distributed with respect to topography of the volcanic edifice, which can be ascribed to the distinct exposure of mountain flanks to sunlight. All SAR observations used in this study were made around sunrise, thus the western flank commonly lies in the shadow of the volcanic edifice, resulting in cooler air masses with a higher refractivity aloft the western flank, which produces a lengthening of the propagation path delay, whereas the air masses above the flat plain southeast of the volcano, the summit region, and the eastern flanks are subjected to diabatic heating due to a more pronounced insolation, which causes refractivity in the overlying air mass to be smaller due to higher air temperature.

Phase delay patterns of the pressure related phase screen (Figure 6f) indicate a lengthening of the delay over several confined steep-sloped areas, which are particularly exposed to westerly winds. The phase delay patterns of the temperature and pressure phase screens (Figure 6e,f) thus have the opposite direction, if compared to the phase delay patterns of the gas plume estimates (Figure 6c,d), resulting in a partial cancellation of the plume related phase delay (Figure 6b). Cancellation of the opposing phase screens hence reflect the negative dependence of volcanic gas emissions on barometric pressure and ambient temperature [77,78]. This is further supported by the spatial distribution of phase delay patterns in the relative humidity related phase screen (Figure 6g), which indicate enhanced relative humidity over exactly the same areas, which show a decreased temperature in the temperature phase screen (Figure 6e).

Appendix F. DEM Error Related Phase Delays

The SRTM-1 DEM that we used as a reference surface for our DInSAR observations is based on data which have been recorded in February 11–22, 2000. Since then several explosive eruptions occurred at Láscar volcano (July 2000, October 2002, December 2005, April 2006, and April 2013), which locally may have resulted in substantial changes in surface altitude that occurred previous to the period considered here. This in turn may have given rise to phase differences in our SAR interferograms, where the ground surface geometries measured by SAR observations and modeled digital elevation are different. The amplitude of such topographical artifacts is largely controlled by the measurement geometry of the DInSAR measurement, and generally increases proportionally to the length of the spatial baseline. As these phase differences can become exceedingly large in interferograms with a long spatial baseline, we instead used very small baseline interferograms for our DInSAR decomposition analysis.

The patterns in the phase delay estimate that we obtained for the spatial baseline *prior* are indeed intimately linked to topographic features (Figure 6h), and thus very likely reflect the development of the land surface during the period following the SRT Mission. The most prominent feature of this phase delay estimate is a pronounced shortening of the phase delay in the summit region of Láscar and on the SW flank of Aguas Calientes, which may be attributed to the deposition of erupted material. Furthermore, a lengthening of the phase delay occurs along morphological depressions, which follow the steep southeastern flank of Láscar and the base of Aguas Calientes, and therefore may be a result of removal or compaction of sedimentary deposits.

Appendix G. Estimation of APS Amplitude using Modeled Atmospheric PWV Contents Obtained from GDAS1 Soundings

The atmospheric phase contributions in our coarsely corrected DInSAR maps (Figure 5b) are mainly caused by the presence of water vapor. Amplitudes of the atmospheric contribution to total DInSAR phase thus to a certain extent may be predicted using the modeled atmospheric PWV contents that were obtained from GDAS1 soundings for the summit region of Láscar volcano (Figure 7b and Table 2, columns 6 and 11).

In order to assess how well these modeled PWV values are suited to reproduce the APS amplitudes we therefore utilized the calculations described in Section 3.4 to determine the differential slant wet delays (dSWDs) that the PWV contents would produce in respective DInSARs, and compared their absolute values to the measured amplitudes of associated phase delays in each of the corresponding DInSAR maps (Figure 5b).

As a first step, we compared these predicted dSWDs (Table A1, column 4) to visually determined estimates of the mean APS amplitude in each of the DInSARs (Table A1, column 5). Using the scale bar ranges of the DInSAR maps in Figure 5 as reference, the mean APS amplitudes in five of the DInSARs (DInSARs 1-3, 5 and 6) were estimated to roughly amount to half of the scales, because a majority of the APS amplitudes does not exceed half of the associated scale, while they rather span 2/3rds of the scale bar in DInSAR 4, and only about 1/4 in DInSAR 7. Comparing the absolute values of the predicted dSWDs with our estimates of the mean APS amplitudes measured in corresponding DInSARs reveals that the latter are surprisingly well represented by the modeled dSWDs ($R^2 = 0.96$; Figure A5a), which at average are offset from measured APS amplitudes by merely 2.1 mm.

In order to obtain more representative values for the measured amplitudes of the atmospheric delays in the DInSAR maps (Figure 5b) we further determined the root-mean-square-deviation (RMSD) of the phase delay amplitudes in each DInSAR map (Table A1, column 6) and compared it to the absolute values of the dSWDs, which the GDAS1 soundings predicted for the acquisition times of corresponding DInSARs. The comparison revealed that the absolute values of the predicted dSWDs are in good agreement with the RMSDs of DInSAR amplitudes ($R^2 = 0.70$; Figure A5b).

Table A1. Absolute values of predicted dSWDs, obtained from GDAS1 soundings, along with estimates of the mean measured APS amplitudes (= scale bar fractions of DInSARs) and RMSDs of phase delay amplitudes in DInSARs.

DInSAR #	Master Scene Date (yyyy-mm-dd)	Slave Scene Date (yyyy-mm-dd)	predicted |dSWD| (mm)	Estimated Mean APS Amplitudes = DInSAR Scale Bar Fraction (mm)	RMSD of DInSAR Amplitudes (mm)
			subset 01		
DInSAR 1	2013-12-12	2013-11-20	8.8	10 = 1/2*20	1.6
DInSAR 2	2013-12-12	2013-12-01	11.3	10 = 1/2*20	1.4
DInSAR 3	2013-12-12	2014-02-05	17.6	15 = 1/2*30	2.3
			subset 02		
DInSAR 4	2013-10-18	2013-12-23	36.2	33.33 = 2/3*50	5.1
DInSAR 5	2013-10-18	2014-01-03	18.7	20 = 1/2*40	2.5
DInSAR 6	2013-10-18	2014-01-14	23.8	20 = 1/2*40	3.8
DInSAR 7	2013-10-18	2014-02-16	8.5	10 = 1/4*40	3.2

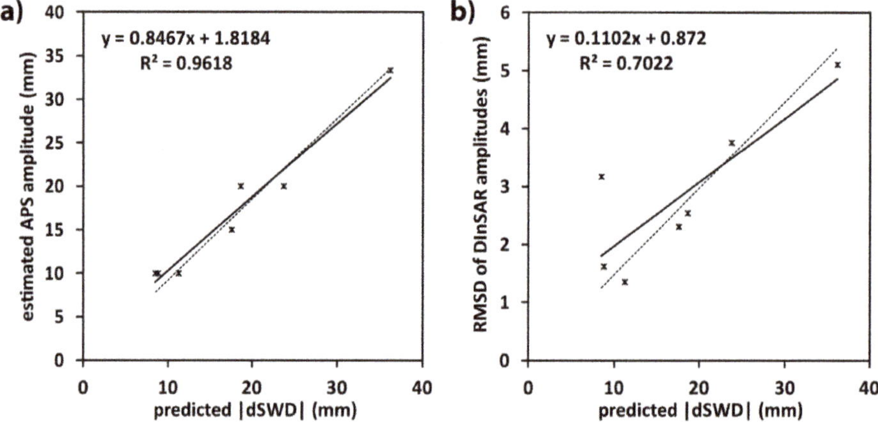

Figure A5. Comparison of the absolute values of the predicted dSWDs with (**a**) estimates of the mean measured APS amplitudes, and with (**b**) RMSDs of phase delay amplitudes in DInSARs. Best fitting linear regression lines (*thick black lines*) are depicted along with their corresponding equations and R-squared values. Additionally, to guide the eye, linear regression lines that are forced through zero (*thin dashed lines*) are given as reference.

Appendix H. Gas Plume Related Phase Delays of DInSAR Time Series Subsets 01 and 02

Interferograms of DInSAR time series *subsets 01* and *02* were decomposed, in order to separately examine the associated gas plume related phase delays. Estimates of gas plume related phase delays were computed for each of the two disjoint time series subsets, i.e., using a limited number of three (*subset 01*), and respectively four temporally interconnected interferograms (*subset 02*). Presented are the results obtained from the WBDD run, that was conducted omitting surface temperature and pressure *priors*, i.e., the phase delays associated to refractivity changes caused by atmospheric pressure and temperature variations were not prevented from leaking into the phase delay estimates obtained for the gas plume. Gas plume estimates of both subsets show a fan-shaped interferometric pattern indicating lengthening of the radar path over the south-eastern flank of Láscar volcano (depicted by yellow to orange colors in Figure A6a,b), which is in good agreement with the common direction of plume transport (see supplementary Figure A1c,d, Figures A2 and A3 in Appendix D, where a detailed description of the wind field is given). Furthermore, this delay lengthening is coherent regarding the expected effect (Equation (1) in Smith and Weintraub, [66]), because the enhanced water vapor content of the volcanic gas plume is expected to increase the refractivity with respect to its surroundings. Negatively correlated phase delay patterns (depicted by cyan to blue colors in Figure A6a,b), which are distributed over the western flanks of Láscar and Aguas Calientes volcanoes, in contrast are incoherent in terms of the expected sign of the phase delay (Equation (1) in Smith and Weintraub, [66]) and thus cannot be attributed to the gas plume. They rather indicate the contribution of another process (or other processes), which was (were) not mitigated from the gas plume estimate. *Subset 02* was additionally processed without the SAR acquisition of 3 January 2014, in order to display the resulting reduction of the phase delay amplitude in the repeating plume related signal (Figure A6c). PWV contents inside the plume and their respective differences were larger for SAR observations of *subset 01*, than they were for observations of *subset 02* (Table 2), suggesting a stronger influence on the gas plume related phase delay signal of *subset 01* (Figure A6a), if compared to *subset 02* (Figure A6b,c). The amplitudes of the estimated phase delays displayed in Figure A6a–c are thus consistent with their respective input dSWD values (Table 3, column 4), and are also in accordance with humidity contrast between plume and atmosphere (Table 2).

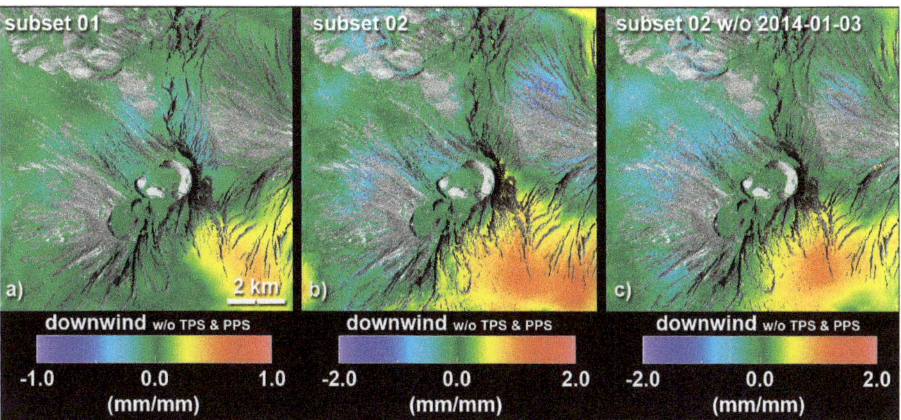

Figure A6. Delay correlation maps depicting estimated interferometric gas plume patterns from (**a**) *subset 01* and (**b**) *subset 02*. (**c**) Estimated interferometric pattern from *subset 02*, where the SAR observation of 03 Januray 2014 was omitted. All three estimates contain phase contributions of the temperature and pressure related phase screens (TPS & PPS not removed). Scales correspond to estimated delay (mm) per theoretical delay (mm). Upper and lower bounds of the scale bar are equal to unity in *subset 01*, but not in *subset 02*, indicating that the estimated delay in *subset 02* was by a factor 2 larger than the theoretical delay.

References

1. Lengliné, O.; Marsan, D.; Got, J.L.; Pinel, V.; Ferrazzini, V.; Okubo, P.G. Seismicity and deformation induced by magma accumulation at three basaltic volcanoes. *J. Geophys. Res. Solid Earth* **2008**, *113*. [CrossRef]
2. Morita, Y.; Nakao, S.; Hayashi, Y. A quantitative approach to the dike intrusion process inferred from a joint analysis of geodetic and seismological data for the 1998 earthquake swarm off the east coast of Izu Peninsula, central Japan. *J. Geophys. Res. Solid Earth* **2006**, *111*. [CrossRef]
3. Watson, I.M.; Oppenheimer, C.; Voight, B.; Francis, P.W.; Clarke, A.; Stix, J.; Miller, A.; Pyle, D.M.; Burton, M.R.; Young, S.R.; et al. The relationship between degassing and ground deformation at Soufriere Hills Volcano, Montserrat. *J. Volcanol. Geotherm. Res.* **2000**, *98*, 117–126. [CrossRef]
4. Kazahaya, R.; Aoki, Y.; Shinohara, H. Budget of shallow magma plumbing system at Asama Volcano, Japan, revealed by ground deformation and volcanic gas studies. *J. Geophys. Res. Solid Earth* **2015**, *120*, 2961–2973. [CrossRef]

5. Girona, T.; Costa, F.; Newhall, C.; Taisne, B. On depressurization of volcanic magma reservoirs by passive degassing. *J. Geophys. Res. Solid Earth* **2014**, *119*, 8667–8687. [CrossRef]
6. Tait, S.; Jaupart, C.; Vergniolle, S. Pressure, gas content and eruption periodicity of a shallow, crystallising magma chamber. *Earth Planet. Sci. Lett.* **1989**, *92*, 107–123. [CrossRef]
7. Sparks, R.S.J. Dynamics of magma degassing. *Geol. Soc. London Spec. Pub.* **2003**, *213*, 5–22. [CrossRef]
8. Green, D.N.; Neuberg, J. Waveform classification of volcanic low-frequency earthquake swarms and its implication at Soufrière Hills Volcano, Montserrat. *J. Volcanol. Geotherm. Res.* **2006**, *153*, 51–63. [CrossRef]
9. Hooper, A.; Zebker, H.; Segall, P.; Kampes, B. A new method for measuring deformation on volcanoes and other natural terrains using InSAR persistent scatterers. *Geophys. Res. Lett.* **2004**, *31*, L23611. [CrossRef]
10. Goldstein, R.M. Atmospheric Limitations to repeat-track radar interferometry. *Geophys. Res. Lett.* **1995**, *22*, 2517–2520. [CrossRef]
11. Zebker, H.; Rosen, P.; Hensley, S. Atmospheric effects in interferometric synthetic aperture radar surface deformation and topographic maps. *J. Geophys. Res.* **1997**, *102*, 7547–7563. [CrossRef]
12. Jung, J.; Kim, D.; Park, S.-E. Correction of Atmospheric Phase Screen in Time Series InSAR using WRF Model for Monitoring Volcanic Activities. *IEEE Trans. Geosci. Remote Sens.* **2014**, *52*, 2678–2689. [CrossRef]
13. Foster, J.; Brooks, B.; Cherubini, T.; Shacat, C.; Businger, S.; Werner, C. Mitigating atmospheric noise for InSAR using a high resolution weather model. *Geophys. Res. Lett.* **2006**, *33*, L16304. [CrossRef]
14. Pichelli, E.; Ferretti, R.; Cimini, D.; Perissin, D.; Montopoli, M.; Marzano, F.S.; Pierdicca, N. Water vapour distribution at urban scale using high-resolution numerical weather model and spaceborne SAR interferometric data. *Nat. Hazard Earth Sys. Sci.* **2010**, *10*, 121–132. [CrossRef]
15. Ulmer, F.-G.; Adam, N. A Synergy Method to Improve Ensemble Weather Predictions and Differential SAR Interferograms. *ISPRS J. Photogramm. Remote Sens.* **2015**, *109*, 98–107. [CrossRef]
16. Ulmer, F.-G.; Adam, N. Characterisation and improvement of the structure function estimation for application in PSI. *ISPRS J. Photogramm. Remote Sens.* **2017**, *128*, 40–46. [CrossRef]
17. Bonforte, A.; Ferretti, A.; Prati, C.; Puglisi, G.; Rocca, F. Calibration of atmospheric effects on SAR interferograms by GPS and local atmosphere models: First results. *J. Atmos. Sol. Terr. Phys.* **2001**, *63*, 1343–1357. [CrossRef]
18. Wadge, G.; Webley, P.W.; James, I.N.; Bingley, R.; Dodson, A.; Waugh, S.; Veneboer, T.; Puglisi, G.; Mattia, M.; Baker, D.; et al. Atmospheric models, GPS and InSAR measurements of the tropospheric water vapour field over Mount Etna. *Geophys. Res. Lett.* **2002**, *29*, 1905. [CrossRef]
19. Rosen, P.A.; Henley, S.; Zebker, H.A.; Webb, F.H.; Fielding, E.J. Surface deformation and coherence measurements of Kilauea volcano, Hawaii, from SIR-C radar interferometry. *J. Geophys. Res.* **1996**, *101*, 23109–23125. [CrossRef]
20. Wadge, G.; Mattioli, G.S.; Herd, R.A. Ground deformation at Soufrière Hills Volcano, Montserrat during 1998–2000 measured by radar interferometry and GPS. *J. Volcanol. Geotherm. Res.* **2006**, *152*, 157–173. [CrossRef]
21. González, P.J.; Bagnardi, M.; Hooper, A.J.; Larsen, Y.; Marinkovic, P.; Samsonov, S.V.; Wright, T.J. The 2014–2015 eruption of Fogo volcano: Geodetic modeling of Sentinel-1 TOPS interferometry. *Geophys. Res. Lett.* **2015**, *42*, 9239–9246. [CrossRef]
22. Wadge, G.; Costa, A.; Pascal, K.; Werner, C.; Webb, T. The variability of refractivity in the atmospheric boundary layer of a tropical island volcano measured by ground-based interferometric radar. *Boundary-Layer Meteorol.* **2016**, *161*, 309–333. [CrossRef]
23. Saastamoinen, J. Introduction to practical computation of astronomical refraction. *B. Géod.* **1972**, *106*, 383–397. [CrossRef]
24. Hanssen, R.F.; Weckwerth, T.M.; Zebker, H.A.; Klees, R. High-resolution water vapour mapping from interferometric radar measurements. *Science* **1999**, *283*, 1295–1297. [CrossRef]
25. Mateus, P.; Nico, G.; Catalão, J. Can spaceborne SAR interferometry be used to study the temporal evolution of PWV? *Atmos. Res.* **2013**, *119*, 70–80. [CrossRef]
26. Burton, M.R.; Oppenheimer, C.; Horrocks, L.; Francis, P.W. Remote sensing of CO_2 and H_2O emission rates from Masaya volcano, Nicaragua. *Geology* **2000**, *28*, 915–918. [CrossRef]
27. Fiorani, L.; Colao, F.; Palucci, A.; Poreh, D.; Aiuppa, A.; Giudice, G. First-time lidar measurement of water vapor flux in a volcanic plume. *Opt. Commun.* **2011**, *284*, 1295–1298. [CrossRef]

28. Bryan, S.; Clarke, A.; Vanderkluysen, L.; Groppi, C.; Paine, S.; Bliss, D.W.; Aberle, J.; Mauskopf, P. Measuring Water Vapor and Ash in Volcanic Eruptions With a Millimeter-Wave Radar/Imager. *IEEE Trans. Geosci. Remote Sens.* **2017**, *55*, 3177–3185. [CrossRef]
29. Kern, C. The Difficulty of Measuring the Absorption of Scattered Sunlight by H_2O and CO_2 in Volcanic Plumes: A Comment on Pering et al. "A Novel and Inexpensive Method for Measuring Volcanic Plume Water Fluxes at High Temporal Resolution," Remote Sensing **2017**, *9*, 146. *Remote Sens.* **2017**, *9*, 534. [CrossRef]
30. Kern, C.; Masias, P.; Apaza, F.; Reath, K.A.; Platt, U. Remote measurement of high preeruptive water vapor emissions at Sabancaya volcano by passive differential optical absorption spectroscopy. *J. Geophys. Res. Solid Earth* **2017**, *122*, 3540–3564. [CrossRef]
31. Gardeweg, M.C.; Sparks, R.S.J.; Matthews, S.J. Evolution of Lascar Volcano, northern Chile. *J. Geol. Soc.* **1998**, *155*, 89–104. [CrossRef]
32. Matthews, S.J.; Sparks, R.S.J.; Gardeweg, M.C. The Piedras Grandes–Soncor eruptions, Lascar volcano, Chile; evolution of a zoned magma chamber in the central Andean upper crust. *J. Petrol.* **1999**, *40*, 1891–1919. [CrossRef]
33. Tamburello, G.; Hansteen, T.H.; Bredemeyer, S.; Aiuppa, A.; Tassi, F. Gas emissions from five volcanoes in northern Chile, and implications for the volatiles budget of the Central Volcanic Zone. *Geophys. Res. Lett.* **2014**, *41*, 4961–4969. [CrossRef]
34. Marín, J.C.; Pozo, D.; Mlawer, E.; Turner, D.D.; Curé, M. Dynamics of Local Circulations in Mountainous Terrain during the RHUBC-II Project. *Mon. Weather Rev.* **2013**, *141*, 3641–3656. [CrossRef]
35. Giovanelli, R.; Darling, J.; Henderson, C.; Hoffman, W.; Barry, D.; Cordes, J.; Eikenberry, S.; Gull, G.; Keller, L.; Smith, J.D.; et al. The optical-infrared astronomical quality of high Atacama sites. II—Infrared characteristics. *Publ. Astron. Soc. Pac.* **2001**, *113*. [CrossRef]
36. Matthews, S.J.; Jones, A.P.; Gardeweg, M.C. Lascar Volcano, northern Chile; evidence for steady-state disequilibrium. *J. Petrol.* **1994**, *35*, 401–432. [CrossRef]
37. De Zeeuw-van Dalfsen, E.; Richter, N.; González, G.; Walter, T.R. Geomorphology and structural development of the nested summit crater of Láscar Volcano studied with Terrestrial Laser Scanner data and analogue modelling. *J. Volcanol. Geotherm. Res.* **2017**, *329*, 1–12. [CrossRef]
38. Matthews, S.J.; Gardeweg, M.C.; Sparks, R.S.J. The 1984 to 1996 cyclic activity of Lascar Volcano, Northern Chile; cycles of dome growth, dome subsidence, degassing and explosive eruptions. *Bull. Volcanol.* **1997**, *59*, 72–82. [CrossRef]
39. Pavez, A.; Remy, D.; Bonvalot, S.; Diament, M.; Gabalda, G.; Froger, J.L.; Julien, P.; Legrand, D.; Moisset, D. Insight into ground deformations at Lascar volcano (Chile) from SAR interferometry, photogrammetry and GPS data: Implications on volcano dynamics and future space monitoring. *Remote Sens. Environ.* **2006**, *100*, 307–320. [CrossRef]
40. Richter, N.; Salzer, J.T.; de Zeeuw-van Dalfsen, E.; Perissin, D.; Walter, T.R. Constraints on the geomorphological evolution of the nested summit craters of Láscar Volcano from high spatio-temporal resolution TerraSAR-X interferometry. *Bull. Volcanol.* **2018**, *80*. [CrossRef]
41. Tassi, F.; Aguilera, F.; Vaselli, O.; Medina, E.; Tedesco, D.; Huertas, A.D.; Poreda, R.; Kojima, S. The magmatic- and hydrothermal-dominated fumarolic system at the active crater of Lascar volcano, northern Chile. *Bull. Volcanol.* **2009**, *71*, 171–183. [CrossRef]
42. Menard, G.; Moune, S.; Vlastélic, I.; Aguilera, F.; Valade, S.; Bontemps, M.; González, R. Gas and aerosol emissions from Lascar volcano (Northern Chile): Insights into the origin of gases and their links with the volcanic activity. *J. Volcanol. Geotherm. Res.* **2014**, *287*, 51–67. [CrossRef]
43. González, C.; Inostroza, M.; Aguilera, F.; González, R.; Viramonte, J.; Menzies, A. Heat and mass flux measurements using Landsat images from the 2000-2004 period, Lascar volcano, northern Chile. *J. Volcanol. Geotherm. Res.* **2015**, *301*, 277–292. [CrossRef]
44. Glaze, L.S.; Francis, P.W.; Rothery, D.A. Measuring thermal budgets of active volcanoes by satellite remote sensing. *Nature* **1989**, *338*, 144–146. [CrossRef]
45. Oppenheimer, C.; Francis, P.W.; Rothery, D.A.; Carlton, R.W.; Glaze, L.S. Infrared image analysis of volcanic thermal features: Lascar Volcano, Chile, 1984–1992. *J. Geophys. Res. Solid Earth* **1993**, *98*, 4269–4286. [CrossRef]
46. Wooster, M.J.; Rothery, D.A. Thermal monitoring of Lascar Volcano, Chile, using infrared data from the along-track scanning radiometer: A 1992–1995 time series. *Bull. Volcanol.* **1997**, *58*, 566–579. [CrossRef]

47. Wooster, M.J. Long-term infrared surveillance of Lascar Volcano: Contrasting activity cycles and cooling pyroclastics. *Geophys. Res. Lett.* **2001**, *28*, 847–850. [CrossRef]
48. Murphy, S.W.; Wright, R.; Oppenheimer, C.; Souza Filho, C.R. MODIS and ASTER synergy for characterizing thermal volcanic activity. *Remote Sens. Environ.* **2013**, *131*, 195–205. [CrossRef]
49. Galle, B.; Johansson, M.; Rivera, C.; Zhang, Y.; Kihlman, M.; Kern, C.; Lehmann, T.; Platt, U.; Arellano, S.; Hidalgo, S. Network for Observation of Volcanic and Atmospheric Change (NOVAC)—A global network for volcanic gas monitoring: Network layout and instrument description. *J. Geophys. Res. Atmos.* **2010**, 115. [CrossRef]
50. Platt, U.; Stutz, J. *Differential Optical Absorption Spectroscopy—Principles and Applications*; Springer: Berlin/Heidelberg, Germany, 2008. [CrossRef]
51. Vandaele, A.C.; Simon, P.C.; Guilmot, J.M.; Carleer, M.; Colin, R. SO_2 absorption cross section measurement in the UV using a Fourier transform spectrometer. *J. Geophys. Res. Atmos.* **1994**, *99*, 25599–25605. [CrossRef]
52. Voigt, S.; Orphal, J.; Bogumil, K.; Burrows, J.P. The temperature dependence (203–293K) of the absorption cross sections of O_3 in the 230–850 nm region measured by Fourier-transform spectroscopy. *J. Photoch. Photobiol. A Chem.* **2001**, *143*, 1–9. [CrossRef]
53. Solomon, S.; Schmeltekopf, A.L.; Sanders, R.W. On the interpretation of zenith sky absorption measurements. *J. Geophys. Res. Atmos.* **1987**, *92*, 8311–8319. [CrossRef]
54. Kurucz, R.L.; Furenlid, I.; Brault, J.; Testerman, L. *Solar Flux Atlas from 296 to 1300 nm*; National Solar Observatory: Sunspot, NM, USA, 1984; 240p.
55. Aiuppa, A.; Moretti, R.; Federico, C.; Giudice, G.; Gurrieri, S.; Liuzzo, M.; Papale, P.; Shinohara, H.; Valenza, M. Forecasting Etna eruptions by real-time observation of volcanic gas composition. *Geology* **2007**, *35*, 1115–1118. [CrossRef]
56. Shinohara, H.; Aiuppa, A.; Guidice, G.; Gurrieri, S.; Liuzzo, M. Variation of H_2O/CO_2 and CO_2/SO_2 ratios of volcanic gases discharged by continuous degassing of Mount Etna volcano, Italy. *J. Geophys. Res. Solid Earth* **2008**, *113*, B09203. [CrossRef]
57. Aiuppa, A.; Bertagnini, A.; Métrich, N.; Moretti, R.; Di Muro, A.; Liuzzo, M.; Tamburello, G. A model of degassing for Stromboli volcano. *Earth Planet. Sci. Lett.* **2010**, *295*, 195–204. [CrossRef]
58. Matsushima, N.; Shinohara, H. Visible and invisible volcanic plumes. *Geophys. Res. Lett.* **2006**, *33*, L24309. [CrossRef]
59. Ebmeier, S.K.; Sayer, A.M.; Grainger, R.G.; Mather, T.A.; Carboni, E. Systematic satellite observations of the impact of aerosols from passive volcanic degassing on local cloud properties. *Atmos. Chem. Phys.* **2014**, *14*, 10601–10618. [CrossRef]
60. Eineder, M.; Adam, N. A flexible system for the generation of interferometric SAR products. In Proceedings of the International Geoscience and Remote Sensing Symposium IGARSS'97. Remote Sensing—A Scientific Vision for Sustainable Development, Singapore, 3–8 August 1997. [CrossRef]
61. Bevis, M.; Businger, S.; Herring, T.A.; Rocken, C.; Anthes, R.A.; Ware, R.H. GPS meteorology: Remote sensing of atmospheric water vapour using the Global Positioning System. *J. Geophys. Res. Atmos.* **1992**, *97*, 15787–15801. [CrossRef]
62. Ulmer, F.-G. On the accuracy gain of electromagnetic wave delay predictions derived by the digital filter initialization technique. *J. Appl. Remote Sens.* **2016**, *10*, 016007:1–016007:7. [CrossRef]
63. Ulmer, F.-G. Cinderella: Method Generalisation of the Elimination Process to Filter Repeating Patterns. In Proceedings of the 2015 IEEE International Conference on Digital Signal Processing (DSP), Singapore, 21–24 July 2015. [CrossRef]
64. Selesnick, I.W.; Baraniuk, R.G.; Kingsbury, N.C. The dual-tree complex wavelet transform. *IEEE Signal Process. Mag.* **2005**, *22*, 123–151. [CrossRef]
65. Fattahi, H.; Amelung, F. DEM error correction in InSAR time series. *IEEE Trans. Geosci. Remote Sens.* **2013**, *51*, 4249–4259. [CrossRef]
66. Smith, E.K.; Weintraub, S. The Constants in the Equation for Atmospheric Refractive Index at Radio Frequencies. *Proc. IRE* **1953**, *41*, 1035–1037. [CrossRef]
67. Oppenheimer, C.; Fischer, T.P.; Scaillet, B. Volcanic degassing: Process and impact. In *Treatise on Geochemistry*, 2nd ed.; Elsevier Ltd.: Amsterdam, The Netherlands, 2014; Volume 4, pp. 111–179. [CrossRef]
68. Battaglia, M.; Troise, C.; Obrizzo, F.; Pingue, F.; De Natale, G. Evidence for fluid migration as the source of deformation at Campi Flegrei caldera (Italy). *Geophys. Res. Lett.* **2006**, *33*. [CrossRef]

69. Samsonov, S.V.; Tiampo, K.F.; Camacho, A.G.; Fernández, J.; González, P.J. Spatiotemporal analysis and interpretation of 1993–2013 ground deformation at Campi Flegrei, Italy, observed by advanced DInSAR. *Geophys. Res. Lett.* **2014**, *41*, 6101–6108. [CrossRef]
70. Ruch, J.; Manconi, A.; Zeni, G.; Solaro, G.; Pepe, A.; Shirzaei, M.; Walter, T.R.; Lanari, R. Stress transfer in the Lazufre volcanic area, central Andes. *Geophys. Res. Lett.* **2009**, *36*. [CrossRef]
71. Werner, C.; Hurst, T.; Scott, B.; Sherburn, S.; Christenson, B.W.; Britten, K.; Cole-Barker, J.; Mullan, B. Variability of passive gas emissions, seismicity, and deformation during crater lake growth at White Island Volcano, New Zealand, 2002–2006. *J. Geophys. Res. Solid Earth* **2008**, *113*. [CrossRef]
72. Chiodini, G.; Caliro, S.; De Martino, P.; Avino, R.; Gherardi, F. Early signals of new volcanic unrest at Campi Flegrei caldera? Insights from geochemical data and physical simulations. *Geology* **2012**, *40*, 943–946. [CrossRef]
73. Dinger, F.; Bobrowski, N.; Warnach, S.; Bredemeyer, S.; Hidalgo, S.; Arellano, S.; Galle, B.; Platt, U.; Wagner, T. Periodicity in the BrO/SO_2 molar ratios in the volcanic gas plume of Cotopaxi and its correlation with the Earth tides during the eruption in 2015. *Solid Earth Discuss.* **2018**, 1–28. [CrossRef]
74. Fujiwara, S.; Rosen, P.A.; Tobita, M.; Murakami, M. Crustal deformation measurements using repeat-pass JERS 1 synthetic aperture radar interferometry near the Izu Peninsula, Japan. *J. Geophys. Res. Solid Earth* **1998**, *103*, 2411–2426. [CrossRef]
75. Knospe, S.; Jonsson, S. Covariance estimation for dInSAR surface deformation measurements in the presence of anisotropic atmospheric noise. *IEEE Trans. Geosci. Remote Sens.* **2010**, *48*, 2057–2065. [CrossRef]
76. Aumento, F. Radon tides on an active volcanic island: Terceira, Azores. *Geofís. Int.* **2002**, *41*, 499–505.
77. Bredemeyer, S.; Hansteen, T.H. Synchronous degassing patterns of the neighbouring volcanoes Llaima and Villarrica in south-central Chile: The influence of tidal forces. *Int. J. Earth Sci.* **2014**, *103*, 1999–2012. [CrossRef]
78. Zimmer, M.; Walter, T.R.; Kujawa, C.; Gaete, A.; Franco-Marin, L. Thermal and gas dynamic investigations at Lastarria volcano, Northern Chile. The influence of precipitation and atmospheric pressure on the fumarole temperature and the gas velocity. *J. Volcanol. Geotherm. Res.* **2017**, *346*, 134–140. [CrossRef]
79. NOAA Air Resources Laboratory: READY Archived Meteorology. Available online: https://ready.arl.noaa.gov/READYamet.php (accessed on 18 September 2018).
80. Edmonds, M.; Herd, R.A.; Galle, B.; Oppenheimer, C.M. Automated, high time-resolution measurements of SO_2 flux at Soufrière Hills Volcano, Montserrat. *Bull. Volcanol.* **2003**, *65*, 578–586. [CrossRef]
81. Global Volcanism Program. Report on Lascar (Chile). In *Bulletin of the Global Volcanism Network*; Venzke, E., Ed.; Smithsonian Institution: Washington, DC, USA, 2015; Available online: https://volcano.si.edu/showreport.cfm?doi=10.5479/si.GVP.BGVN201506-355100 (accessed on 18 September 2018).
82. Giovanelli, R. Optical seeing and infrared atmospheric transparency in the upper Atacama desert. Astronomical Site Evaluation in the Visible and Radio Range. In *Astronomical Site Evaluation in the Visible and Radio Range. ASP Conference Proceedings, 266*; Vernin, J., Benkhaldoun, Z., Muñoz-Tuñón, C., Eds.; Astronomical Society of the Pacific: San Francisco, CA, USA, 2002; p. 366, ISBN 1-58381-106-0.
83. Vuille, M. Atmospheric circulation over the Bolivian Altiplano during dry and wet periods and extreme phases of the Southern Oscillation. *Int. J. Climatol.* **1999**, *19*, 1579–1600. [CrossRef]
84. Messerli, B.; Grosjean, M.; Bonani, G.; Bürgi, A.; Geyh, M.A.; Graf, K.; Ramseyer, K.; Romero, H.; Schotterer, U.; Schreier, H.; Vuille, M. Climate change and natural resource dynamics of the Atacama Altiplano during the last 18,000 years: A preliminary synthesis. *Mt. Res. Dev.* **1993**, 117–127. [CrossRef]
85. Vuille, M.; Ammann, C. Regional snowfall patterns in the high, arid Andes. *Clim. Chang.* **1997**, *36*, 413–423. [CrossRef]
86. Mori, T.; Mori, T.; Kazahaya, K.; Ohwada, M.; Hirabayashi, J.I.; Yoshikawa, S. Effect of UV scattering on SO_2 emission rate measurements. *Geophys. Res. Lett.* **2006**, *33*. [CrossRef]
87. Symonds, R.B.; Gerlach, T.M.; Reed, M.H. Magmatic gas scrubbing: Implications for volcano monitoring. *J. Volcanol. Geotherm. Res.* **2001**, *108*, 303–341. [CrossRef]
88. Rodríguez, L.A.; Watson, I.M.; Edmonds, M.; Ryan, G.; Hards, V.; Oppenheimer, C.M.; Bluth, G.J. SO_2 loss rates in the plume emitted by Soufrière Hills volcano, Montserrat. *J. Volcanol. Geotherm. Res.* **2008**, *173*, 135–147. [CrossRef]

89. Beirle, S.; Hörmann, C.; Penning de Vries, M.; Dörner, S.; Kern, C.; Wagner, T. Estimating the volcanic emission rate and atmospheric lifetime of SO_2 from space: A case study for Kīlauea volcano, Hawaii. *Atmos. Chem. Phys.* **2014**, *14*, 8309–8322. [CrossRef]
90. Samsonov, S.V.; Trishchenko, A.P.; Tiampo, K.; González, P.J.; Zhang, Y.; Fernández, J. Removal of systematic seasonal atmospheric signal from interferometric synthetic aperture radar ground deformation time series. *Geophys. Res. Lett.* **2014**, *41*, 6123–6130. [CrossRef]
91. Johansson, M.E.B. Application of passive DOAS for studies of megacity air pollution and volcanic gas emissions. Ph.D. Thesis, Chalmers University of Technology, Gothenburg, Sweden, 2009, ISBN 978-91-7385-238-8
92. Marquard, L.C.; Wagner, T.; Platt, U. Improved air mass factor concepts for scattered radiation differential optical absorption spectroscopy of atmospheric species. *J. Geophys. Res. Atmos.* **2000**, *105*, 1315–1327. [CrossRef]
93. McClatchey, R.A.; Fenn, R.W.; Selby, J.A.; Volz, F.E.; Garing, J.S. *Optical properties of the atmosphere (No. AFCRL-72-0497)*; Air Force Cambridge Research Laboratories, Hanscom AFB: Bedford, MA, USA, 1972.
94. Wagner, T.; Beirle, S.; Grzegorski, M.; Platt, U. Global trends (1996–2003) of total column precipitable water observed by Global Ozone Monitoring Experiment (GOME) on ERS-2 and their relation to near-surface temperature. *J. Geophys. Res. Atmos.* **2006**, *111*, D12102. [CrossRef]
95. Granger, R.J. An examination of the concept of potential evaporation. *J. Hydrol.* **1989**, *111*, 9–19. [CrossRef]
96. Horrocks, L.A.; Oppenheimer, C.; Burton, M.R.; Duffell, H.J. Compositional variation in tropospheric volcanic gas plumes: Evidence from ground-based remote sensing. *Geol. Soc. Lond. Spec. Pub.* **2003**, *213*, 349–369. [CrossRef]
97. Singh, V.P.; Xu, C.Y. Evaluation and generalization of 13 mass-transfer equations for determining free water evaporation. *Hydrol. Process.* **1997**, *11*, 311–323. [CrossRef]
98. Dalton, J. Experimental essays on the constitution of mixed gases on the force of steam or vapor from water and other liquids in different temperatures, both in a Torricellian vacuum and in air; on evaporation and on the expansion of gases by heat. *Mem. Lit. Philos. Soc. Manch.* **1802**, *5*, 535–602.
99. Penman, H.L. Natural evaporation from open water, bare soil, and grass. *P. Roy. Soc. Lond.* **1948**, *A193*, 120–146. [CrossRef]
100. Van Bavel, C.H.M. Potential evaporation: The combination concept and its experimental verification. *Water Resour. Res.* **1996**, *2*, 455–467. [CrossRef]
101. Tennekes, H. The logarithmic wind profile. *J. Atmos. Sci.* **1973**, *30*, 234–238. [CrossRef]
102. Mason, P.J. The formation of areally-averaged roughness lengths. *Q. J. Roy. Meteorol. Soc.* **1988**, *114*, 399–420. [CrossRef]
103. Magnus, G. Versuche über die Spannkräfte des Wasserdampfs. *Ann. Phys.* **1844**, *137*, 225–247. [CrossRef]
104. Sonntag, D. Important new values of the physical constants of 1986, vapor pressure formulations based on the ITS-90, and psychrometer formulae. *Z. Meteorol.* **1990**, *70*, 340–344.
105. Alduchov, O.A.; Eskridge, R.E. Improved Magnus form approximation of saturation vapor pressure. *J. Appl. Meteorol.* **1996**, *35*, 601–609. [CrossRef]
106. Jones, F.E.; Harris, G.L. ITS-90 density of water formulation for volumetric standards calibration. *J. Res. Natl. Inst. Stand.* **1992**, *97*, 335–340. [CrossRef] [PubMed]
107. Plate, E.J.; Quraishi, A.A. Modeling of Velocity Distributions Inside and Above Tall Crops. *J. Appl. Meteorol.* **1965**, *4*, 400–408. [CrossRef]
108. Hansen, F.V. *Surface Roughness Lengths (No. ARL-TR-61)*; Army Research Lab White Sands Missile Range NM: Adelphi, MD, USA, 1993; pp. 88002–85501.
109. Garrison, J.D.; Adler, G.P. Estimation of precipitable water over the United States for application to the division of solar radiation into its direct and diffuse components. *Sol. Energy* **1990**, *44*, 225–241. [CrossRef]
110. Nann, S.; Riordan, C. Solar spectral irradiance under clear and cloudy skies: Measurements and a semiempirical model. *J. Appl. Meteorol.* **1991**, *30*, 447–462. [CrossRef]

Article

The Contribution of Multi-Sensor Infrared Satellite Observations to Monitor Mt. Etna (Italy) Activity during May to August 2016

Francesco Marchese [1,*], Marco Neri [2], Alfredo Falconieri [1], Teodosio Lacava [1], Giuseppe Mazzeo [1], Nicola Pergola [1] and Valerio Tramutoli [3]

1 Consiglio Nazionale delle Ricerche, Istituto di Metodologie per l'Analisi Ambientale, C. da S. Loja, 85050 Tito Scalo, Italy; alfredo.falconieri@imaa.cnr.it (A.F.); teodosio.lacava@imaa.cnr.it (T.L.); giuseppe.mazzeo@imaa.cnr.it (G.M.); nicola.pergola@imaa.cnr.it (N.P.)

2 Istituto Nazionale di Geofisica e Vulcanologia, Sezione di Catania, Osservatorio Etneo, Piazza Roma 2, 95125 Catania, Italy; marco.neri@invg.it

3 Università della Basilicata, Scuola di Ingegneria, Via dell'Ateneo Lucano, 10, 85100 Potenza, Italy; valerio.tramutoli@unibas.it

* Correspondence: francesco.marchese@imaa.cnr.it; Tel.: +39-0971427225

Received: 8 October 2018; Accepted: 30 November 2018; Published: 4 December 2018

Abstract: In May 2016, three powerful paroxysmal events, mild Strombolian activity, and lava emissions took place at the summit crater area of Mt. Etna (Sicily, Italy). During, and immediately after the eruption, part of the North-East crater (NEC) collapsed, while extensive subsidence affected the Voragine crater (VOR). Since the end of the May eruptions, a diffuse fumarolic activity occurred from a fracture system that cuts the entire summit area. Starting from 7 August, a small vent (of ~20–30 m in diameter) opened up within the VOR crater, emitting high-temperature gases and producing volcanic glow which was visible at night. We investigated those volcanic phenomena from space, exploiting the information provided by the satellite-based system developed at the Institute of Methodologies for Environmental Analysis (IMAA), which monitors Italian volcanoes in near-real time by means of the RST_{VOLC} (Robust Satellite Techniques–volcanoes) algorithm. Results, achieved integrating Advanced Very High Resolution Radiometer (AVHRR) and Moderate Resolution Imaging Spectroradiometer (MODIS) observations, showed that, despite some issues (e.g., in some cases, clouds masking the underlying hot surfaces), RST_{VOLC} provided additional information regarding Mt. Etna activity. In particular, results indicated that the Strombolian eruption of 21 May lasted longer than reported by field observations or that a short-lived event occurred in the late afternoon of the same day. Moreover, the outcomes of this study showed that the intensity of fumarolic emissions changed before 7 August, as a possible preparatory phase of the hot degassing activity occurring at VOR. In particular, the radiant flux retrieved from MODIS data decreased from 30 MW on 4 July to an average value of about 7.5 MW in the following weeks, increasing up to 18 MW a few days before the opening of a new degassing vent. These outcomes, in accordance with information provided by Sentinel-2 MSI (Multispectral Instrument) and Landsat 8-OLI (Operational Land Imager) data, confirm that satellite observations may also contribute greatly to the monitoring of active volcanoes in areas where efficient traditional surveillance systems exist.

Keywords: Mt. Etna; multi-platform satellite observations; RST_{VOLC}

1. Introduction

Several papers have shown that satellite remote sensing may play an important role for studying and monitoring thermal volcanic activity, due to global coverage, continuity, and high frequency of observation, particularly in remote areas where ground-based surveillance systems are often lacking (e.g., [1–5]).

Sensors such as TM (Thematic Mapper) and ASTER (Advanced Spaceborne Thermal Emission and Reflection Radiometer), that have a repeat cycle of 16 days and offer channels in the SWIR (shortwave infrared) and TIR (thermal infrared) bands with a spatial resolution of 30–90 m, were widely used to investigate volcanic thermal anomalies (e.g., lava bodies, fumarole fields) [1,6–8]. HYPHERION, which is a hyperspectral imaging spectrometer providing VNIR (visible, near-infrared)/SWIR data at 220 wavelengths, was profitably used even for characterizing hot magmatic surfaces (e.g., [7,8]). AVHRR (Advanced Very High Resolution Radiometer) and MODIS (Moderate Resolution Imaging Spectroradiometer), acquiring data in the MIR (medium infrared) band and offering a good compromise between spatial and temporal resolution (1.1 km at the nadir; up to 6 h for AVHRR), represented key instruments for monitoring active volcanoes from space (e.g., [9–14]). SEVIRI (Spinning Enhanced Visible and Infrared Imager), like other geostationary satellite sensors, enabled the prompt identification of short-lived eruptive events (e.g., [15–19]), thanks to the high frequency of observation (15 min) and despite the low spatial resolution (3 km at the sub-satellite point).

Data from the above-mentioned sensors were exploited to retrieve the volcanogenic radiant flux, the time variations of which can be used as a proxy of the intensity changes of volcanic eruptions (e.g., [20–24]). This parameter enables the estimation of lava effusion rate, which is a critical parameter for numerical models that aim to predict lava flow paths (e.g., [25]).

In this paper, we present the results of satellite monitoring of Mt. Etna (Sicily, Italy) thermal activity between May and August 2016, integrated with ground-based structural and volcanological data. In particular, we investigate the eruptive events occurring in May and the fumarolic emissions recorded before the opening of a small degassing vent within the Voragine crater (VOR). The latter is one of Mt. Etna summit craters which has opened at the top of the central conduit, see Figure 1 [26,27].

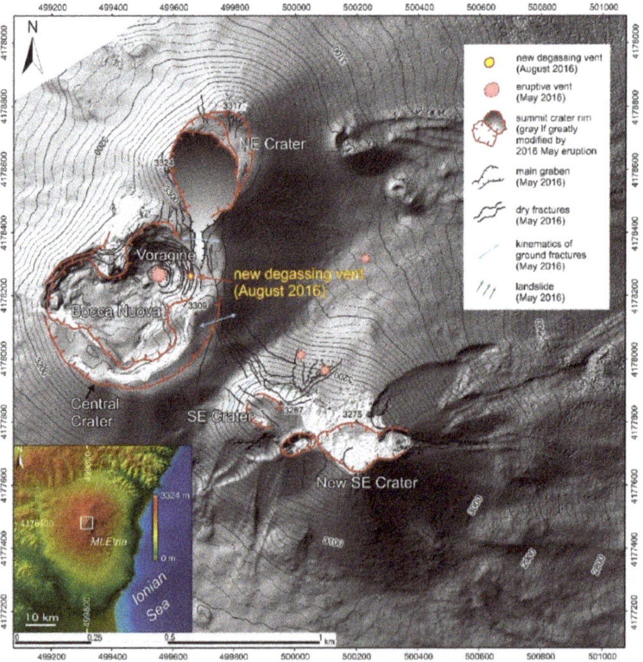

Figure 1. Structural map of the summit area of the Mount Etna volcano, updated in August 2016.

These events were investigated by means of a satellite-based system developed at IMAA (Institute of Methodologies for Environmental Analysis) implementing the RST_{VOLC} algorithm [28]. This monitoring system integrates AVHRR and MODIS observations for monitoring Italian volcanoes in near-real time, generating hotspot products (i.e., JPG, Kml, and ASCII files) a few minutes after the sensing time [29]. The work aims at assessing the contribution that multi-platform satellite observations may provide for better monitoring Mt. Etna, complementing the information provided by traditional surveillance systems.

2. The 2016 Mt. Etna Eruptive Activity

Five months after the early-December 2015 eruptions [30,31], ash emissions resumed at the North-East crater (NEC) during the night of 15–16 May 2016, while a Strombolian activity started the day after [32]. On the morning of 18 May, a 20–30 m long, short-lived (a few minutes) eruptive fissure activated on the northern flank of the NEC displaying weak spattering, immediately followed by violent Strombolian activity which occurred at the VOR crater. At that time, an eruptive fissure also opened feeding a lava flow, which expanded on the high Western flank of Mt. Etna. Within a couple of hours, the Strombolian activity at VOR totally filled this crater and the adjacent Bocca Nuova crater (BN), finally overflowing the BN western rim. The eruptive activity ceased the following night but resumed on early 19 May for ~1 h. On 20 May, a strong explosive activity started again at VOR and a new fracture field opened between VOR and the New South-East crater (NSEC), feeding a lava flow that expanded toward the east in the high Valle del Bove depression. At the same time, a new lava overflow occurred from the western rim of the BN. After a break of about a day, mild Strombolian activity resumed during the night between the 22 and 23 May at NEC, while lava fountaining occurred on the 24–25 May at VOR which eventually caused the total filling and obstruction of NEC, VOR, and BN.

At the end of the 15–25 May eruptions, a ~N-S fractured area characterized the volcano's summit. This fracture field was ~400 m wide and ~2000 m long, extending from the northern flank of the NEC to the eastern flank of the NSEC cone, crossing the eastern rim of the VOR, see Figure 1. This fracture field is bounded towards the east by a graben several tens of meters wide that also caused the collapse of the southern portion of the pyroclastic cone of the NEC, radically changing its morphology. As stated before, in the central crater, which contains VOR and BN, see Figure 1, the eruptive activity emerged at VOR only. After the end of the Strombolian activity that occurred on the 23–25 May, the VOR showed conspicuous subsidence phenomena, evidenced by the formation of numerous sub-circular and concentric fractures (lunar cracks) placed around a weakly degassing vent positioned at the bottom of this crater. Moreover, the BN was totally obstructed by products that erupted in May 2016; however, soon after the end of the eruption a weak subsidence also began to affect this crater.

Late in the evening of 7 August 2016, the top of Etna showed almost continuous flashes. Since that moment, a new 20−30 m wide vent, placed on the inner eastern rim of VOR, emits a pulsating emission of incandescent gases up to over 600 °C, see Figure 2; video footage of this activity can be found as a supplementary material. The new degassing vent fits perfectly into the structural framework inherited from the eruption of May 2016, see Figure 3. It is located in the area where the graben, described before, intercepts the edge of the VOR; i.e., an area subject to open for the "pull" induced by the movement of the eastern flank of the volcano [33–37] and for the subsidence that affects the bottom of the VOR.

(a)

(b)

Figure 2. Mt. Etna activity of 10 August 2016 from the degassing vent opening within the Voragine crater (VOR) crater; (**a**) morning picture; (**b**) afternoon picture with evidence of the volcanic glow (credits: Marco Neri).

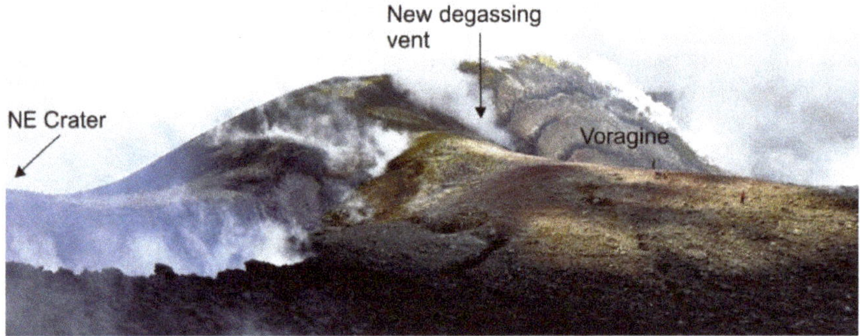

Figure 3. New degassing vent opened on 7 August 2016 and located along the fracture system formed during the May 2016 eruption (see Figure 1 for map view) (credits: Marco Neri).

3. Methods: RST$_{VOLC}$ Algorithm

The RST$_{VOLC}$ algorithm [28] identifies volcanic hotspots by means of two local variation indices defined as:

$$\otimes_{MIR}(x,y,t) = \frac{BT_{MIR}(x,y,t) - \mu_{MIR}(x,y)}{\sigma_{MIR}(x,y)} \tag{1}$$

$$\otimes_{MIR-TIR}(x,y,t) = \frac{\Delta T(x,y,t) - \mu_{\Delta T}(x,y)}{\sigma_{MIR}(x,y)} \tag{2}$$

In Equation (1), $BT_{MIR}(x,y,t)$ is the MIR (medium infrared) brightness temperature measured at the time t for each pixel (x,y), whereas $\mu_{MIR}(x,y)$ and $\sigma_{MIR}(x,y)$ are the relative temporal mean and standard deviation. In Equation (2), $\Delta T(x,y,t) = BT_{MIR}(x,y,t) - BT_{TIR}(x,y,t)$, where $BT_{TIR}(x,y,t)$

is the brightness temperature measured in the TIR (thermal infrared) band at around 11 μm wavelength; $\mu_{\Delta T}(x,y)$ and $\sigma_{\Delta T}(x,y)$ stand for the relative temporal mean and standard deviation. These terms are calculated after processing cloud-free satellite records selected according to specific homogeneity criteria (i.e., same spectral channel/s, calendar month, satellite overpass time). In particular, the OCA (One Channel Cloud-Detection Approach) RST-based method [38] is generally used to filter out cloudy pixels from the scenes. The iterative $k\sigma$ clipping filter, which is also implemented within the RST_{VOLC} process, enables the removal of signal outliers (e.g., extremely hot pixels) [29].

The $\otimes_{MIR}(x,y,t)$ index identifies anomalous signal variations in the MIR band of sensors like AVHRR (channel 3: 3.55–3.93 μm) and MODIS (channels 21/22: 3.929–3.989 μm), where hot magmatic surfaces reach the peak of thermal emissions [39,40]. The $\otimes_{MIR-TIR}(x,y,t)$ index is used jointly with the previous one for minimizing spurious effects associated with non-volcanological signal fluctuations [28]. RST_{VOLC}, combining those indices, is capable of guaranteeing an efficient identification of volcanic thermal anomalies (a cloud-masking procedure is used in the daytime before running the algorithm) under different observational conditions (e.g., [12,18,41–43]).

4. Results

4.1. Monitoring the Paroxysmal Events of May 2016

To investigate the Mt. Etna eruptive events of May 2016, we show, in Figure 4, the curve of the volcanic tremor (top panel) and the time series of the radiant flux (Q_{rad}) retrieved from AVHRR and MODIS data uncorrected for atmospheric effects (bottom panel). In more detail, in Figure 4b, we also show eruption chronology and overcast periods, for better assessing the impact of clouds on the achieved results. We estimated the radiant flux from infrared AVHRR data by outputs of a dual band-three components method [44]; the mean Q_{rad} value calculated from two end-members was considered. The formulation proposed by [45] and amended by [46] was used to retrieve the same parameter from MODIS data.

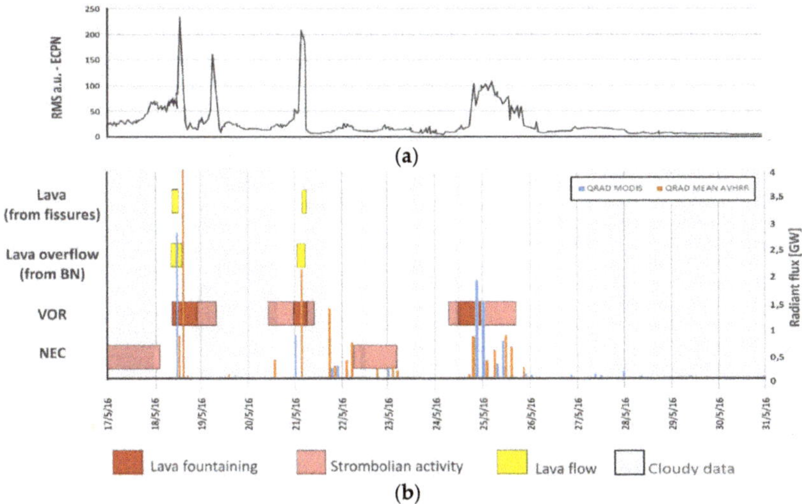

Figure 4. (**a**) Root-mean-square (RMS) tremor amplitude at Etna Cratere del Piano (ECPN) seismic station, a.u. = arbitrary unit; (**b**) radiant flux retrieved from Advanced Very High Resolution Radiometer (AVHRR) (orange bars) and Moderate Resolution Imaging Spectroradiometer (MODIS) (blue bars) data from the 17 to 31 May. Eruption chronology and information about overcast periods (in light grey), provided by the One channel Cloud-Detection Approach (OCA) method, is also reported.

Figure 4b shows that the radiant flux estimated by satellite ranged, during the period of interest, from a few MW up to a few GW. In particular, the paroxysmal event of May 18 was the most intense, leading to Q_{rad} values, retrieved from daytime MODIS and AVHRR data acquired under comparable values of satellite zenith angle (SZA), in the range 2.5–4.0 GW. This paroxysm, occurring with lava fountaining from VOR and a lava overflow from BN, generated the major peak in the volcanic tremor, see Figure 4a. The following Strombolian activity lead to another sudden increase of this parameter but it was undetected by RST_{VOLC} because of clouds, see Figure 4b.

During 20–21 May, another significant increase of Q_{rad}, up to about 2.0 GW, was recorded. This increase of radiant flux was determined by an eruptive activity similar to that of 18 May (see eruption chronology), leading to the third abrupt increment in the volcanic tremor. The latter increased, although in a less significant way than before, on 24 May when another lava fountaining activity occurred. Based on the retrieved Q_{rad} values, this eruptive episode was slightly less intense than the previous one. Figure 4b shows that RST_{VOLC} also identified the Strombolian event of 22–23 May. Moreover, it detected a thermal anomaly in between the Strombolian activities of 21 May and 22 May, as indicated by values of the radiant flux retrieved in that period which were mostly lower than 1 GW. In addition, an abrupt increase of Q_{rad} was recorded in short-time intervals (e.g., in between AVHRR and MODIS observations of 24 May at 20:23 UTC and 21:12 UTC), revealing some abrupt variations in the intensity of the volcanic thermal emissions. However, due to cloud coverage and satellite overpass times, some thermal activities (e.g., see 24 May) were undetected from space. On the other hand, even the discontinuous identification of a low-level thermal anomaly after 25 May, possibly associated to a weak degassing from VOR (leading to a minimum Q_{rad} value of about 7.5 MW), is ascribable to clouds.

4.2. Investigating the Thermal Activity of June–August 2016

In Figure 5, we show the temporal trend of the total MIR radiance retrieved from nighttime AVHRR and MODIS records of June–August 2016. We investigated only data with relatively low values of satellite zenith angle (i.e., SZA < 40°), unlike the eruptive events of May 2016, for minimizing the impact of the satellite viewing geometry on the thermal anomaly detected from space.

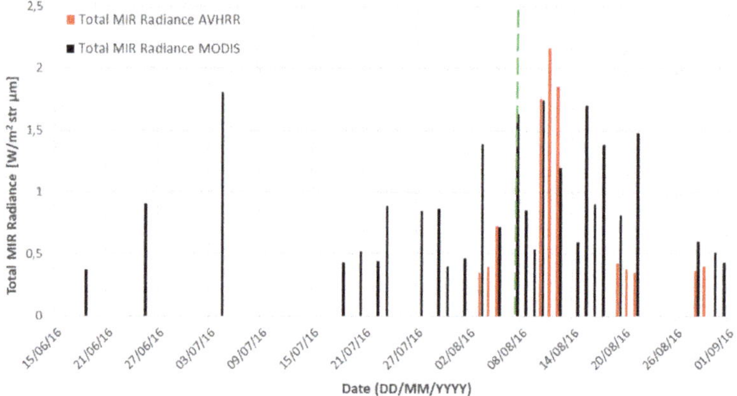

Figure 5. Total medium infrared (MIR) radiance retrieved from hotspot pixels identified by Robust Satellite Techniques–volcanoes (RST_{VOLC}) on nighttime AVHRR (blue bars) and MODIS (red bars) data of June–August 2016. Dotted green line indicates the start of hot degassing activity at VOR. Note that no MODIS data were acquired at Institute of Methodologies for Environmental Analysis (IMAA) during 1−3 July because of antenna problems; AVHRR data acquired in July were unprocessed due to geo-location issues.

During the first week of June, no MODIS data were acquired at IMAA because of some antenna problems. In July, AVHRR records were unprocessed due to some geolocation issues (the system was back to being fully operational in early August). Despite these limitations, reducing the number of available satellite scenes, RST_{VOLC} provided information about changes in Mt. Etna thermal activity.

Indeed, after the identification of a sporadic thermal anomaly in June RST_{VOLC} revealed, since early July, the occurrence of more continuous thermal emissions at the crater area. In more detail, Figure 5 shows that after 4 July, the total MIR radiances decreased, because of a less significant thermal anomaly (see Figure 6), showing small fluctuations until 3 August when an evident increase of this parameter was recorded. During 10–12 August, detected thermal anomaly was more extended in terms of hot spot pixels, as shown by RST_{VOLC} maps which are not reported here, leading to peak of total MIR radiance.

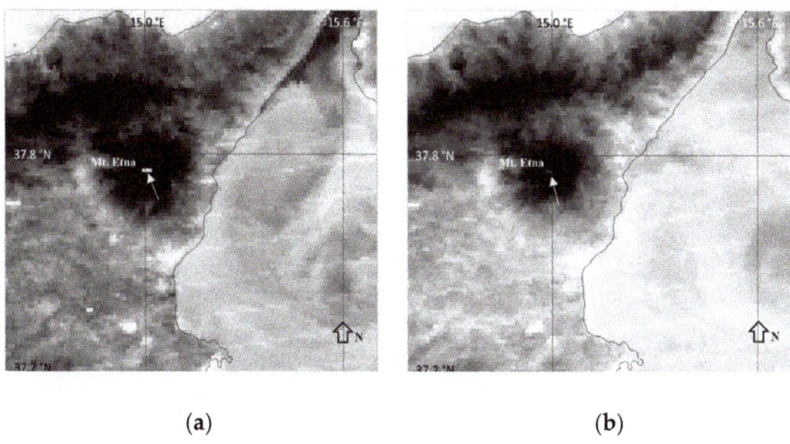

(a) (b)

Figure 6. (**a**) MODIS channel 22 (MIR) images with an indication (see white arrow) of the volcanic thermal anomaly identified by RST_{VOLC} (brighter tones indicate higher brightness temperature values); 4 July 2016 at 21:06 UTC (BT_{22MAX} = 297.65 K); (**b**) 6 July 2016 at 20:53 UTC (BT_{22MAX} = 290.29 K).

To assess RST_{VOLC} detections, as well as changes of thermal activity revealed by Figure 5, we analysed the cloud-free Sentinel 2A-MSI (Multispectral Instrument) data of June–August 2016 made freely available online by the Sentinel Hub.

The MSI has 13 spectral channels centered in the VNIR and SWIR bands having a different spatial resolution [47]. In Figure 7, we show a number of false color composite images, nine in total from 2 June to 28 August, magnified over the Mt. Etna crater area. These RGB (red-green-blue) products, at a 20-m spatial resolution, were generated by using bands 12 (2.19 μm), 8A (0.865 μm) and 4 (0.665 μm). The figure shows that some crater pixels assumed the red color owing to the dominance of the SWIR component.

Since high-temperature magmatic surfaces are highly radiant even in the SWIR region (e.g., [48]), and considering that other features (e.g., clouds) are clearly recognizable in Figure 7, the presence of a thermal anomaly at the Mt. Etna crater area, before and after the start of hot degassing activity at VOR, was confirmed. Besides, it can be noted that some VOR pixels were brighter, especially, on satellite scenes of 2 July (i.e., two days before the evident increase of total MIR radiance revealed by Figure 5) and 21 August (see Figure 7).

The false color composite imagery, at a 30-m spatial resolution, of Figure 8 generated from Landsat 8-OLI (Landsat 8-Operational Land Imager) data of 5 July and 6 August provided by UGSG (United States Geological Survey), further corroborate information provided by RST_{VOLC}. The figure confirms that a thermal anomaly affected the Mt. Etna crater area before the opening of the new degassing vent; see pixels, in red, highly radiant in OLI band 7 (2.1–2.3 μm).

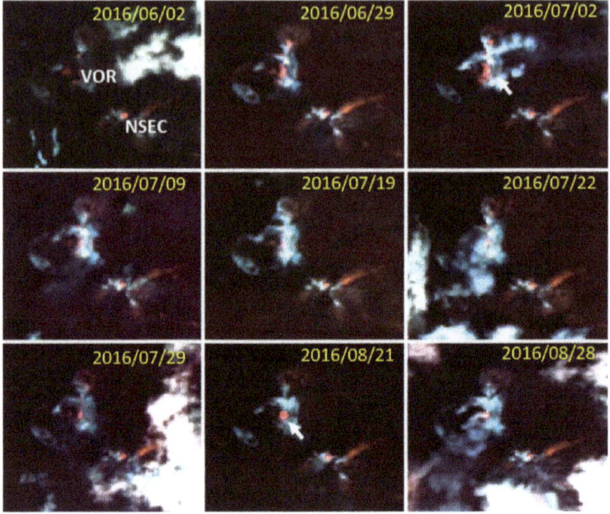

Figure 7. RGB (Red= Band 12, 2.19 μm; Green = Band 8A, 0.865 μm; Blue = Band 4, 0.665 μm) Sentinel-2 products, at 20 m spatial resolution, of June–August 2016 generated from Level 1C data excluding images completely overcast over target area. The white arrow indicates pixels more radiant in the shortwave infrared (SWIR) band at VOR.

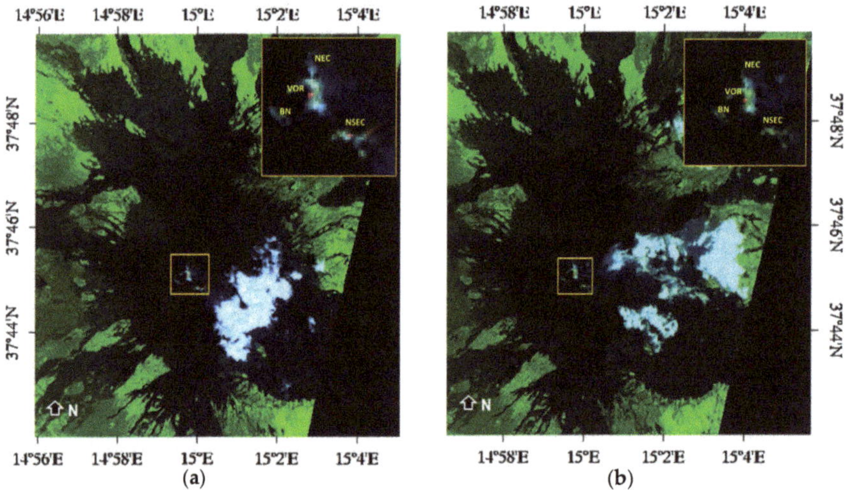

Figure 8. RGB (Red = Band 7, 2.1–2.3 μm; Green = Band 5, 0.845–0.885 μm; Blue = Band 2, 0.450–0.515 μm) Landsat 8-Operational Land Imager (OLI) images, at a 30-m spatial resolution, showing hotspot pixels (in red), magnified at the top-right side of each panel, affecting the Mt. Etna area. (**a**) 5 July 2016; (**b**) 6 August 2016.

To localize, in a more accurate way, the source of the thermal anomaly identified by the satellite, we analyzed the temporal trend of the SWIR radiance. In more detail, we used the dark object subtraction (DOS) method [49] to correct SWIR radiance for effects of atmospheric scattering by subtracting the radiance value of the darkest object from each pixel of the image (e.g., [50]). We focused our analyses on the VOR area since the thermal anomaly detected by RST_{VOLC} generally did not include the NSEC; see AVHRR pixel polygon in green overlapped with the OLI sub-scene in the inset of Figure 9. The latter shows the similar behavior of SWIR curves at nine VOR pixels, apart from 5 July when the SWIR signal decreased only at VOR5 and VOR7.

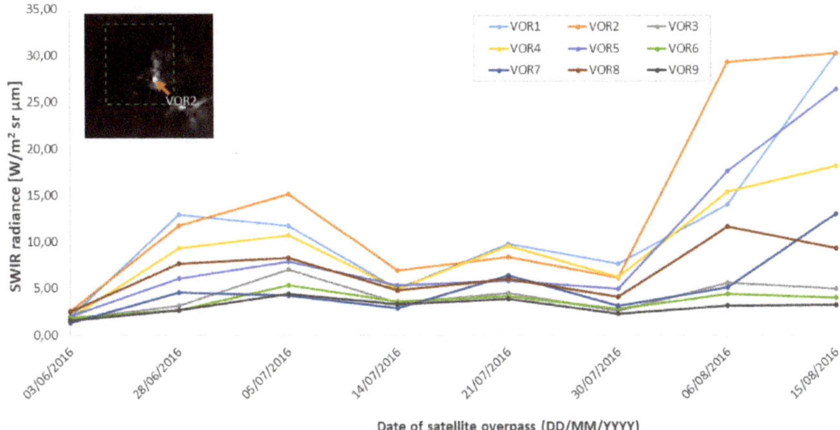

Figure 9. Temporal plot of the OLI-SWIR radiance retrieved, after applying the dark object subtraction (DOS) method, over nine different VOR pixels. In the inset, the OLI sub-image of 6 August, with indication of most radiant pixel (VOR2), and the AVHRR pixel polygon (in dotted green) associated to thermal anomaly flagged by RST_{VOLC} before the opening of the new degassing vent.

In addition, the plot shows that the strongest increase in the SWIR signal was recorded at VOR2 on 6 August, see the orange curve. By the temporal trend of the total SWIR radiance retrieved, pixel by pixel, along the A-B transect region of Figure 10, we found that VOR2, whose geographic coordinates of the center are reported in Figure 9, was the most radiant pixel in OLI band 7 (see value of the analyzed parameter retrieved in correspondence to the dotted red line). Hence, it is reasonable to suppose that the volcanic thermal emissions from the VOR2 area affected, in a more significant way, the thermal anomaly detected by RST_{VOLC} after eruptions of May and before the start of high-temperature degassing activity on 7 August.

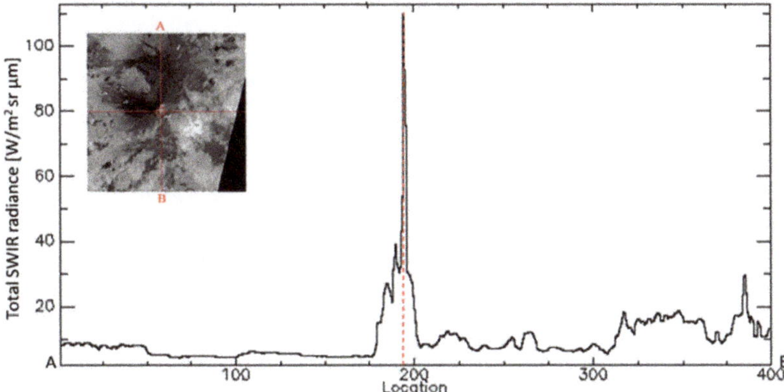

Figure 10. Spatial profile of the total SWIR radiance retrieved along the A–B transect region intersecting the VOR area (see OLI data sub-scene in the inset of the figure). The most radiant pixel, whose location is indicated by the dotted red line, having geographic coordinates of the center 37°45′6.08″N, 14°59′45.08″E, corresponds to VOR2 in Figure 9.

5. Discussion

In this paper, we have investigated the Mt. Etna activity of May–August 2016, exploiting the information provided by the satellite-based monitoring system operating at IMAA (whose products are currently used only for research purposes) implementing the RST_{VOLC} algorithm. The latter runs on both AVHRR and MODIS data, which currently guarantee more than 10 observations per day over Italy, increasing the probability of processing cloud-free satellite scenes [51].

The results of this study confirm that RST_{VOLC} performed in a similar way when detecting thermal anomalies regardless of the satellite data used; i.e., despite the different features of AVHRR and MODIS instruments. Indeed, by combining values of the radiant flux, it was possible to investigate eruptions occurring in May in a more continuous way than using data from a single satellite sensor; although, a different retrieval method was used. Indeed, despite the possible inaccuracy in estimating the radiant flux due to the analysis of satellite records which were uncorrected for atmospheric effects, we observed a good agreement between temporal changes of Q_{rad} and information provided by field observations. High values of radiant flux characterized periods of lava effusion/fountaining, whereas low values of the same parameter were retrieved in the presence of Strombolian eruptions, as shown in Figure 4.

Regarding the few discrepancies with ground-based observations, they were mostly related to clouds. Although recent literature studies performed by means of ground-based hyperspectral imagers have suggested that these features could be accounted for under certain circumstances [45], clouds generally represent a common issue for satellite-based methods developed for monitoring thermal volcanic activity. In this work, clouds had a not negligible impact on the results of the thermal anomaly detection because, on some days, they partially or completely obscured the underlying hot surfaces. Consequently, some Strombolian activities were identified by satellite some hours after the eruption onset; RST_{VOLC} detected, for instance, the first thermal anomaly over the Mt. Etna area on 18 May, on AVHRR overpass of 01:17 UTC (universal time coordinated); although, a Strombolian activity was already in progress the day before, see eruption chronology in Figure 4. Moreover, a few daytime MODIS scenes strongly affected by clouds, which were not completely removed by the OCA method, showed artifacts. These features were recognized and filtered out from the analyzed time series.

Despite the impact of clouds and satellite viewing geometry on the thermal anomaly identification (e.g., as for AVHRR data acquired at 17:05 UTC), this study provides additional information concerning eruptive events occurring in May. Specifically, Figure 2 showed that, in spite of possible lava cooling effects, the eruptive activity of 21 May probably lasted longer than was indicated by field observations. Although, the occurrence of a short-lived eruptive event in the late afternoon of the same day cannot be excluded. The unfavorable meteorological conditions possibly affected the direct/visual observation of these phenomena. Concerning the short-term variations of Q_{rad} also revealed in Figure 2, they require further investigation to be assessed, analyzing, for instance, infrared SEVIRI data.

RST_{VOLC} detections performed after the eruptive events of May revealed that the intensity of fumarolic emissions from the fracture fields cutting the entire summit area changed before 7 August, possibly due to a preparatory phase of hot degassing activity occurring at VOR. This is even more evident from looking at the curve of the radiant flux (in blue), retrieved from nighttime MODIS data, in Figure 11. The plot shows that Q_{rad} was around 30 MW on 4 July, then decreased in the following weeks, when its average value was ~7.5 MW, and increased up to 18 MW on 3 August (see right axis); i.e., four days before the opening of the new degassing vent marked by the dotted black line. After a slight reduction, the radiant flux once again increased on 7 August owing to the high-temperature degassing activity in progress at VOR, reaching its peak of around 33 MW three days later, see Figure 11.

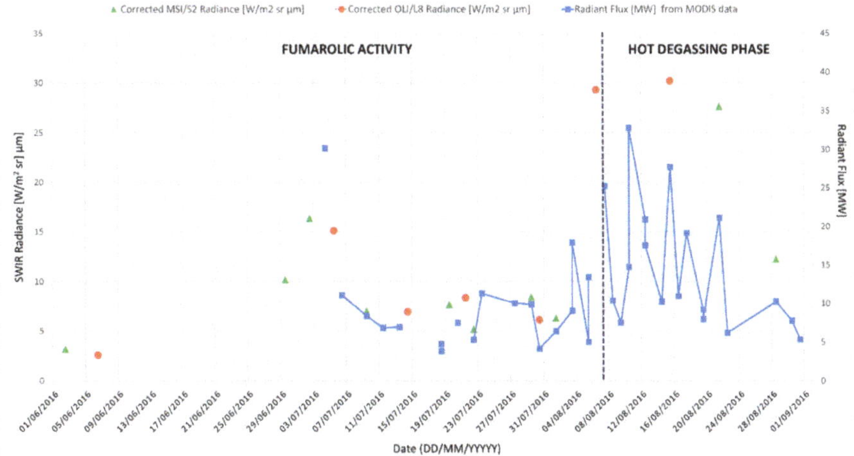

Figure 11. SWIR radiance (left axis), from OLI band 7 (red points) and MSI band 12 (green triangles) data of June–August 2016, measured over VOR2 area and corrected using the DOS method (see Section 4.2). In blue, the temporal trend of radiant flux (right axis) retrieved after filtering MODIS data for values of SZA < 40°. The dotted black line indicates the start of hot degassing activity at VOR.

Those variations of radiant flux were consistent with information independently provided by OLI and MSI data at the VOR2 location, corresponding to the area where the new degassing vent would later open, see the previous section. Figure 11 shows that SWIR radiance measured at the VOR2 crater area was higher in early July compared to mid-end July, and significantly increased on 6 August (see left axis), strengthening the hypothesis formulated above.

The information retrieved by satellite is perfectly compatible with the actual evolution of the system of degassing fractures, which progressively opened in the summit area of Mt. Etna. It should be pointed out that, at the end of May 2016, eruptions from all summit craters had conduits almost totally obstructed by the erupted volcanic products, and, therefore, they were no longer capable of degassing efficiently. Under these conditions, it is very likely that the pressure of the volcanic gases that rise inside the obstructed conduits have exerted a considerable pressure, also causing the heating and the thermal alteration of the rocks in the immediate vicinity of the conduits themselves, looking for alternative outputs. Therefore, the fracture system opening in the summit area could have been an alternative way to allow the volcanic gas to reach the surface. At first, this gradual fracturing process triggered the thermal anomalies observed from space during the first days of July. Finally, the opening of the degassing vent on August 7 close to the VOR's rim could be the culmination of this fracturing process, which eventually drained most of the gas from the obstructed NEC conduit.

This evolution has already been observed several times during the last twenty years at Etna. Starting from January 1998, a system of N-S fractures opened, and then progressively propagated, affecting exactly the same portion of the summit area, generating dry fractures, fumarolic activities, and eruptive fissures [33], similar to what happened in May–August 2016. Thus, the opening of the vent on 7 August was possibly triggered by the intersection of this developing fracture system with the North-East crater (NEC) plumbing system, leading to the draining of gas that had remained blocked inside its conduits after the May eruption. Finally, it should be emphasized that the opening of this fracture system seems to be triggered by an acceleration of the deformations affecting the eastern flank of the volcano [26,27,35–37]); therefore, it is possible that the opening of new degassing or erupting vents will continue to take place in this portion of the summit crater area.

6. Conclusions

This study demonstrates that multi-platform observing systems may also provide a relevant contribution for monitoring active volcanoes in areas where efficient ground-based surveillance systems exist, enabling the identification of low-temperature fumarole fields whose intensity variations may precede new and more significant phases of thermal unrest.

In this context, the performance of the RST_{VOLC} system may be further increased using VIIRS (Visible Infrared Imaging Radiometer Suite), flying onboard Suomi NPP (Suomi National Polar-orbiting Partnership) and JPPS-1 (Joint Polar Satellite System) satellites. VIIRS collects data in 22 different spectral bands, ranging from 0.412 μm (VIS) to 12.01 μm (TIR) and including 17 spectral channels at a 750-m spatial resolution, i.e., the moderate resolution bands (M-bands) and the Day/Night panchromatic band (DNB), and five imaging resolution bands (I-bands), having a spatial resolution of 375 m at the nadir. Among the channels, the I4 band (3.55–3.93 μm) should guarantee further improvements in the identification of weak thermal anomalies (i.e., those of low temperature and/or spatial extent), as indicated by some investigations currently in progress, making RST_{VOLC} even more effective in detecting subtle changes of thermal volcanic activity at Mt. Etna.

Author Contributions: F.M., M.N. and N.P. conceived the work and wrote the paper. All the other authors contributed in analyzing and interpreting results.

Funding: This research received no external funding.

Acknowledgments: Landsat 8-OLI imagery used in this work were provided by the USGS (United States Geological Survey) through the Earth Explorer portal (https://earthexplorer.usgs.gov/). Sentinel-2 data are made free available online by the SENTINEL Hub (http://www.sentinel-hub.com/).

Conflicts of Interest: The authors declare no conflict of interest.

References

1. Dehn, J.; Dean, K.; Engle, K. Thermal monitoring of North Pacific volcanoes from space. *Geology* **2000**, *28*, 755–758.
2. Dean, K.; Servilla, M.; Roach, A.; Foster, B.; Engle, K. Satellite monitoring of remote volcanoes improves study efforts in Alaska. *Eos Trans. Am. Geophys. Union* **1998**, *79*, 413–423. [CrossRef]
3. Coppola, D.; Cigolini, C. Thermal regimes and effusive trends at Nyamuragira volcano (DRC) from MODIS infrared data. *Bull. Volcanol.* **2013**, *75*, 744. [CrossRef]
4. Rothery, D.A.; Francis, P.W.; Wood, C.A. Volcano monitoring using short wavelength infrared data from satellites. *J. Geophys. Res. Solid Earth* **1998**, *93*, 7993–8008. [CrossRef]
5. Oppenheimer, C. Lava flow cooling estimated from Landsat Thematic Mapper infrared data: The Lonquimay eruption (Chile, 1989). *J. Geophys. Res. Solid Earth* **1991**, *96*, 21865–21878. [CrossRef]
6. Pieri, D.; Abrams, M. ASTER observations of thermal anomalies preceding the April 2003 eruption of Chikurachki volcano, Kurile Islands, Russia. *Remote Sens. Environ.* **2005**, *99*, 84–94. [CrossRef]
7. Davies, A.G.; Chien, S.; Baker, V.; Doggett, T.; Dohm, J.; Greeley, R.; Ip, F.; Castan, R.; Cichy, B.; Rabideau, G.; et al. Monitoring active volcanism with the autonomous sciencecraft experiment on EO-1. *Remote Sens. Environ.* **2006**, *101*, 427–446. [CrossRef]
8. Abrams, M.; Pieri, D.; Realmuto, V.; Wright, R. Using EO-1 Hyperion data as HyspIRI preparatory data sets for volcanology applied to Mt Etna, Italy. *IEEE J. Sel. Top. Appl. Earth Obs. Remote Sens.* **2013**, *6*, 375–385. [CrossRef]
9. Higgins, J.; Harris, A. VAST: A program to locate and analyze volcanic thermal anomalies automatically from remotely sensed data. *Comput. Geosci.* **1997**, *23*, 627–645. [CrossRef]
10. Dehn, J.; Dean, K.G.; Engle, K.; Izbekov, P. Thermal precursors in satellite images of the 1999 eruption of Shishaldin Volcano. *Bull. Volcanol.* **2002**, *64*, 525–534. [CrossRef]
11. Wright, R.; Flynn, L.P. On the retrieval of lava-flow surface temperatures from infrared satellite data. *Geology* **2003**, *31*, 893–896. [CrossRef]

12. Marchese, F.; Lacava, T.; Pergola, N.; Hattori, K.; Miraglia, E.; Tramutoli, V. Inferring phases of thermal unrest at Mt. Asama (Japan) from infrared satellite observations. *J. Volcanol. Geotherm. Res.* **2012**, *237*, 10–18. [CrossRef]
13. Lombardo, V. AVHotRR: Near-real time routine for volcano monitoring using IR satellite data. *Geol. Soc. Lond. Spec. Publ.* **2015**, *426*, 73–92. [CrossRef]
14. Miller, P.I.; Harris, A.J. Near-real-time service provision during effusive crises at Etna and Stromboli: Basis and implementation of satellite-based IR operations. *Geol. Soc. Lond. Spec. Publ.* **2016**, *426*, SP426-26. [CrossRef]
15. Harris, A.J.L.; Pilger, E.; Flynn, L.P.; Garbeil, H.; Mouginis-Mark, P.J.; Kauahikaua, J.; Thornber, C. Automated, high temporal resolution, thermal analysis of Kilauea volcano, Hawai'i, using GOES satellite data. *Int. J. Remote Sens.* **2001**, *22*, 945–967. [CrossRef]
16. Marchese, F.; Ciampa, M.; Filizzola, C.; Mazzeo, G.; Lacava, T.; Pergola, N.; Tramutoli, V. On the exportability of Robust Satellite Techniques (RST) for active volcanoes monitoring. *Remote Sens.* **2010**, *2*, 1575–1588. [CrossRef]
17. Ganci, G.; Harris, A.J.L.; Del Negro, C.; Guehenneux, Y.; Cappello, A.; Labazuy, P.; Calvari, S.; Gouhier, M. A year of lava fountaining at Etna: Volumes from SEVIRI. *Geophys. Res. Lett.* **2012**, *39*, L06305. [CrossRef]
18. Marchese, F.; Falconieri, A.; Pergola, N.; Tramutoli, V. A retrospective analysis of the Shinmoedake (Japan) eruption of 26–27 January 2011 by means of Japanese geostationary satellite data. *J. Volcanol. Geotherm. Res.* **2014**, *269*, 1–13. [CrossRef]
19. Kaneko, T.; Takasaki, K.; Maeno, F.; Wooster, M.J.; Yasuda, A. Himawari-8 infrared observations of the June–August 2015 Mt Raung eruption, Indonesia. *Earth Planets Space* **2018**, *70*, 89. [CrossRef]
20. Harris, A.J.; Butterworth, A.L.; Carlton, R.W.; Downey, I.; Miller, P.; Navarro, P.; Rothery, D.A. Low-cost volcano surveillance from space: Case studies from Etna, Krafla, Cerro Negro, Fogo, Lascar and Erebus. *Bull. Volcanol.* **1997**, *59*, 49–64. [CrossRef]
21. Coppola, D.; Piscopo, D.; Staudacher, T.; Cigolini, C. Lava discharge rate and effusive pattern at Piton de la Fournaise from MODIS data. *J. Volcanol. Geotherm. Res.* **2009**, *184*, 174–192. [CrossRef]
22. Wright, R.; Blackett, M.; Hill-Butler, C. Some observations regarding the thermal flux from Earth's erupting volcanoes for the period of 2000 to 2014. *Geophys. Res. Lett.* **2015**, *42*, 282–289. [CrossRef]
23. Wright, R.; Pilger, E. Radiant flux from Earth's subaerially erupting volcanoes. *Int. J. Remote Sens.* **2008**, *29*, 6443–6466. [CrossRef]
24. Bonny, E.; Wright, R. Predicting the end of lava-flow-forming eruptions from space. *Bull. Volcanol.* **2017**, *79*, 52. [CrossRef]
25. Vicari, A.; Ganci, G.; Behncke, B.; Cappello, A.; Neri, M.; Del Negro, C. Near-real-time forecasting of lava flow hazards during the 12–13 January 2011 Etna eruption. *Geophys. Res. Lett.* **2011**, *38*, 13. [CrossRef]
26. Neri, M.; Acocella, V.; Behncke, B.; Giammanco, S.; Mazzarini, F.; Rust, D. Structural analysis of the eruptive fissures at Mount Etna (Italy). *Ann. Geophys.* **2011**, *54*, 464–479. [CrossRef]
27. Cappello, A.; Bilotta, G.; Neri, M.; Del Negro, C. Probabilistic modeling of future volcanic eruptions at Mount Etna. *J. Geophys. Res. Solid Earth* **2013**, *118*, 1925–1935. [CrossRef]
28. Marchese, F.; Filizzola, C.; Genzano, N.; Mazzeo, G.; Pergola, N.; Tramutoli, V. Assessment and improvement of a Robust Satellite Technique (RST) for thermal monitoring of volcanoes. *Remote Sens. Environ.* **2011**, *115–116*, 1556–1563. [CrossRef]
29. Pergola, N.; Coviello, I.; Filizzola, C.; Lacava, T.; Marchese, F.; Paciello, R.; Tramutoli, V. A review of RSTVOLC, an original algorithm for automatic detection and near-real-time monitoring of volcanic hotspots from space. *Geol. Soc. Lond. Spec. Publ.* **2015**, *426*, 55–72. [CrossRef]
30. Corsaro, R.A.; Andronico, D.; Behncke, B.; Branca, S.; Caltabiano, T.; Ciancitto, F.; Cristaldi, A.; De Beni, E.; La Spina, A.; Lodato, L.; et al. Monitoring the December 2015 summit eruptions of Mt. Etna (Italy): Implications on eruptive dynamics. *J. Volcanol. Geotherm. Res.* **2017**, *341*, 53–69. [CrossRef]
31. Neri, M.; De Maio, M.; Crepaldi, S.; Suozzi, E.; Lavy, M.; Marchionatti, F.; Calvari, S.; Buongiorno, F. Topographic Maps of Mount Etna's Summit Craters, updated to December 2015. *J. Maps* **2017**, *13*, 674–683. [CrossRef]
32. Cannata, A.; Di Grazia, G.; Giuffrida, M.; Gresta, S.; Palano, M.; Sciotto, M.; Viccaro, M.; Zuccarello, F. Space-time evolution of magma storage and transfer at Mt. Etna volcano (Italy): The 2015–2016 reawakening of Voragine crater. *Geochem. Geophys. Geosyst.* **2018**, *19*, 471–495. [CrossRef]

33. Neri, M.; Acocella, V. The 2004-05 Etna eruption: Implications for flank deformation and structural behaviour of the volcano. *J. Volcanol. Geotherm. Res.* **2006**, *158*, 195–206. [CrossRef]
34. Neri, M.; Casu, F.; Acocella, V.; Solaro, G.; Pepe, S.; Berardino, P.; Sansosti, E.; Caltabiano, T.; Lundgren, P.; Lanari, R. Deformation and eruptions at Mt. Etna (Italy): A lesson from 15 years of observations. *Geophys. Res. Lett.* **2009**, *36*, L02309. [CrossRef]
35. Siniscalchi, A.; Tripaldi, S.; Neri, M.; Balasco, M.; Romano, G.; Ruch, J.; Schiavone, D. Flank instability structure of Mt Etna inferred by a magnetotelluric survey. *J. Geophys. Res.* **2012**, *117*, B03216. [CrossRef]
36. Acocella, V.; Neri, N.; Behncke, B.; Bonforte, A.; Del Negro, C.; Ganci, G. Why does a mature volcano need new vents? The case of the New Southeast Crater at Etna. *Front. Earth Sci.* **2016**, *4*, 67. [CrossRef]
37. Giammanco, S.; Melián, G.; Neri, M.; Hernández, P.A.; Sortino, F.; Barrancos, J.; López, M.; Pecoraino, G.; Perez, N.M. Active tectonic features and structural dynamics of the summit area of Mt. Etna (Italy) revealed by soil CO2 and soil temperature surveying. *J. Volcanol. Geotherm. Res.* **2016**, *311*, 79–98. [CrossRef]
38. Cuomo, V.; Filizzola, C.; Pergola, N.; Pietrapertosa, C.; Tramutoli, V. A self-sufficient approach for GERB cloudy radiance detection. *Atmos. Res.* **2004**, *72*, 39–56. [CrossRef]
39. Pergola, N.; Marchese, F.; Tramutoli, V. Automated detection of thermal features of active volcanoes by means of infrared AVHRR records. *Remote Sens. Environ.* **2004**, *93*, 311–327. [CrossRef]
40. Zakšek, K.; Hort, M.; Lorenz, E. Satellite and ground based thermal observation of the 2014 effusive eruption at Stromboli volcano. *Remote Sens.* **2015**, *7*, 17190–17211. [CrossRef]
41. Marchese, F.; Filizzola, C.; Mazzeo, G.; Paciello, R.; Pergola, N.; Tramutoli, V. Robust Satellite Techniques for thermal volcanic activity monitoring, early warning and possible prediction of new eruptive events. In Proceedings of the 2009 IEEE International Geoscience and Remote Sensing Symposium, Cape Town, South Africa, 12–17 July 2009; Volume 2, p. II-953.
42. Lacava, T.; Marchese, F.; Arcomano, G.; Coviello, I.; Falconieri, A.; Faruolo, M.; Pergola, N.; Tramutoli, V. Thermal monitoring of Eyjafjöll volcano eruptions by means of infrared MODIS data. *IEEE J. Sel. Top. Appl. Earth Obs. Remote Sens.* **2014**, *7*, 3393–3401. [CrossRef]
43. Lacava, T.; Kervyn, M.; Liuzzi, M.; Marchese, F.; Pergola, N.; Tramutoli, V. Assessing performance of the RSTVOLC multi-temporal algorithm in detecting subtle hot spots at Oldoinyo Lengai (Tanzania, Africa) for comparison with MODLEN. *Remote Sens.* **2018**, *10*, 1177. [CrossRef]
44. Harris, A.J.; Blake, S.; Rothery, D.A.; Stevens, N.F. A chronology of the 1991 to 1993 Mount Etna eruption using advanced very high resolution radiometer data: Implications for real-time thermal volcano monitoring. *J. Geophys. Res. Solid Earth* **1997**, *102*, 7985–8003. [CrossRef]
45. Kaufman, Y.J.; Justice, C.O.; Flynn, L.P.; Kendall, J.D.; Prins, E.M.; Giglio, L.; Ward, D.E.; Menzel, W.P.; Setzer, A.W. Potential global fire monitoring from EOS-MODIS. *J. Geophys. Res.* **1998**, *103*, 32215–32238. [CrossRef]
46. Giglio, L. *MODIS Collection 5 Active Fire Product User's Guide 2003, Version 2.5*; Department of Geographical, University of Maryland: College Park, MD, USA, 2003; 61p.
47. European Space Agency, Sentinel Online. Available online: https://sentinel.esa.int/web/sentinel/user-guides/sentinel-2-msi/resolutions/spatial (accessed on 1 December 2018).
48. Blackett, M.; Wooster, M.J. Evaluation of SWIR-based methods for quantifying active volcano radiant emissions using NASA EOS-ASTER data. *Geomatics Nat. Hazards Risk* **2001**, *2*, 51–78. [CrossRef]
49. Chavez, P.S., Jr. An improved dark-object subtraction technique for atmospheric scattering correction of multispectral data. *Remote Sens. Environ.* **1988**, *24*, 459–479. [CrossRef]
50. Singh, A.; Raju, A.; Pati, P.; Kumar, N. Mapping of coal fire in Jharia coalfield, India: A remote sensing based approach. *J. Indian Soc. Remote Sens.* **2017**, *45*, 369–376. [CrossRef]
51. Marchese, F.; Mazzeo, G.; Filizzola, C.; Coviello, I.; Falconieri, A.; Lacava, T.; Paciello, R.; Pergola, N.; Tramutoli, V. Issues and Possible Improvements in Winter Fires Detection by Satellite Radiances Analysis: Lesson Learned in Two Regions of Northern Italy. *IEEE J. Sel. Top. Appl. Earth Obs. Remote Sens.* **2017**, *10*, 3297–3313. [CrossRef]

Article

The 2014 Effusive Eruption at Stromboli: New Insights from In Situ and Remote-Sensing Measurements

Federico Di Traglia [1,*] Sonia Calvari [2], Luca D'Auria [3,4,5], Teresa Nolesini [6], Alessandro Bonaccorso [2], Alessandro Fornaciai [7], Antonietta Esposito [3], Antonio Cristaldi [2], Massimiliano Favalli [7] and Nicola Casagli [1]

1 Dipartimento di Scienze della Terra, Università degli Studi di Firenze, Via La Pira 4, 50121 Firenze, Italy; nicola.casagli@unifi.it

2 Istituto Nazionale di Geofisica e Vulcanologia, Osservatorio Etneo – Sezione di Catania, Piazza Roma 2, 95125 Catania, Italy; sonia.calvari@ingv.it (S.C.); alessandro.bonaccorso@ingv.it (A.B.); antonio.cristaldi@ingv.it (A.C.)

3 Istituto Nazionale di Geofisica e Vulcanologia, Osservatorio Vesuviano – Sezione di Napoli, Via Diocleziano 328, 80124 Napoli, Italy; ldauria@iter.es (L.D.); antonietta.esposito@ingv.it (A.E.)

4 Instituto Volcanológico de Canarias (INVOLCAN), 38320 San Cristóbal de La Laguna, Tenerife, Canary Islands, Spain

5 Instituto Tecnológico y de Energías Renovables (ITER), 38600 Granadilla de Abona, Tenerife, Canary Islands, Spain

6 Centro per la Protezione Civile, Università degli Studi di Firenze, Piazza San Marco 4, 50121 Firenze, Italy; teresa.nolesini@unifi.it

7 Istituto Nazionale di Geofisica e Vulcanologia, Sezione di Pisa, Via della Faggiola 32, 56126 Pisa, Italy; alessandro.fornaciai@ingv.it (A.F.); massimiliano.favalli@ingv.it (M.F.)

* Correspondence: federico.ditraglia@unifi.it; Tel.: +39-055-275-5981

Received: 15 November 2018; Accepted: 12 December 2018; Published: 14 December 2018

Abstract: In situ and remote-sensing measurements have been used to characterize the run-up phase and the phenomena that occurred during the August–November 2014 flank eruption at Stromboli. Data comprise videos recorded by the visible and infrared camera network, ground displacement recorded by the permanent-sited Ku-band, Ground-Based Interferometric Synthetic Aperture Radar (GBInSAR) device, seismic signals (band 0.02–10 Hz), and high-resolution Digital Elevation Models (DEMs) reconstructed based on Light Detection and Ranging (LiDAR) data and tri-stereo PLEIADES-1 imagery. This work highlights the importance of considering data from in situ sensors and remote-sensing platforms in monitoring active volcanoes. Comparison of data from live-cams, tremor amplitude, localization of Very-Long-Period (VLP) source and amplitude of explosion quakes, and ground displacements recorded by GBInSAR in the crater terrace provide information about the eruptive activity, nowcasting the shift in eruptive style of explosive to effusive. At the same time, the landslide activity during the run-up and onset phases could be forecasted and tracked using the integration of data from the GBInSAR and the seismic landslide index. Finally, the use of airborne and space-borne DEMs permitted the detection of topographic changes induced by the eruptive activity, allowing for the estimation of a total volume of $3.07 \pm 0.37 \times 10^6$ m^3 of the 2014 lava flow field emplaced on the steep Sciara del Fuoco slope.

Keywords: Stromboli volcano; landslides; effusive activity; Ground-Based InSAR; infrared live cam; seismic monitoring; PLEIADES; Digital Elevation Models; optical sensors; volcano remote sensing

---

1. Introduction

Stromboli volcano (Italy; Figure 1), a stratovolcano located at the easternmost end of the Aeolian Archipelago, experienced a flank eruption from August–November 2014 [1–7]. In this paper, in situ and remote-sensing measurements at Stromboli between May and November 2014 are presented. Data comprise videos recorded by the visible and infrared cameras network [8,9] (Figure 1a), ground displacement recorded by the permanent sited Ku-band, Ground-Based Interferometric Synthetic Aperture Radar (GBInSAR) device [10,11], ground displacements in the seismic band (0.02–10 Hz) [12,13] (Figure 1a), and high-resolution Digital Elevation Models (DEMs) reconstructed based on tri-stereo PLEIADES-1 imagery and Light Detection and Ranging (LiDAR) data, respectively. Data allowed us to characterize the precursors and the phenomena that occurred during the eruption, as well as an estimation of the erupted volume.

Stromboli is characterized by persistent Strombolian activity from several vents within a crater terrace area [9,14] (Figure 1b). The volcanic edifice is frequently affected by landslides [15], mainly within the Sciara del Fuoco (SdF; Figure 1b), a collapse depression on the NW flank of the island formed during the last 13 ka [16,17]. Landslides triggered tsunamis on average every 20 years [18], especially during flank eruptions or paroxysmal explosions [12,19,20]. Small to large volcano slope instability characterized the initial phases of the last four flank eruptions (1985–86, 2002–03, 2007, 2014) [21–24], triggering tsunamis only during the 2002–03 event [22,25].

The last flank eruption started at Stromboli on 7 August 2014, preceded by 2 months of increased Strombolian activity and several lava overflows from the craters expanding along the SdF [2,26]. Overflows were often accompanied by landslides along the SdF [7], described as rock-falls and/or gravel slides, evolving down slope to gravel flows (Figure 2a). The onset of the 2014 flank eruption (6–7 August) involved the breaching of the summit cone with emplacement of a landslide along the SdF (Figure 2b), the opening of an eruptive fissure on the NE flank of the cone [24], and the effusion of lava from the crater rim at first and from the eruptive fissure later, feeding the 2014 lava flow field [4,5,7]. The eruption was characterized by the lava effusion in the SdF from a fissure at 650 m above sea level (a.s.l.) that lasted until 13 November 2014 [4].

The 2014 lava flow field was similar to others erupted from high-elevation vents (as described for the 2002–2003 lava flow field by [27]), comprising: (i) a series of tumuli and lava flows around the effusive vent at ~650 m a.s.l. (proximal shield); (ii) a medial zone fed by small flows, and characterized by frequent lava crumbling down slope and producing a debris field; (iii) a basal toe composed of the debris flow field emplaced above the stacked lava delta [28]. The erupted volume has been estimated to be 7.4×10^6 m^3 by [4] and $\sim 5.5 \times 10^6$ m^3 by [5].

2. Materials and Methods

2.1. The Camera Monitoring Network (Istituto Nazionale di Geofisica e Vulcanologia, Osservatorio Etneo - INGV-OE)

The camera monitoring network comprises thermal infrared and visible cameras located at Il Pizzo, at ~918 m elevation and ~250 m from the craters, plus visible and thermal infrared cameras (SQV400 and SQT400, respectively) at 400 m elevation on the N flank of the SdF and ~800 m from the craters (Figure 1a). The SQV400 and SQT400 cameras allow a view from NE of the North-East Crater NEC area (Figure 1a) and of the upper eastern sector of the SdF. An additional thermal camera installed in July 2014, is located at the lower eastern end of the SdF (~190 m a.s.l., Figure 1b) and is focused on the lower portion of the slope between ~400 m and the N coast line. To obtain a description of the eruptive activity, the total number of explosive events that occurred during each day of cloud-free observation was manually counted and plotted as an integer versus time. On average, between 24 July 2014 and 10 August 2014, ~30% of the days were affected by clouds and/or by system failure. In such cases data are lacking.

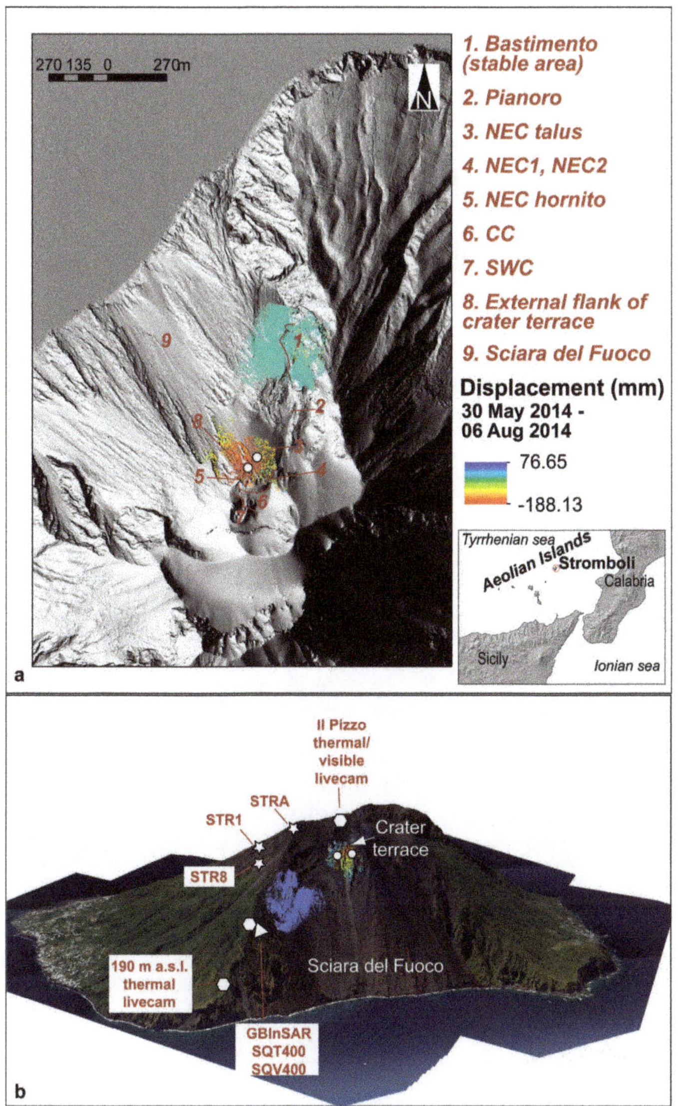

Figure 1. (**a**) Geographic setting of the Island of Stromboli, located at the easternmost end of the Aeolian Archipelago, and names of the summit vents within the crater terrace. NEC: North-East Crater area; CC: Central Crater area; SWC: South-West Crater area (**b**) 3D-view of the investigated sectors observed from north, highlighting the location of the measurement stations used in this work. STR1, STR8, and STRA: broadband seismic stations. SQT400 and SQV400: thermal and visible live cam, respectively. In both (**a**) and (**b**), the line of sight (LOS) displacement map produced by the GBInSAR apparatus between 30 May 2014 and 6 August 2014 is also shown, highlighting the strong inflation of the crater terrace and the stability of the Sciara del Fuoco before the 2014 flank eruption. The measure-areas of the time-series are indicated by white circles. All Digital data were collected in the Projected coordinate system: WGS 1984 UTM zone 33 Projection: Transverse Mercatore. Maps were generated using ESRI ArcGIS CAMPUS (Università degli Studi di Firenze Licence; http://www.siaf.unifi.it/vp-1275-arcgis-licenza-campus.html).

Figure 2. (**a**) Gravel flow associated with the 7 July 2014 overflow. The landslide is composed by mixed breccia from the overflow front and the (mainly fine-grained) debris eroded along the Sciara del Fuoco (photo taken by the Università degli Studi di Firenze personnel during field survey). (**b**) Landslides associated with the collapse of the NEC-hornito, occurred on 6 August 2014 (photo taken by UNIFI personnel during boat survey).

2.2. Seismic Network (Istituto Nazionale di Geofisica e Vulcanologia, Osservatorio Vesuviano - INGV-OV)

During the 2014 eruption the seismic broadband network of Stromboli [29] was operating with 10 broadband seismic stations. In this work, we used data from 3 seismic stations (Figure 1b). STR1 is located on the flank of the volcano at an elevation of about 560 m and has the longest and most complete record of seismological parameters. STRA is located at about 840 m, it is the closest to the summit craters, therefore is particularly suitable to study the dynamics of explosive activity. Lastly, STR8, located at about 570 m, is the closest to the SdF, hence it is the best suited to study seismic signals related to landslides in this area.

The seismological parameters considered here are: the amplitude of volcanic tremor, the amplitude of explosion quakes and the inclination of the seismic polarization in the Very-Long-Period (VLP) band (0.05–0.5 Hz). We also performed the analysis of seismic signals, using neural networks, which allows detecting signals related to landslides occurring along the SdF [30]. Landslides signals have shown to be a reliable precursor of effusive activity at Stromboli. Their occurrence rate increased markedly during the onset of the previous effusive eruption in 2007 [12]. To quantify the intensity of landslide activity, Esposito, et al. [31] defined a landslide index following a fuzzy logic scheme. This parameter varies smoothly between 0 and 1, with 0 indicating a low probability of ongoing landslide activity and, conversely, 1 indicating a high probability of occurrence.

2.3. GBInSAR

InSAR technique has been successfully used to catch slope deformation at active volcanoes [31–33], though Synthetic Aperture Radar (SAR) satellites are no longer able to follow the evolution of very fast slope deformation. Conversely, given their repeat time (minutes), GBInSAR can correctly measure very fast displacements, thus the InSAR technique is ideal for monitoring, surveillance, and early-warning applications [24,34]. At Stromboli, a GBInSAR apparatus is working since 2003, located in a stable area N of the SdF (Figure 1b), and consisting of a transmitting and a receiving antenna moving along a 3 m long rail [10]. The GBInSAR, working in Ku-band, has also the great advantage of penetrating the dust clouds (abundant especially during the collapse events, see Figure 2), and working in every light and atmospheric condition. The GBInSAR measures ground displacement along the line of sight (LOS; Figure 1a,b), by computing via cross correlation the phase differences between the backscattered signals associated with two SAR images. Range and cross-range resolution are on average 2×2 m, with a measured displacement precision lower than 1 mm [10,11]. Due to the short-elapsed time (11 min) between two subsequent measurements, the interferometric displacements are usually smaller than half the wavelength (8.6 mm for the Ku-band), therefore unwrapping procedures are generally not necessary [10]. Moreover, unwrapping is a time-consuming process, avoiding the operational use of the GBInSAR during crisis management. For the Stromboli GBInSAR images, a coherence mask (threshold equal to 0.8) is set to reduce the noisy areas of the interferogram [11]. Displacement maps and time-series are obtained "stacking" the interferogram phase with a displacement measurement precision of 0.5 mm, obtained using 1-h averaged SAR images [35].

2.4. Topographic Data and Co-Registration

Topographic change detection of Stromboli island due to the 2014 eruption was performed by comparing pre- and post- eruptive DEMs, the former generated from airborne LiDAR data and the post-eruption DEM from PLEIADES-1 satellite data.

The pre-2014 LiDAR-DEM was obtained elaborating the 3D point cloud acquired during an airborne survey carried on May 2012 by "BLOM Compagnia Generale Ripreseaeree S.P.A." (www.blomasa.com). The data were acquired using the Leica ADS80 sensor which has instrumental vertical and horizontal accuracy of $\pm$10/20 cm and $\pm$25 cm, respectively (see [36] for 2012 DEM features). The acquired point cloud has a mean point density of 8 pt/m^2.

The post-2014 DEM was derived from the very high-resolution (VHR) tri-stereo optical imagery from the PLEIADES-1 satellites. PLEIADES-1 constellation is composed by two satellites, PLEIADES-1A (PHR1A) and PLEIADES-1B (PHR1B). These satellites can sense three or more synchronous images of the same area, with angles variable between ~6° e ~28°. The stereoscopic triplet is composed of three nearly simultaneously acquired images, one backward looking, one forward looking, plus a third near-nadir image [37–41]. Tri-stereo images are 100% cloud free and were acquired by PHR1A on 27 May 2017, with a total areal coverage of 58 km^2. This acquisition mode permits to obtain a DEM using the photogrammetric processing of three images (tri-stereo mode), with a pixel xy resolution of $1 \text{ m} \times 1 \text{ m}$. The dataset comprised also VHR optical imagery ($0.5 \text{ m} \times 0.5 \text{ m}$ resolution for Panchromatic + $2 \text{ m} \times 2 \text{ m}$ Multispectral data). To assess the accuracy of the heights and their horizontal position in the Pleiades-1 DEM, six Ground-Control Points (GCPs) were collected on the map database (Cartographic XY standard deviation: 0.15 m). Nine tie points were automatically collected in the images. A block adjustment including all the satellite scenes was performed. The block adjustment was validated when the following accuracy was achieved: (i) pixel xy bias smaller than 0.3 pixels; (ii) pixel xy standard deviation smaller than 0.3 pixels; (iii) pixel xy maximum smaller than 2 pixels.

Topographic change detection using multi-temporal DEMs was performed by differencing two DEMs of the same area derived from data taken at different time. This calculation is often affected by errors depending on mismatches between the two DEMs, which lead to artefact Δz [42]. This error can be detected and reduced by measuring and minimizing the DEM differences in areas where the two DEMs are supposed to be equal, i.e., those areas not reasonably affected by relevant natural changes.

The 2012 and 2017 topographic data were here co-registered by minimizing the root mean square (RMS) error between two DEMs. The methods used iteratively vary the three angles of rotation, the translation, and the magnification or reduction factor of one DEM using a custom-made algorithm based on the MINUIT minimization library [43], as described by [44–46]. MINUIT is a tool to find the minimum value of multi-parameter functions and can be freely downloaded (http://www.cern.ch/minuit). The LiDAR-PLEIADES DEMs co-registration was performed by calculating the minimization parameters for the whole volcano with the exclusion of the investigated areas (i.e., the SdF). The RMS error in this case is given by the MINUIT output. The 2012 LiDAR was used as reference for co-registering the PLEIADES DEM. It is worth emphasizing that this RMS error is the root mean square residuals (in elevation) between a 3D point cloud and the reference DEM, rather than a true absolute error. Co-registration decreased the RMS error from the initial 3.37 m to the final 1.27 m.

The differences between the co-registered DEMs were used to detect and outline the extent of the areas affected by topographic changes in the considered time spam, and to calculate the volume and thickness variations inside them. Moreover, the perimeters of observed changes were double-checked using the ortho-rectified PLEIADES-1 images. Field surveys also strengthened data interpretation.

Volume (V) emplaced or lost between two acquisitions was calculated from the DEM difference according to the equation:

$$V = \sum_i \Delta x^2 \Delta z_i, \quad (1)$$

(see [42] and references therein), where x is the grid step and z_i is the height variation within the grid cell i. These values were then summed up for all the cells in the selected areas in which we want to calculate volume changes.

The residual volume errors were calculated as:

$$\text{Err}_{(\text{V,high})} = \text{A}\sigma_{\Delta z}, \quad (2)$$

where A is the investigated area and $\sigma_{\Delta z}$ is the co-registration residual RMS error between the two DEMs. Following [42], this formula assumes that the errors were completely correlated, and thus the volume error calculated as above is overestimated.

3. Results

Explosive activity at Stromboli from the crater terrace was at an average of ~10 explosions h^{-1} during the first four months of 2014 (INGV Reports, available at www.ct.ingv.it). Between end of June–early July, the explosion number peaked at ~25 explosions h^{-1} (INGV Reports, available at www.ct.ingv.it), when also several lava flows poured out from the South-West crater (SWC, Figure 1a), and from two of the three vents located in the NEC area (NEC1 and NEC-hornito, Figure 1a, Table 1, Supplementary Materials), spreading along the SdF. Between 22 June and 5 August 2014, several powerful explosions and small landslides along the SdF occurred (Supplementary Materials).

The GBInSAR recorded an increase in the LOS displacement rate directed towards the sensor in the crater terrace area, indicating inflation of the summit plumbing system, starting from 30 May 2014 (Figure 3B,C). The following months (June–early August 2014) were characterized by slightly fluctuating displacement rate, with values always above the threshold that distinguishes the anomalous from the persistent activity (>0.05 mm/h directed towards the sensor; as defined by [11]). A small deflation of the crater terrace occurred around 24 July 2014, after the output of several small overflows, then a more rapid inflation of the crater terrace, and a week later also of the NEC-talus, was observed (Figure 3C).

Table 1. Summary of the eruptive activity during the run-up and onset phases of the 2014 flank eruption, as derived from the camera monitoring network.

Date	Live-Cams Observations
30 May 2014–31 Jul 2014	Frequent explosive activity (~15 explosions/hour) Overflows Landslides
01 Aug 2014–05 Aug 2014	Increased explosion frequency (up to 30 explosions/hour)
06 Aug 2014 at 08:50	Overflow (between NEC2 and NEC-hornito)
06 Aug 2014 at 11:00	Arcuate fractures on the NE crater rim between NEC1 and NEC-hornito
06 Aug 2014 at 12:22	First incandescent blocks from the NEC2/NEC-hornito into the sea
06 Aug 2014 at 12:29	Overflow (between NEC-hornito and SWC)
06 Aug 2014 12:32–13:00	Landslide (N flank of the crater terrace)
06 Aug 2014 12:35–13:00	Incandescent blocks accumulation along the coast
06 Aug 2014 at 13:08	Overflow from NEC-hornito reached the coast
06 Aug 2014 14:05–14:08	Three hot "gravel flows" from the NEC-hornito along the SdF, reached the coast and went on spreading along the sea surface for several tens of meters
06 Aug 2014 at 14:50	Overflow from NEC-hornito reached the coast
06 Aug 2014 at 15:46	Overflow from NEC-hornito reached the coast
06 Aug 2014 at 16:02	Hot "gravel flows" from the NEC-hornito along the SdF
06 Aug 2014 at 16:08	Overflow from NEC-hornito reached the coast
07 Aug 2014 ~02:30	Hot "gravel flows" from the NEC-hornito along the SdF
07 Aug 2014 ~03:00	- NEC-hornito lava decreased - Explosive activity from NEC increased - Increased size of the hot avalanche deposit
07 Aug 2014 at 03:40	Hot "gravel flows" from the NEC onto the Pianoro flat area
07 Aug 2014 04:01	NE flank of NEC1 started to collapse Lava flow from the NEC towards the Pianoro flat area Hot "gravel flows" from the NEC-talus onto the Pianoro flat area and along the SdF
07 Aug 2014 05:01	Downslope curved fracture opened on the flank of the cone
07 Aug 2014 05:04–05:16	V1 opened at 05:04 and V2 at 05:16, both at ~650 m a.s.l. Landslides onto the Pianoro flat area and along the SdF Lava flow onto the Pianoro flat area
07 Aug 2014 05:30	Lava flow along the SdF
07 Aug 2014 ~06:02	Hot "gravel flows" from the NEC-talus reached the coast
07 Aug 2014 06:24	Lava flow from V1 reached the coast

One of the most striking features in the seismological parameters during the 2014 eruption was the progressive, but rapid increase in both volcanic tremor and explosion-quake amplitude, starting approximately on 2 August 2014 (Figure 3D,E). Both quantities showed a steady increment, followed by a sudden drop in the amplitudes occurring about on 6 August 2014 at 12:00 (Figure 3D,E). This event was associated with the onset of the effusive activity on 6 August 2014, clearly evidenced by the landslide index, approaching the value of 1 in the same period (Figure 3A). However, a detailed analysis of the VLP polarization (Figure 3F), which is a good indicator of changes in the level of the magmatic column within the conduits, revealed that the most significant drop in the magma column level occurred only at about 04:00 of 7 August 2014, just before the effusive vent opened at 650 m a.s.l. (V1 and V2; see Table 1, Figure 4, and Supplementary Materials for further details).

The displacement rate suddenly increased on 5 August 2014 for both the external flank of the crater terrace and the NEC-talus, reaching their maximum on 6 August 2014 (Figure 3B,C). An increase in the explosion rate and intensity was recorded on 6 August 2014 morning, especially from the NEC. Several overflows occurred since 08:50 (flows F1 to F3 in Figure 4), and the rim of the crater terrace started to be clearly unstable since 11:00, when three arcuate fractures on the NE crater rim formed between NEC1 and NEC-hornito (Table 1, Supplementary Materials).

The GBInSAR recorded an increase in the displacement rate (since 10:00 of 6 August 2014) in both the external flank of the crater terrace and the NEC-talus. The maximum displacement was recorded mainly in the NEC-hornito area (Figure 5a). Rock-falls started to occur at 12:22, and soon after a lava flow erupted from the saddle between the NEC-hornito and the SWC (Flow F2A in Figure 4a), expanding N along the SdF (12:32). Between 13:00 and 14:21, the GBInSAR recorded a large loss in the SAR coherence in the NEC-hornito, consistent with the occurrence of fast-moving material (landslides and/or overflows) from the NEC-hornito (Figure 5b). Three landslides (starting at 14:05, 14:06, and 14:07, Figure 6), described as rock-falls/gravel slides evolving in rapid gravel flows, were directly observed (Figure 2b). The collapse scar observed on a photo taken from helicopter (Figure 2a) had an estimated size of ~ 40 m width, 30 m high and 20 m depth, with a total collapsed volume estimated in ~ 24,000 m^3. Effusive vents moved northward from the NEC-hornito towards the NEC1. Afterwards, the displacement towards the sensor (up to 236 mm h^{-1}) shifted from the NEC-hornito towards the NEC-talus (Figure 5c). Between 01:29 and 01:32 of 7 August 2014 (Figure 5c), the GBInSAR detected the formation of a fracture upslope of the NEC-talus. At the same time, the GBInSAR recorded an inversion in the LOS displacement rate (movements directed away from the sensor) in the crater terrace area, coherent with the deflation of the shallow conduit (Figure 5d).

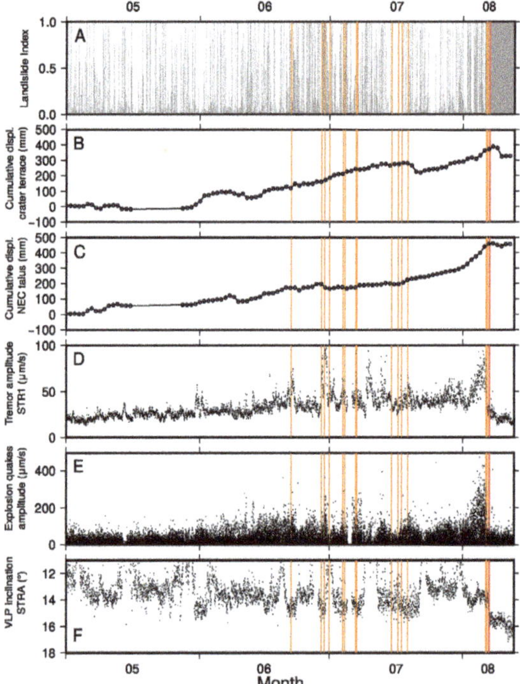

Figure 3. Relevant geophysical time-series from May to August 2014. (**A**) Landslide index derived from seismic data; GBInSAR cumulative displacement data for (**B**) the crater terrace and (**C**) for the NEC-talus. (**D**) Volcanic tremor amplitude at station STR1. (**E**) Explosion-quake amplitudes at station STR1. (**F**) VLP polarization inclination at station STRA.

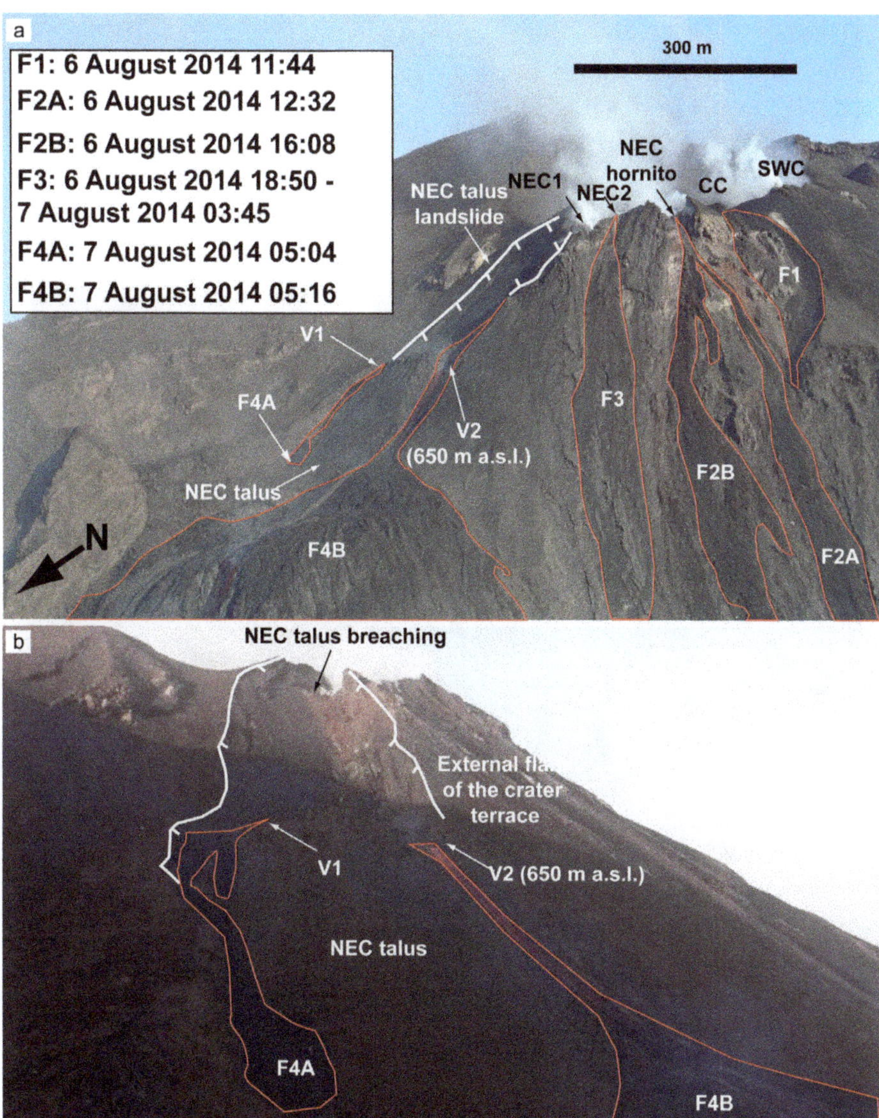

Figure 4. (**a**) The external flank of the crater terrace after the onset of the 2014 flank eruption (photo taken by INGV personnel on 11 August 2014 from helicopter survey). Lava overflows emitted during the run-up phase (F1 to F3) and lava flow produced during and after the onset (F4A and F4B) are also reported. (**b**) Detail of the NEC-talus area highlighting the scar produced during the onset of the eruption (7 August 2014), the first lava flow after crater breaching (F4A, erupted from vent V1), the upper portion of the 2014 lava flow field (F4B), and the vent of the 2014 flank eruption (V2) (photo taken by UNIFI personnel on 8 October 2014 from field survey). It is interesting to note the internal stratification in the NEC-talus remnants (consisting of volcaniclastic levels) that are continuous in all the inner walls of the landslide scar.

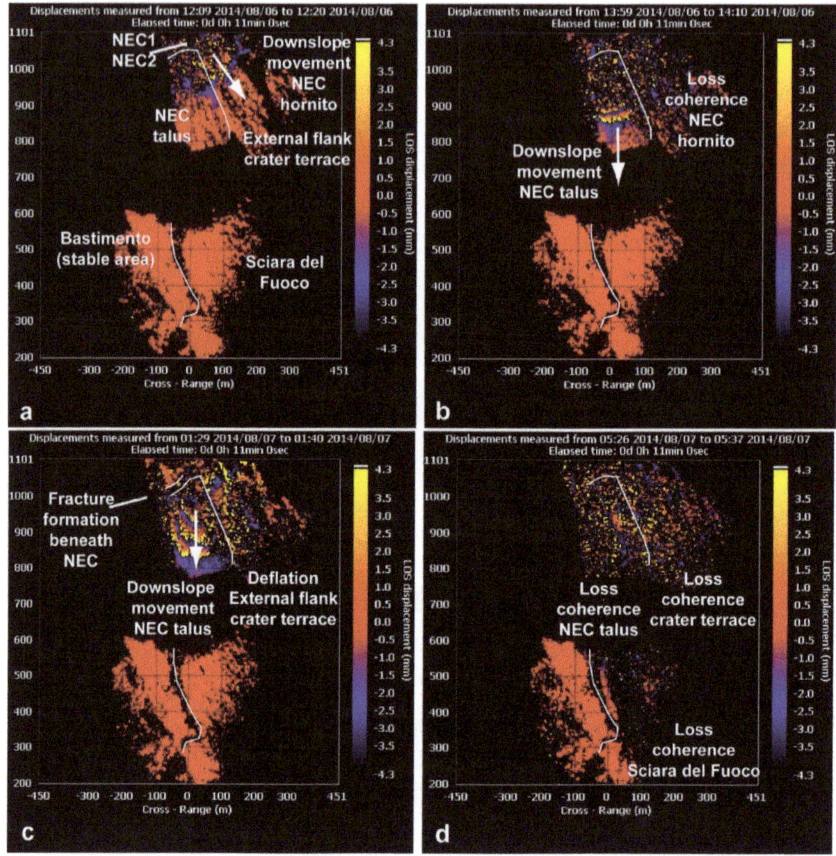

Figure 5. GBInSAR interferograms formed using a pair of images recorded between (**a**) the 12:09 and 12:20 on 6 August 2014, evidencing the instability of the NEC-hornito; (**b**) the 13:59 and 14:10 on 6 August 2014, evidencing the instability of the NEC-talus; (**c**) the 01:29 and 01:40 on 7 August 2014, evidencing the concomitant instability of the NEC-talus and the deflation of the crater terrace, coherent with the sliding of the NEC-talus induced by the NE-ward magma migration from the crater terrace; (**d**) the 05:26 and 05:37 on 7 August 2014, characterized by loss in SAR coherence due to fast movement both in the crater area (sliding processes) and along the SdF (lava flow emplacement). Interferograms in (**a–c**) are affected by phase wrapping.

At 05:04 V1 opened at ~650 m a.s.l. (Figure 4) and at 05:08 the NE flank of NEC1 started collapsing, as evidenced by the associated high-frequency seismic signal (Figure 7). At 05:01, the live cam detected a downslope curved fracture opened on the flank of the NEC cone, followed at 05:16 by the opening of the effusive vent V2 at ~650 m a.s.l. (Table 1 and Figure 4). The onset of the increased landslide activity occurred at 05:08, and its amplitude reached a peak at ~05:24, i.e., during the opening of V1 and V2 (Table 1 and Figure 7). With the shift of the active vent from V1 to V2, a lava flow started spreading along the Pianoro (Supplementary Materials) and the flank eruption started.

We analyzed in detail 24 h of high-frequency seismic signals related to landslide activity along the SdF starting from 10:00 on 6 August 2014 (Figure 7B). Between 10:00 and 18:00 on 6 August 2014, the repeated lava overflows triggered a moderate landslide activity along SdF (Figure 7A). Before the opening of vent V1, at 04:01 on 6 August, we observed a progressive increase in the landslides, which dropped just after this event. However, the clear onset of the increased landslide activity occurred at 05:08 and the amplitude reached a peak at around 05:24, a few minutes after the opening of vent V2 at 05:16. After a pause of about 8 min, the landslide activity resumed at 05:32 and lasted for about 1 h, before reaching a stationary level (slightly higher than the value observed before 05:08) and possibly related to the ongoing lava flow (Figure 7C).

The GBInSAR recorded a large loss in coherence at the NEC-talus and the SdF area, consistent with the onset of the effusion (Figure 5D). At 05:29 the flow had travelled the Pianoro flat surface and at 05:30 started to expand along the SdF steep slope (Supplementary Materials). Here it formed at least 5 lava flow branches (Supplementary Materials), reaching the coast with a hot avalanche 30 min later (Table 1, Supplementary Materials). The landslide activity resumed at 05:32, and the collapse lasted for about 1 h, coherent with several landslides' pulses (Figure 8). By 06:24 several lava branches reached the sea, forming a lava fan ~500 m wide (Table 1, Supplementary Materials). Effusive and explosive activity declined between 7 and 8 August 2014, with only one lava flow active along the coast and deeper magma level within the feeder conduit, evidenced by more collimated jets produced during the explosive activity at the summit vents (Table 1, Supplementary Materials).

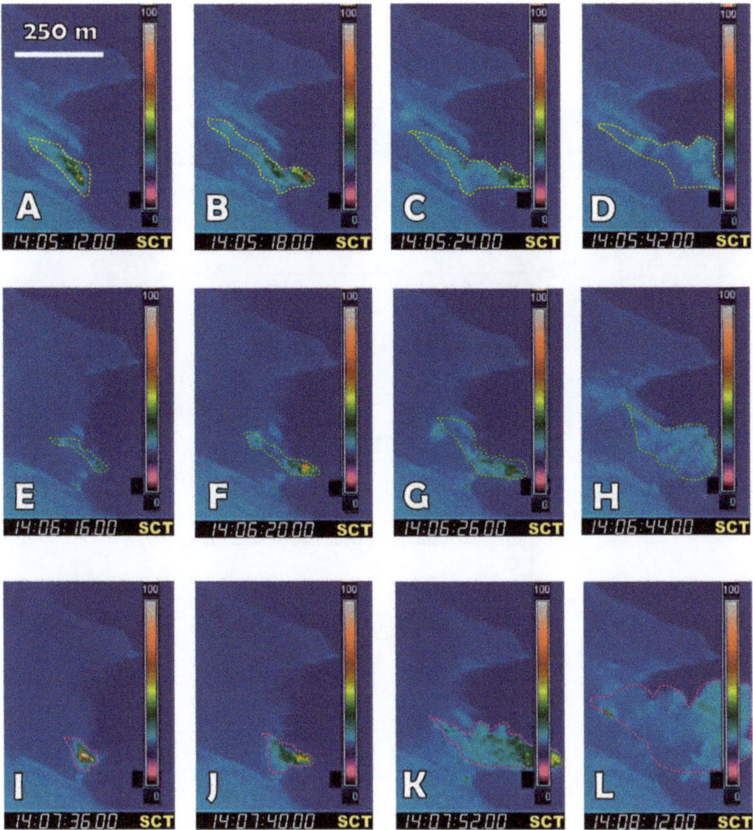

Figure 6. Thermal images recorded by the SCT camera located at Labronzo, ~100 m a.s.l. (see Figure 1b for location), showing the hot "gravel flows" that reached the sea on 6 August 2014, at 14:05 (**A–D**), 14:06 (**E–H**) and 14:07 (**I–L**), spreading on the sea surface. Temperature is in °C. The velocities of the three landslides of hot blocks in the frontal region were 9.8 m s^{-1}, 5.9 m s^{-1} and 9.8 m s^{-1}, respectively, based on web-cameras measurements.

Loss in SAR coherence persisted until 8 August 2014 on the whole scene illuminated by the GBInSAR device. In the crater terrace, the coherence increased again since 9 August 2014 when no explosive activity was observed from the monitoring cameras, and the GBInSAR recorded displacement away from the sensor until late September indicating deflation of the summit.

The images recorded by the thermal camera located at Il Pizzo (SPT; see Figure 1B for location) showed significant morphology changes of the summit craters since 9 August 2014, showing that the apex of cones and hornitos within NEC1, NEC-hornito and SWC had been removed by the several collapses and hot "gravel flows" of the previous days. The thermal images displayed also a much lower temperature of the whole crater area, indicating a decline in the explosive activity. Likewise, the rate of effusion decreased significantly from the previous day, with only one wider lava flow starting from the eruptive fissure and replacing the 4–5 branches of the previous day. This flow was reaching the coast forming a narrow channel apparently cooling down, with overflows sometimes spilling over the channel levees.

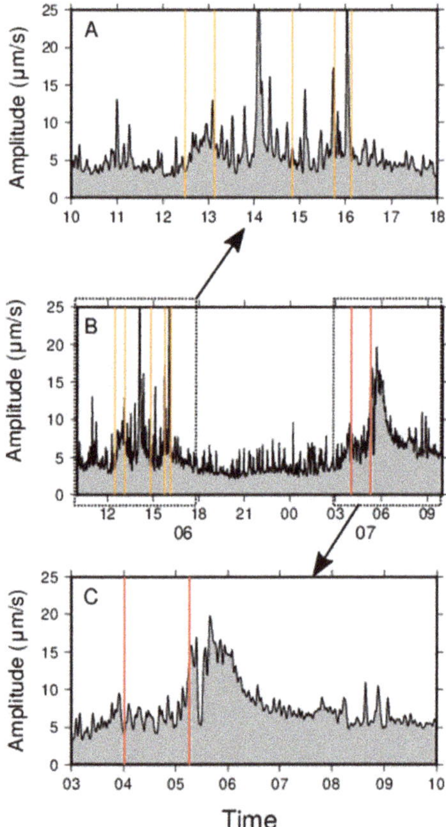

Figure 7. Plot of the high-frequency (>10 Hz) seismic amplitudes averaged over 10 s windows (station STR8, EW seismic component) for: (**A**) details referred to the time interval 10:00–18:00 of 6 August 2014, highlighting the signals associated with landslides that occurred soon after the onset of the overflows, whose onset is marked by orange vertical lines; (**B**) the time interval 10:00 on 6 August 2014–10:00 on 7 August 2014; (**C**) details referred to the time interval 03:00–10:00 of 7 August 2014, highlighting the signals associated with landslides that occurred at the opening V1 and V2 vents (red vertical lines).

Until 15 October 2014 the GBInSAR recorded displacement towards the sensor, indicating inflation of the crater area. Afterwards, very low displacement rate (<0.05 mm/h) oscillating between away and towards the sensor, have been recorded. In the SdF area, after the onset of the 2014 flank eruption, the radar recorded a very low coherence for three days; this was related to the initial fast-moving lava flow. Since 10 August 2014, low coherence zones were related only to a small part of the monitored scenario, corresponding to active lava tongues, while the rest of the monitored SdF showed variable displacement rates (1–10 mm/h).

The seismic signals related to minor landslides along SdF continued to be recorded after the development of the lava flow on SdF.

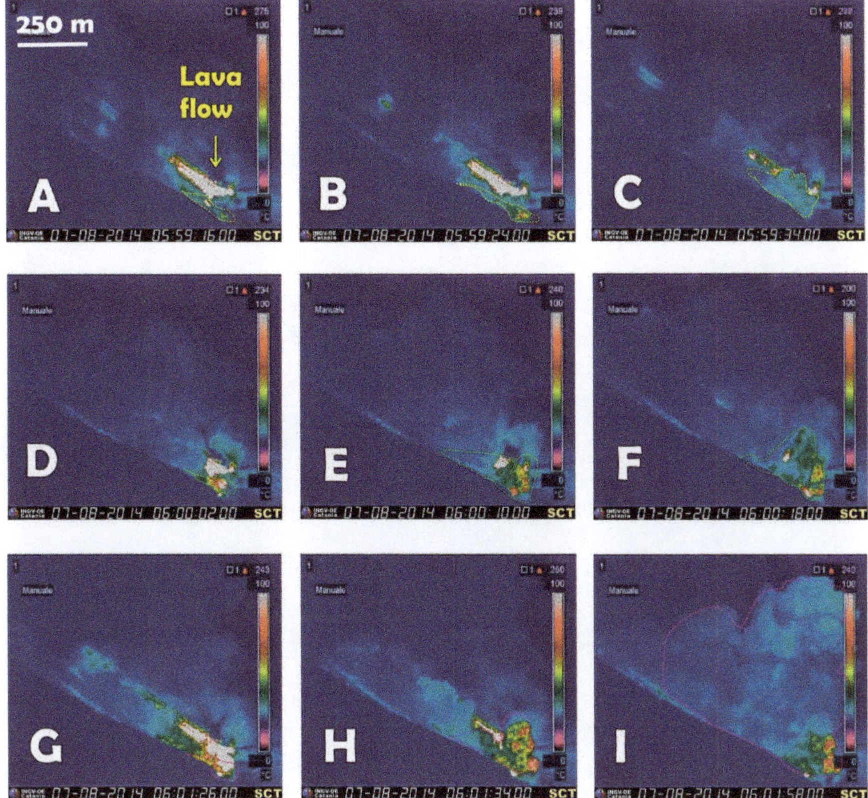

Figure 8. Lava flows (white) and hot avalanches (yellow, green, and purple dotted outline) spreading along the lower slope of Sciara del Fuoco and to the sea, as observed by the thermal images recorded by the SCT camera located at Labronzo, ~100 m a.s.l. on the east side of the Sciara on 7 August 2014. See Figure 1b for camera location. (**A–C**): lava flow spreading to the coast, flanked by a hot avalanche (yellow dotted outline) at 5:59; (**D–F**): another hot avalanche spreading at 6:00 (yellow dotted line); (**G–I**): a much bigger hot avalanche spreading at 6:01 (red dotted outline) forming a conspicuous dust cloud).

In the period between the acquisitions of the two DEMs (May 2012–May 2017), large topographic changes occurred. The areas interested in volume gain and loss are shown in Figure 9, which includes the crater terrace and the SdF depression. The largest variations were related to the emplacement of the 2014 lava flow field in the NE part of the SdF (Figure 9). The total emplaced volume is

$3.07 \pm 0.37 \times 10^6$ m^3, calculated inside the black dashed line. The average thickness of lava was 10.45 ± 1.27 m, with the maximum value reaching 35 m (Figure 9, profile 1).

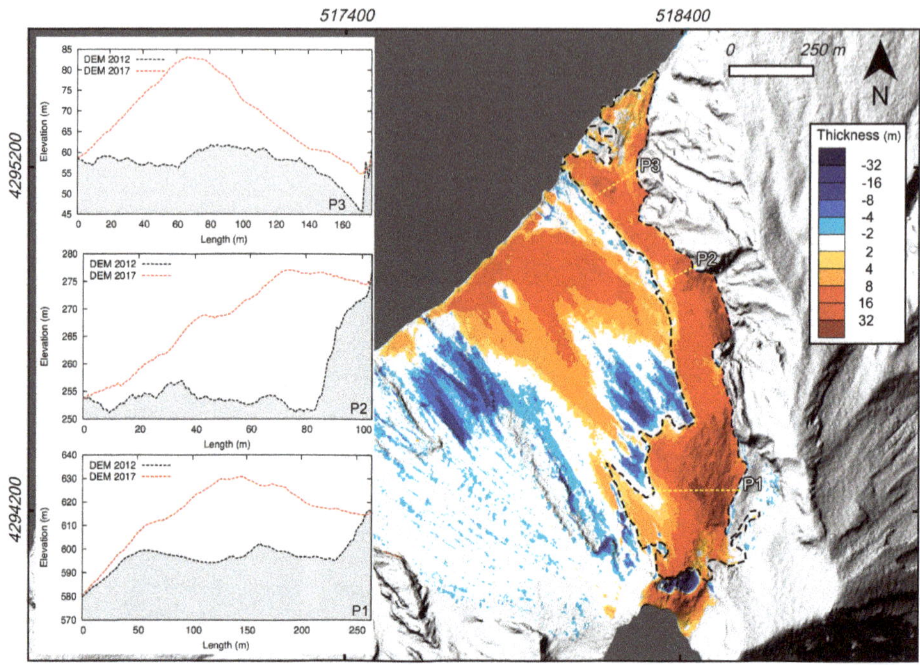

Figure 9. Stromboli topographic changes detected comparing 2012 and 2017 DEMs. Dashed black line encloses the area covered by the 2014 lava flow. Dashed yellow lines, labelled with P1, P2 and P3, show the position of the profiles along the flow represented in the insets.

In the central part of the SdF, between 0 and ~518 m a.s.l., an increased volume of $1.70 \pm 0.30 \times 10^6$ m^3 has been estimated, with an average thickness of 7.11 ± 1.27 m. This volume is related to the accumulation of volcaniclastic material derived from the erosion of the SdF, from the overflow-induced landslides, and from the NEC-talus collapse [7].

4. Discussion

The run-up phase of the 2014 flank eruption was characterized by very vigorous explosive activity and episodic lava overflows from the crater terrace. This activity was preceded by an increase in the displacement rate towards the sensor of the crater terrace at anomalous level since the end of May 2014, and by frequent landslides along the SdF. Ground deformation of the crater terrace strongly increased on 26 July 2014 and was followed by increase of the volcanic tremor and explosion-quake amplitude, starting approximately on 2 August 2014. The explosive activity peaked on 5–6 August 2014, anticipating by ~1 day the increase in the displacement rate towards the sensor (inflation of the crater terrace).

Different authors [1,3,6] suggested that the 2014 eruption was triggered by a gradual recharge of the shallow plumbing system and upper conduits. The SO_2 flux in the persistent Stromboli plume has been measured for two months before the onset of the eruption, detecting a strong increase above normal activity in the SO_2 pulses (puffing and explosions) before the effusion onset. This is consistent with the increase in gas bubble supply and magma transport rate feeding the uppermost storage system at Stromboli [47]. Ground displacement is consistent with the presence of a very shallow reservoir below the crater terrace [2]. Long-term (month to weeks) increase in the ground deformation recorded by the GBInSAR in the crater terrace area, as well as the long-term rise of the VLP seismic source towards the surface [5,6], are associated with the accumulation of magma in the uppermost storage system at Stromboli [2]. Contrarily, the detected increase in short-term (days to hours) ground deformation and amplitude of volcanic tremor are associated with the impulsive bulk degassing from the shallow plumbing system [35,48], whereas the increase in the explosion number and intensity and amplitude of explosion quakes are indicative of the increase in the release rate and overpressure intensity of individual explosions.

One of the most interesting aspects of the data presented here is the relationship between the instability events and the effusive activity at Stromboli volcano (Figure 10). Other authors have already shown the critical state of Stromboli during the very early stages of effusive eruptions [2,5,11,22,49]. During the run-up to the 2014 flank eruption, gravel flows consisting of mixed breccia and loose volcaniclastic deposits were generated mainly in two ways: (i) by the crumbling of the overflows, or (ii) by the collapse of some portions of the external part of the crater terrace and apex of the hornitos (Figure 2). The second type of collapse is a recurring phenomenon at Stromboli, occurring when there is a large accumulation of magma below the crater terrace (causing overflows, increase in the frequency of explosions, spattering activity, inflation of the whole summit). Furthermore, weakness of the crater wall (e.g., induced by the strong explosive activity on 5 August 2014), and mechanical erosion caused by vent opening and by lava fingering, may have triggered the NEC-hornito collapse on 6 August 2014 [48]. Here we have documented for the first time that, after the NEC-hornito collapse, eruptive vents migrated from south to north (Figure 10a). Overflows and mass-wasting in the NEC-hornito area preceded the formation of fracture below the NEC1 (Figure 10b) by ~11 h, and the onset of the flank eruption and NEC1 collapse by ~15 h, suggesting that 11–15 h before the onset of the effusive eruption, the crater terrace was full of magma. Then, the location of effusive vents displayed a clockwise rotation along the rim of the crater terrace, from the NEC-hornito towards NEC2 and NEC1 (Figure 2), and finally magma migrated towards the NEC-talus, inducing fracture opening and landslides occurrence. This process suggested a gradual pressurization and/or expansion of the feeder dike along a SW-NE direction, which is also the main tectonic trend of the island [49,50]. Lava started to outflow from a first vent (V1; Figure 10c) before the NEC-talus collapse and shifted at vent V2 simultaneously with the beginning of the NEC-talus landslide (Figure 10d). There could be different explanations for this vent shift:

1. a structural barrier generated by the presence of a buried structure [51,52] or different stratified material (Figure 2b), which in turn has caused the magma-filled fracture to re-orientate [53,54];
2. magma to flow from V1 at V2 in response to the tensile stress occurring in the talus produced by the downwards displacement observed by GBInSAR.

It is important to emphasize that this shift has prevented the dike from intruding within the SdF, as happened in the early stages of past eruptions (e.g., the 2007 eruption) [8]. Instead during the first days of the 2002–03 flank eruption, the intrusion of magma propagating laterally from the conduit towards a lower altitude was considered the cause that destabilized the SdF slope, with the triggering of tsunamigenic landslides on 30 December 2002 [22–25].

The flank eruption onset produced the drainage of a superficial magma reservoir [5,49], inducing ground deflation [55], a deepening of the VLP seismic source [5], and the cessation of summit explosive activity [5].

The volume of the drained magma can have critical implications on the eruption style and on its transition from effusive to explosive [56]. A previous author [4] calculated a total bulk volume of 7.4×10^6 m^3 using data derived by the new satellite Technology Experiment Carrier-1 (TET-1) by the German Aerospace Center (DLR), whereas others [5] evaluated a total bulk volume of 5.5×10^6 m^3 by thermal images from satellites using the Moderate Resolution Imaging Spectroradiometer (MODIS) sensor on-board the Terra and Aqua satellites. This last value was also confirmed by [57]. In MODIS approach, the error is usually estimated at 50% [58].

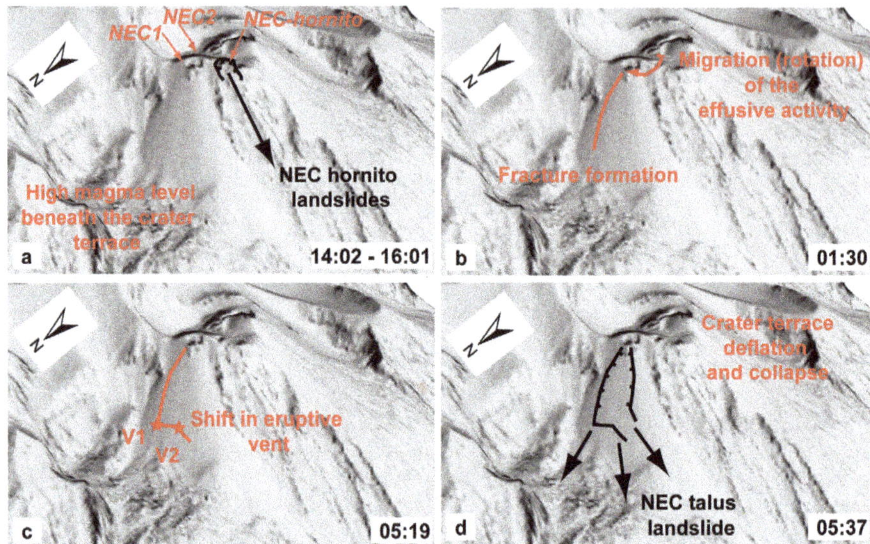

Figure 10. Schematic sketches representing the onset of the 2014 flank eruption as reconstructed on the basis of the new data presented in this paper. (**a**) The run-up phase of the 2014 flank eruption was characterized by high magma level beneath the crater terrace, which peaked on 6 August 2014 morning, with the occurrence of a large overflow from the NEC-hornito, triggering its collapse. Then the activity migrated from the NEC-hornito towards the NEC1. (**b**) The onset of the eruption coincided with the lateral dike propagation from the crater terrace along a NE-SW direction, triggering the instability of the NEC-talus, as evidenced by the increase in displacement rate at NEC-talus (until the loss of coherence in the GBInSAR data, compare with Figure 5) and landslides number (Figure 7). (**c**) Then the shift in the effusive vent from V1 to V2 occurred, (**d**) closely related to the NEC-talus collapse. Magmatic and mass-wasting phenomena are represented in red and in black, respectively.

To improve the volume estimation, we calculated the topographic change by comparing the 2012 and 2017 DEMs. This approach considers the volume variations due to erosion and accumulation occurred between the two acquisitions, therefore also the topographic changes occurred before or after the eruption and the volume of the lava flow emplaced below sea level during the first days of the eruption. However, in this area there was no evidence of significant erosive phenomena [7], and moreover the evidence of large accumulation (i.e., more than 1–2 $\times$ 10^6 m^3) of lava or a debris flow under the sea is still missing [28].

We estimated that during the 2014 flank eruption, a volume of $3.07 \pm 0.37 \times 10^6$ m^3 of lava was emplaced on the steep SdF slope, i.e., most of the volume emplaced in the subaerial part of the volcano. This value is close to the previous estimations of 5.5×10^6 m^3 [5,57], which however comprises also the minor portion emplaced below sea level.

Some authors [57] proposed that the critical factor for triggering the most violent explosive events (paroxysms, [59–61]) is the decompression rate, suggesting that rates >10 Pa s^{-1} can potentially favor the ascent of gas-rich magma batches responsible for the paroxysmal explosions that occurred at Stromboli during the 2002–03 and 2007 flank eruptions.

A critical aspect is that the paroxysmal explosions of both 5 April 2003 and 15 March 2007 occurred while effusive activity was still going on, and when the lava effusion rate (related to the magma decompression rate) was decreasing with respect to the initial phases of the eruptions. Moreover, the 2002–03 and 2014 flank eruptions were characterized by the same mean effusion rate [57] but, differently from the 2002–03, the 2014 eruption did not produce any paroxysm during the effusive activity.

A different approach to understand the transition from effusive to explosive style can be obtained by estimating the cumulated effusive volume. In fact, a previous author [56] proposed that the erupted volume during flank eruptions at Stromboli is very important for hazard estimation, because it is considered a key factor for the triggering of paroxysmal explosive activity. Calvari et al. [56] proposed that there is an erupted magma bulk volume limit of $6.5 \pm 1 \times 10^6$ m^3. Once that this volume is poured out, for example by drainage due to the opening of a flank fissure, the decompression induced on the shallow storage can trigger paroxysmal activity, by bringing at shallow levels the gas-rich magma that is in the lower conduit [62–64]. Calvari et al. [56] point out that the rate of effusion, and consequently the decompression rate, is less crucial to trigger powerful explosive activity. This interesting aspect has also been observed trough laboratory experiments allowing direct observations of explosive expansion of a slowly decompressed magma analog [65]. The analog experiments highlighted that under these conditions an explosive eruption would only occur when sufficient magma was spilled from the conduit, in agreement with the considerations of [56]. The inferred volume of 2014 emitted lava ($3.07 \pm 0.37 \times 10^6$ m^3), even adding 1–2 $\times$ 10^6 m^3 of underwater volume not measurable with our DEMs approach, is below the limit of $6.5 \pm 1 \times 10^6$ m^3 suggested by [56] and, therefore, consistent with the absence of the paroxysmal explosion. This consideration also explains the different behavior of the 2014 eruption (with absence of paroxysmal activity), respect to the 2002–03 eruption (with paroxysmal explosion). In fact, although the two eruptions had the same mean effusion rate, the 2014 emitted volume did not reach the critical limit (6.5×10^6 m^3) that was instead overcome by the 2002–03 eruption.

5. Conclusions

Integration of in situ and remote-sensing measurements can strongly support the management of eruptive crises at active volcanoes. In this paper, data from the in situ (visible and infrared) live-cam and seismic monitoring network, measuring the ground motion in the seismic band (0.02–10 Hz), have been integrated with ground displacement recorded by a GBInSAR device. Furthermore, to constrain the erupted lava volume, DEMs from LiDAR and PLEIADES-1 tri-stereo were compared, also supported by the high-resolution PLEIADES-1 optical images, supporting the areal mapping of the 2014 lava flow field. The main findings are:

- live-cam and explosion-quake data revealed that the explosive activity peaked between 5 and 6 August 2014, whereas the GBInSAR device recorded a drastic increase in the displacement rate since the morning of 6 August, consistent with a strong inflation of the crater terrace;
- ground displacement started to show evidence of sliding in the crater terrace after the 6 August 2014 evening, as also evidenced by seismic signals;
- the onset of the 2014 flank eruption involved the breaching of the summit cone with emplacement of a landslide along the SdF (anticipated by the GBInSAR measurements, observed by the live cam, and recorded by the seismic data), the opening of an eruptive fissure on the NE flank of the cone (observed by the live cam), and the effusion of lava from the crater rim at first and from the eruptive fissure later, feeding the 2014 lava flow field;
- the eruption was characterized by the lava effusion along the SdF from a fissure at 650 m above sea level (a.s.l.) that lasted until 13 November 2014, with a total volume of $3.07 \pm 0.37 \times 10^6$ m^3 of lava emplaced on the steep SdF slope;
- this volume is below the limit of $6.5 \pm 1 \times 10^6$ m^3 expected for triggering a paroxysmal explosion.

This work highlights the importance of considering data from in situ sensors and remote-sensing platforms in monitoring active volcanoes. Comparison of data from live-cams, seismic monitoring, and ground displacements recorded by GBInSAR devices provide information about the eruptive activity, nowcasting the shift in eruptive style from effusive to explosive. At the same time, the landslide activity during the run-up and onset phases could be forecasted and tracked using the integration of data from the GBInSAR and the seismic landslide index. Finally, the use of airborne and space-borne DEMs permitted the detection of topographic changes induced by the eruptive activity, allowed for the estimation of the total erupted volumes.

Supplementary Materials: The following are available online at http://www.mdpi.com/2072-4292/10/12/2035/s1, Text 1. Supporting information to the main text.

Author Contributions: Conceptualization: F.D.T., S.C., A.B., L.D., T.N., A.F.; Live-cam data analysis: S.C., A.C.; Seismic data analysis: L.D., A.E.; GBInSAR data analysis: F.D.T., T.N., N.C.; DEMs comparison: A.F., M.F.; Writing—Original Draft Preparation: F.D.T., S.C., A.B., L.D., A.F.; Writing—Review & Editing: T.N., A.E., M.F., N.C.

Funding: This work has been financially supported by the "Presidenza del Consiglio dei Ministri – Dipartimento della Protezione Civile" (Presidency of the Council of Ministers – Department of Civil Protection) within the framework of the InGrID2015-2016 and SAR.NET2017 projects (Scientific Responsibility: NC); this publication, however, does not reflect the position and the official policies of the Department. FDiT has been supported by a post-doc fellowship founded by the "Università degli Studi di Firenze – Ente Cassa di Risparmio di Firenze" (D.R. n. 127804 (1206) 2015; "*Volcano Sentinel*" project). This work has been financially supported by "*Volcano Sentinel—extension*" project (Call: "Settore ricerca scientifica e innovazione tecnologica"; founded by: Ente Cassa di Risparmio di Firenze. Scientific Responsibility: FeDiT).

Acknowledgments: For scientific, technical, logistic support during the 2014 Stromboli flank eruption, we thank: (UNIFI staff) Federica Bardi, Giuseppe De Rosa, Matteo Del Soldato, Giulia Dotta, Emanuele Intrieri, Teresa Salvatici, Lorenzo Solari, Luca Tanteri; (INGV-OE Staff) The technicians of INGV-OE are thanked for the maintenance of the camera monitoring network; (INGV-OV Staff) The technicians of INGV-OV are thanked for the maintenance of the seismic monitoring network; (DPC staff) Chiara Cardaci, Stefano Ciolli, Luigi Coppola, Domenico Mangione, Damiano Piselli, Antonio Ricciardi, Salvatore Zaia. F.D.T. would like to thank Carmine Allocca, Diego Coppola, Dario Delle Donne, Riccardo Genco, Giorgio Lacanna, Marco Laoiolo, Marco Pistolesi, Sébastien Valade for their constructive discussions during and after the 2014 flank eruption.

Conflicts of Interest: The authors declare no conflict of interest.

References

1. Rizzo, A.L.; Federico, C.; Inguaggiato, S.; Sollami, A.; Tantillo, M.; Vita, F.; Bellomo, S.; Longo, M.; Grassa, F.; Liuzzo, M. The 2014 effusive eruption at Stromboli volcano (Italy): Inferences from soil CO_2 flux and 3He/4He ratio in thermal waters. *Geophys. Res. Lett.* **2015**, *42*, 2235–2243. [CrossRef]
2. Di Traglia, F.; Battaglia, M.; Nolesini, T.; Lagomarsino, D.; Casagli, N. Shifts in the eruptive styles at Stromboli in 2010–2014 revealed by ground-based InSAR data. *Sci. Rep.* **2015**, *5*, 13569. [CrossRef] [PubMed]
3. Inguaggiato, S.; Vita, F.; Cangemi, M.; Mazot, A.; Sollami, A.; Calderone, L.; Morici, S.; Paz, M.P.J. Stromboli volcanic activity variations inferred from observations of fluid geochemistry: 16 years of continuous monitoring of soil CO_2 fluxes (2000–2015). *Chem. Geol.* **2017**, *469*, 69–84. [CrossRef]
4. Zakšek, K.; Hort, M.; Lorenz, E. Satellite and ground based thermal observation of the 2014 effusive eruption at Stromboli volcano. *Remote Sens.* **2015**, *7*, 17190–17211. [CrossRef]
5. Valade, S.; Lacanna, G.; Coppola, D.; Laiolo, M.; Pistolesi, M.; Delle Donne, D.; Genco, R.; Marchetti, E.; Ulivieri, G.; Allocca, C.; et al. Tracking dynamics of magma migration in open-conduit systems. *Bull. Volcanol.* **2016**, *78*, 78. [CrossRef]
6. Liotta, M.; Rizzo, A.L.; Barnes, J.D.; D'Auria, L.; Martelli, M.; Bobrowski, N.; Wittmer, J. Chlorine isotope composition of volcanic rocks and gases at Stromboli volcano (Aeolian Islands, Italy): Inferences on magmatic degassing prior to 2014 eruption. *J. Volcanol. Geotherm. Res.* **2017**, *336*, 168–178. [CrossRef]
7. Di Traglia, F.; Nolesini, T.; Ciampalini, A.; Solari, L.; Frodella, W.; Bellotti, F.; Fumagalli, A.; De Rosa, G.; Casagli, N. Tracking morphological changes and slope instability using spaceborne and ground-based SAR data. *Geomorphology* **2018**, *300*, 95–112. [CrossRef]
8. Calvari, S.; Lodato, L.; Steffke, A.; Cristaldi, A.; Harris, A.J.L.; Spampinato, L.; Boschi, E. The 2007 Stromboli flank eruption: Chronology of the events, and effusion rate measurements from thermal images and satellite data. *J. Geophys. Res. Solid Earth* **2010**, *115*, B04201. [CrossRef]
9. Calvari, S.; Bonaccorso, A.; Madonia, P.; Neri, M.; Liuzzo, M.; Salerno, G.G.; Behncke, B.; Caltabiano, T.; Cristaldi, A.; Giuffrida, G.; et al. Major eruptive style changes induced by structural modifications of a shallow conduit system: The 2007–2012 Stromboli case. *Bull. Volcanol.* **2014**, *76*, 841. [CrossRef]
10. Antonello, G.; Casagli, N.; Farina, P.; Leva, D.; Nico, G.; Sieber, A.J.; Tarchi, D. Ground-based SAR interferometry for monitoring mass movements. *Landslides* **2004**, *1*, 21–28. [CrossRef]

11. Di Traglia, F.; Nolesini, T.; Intrieri, E.; Mugnai, F.; Leva, D.; Rosi, M.; Casagli, N. Review of ten years of volcano deformations recorded by the ground-based InSAR monitoring system at Stromboli volcano: A tool to mitigate volcano flank dynamics and intense volcanic activity. *Earth Sci. Rev.* **2014**, *139*, 317–335. [CrossRef]
12. Martini, M.; Giudicepietro, F.; D'Auria, L.; Esposito, A.M.; Caputo, T.; Curciotti, R.; Raffaele Curciotti De Cesare, W.; Orazi, M.; Scarpato, G.; Caputo, A.; et al. Seismological monitoring of the February 2007 effusive eruption of the Stromboli volcano. *Ann. Geophys.* **2007**, *50*, 775–788.
13. Martini, M.; D'Auria, L.; Caputo, T.; Giudicepietro, F.; Peluso, R.; Caputo, A.; De Cesare, W.; Esposito, A.M.; Orazi, M.; Scarpato, G. Seismological insights on the shallow magma system. In *The Stromboli Volcano: An Integrated Study of the 2002–2003 Eruption*; American Geophysical Union: Washington, DC, USA, 2008; pp. 279–286.
14. Marotta, E.; Calvari, S.; Cristaldi, A.; D'Auria, L.; Di Vito, M.A.; Moretti, R.; Peluso, R.; Spampinato, L.; Boschi, E. Reactivation of Stromboli's summit craters at the end of the 2007 effusive eruption detected by thermal surveys and seismicity. *J. Geophys. Res.* **2015**, *120*, 7376–7395. [CrossRef]
15. Di Traglia, F.; Bartolini, S.; Artesi, E.; Nolesini, T.; Ciampalini, A.; Lagomarsino, D.; Martí, J.; Casagli, N. Susceptibility of intrusion-related landslides at volcanic islands: The Stromboli case study. *Landslides* **2018**, *15*, 21–29. [CrossRef]
16. Kokelaar, P.; Romagnoli, C. Sector collapse, sedimentation and clast population evolution at an active island-arc volcano: Stromboli, Italy. *Bull. Volcanol.* **1995**, *57*, 240–262. [CrossRef]
17. Tibaldi, A. Multiple sector collapses at Stromboli volcano, Italy: How they work. *Bull. Volcanol.* **2001**, *63*, 112–125. [CrossRef]
18. Maramai, A.; Graziani, L.; Tinti, S. Tsunamis in the Aeolian Islands (southern Italy): A review. *Mar. Geol.* **2005**, *215*, 11–21. [CrossRef]
19. Barberi, F.; Rosi, M.; Sodi, A. Volcanic hazard assessment at Stromboli based on review of historical data. *Acta Vulcanol.* **1993**, *3*, 173–187.
20. Rosi, M.; Pistolesi, M.; Bertagnini, A.; Landi, P.; Pompilio, M.; Di Roberto, A. Stromboli volcano, Aeolian Islands (Italy): Present eruptive activity and hazards. *Geol. Soc. Lond. Mem.* **2013**, *37*, 473–490. [CrossRef]
21. De Fino, M.; La Volpe, L.; Falsaperla, S.; Frazzetta, G.; Neri, G.; Francalanci, L.; Rosi, M.; Sbrana, A. The Stromboli eruption of 6 December 1985–25 April 1986: Volcanological, petrological and seismological data. *Rend. Della Soc. Italiana Mineral. Petrol.* **1988**, *43*, 1021–1038.
22. Bonaccorso, A.; Calvari, S.; Garfì, G.; Lodato, L.; Patanè, D. Dynamics of the December 2002 flank failure and tsunami at Stromboli volcano inferred by volcanological and geophysical observations. *Geophys. Res. Lett.* **2003**, *30*. [CrossRef]
23. Casagli, N.; Tibaldi, A.; Merri, A.; Del Ventisette, C.; Apuani, T.; Guerri, L.; Fortuny-Guasch, J.; Tarchi, D. Deformation of Stromboli Volcano (Italy) during the 2007 eruption revealed by radar interferometry, numerical modelling and structural geological field data. *J. Volcanol. Geotherm. Res.* **2009**, *182*, 182–200. [CrossRef]
24. Carlà, T.; Intrieri, E.; Di Traglia, F.; Nolesini, T.; Gigli, G.; Casagli, N. Guidelines on the use of inverse velocity method as a tool for setting alarm thresholds and forecasting landslides and structure collapses. *Landslides* **2017**, *14*, 517–534. [CrossRef]
25. Tinti, S.; Pagnoni, G.; Zaniboni, F. The landslides and tsunamis of the 30th of December 2002 in Stromboli analysed through numerical simulations. *Bull. Volcanol.* **2006**, *68*, 462–479. [CrossRef]
26. Carlà, T.; Intrieri, E.; Di Traglia, F.; Casagli, N. A statistical-based approach for determining the intensity of unrest phases at Stromboli volcano (Southern Italy) using one-step-ahead forecasts of displacement time series. *Nat. Hazards* **2016**, *84*, 669–683. [CrossRef]
27. Acocella, V.; Neri, M.; Scarlato, P. Understanding shallow magma emplacement at volcanoes: orthogonal feeder dikes during the 2002–2003 Stromboli (Italy) eruption. *Geophys. Res. Lett.* **2006**, *33*. [CrossRef]
28. Lodato, L.; Spampinato, L.; Harris, A.J.L.; Calvari, S.; Dehn, J.; Patrick, M. The Morphology and Evolution of the Stromboli 2002–2003 Lava Flow Field: An Example of Basaltic Flow Field Emplaced on a Steep Slope. *Bull. Volcanol.* **2007**, *69*, 661–679. [CrossRef]
29. Di Traglia, F.; Nolesini, T.; Solari, L.; Ciampalini, A.; Frodella, W.; Steri, D.; Allotta, B.; Rindi, A.; Monni, N.; Marini, L.; et al. Lava delta deformation as a proxy for submarine slope instability. *Earth Planet. Sci. Lett.* **2018**, *488*, 46–58. [CrossRef]

30. De Cesare, W.; Orazi, M.; Peluso, R.; Scarpato, G.; Caputo, A.; D'Auria, L.; Giudicepietro, F.; Martini, M.; Buonocunto, C.; Capello, M.; Esposito, A.M. The broadband seismic network of Stromboli volcano, Italy. *Seismol. Res. Lett.* **2009**, *80*, 435–439. [CrossRef]
31. Esposito, A.M.; D'Auria, L.; Giudicepietro, F.; Peluso, R.; Martini, M. Automatic recognition of landslides based on neural network analysis of seismic signals: An application to the monitoring of Stromboli volcano (Southern Italy). *Pure Appl. Geophys.* **2013**, *170*, 1821–1832. [CrossRef]
32. Muller, C.; del Potro, R.; Biggs, J.; Gottsmann, J.; Ebmeier, S.K.; Guillaume, S.; Cattin, P.-H.; Van der Laat, R. Integrated velocity field from ground and satellite geodetic techniques: Application to Arenal volcano. *Geophys. J. Int.* **2014**, *200*, 863–879. [CrossRef]
33. Schaefer, L.N.; Lu, Z.; Oommen, T. Dramatic volcanic instability revealed by InSAR. *Geology* **2015**, *43*, 743–746. [CrossRef]
34. Bonforte, A.; Guglielmino, F. Very shallow dyke intrusion and potential slope failure imaged by ground deformation: The 28 December 2014 eruption on Mount Etna. *Geophys. Res. Lett.* **2015**, *42*, 2727–2733. [CrossRef]
35. Frodella, W.; Ciampalini, A.; Bardi, F.; Salvatici, T.; Di Traglia, F.; Basile, G.; Casagli, N. A method for assessing and managing landslide residual hazard in urban areas. *Landslides* **2018**, *15*, 183–197. [CrossRef]
36. Di Traglia, F.; Cauchie, L.; Casagli, N.; Saccorotti, G. Decrypting geophysical signals at Stromboli Volcano (Italy): Integration of seismic and Ground-Based InSAR displacement data. *Geophys. Res. Lett.* **2014**, *41*, 2753–2761. [CrossRef] [PubMed]
37. Gleyzes, M.A.; Perret, L.; Kubik, P. Pleiades system architecture and main performances. *Int. Arch. Photogramm. Remote Sens. Spat. Inf. Sci.* **2012**, *39*, 537–542. [CrossRef]
38. Perko, R.; Raggam, H.; Gutjahr, K.; Schardt, M. Assessment of the mapping potential of Pléiades stereo and triplet data. *ISPRS Ann. Photogramm. Remote Sens. Spat. Inf. Sci.* **2014**, *2*, 103. [CrossRef]
39. Zhou, Y.; Parsons, B.; Elliott, J.R.; Barisin, I.; Walker, R.T. Assessing the ability of Pleiades stereo imagery to determine height changes in earthquakes: A case study for the El Mayor-Cucapah epicentral area. *J. Geophys. Res. Solid Earth* **2015**, *120*, 8793–8808. [CrossRef]
40. Bagnardi, M.; González, P.J.; Hooper, A. High-resolution digital elevation model from tri-stereo Pleiades-1 satellite imagery for lava flow volume estimates at Fogo Volcano. *Geophys. Res. Lett.* **2016**, *43*, 6267–6275. [CrossRef]
41. Rieg, L.; Klug, C.; Nicholson, L.; Sailer, R. Pléiades Tri-Stereo Data for Glacier Investigations—Examples from the European Alps and the Khumbu Himal. *Remote Sens.* **2018**, *10*, 1563. [CrossRef]
42. Favalli, M.; Fornaciai, A.; Mazzarini, F.; Harris, A.; Neri, M.; Behncke, B.; Pareschi, M.T.; Tarquini, S.; Boschi, E. Evolution of an active lava flow field using a multitemporal LIDAR acquisition. *J. Geophys. Res. Solid Earth* **2012**, *115*. [CrossRef]
43. James, F.; Roos, M. *MINUIT Computer Code*; Program D-506; CERN: Geneva, Switzerland, 1977.
44. Kolzenburg, S.; Favalli, M.; Fornaciai, A.; Isola, I.; Harris, A.J.L.; Nannipieri, L.; Giordano, D. Rapid updating and improvement of airborne LIDAR DEMs through ground-based SfM 3-D modeling of volcanic features. *IEEE Trans. Geosci. Remote Sens.* **2016**, *54*, 6687–6699. [CrossRef]
45. Richter, N.; Favalli, M.; de Zeeuw-van Dalfsen, E.; Fornaciai, A.; Fernandes, R.M.D.S.; Pérez, N.M.; Levy, J.; Silva Victória, S.; Walter, T.R. Lava flow hazard at Fogo Volcano, Cabo Verde, before and after the 2014-2015 eruption. *Nat. Hazards Earth Syst. Sci.* **2016**, *16*, 1925–1951. [CrossRef]
46. Favalli, M.; Fornaciai, A.; Nannipieri, L.; Harris, A.; Calvari, S.; Lormand, C. UAV-based remote sensing surveys of lava flow fields: A case study from Etna's 1974 channel-fed lava flows. *Bull. Volcanol.* **2018**, *80*, 29. [CrossRef]
47. Delle Donne, D.; Tamburello, G.; Aiuppa, A.; Bitetto, M.; Lacanna, G.; D'Aleo, R.; Ripepe, M. Exploring the explosive-effusive transition using permanent ultraviolet cameras. *J. Geophys. Res. Solid Earth* **2017**, *122*, 4377–4394. [CrossRef]
48. Calvari, S.; Intrieri, E.; Di Traglia, F.; Bonaccorso, A.; Casagli, N.; Cristaldi, A. Monitoring crater-wall collapse at active volcanoes: A study of the 12 January 2013 event at Stromboli. *Bull. Volcanol.* **2016**, *78*, 39. [CrossRef]
49. Neri, M.; Lanzafame, G. Structural features of the 2007 Stromboli eruption. *J. Volcanol. Geotherm. Res.* **2009**, *182*, 137–144. [CrossRef]
50. Bonaccorso, A.; Bonforte, A.; Gambino, S.; Mattia, M.; Guglielmino, F.; Puglisi, G.; Boschi, E. Insight on recent Stromboli eruption inferred from terrestrial and satellite ground deformation measurements. *J. Volcanol. Geotherm. Res.* **2009**, *182*, 171–181. [CrossRef]

51. Finizola, A.; Sortino, F.; Lénat, J.F.; Valenza, M. Fluid circulation at Stromboli volcano (Aeolian Islands, Italy) from self-potential and CO_2 surveys. *J. Volcanol. Geotherm. Res.* **2002**, *116*, 1–18. [CrossRef]
52. Giordano, G.; Porreca, M. Field observations on the initial lava flow and the fracture system developed during the early days of the Stromboli 2007 eruption. *J. Volcanol. Geotherm. Res.* **2009**, *182*, 145–154. [CrossRef]
53. Gudmundsson, A. Deflection of dykes into sills at discontinuities and magma-chamber formation. *Tectonophysics* **2011**, *500*, 50–64. [CrossRef]
54. Gudmundsson, A. Strengths and strain energies of volcanic edifices: Implications for eruptions, collapse calderas, and landslides. *Nat. Hazards Earth Syst. Sci.* **2012**, *12*, 2241–2258. [CrossRef]
55. Bonaccorso, A.; Gambino, S.; Guglielmino, F.; Mattia, M.; Puglisi, G.; Boschi, E. Stromboli 2007 eruption: Deflation modeling to infer shallow-intermediate plumbing system. *Geophys. Res. Lett.* **2008**, *35*. [CrossRef]
56. Calvari, S.; Spampinato, L.; Bonaccorso, A.; Oppenheimer, C.; Rivalta, E.; Boschi, E. Lava effusion—A slow fuse for paroxysms at Stromboli volcano? *Earth Planet. Sci. Lett.* **2011**, *301*, 317–323. [CrossRef]
57. Ripepe, M.; Pistolesi, M.; Coppola, D.; Delle Donne, D.; Genco, R.; Lacanna, G.; Laiolo, M.; Marchetti, E.; Ulivieri, G.; Valade, S. Forecasting Effusive Dynamics and Decompression Rates by Magmastatic Model at Open-vent Volcanoes. *Sci. Rep.* **2017**, *7*, 3885. [CrossRef]
58. Coppola, D.; Laiolo, M.; Piscopo, D.; Cigolini, C. Rheological control on the radiant density of active lava flows and domes. *J. Volcanol. Geotherm. Res.* **2013**, *249*, 39–48. [CrossRef]
59. Calvari, S.; Spampinato, L.; Lodato, L.; Harris, A.J.; Patrick, M.R.; Dehn, J.; Burton, M.; Andronico, D. Chronology and complex volcanic processes during the 2002–2003 flank eruption at Stromboli volcano (Italy) reconstructed from direct observations and surveys with a handheld thermal camera. *J. Geophys. Res. Solid Earth* **2015**, *110*. [CrossRef]
60. Calvari, S.; Spampinato, L.; Lodato, L. The 5 April 2003 vulcanian paroxysmal explosion at Stromboli volcano (Italy) from field observations and thermal data. *J. Volcanol. Geotherm. Res.* **2006**, *149*, 160–175. [CrossRef]
61. Rosi, M.; Bertagnini, A.; Harris, A.J.L.; Pioli, L.; Pistolesi, M.; Ripepe, M. A case history of paroxysmal explosion at Stromboli: Timing and dynamics of the 5 April 2003 event. *Earth Planet. Sci. Lett.* **2006**, *243*, 594–606. [CrossRef]
62. Bertagnini, A.; Métrich, N.; Landi, P.; Rosi, M. Stromboli volcano (Aeolian Archipelago, Italy): An open window on the deep-feeding system of a steady state basaltic volcano. *J. Geophys. Res.* **2003**, *108*, 2336. [CrossRef]
63. Métrich, N.; Bertagnini, A.; Landi, P.; Rosi, M.; Belhadj, O. Triggering mechanism at the origin of paroxysms at Stromboli (Aeolian Archipelago, Italy): The 5 April, 2003 eruption. *Geophys. Res. Lett.* **2005**, *32*, L10305. [CrossRef]
64. Bonaccorso, A.; Calvari, S.; Linde, A.; Sacks, S.; Boschi, E. Dynamics of the shallow plumbing system investigated from borehole strainmeters and cameras during the 15 March 2007 Vulcanian paroxysm at Stromboli volcano. *Earth Planet. Sci. Lett.* **2012**, *357–358*, 249–256. [CrossRef]
65. Rivalta, E.; Pascal, K.; Phillips, J.; Bonaccorso, A. Explosive expansion of a slowly decompressed magma analogue: Evidence for delayed bubble nucleation. *Geochem. Geophys. Geosyst.* **2013**, *14*, 3067–3084. [CrossRef]

Article

Multi-Sensor SAR Geodetic Imaging and Modelling of Santorini Volcano Post-Unrest Response

Elena Papageorgiou [1], Michael Foumelis [2,*], Elisa Trasatti [3], Guido Ventura [3], Daniel Raucoules [2] and Antonios Mouratidis [1]

1 School of Geology, Aristotle University of Thessaloniki (AUTh), 54124 Thessaloniki, Greece; elenpapageo@geo.auth.gr (E.P.); amourati@auth.gr (A.M.)

2 BRGM—French Geological Survey, 3 Claude-Guillemin, 45060 Orléans, France; d.raucoules@brgm.fr

3 Istituto Nazionale di Geofisica e Vulcanologia (INGV), 605 Via di Vigna Murata, 00143 Roma, Italy; elisa.trasatti@ingv.it (E.T.); guido.ventura@ingv.it (G.V.)

* Correspondence: m.foumelis@brgm.fr; Tel.: +33-023-868-3226

Received: 25 December 2018; Accepted: 24 January 2019; Published: 28 January 2019

Abstract: Volcanic history of Santorini over recent years records a seismo-volcanic unrest in 2011–12 with a non-eruptive behavior. The volcano deformation state following the unrest was investigated through multi-sensor Synthetic Aperture Radar Interferometry (InSAR) time series. We focused on the analysis of Copernicus Sentinel-1, Radarsat-2 and TerraSAR-X Multi-temporal SAR Interferometric (MT-InSAR) results, for the post-unrest period 2012–17. Data from multiple Sentinel-1 tracks and acquisition geometries were used to constrain the E-W and vertical components of the deformation field along with their evolution in time. The interpretation of the InSAR observations and modelling provided insights on the post-unrest deformation pattern of the volcano, allowing the further re-evaluation of the unrest event. The increase of subsidence rates on Nea Kameni, in accordance with the observed change of the spatial deformation pattern, compared to the pre-unrest period, suggests the superimposition of various deformation sources. Best-fitting inversion results indicate two deflation sources located at southwestern Nea Kameni at 1 km depth, and in the northern intra-caldera area at 2 km depth. A northern sill-like source interprets the post-unrest deflation attributed to the passive degassing of the magma intruded at 4 km during the unrest, while an isotropic source at Nea Kameni simulates a prevailing subsidence occurring since the pre-unrest period (1992–2010).

Keywords: volcano deformation; SAR interferometry; post-unrest deflation; inversion modelling; Santorini

---

1. Introduction

Santorini volcano is considered to belong in caldera-forming systems undergoing long-term periods of quiescence, of approximately 20,000 years (Figure 1). Although its last big eruption dates back 3600 years, its volcanic activity up to the most recent eruption in 1950 [1] was intertwined with the building of the intra-caldera islets of Palea and Nea Kameni. The latest volcano's reactivation was followed by the restless period of 2011, which did not culminate in an eruption. Increased microseismic activity [2] and significant ground uplift [3] reaching 14 cm from March 2011 to March 2012 at Cape Skaros (Figure 1), and 9 cm at Nea Kameni islet [4], underlined the seismo-volcanic unrest. Most studies based on geodetic data interpret the episode with a single inflation source due to magma intrusion, at the northern part of the caldera, estimated within 3–4 km depth [4–8]. An alternative model based on GPS data [9], proposes two inflationary magmatic sources that relocate in depth and geographic location throughout the unrest. The evolution of the shallow magma body beneath Santorini appears to have been regulated by the episodic rapid magma supply from a deeper magmatic system, while a significant volume of magma was intruded in short pulses between January 2011 and

April 2012 [6,10]. Since the ending of the 2011–12 unrest, the volcano has presented no further activity, as confirmed by geodetic measurements [4,11] and seismicity [2]. Similar behavior, with extended periods of quiescence, interrupted by short non-eruptive activity, is also observed in other caldera volcanic systems [12–14].

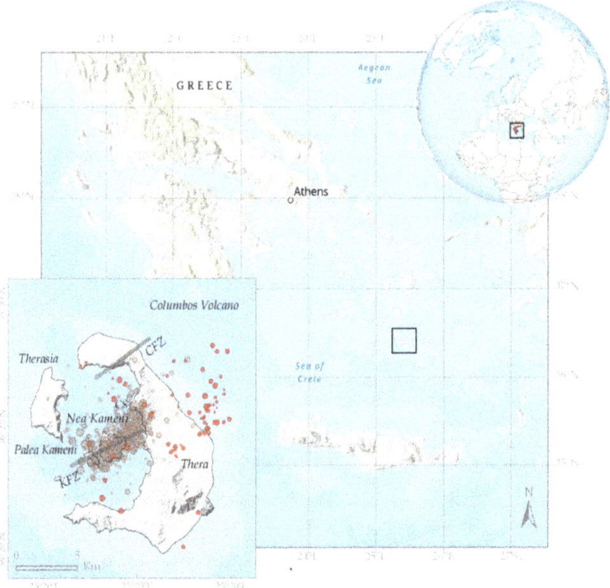

Figure 1. Location map of Santorini volcano and seismicity distribution for the period 2011–17 [2]. Earthquake foci in gray color represent the unrest period between January 2011 and May 2012 (ML $\leq$ 3.3), whereas epicenters in red correspond to the post-unrest period from September 2012 to August 2017 (ML $\leq$ 2.3). KFZ and CFZ represent the Kameni and Columbos Fault Zones, respectively. CS indicates Cape Skaros undergoing the maximum uplift (14 cm) during the 2011–12 unrest episode [4].

Recent advances in geodetic imaging techniques and the systematic availability of Synthetic Aperture Radar (SAR) data from spaceborne missions provided the necessary tools and data, in order to monitor the deformation field of Santorini volcano over the post-unrest period, and to re-evaluate the deformation mechanism during the unrest. MT-InSAR techniques were employed to generate and analyze the SAR deformation time series. A comprehensive analysis of multi-sensor SAR acquisitions of Copernicus Sentinel-1, Radarsat-2 and TerraSAR-X satellite missions was performed to investigate the evolution of ground deformation from 2012 to 2017. The Copernicus Sentinel-1 and Radarsat-2 measurements were further analyzed in an inversion framework to estimate the source parameters that can best resolve the observed displacements, while TerraSAR-X data provided an additional source of information to verify the deformation pattern. Finally, insights on the post-unrest response of the volcano and the interrelationship between the post-unrest interval and the unrest event were investigated, while ERS-1 and -2 and ENVISAT data from 1992 to 2010 [3,4] were additionally used to interpret the similarity between the pre- and post-unrest volcano's deformation.

2. Santorini Volcanic Setting

Santorini volcano is part of the Hellenic Volcanic Arc in the Southern Aegean Sea (Figure 1), and it is partly situated on a SW-NE trending tectonic horst, the Amorgos Ridge, whereas the northwestern half of the volcanic field lies within the Anydros Basin [15]. Extensional tectonics seems to have a profound effect on Santorini volcano. Regional fault systems, namely the NE-SW striking Kameni and

Columbos Fault Zones (KFZ, CFZ) [16–18] mark the alignment of several eruptive vents [16,19] and have been interpreted as major normal faults moving in response to a NW-SE extension [20].

Volcanism in Santorini evolved with time from an early formation of volcanic cones, two successive explosive cycles of pyroclastic eruptions, and the later collapse of the caldera with the development of shield volcanoes within the caldera [16]. Since the caldera-forming Minoan eruption in the late 1600s BC, a peculiar volcanic configuration was finally set, with two active volcanoes, the Nea Kameni volcano located at the center of the caldera and the Columbos submarine volcano located almost 7 km NE of Thera [21–23] (Figure 1). The largest volcanic eruptions date to 197 BC, 1866, 1925 and 1949–50 and are all associated with Nea Kameni volcano. However, an eruption in 1650 AD took place offshore at Columbos volcano [24,25]. Aseismic and small-scale intrusions are inferred to have been emplaced on the northern caldera between 1994 and 1999 [26]. Marine geophysical surveys in the same area provide evidence of shallow intrusions of the post-Minoan activity [27]. The most recent activity of the volcano was recorded at the beginning of 2011 lasting up to the first half of 2012 [3–9]. This activity was marked by an increase of low magnitude (ML < 3.2) earthquakes, mainly aligned along the Kameni fault [28], deformation and changes in the geochemical parameters of the gas emissions. These unrest signals were concentrated within the caldera sector surrounding Nea Kameni and Thera [2,29]. Since the ending of the unrest, seismicity associated with caldera volcanic activity has decreased (Figure 1).

3. Materials and Methods

3.1. SAR Interferometric Processing

Differential Interferometric SAR (DInSAR) techniques are widely used to detect and monitor subtle ground displacements associated with volcanoes, such as dome growth and subsidence [30,31]. By combining SAR observations from different sensors and viewing geometries, over common acquisition periods, a proper characterization of the spatial and temporal deformation patterns can be robustly achieved. Advanced Multi-Temporal InSAR (MT-InSAR) time series analysis techniques offer the capability to monitor the temporal evolution of ground deformation phenomena. These techniques lie on the capability to utilize large series of SAR imagery to measure small deformation signals, to millimeter level of accuracy, on individual Persistent Scatterers (PS) point targets (human infrastructures, natural scatterers). Several MT-InSAR techniques have been proposed (e.g., References [32–45]), each following different approaches for the identification of point scatterers and resolving displacement histories. A detailed review of proposed MT-InSAR techniques in the literature is presented in Reference [46].

In this study, the MT-InSAR approach proposed by Reference [47] was adopted, by using the GAMMA s/w packages. The implemented Terrain Observation by Progressive Scans (TOPS) acquisition mode [48] on Sentinel-1 mission required specific/particular interferometric handling to ensure proper coregistration of burst-type data compared to standard stripmap acquisitions [49]. Sentinel-1 captures 250 km in three sub-swaths in range, each divided into nine bursts in azimuth. Burst synchronization ensures the interferometric capability of the mission. Since accuracy at 0.005 pixels is required [50], coregistration is performed on a pixel basis using initial orbital information and DEM-assisted cross-correlation methods, while an iterative refinement procedure is followed, by using the Enhanced Spectral Diversity (ESD) technique [49,51]. The coregistration scheme was further adapted to avoid temporal decorrelation effect on ESD estimates, especially since the burst overlap areas for Santorini are spatially limited over land. The estimation of ESD offsets for each scene were based on the temporally closest already coregistered slave, in such a way that coregistration proceeds successively, starting from the scenes closest to the master to the most temporally distant ones.

Interferometric processing with multi-looking factors of 6x2 in range and azimuth, respectively, was considered in order to mitigate signal variability and to obtain comparable pixel spacing to previous ERS-ENVISAT results. Topographic phase was simulated and subtracted based on a 20 m resolution Digital Elevation Model (DEM) generated from 1/50.000 scale topographic map of the Hellenic Military Geographical Service (HGMS).

Since no variability of deformation within a 6-day interval was expected, only acquisitions from Sentinel-1A satellite were considered (12-days repeat cycle), offering a sufficiently dense temporal stack, to minimize unwrapping artifacts due to decorrelation effects, and characterize the spatio-temporal behavior of the deformation signal. More specifically, three Sentinel-1 orbit geometries were used (Table 1), covering the period from October 2014 up to December 2017. Data used for the ascending relative orbit A029 amounts to 91 acquisitions, while for the descending relative orbits D109 and D036 correspond to 93 and 92 acquisitions, respectively (Figure 2).

Table 1. Overview of SAR data in-use.

Mission	Orbit	Track	Acquisition Mode	Incidence	Observation Period	No. Scenes
Sentinel-1A	Ascending	029	TOPS IW1	33.9	2014–17	91
Sentinel-1A	Descending	109	TOPS IW1	33.9	2014–17	93
Sentinel-1A	Descending	036	TOPS IW3	43.9	2014–17	92
Radarsat-2	Descending	-	Stripmap	33.5	2012–16	20
TerraSAR-X	Descending	-	Stripmap	27.2	2012–13	25

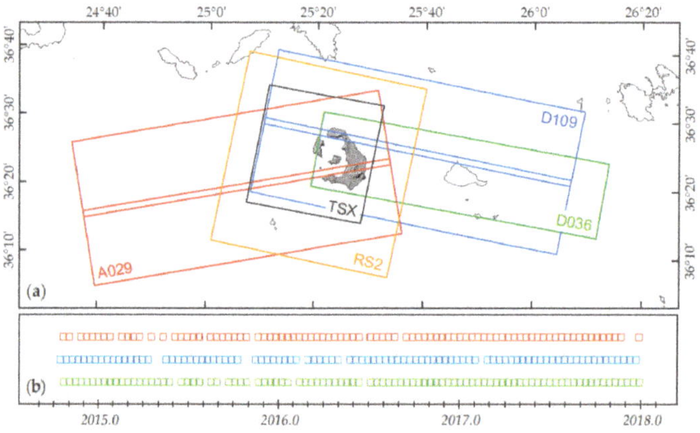

Figure 2. (**a**) Sentinel-1 IW burst coverage over Santorini volcano for the Sentinel-1 ascending 029 (A029), and descending orbits 109 (D109) and 036 (D036), and for Radarsat-2 (RS2) and TerraSAR-X (TSX) frames; (**b**) Temporal distribution of Sentinel-1 data for the A029, D109 and D036 acquisition geometries.

A key point of the processing, given the size of the data stack, was to derive a redundancy of all possible interferometric pairs. The final selection of the interferograms was done by limiting the temporal baselines, as well as by using an upper threshold on the maximum normal baseline value. Although the orbital tube of Sentinel-1 is well-tuned for interferometric applications (within 120 m radius), the presence of steep slopes along the caldera walls, of height differences up to approx. 300 m, led to the further control of the baselines. Interferometric stacks of 160, 160 and 242 pairs were generated for the A029, D109 and D036 orbits accordingly, for the final MT-InSAR configuration (Table 2).

Table 2. Interferometric pairs considered in Sentinel-1 MT-InSAR processing.

Mission	Orbit	Track	Temporal Separation (Days)	Normal Baseline (m)	No. of Pairs
Sentinel-1A	Ascending	029	$120 \leq dt \leq 240$	$Bp \leq 20$	160
Sentinel-1A	Descending	109	$120 \leq dt \leq 240$	$Bp \leq 20$	160
Sentinel-1A	Descending	036	$90 \leq dt \leq 270$	$Bp \leq 20$	242

The Singular Value Decomposition (SVD) approach was used to obtain a solution for the phase time series, based on a set of multi-reference point differential interferograms. Deformation phase time series were estimated using a weighted least-squares algorithm that minimized the sum of squared weighted residual phases [36]. The phase model included a height related term proportional to the derivative of the interferometric phase with respect to height, allowing the update of PS heights during the interferometric processing. To mitigate the atmospheric contribution, filtering based on assumptions of atmospheric statistics [33] was applied in both temporal and spatial domains, in a way to minimize the effect of spatially correlated and temporally uncorrelated signal.

Redundancy of interferometric pairs reduces uncertainties in the time series mainly due to phase errors introduced by decorrelation and residual topography. With respect to the selected pairs (Figure S1), few scenes per orbit were excluded from the analysis; still temporal sampling was sufficiently dense to examine temporal variability in PS displacement histories. For each solution the linear displacement phase rate, the standard deviation of the phase time series relative to the linear fit, as well as the unwrapped phases for each PS target were stored.

3.2. Hosted Processing on Geohazards Exploitation Platform

The Geohazards Lab initiative, originated by the European Space Agency (ESA) with the support of several other space agencies of the Committee on Earth Observation Satellites (CEOS), is based on a group of interoperable platforms with federated resources providing Earth Observation (EO) data access, hosted processing and e-collaboration capabilities to animate and support the geohazards user community [52]. One of its precursors is the Geohazards Exploitation Platform (GEP) (https://geohazards-tep.eu), developed in the framework of the ESA Thematic Exploitation Platforms (TEP) initiative and has been available since 2016. Today, GEP has primary focus on mapping hazard-prone land surfaces and monitoring terrain deformation. The platform has been expanded to include a broad range of products and services, currently available or under development on cloud processing resources like the GEP, to support experts and users to better understand geohazards and relevant risks.

In the framework of the ESA funded project "Disaster Risk Reduction using innovative data exploitation methods and space assets", ground deformation maps were produced on GEP for the Santorini volcano. Processing was based on a customized implementation of the StaMPS Persistent Scatterer Interferometry (PSI) software [40]. For the period 2012–16 average Line-Of-Sight (LOS) velocities were based on 20 descending Radarsat-2 images, while TerraSAR-X data (25 scenes in descending orbit) were used for the 2012–13 period (Table 1). Datasets are available via Zenodo repository [53].

GEP PSI results are provided as raw velocities generated from automatic processors and have not undergone any post-processing. The adjustment of the reference point was addressed to ensure compatibility with Sentinel-1 results obtained herein, and visual inspections were performed to exclude regions potentially affected by isolated unwrapping errors.

4. Interferometric Results

A multi-temporal interferometric analysis approach was implemented to explore the dense temporal series of Sentinel-1 data, and to provide the displacement time series for each identified PS target. The MT-InSAR results are shown in Figure 3, presenting the measured LOS displacement rates for each of the Sentinel-1 acquisition geometries. Common deformation patterns are retrieved, indicating LOS motion rates up to −7 and −9 mm/yr over Nea Kameni islet. Variations of maximum observed motion could be partly attributed to the different incident angles between the acquisition geometries. Despite the large amount of data, the relatively short observation period (3.2 years) does not permit substantial decrease of the measurement uncertainties, though they were maintained at a level of 1.5–2.0 mm/yr (Figure S2). Additionally, the relatively high coherence levels, due to the short span of the interferometric pairs used, especially over Nea Kameni, underline the robustness of the obtained results.

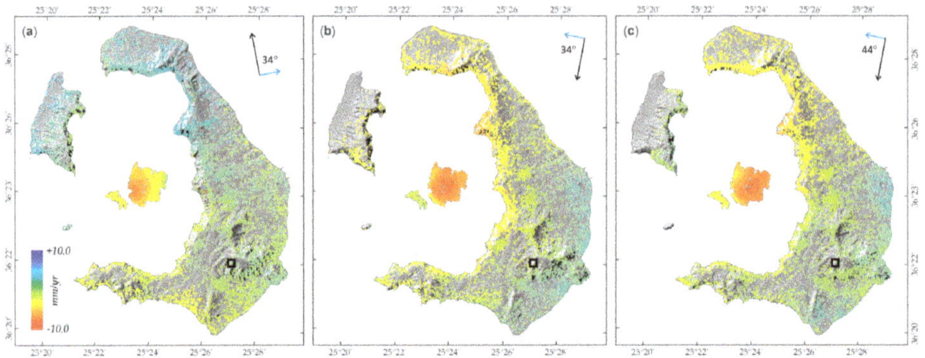

Figure 3. Sentinel-1 MT-InSAR LOS displacement rates of Santorini volcano for the period 2014–17 (**a**) Ascending orbit 029; (**b**) Descending orbit 109; (**c**) Descending orbit 036. Displacement rates are given relative to the reference point, marked by a rectangle. The line-of-sight and azimuth directions of the satellite are displayed by blue and black arrows, respectively. LOS velocity is negative away from the satellite. Values in degrees correspond to the incidence angles.

It is worth mentioning that deformation does not seem to follow exactly the same pattern among the different orbits (Figure 3). Therefore, the southwestern part of Nea Kameni seems to be most affected in the ascending orbit, while in the descending ones, maximum deformation is shifted to the central and northern parts of the islet. This spatial variability implies the presence of horizontal motion, also mentioned in previous study [4].

The displacement time series for a PS target over Nea Kameni islet is shown in Figure 4. It is possible to observe the presence of a periodic (annual) component of the displacement, identical both in terms of period and amplitude for all Sentinel-1 datasets (Figure 5). The amplitudes of the oscillations (up to 5 mm) are not negligible with respect to the expected linear deformations rates. Such cyclic motions can therefore significantly affect, by under- or over-estimation, the deformation rates calculated based on linear regression of the displacement time series over a few years' time-spans.

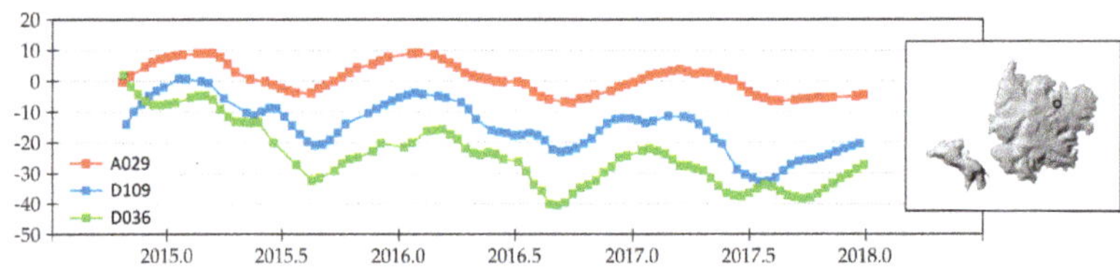

Figure 4. Temporal evolution of LOS displacements over selected PS target located on Nea Kameni, for the three Sentinel-1 acquisition geometries shown in Figure 2.

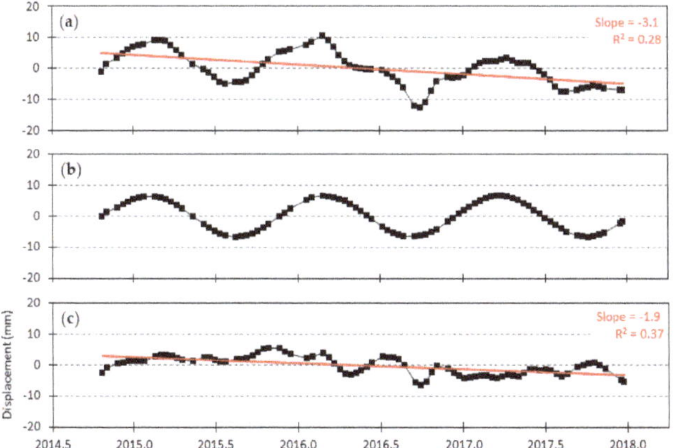

Figure 5. (**a**) Time series of LOS displacements at a selected point (see Figure 4); (**b**) Seasonal component obtained by Fourier transform; (**c**) Time series corrected for the seasonal oscillations.

To compensate for this source of error, a linear interpolation to the time series was initially applied to produce a regular 6-day sampling, in order to fill gaps in the temporal series. To quantify quasi-annual motion, we then calculated a Fourier transform of the time series by only selecting periods between 200 and 500 days. Other frequency intervals were set to zero. The maximum amplitude of the resulting spectrum was then identified. The corresponding cyclic function (d) of time (t) was retrieved on the basis of the amplitude (A), period (T) and phase (φ) of the Fourier transform for the frequency of the spectrum's maximum (Equation (1)):

$$d = A\cos(2\pi \frac{t}{T} + \varphi) \quad (1)$$

The obtained cyclic component was finally removed from the initial time series and the LOS displacement rates for each PS were re-computed as the slopes of the corrected displacement time series (Figure 5). Such periodic signal can be attributed to the topography-dependent atmospheric component of the SAR, also reported for several other volcanoes [54,55].

By compensating for the seasonal signal in the time series, a noteworthy decrease in the displacement rates is apparent for the major part of the volcano (Figure 6). The new estimation of the displacement rates provided the actual deformation that was used to model the source dynamics. It was not in fact possible for such correction to be derived from Radarsat-2 PSI results, as data provided via GEP platform [53] concerned only LOS velocities and not actual time series.

Furthermore, LOS measurements were decomposed into vertical and E-W motion components to facilitate the better interpretation of motion patterns. The E-W and vertical components were calculated using all three Sentinel-1 geometries in a comprehensive decomposition scheme, considering as well the variability of the incidence angles within the scenes (Figure 7). The procedure involved rasterization of MT-InSAR point vectors and calculation of the two components on common pixels. Then, spline interpolation was applied to fill gaps and obtain a more visually appealing result. It is worth noting that, for the comparison against modelling results and the calculation of misfits, only pixels presenting real and not interpolated data were considered.

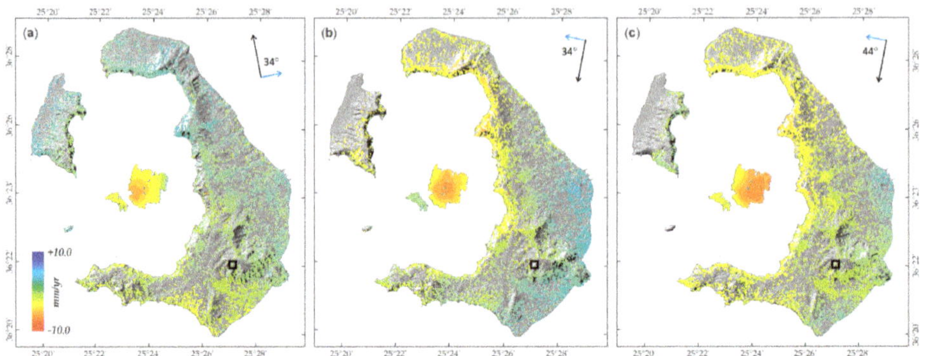

Figure 6. Sentinel-1 MT-InSAR LOS displacement rates after the correction for seasonal effects (**a**) Ascending orbit 029; (**b**) Descending orbit 109; (**c**) Descending orbit 036. Displacement rates are given relative to the reference point, marked by a rectangle. The line-of-sight and azimuth directions of the satellite are displayed by blue and black arrows, respectively. Values in degrees correspond to the incidence angles.

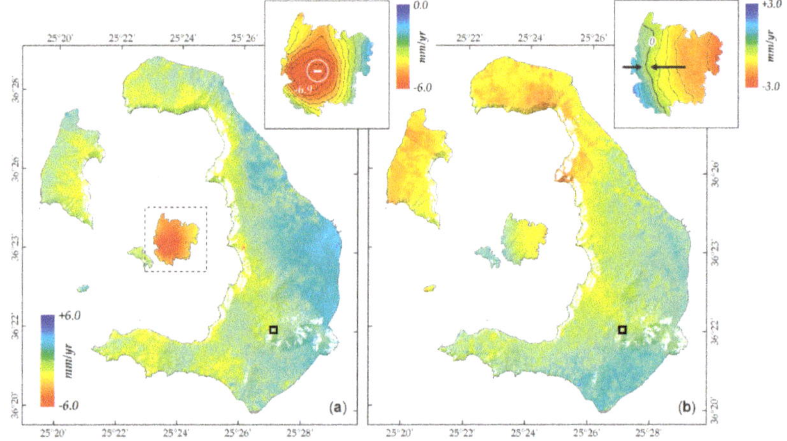

Figure 7. (**a**) Vertical and (**b**) East-West displacement rate maps of Santorini volcano for the period 2014–17, decomposed and interpolated, by combining Sentinel-1 LOS observations (see Figure 2). Displacement rates are given relative to the reference point, marked by a rectangle.

Prevalent subsidence for the entire Nea Kameni is indicated by the vertical motion map of Figure 7, reaching −7 mm/yr. The observed concentric deformation pattern presents an opening towards northern parts of the islet, where high deformation rates are also evident. Motion for the rest of the area has also been reduced after compensating for the seasonal signal, with overall rates (a couple of mm/yr), mostly within the uncertainties of the measurements. Concerning horizontal E-W motion, as expected, the N-S extending zero deformation zone on Nea Kameni is well aligned to the local concentric subsidence pattern, showing opposite motions on both sides and towards the area of maximum deformation. For the rest of the volcano, of significant interest is the relatively high westward motion of Cape Skaros, an area that accumulated the highest inflation rates during the 2011–12 unrest [4]. On the basis of the SAR Interferometry results, Nea Kameni remains the area exhibiting the highest ground deformation during the post-unrest period, with higher rates compared to the pre-unrest period (average subsidence rate of -5 ± 0.6 mm/yr for the 1992–2010 period) [3,4,56].

In addition to this, the ongoing deformation at Cape Skaros on the main island indicates that the volcano has not fully recovered yet.

Finally, regarding GEP PSI results and specifically those of Radarsat-2 (2012–16), having major overlap with the Sentinel-1 observations (2014–17), they are showing comparable motion in terms of both spatial pattern and magnitude (Figure 8a). Providing a complementary LOS measurement angle and sufficiently large time span for obtaining a robust solution, they were considered in the source modelling. On the contrary, for TerraSAR-X PSI (Figure 8b), and mainly due to the difference in the temporal coverage (2012–13), as well as the level of noise in the data, a decision was made not to include them in the inversion modelling. Nonetheless, they provided valuable information to underline the higher deflation rates observed during the early post-unrest phase. LOS displacement rates up to –12 mm/yr were observed in the center of the Nea Kameni islet with a visible concentric deformation pattern, whereas a similar localized deformation is denoted at the northern part of the islet. Apart from the higher deformation of TerraSAR-X data, due to the temporal proximity to the unrest, spatial deformation patterns are consistent with Sentinel-1 data, indicating the persistence of the deformation maxima areas since 2012.

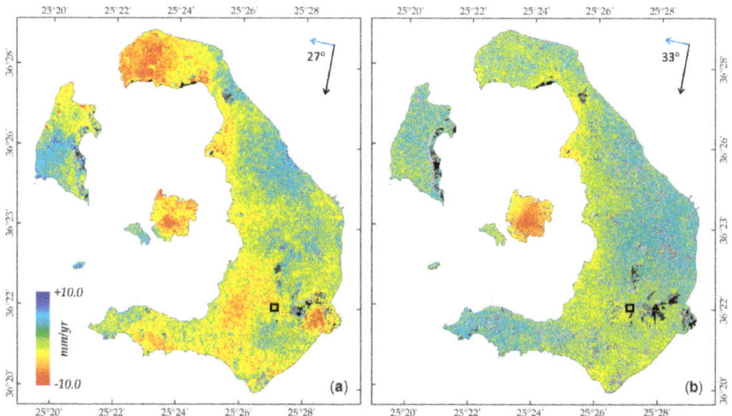

Figure 8. PSI average LOS velocities for (**a**) Radarsat-2 data (2012–2016), and (**b**) TerraSAR-X data (2012–2013), obtained from GEP [53]. The line-of-sight and azimuth directions of the satellite are displayed by blue and black arrows, respectively. Values in degrees correspond to the incidence angles.

5. Source Modelling

InSAR data show subsidence of the inner islets and, partially in the main island of Thera, across the inner walls (Figure 6). This pattern is quite new with respect to the pre-unrest, since 1992–2010 InSAR results show a subsidence pattern limited to the Kameni islets, and negligible deformation for the rest of the caldera [3,4,56,57]. Furthermore, the subsidence on the Kameni islets after the unrest extends to a wider area with respect to the pre-unrest period. During the unrest, higher uplift was concentrated on the main island at Cape Skaros, whereas Nea Kameni demonstrated comparable uplift rates of the northern part, with an observed tilting towards the south.

SAR modelling of a single deflating source attributed to the caldera-wide deformation yielded residuals in Nea Kameni justifying the use of a second source for the inversion (Figure S3). In order to take into account the localized subsidence at Nea Kameni and the caldera-wide deformation, two separated sources of deformation were considered. A joint inversion of mean velocity data was performed from ascending and descending Sentinel-1, together with descending Radarsat-2. The datasets were pre-processed by masking the areas undergoing layover issues, affecting in particular the inner cliffs in the ascending orbit of Sentinel-1 data. Afterwards, the four datasets were subsampled with a step of 60 m in the inner islets and 200–300 m in the remaining islands of Santorini.

The non-linear geodetic inversion was performed considering a 2-steps procedure based on the Neighbourhood Algorithm [59,60]. The VSM (Volcano and Seismic source Model) is a FORTRAN code retrieving not only the best-fit model, but also performing a second step with a Bayesian inference in the generated model ensemble. For the reasons described above, two sources were adopted, one for the caldera-wide deformation, and another one for the localized subsidence at Kameni islets. After several attempts aiming to minimize the residuals (a chi-square function) between data and models, we obtained a chi-square equal to 1.2, and the preferred source models are to be a sill-like source [58] for the caldera-wide deformation, and an isotropic source for the localized deformation in Nea Kameni [61,62].

Results are shown in Figures 9 and 10 by combining Sentinel-1 data for the East-West and vertical components. The modelled data is simply the East-West and vertical component of the mean model, without further interpolation. Detailed comparisons for all the datapoints jointly inverted are reported in Figure S4, while the posterior probability density functions are shown in Figures S5 and S6. The caldera-wide source is located below the sea, north of Kameni, at a depth of 2 km. It has a radius of a few hundred meters and underwent a volume variation rate of $-12{\cdot}10^4$ m^3 yr^{-1} during 2012–17 (Table 3). The volume variation is not inverted, but it is computed from the sill parameters, considering the medium as Poissonian. The second source is located below the area of maximum subsidence at Nea Kameni, at a depth of about 1 km, and characterized by a volume variation rate of $-1.4{\cdot}10^4$ m^3 yr^{-1}. The surface projection of the source center coincides with the change of direction of the East-West component, as expected for isotropic sources. The residuals show a trend of underestimation in the northern islands of Santorini for the horizontal component. However, the horizontal displacement is affected by larger errors than the vertical, amounting to few mm/yr. The vertical component shows some residuals across the northern cliffs, but as that area was affected by layover issues in the ascending orbit, it is therefore characterized by a larger uncertainty than the rest of the dataset.

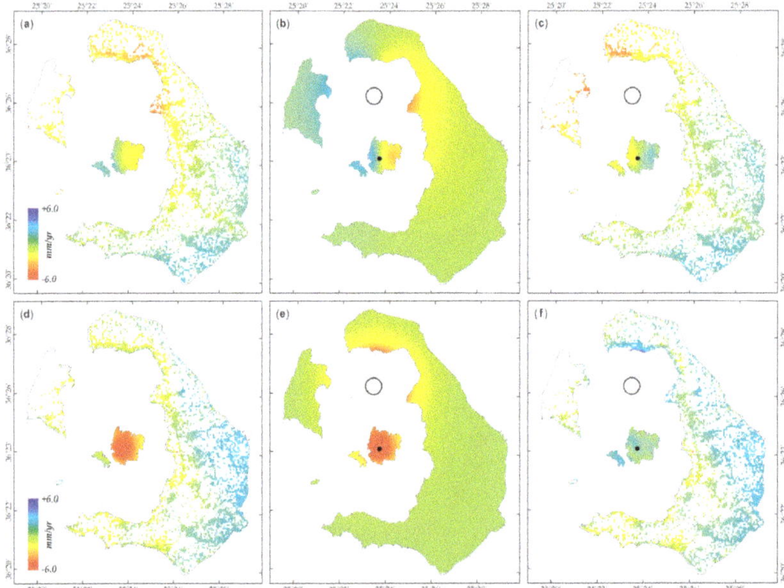

Figure 9. Results from the VSM (Volcano and Seismic source Model) inversion of the four datasets reported in Table 3. Data comparisons concern East-West and vertical components. In the first column are the East-West (**a**) and vertical (**d**) data, in the second column (**b**,**e**) the model, and in the third column (**c**,**f**) the residuals, respectively. Black open circle and black dot are the surface projections of the center of the caldera-wide source and the Nea Kameni source, respectively.

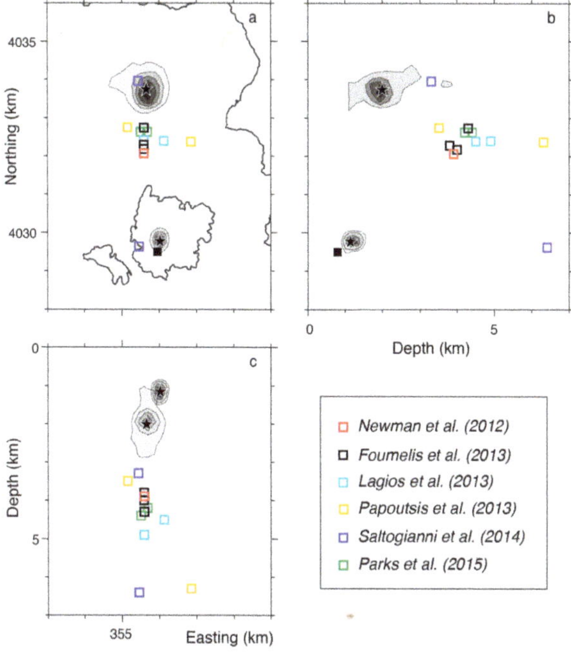

Figure 10. (**a–c**) Distribution of the deformation source location presented in this study, compared to previously published solutions. 2D posterior probability density functions of the two sources inverted are reported with 20% confidence contour. The mean models of this study are marked by a black star. Previous sources are reported as empty squares if referred to the unrest (inflation), filled squares if referred to the slow pre-unrest subsidence.

Table 3. Mean values of the inverted source parameters.

Model	E [a] km	N [a] km	Depth km	ΔV/t 10^4 m^3 yr^{-1}	Radius m	ΔP/μ/t 10^{-4} yr^{-1}
Sill (caldera—wide)	355.7 ± 0.3	4033.7 ± 0.3	2.0 ± 0.3	-12 ± 6 [b]	520 ± 200	-4 ± 3
Mogi (Kameni)	356.0 ± 0.1	4029.8 ± 0.2	1.1 ± 0.2	-1.4 ± 0.3	n.a.	n.a.

[a] Easting and Northing are in Universal Transverse Mercator (UTM) projection, zone 35N. [b] Volume variation rate not inverted, but computed from Reference [58].

6. Discussion

The temporal evolution of ground deformation at Santorini, after the unrest episode, seems to be controlled by a set of two shallow sources, favoring a more complex scenario of the volcano dynamics (Figure 11). The model proposed here, probed with four different C-band SAR observations from Sentinel-1 and Radarsat-2 missions, led to better constrain the deformation sources acting during the post-unrest period. The two-source model considers two deflating sources with a 4 km distance in between. The first source is a deflating, Mogi-like source located at about 1 km depth and centered below Kameni island. The second source is a sill-like body at 2 km depth located just above the deeper (about 4 km) source of the 2011–12 unrest episode [4–9]. According to the available geochemical data, an increase of CO_2 concentration has been observed after the unrest period [63]. This increase has been interpreted as an increase in the permeability of the volcanic pile along the Kameni fault resulting from the seismic activity during the unrest. According to these data, and taking into account the results of recent models of degassing from magma chambers [64], we propose that dynamics of our sill-like deflating source is related to the passive degassing (second boiling) of the magma intruded at 4–5 km depth during the unrest. This interpretation is consistent with the shallower depth (about 2 km) and geometry (sill-like) of the top of the magma reservoir, in which a volatile saturated zone is formed at the top by accumulation of gases associated to initial states of fractional crystallization and cooling [64–66]. This interpretation is also supported by the available geochemical data, which show a significant increase of CO_2 concentration values from the unrest, uplift period (May 2010–February 2012, CO_2 = 400 mmol/mol) to the post-unrest (deflation) period (March–July 2012; CO_2 = 800 mmol/mol) [62]. Such increase excludes the occurrence of drain-back of magma because, in that case, a decrease in the CO_2 concentration values should be observed.

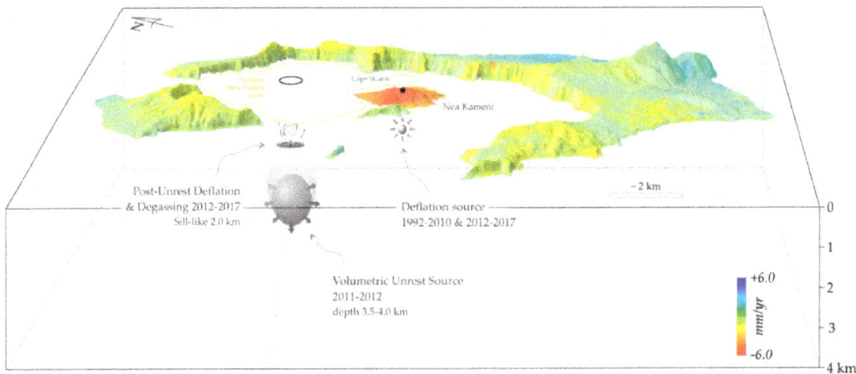

Figure 11. Schematic diagram summarizing SAR geodetic results presented herein and in Reference [4]. A shallow caldera-wide source (sill) at around 2 km depth reflects the post-unrest deflation of the deeper magma chamber responsible for the 2011–12 unrest episode. The volumetric spherical/spheroidal unrest source currently seems to undergo a depressurization due to degassing from its upper parts. A shallower deflating source (Mogi), at around 1 km depth, is also shown below Nea Kameni, exhibiting similar behavior for both the 1992–2010 pre-unrest and 2012–17 post-unrest periods.

Therefore, the degassing from this shallower source, which easily develops along the preferred pathway of the Kameni fault zone (e.g., References [28,63]) is responsible for the observed deflation of the 2 km deep sill-like source. The volume variation involved, as computed from the elastic models is in the order of 10^5 m^3/yr, which is 1/100 of the volume variation involved during the 2011–12 unrest. Regarding the nature of the degassing magma, Reference [67] suggest that a ^{3}He- and volatile-rich mafic magma intruded during the unrest. According to the conceptual model proposed above, such volatiles migrated in the uppermost portion of the reservoir and were passively released to the surface along the Kameni line. A component of deformation related to magma cooling can also be invoked for the observed subsidence, however, according to Reference [64], it plays a minor role.

Prior studies based on GPS and SAR observations for the unrest period [4–9], yielded a single inflating magma source in the northern intra-caldera area. While nearly all solutions follow almost N-S dispersion, they still indicate the same unrest center, if we take into consideration their location uncertainties (Figure 10). On the contrary, Reference [9] using GPS time series determined two source locations, one offshore, about 1 km northern of the N-S cluster solutions, with comparable depth, and a deeper one (7–8 km) on Nea Kameni islet. Our post-unrest deflating sill at the northern intra-caldera area is in good agreement with the unrest Mogi source of Reference [9], for their better constrained solution (period P1, Jan–Jun 2011), in terms of number of GPS stations and larger deformation gradients. On the whole, the inefficacy to well constrain magma sources using single SAR LOS observations, or GPS measurements that suffer from poor spatial sampling should be underlined. The utilization of multiple geodetic assets is thus beneficial and should be certainly taken into account when designing volcanic monitoring systems.

The InSAR results and modelling for the Nea Kameni deformation source of this study demonstrate a comparable pattern to previous ERS and ENVISAT observations covering the 1992–2010 pre-unrest period [3,4,56] (Figure 12a). The source's identical location and volume variation rate between the two intervals points to the hypothesis of a steady deformation signal. By assuming the pre-unrest deformation rate (ERS/ENVISAT) as background motion, and subtracting it from the Sentinel-1 post-unrest deformation rates, we were able to reveal the post-unrest response signal (Figure 12). Thereafter, to simplify the comparison between different LOS incident angles, we only considered the vertical component of the displacement.

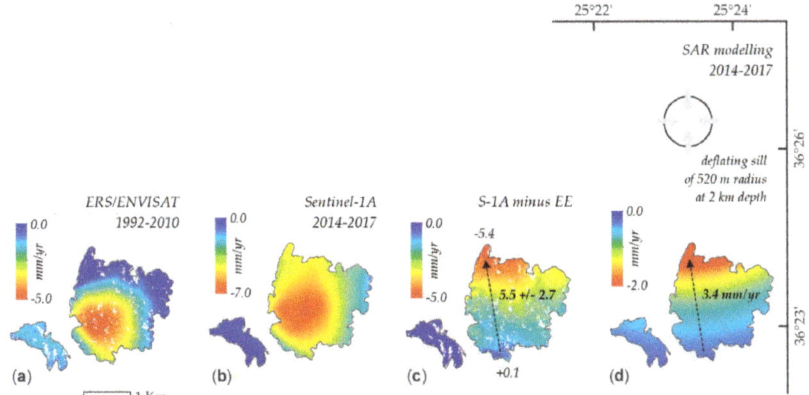

Figure 12. Vertical motion on Kameni islets derived from (**a**) ERS-ENVISAT observations for the 1992-2010 pre-unrest period [4]; (**b**) Sentinel-1 MT-InSAR observations for the 2014–17 post-unrest period; (**c**) Differences between Sentinel-1 post-unrest and ERS-ENVISAT pre-unrest motion rates; (**d**) Forward modelling results considering only the deflating 2014–17 post-unrest caldera-wide source. By compensating for the steady pre-unrest subsidence (**a**), Nea Kameni seems to undergo a tilt towards the north during the post-unrest period (**c**), fully compatible with the deformation pattern induced by the deflating caldera-wide source (**d**).

The outcome specifies a ramp tilting towards the north with rate of 5.5 $\pm$ 2.7 mm/yr. Modelling results adopting only the sill (the caldera-wide source) show an identical tilt at Nea Kameni (3.4 mm/yr), due to the deflation of the sill source at the northern caldera (Figure 12d). The computed tilt rate is within the uncertainty of SAR estimates. This demonstrates that local subsidence at Nea Kameni can be considered of the same amount during 1992–2010 and 2014–17, while the residual tilt of 2014–17 at Nea Kameni is attributed to the caldera-wide deflating source of the post-unrest period.

The existence of the local post-unrest shallow deformation source at Nea Kameni, having the same behavior as in the pre-unrest period in terms of both deformation rate and spatial pattern, suggests its continuous action during the unrest. Modelling misfits over Nea Kameni for the unrest period indicate a systematic over-estimation of both SAR and GPS data, as demonstrated in other studies only considering a single offshore magma source [4–6,9]. Therefore, the same volcanic activity is supposed to occur at Nea Kameni during the unrest, consistent with the pre-/post-unrest steady subsidence of -5 mm/yr [3,4]. However, such deformation cannot be isolated in the uplift data, since a total amount of 9 cm was measured at Nea Kameni during the unrest. In order to quantify the effect of the local source to the main source of the unrest, we ran a model with two sources. The first one was the unrest source as in Reference [4], and the other one was the Nea Kameni source as constrained in this work. The percentual change of the volume variation rate in the case of a single active source, as opposed to the case where both sources are active, amounts to 0.2%. So, in case the deflation at Nea Kameni is taken into account during the unrest, the inflating magma source is still located at the same position and depth (3.7 km), undergoing an increase of only 0.2% of the volume variation rate.

This is not surprising, since the unrest generated large displacements that were actually masking out the deformation at Nea Kameni. In our opinion this prevents an independent constraint of the Nea Kameni source from data, since its action is partially concealed by the unrest. The unrest source has a volume variation of the order of 10^7 m^3/yr, while the Nea Kameni source amounts to 10^4 m^3/yr, so it fits the effect of the Nea Kameni source on the unrest source being of the order of 10^{-3}.

The Nea Kameni deformation source is suggested by Reference [4] as the effect of variations within the shallow hydrothermal system, the existence of which is reported by Reference [68] at 800–1000 m depth. A different mechanism is suggested by Reference [57], whereas slow subsidence could reflect the thermal cooling and the load-induced relaxation of the substrate due to lava flows emitted between 1866 and 1870. In view of our SAR results, and the identification of the same slow subsidence signal before and after the unrest, this latter scenario could not also be ruled out. Based on geodetic data [9] and petrological studies [69,70], deeper volcanic-related processes beneath Nea Kameni are also proposed. If, over and above, we take into account that historic lava flows built up the intra-caldera islets of Palea and Nea Kameni and that the last lava emplacement took place in 1950, the involvement of still-ongoing deep processes beneath Kameni islets could support this hypothesis.

As a final remark, our results show that a change (decrease) in the subsidence rate at Nea Kameni could reflect the onset of a renewed unrest, and as a consequence, the geodetic monitoring of ground deformation represents a key to deciphering the dynamics of the Santorini shallower plumbing system, even for periods not accompanied by seismic activity or increased degassing. The Kameni deformations are therefore of primary importance to correctly interpret future unrest episodes and evaluate the volcanic hazard.

7. Conclusions

The MT-InSAR analysis of the dense temporal Sentine-1 data series allowed us to measure the post-unrest ground deformation of the Santorini volcano. The results provide additional insights on the recognition of an annual, seasonal periodicity in the displacement time series, affecting the accurate estimation of the deformation gradients. Whether this oscillation pre-existed during the unrest period, or even before, cannot be ascertained due to the non-systematic temporal coverage of former SAR acquisitions. Hence, this new estimation provides an important step for a more reliable assessment of the volcano deformation.

The new volcano state after the unrest period is confirmed by the geodetic analysis of multiple SAR sensor data (Sentinel-1, Radarsat-2 and TerraSAR-X) for the period 2012–17. The post-unrest response to the 2011–12 inflation episode is well explained by a shallow sill-like source at 2 km depth. This source is located just above the ~4 km deep inflation source responsible for the 2011–12 uplift. The wide observed deflation extending up to Thera Is. (Cape Skaros and northern caldera walls) provides evidence that the volcano apparently remains still under a recovery state. According to the geochemical and isotopic data on gas emissions, this deflation reflects the passive degassing of the shallow (top) portion of an intrusion of relatively poorly evolved magma emplaced during the unrest. The degassing pathways could be represented by the Kameni fault and fractures, which were re-activated by earthquakes during the unrest period, allowing an increase in local permeability.

The presence of a steady subsidence source at Nea Kameni, in accordance with the pre-unrest period, led to the re-evaluation of the 2011–12 unrest. Our interpretation model suggests the co-existence of the Kameni source during the unrest, having however a lower impact compared to the larger deformations induced by the inflation source.

Supplementary Materials: The following are available online at http://www.mdpi.com/2072-4292/11/3/259/s1, Text 1: Supplementary material of the manuscript. Figure S1: Temporal separation versus normal baseline plots displaying the InSAR pairs (connecting lines) considered in the MT-InSAR processing. Sentinel-1 scenes in blue color were not considered in the solutions. Figure S2: Sentinel-1 MT-InSAR LOS displacement rate uncertainties for (a) Ascending orbit 029; (b) Descending orbit 109; (c) Descending orbit 036. Uncertainties of estimated rates (see Figure 3) are given relative to the reference point, marked by a rectangle. The line-of-sight and azimuth directions of the satellite are displayed by blue and black arrows, respectively. Values in degrees correspond to the incidence angles. Figure S3: Comparison between subsampled data (first column) and modeled data (second column) for the single source model of the post-unrest. The third column reports the residuals (observed data minus modeled data). (a–c) Ascending orbit track 029 (a029); (d–f) Descending orbit track 036 (d036); (g–i) Descending orbit track 109 (d109); (j–l) Descending radarsat-2 data (drs2). The shaded circle is the surface projection of the sill-like source with its actual radius (about 2000 m). The chi-square for this model is 3.5. Figure S4: Comparison between subsampled data (first column) and modeled data (second column). The third column reports the residuals (observed data minus modeled data). (a–c) Ascending orbit track 029 (a029); (d–f) Descending orbit track 036 (d036); (g–i) Descending orbit track 109 (d109); (j–l) Descending radarsat-2 data (drs2). The star is the center of the Kameni source, the shaded circle is the surface projection of the sill-like source with its actual radius. Figure S5: Posterior Probability Density (PPD) functions retrieved by the non-linear joint inversion of the two sources. (a,b) Coordinates of the centre of the sill-like source (S), (c) its depth, (d) radius, (e) potency, i.e., overpressure ΔP vs rigidity μ; (f,g) Coordinates of the spherical source (Mogi source, M), (h) its depth, (i) its volume variation. Easting and Northing are in UTM-WGS84 projection, zone 35. The dashed line is the mean model. Figure S6: Two-dimensional PPD distributions. Each panel represents the 2D distribution of the inverted parameters. Contour lines every 10% confidence. The red cross is the mean model listed in Table 3. Each panel reports the numbering of two parameters for which the 2D-distribution is computed, considering the exact order as reported in Figure S5 and Table 3, i.e., a) = 1, b) = 2, c) = 3, d) = 4, e) = 5, f) = 6, g) = 7, h) = 8, i) = 9.

Author Contributions: Conceptualization, E.P. and M.F.; methodology, E.P., M.F. and E.T.; formal analysis, M.F. and D.R.; writing—original draft preparation, E.P.; writing—review and editing, E.P., M.F., E.T., G.V., D.R. and A.M. All the authors contributed in analyzing and interpreting the results.

Funding: This research was realized in the framework of the Operational Programme "Human Resources Development Program, Education and Lifelong Learning" of the State Scholarships Foundation, within the National Strategic Reference Framework (2014–20) for the Action "Strengthening Post Doctoral Research", co-financed by the European Social Fund—ESF and the Greek government.

Acknowledgments: The authors would like to thank Professor Costas Papazachos from the Aristotle University of Thessaloniki for the useful discussions over the seismicity of the study area. M. Parks is acknowledged for helpful insights. The suggestions and recommendations by anonymous reviewers that contributed to improving the manuscript are also acknowledged.

Conflicts of Interest: The authors declare no conflicts of interest.

References

1. Georgalas, G. L' éruption du volcan de Santorini en 1950. *Bull. Volcanol.* **1953**, *13*, 39–55. [CrossRef]
2. Permanent Regional Seismological Network operated by the Aristotle University of Thessaloniki. Available online: http://geophysics.geo.auth.gr/ss/station_index_en.html (accessed on 20 December 2018).
3. Papageorgiou, E.; Foumelis, M.; Parcharidis, I. Long- and Short-Term Deformation Monitoring of Santorini Volcano: Unrest Evidence by DInSAR Analysis. *IEEE JSTARS* **2012**, *5*, 1531–1537. [CrossRef]
4. Foumelis, M.; Trasatti, E.; Papageorgiou, E.; Stramondo, S.; Parcharidis, I. Monitoring Santorini volcano (Greece) breathing from space. *Geophys. J. Int.* **2013**, *193*, 161–170. [CrossRef]
5. Newman, A.V.; Stiros, S.; Feng, L.; Psimoulis, P.; Moschas, F.; Saltogianni, V.; Jiang, Y.; Papazachos, C.; Panagiotopoulos, D.; Karagianni, E.; et al. Recent geodetic unrest at Santorini Caldera, Greece. *Geophys. Res. Lett.* **2012**, *39*, L06309. [CrossRef]
6. Pyle, D.M.; Raptakis, C.; Zacharis, V. Evolution of Santorini Volcano dominated by episodic and rapid fluxes of melt from depth. *Nat. Geosci.* **2012**, *5*, 749–754.
7. Papoutsis, I.; Papanikolaou, X.; Floyd, M.; Ji, K.H.; Kontoes, C.; Paradissis, D.; Zacharis, V. Mapping inflation at Santorini volcano, Greece, using GPS and InSAR. *Geophys. Res. Lett.* **2013**, *40*, 267–272. [CrossRef]
8. Lagios, E.; Sakkas, V.; Novali, F.; Bellotti, F.; Ferretti, A.; Vlachou, K.; Dietrich, V. SqueeSAR^{TM} and GPS ground deformation monitoring of Santorini Volcano (1992–2012): Tectonic Implications. *Tectonophysics* **2013**, *594*, 38–59. [CrossRef]
9. Saltogianni, V.; Stiros, S.C.; Newman, A.V.; Flanagan, K.; Moschas, F. Time-space modeling of the dynamics of Santorini volcano (Greece) during the 2011–2012 unrest. *J. Geophys. Res. Solid Earth* **2014**, *119*. [CrossRef]
10. Hooper, A. A volcano's sharp intake of breath. *Nat. Geosci.* **2012**, *5*, 686–687. [CrossRef]
11. Papageorgiou, E.; Foumelis, M.; Mouratidis, A.; Papazachos, C. Sentinel-1 Monitoring of Santorini volcano post-unrest state. In Proceedings of the IGARSS 2018, Valencia, Spain, 23–27 July 2018.
12. Del Gaudio, C.; Aquino, I.; Ricciardi, G.P.; Ricco, C.; Scandone, R. Unrest episodes at Campi Flegrei: A reconstruction of vertical ground movements during 1905–2009. *J. Volcanol. Geotherm. Res.* **2010**, *195*, 48–56. [CrossRef]

13. Aiuppa, A.; Tamburello, D.; Di Napoli, R.; Cardellini, C.; Chiodini, G.; Giudice, G.; Grassa, F.; Pedone, M. First observations of the fumarolic gas output from a restless caldera: Implications for the current period of unrest (2005–2013) at Campi Flegrei. *Geochem. Geophys. Geosyst.* **2013**, *14*, 4153–4169. [CrossRef]
14. Wicks, C.W.; Thatcher, W.; Dzurisin, D.; Svarc, J. Uplift, thermal unrest and magma intrusion at Yellowstone caldera. *Nature* **2006**, *440*, 72–75. [CrossRef] [PubMed]
15. Perissoratis, C. The Santorini volcanic complex and its relation to the stratigraphy and structure of the Aegean Arc, Greece. *Marine Geology* **1995**, *128*, 37–58. [CrossRef]
16. Druitt, T.H.; Mellors, R.A.; Pyle, D.M.; Sparks, R.S.J. Explosive volcanism on Santorini, Greece. *Geol. Mag.* **1989**, *126*, 95–126. [CrossRef]
17. Fytikas, M.; Kolios, N.; Vougioukalakis, G.E. Post-Minoan volcanic activity on the Santorini volcano. Volcanic hazard and risk. Forecasting possibilities. In *Thera and the Aegean World III*; Hardy, D.A., Keller, J., Galanopoulos, V.P., Flemming, N.C., Druitt, T.H., Eds.; The Thera Foundation: London, UK, 1990; Volume 2, pp. 183–198.
18. Sakellariou, D.; Sigurdsson, H.; Alexandri, M.; Carey, S.; Rousakis, G.; Nomikou, P.; Georgiou, P.; Ballas, D. Active tectonics in the Hellenic volcanic Arc: The Kolumbo submarine volcanic zone. *Bull. Geol. Soc. Greece* **2011**, *2*, 1056–1063. [CrossRef]
19. Pyle, D.M.; Elliott, J. Quantitative morphology, recent evolution, and future activity of the Kameni Islands volcano, Santorini, Greece. *Geosphere* **2006**, *2*, 253–268. [CrossRef]
20. Pe-Piper, G.; Piper, D.J.W. The SouthAegean active volcanicArc: Relationships between magmatism and tectonics. *Develop. Volc.* **2005**, *7*, 113–133.
21. Pichler, H.; Kussmaul, S. Comments on the geological map of the Santorini islands. In *Thera and the Aegean World II*; Doumas, C., Ed.; The Thera Foundation: London, UK, 1980; pp. 413–427.
22. Vougioukalakis, G.; Mitropoulos, D.; Perissoratis, C.; Andrinopoulos, A.; Fytikas, M. The submarine volcanic centre of Coloumbo, Santorini, Greece. *Bull Soc Geol Greece XXX* **1994**, *3*, 351–360.
23. Nomikou, P.; Carey, S.; Papanikolaou, D.; Croff Bell, K.; Sakellariou, D.; Alexandri, M.; Bejelou, K. Submarine volcanoes of Kolumbo volcanic zone NE of Santorini Caldera, Greece. *Glob. Planet. Chang.* **2012**, *90*, 135–151. [CrossRef]
24. Fouqué, F. *Santorin et ses Eruptions*; Masson Et Cie: Paris, France, 1879; pp. 1–440.
25. Vougioukalakis, G.E.; Francalanci, L.; Sbrana, A.; Mitropoulos, D. The 1649–1650 Coloumbo submarine volcano activity, Santorini, Greece. In Proceedings of the European Laboratory Volcanoes; Barberi, F., Casale, R., Fratta, M., Eds.; European Commission, European Science Foundation: Luxembourg, 1995; pp. 189–192.
26. Stiros, S.C.; Psimoulis, P.; Vougioukalakis, G.; Fyticas, M. Geodetic evidence and modeling of slow, small-scale inflation episode in the Thera (Santorini) volcano caldera, AegeanSea. *Tectonophysics* **2010**, *494*, 180–190. [CrossRef]
27. Sakellariou, D.; Rousakis, G.; Sigurdsson, H.; Nomikou, P.; Katsenis, I.; CroffBell, K.; Carey, S. Seismic stratigraphy of Santorini's caldera: A contribution to the understanding of the Minoan eruption. In Proceedings of the 10th Symposium on Oceanography and Fisheries, Athens, Greece, 7–11 May 2012.
28. Papazachos, C.B.; Panagiotopoulos, D.; Newman, A.V.; Stiros, S.; Vougioukalakis, G.; Fytikas, M.; Laopoulos, T.; Albanakis, K.; Vamvakaris, D.; Karagianni, E.; et al. Quantifying the current unrest of the Santorini volcano: Evidence from a multiparametric dataset, involving seismological, geodetic, geochemical and other geophysical data. In Proceedings of the EGU General Assembly 2012, Vienna, Austria, 22–27 April 2012.
29. Hellenic Seismic Network (HL), Institute of Geodynamics, National Observatory of Athens (NOA). Available online: http://bbnet.gein.noa.gr (accessed on 20 December 2018).
30. Lu, Z.; Masterlark, T.; Power, J.A.; Wicks, C.; Dzurisin, D.; Thatcher, W. Subsidence at Kiska volcano, western Aleutians, detected by satellite radar interferometry. *Geophys. Res. Lett.* **2002**, *29*, 1855. [CrossRef]
31. Dzurisin, D. A comprehensive approach to monitoring volcano deformation as a window on the eruption cycle. *Rev. Geophys.* **2003**, *41*, 1001. [CrossRef]
32. Ferretti, A.; Prati, C.; Rocca, F. Nonlinear subsidence rate estimation using permanent scatterers in differential SAR interferometry. *IEEE TGRS* **2000**, *38*, 2202–2212. [CrossRef]
33. Ferretti, A.; Prati, C.; Rocca, F. Permanent scatterers in SAR interferometry. *IEEE TGRS* **2001**, *39*, 8–20. [CrossRef]

34. Berardino, P.; Fornaro, G.; Lanari, R.; Sansosti, E. A new algorithm for surface deformation monitoring based on small baseline differential SAR interferograms. *IEEE TGRS* **2002**, *40*, 2375–2383. [CrossRef]
35. Mora, O.; Mallorqui, J.J.; Broquetas, A. Linear and nonlinear terrain deformation maps from a reduced set of interferometric SAR images. *IEEE TGRS* **2003**, *41*, 2243–2253. [CrossRef]
36. Schmidt, D.A.; Bürgmann, R. Time-dependent land uplift and subsidence in the Santa Clara valley, California, from a large interferometric synthetic aperture radar data set. *J. Geophys. Res. Solid Earth* **2003**, *108*. [CrossRef]
37. Werner, C.; Wegmüller, U.; Strozzi, T.; Wiesmann, A. Interferometric point target analysis for deformation mapping. In Proceedings of the IGARSS 2003, Toulouse, France, 21–25 July 2003.
38. Duro, J.; Inglada, J.; Closa, J.; Adam, N.; Arnaud, A. High resolution differential interferometry using time series of ERS and ENVISAT SAR data. In Proceedings of the FRINGE 2003 Workshop, Frascati, Italy, 1–5 December 2003.
39. Lanari, R.; Mora, O.; Manunta, M.; Mallorquí, J.J.; Berardino, P.; Sansosti, E. A small-baseline approach for investigating deformations on full-resolution differential SAR interferograms. *IEEE TGRS* **2004**, *42*, 1377–1386. [CrossRef]
40. Hooper, A.; Zebker, H.; Segall, P.; Kampes, B. A new method for measuring deformation on volcanoes and other natural terrains using InSAR persistent scatterers. *Geophys. Res. Lett.* **2004**, *31*, L23611. [CrossRef]
41. Crosetto, M.; Biescas, E.; Duro, J.; Closa, J.; Arnaud, A. Generation of advanced ERS and Envisat interferometric SAR products using the stable point network technique. *Photogram. Eng. Remote Sens.* **2008**, *74*, 443–450. [CrossRef]
42. Ferretti, A.; Fumagalli, A.; Novali, F.; Prati, C.; Rocca, F.; Rucci, A. A new algorithm for processing interferometric data-stacks: SqueeSAR. *IEEE TGRS* **2011**, *49*, 3460–3470. [CrossRef]
43. Perissin, D.; Wang, T. Repeat-pass SAR interferometry with partially coherent targets. *IEEE TGRS* **2012**, *50*, 271–280. [CrossRef]
44. Van Leijen, F. Persistent Scatterer Interferometry Based on Geodetic Estimation Theory. Doctoral Dissertation, Delft University of Technology, Delft, The Netherlands, 2014.
45. Goel, K.; Adam, N. A distributed scatterer interferometry approach for precision monitoring of known surface deformation phenomena. *IEEE TGRS* **2014**, *52*, 5454–5468. [CrossRef]
46. Crosetto, M.; Monserrat, O.; Cuevas-González, M.; Devanthéry, N.; Crippa, B. Persistent Scatterer Interferometry: A review. *ISPRS J. Photogramm. Remote Sens.* **2016**, *115*, 78–89. [CrossRef]
47. Wegmüller, U.; Werner, C.; Strozzi, T.; Wiesmann, A.; Frey, O.; Santoro, M. Sentinel-1 Support in the GAMMA Software. *Procedia Comput. Sci.* **2016**, *100*, 1305–1312. [CrossRef]
48. De Zan, F.; Monti Guarnieri, A. TOPSAR: Terrain observation by progressive scans. *IEEE TGRS* **2006**, *44*, 2352–2360. [CrossRef]
49. Yague-Martinez, N.; Prats-Iraola, P.; Rodriguez Gonzalez, F.; Brcic, R.; Shau, R.; Geudtner, D.; Eineder, M.; Bamler, R. Interferometric Processing of Sentinel-1 TOPS Data. *IEEE TGRS* **2016**, *54*, 2220–2234. [CrossRef]
50. Prats-Iraola, P.; Nannini, M.; Yague-Martinez, N.; Scheiber, R.; Minati, F.; Vecchioli, F.; Costantini, M.; Borgstrom, S.; De Martino, P.; Siniscalchi, V.; et al. Sentinel-1 TOPS interferometric time series results and validation. In Proceedings of the IGARSS 2016, Beijing, China, 10–15 July 2016.
51. Scheiber, R.; Moreira, A. Coregistration of interferometric SAR images using spectral diversity. *IEEE TGRS* **2000**, *38*, 2179–2191. [CrossRef]
52. Geohazards Lab Intitiave. Available online: http://ceos.org/ourwork/workinggroups/disasters/geohazards-lab/ (accessed on 20 December 2018).
53. El Saer, A.; Diakogianni, G.; Papoutsis, I. Systematic PSI processing over Santorini volcano: The 2012–2016 post unrest period [Data set]. *Zenodo* **2017**. [CrossRef]
54. Samsonov, S.V.; Trishchenko, A.P.; Tiampo, K.; González, P.J.; Zhang, Y.; Fernández, J. Removal of systematic seasonal atmospheric signal from interferometric synthetic aperture radar ground deformation time series. *Geophys. Res. Lett.* **2014**, *41*, 6123–6130. [CrossRef]
55. Hu, Z.; Mallorqui, J.J. Direct method to estimate atmospheric phase delay for InSAR with global atmospheric models. In Proceedings of the IGARSS 2018, Valencia, Spain, 23–27 July 2018.
56. Papageorgiou, E.; Foumelis, M.; Parcharidis, I. SAR Interferometric analysis of ground deformation at Santorini Volcano (Greece). In Proceedings of the FRINGE 2011 Workshop, Frascati, Italy, 19–23 September 2011.

57. Parks, M.M.; Moore, J.D.P.; Papanikolaou, X.; Biggs, J.; Mather, T.A.; Pyle, D.M.; Raptakis, C.; Paradissis, D.; Hooper, A.; Parsons, B.; et al. Evolution From quiescence to unrest: 20 years of satellite geodetic measurements at Santorini volcano, Greece. *J. Geophys. Res. Solid Earth* **2015**, *120*, 1309–1328. [CrossRef]
58. Fialko, Y.; Khazan, Y.; Simons, M. Deformation due to a pressurized horizontal circular crack in an elastic half-space, with applications to volcano geodesy. *Geophys. J. Int.* **2001**, *146*, 181–190. [CrossRef]
59. Sambridge, M. Geophysical inversion with a neighbourhood algorithm-I. Searching a parameter space. *Geophys. J. Int.* **1999**, *138*, 479–494. [CrossRef]
60. Sambridge, M. Geophysical inversion with a neighbourhood algorithm-II. Appraising the ensemble. *Geophys. J. Int.* **1999**, *138*, 727–746. [CrossRef]
61. Mogi, K. Relations between the eruptions of various volcanoes and the deformations of the ground surfaces around them. *Bull. Earthq. Res. Inst. Univ. Tokyo* **1958**, *36*, 99–134.
62. McTigue, D.F. Elastic stress and deformation near a finite spherical magma body: Resolution of the point source paradox. *J. Geophys. Res.* **1987**, *92*, 931–940. [CrossRef]
63. Tassi, F.; Vaselli, O.; Papzachos, C.B.; Giannini, L.; Chiodini, G.; Vougioukalakis, G.E.; Karagianni, E.; Vamvakaris, D.; Panagiotopoulos, D. Geochemical and isotopic changes in the fumarolic and submerged gas discharges during the 2011–2012 unrest at Santorini caldera (Greece). *Bull. Volcanol.* **2013**, *75*, 1–15. [CrossRef]
64. Edmonds, M.; Woods, A. Exsolved volatiles in magma reservoirs. *J. Volcanol. Geotherm. Res.* **2018**, *368*, 13–30. [CrossRef]
65. Lowenstern, J.B.; Sinclair, D.W. Exsolved magmatic fluid and its role in the formation of comb-layered quartz at the Cretaceous Logtung W-Modeposit, Yukon Territory, Canada. *Trans. R. Soc. Edinburgh Earth Sci.* **1996**, *87*, 291–303. [CrossRef]
66. Candela, P.A. A review of shallow, ore-related granites: Textures, volatiles, and ore metals. *J. Petrol.* **1997**, *38*, 1619–1633. [CrossRef]
67. Rizzo, A.L.; Barberi, F.; Carapezza, M.L.; Di Piazza, A.; Francalanci, L.; Sortino, F.; D'Alessandro, W. New mafic magma refilling a quiescent volcano: Evidence from He-Ne-Ar isotopes during the 2011–2012 unrest at Santorini, Greece. *Geochem. Geophys. Geosyst.* **2015**, *16*. [CrossRef]
68. Fytikas, M.; Karydakis, G.; Kavouridis, T.H.; Kolios, N.; Vougioukalakis, G. Geothermal research on Santorini. In *Thera and the Aegean World III*; Hardy, D.A., Keller, J., Galanopoulos, V.P., Flemming, N.C., Druitt, T.H., Eds.; The Thera Foundation: London, UK, 1990; Volume 2, pp. 241–249.
69. Cadoux, A.; Scaillet, B.; Druitt, T.H.; Deloule, E. Magma Storage Conditions of Large Plinian Eruptions of Santorini Volcano (Greece). *J. Petrol.* **2014**, *55*, 1129–1171. [CrossRef]
70. Browning, J.; Drymoni, K.; Gudmundsson, A. Forecasting magma-chamber rupture at Santorini volcano, Greece. *Sci. Rep.* **2015**, *5*, 15785. [CrossRef] [PubMed]

Article

The 2014–2015 Lava Flow Field at Holuhraun, Iceland: Using Airborne Hyperspectral Remote Sensing for Discriminating the Lava Surface

Muhammad Aufaristama [1,*], Armann Hoskuldsson [1], Magnus Orn Ulfarsson [2], Ingibjorg Jonsdottir [1,3] and Thorvaldur Thordarson [1,3]

1 Institute of Earth Sciences, University of Iceland, Sturlugata 7, 101 Reykjavík, Iceland; armh@hi.is (A.H.); ij@hi.is (I.J.); torvth@hi.is (T.T.)
2 Faculty of Electrical and Computer Engineering, University of Iceland, Hjardarhagi 2-7, 107 Reykjavik, Iceland; mou@hi.is
3 Faculty of Earth Sciences, University of Iceland, Sturlugata 7, 101 Reykjavík, Iceland
* Correspondence: mua2@hi.is; Tel.: +354-855-1242

Received: 31 January 2019; Accepted: 18 February 2019; Published: 26 February 2019

Abstract: The Holuhraun lava flow was the largest effusive eruption in Iceland for 230 years, with an estimated lava bulk volume of ~1.44 km^3 and covering an area of ~84 km^2. The six month long eruption at Holuhraun 2014–2015 generated a diverse surface environment. Therefore, the abundant data of airborne hyperspectral imagery above the lava field, calls for the use of time-efficient and accurate methods to unravel them. The hyperspectral data acquisition was acquired five months after the eruption finished, using an airborne FENIX-Hyperspectral sensor that was operated by the Natural Environment Research Council Airborne Research Facility (NERC-ARF). The data were atmospherically corrected using the Quick Atmospheric Correction (QUAC) algorithm. Here we used the Sequential Maximum Angle Convex Cone (SMACC) method to find spectral endmembers and their abundances throughout the airborne hyperspectral image. In total we estimated 15 endmembers, and we grouped these endmembers into six groups; (1) basalt; (2) hot material; (3) oxidized surface; (4) sulfate mineral; (5) water; and (6) noise. These groups were based on the similar shape of the endmembers; however, the amplitude varies due to illumination conditions, spectral variability, and topography. We, thus, obtained the respective abundances from each endmember group using fully constrained linear spectral mixture analysis (LSMA). The methods offer an optimum and a fast selection for volcanic products segregation. However, ground truth spectra are needed for further analysis.

Keywords: hyperspectral; FENIX; lava field; SMACC; LSMA

1. Introduction

Lava flow emplacement is an important constructive geological process that contributes to reshaping natural landscapes [1–3]. To assess the hazards and long-term impacts posed by lava flows, it is vital to understand aspects such as the return period of effusive eruptions, to map the areas covered by eruptions in the past and to characterize the evolution of lava flow surfaces after emplacement [4,5]. In high eruption frequency areas, lava flows often overlap each other. If the overlapping lava flows erupt within a short time span and have similar chemical and surface characteristics, discrimination will be further complicated by their similar spectral signatures. Spectral reflectance plays an important role in visible and shortwave infrared (VIS-SWIR) remote sensing. Each material absorbs and reflects the incoming radiation in a characteristic way. In the 400–2500 nm range, minerals display absorption features due to the interaction of light with cations (Fe, Mg, Al) and anions (OH, CO_3) [6]. Reflectance spectra provide information about the specific material and their composition. They are used for different applications such as classification of remotely sensed data, identification of mineral features of rock, and environmental assessment [7,8]. The interest in reflectance spectra of volcanic rocks has increased recently as they can play an important role as planetary analogues. In fact, these spectra can be used to identify compounds by data acquired by ongoing solar system exploration missions [9,10].

Characterization of surface spectral reflectance by satellite remote sensing is constrained by the spectral range and resolution (i.e., number of spectral bands) as well as by the spatial resolution of the imagery. Whereas multispectral imagery can be acquired at very high spatial resolution (e.g., WorldView [11,12]); the spatial resolution of hyperspectral satellite data remains low (e.g., EO-1 Hyperion with a ground resolution of 30 m x 30 m); and spectral mixing is thus a major issue [13]. The spectral reflectance of lava of different compositions has also been documented using laboratory spectrometry with decimeter-size samples [14]. For accessible volcanic terrains, field spectrometry offers a useful alternative approach for characterizing the spectral reflectance of contrasted lava surfaces and for documenting its spatial variation at different spatial scales [5,14]. The great variety of morphologies observed in the 2014–2015 Holuhraun lava flows [1,15] encouraged a detailed study of their spectral characteristics, to obtain information about lava composition and detect possible differences in the spectra of the flow. In spectroscopy, the identification of the mineral constituents of major rock types is typically approached using spectral unmixing methods [5,16]. Usually, in the visible and near-infrared spectral range, mafic rocks are characterized by very low reflectance due to the presence of large amounts of dark mafic minerals [14]. The 2014–2015 lava flow at Holuhraun in NE Iceland offers an excellent diverse surface environment for investigating and characterizing lava deposits. Its intense volcanic activity [1,17–19], geomorphological complexity [20], and well-documented flank eruptions [1] perplex the remote sensing monitoring of the bulk volcanic edifice. However, the detailed field mapping of lithologies is frequently obstructed by difficulties in accessibility, the scale of lava flow fields, topography, while remote sensing has become increasingly important in mapping volcanic terrains and specifically in mapping lava flows. Mapping individual lava flows using satellite remote sensing is challenging for at least three reasons: vegetation cover, spatial overlapping, and spectral similarity [3,4]. Moreover, a high eruption frequency often leads to lava flows overlapping each other. If the overlapping lava flows are erupted within a short period and have similar chemical and surface characteristics, discrimination will be further complicated by their similar spectral signatures.

Hyperspectral remote sensing provides information on hundreds of distinct and contiguous channels of the electromagnetic spectrum, thus enabling the identification of multiple ground objects through their detailed spectral profiles. However, restrictions on the spatial resolution of hyperspectral data, the multiple scattering of the incident light between objects, and microscopic material mixing form the mixed pixel problem. Pixels are identified as mixed when they are composed of the spectral signatures of more than one ground object. Therefore, we adopted linear spectral mixture analysis (LSMA) techniques [8,21], which model the pixel spectra as a combination of pure components (endmembers) weighted by the fractions (abundances) that contribute to the total reflectance of the mixed pixel [22]. Ideally, each selected endmember from the hyperspectral image under study has the maximum possible abundance of a single physical material present and minimum abundance of the rest of the physical materials. Spectral unmixing typically consists of two main substages: (a) endmember extraction; and (b) abundance estimation [22]. In this paper, we focus on both endmember extraction and estimation of fractional abundances of the lava field products on 2014–2015 Holuhraun lava fields. For this purpose, an airborne hyperspectral image with an AisaFENIX sensor on board a NERC Airborne Research Facility (Natural Environment Research Council Airborne Research Facility) campaign was acquired at Holuhraun after the eruption and for the sub-pixel analysis we used the sequential maximum angle convex cone (SMACC) algorithm to identify the spectral image endmembers while the LSMA method was employed to retrieve the abundances. Our approach was narrowed to the eruptive fissure vent part since it is considered to have a more diverse surface. The resulting abundances from the LSMA method were both quantitatively and qualitatively compared with the spectral indices technique, aerial and field photographs, respectively. The objective was to retrieve the main lava surface type contributing to the signal recorded by airborne hyperspectral at the very top surface of Holuhraun.

2. The 2014–2015 Eruption at Holuhraun

The eruption took place in the tectonic fissure swarm between the Bárðarbunga-Veiðivötn and the Askja volcanic systems (Figure 1a). It lasted about six months (31 August 2014 to 27 February 2015) and produced a bulk volume ~1.44 km^3 of basaltic lava [1]. Lava effusion rates during the eruption period range from 320 to 10 m^3/s. Averaged values are ∼250, 100, and 50 m^3/s during the initial (August–September 2014), intermediate (October–December 2014) and final phase (December 2014 to February 2015), respectively [1,17] (Figure 1b). The lava was emplaced on the sandur plains (glacial outwash sediment plains) north of the Vatnajökull/Dyngjujökull glacier, partially covering the previous two Holuhraun lava flow fields south of the Askja caldera [1]. The area is gently sloping (average inclination <0.5%; i.e., ∼0.3°) to the east-northeast. The shallow gradient resulted in low topographic forcing of the flow and, therefore, rather slow lava flow advance. During its emplacement history, the lava field was initially dominated by channels and horizontal expansion. Then it transitioned to grow in volume primarily by inflation, tube-fed flow (i.e., transport of lava through roofed over partially or filled channels) and vertical stacking of lava-lobes. The 2014–2015 effusive eruption products originate from intense activity in the vent, in which high oxidation occurs in this area. The main lava channel shows significant inflation (5–10 m). Lava advancement rates were generally low ∼0.0167 m/s during the initial eruption phase [1] and dropped to ∼0.0017 m/s during the middle of November 2014 [23]. The six-month-long effusive eruption features diverse surface structures and morphologies. The 2014–2015 lava flow at Holuhraun in NE Iceland offers an excellent diverse surface environment to investigate and characterize lava deposits.

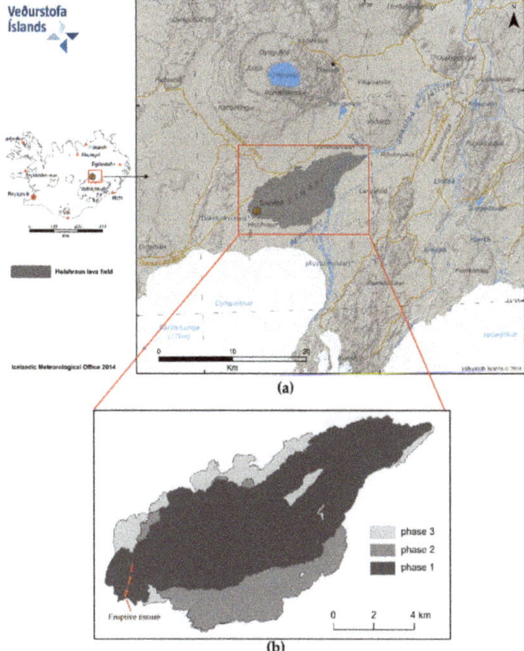

Figure 1. Bárðarbunga volcano and the Holuhraun lava flow field. (**a**) geological setting by the Icelandic Meteorological Office (after modification) [24], (**b**) coverage of the three main phases after Pedersen et al [1].

3. Spectral Unmixing on Lava

Various spectroscopy studies [2,5,7,14,25] over the volcanic area have examined the mineralogical composition of the extensive lava fields. Usually, in the visible (VIS) and near-infrared (NIR) spectral range, mafic rocks are characterized by very low reflectance due to the presence of large amounts of dark mafic minerals [14]. Spectral indices provide the first efficient way to emphasize subtle spectral variations at the surface [26]. More elaborate methods have been developed to discriminate and quantify mixtures of mafic minerals. They have been used to derive composition maps of mafic minerals [27–29]. However, some lava flows can have a similar chemical/mineralogical composition but dissimilar spectral behaviour due to the different grain size, surface texture, and presence of weathering [13,14]. The main components of igneous rocks do not display any peculiar spectral features in the visible and near infrared spectral range. In the case of basalts, the only spectral feature commonly found is an absorption peak, due to iron, located around 1000 nm [26]. However, in the case of hydrothermal alteration, hydroxyl bearing minerals show distinctive absorption features in the 2000–2500 nm spectral region [30]. Because of the heterogeneity of the lava surface, mixed pixels are very common which is illustrated in Figure 2a,b.

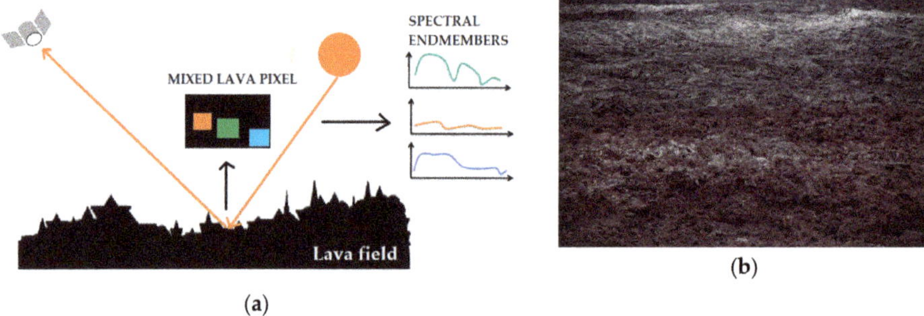

Figure 2. Illustration of (**a**) the mixed pixel in the lava surface caused by the presence of small, sub pixel targets within the area; (**b**) variability of lava surfaces in Holuhraun lava field which include the oxidizing surface, sulfate mineral, and lava.

Spectral Mixing Analysis (SMA) has been specifically developed to account for mixtures [10]. Analysis of the data sample can simply be performed on these abundance fractions rather than the sample itself. This method is well-suited for spectroscopic analysis because most of the spectral shapes are due to different materials. The signal detected by a sensor at a single pixel is frequently a combination of numerous disparate signals. Unmixing techniques were applied to the volcano of Nyamuragira for discriminating lava flows of different ages by Li et al. [5]. The most recent study by Daskalopoulou et al. [16], used unmixing techniques to segregate lava flows and related products from the historical Mt. Etna. Nonetheless, there are no findings concerning lava flow delineation through unmixing in Iceland.

4. Data Acquisitions and Methods

4.1. Airborne Hyperspectral Data Acquisitions

Airborne hyperspectral data were acquired on 4 September 2015 between 16.56 and 17.58 (local time) with an AisaFENIX sensor (Specim, Spectral Imaging Ltd, http://www.specim.fi) [31] on board a NERC Airborne Research Facility (Natural Environment Research Council Airborne Research Facility http://www.bas.ac.uk/nerc-arf) aircraft [32]. Pushbroom VNIR and SWIR sensor, are two separate detectors with common fore-optics. The hyperspectral data contain 622 channels with spectral range from ~400 nm to 2500 nm (break at ~970 nm). The pixel size of this data is explained in Section 4.2.2.

In total, eight flights were acquired at the Holuhraun lava flow during this period with an average altitude of 2.4 km (Figure 3a). The data are delivered as level 1b ENVI BIL format files which means that radiometric calibration algorithms have been applied and navigation information has been synced to the image data (Figure 3b). In this study, we subset the data to focus on the area around the eruptive fissures vent (Figure 3c) which is thought to have a diverse surface and has field photographs. Very high-resolution aerial photographs of the lava field (0.5 m spatial resolution) from Loftmyndir ehf (http://www.loftmyndir.is/) [33] were used for comparison and validation of the unmixing results.

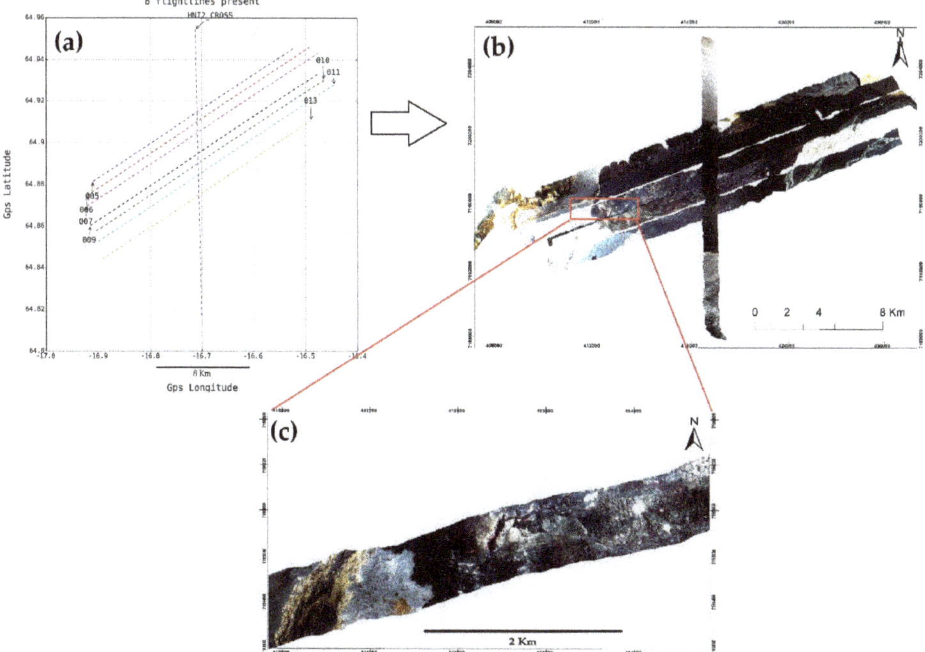

Figure 3. (**a**) Map showing line acquisition of FENIX hyperspectral image in the Holuhraun lava field; (**b**) Image mosaic from eight FENIX lines collected during the campaign (red box shows the image subset location); (**c**) Image subset of the focusing study area in the eruptive fissure vent of Holuhraun.

4.2. Spectral Unmixing and Abundance Retrieval

The processing workflow towards unmixing and generating abundance consists of four steps: (1) Atmospheric correction to retrieve surface reflectance; (2) Data masking, geocorrection, reprojection, and resampling; (3) An endmember selection algorithm was adopted to select the endmembers; then a linear spectral mixing analysis method was employed to retrieve the abundance (Figure 4).

4.2.1. Atmospheric Correction

Remote-sensing applications require removing the atmospheric effect from the imagery, to retrieve the spectral reflectance of the surface materials. In this study, the data were atmospherically corrected using the quick atmospheric correction (QUAC) algorithm [34,35], since we had no prior knowledge to perform empirical calibration [36,37]. QUAC is an in-scene approach, requiring only an approximate specification of sensor band locations (i.e., central wavelengths) and their radiometric calibration; no additional metadata is required [35]. QUAC does not involve first principles radiative transfer calculations, and therefore it is significantly faster than physics-based methods; however, it is also more approximate [35].

4.2.2. Data Masking, Geocorrection, Reprojection, and Resampling

In this study, we use the Airborne Processing Library (APL) software for processing the data [38]. The first step of the APL processing is to apply the mask of bad channels to atmospherically corrected data, creating a new file with bad channels set to zero (Appendix A on Figure A1). The next step uses the navigation file, the view vector file, and the digital elevation file (DEM) to calculate the ground position for each pixel then change the projection to UTM (Universal Transverse Mercator) Zone 28N [38]. We used satellite-based ASTER sensor for the DEM. In the final step we resampled output pixel size to ~3.5 m according to the height above ground level (AGL) that is given by the theoretical pixel size chart that can be found in Appendix A on Figure A2 (https://nerc-arf-dan.pml.ac.uk/trac/wiki/Processing/PixelSize) [39].

4.2.3. Endmembers Selection

The conventional image-based endmember selection approach based on scatterplots of the image bands may not be effective in identifying a sufficient number of endmembers. In this paper, we employed the sequential maximum angle convex cone (SMACC) algorithm [34] to identify spectral image endmembers. Endmembers are spectra that represent pure surface materials in a spectral image. The extreme points were used to determine a convex cone, which defined the first endmember. A constrained oblique projection was applied to the existing cone to derive the next endmember. The cone was then increased to include a new endmember [8,40]. This process was repeated until a projection derived an endmember that already existed within the convex cone, or until a specified number of endmembers was satisfied [21]. When implemented with SMACC, the output endmember number was set as 5, 10, 15, 20, and 30 respectively. Better endmembers could be identified easily from the 15 endmembers output (more detail in Section 6.2). Then, we used the selected 15 endmembers for deriving the abundance.

4.2.4. Linear Spectral Mixture Analysis

The linear spectral mixture analysis (LSMA) approach was adopted to calculate the abundance of endmembers for each pixel. LSMA assumes that the spectrum measured by a sensor is a linear combination of the spectra of all components (endmembers) within the pixel, and the spectral proportions of the endmembers (i.e., their abundance) reflect the proportion of area covered by distinct features on the ground [8,21]. The general equation for linear spectral mixing can be expressed as:

$$\boldsymbol{R}_{ij,\lambda} = \sum_{n=1}^{N} \boldsymbol{p}_{ij,n}\boldsymbol{R}_{n,\lambda} + \boldsymbol{E}_{\lambda} \quad (1)$$

where $\boldsymbol{R}_{ij,\lambda}$ is the measured reflectance at wavelength $\boldsymbol{\lambda}$ for pixel $\boldsymbol{ij}$, where $\boldsymbol{i}$ is the column pixel number, and $\boldsymbol{j}$ is the line pixel number; $\boldsymbol{p}_{ij,n}$ is the fraction of endmembers $\boldsymbol{n}$ contributing to the image spectrum of pixel $\boldsymbol{ij}$; $\boldsymbol{N}$ is the total number of endmembers; $\boldsymbol{R}_{n,\lambda}$ is the reflectance of endmember $\boldsymbol{n}$ at wavelength $\boldsymbol{\lambda}$; and $\boldsymbol{E}_{\lambda}$ is the error at wavelength $\boldsymbol{\lambda}$ of the fit of $\boldsymbol{N}$ spectral endmembers. The fraction $\boldsymbol{p}_{ij,n}$ can be solved using a least-square method with fully constrained unmixing. Fully constrained unmixing means that the sum of the endmember fractional (abundance) values for each pixel must equal unity, which requires a complete set of endmembers. Therefore, it should meet the following two conditions:

$$0 \leq \boldsymbol{p}_{ij,n} \leq 1 \quad (2)$$

$$\sum_{n=1}^{N} \boldsymbol{p}_{ij,n} = 1 \quad (3)$$

In the majority of cases, the unmixing is only partially constrained because the extracted endmember set is incomplete for the image and only term (2) (i.e., Equation (2)) is satisfied. In this study, fully constrained LSMA were applied to the FENIX image to obtain the abundance result and both SMACC and LSMA were executed by ENVI 5.3 and IDL 8.5 language programming.

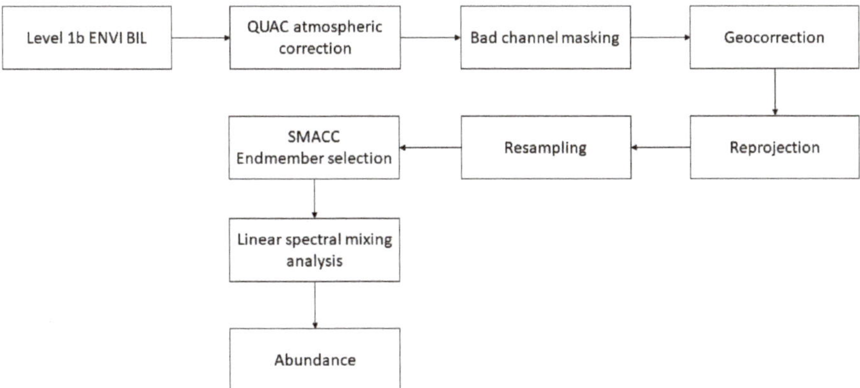

Figure 4. The workflow processing to derive an abundance map from the FENIX hyperspectral data

5. Results

5.1. Endmember Groups

The approximate locations of the 15 endmembers selected are shown in Figure 5a. SMACC first finds the brightest spectral in the image and defines it as the first endmember. In this study, the first endmember (endmember 1) represented saturated hot material. We grouped these 15 endmembers into six groups; (1) basalt; (2) hot material; (3) oxidized surface; (4) sulfate mineral; (5) water; and (6) noise (Figure 5b–g). These groups were based on the similar shape of the endmembers with the USGS spectral library; however, the amplitude of the endmembers within a group vary due to illumination conditions, spectral variability, and topography. We added up the abundances within the group to derive the abundance according to this endmembers group.

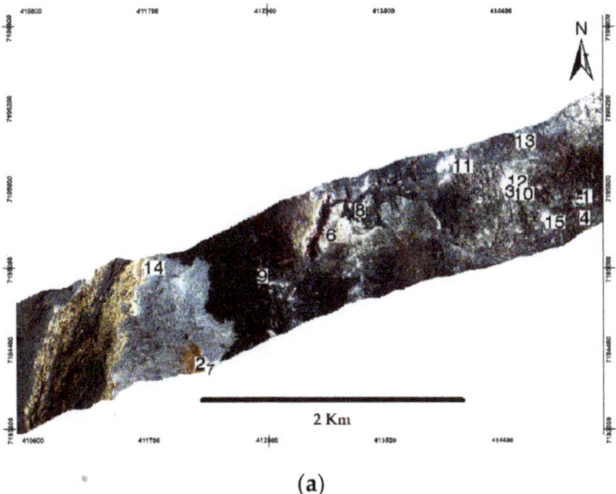

(a)

Figure 5. *Cont.*

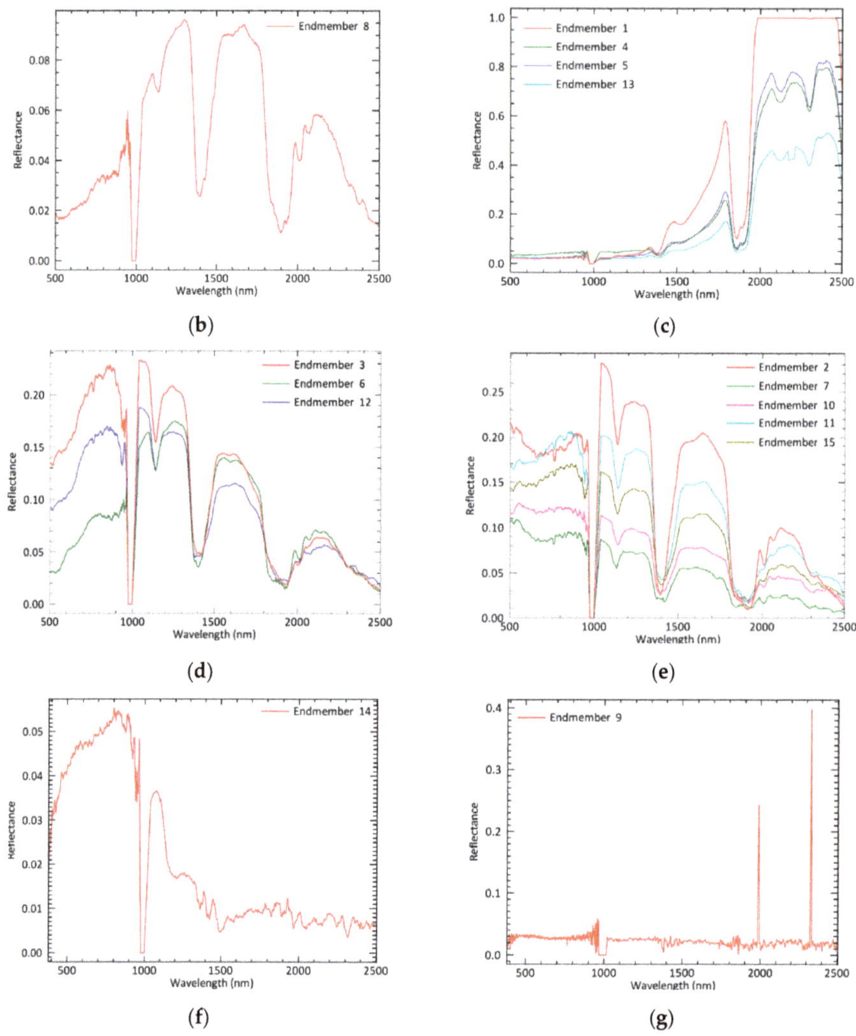

Figure 5. (**a**) The spatial distribution of 15 endmembers extracted by SMACC; The numbers on the image indicate the approximate location of the pixels selected as the represented endmembers of (**b**) basalt; (**c**) hot material; (**d**) oxidized surface; (**e**) sulfate mineral; (**f**) water; and (**g**) noise, extracted by SMACC.

5.2. Basalt Abundance

Figure 6a indicates the presence of the dominant basalt abundance pixel throughout the image. This abundance is associated with endmember 8 which is characterized by very low reflectance (Figure 5b) due to the presence of large amounts of dark mafic rock since the study area is dominated by basaltic lava (Figure 6b).

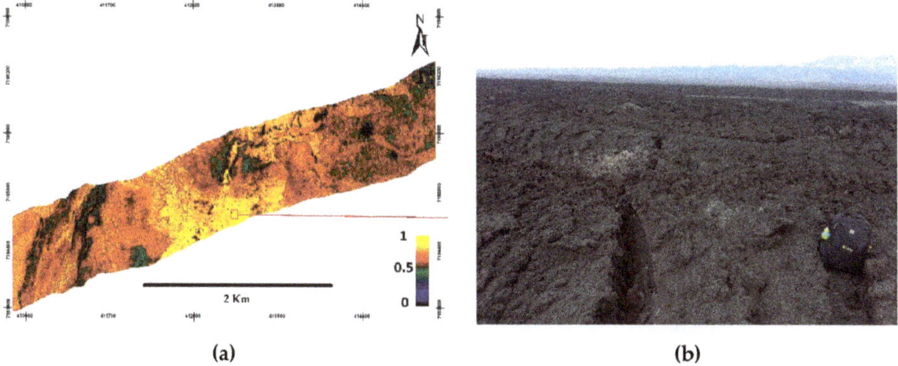

(a) (b)

Figure 6. (**a**) The abundance map for basalt endmember, yellow areas indicate the highest fraction of basalt meanwhile the black areas indicate the lowest fraction of basalt (the red box shows the approximate location of the field photo); (**b**) field photograph of basaltic lava field of the Holuhraun.

5.3. Hot Material Abundance

As shown in Figure 7a, the hot material abundance map is very sparse. This abundance is described as blends of the endmember 1, 4, 5, and 13 which are characterized by very high reflectance in the SWIR due to the presence of hot material (Figure 5c). Figure 5a shows that endmembers 1, 4, 5, and 13 are located in the lower right corner and the upper part of the image, Figure 7b shows a false color (NIR-SWIR) image which agrees with the abundance map, i.e., some patches of hot material (red-yellow color) exist in the area. The false color image is created by stacking R: 2200 nm; G: 1600 nm, and B: 896 nm. This indicates that the lava field is still emitting hot material during the data acquisition.

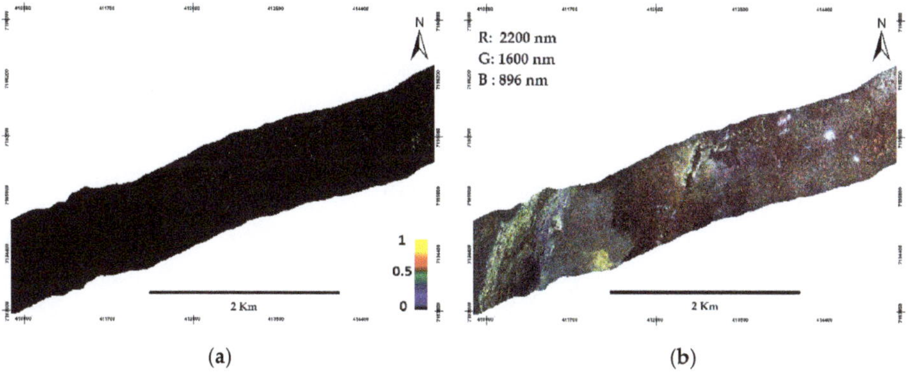

(a) (b)

Figure 7. (**a**) Abundance map for the hot material endmember, yellow areas indicate the highest fraction of hot material meanwhile the black areas indicate the lowest fraction of incandescent lava; (**b**) The false color (NIR-SWIR) image show that hot material (red-yellow color) exists in the area.

5.4. Oxidized Surface Abundance

The oxidized surface endmembers (3, 6, and 12) have the highest abundance fraction at the vent as shown in Figure 8a. This agrees with a field observation shown in Figure 8b which highlights the matching dominant oxidized surface at the vent wall.

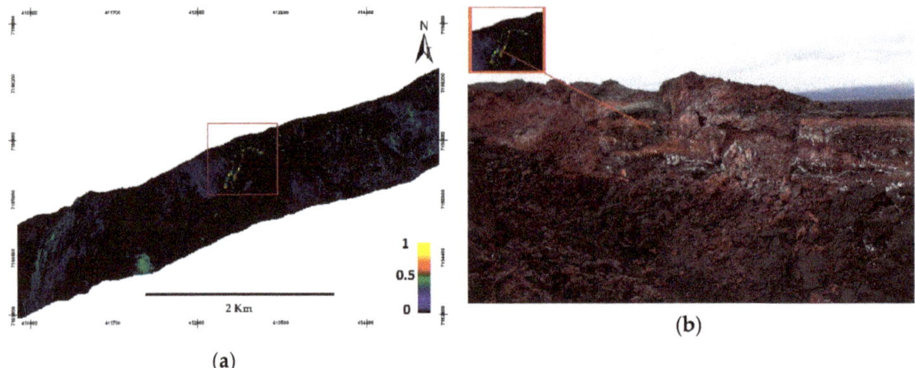

Figure 8. (**a**) Abundance map oxidized surface endmember; yellow areas indicate the highest fraction of oxidized surface meanwhile the black areas indicate the lowest fraction of oxidized surface; (**b**) Field photograph of an oxidized surface of the vent wall (red box and the line shows the approximate location of the field photograph)

5.5. Sulfate Mineral Abundance

The sulfate mineral endmembers (2, 7, 10, 11, and 15) have the highest abundance fraction around the lava pond and there are four most prominent areas for the sulfate (Figure 9a). This surface mineral looked as if it had been dusted by snow (white color) commonly identified as *thernadite* (Na_2SO_4) [41]. This can be directly seen from a true color image. This mineral formed as the flow cooled, a thin sublimate coating formed on the surface of the lava [41]. Figure 9b,c shows the *thernadite* formed in surface lava at Holuhraun.

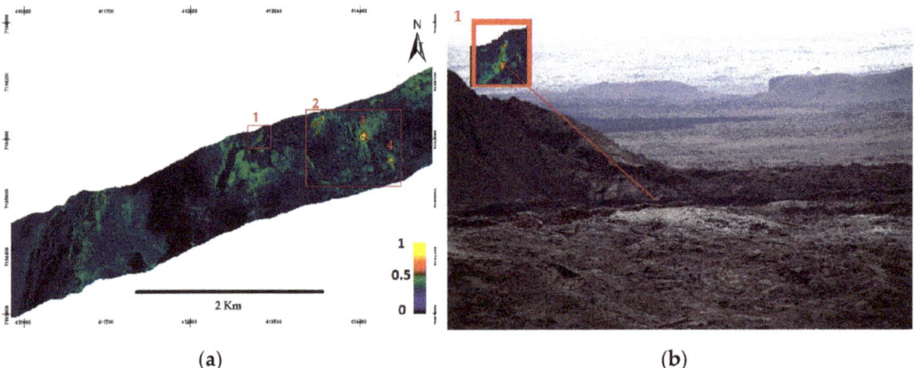

Figure 9. *Cont.*

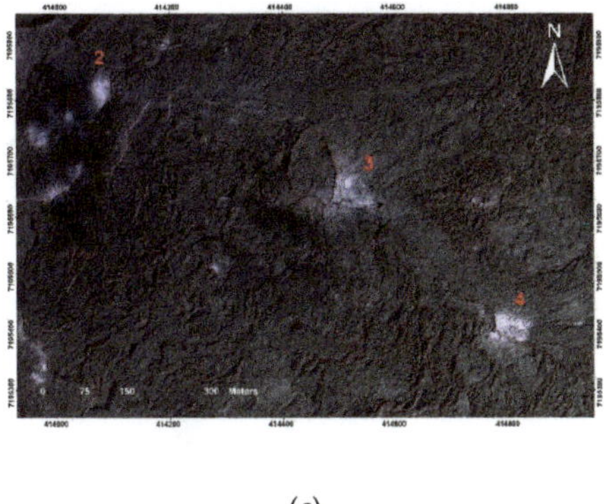

(c)

Figure 9. (**a**) The Abundance map for the sulfate mineral endmember, the yellow areas indicate the highest fraction of sulfate mineral meanwhile the black areas indicate the lowest fraction of the sulfate mineral; (**b**) Field photograph of sulfate mineral (white surface) formed on the surface of lava (the red boxes and lines show the approximate location of the field and aerial photo respectively); (**c**) aerial photograph of sulfate mineral (white surface) formed on the surface of lava (The numbers on the image indicate the approximate location of the sulfate for both the abundance and photograph).

5.6. Water Abundance

The water abundance (Figure 10a) has the highest abundance fraction at the location mainly recognized as a glacial river (Figure 10b). Endmember 14 represents water which is characterized by a relatively low reflectance and has the highest reflectance in the blue wavelength. Water has high absorption and virtually no reflectance in the NIR-SWIR wavelengths range (Figure 5f).

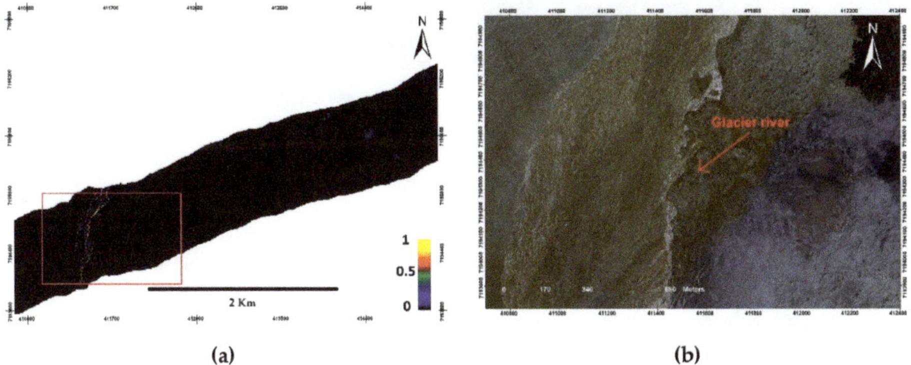

(a) (b)

Figure 10. (**a**) The abundance map for water endmember, the highest abundance fraction indicated by a yellow color, and the lowest abundance fraction indicated by a black red box shows the approximate location of the aerial photograph); (**b**) aerial photograph of the glacial river.

5.7. Noise Abundance

Figure 11 shows the abundance map corresponding to endmember 9. We consider this endmember as representing noise due to an unrecognized spectral signature since this spectrum is characterized by saturated reflectance in channels ~2000 nm and ~2400 nm (Figure 5g). The saturated reflectance could be due to corrupted bands in some pixels.

Figure 11. The abundance map for the noise endmember, the highest abundance fraction is indicated by the yellow color, and the lowest abundance fraction is indicated by the black color.

5.8. False Color Abundance

The abundance results depicted as false color (R: Oxidized surface; G: Sulfate mineral; B: Basalt) images show that the majority of rocks or minerals in the study area are dominated by basalt as shown in the blue color in Figure 12a. The other colors such as magenta and yellow indicate a mixture. The mixture phenomenon is illustrated in Figure 12b, as the surface has 0.25 oxidized surface mix with 0.75 basalt resulting in the magenta color; and 0.25 oxidized surface mix with 0.75 sulfate mineral resulting in the yellow color pixel.

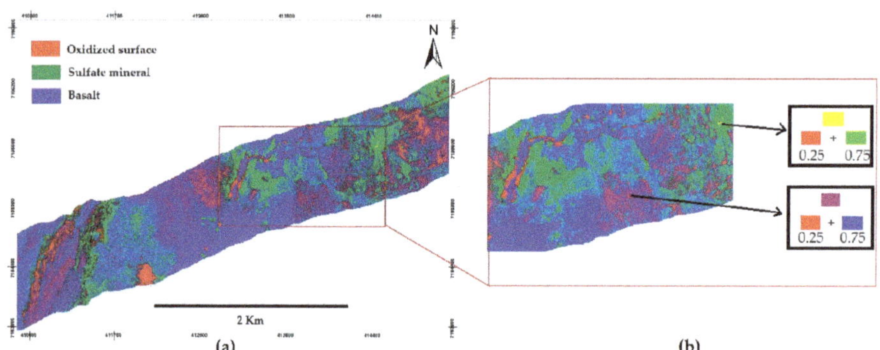

Figure 12. (**a**) False color of abundance highlighting for R: oxidized surface; G: sulfate mineral; and B: Basalt; (**b**) Illustration of the mixed pixels in the area, 0.25 oxidized surface mix with 0.75 basalt resulting in the magenta color; and 0.25 oxidized surface mix with 0.75 sulfate mineral resulting in the yellow color pixel.

5.9. Validation

The very high-resolution aerial photograph was used for ground truth. The aerial photograph was classified into oxidized surface, sulfate, basalt, and water using visual image interpretation and used for validation of the unmixing results. We only validate three endmembers for basalt—oxidized, sulfate, and water—since the noise and hot material cannot be detected based on visual interpretation. We classified the endmembers that have fractional abundance > 0.5. Validation was based on 150 randomly generated point samples within each class. Table 1 show the validation results, with a resulting mean overall accuracy 79% and mean Kappa index of 0.73. This result shows that the abundances have moderate agreement with the sample points.

Table 1. Validation of the endmembers that have abundance > 0.5.

Class	Overall Accuracy	Kappa Index	Mean Overall Accuracy	Mean Kappa Index
Basalt/Non-Basalt	70%	0.62	79%	0.73
Sulfate/Non-Sulfate	93%	0.89		
Oxidized/Non-Oxidized	77%	0.72		
Water/Non-Water	76%	0.70		

6. Discussion

6.1. Comparison with the Existing Spectral Index Technique

The correlation between the spectral index images and the abundance image was analyzed. We only correlated the three endmembers since there are no reference spectral indices for sulfate mineral, hot material, and noise. Here we compared the basalt, oxidized, and water abundance images with the mafic, oxidized, and water index images proposed by Inzana et al., Podwysocki et al. and Xu respectively [42–44] (Appendix B). We applied these indices to the hyperspectral image and compared them with the result from each abundance. Figure 13a–c shows the scatter plots results. The R^2 values were 0.46, 0.91, and 0.77 for the basalt, oxidized surface, and water, respectively. The oxidized surface and water indicate a good correlation with the indices (Figure 13b,c). This suggests that both oxidation and water generated from a spectral index are properly validated [2,44]. Meanwhile, basalt shows a low correlation with the mafic index (Figure 13a) suggesting that the estimates of the basalt surface from the unmixing technique is an overestimation, since the basalt abundance shows the older lava flows as mafic with a relatively high fraction compared to the mafic index that only showed for fresh lava flow. This being due to a full spectrum of hyperspectral can easily differentiate between basalt surface and non-basalt.

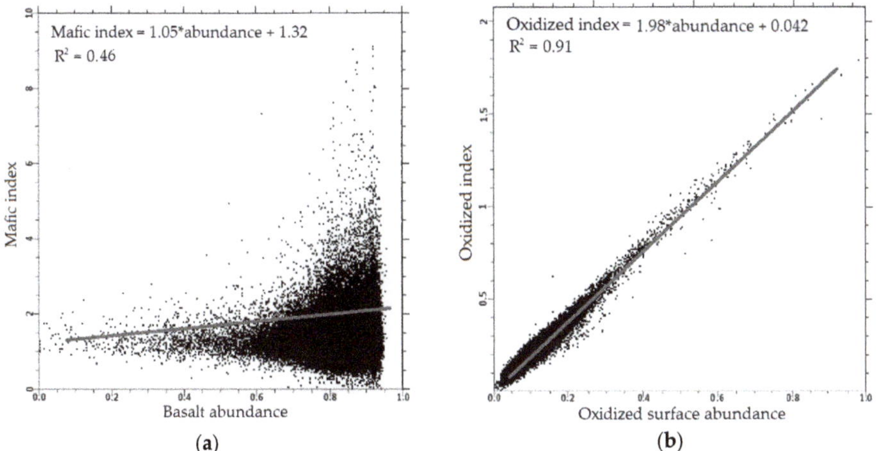

(a) (b)

Figure 13. *Cont.*

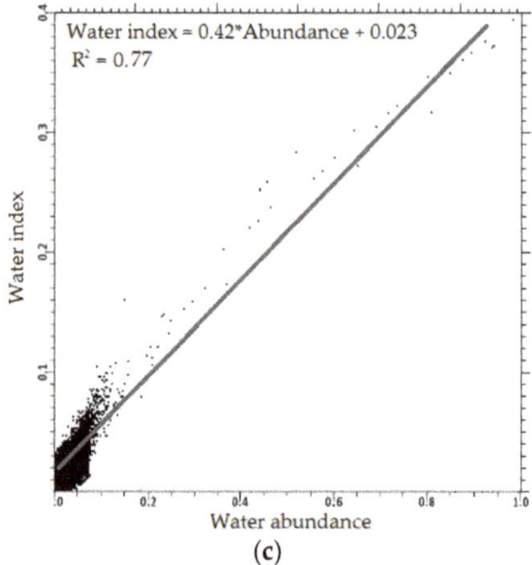

Figure 13. Linear regression analysis between the spectral index images and the (**a**) basalt abundance; (**b**) oxidized surface abundance; and (**c**) water index.

6.2. Number of Endmembers

The determination of the number of endmembers is critical, since underestimation may result in a poor representation of the mixed pixels, whereas overestimation may result in an overly segregated area [16]. Table 2 shows the relationship between the number of endmember and the number of pixels that have fractional abundance > 0.5 and the mean correlation with mafic, oxidized, and water index. We considered abundance >0.5 as high abundance. As the number of endmembers increase, the number of pixels also increases for an oxidized surface, sulfate mineral, water, and noise abundances, respectively. This is due to an increase of endmembers that is detected for each group. Meanwhile, the basalt abundance shows the opposite, as the endmembers increase the number of pixels with abundance >0.5 decreases. These results show that as more endmembers are considered the mixing of basalt with other endmembers increases resulting in a decrease of the fractional abundance of basalt. According to the results, we considered the 15 endmembers as an optimum number for this study since they have the highest mean correlation with mafic, oxidized, and water index. Clearly, the selection of appropriate endmembers in such a diverse volcanic environment, considering the particularities of the FENIX dataset, is of great importance in order to obtain accurate unmixing results. In addition, since only a small number of the available materials spectra are expected to be present in a single pixel, the abundance vectors are often sparse [45].

Table 2. Comparison number of pixels that have abundance >0.5, $\overline{R^2}$ and number of endmembers.

Number of Endmembers	Number of Pixels Abundance						$\overline{R^2}$
	Oxidized Surface	Sulfate Mineral	Hot Material	Water	Noise	Basalt	
5	19	57	19	0	0	522481	0.27
10	86	115	19	0	2	522406	0.35
15	91	215	34	373	2	522266	0.71
20	95	232	36	373	5	522046	0.67
30	97	250	40	373	7	521707	0.69

6.3. Size of Lava Field Area

As the methods were only tested on a subset area of the lava field vent, to apply the methods for the entire lava flow is challenging for several reasons. (1) The high spatial heterogeneity typically gives rise to mixed pixels containing multiple materials and it will increase the number of endmembers detected by SMACC [40]. (2) Different illumination occurs within the different flight lines for the entire lava flow (Figure 3b) since the data acquisition time is acquired between 16.56 and 17.58 local times which results from the very low sun angle during the acquisition. This problem can be approached by collecting ground truth spectra, extensive calibration, and atmospheric correction using simultaneous and constrained calibration of multiple hyperspectral images through a new generalized empirical line model purposed by Kizel et al. [37]. (3) The computation time to perform unmixing also must be considered for the entire lava field since the area is relatively large (84 km^2) and the hyperspectral data contains 622 channels with a 3.5-meter spatial resolution. In order to process the full set of data we need to consider using high performance computing (HPC) [46].

6.4. Using Full Optical Region for Mapping Recent Lava Flow (VIS-SWIR-TIR)

Hyperspectral VIS-SWIR image data is effective for discrimination mafic, oxidation, sulfate etc. However, not all the minerals and surface type are always mapped uniquely with VIS-SWIR hyperspectral data. A typical surface such as rock forming minerals associated with unaltered rocks and alteration minerals associated with altered rocks can be identified with TIR (Thermal Infrared) data [47–49]. Image processing methods that have become standard for hyperspectral VNIR/SWIR data analysis also work for hyperspectral TIR data [47]. Vaughan et al [47] showed that pixel classification techniques based on spectral variability within the scene and mineral libraries for matching spectral emissivity features can be used for TIR-derived mineral maps using SEBASS hyperspectral TIR image data. Hyperspectral TIR instruments operational for airborne surveys are also available in the NERC Airborne Research Facility with a Specim AisaOWL sensor [48]. A synergistic use of airborne data from both FENIX (VIS-SWIR) and OWL (TIR) allows great potential for lava discrimination in future study due to the complementary nature of the reflective (VIS-SWIR) and emissive (TIR) spectral regions. This might significantly improve our understanding of physical lava surface properties. Specifically, VIS-SWIR imaging spectrometers can discriminate surface materials and TIR data acquisitions can help to identify the thermal characteristics of different materials [47–49]. For instance, combining emissivity spectra with reflectance spectra in a mixing model would improve discriminating lava from surfaces [50–52].

7. Conclusions

In this study, an application of potential spectral unmixing methods on 2014–2015 Holuhraun lava flow field was presented. In total, we acquired fifteen spectral endmembers and their abundances. The first endmember was chosen as the brightest pixel which represented saturated incandescent lava. We grouped these 15 endmembers into six groups (basalt, oxidized surface, sulfate mineral, hot material, water, and noise) based on the shape of the endmembers since the amplitude varies due to illumination conditions, spectral variability, and topography. The endmembers represent pure surface materials in a hyperspectral image. We concluded that the selection of appropriate endmembers in such a diverse volcanic environment, considering the particularities of the FENIX dataset, is of great importance in order to obtain accurate unmixing results. Combination of SMACC and LSMA methods offers an optimum and a fast selection for volcanic products segregation However, ground-truthing spectra are recommended for further analysis. A synergistic use of airborne data from both FENIX (VIS-SWIR) and OWL (TIR) gives a great potential for lava discrimination in future study due to the complementary nature of the reflective (VIS-SWIR) and emissive (TIR) spectral regions. This might significantly improve our understanding of physical lava surface properties.

Author Contributions: Conceptualization, M.A.; Supervision, A.H., M.O.U., I.J. and T.T.

Funding: The first author was supported by the Indonesia Endowment Fund for Education (LPDP) Grant No. 20160222025516, European Network of Observatories and Research Infrastructures for Volcanology (EUROVOLC), The European Facility for Airborne Research (EUFAR) and Vinir Vatnajökuls during his Ph.D. project.

Acknowledgments: Authors would like to thank Robert Askew and Catherine Gallagher from the Institute of Earth Sciences, University of Iceland for the fieldwork photos around the lava field. Authors also would also like to thank anonymous reviewers for their constructive comments for the manuscript.

Conflicts of Interest: The authors declare no conflict of interest.

Appendix A

The bad channels in this data are located at 968 nm and 1014 nm. Figure A1 show the spectral reflectance before masking (Figure A1A) and after channel masking (Figure A1B).

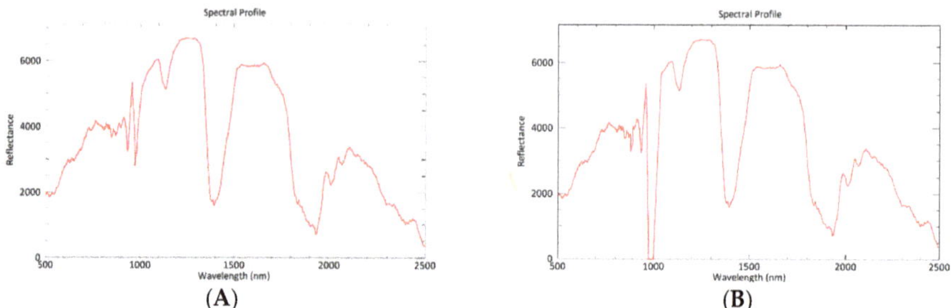

Figure A1. Spectral reflectance of material (**A**) before masking; (**B**) after channel masking.

Figure A2. This shows the theoretical pixel size at the nadir for Fenix. The pixel size will be larger at the edges of the swath, in this study, the AGL is ~ 2400 m so according to the graph the optimal pixel size resample for the FENIX is ~3.5 m.

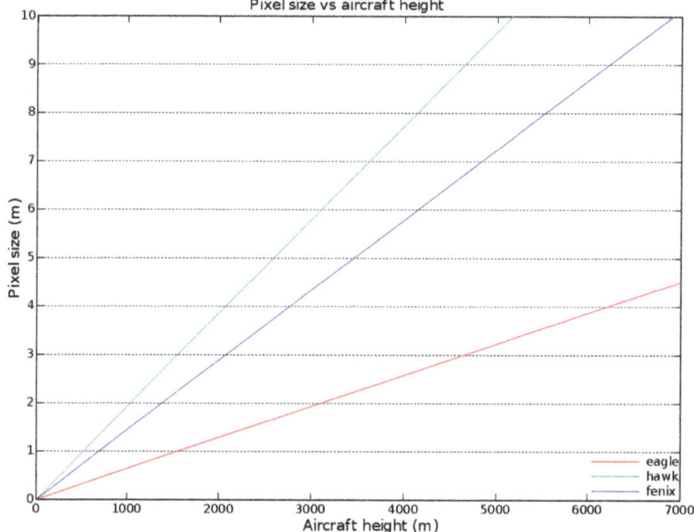

Figure A2. The theoretical pixel size at the nadir for FENIX, EAGLE, and HAWK. In this study we used FENIX airborne for data acquisition [39].

Appendix B

The mafic indices originated, developed by Inzana et al. [42] to distinguish mafic from non-mafic rocks are from Landsat TM image, expressed as follows:

$$Mafic\ index = \frac{\rho_{1600\text{nm}}}{\rho_{860\text{nm}}} * \frac{\rho_{640\text{nm}}}{\rho_{860\text{nm}}} \tag{A1}$$

where $\rho_{1600\text{nm}}$ is the measured reflectance at wavelength 1600 nm, $\rho_{640\text{nm}}$ is the measured reflectance at wavelength 640 nm, and $\rho_{860\text{nm}}$ is the measured reflectance at wavelength 860 nm.

The oxidized index originated designed any multispectral sensor with bands that fall within the red channel and blue channel [43], expressed as follows:

$$Oxidized\ index = \frac{\rho_{640\text{nm}}}{\rho_{500\text{nm}}} \tag{A2}$$

where $\rho_{500\text{nm}}$ is the measured reflectance at wavelength 500 nm.

We calculated the water index using the Modified Normalized Difference Water Index (MNDWI) [44]. This index enhances open water features while suppressing noise from built-up land, vegetation, and soil. This is expressed as follows:

$$Water\ index = \frac{\rho_{600\text{nm}} - \rho_{1600\text{nm}}}{\rho_{600\text{nm}} + \rho_{1600\text{nm}}}$$

where $\rho_{600\text{nm}}$ is the measured reflectance at wavelength 600 nm.

References

1. Pedersen, G.B.M.; Höskuldsson, A.; Dürig, T.; Thordarson, T.; Jónsdóttir, I.; Riishuus, M.S.; Óskarsson, B.V.; Dumont, S.; Magnusson, E.; Gudmundsson, M.T.; et al. Lava field evolution and emplacement dynamics of the 2014–2015 basaltic fissure eruption at Holuhraun, Iceland. *J. Volcanol. Geotherm. Res.* **2017**. [CrossRef]
2. Li, L.; Solana, C.; Canters, F.; Chan, J.; Kervyn, M. Impact of environmental factors on the spectral characteristics of lava surfaces: field spectrometry of basaltic lava flows on Tenerife, Canary Islands, Spain. *Remote Sens.* **2015**, *7*, 16986–17012. [CrossRef]
3. Li, L.; Solana, C.; Canters, F.; Kervyn, M. Testing random forest classification for identifying lava flows and mapping age groups on a single Landsat 8 image. *J. Volcanol. Geotherm. Res.* **2017**, *345*, 109–124. [CrossRef]
4. Head, E.M.; Maclean, A.L.; Carn, S.A. Mapping lava flows from Nyamuragira volcano (1967-2011) with satellite data and automated classification methods. *Geomatics, Nat. Hazards Risk* **2013**, *4*, 119–144. [CrossRef]
5. Li, L.; Canters, F.; Solana, C.; Ma, W.; Chen, L.; Kervyn, M. Discriminating lava flows of different age within Nyamuragira's volcanic field using spectral mixture analysis. *Int. J. Appl. Earth Obs. Geoinf.* **2015**, *40*, 1–10. [CrossRef]
6. Amici, S.; Piscini, A.; Neri, M. Reflectance Spectra Measurements of Mt. Etna: A Comparison with Multispectral / Hyperspectral Satellite. *Adv. Remote Sens.* **2014**, *3*, 235–245. [CrossRef]
7. Graettinger, A.H.; Ellis, M.K.; Skilling, I.P.; Reath, K.; Ramsey, M.S.; Lee, R.J.; Hughes, C.G.; McGarvie, D.W. Remote sensing and geologic mapping of glaciovolcanic deposits in the region surrounding Askja (Dyngjufjöll) volcano, Iceland. *Int. J. Remote Sens.* **2013**, *34*, 7178–7198. [CrossRef]
8. Zhang, X.; Shang, K.; Cen, Y.; Shuai, T.; Sun, Y. Estimating ecological indicators of karst rocky desertification by linear spectral unmixing method. *Int. J. Appl. Earth Obs. Geoinf.* **2014**, *31*, 86–94. [CrossRef]
9. Combe, J.P.; Le Mouélic, S.; Sotin, C.; Gendrin, A.; Mustard, J.F.; Le Deit, L.; Launeau, P.; Bibring, J.P.; Gondet, B.; Langevin, Y.; et al. Analysis of OMEGA/Mars Express data hyperspectral data using a Multiple-Endmember Linear Spectral Unmixing Model (MELSUM): Methodology and first results. *Planet. Space Sci.* **2008**, *56*, 951–975. [CrossRef]
10. Adams, J.B.; Smith, M.O.; Johnson, P.E. Spectral mixture modeling: A new analysis of rock and soil types at the Viking Lander 1 Site. *J. Geophys. Res.* **1986**. [CrossRef]
11. Kruse, F.A.; Perry, S.L. Mineral mapping using simulated worldview-3 short-wave-infrared imagery. *Remote Sens.* **2013**, *6*, 2688–2703. [CrossRef]
12. Sun, Y.; Tian, S.; Di, B. Extracting mineral alteration information using WorldView-3 data. *Geosci. Front.* **2017**. [CrossRef]
13. Aufaristama, M.; Höskuldsson, Á.; Jónsdóttir, I.; Ólafsdóttir, R. Mapping and Assessing Surface Morphology of Holocene Lava Field in Krafla (NE Iceland) Using Hyperspectral Remote Sensing. *IOP Conf. Ser. Earth Environ. Sci.* **2016**, *29*, 1–6. [CrossRef]
14. Spinetti, C.; Mazzarini, F.; Casacchia, R.; Colini, L.; Neri, M.; Behncke, B.; Salvatori, R.; Buongiorno, M.F.; Pareschi, M.T. Spectral properties of volcanic materials from hyperspectral field and satellite data compared with LiDAR data at Mt. Etna. *Int. J. Appl. Earth Obs. Geoinf.* **2009**, *11*, 142–155. [CrossRef]

15. Kolzenburg, S.; Jaenicke, J.; Münzer, U.; Dingwell, D.B. The effect of inflation on the morphology-derived rheological parameters of lava flows and its implications for interpreting remote sensing data - A case study on the 2014/2015 eruption at Holuhraun, Iceland. *J. Volcanol. Geotherm. Res.* **2018**, *357*, 200–212. [CrossRef]
16. Daskalopoulou, V.; Sykioti, O.; Karagiannopoulou, C. Application of Spectral Unmixing on Hyperspectral data of the Historic volcanic products of Mt. Etna (Italy). *Multidiscip. Digit. Publ. Inst. Proc.* **2018**, *2*, 329. [CrossRef]
17. Coppola, D.; Ripepe, M.; Laiolo, M.; Cigolini, C. Modelling satellite-derived magma discharge to explain caldera collapse. *Geology* **2017**, *45*, 523–526. [CrossRef]
18. Aufaristama, M.; Hoskuldsson, A.; Jonsdottir, I.; Ulfarsson, M.; Thordarson, T. New Insights for Detecting and Deriving Thermal Properties of Lava Flow Using Infrared Satellite during 2014–2015 Effusive Eruption at Holuhraun, Iceland. *Remote Sens.* **2018**, *10*, 151. [CrossRef]
19. Rossi, C.; Minet, C.; Fritz, T.; Eineder, M.; Bamler, R. Temporal monitoring of subglacial volcanoes with TanDEM-X - Application to the 2014–2015 eruption within the Bardarbunga volcanic system, Iceland. *Remote Sens. Environ.* **2016**, *181*, 186–197. [CrossRef]
20. Dirscherl, M.; Rossi, C. Geomorphometric analysis of the 2014–2015 Bárðarbunga volcanic eruption, Iceland. *Remote Sens. Environ.* **2018**. [CrossRef]
21. Adams, J.B.; Sabol, D.E.; Kapos, V.; Almeida Filho, R.; Roberts, D.A.; Smith, M.O.; Gillespie, A.R. Classification of multispectral images based on fractions of endmembers: Application to land-cover change in the Brazilian Amazon. *Remote Sens. Environ.* **1995**. [CrossRef]
22. Quintano, C.; Fernández-Manso, A.; Shimabukuro, Y.E.; Pereira, G. Spectral unmixing. *Int. J. Remote Sens.* **2012**. [CrossRef]
23. Kolzenburg, S.; Giordano, D.; Thordarson, T.; Höskuldsson, A.; Dingwell, D.B. The rheological evolution of the 2014/2015 eruption at Holuhraun, central Iceland. *Bull. Volcanol.* **2017**, *79*, 45. [CrossRef]
24. Icelandic Meteorological Office Holuhraun. Available online: http://en.vedur.is/earthquakes-and-volcanism/articles/nr/3122 (accessed on 11 May 2017).
25. Tayebi, M.H.; Tangestani, M.H.; Vincent, R.K.; Neal, D. Spectral properties and ASTER-based alteration mapping of Masahim volcano facies, SE Iran. *J. Volcanol. Geotherm. Res.* **2014**, *287*, 40–50. [CrossRef]
26. Clark, R.N.; Roush, T.L. Reflectance spectroscopy: quantitative analysis techniques for remote sensing applications. *J. Geophys. Res.* **1984**. [CrossRef]
27. Zhang, J.; Rivard, B.; Sánchez-Azofeifa, A. Spectral unmixing of normalized reflectance data for the deconvolution of lichen and rock mixtures. *Remote Sens. Environ.* **2005**. [CrossRef]
28. Zhang, J.; Rivard, B.; Sanchez-Azofeifa, A. Derivative spectral unmixing of hyperspectral data applied to mixtures of lichen and rock. *IEEE Trans. Geosci. Remote Sens.* **2004**.
29. Rowan, L.C.; Mars, J.C.; Simpson, C.J. Lithologic mapping of the Mordor, NT, Australia ultramafic complex by using the Advanced Spaceborne Thermal Emission and Reflection Radiometer (ASTER). *Remote Sens. Environ.* **2005**. [CrossRef]
30. Hellman, M.J.; Ramsey, M.S. Analysis of hot springs and associated deposits in Yellowstone National Park using ASTER and AVIRIS remote sensing. *J. Volcanol. Geotherm. Res.* **2004**. [CrossRef]
31. Hyperspectral Imaging Cameras And Systems - Specim. Available online: http://www.specim.fi/ (accessed on 1 December 2018).
32. NERC Airborne Research Facility - British Antarctic Survey. Available online: https://www.bas.ac.uk/polar-operations/sites-and-facilities/facility/nerc-airborne-research-facility-2/ (accessed on 3 December 2018).
33. Loftmyndir ehf. Available online: http://www.loftmyndir.is/ (accessed on 12 February 2019).
34. Bernstein, L.S.; Adler-Golden, S.M.; Sundberg, R.L.; Levine, R.Y.; Perkins, T.C.; Berk, A.; Ratkowski, A.J.; Felde, G.; Hoke, M.L. A new method for atmospheric correction and aerosol optical property retrieval for VIS-SWIR multi- and hyperspectral imaging sensors: QUAC (QUick Atmospheric Correction). *IGRSS* **2005**, *5*. [CrossRef]
35. Bernstein, L.S. Quick atmospheric correction code: algorithm description and recent upgrades. *Opt. Eng.* **2012**, *51*, 111719. [CrossRef]
36. Karpouzli, E.; Malthus, T. The empirical line method for the atmospheric correction of IKONOS imagery. *Int. J. Remote Sens.* **2003**. [CrossRef]

37. Kizel, F.; Benediktsson, J.A.; Bruzzone, L.; Pedersen, G.B.M.; Vilmundardottir, O.K.; Falco, N. Simultaneous and constrained calibration of multiple hyperspectral images through a new generalized empirical line model. *IEEE J. Sel. Top. Appl. Earth Obs. Remote Sens.* **2018**. [CrossRef]
38. Warren, M.A.; Taylor, B.H.; Grant, M.G.; Shutler, J.D. Data processing of remotely sensed airborne hyperspectral data using the Airborne Processing Library (APL): Geocorrection algorithm descriptions and spatial accuracy assessment. *Comput. Geosci.* **2014**, *64*, 24–34. [CrossRef]
39. Processing/PixelSize. Available online: https://nerc-arf-dan.pml.ac.uk/trac/wiki/Processing/PixelSize (accessed on 31 December 2018).
40. Gruninger, J.H.; Ratkowski, A.J.; Hoke, M.L. The sequential maximum angle convex cone (SMACC) endmember model. *SPIE* **2004**, *5425*, 1.
41. Moore, R.B.; Clague, D.A.; Rubin, M.; Bohrson, W.A. Volcanism in Hawaii. In *U.S. Geological Survey Professional Paper 1350*; USGS: Reston, VA, USA, 1987; p. 557. ISBN 3663537137.
42. Inzana, J.; Kusky, T.; Higgs, G.; Tucker, R. Supervised classifications of Landsat TM band ratio images and Landsat TM band ratio image with radar for geological interpretations of central Madagascar. *J. African Earth Sci.* **2003**, *37*, 59–72. [CrossRef]
43. Podwysocki, M.H.; Segal, D.B.; Abrams, M.J. Use of multispectral scanner images for assessment of hydrothermal alteration in the Marysvale, Utah, mining area. *Econ. Geol.* **1983**. [CrossRef]
44. Xu, H. Modification of normalised difference water index (NDWI) to enhance open water features in remotely sensed imagery. *Int. J. Remote Sens.* **2006**. [CrossRef]
45. Giampouras, P.V.; Themelis, K.E.; Rontogiannis, A.A.; Koutroumbas, K.D. Simultaneously Sparse and Low-Rank Abundance Matrix Estimation for Hyperspectral Image Unmixing. *IEEE Trans. Geosci. Remote Sens.* **2016**. [CrossRef]
46. Plaza, A.; Du, Q.; Chang, Y.; King, R.L. High Performance Computing for Hyperspectral Remote Sensing. *IEEE J. Sel. Top. Appl. Earth Obs. Remote Sens.* **2011**, *4*, 528–544. [CrossRef]
47. Vaughan, R.G.; Calvin, W.M.; Taranik, J.V. SEBASS hyperspectral thermal infrared data: Surface emissivity measurement and mineral mapping. *Remote Sens. Environ.* **2003**, *85*, 48–63. [CrossRef]
48. Schlerf, M.; Rock, G.; Lagueux, P.; Ronellenfitsch, F.; Gerhards, M.; Hoffmann, L.; Udelhoven, T. A hyperspectral thermal infrared imaging instrument for natural resources applications. *Remote Sens.* **2012**, *4*, 3995–4009. [CrossRef]
49. Riley, D.; Hecker, C. Mineral Mapping with Airborne Hyperspectral Thermal Infrared Remote Sensing at Cuprite, Nevada, USA. In *Thermal Infrared Remote Sensing: Sensors, Methods, Applications*; Kuenzer, C., Dech, S., Eds.; Springer: Heidelberg, The Netherlands, 2013; Vol. 17, pp. 495–514. ISBN 978-94-007-6639-6.
50. Ball, M.; Pinkerton, H.; Harris, A.J.L. Surface cooling, advection and the development of different surface textures on active lavas on Kilauea, Hawai'i. *J. Volcanol. Geotherm. Res.* **2008**, *173*, 148–156. [CrossRef]
51. Ramsey, M.S.; Harris, A.J.L.; Crown, D.A. What can thermal infrared remote sensing of terrestrial volcanoes tell us about processes past and present on Mars? *J. Volcanol. Geotherm. Res.* **2016**, *311*, 198–216. [CrossRef]
52. Harris, A. *Thermal Remote Sensing of Active Volcanoes: A User's Manual*; Cambridge University Press: Cambridge, UK, 2013; ISBN 9781139029346.

Technical Note

Processing Thermal Infrared Imagery Time-Series from Volcano Permanent Ground-Based Monitoring Network. Latest Methodological Improvements to Characterize Surface Temperatures Behavior of Thermal Anomaly Areas

Fabio Sansivero * and Giuseppe Vilardo

Istituto Nazionale di Geofisica e Vulcanologia, Osservatorio Vesuviano, Via Diocleziano 328, 80124 Napoli, Italy; giuseppe.vilardo@ingv.it

* Correspondence: fabio.sansivero@ingv.it; Tel.: +39-081-6108217

Received: 29 January 2019; Accepted: 4 March 2019; Published: 6 March 2019

Abstract: In this technical paper, the state-of-art of automated procedures to process thermal infrared (TIR) scenes acquired by a permanent ground-based surveillance system, is discussed. TIR scenes regard diffuse degassing areas at Campi Flegrei and Vesuvio in the Neapolitan volcanic district (Italy). The processing system was developed in-house by using the flexible and fast processing Matlab© environment. The multi-step procedure, starting from raw infrared (IR) frames, generates a final product consisting mainly of de-seasoned temperatures and heat fluxes time-series as well as maps of yearly rates of temperature change of the IR frames. Accurate descriptions of all operational phases and of the procedures of analysis are illustrated; a Matlab© code (Natick, MA, USA) is provided as supplementary material. This product is ordinarily addressed to study volcanic dynamics and improve the forecasting of the volcanic activity. Nevertheless, it can be a useful tool to investigate the surface temperature field of any areas subjected to thermal anomalies, both of natural and anthropic origin.

Keywords: volcano monitoring; thermal imaging; time series; Seasonal-Trend Decomposition; heat flux

---

1. Introduction

Thermal infrared (TIR) ground-based observations are largely used in volcanology, both in research and in surveillance activities, to investigate volcanic plumes and gases, lava flows, lava lakes and fumarole fields [1–17]. Generally, the observations were made during a limited time span such as eruption phases or field campaigns with temporarily installed TIR stations or handheld cameras. In the last years the number of surveillance and research activities aimed to undertake TIR continuous observations of volcanic areas have increased [12–20]. Improvements in monitoring tools and analysis techniques of long TIR time-series of infrared (IR) scenes of volcanic areas are becoming matter of great interest since they give the opportunity to track changes of surface thermal anomalies that may reveal a renewal of eruptive activity. Several works identified thermal precursors before eruptions by using TIR observations [21–24] and these insightful results, also provided by field campaigns, have suggested planning permanent fixed installations of ground TIR stations at active volcanoes in the world.

At present time, few commercial software packages, based on general-purpose procedures, are available to process TIR time-series and they are not aimed for a near real-time automated analysis of large dataset. Generally, they involve manual processing steps and cannot be used in daily continuous automated volcano monitoring activity. Recently, [20] introduced automated analysis techniques of long TIR time-series of images acquired inside the Campi Flegrei volcanic area and previously [19] discussed about analysis techniques applied to TIR scenes inside the Vesuvius crater. The studied zones of these works are diffuse degassing areas of quiescent volcanoes characterized by low temperatures of released gas fluxes.

In this work, recent developments of processing methodologies of several-years long TIR time-series of volcanic areas from a permanent surveillance network are discussed in detail. Additionally, the automation of the processes is discussed. Step-by step descriptions of all operational phases and of the theoretical basis are reported in order to provide a clear explanation of the applied procedures. The main final results are trends of temperatures, heat fluxes and yearly rate of temperature change of the studied areas. In particular, a detailed study with a focus on seasonal component removal and on pixel alignment of IR frames (co-registration) was carried out. The code of fully-automated Matlab© application (ASIRA, Automated System of InfraRed Analysis) used to process the IR data is provided as Supplementary Materials.

2. The Study Areas

The TIR frames time-series, used to develop and test the methodologies described in this work, were acquired by stations of TIRNet (Thermel InfraRed Network), a surveillance network operated by the Osservatorio Vesuviano, section of National Institute of Geophysics and Volcanology (INGV), consisting of six permanent ground stations installed at Campi Flegrei caldera and Vesuvius crater (Figure 1). Campi Flegrei (CF) is an active volcanic field including part of the city of Napoli (Italy). Nowadays, although quiescent and the last eruption occurred in 1538 (Monte Nuovo; [25]), the CF area is affected by significant ground deformation (Bradyseism), low to moderate seismic activity, hot fumaroles fields and diffuse degassing zones. The target areas acquired by TIR cameras in the Solfatara crater and its surroundings are shown in (Figure 1a). The monitored area represents the main surface expression of the CF caldera hydrothermal system with gases emissions originated by interaction between fluids of magmatic and meteoric origin [26–28]. The Somma–Vesuvius volcanic complex, located east of the city of Naples, is one of most dangerous volcanoes in the world and the latest eruption occurred on 1944 [29]. The recent dynamic of the Vesuvius is characterized by low-level shallow seismicity and by low temperature fumarolic activity mainly concentrated in the crater area [30–32]. TIR scenes are from low-temperatures surface thermal anomaly on the western inner slope of the Vesuvius crater (Figure 1b).

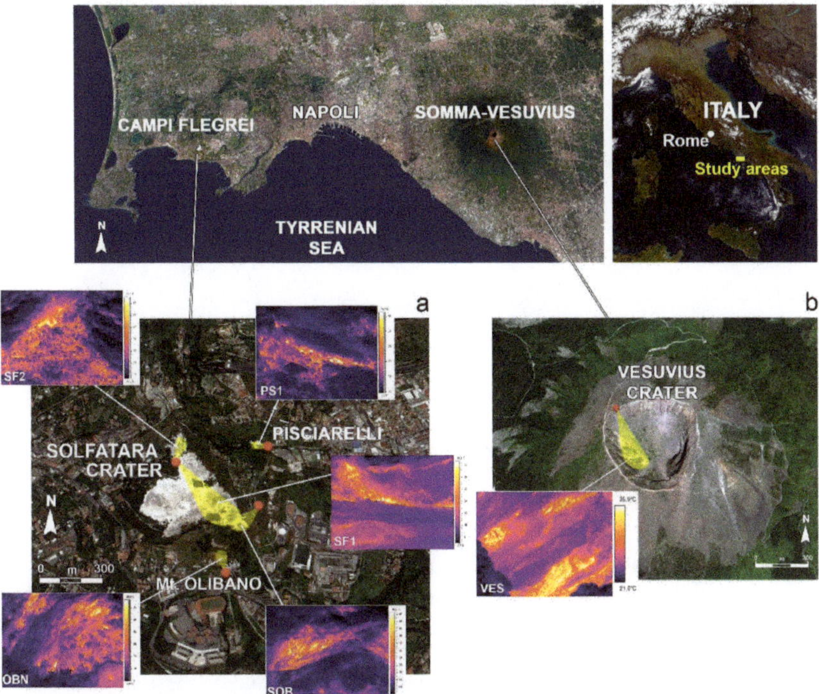

Figure 1. The Solfatara area (**a**) and Vesuvius crater (**b**) acquired by Thermel InfraRed Network (TIRnet) cameras. Red points are infrared (IR) stations locations and yellow regions represent the framed areas.

3. Materials and Methods

3.1. The IR Sensors and Data Acquisition

TIRNet stations were equipped with FLIR System, Inc. IR cameras, which acquire IR frames in the 7.5–13 μm waveband. The IR sensor installed at Campi Flegrei caldera is the FLIR SC655 and at Vesuvius is the FLIR A40 M, both with a focal plane array (FPA) uncooled microbolometer detector, of which the resolution was, respectively, 640 × 480 and 320 × 240 pixels. Accuracy was ±2 °C (SC655 and A40 M) and thermal sensitivity at 50/60 Hz was <30 mK (SC655) and 80 mK @ +25°C (A40 M). All IR cameras were set to a −40° to 120 °C temperature range. The optics used depended both on the distance sensor-target and type of IR camera and varied from 24.6 mm (FoV 25° × 19°) of SC655 camera to 36 mm (FoV 24° × 23.4°) of A40 M camera. The technical specifications of FLIR cameras and the features of target areas are reported in Table 1.

The IR stations acquired three IR frames of the target area every day at night-time. As solar heating can drastically decrease the thermal contrast between fumarole anomaly and the heated surrounding rocks [33] and references therein], the acquisitions of TIR frames were carried out at night (00:00, 02:00, 04:00 AM) in order to minimize diurnal heating effects.

After IR frames acquisition, WiFi radio or UMTS (Universal Mobile Telecommunications Service) modem transmits TIR data to the INGV-Osservatorio Vesuviano server of TIRNet in order to process them and to display the results in the surveillance room.

Table 1. Technical specifications of remote stations, FLIR infrared cameras and target areas details.

Remote Station	Camera Model	Resolution (pixel)	FoV	Data Transmission	Station UTM Coordinates (m)	Sensor-Target Average Distance (m)	Average Pixel Size (cm)
SF1	FLIR A655SC	640 × 480	25° × 19°	WiFi	X: 427.460 Y: 4.520.154	340	23.1
SF2	FLIR A645SC	640 × 480	15° × 11.9°	WiFi	X: 427.460 Y: 4.520.154	114	4.6
PS1	FLIR A645SC	640 × 480	15° × 11.9°	UMTS	X: 428.081 Y: 4.520.117	140	5.6
OBN	FLIR A645SC	640 × 480	25° × 19°	WiFi	X: 427.695 Y: 4.519.530	65	2.9 ÷ 5.4
SOB	FLIR A655SC	640 × 480	25° × 19°	WiFi	X: 427.810 Y: 4.519.878	90	5.5 ÷ 6.7
VES	FLIR A40	320 × 240	24° × 18°	WiFi	X: 451.325 Y: 4.519.281	225	30

As temperature values of TIR scenes are influenced by the atmospheric conditions (e.g., air temperature and humidity; [10]) and by the emissivity of target area, atmospheric correction was necessary. A probe of the IR station detected the values of air temperature and humidity and these values were transferred to the FLIR camera, which then applied the internal algorithm (LOWTRAN; [33]). This algorithm performed the atmospheric correction to the acquired IR frame in function of detector-target distance, emissivity of the target, air temperature and air relative humidity. The emissivity of the volcanic terrains (thermally altered pyroclasts), which characterize the target areas, was assumed to be 0.9 [34]).

The accuracy of the temperature measurements also depended on the orientation of the field of view, which should be as parallel as possible to the target. Generally, despite the calibration and correction of camera parameters, the detected IR temperatures were underestimated due to extrinsic field conditions mainly influenced by the presence of condensed water in fumarole gases which can partially hide the hot areas [14,35,36]. Therefore, the measured IR temperatures are to be considered apparent temperatures values that can differ from the real surface temperatures of the target area [1,37].

The resolution of FLIR cameras and the small distances between sensors and target areas allowed to detect correctly small thermal anomalies, and moreover, to minimize the attenuation of radiated energy of those non-homogeneous pixels which integrate both hot and cold areas [33]. In addition, the limitations in the calculation of real temperature were deemed not critical when the purpose was to investigate relative spatio-temporal variations of surface temperature field in volcanic areas [38].

3.2. Data Processing Procedures

The IR frames acquired by TIRNet stations were processed according to a multi-step procedure consisting of five main steps (Figure 2). The entire process is accomplished by the fully automated Matlab© software ASIRA (Natick, MA, USA), which was developed starting from the initial structure described by [19] and then by [20]. A detailed explanation of operative procedures is described in the following paragraphs. In the Appendix A, synthetic technical sheets of Matlab© code are reported.

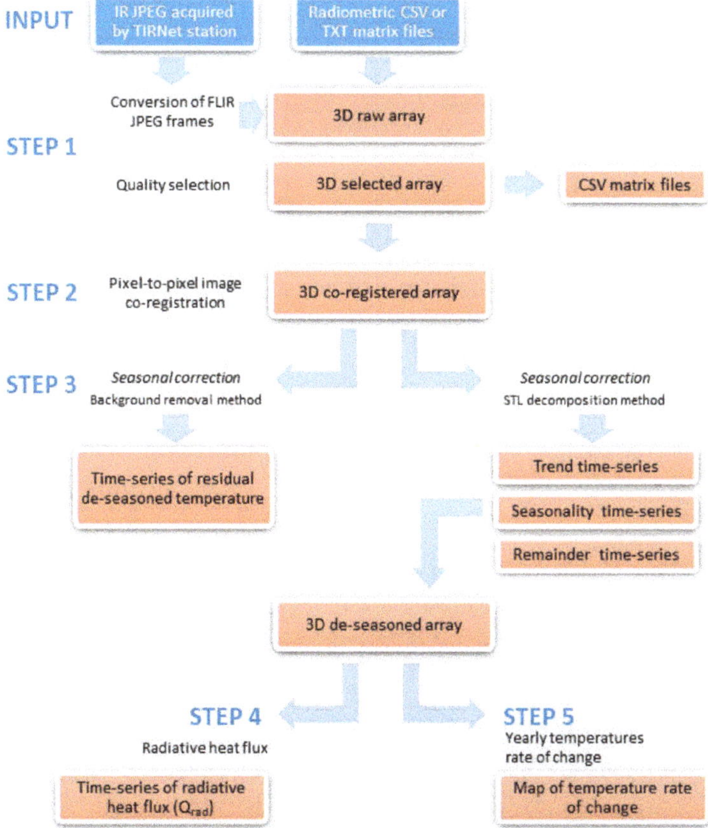

Figure 2. Block diagram of IR images processing steps. 3D: three-dimensional.

3.2.1. Step 1—IR Files Conversion, Archiving and Image Quality Selection

The FLIR IR raw files (radiometric JPEG), transmitted by remote TIRNet stations to the acquisition server, were imported in the Matlab© environment, then saved in appropriate storage folders both as a single CSV file and in a Matlab© three-dimensional (3D) matrix (Matlab© function: 'step01.m'). Occasionally, the presence of wide blurred areas, due to the condensation of water vapor from the fumaroles plume and the occurrence of heavy rain, caused the homogenization of the IR temperatures [18–20] and generated low quality IR frames. With the aim of removing low quality data, only the IR scenes that satisfied the following condition were selected:

$$\sigma F_i > m\sigma - c * \sigma F_\sigma \tag{1}$$

where σF_i is the Standard Deviation (SD) of the i-th IR frame, $m\sigma$ is the median of SD values of all IR frames of the station time-series, σF_σ is the Standard Deviation of all Standard Deviations of IR frames of the station time-series, and c is a user-defined coefficient depending on the statistical distribution of data (Matlab© function: 'step01.m'). We found c = 1 a suitable value to obtain a homogeneous data set by excluding very low-quality images.

This step converted input data (FLIR radiometric JPEG, CSV or TXT IR matrix) into Matlab© 3D arrays [resY, resX, n], where (resY, resX) is the image resolution and n is the number of IR collected frames.

3.2.2. Step 2—IR Frames Co-registration

The accurate alignment of all the IR frames related to a station time-series was necessary to proceed to further analysis. Since the IR framed area can vary in time, due to ground movements affecting volcanic areas or simply to maintenance services, a correction of IR frames position in respect of a reference IR frame was carried out (co-registration). This correction performed the alignment of the same pixels, of all IR frames belonging to the same station, by using the flow-based, image registration Matlab© algorithm, SIFT flow [39]. The SIFT flow algorithm matches pixel-to-pixel correspondences between two images and it is able to find dense scene correspondence despite substantial differences in spatial arrangement of compared images (Matlab© function: 'step02.m').

3.2.3. Step 3—Seasonal Component Removal

A simple plot of the time-series of temperature values evidenced a typical recurring pattern due to the seasonal influence over the surface temperatures (background raw maximum temperature plot in Figure 3). The temperature time-series of raw IR frames were representative of both exogenous (e.g., seasonal) and endogenous (thermal anomaly) components. Therefore, in order to highlight the possible spatio-temporal variation of thermal anomalies, it was necessary to remove the seasonal component in the raw temperature time-series (de-seasoned time-series).

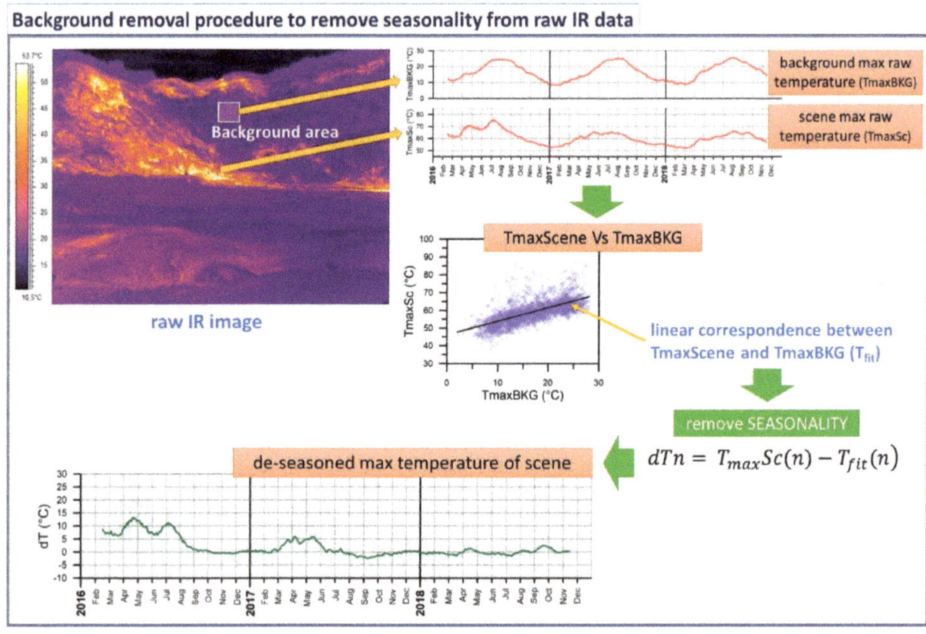

Figure 3. Processing scheme of background removal procedure (BKGr).

Different methods to remove seasonal component in time-series were previously tested to TIRNet data [18–20] and two methods demonstrated to be effective to perform seasonal adjustment: the background removal (BKGr) and the STL decomposition (STLd, Seasonal-Trend decomposition based on Loess) [40]. The effectiveness of these two different methodologies depends on the time-length of the dataset. BKGr is applied on time-series shorter than two years that cannot be processed by STLd as it requires several-years-long time-series. The BKGr removes the seasonality only to maximum and average temperatures of IR time-series and does not perform the seasonal adjustment to all the pixels of IR frame. Diversely, STLd can remove the seasonal component to all the pixels of IR frame, allowing to perform deeper analysis to the IR dataset.

The Background Removal Procedure (BKGr)

The BKGr procedure [18–20] consisted of the removal of background temperature time-series to raw IR frames time-series. Background temperatures were detected in a background area of the IR scene not influenced by thermal anomaly. The procedure was based on the evidence that a linear correspondence is between maximum (or mean) temperature of background area ($T_{max}BKG$) and maximum (or mean) temperature of IR scene (TmaxSc), as previously reported by [19,20] and illustrated in Figure 3 (TmaxSc vs TmaxBKG plot). This correspondence allows the application of the following equation:

$$dTn = T_{max}Sc(n) - T_{fit}(n) \tag{2}$$

were $dT(n)$ is the residual de-seasoned temperature value, $T_{max}Sc(n)$ is the maximum temperature of the n IR scene and $Tfit(n)$ is the value of $T_{max}Sc(n)$ in correspondence of $T_{max}BKG(n)$ according to the linear fitting equation of the two variables (Figure 3; Matlab© function: 'step03.m').

The accurate selection of the background area (BKG) was crucial as it strongly influenced the efficiency of this procedure. BKG had to be outside the region of the IR frame affected by thermal anomaly and also characterized by similar lithology of the anomaly area, without vegetation and any kind of anthropic object. An efficient way to test the quality of the chosen BKG was to perform a linear regression to time-series of average temperature values of BKG. A suitable BKG must have the slope of the linear regression equation near to zero (Figure 4a).

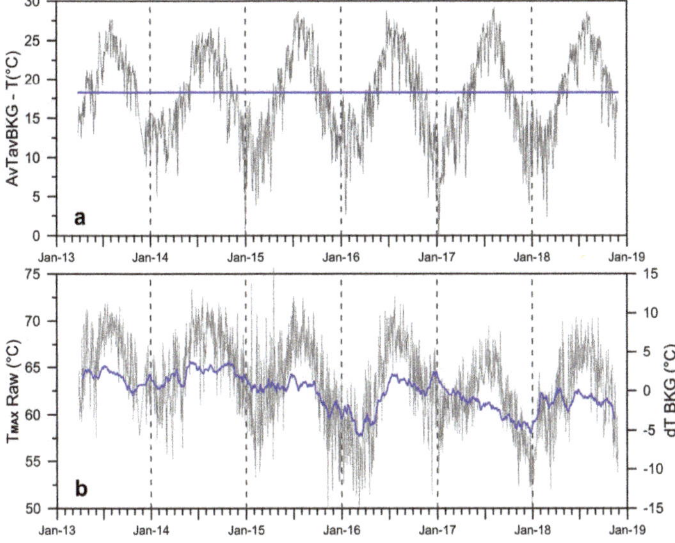

Figure 4. (**a**) Time-series of average temperature values of Pisciarelli background area (grey color) and linear regression fit (blue color); (**b**) the results of the background removal procedure applied to Pisciarelli station: RAW maximum temperature of IR scene (grey color) and residual temperature value dT (blue color).

The main advantage of BKGr method was the possibility to apply seasonal correction to short temperature time-series; nevertheless, the results are expressed in terms of temperature residuals and not as absolute temperatures (Figure 4b).

The STL Decomposition Method (STLd)

STL is a flexible, iterative non-parametric and robust method developed by [40] to decompose time-series into three components, according to an additive model:

$$TS_i = T_i + S_i + R_i \quad (3)$$

where TS_i is the time-series of i-th pixel of IR frame, T_i is the Trend, which represents a general tendency of data to move in a certain direction, S_i is the Seasonality, which is a repetitive pattern over time due to exogenous causes, and R_i is the remainder, e.g., TS_i removed of Trend and Seasonality components (Matlab© function: 'step04.m').

The stl() function is available in the R statistical programming language [41]. STL is an acronym for "Seasonal and Trend decomposition using Loess", where Loess is a method for estimating nonlinear relationships. The Loess (LOcal regrESSion) algorithm performs smooth estimate $g(t)$ for temperature T at all times t, not just at time t_i for which T has been observed. There are several parameters to set in the STL algorithm [40]. The main parameters are the number of observations *n.p* per seasonal cycle, the trend window (*t.window*) and the seasonal window (*s.window*). These last two parameters specify how quickly the trend and seasonal components can change. In different words, *t.window* is the number of consecutive observations to be used when estimating the trend; *s.window* is the number of consecutive years to be used in estimating each value in the seasonal component.

The 'standard' use of STL function in R is: *stl(time-series, s.window = "periodic").* By using the setting *s.window = "periodic"*, Loess smoothing is effectively replaced by the mean of the seasonal sub-series. This way, STL assumes the same seasonal cycle for each year of the time-series; therefore, the seasonal component for January is simply the mean of all January values and similarly for the other months.

STL can be set to be robust to outliers, so that occasional uncommon observations will not affect the trend and seasonal components, but only the remainder component.

As the STL algorithm was developed in R language only, a specific script was created to integrate the STL function into the Matlab© processing procedure. The script ('step04.m') consisted of two parts: a) Matlab© code which calls b) R code by using a Matlab toolbox (*RunRcode*, Matlab File Exchange).

When calling the STL function ('STLIR.R'), the *s.window* parameter was set to *'periodic'* and the *t.degree* was set to 0. This last one parameter is the degree of locally-fitted polynomial in trend extraction. Moreover, it was important to set the periodicity when creating the temperature time-series in R script. For TIRNet temperature data, the periodicity was set to 365.

STL needs at least a two-year long, continuous time-series; otherwise, it does not process the dataset. If the dataset is not continuous, due to data lack in some periods, it has to be resampled daily. In case of shorter dataset, only BKGr method can be applied.

The STLd procedure can be simply applied to statistical time-series (e.g., raw maximum temperatures time-series; Figure 5) or applied to time-series of all pixel temperatures of the IR frames by using the processing scheme reported in Figure 6.

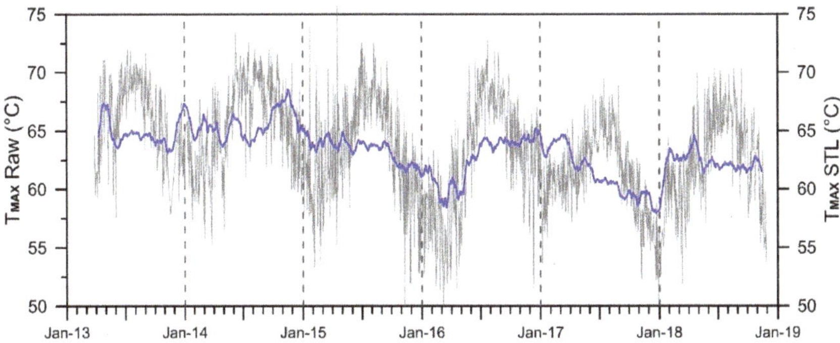

Figure 5. STL decomposition procedure applied to Pisciarelli station: RAW maximum temperature of IR scene (grey color) and de-seasoned maximum temperature of IR scene (blue color).

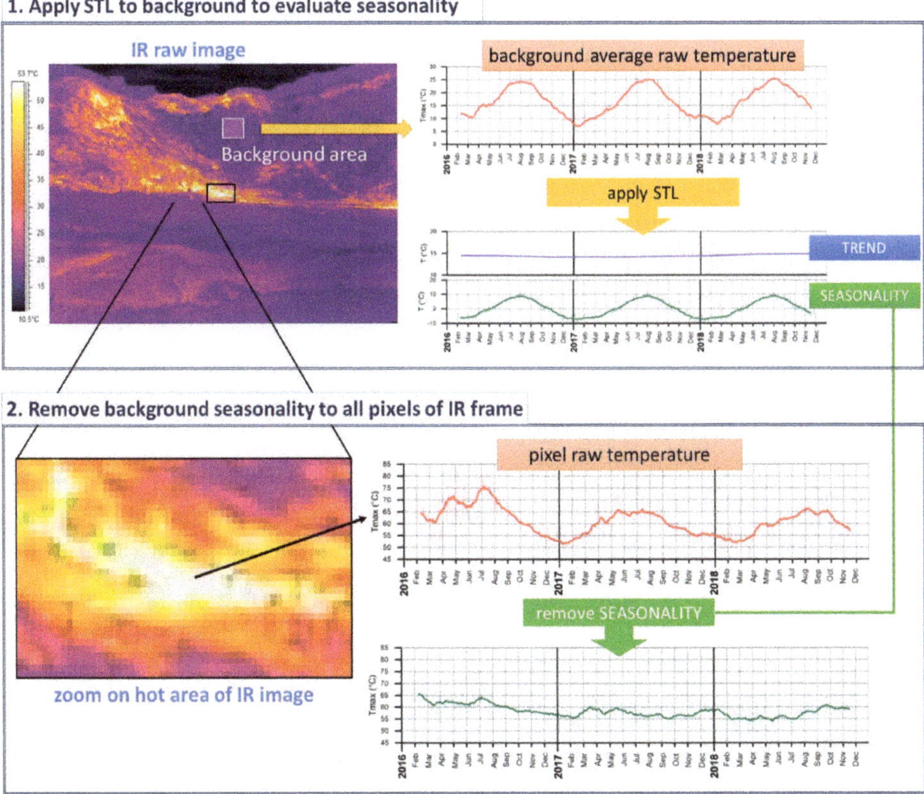

Figure 6. Processing scheme used to remove seasonal component of TIRNet data by using R script including STL procedure ('STLIR.R') and Matlab code. 1) Evaluation of seasonal component (S) by applying STL to background area. 2) Removal of S to a pixel raw temperature in order to get the de-seasoned values.

The STLd procedure used to remove seasonal component of all pixels from IR time-series (Figure 6) required, as a first step, the evaluation of the average temperatures time-series of BKG (TavBkgTS). The STL function was then applied to the TavBkgTS time-series to decompose it into three components:

$$\text{TavBkgTS} = \text{TrendBkgTS} + \text{SeasonBkgTS} + \text{RemBkgTS} \tag{4}$$

where TrendBkgTS, SeasonBkgTS and RemBkgTS are, respectively, Trend, Seasonality and Remainder time-series of TavBkgTS.

As the background area is not influenced by thermal anomaly, the SeasonBkgTS can be assumed to be representative of the seasonal component affecting all pixels of the frames acquired by an IR station. This assumption makes it possible to apply the following relation:

$$\text{T}_{\text{des}}\text{TS}_{\text{i}} = \text{T}_{\text{i}}\text{TS - SeasonBkgTS} \tag{5}$$

where $\text{T}_{\text{des}}\text{TS}_{\text{i}}$ is the de-seasoned time-series of ith pixel of IR frame and $\text{T}_{\text{i}}\text{TS}$ is the time-series of raw temperatures of the same pixel.

In brief, STLd procedure removed the seasonal component to temperature time-series from all the pixels of IR frames, acquired by a IR station, by subtracting the seasonal component of BKG (SeasonBkgTS). Additionally, in the STLd method, the correct choice of the BKG is fundamental. A direct control of BKG quality is to plot the values of TrendBkgTS: they must be without significant variations (Trend plot in Figure 6).

The final result is a Matlab© 3D array representative of IR frames with de-seasoned temperature values. These arrays are relevant to perform advanced pixel-to-pixel processing methods, needing de-seasoned IR data, which are reported in the next steps.

Additional output of processing step 5 is a map showing locations of maximum temperatures values detected in all IR frames.

3.2.4. Step 4—Radiative Heat Flux (Q_{rad})

The estimation of radiative heat flux (Q_{rad}) from an area of IR frame mainly characterized by thermal anomaly (Region of Anomaly, *RoA*) is a newly proposed processing technique that can offer an interesting contribution to the investigation of possible variations of radiative thermal emissions.

In order to estimate Q_{rad}, which is the thermal energy emitted per unity of area in a unity of time, the *RoA* has to include pixels whose temperatures are representative of the main thermal anomaly. Nevertheless, the *RoA* is usually not homogeneous and it is characterized by the presence of both high temperature sources (fumaroles) and low temperature sources (surrounding emission-free rocks). In addition, when sensor-target distance is more than approximately 10 m, the pixels of *RoA* can be several centimeters large, and therefore, some temperatures are underestimated if their pixels integrate both high and low temperatures [42,43]. Consequently, the variations trend of Q_{rad} can be sensibly flattened. A solution to this problem is to calculate the Standard Deviation (SD) of pixels' temperatures of a specific *RoA* and then to use only temperature values ($T_{ROA}H$) greater than 2SD to estimate Q_{rad}.

Finally, the Q_{rad} of a specific *RoA* (W/m^2) is calculated by using the Stefan-Boltzmann equation:

$$Q_{rad}RoA = A\sum_{i=1}^{n}\sigma\varepsilon(T_{RoA}H_i)^4 \tag{6}$$

where σ is the Stefan-Boltzmann constant, ε is the emissivity (for pyroclastic rocks is assumed to be 0.9) and *A* is the investigated area size (m^2) obtained by multiplying pixel area and length n of $T_{ROA}H$ time-series.

The detection of any possible change of Q_{rad} trends, even though related to a specific *RoA*, allows to better characterize thermal behavior of the studied area if *RoA* is representative of the main thermal anomaly.

The use of de-seasoned time-series of temperature values ($T_{des}TS$) is essential in order to evaluate Q_{rad} changes due to endogenous sources only. This means that only the dataset processed with STLd can be used.

The Matlab© code performing *Qrad* ('step04.m') is available as Supplementary Materials and its functionalities are illustrated in the Appendix A.

3.2.5. Step 5—Yearly Rate of Temperatures Change (YRTC)

The thermal variations, in a defined time interval, of every single pixels of IR frame, can be evidenced by evaluating the yearly rate of temperatures change (YRTC). This kind of elaboration produces a map of the IR frame, according to a color scale, of yearly rate of change of pixels' temperature. The yearly rate of temperatures change is represented by the values of slope coefficients of the linear fit of time-series temperatures of every pixel. On the other hand, the selected time interval has to be characterized by a progressive increase or decrease of maximum temperatures of IR frame, according to a correspondence as linear as possible. This needs a preliminary investigation of the temperatures trend over time.

The YRTC map is created by overlapping the values of slope coefficients on a picture (in the visible range) of the framed area. In order to show the yearly rate of change values of pixels whose temperature time-series best fit a linear model, a mask was applied. This mask allowed the display of values related to pixels whose linear regressions of temperature time-series had coefficients of determination (R^2) higher than a user-defined threshold value.

The YRTC map gives the opportunity to evidence possible connections between temperature increase/decrease and geological features of the monitored site (Figure 7).

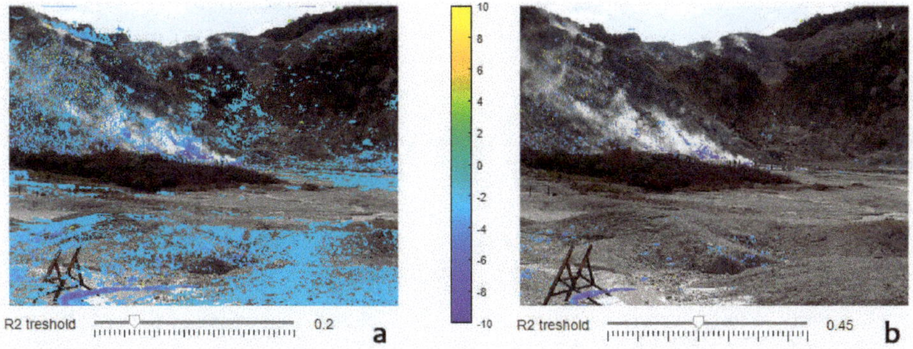

Figure 7. Yearly rate of temperature change maps of SF1 in the period 2016.02.01–2016.11.30. Map (**a**) has R^2 threshold value = 0.2. Map (**b**) has and R^2 threshold value = 0.45. In this time-interval SF1 maximum temperatures decreased of about 10 °C as evidenced by the color map.

The Matlab© code performing YRTC data ('step06.m') is available as Supplementary Materials and its functionalities are illustrated in the Appendix A.

3.3. *System Automation and Graphic Interface*

The above-described methodologies were performed as steps by Matlab© functions, which can be executed with a command line or managed by a user-friendly graphic interface (GUI). Settings can be saved in user-defined configuration files. Due to the modular structure of the processing steps, they can be performed singularly or grouped in an automated sequence in order to execute the whole procedure at defined time by using the GUI that integrates the automation code. Automation is necessary if IR data processing is aimed to surveillance purposes.

The GUI Matlab code (asira_gui.m) is available as Supplementary Materials and its functionalities are illustrated in the Appendix A.

4. Results and Discussion

In order to discuss the advantages and the limits of the above presented processing methodology, the results obtained by applying the five processing steps are reported. Two datasets were processed: (1) the first consisted of 2.901 IR JPEG frames acquired in the period 2016.01.27–2019.01.13 at Solfatara 1 (SF1) station; (2) the second consisted of 5.850 IR JPEG frames acquired in the period 2013.03.26–2019.01.13 at Pisciarelli (PS1) station.

4.1. Data Quality Selection

The relation (1), discussed in §3.2.1, was used to remove low-quality IR frames before starting the analysis of data. The efficiency of this procedure depended on the choice of the coefficient *c* which was influenced by the statistical distribution of data. Low values of Standard Deviation of IR frames temperatures were an indicator of low quality data and the lower the coefficient *c*, the higher the number of IR frames discarded as low quality ones. The analysis of data acquired by SF1 suggested c = 1 as an appropriate value (Figure 8), as the visual inspection of discarded frames (about 11% of total frames) confirmed that they were mainly low-quality ones. This kind of preliminary analysis had to be made to every dataset from different stations as the coefficient *c* can be different depending on the physical and geometrical characteristics of framed area and IR sensor.

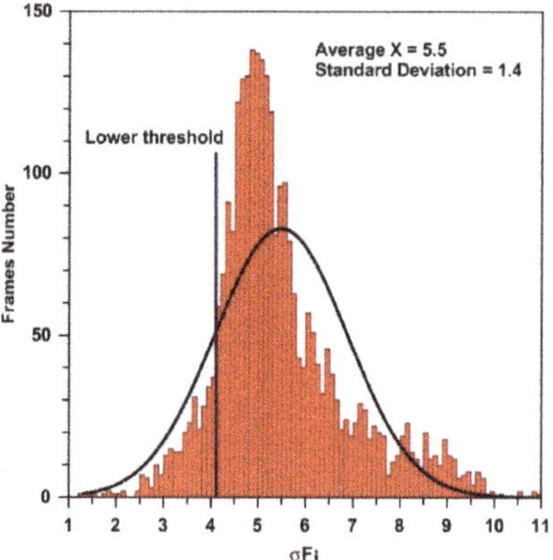

Figure 8. Frequency distribution of Standard Deviation values of IR frames temperatures (σF) acquired at SF1 station. The line 'Lower Threshold', which is defined by the relation (1) with *c* parameter equal to 1, splits good-quality frames (on the right of the line) and low-quality frames (on the left of the line).

4.2. Seasonal Component Removal

Two different methodologies of seasonal component removal are used in order to process IR datasets having different time-length. The background removal procedure (BKGr), previously proposed to seasonal correction [19,20], is suitable to very short datasets even though it has some limitations in the final output. The main limit was that the removal of seasonal component produces only residuals of maximum or median values of temperatures instead of absolute temperature values. Although this kind of analysis does not take full advantage of all the intrinsic information contained inside the IR frames, the BKGr method generated trends of temperature residuals which provide adequate information to characterize the thermal behavior of studied area.

The STL decomposition method (STLd), proposed for the first time in this work to remove seasonality to IR temperature time-series, needed a nearly two-year long datasets due to the statistical approach of the robust and widely applied algorithm. By using STLd method, it was possible to estimate the Seasonality as a separate component of temperature time-series. This feature allows the removal of seasonality to all pixels of IR frames, giving the opportunity to apply further analysis methods (e.g., radiative Heat Flux estimate), which needed the whole frame to be de-seasoned. In Figure 9, the background area boundaries (Figure 9a) inside the SF1 IR frame and the plot of Trend component of background area, obtained by applying STLd method are reported (Figure 9b). The constant and flat temperature values of background Trend confirmed the appropriate choice of this area.

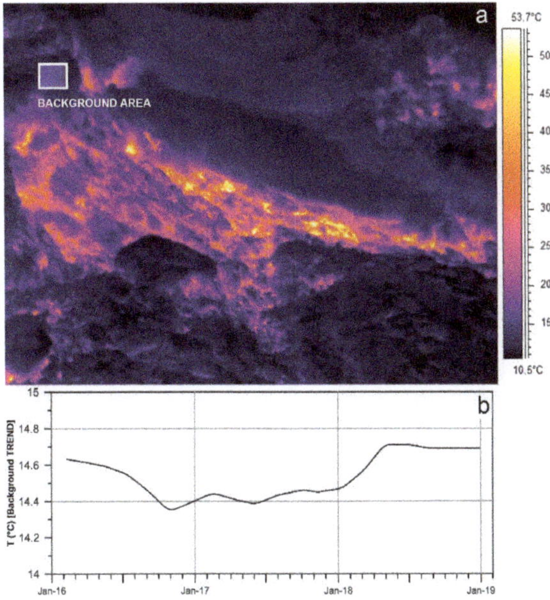

Figure 9. Background area boundaries inside the SF1 IR frame (**a**) and plot of Trend component of background area which is obtained by applying STLd method (**b**). Trend values varie between 14.35 and 14.7 °C.

The Trend component evaluated by the STLd method was useful to estimate the long-term thermal behavior of the studied area even though it was not suitable for short-term observations. In order to describe short-term thermal behavior, aimed to surveillance purpose, it was necessary to merge both Trend and Reminder components to obtain T + R plots (Figure 10, blue line plot).

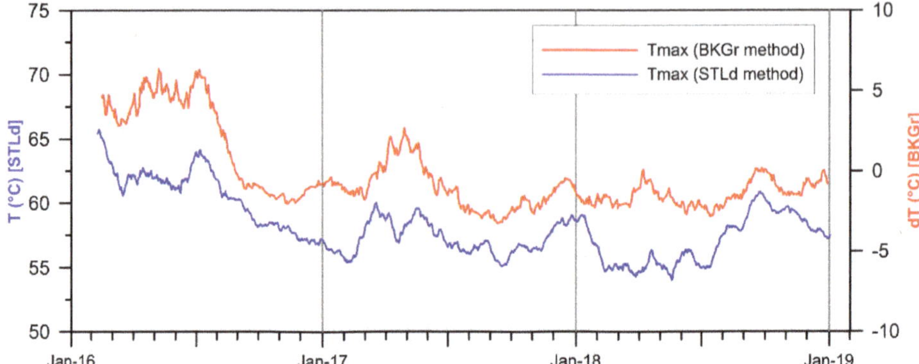

Figure 10. Plots of temperatures acquired at SF1 station removed of seasonal component. Red line = temperature residuals obtained by applying BKGr method; blue line = Trend+Reminder values obtained by applying STLd method.

Despite the reported limits of BKGr method, the comparison between temperature residuals plots with the BKGr method and T+R plots by STLd method of SF1 IR frames (Figure 10) showed a close similarity of data trends. This similarity confirms the effectiveness of the BKGr method to process datasets shorter than two years.

4.3. Radiative Heat Flux Estimate

The computation of radiative heat flux was available on IR frames where seasonality was removed by applying the STLd method. In order to obtain the correct trend of radiative heat flux of a definite area, a correct selection of area boundaries was necessary. The heat flux computation strongly depended on the Hpix BKGpix ratio of the selected area, where Hpix is the number of pixels related to thermal anomaly and BKGpix is the number of pixels related to emission-free rocks. The higher this ratio is, the more accurate the radiative heat flux estimation. This way, the choice of boundaries of processed areas had to be made in order to include as many Hpix as possible. The solution to attenuate the underestimate the heat flux due to the presence of BKGpix, proposed in the §3.2.4, was to select pixels whose temperatures were greater than 2σ of the frequency distribution of temperatures from selected area. The plots reported in Figure 11 show how efficient this kind of solution was. In this figure, plot a) reports heat flux time-series of Areas 1, 2 and 3 evaluated by selecting all the pixels inside each area; plot b) reports heat flux time series of the same areas, evaluated by applying the selection of pixels greater than 2σ of temperatures frequency distribution. The blue line plot is from Area 1, which only includes the major thermal anomaly of the SF1 frame, characterized by the higher Hpix BKGpix ratio. Red line plots of Area 2 and black line plots of Area 3 are representative of lower Hpix BKGpix ratios due to higher number of BKGpix included in the selected areas.

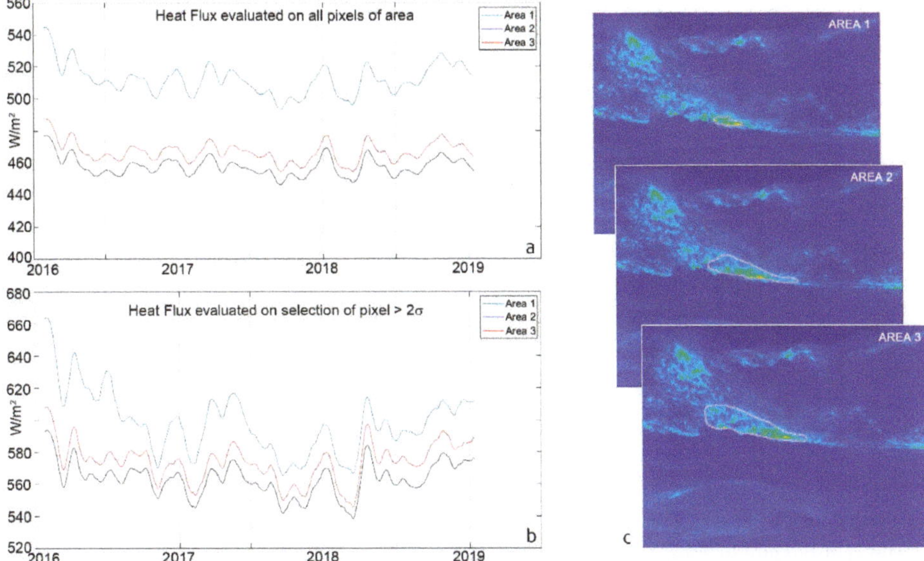

Figure 11. Heat flux plots of selected areas inside SF1 frames. Area 1 (c) includes the mayor thermal anomaly, Area 2 and 3 (c) include emission-free rocks. Plot a) = heat flux trends (smoothed with window = 29) of Areas 1, 2 and 3 evaluated by selecting all the pixels inside each area. Plot b) = heat flux trends (smoothed with window = 29) of Areas 1, 2 and 3 evaluated by applying the selection of pixels greater than 2 s of temperatures frequency distribution.

The comparison between Figure 11a,b evidences an underestimate of heat flux values when the computation includes all the pixels inside the selected areas (Figure 11a). Figure 11b was obtained selecting only the pixels whose temperatures values were greater than 2σ and showed a remarkable decrease of heat flux underestimate, better evidencing trend variations.

4.4. Yearly Rate of Temperature Change Estimate

As reported in §3.2.5, the final product of this processing step was a color scale map of the yearly rate of temperature change (YRTC) values overlapped to a picture (in the visible range) of the framed area. YRTC data were filtered according to a threshold value of the coefficient of determination (R^2) of the linear regressions of pixels' temperature time-series. Two different examples of yearly temperature rate of change maps of PS1 area in the same time-interval (2016.03.10–2016.07.10) are reported in Figure 12. In this time-interval the PS1 temperatures were subjected to an increase of about 10 °C. The maps of Figure 12 only differ in the choice of R^2 threshold value; hence, a correct choice of these parameter is critical to produce a map that is easy to comprehend. Map **b** (R^2 = 0.7) shows better evidence of pixels whose temperatures rate of change time-series values best fit a linear model than map **a** (R^2 = 0.5). The ASIRA code allows the user to select both color scale limits and different values of R^2 threshold by using a user-friendly GUI in order to achieve the right balance between optimal visual result and reliability of data visualized.

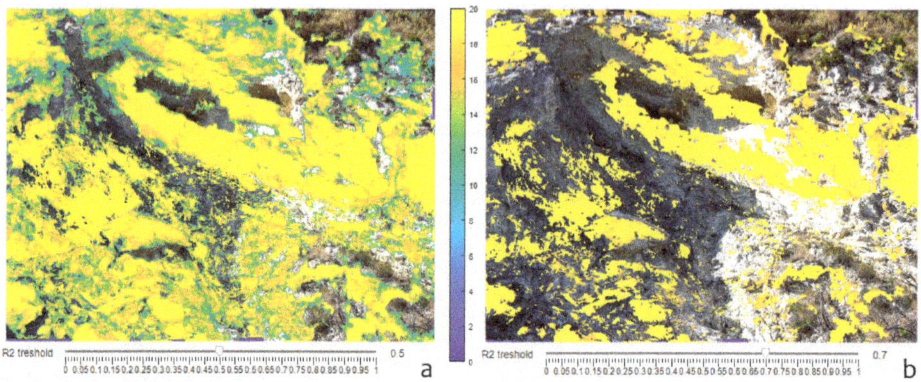

Figure 12. Yearly rate of temperature change maps of PS1 in the period 2016.03.10–2016.07.10. Map (**a**) has R^2 threshold value = 0.5. Map (**b**) has and R^2 threshold value = 0.7.

5. Conclusions

Relevant contribution to the surveillance of volcanic areas affected by thermal anomalies can be provided by monitoring the spatio-temporal evolution of surface temperatures field. The acquisition of IR image data by ground-based monitoring network is an effective tool to perform this task. However, the analysis of IR data time-series is not easy to accomplish due to the influence over IR temperatures of both exogenous and endogenous processes.

In this paper, we have presented a unique operational processing chain developed in Matlab© environment which allows the detection and quantification of possible changes in time and space of the ground-surface thermal features. This application (ASIRA, Automated System of InfraRed Analysis) performed a multi-step procedure that generated both trends of temperatures and heat fluxes as well as maps of yearly rate of temperatures change. The procedure implemented new algorithms based on improvements of previously proposed methods and also original techniques aimed to effectively remove seasonal component of IR temperature time-series and to evaluate radiative heat fluxes of thermal anomaly areas.

ASIRA can be performed as separate steps or executed in a fully-automated way by using a user-friendly graphic interface. The Matlab© code of ASIRA and the Operative Manual are included as Supplementary Materials.

The ASIRA code was applied to process IR data acquired by stations of TIRNet surveillance network operated by the Osservatorio Vesuviano, section of National Institute of Geophysics and Volcanology (INGV) at Campi Flegrei volcanic area (Italy). The results show the effectiveness of this method to provide a valuable contribution to the continuous monitoring of thermal anomalies related to studied areas.

This operative tool has been conceived for volcanic surveillance of diffuse degassing areas and low-temperature fumarole fields which variations may precede significant phases of volcanic unrest. Notwithstanding, the procedure can be applied to monitor different volcanic scenarios (i.e., lava-flows, active volcanic vents and eruptive fractures) but also different natural and environmental hazards (fires, waste-disposal sites, pollution discharges, landslides, etc.).

Supplementary Materials: the Matlab© code of A.S.I.R.A. (Automated System of InfraRed Analysis) which is described in Appendix A), and the Operative Manual (pdf file) are provided at the following link: http://www.mdpi.com/2072-4292/11/5/553/s1.

Author Contributions: F.S. was responsible for the ASIRA design and implementation in the Matlab© software, data acquisition and analysis, and writing the manuscript. G.V. was responsible for the research design, data analysis and writing the manuscript. Both authors contributed to the software validation.

Funding: TIRNet monitoring network was partially funded by the 2000–2006 National Operating Program (NOP) and by SISTEMA project, which has been developed in the framework of the Campania Regional Operating Program (ROP) FESR 2007-2013.

Conflicts of Interest: The authors declare no conflict of interest.

Appendix A

The Matlab application ASIRA: operational structure and technical notes.

ASIRA is an acronym of Automated System of InfraRed Analysis and consists of Matlab© code subdivided into five independent processing steps (step01.m, step02.m, step03.m, step04.m, step05.m) that can be easily managed by a graphic user interface (asira_gui.m). Moreover, additional Matlab© scripts and libraries are needed to ASIRA functionalities. Figure A1 shows screen-captures of different tabs of the graphics interface representative of five processing steps and automation settings.

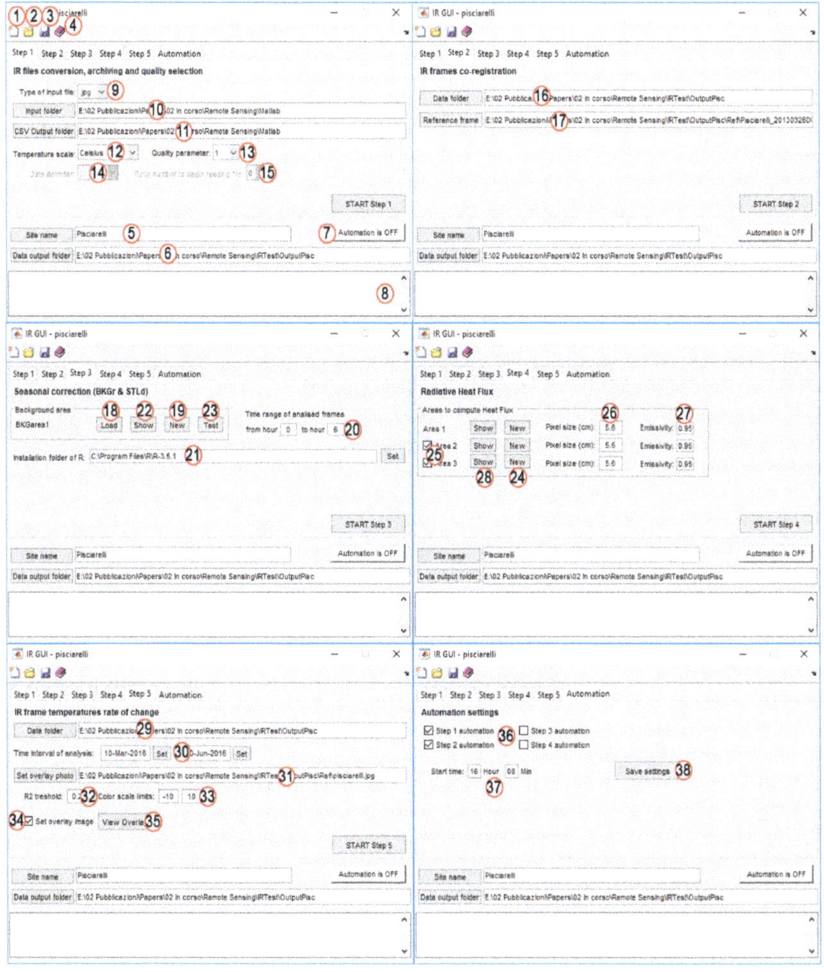

Figure A1. Tabs of the Automated System of InfraRed Analysis (ASIRA) graphics interface representative of five processing steps and automation settings. Numbers inside the red circles are the IDs reported in the technical sheets of Appendix A.

Below are reported synthetic technical sheets of functionalities and type of input and output data of the different processing steps and of the graphic user interface.

Table A1. Technical specifications of functionality and input/output data type of graphic interface.

Script Name		
asira_gui.m		
Functionality		
Graphic user interface (GUI) with management of configuration file		
Inputs description	**Inputs type**	**ID**
New configuration file	File name in common dialog window by pressing toolbar button	1
Open configuration file	File name in common dialog window by pressing toolbar button	2
Save current configuration file	Toolbar button	3
Open Operative Guide	Toolbar button	4
Site name (study area)	String inserted by edit window	5
Output folder of processed data (common to all steps)	Folder path inserted by common dialog window	6
Automation button (activate/deactivate automation)	Button	7
Enable/disable automation of processing step	Check box selection	36
Time to start automation process	Text boxes to input Hour and Minutes	37
Save automation settings	Button	38
Outputs description	**Output type**	**ID**
Log window showing processing messages	Text displayed in box area	8

Table A2. Technical specifications of functionality and input/output data type of STEP 1.

Script Name		
step01.m		
Functionality		
IR files conversion, archiving and quality selection (tab 'Step 1' in GUI)		
Inputs description	**Inputs type**	**ID**
Type of input file	'.jpg/.csv/.txt' inserted by drop-down menu	9
Data input folder	Folder path in common dialog window by pressing button	10
Output folder of CSV files [1]	Folder path in common dialog window by pressing button	11
Temperature scale	'Celsius/Fahrenheit' inserted by drop-down menu	12
Quality selection parameter	'05/1/1.5/2' inserted by drop-down menu	13
Data delimiter of csv/txt input files [1]	',/;/TAB/SPACE' inserted by drop-down menu	14
Row number to begin reading data in csv/txt file	Integer inserted by drop-down menu	15
Outputs description	**Output type**	**ID**
Log window showing processing messages	Text displayed in box area	8
CSV files of quality selected IR frames	Matrix CSV files of temperature values from IR scenes	
Arrays of quality selected IR data, yearly split	Matlab (.mat) archives in output folder	

Table A3. Technical specifications of functionality and input/output data type of STEP 2.

Script Name		
step02.m		
Functionality		
IR frames co-registration (tab 'Step 2′ in GUI)		
Inputs description	**Inputs type**	**ID**
Data input folder (containing .mat archives of Step 1)	Folder path in common dialog window by pressing button	16
Reference IR frame	File name & path in common dialog window by pressing button	17
Outputs description	**Output type**	**ID**
Log window showing processing messages	Text displayed in box area	8
Arrays of co-registered IR data, yearly split	Matlab (.mat) archives in output folder	

Table A4. Technical specifications of functionality and input/output data type of STEP 3.

Script Name		
step03.m		
Functionality		
Seasonal correction with BKGr and STLd methods (tab 'Step 3′ in GUI)		
Inputs description	**Inputs type**	**ID**
Load background area	File name & path in common dialog window by pressing button	18
New background area	File name & path in common dialog window by pressing button and selection of area over IR image	19
Daily time range of IR frames	Integers (hours) in text boxes	20
Installation folder of R statistical package (STL)	Folder path in common dialog window by pressing button	21
Outputs description	**Output type**	**ID**
Log window showing processing messages	Text displayed in box area	8
Show background area image	JPEG image of background area	22
Test background area image	Plots of Tmax and STL Trend of background area (by choice)	23
Array of de-seasoned IR data	Matlab (.mat) archive in output folder	
Data sheets of processed temperatures of IR frames	Excel file in output folder	

Table A5. Technical specifications of functionality and input/output data type of STEP 4.

Script Name		
step04.m		
Functionality		
Radiative heat flux estimation (tab 'Step 4′ in GUI)		
Inputs description	**Inputs type**	**ID**
New heat flux areas (Area 1, 2, 3)	Selection of heat flux area over IR image by pressing button	24
Enable/disable heat flux areas to process (Area 2, 3)	Check box selection	25
Pixel size of heat flux areas (Area 1, 2, 3)	Numeric values in text box	26
Emissivity of heat flux areas (Area 1, 2, 3)	Numeric values in text box	27
Outputs description	**Output type**	**ID**
Log window showing processing messages	Text displayed in box area	8
Show heat flux areas (areas 1, 2, 3)	JPEG images by pressing button	28
Arrays of heat flux data	Matlab (.mat) archive in output folder	
Data sheets of heat fluxes of IR frames	Excel file in output folder	

Table A6. Technical specifications of functionality and input/output data type of STEP 5.

Script Name		
step05.m		
Functionality		
Temperature rate of change during selected time-period (tab 'Step 5′ in GUI)		
Inputs description	**Inputs type**	**ID**
Data input folder (containing .mat output files of previous Steps)	Folder path in common dialog window by pressing button	29
Time interval of analysis	Dates picked over calendar	30
Photo of studied area to use in data overlay	File name & path in common dialog window by pressing button	31
Threshold value of R^2 extracted from linear regressions of pixels time-series	Numeric values in text box	32
Limits of color scale to use in temperature rate of change map	Numeric values in text box	33
Enable/disable data overlay on photo of studied area	Check box selection	34
Outputs description	**Output type**	**ID**
Log window showing processing messages	Text displayed in box area	8
Show map of temperature rate of change	JPEG image by pressing button	35
Arrays of temperature rate of change data	Matlab (.mat) archive in output folder	
Data sheets of temperature rate of change data	Excel file in output folder	

Although the processing steps can be managed separately, the processing chain needs data to be analyzed by the first three steps in sequential way. The Operative Manual of ASIRA is available as Supplementary Materials together with Matlab© scripts and Open Source Toolboxes and functions. Matlab© scripts code is widely commented in order to understand the features and the functionality.

References

1. Ball, M.; Pinkerton, H. Factors affecting the accuracy of thermal imaging cameras in volcanology. *J. Geophys. Res.* **2006**, *111*, B11203. [CrossRef]
2. Lagios, E.; Vassilopoulou, S.; Sakkas, V.; Dietrich, V.; Damiata, B.N.; Ganas, A. Testing satellite and ground thermal imaging of low-temperature fumarolic fields: The dormant Nisyros Volcano (Greece). *ISPRS J. Photogramm. Remote Sens.* **2007**, *62*, 447–460. [CrossRef]
3. Stevenson, J.A.; Varley, N. Fumarole monitoring with a handheld infrared camera: Volcan de Colima, Mexico, 2006–2007. *J. Volcanol. Geotherm. Res.* **2008**, *177*, 911–924. [CrossRef]
4. Oppenheimer, C.; Lomakina, A.S.; Kyle, P.R.; Kingsbury, N.G.; Boichu, M. Pulsatory magma supply to a phonolite lava lake. *Earth Planet. Sci. Lett.* **2009**, *284*, 392–398. [CrossRef]
5. Calvari, S.; Lodato, L.; Steffke, A.; Cristaldi, A.; Harris, A.J.L.; Spampinato, L.; Boschi, E. The 2007 Stromboli eruption: Event chronology and effusion rates using thermal infrared data. *J. Geophys. Res.* **2010**, *115*, B04201. [CrossRef]
6. Schöpa, A.; Pantaleo, M.; Walter, T.R. Scale-dependent location of hydrothermal vents: Stress field models and infrared field observations on the Fossa Cone, Vulcano Island, Italy. *J. Volcanol. Geotherm. Res.* **2011**, *203*, 133–145. [CrossRef]
7. Spampinato, L.; Calvari, S.; Oppenheimer, C.; Boschi, E. Volcano surveillance using infrared cameras. *Earth Sci. Rev.* **2011**, *106*, 63–91. [CrossRef]
8. Vaughan, R.G.; Keszthelyi, L.P.; Lowenstern, J.B.; Jaworowski, C.; Heasler, H. Use of ASTER and MODIS thermal infrared data to quantify heat flow and hydrothermal change at Yellowstone National Park. *J. Volcanol. Geotherm. Res.* **2012**, *233–234*, 72–89. [CrossRef]
9. Spampinato, L.; Ganci, G.; Hernández, P.A.; Calvo, D.; Tedesco, D.; Pérez, N.M.; Calvari, S.; Del Negro, C.; Yalire, M.M. Thermal insights into the dynamics of Nyiragongo lava lake from ground and satellite measurements. *J. Geophys. Res. Solid Earth* **2013**, *118*, 5771–5784. [CrossRef]

10. Zaksek, K.; Shirzaei, M.; Hort, M. Constraining the uncertainties of volcano thermal anomaly monitoring using a K alman filter technique. In *Remote Sensing of Volcanoes and Volcanic Process*; Pyle, D.M., Mather, T.A., Biggs, J., Eds.; Geological Society of London Special Publications: London, UK, 2013; Volume 380, pp. 139–160.
11. Vaughan, R.G.; Heasler, H.; Jaworowski, C.; Lowenstern, J.B.; Keszthelyi, L.P. Provisional maps of thermal areas in Yellowstone National Park, based on satellite thermal infrared imaging and field observations. *US Geol. Surv. Sci. Investig. Rep.* **2014**, *5137*, 22.
12. Patrick, R.M.; Orr, T.; Antolik, L.; Lopaka, L.; Kamibayashi, K. Continuous monitoring of Hawaiian volcanoes with thermal cameras. *J. Appl. Volcanol.* **2014**, *3*, 1. [CrossRef]
13. Cerminara, M.; Esposti Ongaro, T.; Valade, S.; Harris, A.J.L. Volcanic plume vent conditions retrieved from infrared images: A forward and inverse modeling approach. *J. Volcanol. Geotherm. Res.* **2015**, *300*, 129–147. [CrossRef]
14. Lewis, A.; Hilley, G.E.; Lewicki, J.L. Integrated thermal infrared imaging and structure-from-motion photogrammetry to map apparent temperature and radiant hydrothermal heat flux at Mammoth Mountain, CA, USA. *J. Volcanol. Geotherm. Res.* **2015**, *303*, 16–24. [CrossRef]
15. Bombrun, M.; Jessop, D.; Harris, A.; Barra, B. An algorithm for the detection and characterisation of volcanic plumes using thermal camera imagery. *J. Volcanol. Geotherm. Res.* **2018**, *352*, 26–37. [CrossRef]
16. Platt, U.; Bobrowski, N.; Butz, A. Ground-Based Remote Sensing and Imaging of Volcanic Gases and Quantitative Determination of Multi-Species Emission Fluxes. *Geosciences* **2018**, *8*, 44. [CrossRef]
17. Valade, S.; Ripepe, M.; Giuffrida, G.; Karume, K.; Tedesco, D. Dynamics of Mount Nyiragongo lava lake inferred from thermal imaging and infrasound array. *Earth Planet. Sci. Lett.* **2018**, *500*, 192–204. [CrossRef]
18. Chiodini, G.; Vilardo, G.; Augusti, V.; Granieri, D.; Caliro, S.; Minopoli, C.; Terranova, C. Thermal monitoring of hydrothermal activity by permanent infrared automatic stations: Results obtained at Solfatara di Pozzuoli, Campi Flegrei (Italy). *J. Geophys. Res.* **2007**, *112*, B12206. [CrossRef]
19. Sansivero, F.; Scarpato, G.; Vilardo, G. The automated infrared thermal imaging system for the continuous long-term monitoring of the surface temperature of the Vesuvius crater. *Ann. Geophys.* **2013**, *56*, S0454. [CrossRef]
20. Vilardo, G.; Sansivero, F.; Chiodini, G. Long-term TIR imagery processing for spatiotemporal monitoring of surface thermal features in volcanic environment: A case study in the Campi Flegrei (Southern Italy). *J. Geophys. Res. Solid Earth* **2015**, *120*, 812–826. [CrossRef]
21. Kieffer, H.H.; Frank, D.; Friedman, J.D. Thermal infrared surveys at Mount St. Helens—observations prior to the eruption of May 18. In: Lipman P.W., Mullineaux D.R. (Eds). The 1980 eruptions of Mount St. Helens, Washington. *USGS Prof. Pap.* **1981**, *1250*, 257–278.
22. Bonaccorso, A.; Calvari, S.; Garfì, G.; Lodato, L.; Patanè, D. Dynamics of the December 2002 flank failure and tsunami at Stromboli volcano inferred by volcanological and geophysical observations. *Geophys. Res. Lett.* **2003**, *30*, 1941–1944. [CrossRef]
23. Hernández, P.A.; Pérez, N.M.; Varekamp, J.C.; Henriquez, B.; Hernández, A.; Barrancos, J.; Padron, E.; Calvo, D.; Melian, G. Crater lake temperature changes of the 2005 eruption of Santa Ana Volcano, El Salvador, Central America. *Pure. App. Geophys.* **2007**, *164*, 2507–2522. [CrossRef]
24. Yokoo, A. Continuous thermal monitoring of the 2008 eruptions at Showa crater of Sakurajima volcano, Japan. *Earth Planets Space* **2009**, *61*, 1345–1350. [CrossRef]
25. Di Vito, M.A.; Acocella, V.; Aiello, G.; Barra, D.; Battaglia, M.; Carandente, A.; Del Gaudio, C.; de Vita, S.; Ricciardi, G.P.; Ricco, C.; Scandone, R.; Terrasi, F. Magma transfer at Campi Flegrei caldera (Italy) before the 1538 AD eruption. *Sci. Rep.* **2016**, *6*, 32245. [CrossRef] [PubMed]
26. Caliro, S.; Chiodini, G.; Moretti, R.; Avino, R.; Granieri, D.; Russo, M.; Fiebig, J. The origin of the fumaroles of La Solfatara (Campi Flegrei, south Italy). *Geochim. Cosmochim. Acta* **2007**, *71*, 3040–3055. [CrossRef]
27. Chiodini, G.; Vandemeulebrouck, J.; Caliro, S.; D'Auria, L.; De Martino, P.; Mangiacapra, A.; Petrillo, Z. Evidence of thermal-driven processes triggering the 2005–2014 unrest at Campi Flegrei caldera. *Earth Planet. Sci. Lett.* **2015**, *414*, 58–67. [CrossRef]
28. Montanaro, C.; Scheu, B.; Mayer, K.; Orsi, G.; Moretti, R.; Isaia, R.; Dingwell, D.B. Experimental investigations on the explosivity of steam-driven eruptions: A case study of Solfatara volcano (Campi Flegrei). *J. Geophys. Res. Solid Earth* **2016**, *121*, 7996–8014. [CrossRef]

29. Cubellis, E.; Marturano, A.; Pappalardo, L. The last Vesuvius eruption in March 1944: Reconstruction of the eruptive dynamic and its impact on the environment and people through witness reports and volcanological evidence. *Nat. Hazards* **2016**, *82*, 95. [CrossRef]
30. Caliro, S.; Chiodini, G.; Avino, R.; Cardellini, C.; Frondini, F. Volcanic degassing at Somma-Vesuvio (Italy) inferred by chemical and isotopic signatures of groundwater. *Appl. Geochem.* **2005**, *20*, 1060–1076. [CrossRef]
31. De Lorenzo, S.; Di Rienzo, I.; Civetta, L.; D'antonio, M.; Gasparini, P. Thermal model of the Vesuvius magma chamber. *Geophys. Res. Lett.* **2006**. [CrossRef]
32. Caliro, S.; Chiodini, G.; Avino, R.; Minopoli, C.; Bocchino, B. Long time-series of chemical and isotopic compositions of Vesuvius fumaroles: Evidence for deep and shallow processes. *Ann. Geophys.* **2011**, *54*, 137–149. [CrossRef]
33. Spampinato, L.; Calvari, S.; Oppenheimer, C.; Boschi, E. Volcano surveillance using infrared cameras. *Earth-Sci. Rev.* **2011**, *106*, 63–91. [CrossRef]
34. Merucci, L.; Bogliolo M., P.; Buongiorno M., F.; Teggi, S. Spectral emissivity and temperature maps of the Solfatara crater from DAIS hyperspectral images. *Ann. Geophys.* **2006**, *49*, 235–244. [CrossRef]
35. Sawyer, G.M.; Burton, M.R. Effects of a volcanic plume on thermal imaging data. *Geophys. Res. Lett.* **2006**, *33*, L14311. [CrossRef]
36. Seward, A.; Salman, S.; Robert Reeves, R.; Chris Bromley, C. Improved environmental monitoring of surface geothermal features through comparisons of thermal infrared, satellite remote sensing and terrestrial calorimetry. *Geothermics* **2018**, *73*, 60–73. [CrossRef]
37. Gaudin, D.; Beauducel, F.; Allemand, P.; Delacourt, C.; Finizola, A. Heat flux measurement from thermal infrared imagery in low-flux fumarolic zones: Example of the Ty fault (La Soufrière de Guadeloupe). *J. Volcanol. Geotherm. Res.* **2013**, *267*, 47–56. [CrossRef]
38. Pantaleo, M.; Walter, T.R. The ring-shaped thermal field of Stefanos crater, Nisyros Island: A conceptual model. *Solid Earth* **2014**, *5*, 183–198. [CrossRef]
39. Liu, C.; Yuen, J.; Torralba, A. SIFT Flow: Dense Correspondence across Scenes and its Applications. *IEEE Trans. Pattern Anal. Mach. Intell.* **2011**, *33*, 978–994. [CrossRef]
40. Cleveland, R.B.; Cleveland, W.S.; McRae, J.E.; Terpenning, I.J. STL: A seasonal-trend decomposition procedure based on loess. *J. Off. Stat.* **1990**, *6*, 3–73.
41. R Core Team (2018) R: A Language and Environment for Statistical Computing. R Foundation for Statistical Computing, Vienna. Available online: https://www.R-project.org (accessed on 22 January 2019).
42. Dozier, J. A method for satellite identification of surface temperature fields of subpixel resolution. *Remote Sens. Environ.* **1981**, *11*, 221–229. [CrossRef]
43. Harris, A.J.L.; Lodato, L.; Dehn, J.; Spampinato, L. Thermal characterization of the Vulcano fumarole field. *Bull. Volcanol.* **2009**, *71*, 441–458. [CrossRef]

Article

Role of Emissivity in Lava Flow 'Distance-to-Run' Estimates from Satellite-Based Volcano Monitoring

Nikola Rogic [1,*], Annalisa Cappello [2] and Fabrizio Ferrucci [1,3]

1 School of Environment, Earth and Ecosystem Sciences, The Open University, Milton Keynes MK7 6AA, UK; fabrizio.ferrucci@unical.it

2 Istituto Nazionale di Geofisica e Vulcanologia, Osservatorio Etneo, 95125 Catania, Italy; annalisa.cappello@ingv.it

3 Department of Environmental and Chemical Engineering, University of Calabria, 87036 Rende (CS), Italy

* Correspondence: nikola.rogic@open.ac.uk; Tel.: +44-077-313-02424

Received: 30 January 2019; Accepted: 14 March 2019; Published: 19 March 2019

Abstract: Remote sensing is an established technological solution for bridging critical gaps in volcanic hazard assessment and risk mitigation. The enormous amount of remote sensing data available today at a range of temporal and spatial resolutions can aid emergency management in volcanic crises by detecting and measuring high-temperature thermal anomalies and providing lava flow propagation forecasts. In such thermal estimates, an important role is played by emissivity—the efficiency with which a surface radiates its thermal energy at various wavelengths. Emissivity has a close relationship with land surface temperatures and radiant fluxes, and it impacts directly on the prediction of lava flow behavior, as mass flux estimates depend on measured radiant fluxes. Since emissivity is seldom measured and mostly assumed, we aimed to fill this gap in knowledge by carrying out a multi-stage experiment, combining laboratory-based Fourier transform infrared (FTIR) analyses, remote sensing data, and numerical modeling. We tested the capacity for reproducing emissivity from spaceborne observations using ASTER Global Emissivity Database (GED) while assessing the spatial heterogeneity of emissivity. Our laboratory-satellite emissivity values were used to establish a realistic land surface temperature from a high-resolution spaceborne payload (ETM+) to obtain an instant temperature–radiant flux and eruption rate results for the 2001 Mount Etna (Italy) eruption. Forward-modeling tests conducted on the 2001 'aa' lava flow by means of the MAGFLOW Cellular Automata code produced differences of up to ~600 m in the simulated lava flow 'distance-to-run' for a range of emissivity values. Given the density and proximity of urban settlements on and around Mount Etna, these results may have significant implications for civil protection and urban planning applications.

Keywords: emissivity; lava flow modeling; remote sensing; volcano monitoring

---

1. Introduction

As less than 10% of the ~1500 active subaerial volcanoes around the world are monitored regularly on the ground, remote sensing (RS) provides an opportunity to increase coverage. A combination of laboratory-based analyses, RS data, and numerical modeling could bridge critical gaps in volcanic hazard assessment and risk mitigation.

The prediction of lava flow 'distance-to-run' (ultimate length) is viewed as the key activity in support of risk mapping and planning the emergency response and crisis management of effusive volcanic events. The impact of volcanic eruptions and distances to which erupted lava will flow depend on several physical and chemical parameters [1–4].

It is widely recognized that RS data can be integrated with ground-based observations during volcanic crisis to facilitate the estimation of thermal anomalies and—depending on spatial and temporal resolutions—forecast the geographic extent of active lava flows. However, a developing lava flow is a complex surface to observe using remote techniques, due to the moving material exhibiting a range of temperatures, textures, vesicularities [5], and thicknesses. Furthermore, the evolution of thermal anomalies may involve continuous changes in energy emitted as surfaces cool, as well as variations that depend on viewing angles [6].

Several automated processes for the detection and measurement of volcanic 'hot-spots'—such as VAST [7], MODVOLC [8,9], RAT [10], MyVOLC and MyMOD [11], among others—have been developed, tested, and run to date. In particular, three projects have marked the development and awareness for a complete and global monitoring capacity: (i) the European Space Agency's (ESA) pilot project GLOBVOLCANO (2008–2011), using high-spatial resolution RS; (ii) the European Commission's European Volcano Observatory Space Services (EVOSS, 2010–2014), centered on high to very-high temporal resolutions, and (iii) the Disaster Risk Management volcano pilot project of the Committee on Earth Observation Satellite (CEOS), focusing on the continuous monitoring of volcanic activity in the whole of Latin America and the Caribbean. These projects, among others, have demonstrated how access to RS data over volcanic regions can enhance the understanding of volcanic activity, enabling hazard mitigation and the identification of developing trends in volcanic activity [12,13].

A focus of this study is emissivity, which is defined as the efficiency of radiating thermal energy at a specific wavelength. Although it is a critical variable in the interpretation of passive RS for accurate surface temperature estimation, emissivity is seldom measured and mostly assumed or estimated to be a fixed value, which does not change as a function of temperature and wavelength.

The fundamental laws of Planck, Stefan-Boltzmann, and Wien demonstrate that surfaces radiate in different regions of the electromagnetic spectrum, depending on their temperature. Planck's radiation law defines the radiation released to be that of a perfect radiator, a blackbody: however, very few terrestrial surfaces would act as near-perfect blackbodies. The ability of non-blackbody surfaces to emit radiation is defined by their emissivity. Emissivity may be defined as the ratio of the radiation emittance of a sample relative to that of a blackbody at the same temperature.

Volcanic surfaces, such as the basalts analyzed here, may be characterized as selective radiators, since their emissivity varies with wavelength and would be less than unity (blackbody). A variety of approaches have been used to derive surface temperatures from RS data, where emissivity is either assumed to be unity, or a fixed 'look up' value. This value for basaltic surfaces is estimated to be 0.90 to 0.95 [14], or it is estimated directly from reflectance data [15] (e.g., 0.80). The motivation of this study is to establish the significance of emissivity variations in the derivation of lava surface temperatures from RS data, and assess the impact of such variations on the estimation of eruption rates and the prediction of lava flow 'distance-to-run' using high spatial resolution spaceborne data.

2. Materials and Methods

2.1. Rock Samples

Due to its persistent activity, Mount Etna (Italy) is frequently targeted for studies involving the application of RS data to detect high-temperature thermal features and measure eruptive products [9,16]. The 2001 eruption presents three main features: (i) despite lasting only 23 days (18 July to 09 August 2001), it gave rise to an outstanding pattern of seven different (Figure 1) fast-developing lava flows [17]; (ii) the total lava flow volume is significant in the recent eruptive history of Etna, and (iii) this eruption could be observed by three high spatial resolution multispectral payloads (TM onboard Landsat 5, ETM+ onboard Landsat 7, and ASTER onboard Terra).

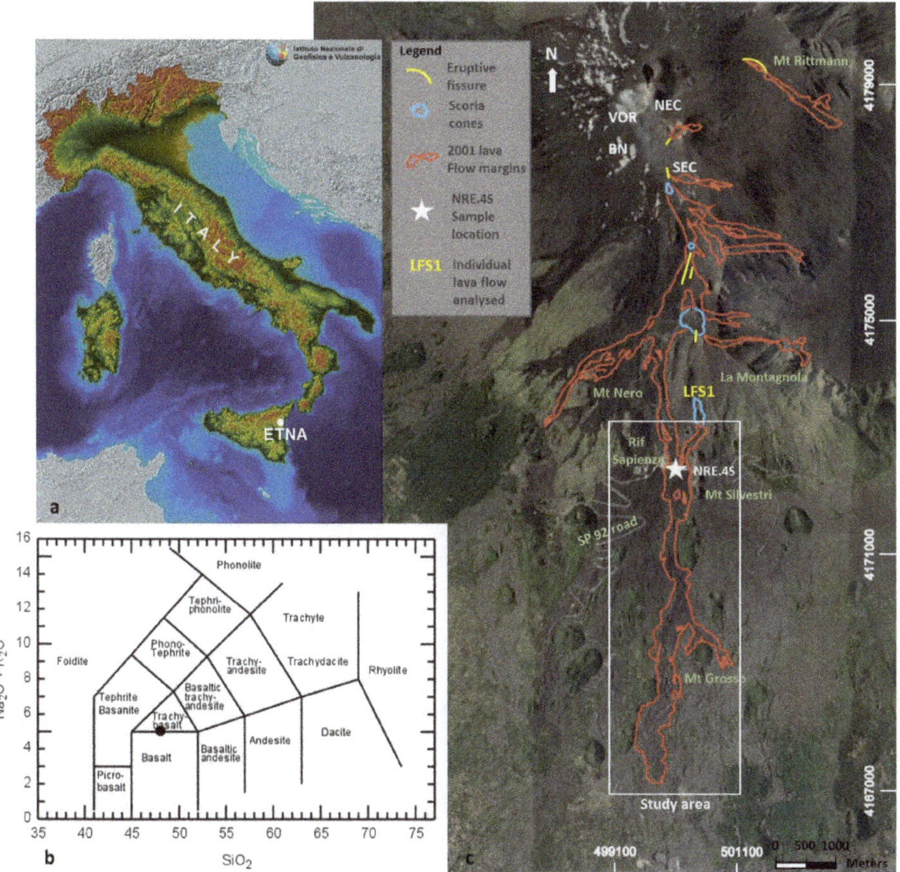

Figure 1. (**a**) A geological map of Italy and the location of Mount Etna [18]; (**b**) Derived silica versus alkalis, superimposed on the chemical classification scheme for NRE.4S; (**c**) The 2001 eruption lava flow margins are highlighted in red and superimposed on the target area in Google Earth. Eruptive fissures are marked in yellow, scoria cones in light blue (after [17]) and within the individual flow study area (LFS1), the approximate sample locations are indicated by the star symbol.

To measure the emissivity of exposed lavas, we collected 10 rock samples termed the NRE.4 Series (NRE.4S) across the main flow (LFS1) (Figure 1), in a grid scaled to the spatial resolution of the ASTER TIR bands (~90 m). Samples were initially investigated using Fourier transform infrared (FTIR) spectroscopy to derive emissivity from both reflectance and radiance data at ambient/low and high temperatures. Existing spaceborne data (ASTER Global Emissivity Database) and numerical modeling (MAGFLOW) are used for data validation.

The chemical composition of the samples (average values for the entire series) and their approximate locations are provided in Table 1 and shown in Figure 1. Previously published information on volumes and effusion rates for the 2001 Mount Etna eruption [17] are also provided in Table 2.

Table 1. X-Ray Fluorescence major elements content, as a component oxide weight percent (wt%).

SiO_2	TiO_2	Al_2O_3	Fe_2O_3	MnO	MgO	CaO	Na_2O	K_2O	P_2O_5	SO_3	LOI	Total
48.15	1.53	16.49	11.19	0.17	5.71	10.49	3.52	1.70	0.53	0.005	−0.30	99.18

Table 2. Volumes and effusion rates for the LFS1 2001 Mount Etna eruption (from [17]).

Acquisition Date	Local Time	Eruption Day	Acquisition Time (s)	Cumulative Volume ($\times 10^6$ m^3)	Time Span (s)	Partial Volume ($\times 10^6$ m^3)	Daily Effusion Rate (m^{3}s^{-1})
18/07/2001	03:00	0	0	0.00	0	0.00	0.00
18/07/2001	13:00	1	36,000	0.37	36,000	0.37	10.28
19/07/2001	16:00	2	133,200	1.70	97,200	1.33	13.68
20/07/2001	13:00	3	208,800	3.50	75,600	1.80	22.81
22/07/2001	11:00	5	374,400	8.58	165,600	5.08	30.68
26/07/2001	12:00	9	723,600	14.98	349,200	6.40	18.33
28/07/2001	16:00	11	910,800	16.99	187,200	2.01	10.74
30/07/2001	11:00	13	1,065,600	18.35	154,800	1.37	8.85
02/08/2001	10:00	16	1,321,200	19.82	255,600	1.47	5.75
04/08/2001	07:00	18	1,483,200	20.62	162,000	0.80	4.94
06/08/2001	11:00	20	1,670,400	21.21	187,200	0.59	3.15
07/08/2001	07:00	21	1,742,400	21.32	72,000	0.11	1.53
09/08/2001	10:00	23	1,926,000	21.40	183,600	0.08	0.44

2.2. Laboratory-Based Data Acquisition

We carried out laboratory Fourier transform infrared (FTIR) analyses on solidified volcanic rock samples from the major 2001 eruption of Mount Etna to derive emissivity at a range of wavelengths and temperatures.

2.2.1. Emissivity from Surface Reflectance Spectra

We collected reflectance spectra of samples measured at an ambient temperature (~295 K) at the Planetary Emissivity Laboratory (DLR, Germany) by the Bruker Vertex 80v FTIR spectrometer, using a gold integration sphere hemispherical reflectance accessory.

The experimental setup [19] measures the reflectance of samples in the visible, the near infrared bands, and medium infrared (MIR) ranges. Reflectances are converted into approximate emissivity. For MIR measurements, a wide-range Mercury Cadmium Telluride (MCT) detector is used (1000–400 cm^{-1}) in tandem with a wide-range germanium (Ge) on potassium bromide (KBr) beam splitter (12,500–420 cm^{-1}). For V-NIR measurements, conversely, an InGaAs Diode detector was used (12,500–5800 cm^{-1}) in tandem with a silicon (Si) on calcium fluoride (CaF_2) beam splitter (15,000–1200 cm^{-1}).

Samples (grain size 500–1000 μm) were placed into individual sample cups, which were placed on the hemispherical reflectance accessory, and aligned. Prior to measuring samples, a gold reference target was used to calibrate the instrument. Finally, individual sample spectra were normalized to the gold reference target spectrum results to obtain reflectance values.

The apparent emissivity (ε') values were derived using the measured reflectance (R) data using Kirchhoff's law (Equation (1)). This approach provides an expected result precision of 0.005 [20].

$$\varepsilon' = 1 - R \quad (1)$$

It is important to note that Kirchhoff's law (1) is only valid for hemispherical reflectance measurements, and is used to approximate emissivity from reflectance data [21]; thus, the term 'apparent' emissivity (ε') is used in contrast to the 'true' emissivity (ε).

2.2.2. Emissivity from Surface Radiance Spectra

The thermal emission spectra of the samples were measured in a vacuum (0.7 mbar) at temperature and wavelength ranges of 400 to 900 K and 5.0 to 16.0 μm, respectively. The experimental setup [19] used an external simulation chamber attached to the FTIR spectrometer measuring the emissivity of solid samples (grain size 100–3000 μm). The emissivity chamber was equipped with an internal web-cam, and several temperature sensors to measure the sample/cup temperature, monitor the equipment, and record chamber temperatures. Both the cup and the sample are heated uniformly by induction, and the temperature of the emitting surface is measured using a thermophile sensor in contact with the surface. The resulting data are calibrated using the emissivity spectrum of the blackbody material [19] to provide the set of 'true' emissivity (ε) data.

2.3. Emissivity from High-Spatial Resolution Satellite Data

The Global Emissivity Database

The Global Emissivity Database (GED v.3) built by NASA's Jet Propulsion Laboratory (JPL) [22] is the most detailed emissivity product available for Earth's land surface. Emissivity, rescaled to 100 m from the original 90-m ASTER TIR pixels, is an average of data acquired at five TIR central wavelengths (8.30 μm, 8.65 μm, 9.10 μm, 10.60 μm, and 11.30 μm) every 16 days, from 2000 to 2008. It was obtained by NASA JPL by combining temperature emissivity separation (TES) algorithms and water vapor scaling (WVS) atmospheric corrections coincident with MODIS MOD07 atmospheric profiles and the MODTRAN 5.2 radiative transfer code [23].

In this study, we examined 12 $1^{\circ} \times 1^{\circ}$ ASTER GEDv3 data 'tiles' downloaded from the NASA EOSDIS Land Processes DAAC [24], centered on Sicily (Italy) and Mount Etna.

3. Emissivity Results

3.1. Emissivity from Reflectance

As a preliminary estimate of surface emissivity, we used FTIR reflectance data measured at ambient temperature. The spectral signatures were consistent with those from previous research on basaltic rocks [14,25], confirming that representative emissivity data can be obtained this way (Figure 2). To extend the observable spectral range, two detectors (KBr at 0.66 to 2.50 μm and MCT at 2.50 to 16.00 μm) were used, so that the data could be merged at 2.63 μm to provide the best signal-to-noise (STN) ratio result for the entire range from visible near-infrared (VNIR) to TIR wavelength. The maximum difference in emissivity at any wavelength between the NRE.4 series samples was 0.02.

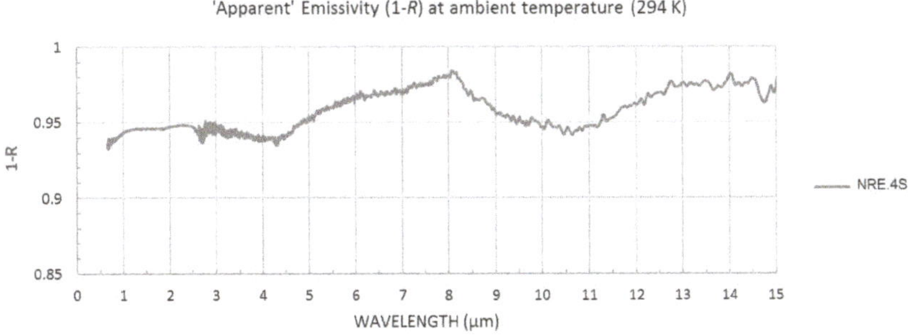

Figure 2. 'Apparent' emissivity (ε') spectral signatures measured and derived from reflectance Fourier transform infrared (FTIR) data (1-R) at ambient temperature for NRE.4S.

3.2. Emissivity from Radiance

The spectral signatures for samples analyzed using thermal emission analysis at 400 K (Figure 3a) display emissivity values consistent with the preliminary reflectance data (Figure 2) in the TIR region (8.0–15.0 μm), with a significantly improved STN ratio and optimal error range (<0.01) for NRE.4 Series. In contrast to ambient/low-temperature data, high-temperature results (Figure 3b) show a steady decrease in emissivity with every temperature increase step (400–900 K). However, this trend could not be observed between 5.0–6.0 μm. This is because of the instrument sensitivity limitations and for this reason, only the results in TIR wavelengths (8.0–15.0 μm) should be used in further analyses. An additional 'cooling test' was performed by measuring the emissivity of the same series in the opposite direction (cooling), by decreasing temperature steps (i.e., 900–400 K), maintaining consistent sample conditions. The deviance in emissivity values during the temperature increase (heating) is shown in Figure 3b and the temperature decrease (cooling) was ≤0.02 with no hysteresis deviation trend in either direction.

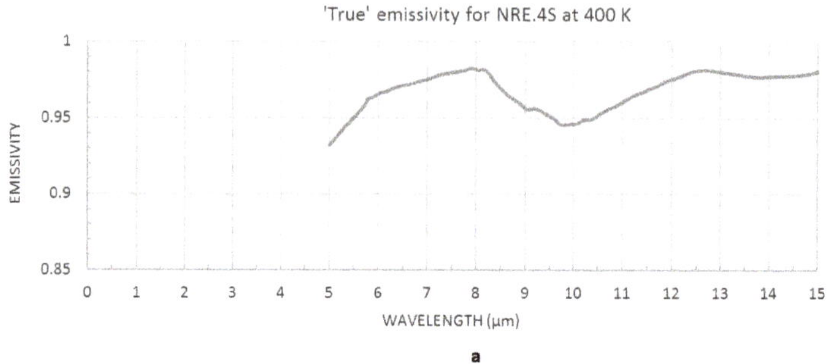

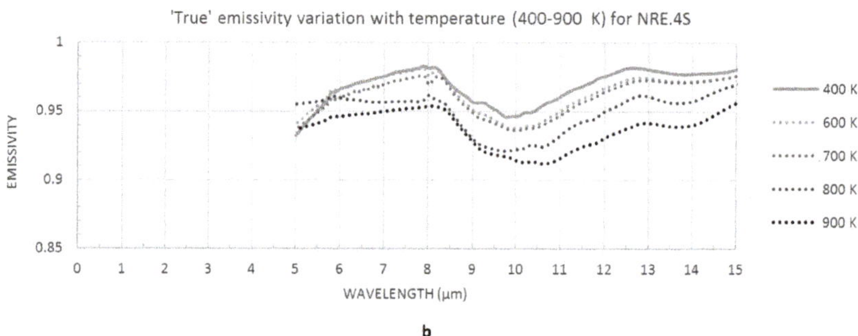

Figure 3. (**a**) 'True' emissivity (ε) spectral signatures at 400 K using thermal emission FTIR for NRE.4S; and (**b**) 'true' emissivity (ε) spectral signature variation with temperature change between 400–900 K.

3.3. Comparison with Emissivity from Satellite Data

The emissivity map (Figure 4) created using existing spaceborne ASTER GEDv3 data [22] over Sicily (Italy) and Mount Etna, displays a mean emissivity variation of the NRE.4S targets analyzed. The highest emissivities are shown in dark blue; these correspond to the Mt Etna region, which is consistent with the emissivity signatures of basaltic volcanic surfaces. Sicily is geologically complex due to its regional tectonics; thus, green and red areas on the map with lower emissivities would represent compositionally different geological units of non-volcanic origin.

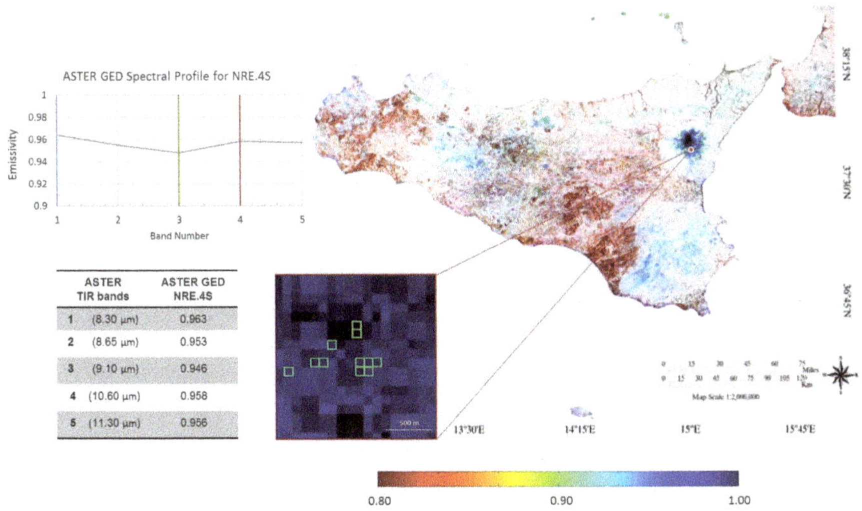

ASTER TIR bands	ASTER GED NRE.4S
1 (8.30 μm)	0.963
2 (8.65 μm)	0.953
3 (9.10 μm)	0.946
4 (10.60 μm)	0.958
5 (11.30 μm)	0.956

Figure 4. (Map) ASTER Global Emissivity Database (ASTER GED) over Sicily, Italy at 100-m pixel resolution at 10.60 μm and 4-3-1 band red–green–blue (RGB) view. The color ramp indicates the mean emissivity values (0.8–1.0). Inset shows samples' pixel locations for NRE.4S. Emissivity spectral profile plot and the table show emissivity variation with wavelength, extracted from ASTER GED for NRE.4S.

For a direct comparison, spaceborne emissivity data (Figures 4 and 5a) and laboratory FTIR results (Figures 2 and 3a) of the same area targets (NRE.4S) are shown at ASTER TIR operating central-wavelength bands (8.30 μm, 8.65 μm, 9.10 μm, 10.60 μm, and 11.30 μm). A relatively comparable trend can be observed (Figure 5b), exhibiting the best data fit at 9.10 μm ($\leq$0.01) and data range/error of $\leq$0.03 at other wavelengths. Accounting for different methodologies used in this study and their limitations (discussed in Section 5), the emissivity range/error is expected. This may suggest that, despite being a nine-year average (2000–2008), the ASTER GED mean emissivity data for the 2001 lava flow area can fit reasonably well with our results obtained using ambient/low-temperature FTIR laboratory measurements. However, the high-temperature FTIR results (Figure 3b) indicate a trend (Figure 5c,d) that suggest an emissivity decrease with temperature increase.

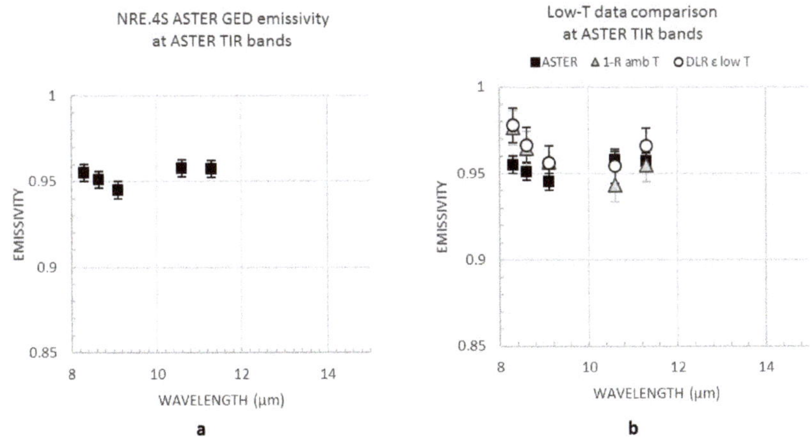

Figure 5. *Cont.*

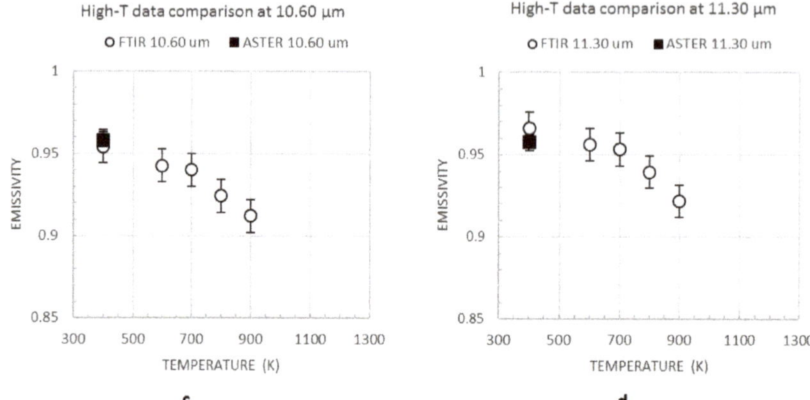

Figure 5. (**a**) ASTER GED mean emissivity values (black squares) plotted at ASTER TIR bands for NRE.4S; (**b**) ASTER GED mean emissivity values (black squares) compared to the 'apparent' (1-*R*) emissivity (grey triangles) and 'true' low-temperature (400 K) emissivity data (empty circles) for NRE.4S; (**c**,**d**) ASTER GED mean emissivity (static) for NRE.4S (black square) at 10.60 μm and 11.30 μm (ASTER TIR bands 4 and 5 respectively), superimposed on the emissivity/temperature trend 400–900 K (empty circles). Error bars represent maximum emissivity difference between NRE.4S samples measured.

4. Emissivity versus Effusion Rate and 'Distance-to-Run'

4.1. From Spaceborne Data

Varying the emissivity and wavelength will have an impact on the computation of integrated temperatures and radiant fluxes, and hence on the estimation of lava effusion rates and 'distance-to-run'. To perform a quantitative evaluation, we selected the best-quality night-time image acquired during the 2001 eruption, aimed to avoid both pixel saturation and the reflected radiances of daytime images in SWIR.

This eruption was observed by three high-spatial resolution payloads on Landsat 5 (ETM), Landsat 7 (ETM+), and Terra (ASTER). The selected image data presented here was acquired by ETM+ on 5 August 2001 at 20:34 (Figure 6a). ETM+ is the multispectral scanning radiometer onboard Landsat 7, providing high-spatial resolution data (30 m in V-NIR-SWIR and 60 m in TIR) in repeat cycles of 16 days. Launched in 1999 and still active at the date of writing, ETM+ provided very high-quality images until 2003, when the linear scan compensator developed a permanent fault affecting the whole image (black stripes).

To assess the sensitivity to variations in emissivity, we selected three values. One end-member is a blackbody (1.0), and the second value (0.80) has been used in published research [15]. The third value (0.93) that was selected for our assessment is the minimum emissivity value at 10.60 μm from our reflectance data (Figure 2), which is also a mean emissivity estimate for basalt [14].

The image was processed for all of the radiant pixels in SWIR and TIR using the three emissivities (1.0, 0.93, and 0.80), following previously established procedures [26]. After radiometric and atmospheric correction [27,28], integrated temperatures are calculated and sub-resolutions were solved in SWIR (1.65 μm and 2.20 μm) to obtain the total radiant flux Q_{r_calc}.

$$Q_{r\_calc} = A\varepsilon\sigma\tau\Phi'(T_e^4 - T_a^4) \tag{2}$$

from which the lava effusion rate [16,29,30] E_{r_calc} is estimated:

$$E_{r\_calc} = \frac{Q_{r\_calc}}{\rho[C_P\Delta T + \Phi\, C_L]} \tag{3}$$

In Equation (2), A is the pixel surface area, ε is the emissivity, σ is the Stefan–Boltzmann's constant, τ is atmospheric transmissivity, and Φ' is the shape of the radiating surface. Note that the symbol Φ' in Equation (2) is not the same symbol representing the percentage of crystals grown (Φ) in Equation (3) or the fitness function (ϕ) in Table 3. The equation for radiative heat transfer has term $T_e{}^4 - T_a{}^4$, where $T_e{}^4$ is the effective temperature to the fourth, power and $T_a{}^4$ is the ambient temperature to the fourth power. The difference in temperature from the hot material to the ambient temperature has a control on the rate of transfer. As we are dealing with very small changes in emissivity, this small difference in temperature will play an important role.

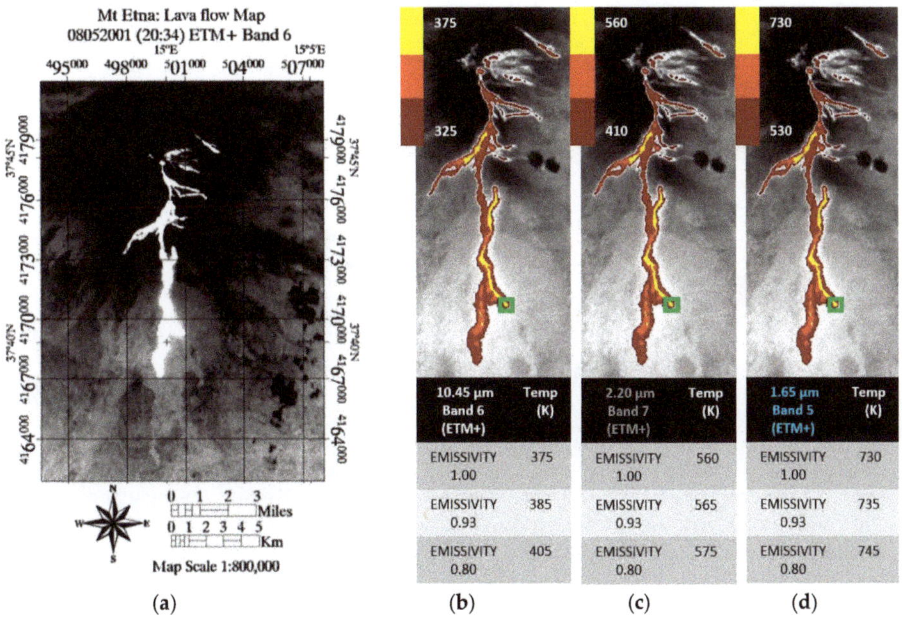

Figure 6. (**a**) Mount Etna high-temperature thermal anomaly scene acquired on 5 August 2001 by Landsat 7 (ETM+); (**b–d**) are all radiant pixel integrated temperatures of the same target area as in (**a**) TIR Band 6 and SWIR bands 7 and 5 for emissivity of 1.0; The green square in (**b–d**) marks the location of the pixel (37°39′28″N 14°59′48″E) used to derive temperature with emissivity variation given below (**b–d**).

In Equation (3), ρ is the lava density (2600 kg m^{-3}); C_P the specific heat capacity (1150 J kg^{-1} K^{-1}); ΔT is the average temperature difference throughout the active flow (100 to 200 K), which is a significant parameter in estimating eruption rate; Φ is the average mass fraction of crystals (0.4 to 0.5) grown in cooling through ΔT, and C_L is the latent heat of crystallization (2.9×10^5 J kg^{-1}).

Note that while Equation (2) includes only observables and variable emissivity, the values used to solve Equation (3) are average values taken from various literature sources, which are specific for this type of lava and Mount Etna [31,32].

We observe that the TIR (Figure 6b) and two SWIR channels (Figure 6c,d) display significantly different integrated pixel temperatures on the active flow, ranging from as low as 325 K (brightness temperature ε = 1 at 10.45 μm TIR wavelength) to as high as 745 K (ε = 0.8 and 1.6 μm SWIR). Overall, 20% emissivity change may give rise to integrated pixel temperature differences in the order of 15 K in SWIR and 25 K in TIR (shown in 6b-d), which is consistent with previous research [15].

We used the instantaneous effusion rate E_{r_calc} derived from high-spatial resolution RS data to attempt a rapid estimation of the maximum lengths that individual lava flow can reach (Figure 7).

By implementing the empirical law $L_{max} = 2E_r^{\frac{1}{2}}$ [33,34] we observe that a moderate difference in emissivity will influence calculated eruption rates and may impact the L_{max} 'distance-to-run' by as much as ±300 m.

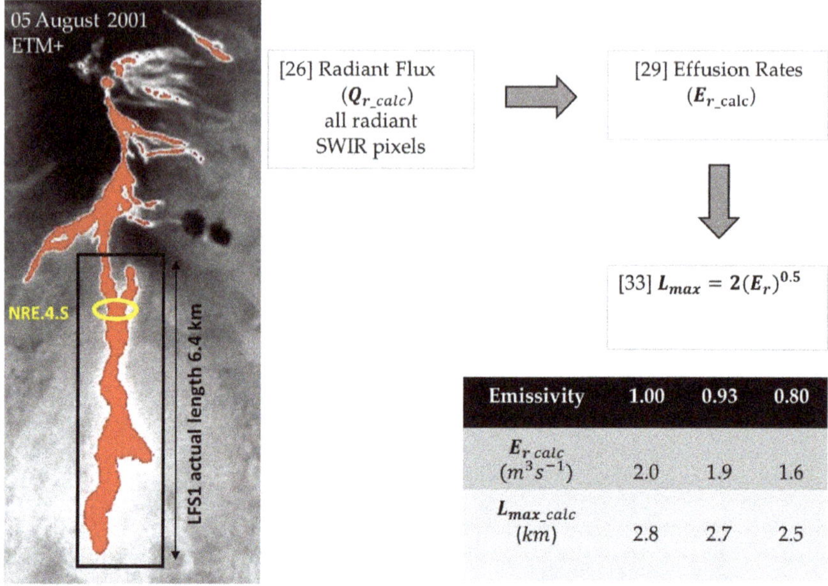

Emissivity	1.00	0.93	0.80
$E_{r\ calc}$ (m^3s^{-1})	2.0	1.9	1.6
L_{max_calc} (km)	2.8	2.7	2.5

Figure 7. ETM+ scene obtained on 5 August 2001 highlighting the high-temperature thermal anomaly, focusing on an individual LFS1 flow analyzed in this study and detailing a flow chart of procedures followed to obtain the maximum lengths that LFS1 lava flow can reach using the empirical approach [33,34]. The table shows the difference in results using various emissivity values (0.80, 0.93, and 1.0).

Often, the effusion rate can be calculated by exploiting ground-based observations, where the information on how the volume of an individual flow has changed in a given time interval can be quite accurate. This is achieved by using information on the rate of advance, if the cross-sectional area of the flow front is known (i.e., width × thickness × rate of advance) or if the volume and time interval are known (length × width × thickness ÷ time). The calculated eruption rate is used in the L_{max} equation to estimate the maximum length that the lava flow can achieve. This approach provides an estimated maximum lava flow length, and is not intended to forecast the exact final length of the flow.

The individual lava flow LFS1 analyzed here reached its maximum length on or around 25 July 2001 with an estimated eruption rate of 18.33 m^3s^{-1} [17]. According to field estimates, after that date, the effusion rate dropped, so lava did not extend along the whole flow length. Our ETM+ data from 5 August 2001 produced an effusion rate of ~2 m^3s^{-1}, which is in line with the rate drop observed for that time period, and corresponds well with field estimates of ~3 m^3s^{-1} on 6 August 2001 (Table 2).

The total volume [17] up to the end of the eruption (Table 2) would suggest an average effusion rate (i.e., cumulative volume divided by the acquisition time in seconds) of 11.1 m^3s^{-1} and L_{max} = 6.7 km, which is slightly greater than and consistent with the observed $L_{max} = 6.4\ km$. Nonetheless, the L_{max} estimation is most needed at the start of the eruption. Our ETM+ calculated effusion rate for 5 August 2001 also highlighted an increase in L_{max} of 5% between emissivity end-members (i.e., 0.80 and 1.0), which may play a role in hazard mitigation at densely populated areas in close proximity to an active volcano.

4.2. From Straightforward Modelling

To reproduce the lava flow path for the 2001 Mount Etna eruption, we used the MAGFLOW cellular automaton propagator [35,36], which uses a physical model accounting for both thermal and rheological evolution of flowing lavas. It is argued that this model has the potential to significantly improve understanding of the dynamics of lava flow emplacement [37] and assist with related hazard assessment and mitigation [38–42]. MAGFLOW sensitivity analyses show that the main controlling factors are topography, rheology, and vent location, while temporal changes in effusion rates will also strongly influence the accuracy of the predictive lava-flow modeling [43,44].

To model the lava flow path for the 2001 Etna eruption, MAGFLOW was run on a pre-eruptive Digital Elevation Model (DEM) using field-derived effusion rates [17] and the typical properties of basaltic rocks (density = 2600 kg/m^3; specific heat capacity = 1150 J kg^{-1} K^{-1}; solidification temperature = 1173 K; extrusion temperature = 1360 K), and we varied the emissivity.

To evaluate its impact on the simulated ultimate lava flow lengths, we carried out a sensitivity test where three different emissivity values (i.e., 0.80, 0.93, and 1.00) were introduced to the model, maintaining constant emissivity, unaffected by temperature changes throughout the simulation. This method was validated using the actual lava flow extent of the 2001 Mount Etna eruption.

The MAGFLOW simulations obtained using the various emissivity values (i.e., 0.80, 0.93, and 1.00) are shown in Figure 8. There is good overall agreement between the actual and the simulated lava flows, but major discrepancies occurred for all simulations and reality due to the neglecting of the ephemeral vent opening that fed the minor, southeastern branch of the flow [17].

Figure 8. MAGFLOW simulation results with changing emissivity (i.e., 0.80, 0.93, and 1.00), showing a difference of up to 600 m in lava flow length.

The quality of fit was quantified using a fitness function (ϕ) computed as the square root of A (sim ∩ real)/A (sim ∪ real), where A (sim ∩ real) and A (sim ∪ real) are the areas of the intersection and union between the simulated and actual lava flows, respectively. The lower and upper limits for ϕ are zero and one, with zero indicating the maximum error (i.e., lack of common areas between the simulated and actual lava flows), and one corresponding to a complete overlap (i.e., the simulated area coincides totally with the actual lava flow field).

The main morphological features (area, length, and average thickness) of actual and simulated lava flows, as well as the results of the fitness function (ϕ) are reported in Table 3.

Table 3. Comparison between the morphological properties (length, area, and average thickness) of the 2001 Etna lava flow derived by field measurements [17] and numerical simulations.

	Length [km]	Area [km^2]	Average Thickness [m]	Fitness [ϕ]
Actual lava flow	6.4	1.95	11	-
Simulation with ε = 0.8	6.9	1.94	12.4	0.76
Simulation with ε = 0.93	6.5	1.83	13.1	0.78
Simulation with ε = 1.0	6.3	1.78	13.5	0.77

The closest approximate values for the area covered and the average distribution were obtained by the MAGFLOW simulation run with an emissivity of 0.8 (1.94 km^2 and 12.4 m, respectively). However, having almost the same extent does not guarantee a best fit (areas may not overlap wholly or in part). This is confirmed by the values of the fitness function ϕ. Whilst the actual lava flow field is quite well reproduced by the MAGFLOW model, with ϕ always higher than 0.75, the maximum value (0.78) is obtained by the simulation run with ε = 0.93, demonstrating that it better reflects the actual field. The MAGFLOW simulation run with ε = 0.93 also reaches the closest flow length (6.5 versus 6.4 km), which is the most critical factor for hazard analysis.

5. Discussion

The land surface temperature derivation and the estimation of eruption rates from spaceborne data rely on assumptions of lava flow emissivity. The majority of research on emissivity to date has been carried out on solid lava at ambient temperatures [14], and it is anticipated that under certain conditions, target radiation emission in the TIR region of the electromagnetic spectrum is inversely proportional to its reflectance [45]. However, there are several drawbacks in using reflectance to derive emissivity values, as the temperature of the sample is not taken into account, and its spatial variation is not recorded. Nonetheless, reflectance data (1-*R*) can be used to provide a first approximation estimate in the absence of 'true emissivity' information.

The 'apparent' emissivity data for the 2001 Mount Etna eruption, which was derived from reflectance data at ambient temperature, is comparable to that of the low-temperature (400 K) emission FTIR data ('true' emissivity), with a range/error of emissivity $\leq$0.03, which is consistent with previous research on basaltic rock spectral signatures [25]. A certain amount of spectral contrast is observed and can be attributed to the instrument's sensitivity and/or the methodology used, which has been acknowledged in previous research [46].

The spaceborne ASTER GED mean emissivity data obtained for the same target area was also relatively consistent with our FTIR results at ambient/low temperatures, producing an emissivity range/error of $\leq$0.03. However, using this satellite-based approach, incorrect emissivity as an input could result in an error in retrieving lava flow surface temperatures, which will have an inherent impact on the computation of the radiant flux, estimates of the mass eruption rate, and lava flow 'distance-to-run' forecasts. Our 5 August 2001 example for the ETM+ satellite scene was used to compute radiant flux show variances in the calculated eruption rate, which has an impact of ~300 m in derived 'distance-to-run' between the emissivity end members (0.8 and 1.0). Similarly, a simple emissivity assessment using numerical forward modeler (MAGFLOW) by means of varying its value (i.e., 0.80, 0.93, and 1.0) showed that emissivity has an impact on simulated lava flow 'distance-to-run' results of up to ~600 m between the emissivity end-members (0.8 and 1.0).

The computation of Q_{r_calc} term in [3] using spaceborne data also involves convection (Q_c) and conduction (Q_k) of the flow. It has been argued that for flows that are not crusted over, Q_{r_calc} is much larger than Q_c and Q_k. However, in case of flows with crusts, Q_c can have as large a heat loss effect as Q_{r_calc} [47]. Etnean 'aa' lava flows, which have been discussed here, have crusts and very rough surfaces as they advance away from their vents and channels; thus, their heat transfer through convection can be as large as that of radiation for the medial and distal parts of the flow. Since the crust may act as an insulator, it limits convection and radiation by the amount of heat conducted from the flow through the crust to the outer surface. This effect may be considered as the most likely contributor to the longer flow lengths. Therefore, emissivity variation may indicate flow phase changes by insulation, allowing it to flow father than a simple effusion rate model would predict. Thus, a model that includes all of the thermal components may be able to account for changes in the thermal conditions (and rheology of the flow). Similarly, if emissivity variation is considered when producing the best dual-band solution using the multi-component approach [48] from spaceborne data, it should also involve applying several different emissivities within a single pixel.

The emissivity variation may also be seen as a proxy for crusting the lava flow, which will insulate the flow, allowing it to flow for longer, and thus further. This approach may provide a means of identifying changes in the insulation properties of an active lava flow by using apparent emissivity. Therefore, it can be argued that a change in the apparent emissivity of the surface of the flow does have an indirect effect on the 'distance-to-run'.

The emissivity data derived from reflectance, low-temperature emission FTIR analyses and/or ASTER GED provide 'static' emissivity values, which may be related to the solidified (cooled) product. However, the high-temperature thermal anomaly observed on Mount Etna has an extrusion temperature of ~1360 K, so there is a need to account for emissivity changes with temperature, as evident from our high-temperature FTIR results. This is consistent with several thermal emission studies of silicate glasses and basaltic lavas, which suggest that the emissivity of molten material is significantly lower than that of the same material in a solid state [49]. Therefore, our preliminary results from high-temperature data imply that it is essential to expand this study to assess the role and significance of emissivity, not only as a 'static' and uniform value across all wavelengths and temperatures, but also taking its response to thermal gradient into account. This will determine the emissivity variation with temperature change, and will provoke further investigation into the role and impact of emissivity in lava flow dynamic modeling and hazard mitigation.

6. Conclusions

Our reflectance and emission FTIR results at ambient/low temperature indicate that emissivity is wavelength-dependent. Both laboratory (FTIR) and spaceborne (ASTER GED) data correspond well for the same target area, and show good correlation at specific TIR wavelengths by exhibiting an emissivity range/error of $\leq$0.03. However, this emissivity information is 'static', relating to the solidified (cooled) product, not reflecting the range of temperatures involved at an active lava flow or the emissivity/temperature trend seen in our high-temperature FTIR results. Furthermore, the theoretical empirical approaches and modeling used in this study indicate that a variation of 0.2 in emissivity may result in significant changes to the prediction of lava flow 'distance-to-run' estimates.

A reliable and exploitable predictive emissivity trend is needed for both modeling and spaceborne applications for a range of temperatures and wavelengths to improve our understanding of the variation of emissivity with temperature. Further, a better understanding of the impact of emissivity on deduced temperature during active lava flow propagation (and cooling) is needed to improve spaceborne data interpretation, which relies on emissivity as an input in computations of lava surface temperatures, radiant fluxes, and ultimately lava flow length estimations.

Author Contributions: F.F. and N.R. led the conception and design of the work. N.R. and A.C. acquired the data. N.R., A.C., F.F. were responsible for analysis, modelling and interpretation of data. All authors participated in drafting and revising the article and have given final approval of the submitted and revised versions.

Funding: This research (and the APC) was fully funded by the Open University, Milton Keynes, U.K.

Acknowledgments: This project and manuscript were improved through invaluable guidance and support of the Open University, School of Environment, Earth and Ecosystem Sciences, Milton Keynes, U.K. (H. Rymer and S. Blake). The authors are grateful to the Natural Environment Research Council (NERC) Field Spectroscopy Facility, University of Edinburgh, U.K. (C. MacLellan and A. Gray) and Clarendon Laboratory, University of Oxford, U.K. (N. Bowles) for accommodating preliminary FTIR pilot studies. Special thanks to the DLR, Institute of Planetary Research, Planetary Emissivity Laboratory, Berlin, Germany (A. Maturilli) for accommodating FTIR measurements and helpful discussions. MAGFLOW simulations were performed within the framework of Tecnolab, the Laboratory for Technological Advance in Volcano Geophysics of the INGV in Catania, Italy. The editor Arlene Cui and the three anonymous reviewers are acknowledged for their constructive and supportive comments, which helped improve this manuscript.

Conflicts of Interest: The authors declare no conflict of interest. The funders had no role in the design of the study; in the collection, analyses, or interpretation of data; in the writing of the manuscript, or in the decision to publish the results.

References

1. Garel, F.; Kaminski, E.; Tait, S.; Limare, A. An experimental study of the surface thermal signature of hot subaerial isoviscous gravity currents: Implications for thermal monitoring of lava flows and domes. *J. Geophys. Res. Solid Earth* **2012**, *117*, B02205. [CrossRef]
2. Vicari, A.; Alexis, H.; Del Negro, C.; Coltelli, M.; Marsella, M.; Proietti, C. Modeling of the 2001 lava flow at Etna volcano by a Cellular Automata approach. *Environ. Model. Softw.* **2007**, *22*, 1465–1471. [CrossRef]
3. McGuire, W.J.; Kilburn, C.R.J.; Murray, J.B. *Monitoring Active Volcanoes: Strategies, Procedures, Techniques*; University of London Press: London, UK, 1995; p. 421.
4. Harris, A.J.L. *Thermal Remote Sensing of Active Volcanoes: A User's Manual*; Cambridge University Press: Cambridge, UK, 2013; p. 736.
5. Ramsey, M.S.; Fink, J.H. Estimating silicic lava vesicularity with thermal remote sensing: A new technique for volcanic mapping and monitoring. *Bull. Volcanol.* **1999**, *61*, 32–39. [CrossRef]
6. Ball, M.; Pinkerton, H. Factors affecting the accuracy of thermal imaging cameras in volcanology. *J. Geophys. Res. Solid Earth* **2006**, *111*, B11203. [CrossRef]
7. Higgins, J.; Harris, A. VAST: A program to locate and analyse volcanic thermal anomalies automatically from remotely sensed data. *Comput. Geosci.* **1997**, *23*, 627–645. [CrossRef]
8. Wright, R.; Flynn, L.; Garbeil, H.; Harris, A.; Pilger, E. Automated volcanic eruption detection using MODIS. *Remote Sens. Environ.* **2002**, *82*, 135–155. [CrossRef]
9. Wright, R.; Flynn, L.P.; Garbeil, H.; Harris, A.J.; Pilger, E. MODVOLC: Near-real-time thermal monitoring of global volcanism. *J. Volcanol. Geotherm. Res.* **2004**, *135*, 29–49. [CrossRef]
10. Di Bello, G.; Filizzola, C.; Lacava, T.; Marchese, F.; Pergola, N.; Pietrapertosa, C.; Piscitelli, S.; Scaffidi, I.; Tramutoli, V. Robust satellite techniques for volcanicand seismic hazards monitoring. *Ann. Geophys.* **2004**, *47*. [CrossRef]
11. Hirn, B.R.; Di Bartola, C.; Laneve, G.; Cadau, E.; Ferrucci, F. SEVIRI Onboard Meteosat Second Generation, and the Quantitative Monitoring of Effusive Volcanoes in Europe and Africa. In Proceedings of the IGARSS 2008, 2008 IEEE International Geoscience and Remote Sensing Symposium, Boston, MA, USA, 7–11 July 2008; pp. 374–377.
12. Wooster, M.J.; Rothery, D.A. Time-series analysis of effusive volcanic activity using the ERS along track scanning radiometer: The 1995 eruption of Fernandina volcano, Galápagos Islands. *Remote Sens. Environ.* **1997**, *62*, 109–117. [CrossRef]
13. Coppola, D.; Macedo, O.; Ramos, D.; Finizola, A.; Delle Donne, D.; Del Carpio, J.; White, R.; McCausland, W.; Centeno, R.; Rivera, M.; et al. Magma extrusion during the Ubinas 2013–2014 eruptive crisis based on satellite thermal imaging (MIROVA) and ground-based monitoring. *J. Volcanol. Geotherm. Res.* **2015**, *302*, 199–210. [CrossRef]
14. Harris, A.J.L. Thermal remote sensing of active volcanism: Principles. In *Thermal Remote Sensing of Active Volcanoes: A User's Manual*; Cambridge University Press: Cambridge, UK, 2013; pp. 70–112.
15. Rothery, D.A.; Francis, P.W.; Wood, C.A. Volcano monitoring using short wavelength infrared data from satellites. *J. Geophys. Res.* **1988**, *93*, 7993–8008. [CrossRef]
16. Harris, A.J.; Butterworth, A.L.; Carlton, R.W.; Downey, I.; Miller, P.; Navarro, P.; Rothery, D.A. Low- cost volcano surveillance from space: Case studies from Etna, Krafla, Cerro Negro, Fogo, Lascar and Erebus. *Bull. Volcanol.* **1997**, *59*, 49–64. [CrossRef]
17. Coltelli, M.; Proietti, C.; Branca, S.; Marsella, M.; Andronico, D.; Lodato, L. Analysis of the 2001 lava flow eruption of Mt. Etna from three-dimensional mapping. *J. Geophys. Res. Earth Surf.* **2007**, *112*, F02029. [CrossRef]
18. Tarquini, S.; Isola, I.; Favalli, M.; Mazzarini, F.; Bisson, M.; Pareschi, M.T.; Boschi, E. TINITALY/01: A new triangular irregular network of Italy. *Ann. Geophys.* **2007**, *50*, 407–425.

19. Maturilli, A.; Helbert, J.; D'Amore, M.; Varatharajan, I.; Ortiz, Y.R. The Planetary Spectroscopy Laboratory (PSL): Wide spectral range, wider sample temperature range. In Proceedings of the SPIE 10765: Infrared Remote Sensing and Instrumentation, San Diego, CA, USA, 18 September 2018.
20. Korb, A.R.; Salisbury, J.W.; Aria, D.M. Thermal-infrared remote sensing and Kirchhoff's law 2. Field measurements. *J. Geophys. Res. Solid Earth* **1999**, *104*, 15339–15350. [CrossRef]
21. Maturilli, A.; Helbert, J.; Witzke, A.; Moroz, L. Emissivity measurements of analogue materials for the interpretation of data from PFS on Mars Express and MERTIS on Bepi-Colombo. *Planet. Space Sci.* **2006**, *54*, 1057–1064. [CrossRef]
22. NASA; JPL. *ASTER Global Emissivity Dataset 100-meter HDF5[37/38/39.012-0-15.0001]*; NASA EOSDIS Land Processes DAAC: Sioux Falls, SD, USA, 2014.
23. Hulley, G.C.; Hook, S.J.; Abbott, E.; Malakar, N.; Islam, T.; Abrams, M. The ASTER Global Emissivity Dataset (ASTER GED): Mapping Earth's emissivity at 100 meter spatial scale. *Geophys. Res. Lett.* **2015**, *42*, 7966–7976. [CrossRef]
24. NASA; JPL. LP DAAC: Land Processes Distributed Active Archive Center. 2014. Available online: https://lpdaac.usgs.gov/dataset_discovery/community/community_products_table/ag100_v003 (accessed on 1 June 2018).
25. Wyatt, M.B.; Hamilton, V.E.; McSween, H.Y.; Christensen, P.R.; Taylor, L.A. Analysis of terrestrial and Martian volcanic compositions using thermal emission spectroscopy: 1. Determination of mineralogy, chemistry, and classification strategies. *J. Geophys. Res.* **2001**, *106*, 14711–14732. [CrossRef]
26. Hirn, B.; Di Bartola, C.; Ferrucci, F. Spaceborne monitoring 2000-2005 of the Pu'u 'O'o-Kupaianaha (Hawaii) eruption by synergetic merge of multispectral payloads ASTER and MODIS. *IEEE Trans. Geosci. Remote Sens.* **2008**, *46*, 2848–2856. [CrossRef]
27. Barsi, J.A.; Barker, J.L.; Schott, J.R. An Atmospheric Correction Parameter Calculator for a single thermal band earth-sensing instrument. In Proceedings of the IGARSS 2003, 2003 IEEE International Geoscience and Remote Sensing Symposium, Toulouse, France, 21–25 July 2003; pp. 3014–3016.
28. Barsi, J.A.; Barker, J.L.; Schott, J.R. Atmospheric Correction Parameter Calculator. 2003 IGARSS03. Available online: https://atmcorr.gsfc.nasa.gov/ (accessed on 1 June 2018).
29. Pieri, D.C.; Baloga, S.M. Eruption rate area, and length relationships for some Hawaiian lava flows. *J. Volcanol. Geotherm. Res.* **1986**, *30*, 29–45. [CrossRef]
30. Wright, R.; Blake, S.; Harris, A.J.; Rothery, D.A. A simple explanation for the space-based calculation of lava eruption rates. *Earth Planet. Sci. Lett.* **2001**, *192*, 223–233. [CrossRef]
31. Harris, A.J.L.; Murray, J.B.; Aries, S.E.; Davies, M.A.; Flynn, L.P.; Wooster, M.J.; Wright, R.; Rothery, D.A. Effusion rate trends at Etna and Krafla and their implications for eruptive mechanisms. *J. Volcanol. Geotherm. Res.* **2000**, *102*, 237–269. [CrossRef]
32. Harris, A.; Dehn, J.; Calvari, S. Lava effusion rate definition and measurement: A review. *Bull. Volcanol.* **2007**, *70*, 1–22. [CrossRef]
33. Kilburn, C.R.J.; Pinkerton, H.; Wilson, L. Forecasting the behaviour of lava flows. In *Monitoring Active Volcanoes*; McGuire, W.J., Kilburn, C.R.J., Murray, J., Eds.; UCL Press: Londin, UK, 1995; pp. 346–368.
34. Kilburn, C.R.J.; Ballard, R.D. *Lava Flows and Flow Fields*; Sigurdsson, H., Ed.; Academic Press: San Diego, CA, USA, 2000; pp. 291–305.
35. Herault, A.; Vicari, A.; Ciraudo, A.; Del Negro, C. Forecasting lava flow hazards during the 2006 Etna eruption: Using the MAGFLOW cellular automata model. *Comput. Geosci.* **2009**, *35*, 1050–1060. [CrossRef]
36. Cappello, A.; Hérault, A.; Bilotta, G.; Ganci, G.; Del Negro, C. MAGFLOW: A physics-based model for the dynamics of lava-flow emplacement. *Geol. Soc. Spec. Publ.* **2016**, *426*, 357–373. [CrossRef]
37. Kereszturi, G.; Cappello, A.; Ganci, G.; Procter, J.; Németh, K.; Del Negro, C.; Cronin, S.J. Numerical simulation of basaltic lava flows in the Auckland Volcanic Field, New Zealand—Implication for volcanic hazard assessment. *Bull. Volcanol.* **2014**, *76*, 1–17. [CrossRef]
38. Del Negro, C.; Cappello, A.; Neri, M.; Bilotta, G.; Hérault, A.; Ganci, G. Lava flow hazards at Mount Etna: Constraints imposed by eruptive history and numerical simulations. *Sci. Rep.* **2013**, *3*, 3493. [CrossRef]
39. Cappello, A.; Geshi, N.; Neri, M.; Del Negro, C. Lava flow hazards—An impending threat at Miyakejima volcano, Japan. *J. Volcanol. Geotherm. Res.* **2015**, *308*, 1–9. [CrossRef]
40. Cappello, A.; Zanon, V.; Del Negro, C.; Ferreira, T.J.; Queiroz, M.G. Exploring lava-flow hazards at Pico Island, Azores Archipelago (Portugal). *Terra Nova* **2015**, *27*, 156. [CrossRef]

41. Pedrazzi, D.; Cappello, A.; Zanon, V.; Del Negro, C. Impact of effusive eruptions from the Eguas–Carvão fissure system, São Miguel Island, Azores Archipelago (Portugal). *J. Volcanol. Geotherm. Res.* **2015**, *291*, 1–13. [CrossRef]
42. Cappello, A.; Ganci, G.; Calvari, S.; Pérez, N.M.; Hernández, P.A.; Silva, S.V.; Cabral, J.; Del Negro, C. Lava flow hazard modeling during the 2014–2015 Fogo eruption, Cape Verde. *J. Geophys. Res. Solid Earth* **2016**, *121*, 2290–2303. [CrossRef]
43. Bilotta, G.; Cappello, A.; Hérault, A.; Vicari, A.; Russo, G.; Del Negro, C. Sensitivity analysis of the MAGFLOW Cellular Automaton model for lava flow simulation. *Environ. Model. Softw.* **2012**, *35*, 122–131. [CrossRef]
44. Bilotta, G.; Cappello, A.; Hérault, A.; Del Negro, C. Influence of topographic data uncertainties and model resolution on the numerical simulation of lava flows. *Environ. Model. Softw.* **2019**, *112*, 1–15. [CrossRef]
45. Rolim, S.B.A.; Grondona, A.; HACKMANN, C.; Rocha, C. A Review of Temperature and Emissivity Retrieval Methods: Applications and Restrictions. *Am. J. Environ. Eng.* **2016**, *6*, 119–128.
46. Sabol, D.E., Jr.; Gillespie, A.R.; Abbott, E.; Yamada, G. Field validation of the ASTER Temperature– Emissivity Separation algorithm. *Remote Sens. Environ.* **2009**, *113*, 2328–2344. [CrossRef]
47. Patrick, M.R.; Dehn, J.; Dean, K. Numerical modeling of lava flow cooling applied to the 1997 Okmok eruption: Approach and analysis. *J. Geophys. Res. Solid Earth* **2004**, *109*, B03202. [CrossRef]
48. Wright, R.; Flynn, L.P. On the retrieval of lava-flow surface temperatures from infrared satellite data. *Geology* **2003**, *31*, 893–896. [CrossRef]
49. Lee, R.J.; Ramsey, M.S.; King, P.L. Development of a new laboratory technique for high-temperature thermal emission spectroscopy of silicate melts. *J. Geophys. Res. Solid Earth* **2013**, *118*, 1968–1983. [CrossRef]

Article

Eruptive Styles Recognition Using High Temporal Resolution Geostationary Infrared Satellite Data

Valerio Lombardo *, Stefano Corradini, Massimo Musacchio, Malvina Silvestri and Jacopo Taddeucci

Istituto Nazionale di Geofisica e Vulcanologia, 00143 Roma, Italy; stefano.corradini@ingv.it (S.C.); massimo.musacchio@ingv.it (M.M.); malvina.silvestri@ingv.it (M.S.); jacopo.taddeucci@ingv.it (J.T.)

* Correspondence: valerio.lombardo@ingv.it; Tel.: +39-06-51860-508

Received: 30 January 2019; Accepted: 12 March 2019; Published: 19 March 2019

Abstract: The high temporal resolution of the Spinning Enhanced Visible and InfraRed Imager (SEVIRI) instrument aboard Meteosat Second Generation (MSG) provides the opportunity to investigate eruptive processes and discriminate different styles of volcanic activity. To this goal, a new detection method based on the wavelet transform of SEVIRI infrared data is proposed. A statistical analysis is performed on wavelet smoothed data derived from SEVIRI Mid-Infrared(MIR) radiances collected from 2011 to 2017 on Mt Etna (Italy) volcano. Time-series analysis of the kurtosis of the radiance distribution allows for reliable hot-spot detection and precise timing of the start and end of eruptive events. Combined kurtosis and gradient trends allow for discrimination of the different activity styles of the volcano, from effusive lava flow, through Strombolian explosions, to paroxysmal fountaining. The same data also allow for the prediction, at the onset of an eruption, of what will be its dominant eruptive style at later stages. The results obtained have been validated against ground-based and literature data.

Keywords: volcanic eruption interpretation; eruption forecasting; MSG SEVIRI; wavelet; remote sensing; thermal measurements; lava fountain; lava flow; Mt.Etna; eruptive style

---

1. Introduction

Determining the beginning and end of an eruption, and forecasting what will be the dominant eruption style, are equally important objectives, crucial for hazard assessment.

Satellite sensors have been increasingly employed for operational monitoring of volcanic thermal features as for example the geostationary Spinning Enhanced Visible and InfraRed Imager (SEVIRI) instrument on board the geostationary Meteosat Second Generation (MSG) satellite [1–4] and the Geostationary Operational Environmental Satellites (GOES) [5], or the polar Moderate Resolution Imaging Spectroradiometer (MODIS) [6], the Along Track Scanning Radiometer (ATSR) [7] and the advanced very-high-resolution radiometer (AVHRR) [3,8–12], on board respectively of the NASA Terra/Aqua, the ESA-ENVISAT, and the NOAA satellites.

In spite of its coarse spatial resolution, the high temporal rate of the geostationary systems represents an important tool for volcano monitoring (e.g., [13–15]). In particular, SEVIRI has 12 spectral channels from visible to thermal infrared, a spatial resolution of 3 km at nadir and a high temporal resolution that ranges from 15 min (earth full disk (EFD)) to 5 min (rapid scan mode (RSM) over Europe and Northern Africa). Exploiting these characteristics, SEVIRI allows a precise timing of the early phase of an eruption and almost a continuous monitoring of volcanic activity from the source to the atmosphere [4,16]. The SEVIRI data are collected in real time from a Meteosat-8 ground station antenna operating from 2010 at Istituto Nazionale di Geofisica e Vulcanologia (INGV) in Rome (Italy) [17,18].

In this work, the SEVIRI measurements have been used to investigate the eruptive processes and to discriminate different styles of volcanic activity by applying a novel procedure based on the correlation between the radiance growing rate at the beginning of the eruption with the specific type of volcanic activity. As test case, the 2013, 2015 and 2017 Etna events have been considered, and the results of the remote sensing analyses have been validated using ground-based information and literature references.

The paper is organized as follows: Section 2 outlines eruptions on Mt. Etna in the 2011–2015 period while Section 3 describes the method developed for the discrimination of the different styles of volcanic activities. In Sections 4 and 5 the results are presented and discussed, respectively. Final conclusions are drawn in Section 6.

2. The 2011–2015 and 2017 Etna Activities

In the 2011–2015 period, the eruptive style of Mt. Etna changed from predominantly effusive to more explosive activity. Strombolian events became more frequent and intense and were often associated with strong explosions, ash emissions, and lava fountaining paroxysmal episodes (e.g., [16]. The duration of these "paroxysmal" episodes varies from minutes to hours [4,19–22]. Between 2 and 8 December 2015, all four summit craters of Mt. Etna generated an extraordinary sequence of eruptive events. The activity consisted of high eruption columns, Strombolian explosions, lava flows, and widespread ash falls [23,24].

In February 2017, a vigorous strombolian activity started from the South-East (SE) crater producing a small lava flow. The direction of the lava flow was towards the southwest flank and flowed down slowly to an elevation of about 2850 m asl. After having lasted approximatively 48 h, the intense explosive-effusive activity began to weaken quickly. On 15 March new lava flows erupted from a fissure vent at the southern base of the SE crater complex. These lava flows descended into the *Valle del Bove* at several locations on the rim between 2500–2700 m elevation. On their descend down the steep western headwall, the lava flow fronts suffered partial collapses and interacted explosively with snow, both leading to the generation of small pyroclastic density currents that produced tall brown ash plumes [25].

3. Materials and Methodology

Statistical analysis was performed for Etnean eruptions using the 3.9 μm SEVIRI channel (Medium InfraRed-MIR channel) acquired from 2011 to 2017. Most of these eruptions were paroxysmal events, including Strombolian explosive activity, often associated with ash/gas emissions, and lava fountaining. Unfortunately, the statistics on lava flows are reduced to two episodes: the two-day long lava flow in December 2015 and the intense effusive activity that occurred from February to March 2017. The analysis of different eruptions has highlighted different radiance trends at the early stages of eruptions.

The December 2015 Etna eruption was characterized by the succession of several eruptive styles in a matter of days. Figure 1 shows the SEVIRI MIR radiance time series, measured from a single SEVIRI pixel centered over the Mt. Etna summit area and collected from 2 to 8 December 2015. The most powerful episode occurred between 2 and 3 December 2015 from the Voragine crater (Ep 1) when high lava fountains were produced without lava flows emitted. The MIR radiance drop that suddenly appeared in the MIR signal on 3 December, clearly indicates the absorption due to the ash emitted during the event [4]. From 4 to 5 December further powerful episodes occurred at the Voragine with strong emission of ash and gases (Ep 2). In this case, the scene was extremely cloudy and the radiances appear low compared to Ep 1 and its highly variable trend [4]. Finally, the activity shifted to the New Southeast Crater on 6 December 2015 (Ep 3), where Strombolian activity and lava flow emission lasted for two days and were fed by the most primitive magma of the study period, that is magma that underwent minimal igneous differentiation from the original composition [24]. Together with the MIR radiance trend, in Figure 1 are also drawn the growing rates for both the lava fountain (red dashed curve) and the lava flow (blue dashed curve). The trend of the 2 and 3 December lava fountain shows a gradual rise to a peak in about 24 h with radiance growing at an exponential rate. The maximum intensity lasts less than one hour, after which the radiance quickly starts to decrease at an exponential rate. From 4 to 5 December, two further potent episodes occurred at the Voragine, characterized by ash emissions and weak Strombolian activity. Radiance trend shows a high variability with the presence of peaks in correspondence of paroxysmal episodes. However, the thick cloud coverage present on those days significantly influences the signal detected at the sensor. Therefore, it is not reliable to characterize this activity based on radiance changes [4]. The trend of the 6–8 December lava flow suddenly changes from steady to rising and the maximum radiance is reached after a few minutes from the beginning of the eruption. Then the radiance is maintained at the maximum value for the duration of the emissions.

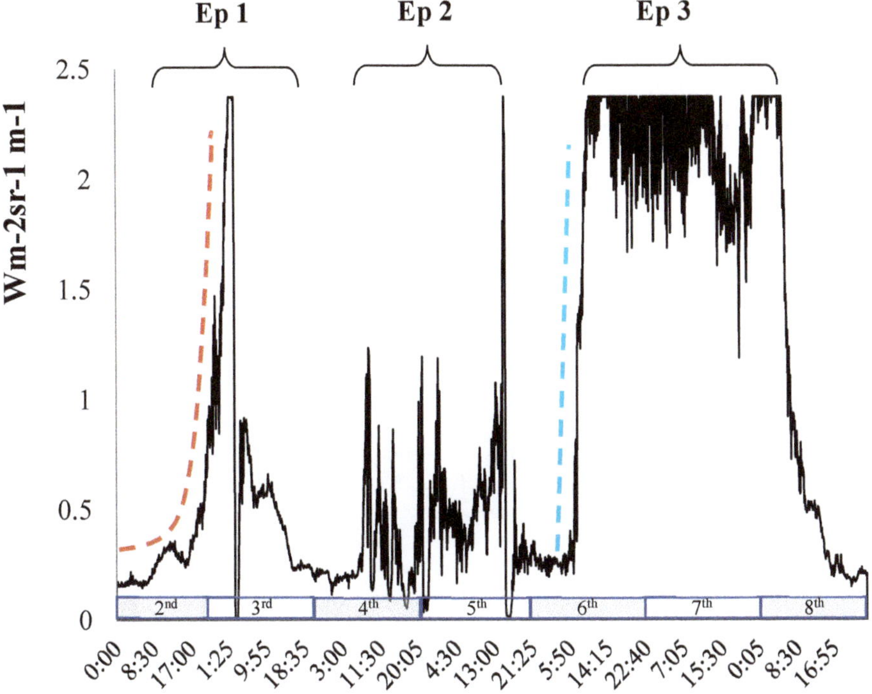

Figure 1. Spinning Enhanced Visible and InfraRed Imager (SEVIRI) Mid-Infrared (MIR) radiance acquired during the 2–8 December 2015 Mt. Etna eruption. Three different eruptive episodes occurred: lava fountaining (**Ep 1**), ash emission (**Ep 2**), and lava flow (**Ep 3**). Different episodes clearly show different trends that can be recognized a posteriori. However, differences are detectable even at the early stages of the eruption. An exponential trend and a sub-vertical trend seem to best fit the initial radiance increment for Ep1 (**red dashed line**) and Ep3 (**blue dashed line**) respectively.

The qualitative analysis of these three distinct episodes clearly shows strong differences in the radiance trends. It is quite easy to distinguish the three episodes knowing the overall trend retrospectively. However, for prevention reasons, it would be important to recognize the type of volcanic activity at the earliest stages of the eruption.

The basic idea is that when an eruption begins, the SEVIRI MIR radiance increases as a function of the temperature and the volume rate of the erupted products. There are two mechanisms that mainly contribute to defining the speed at which the integrated pixel temperature increases. On the one hand, the type of volcanic products generated by the eruption and their eruptive rates. For example, mixtures of ash, lava blebs and broiling gas at Mt. Etna are normally at temperatures between 200 °C and 700 °C, but they can exceed temperatures of 1000 °C, while the measured eruptive temperatures of Etnean lava flows range between 1080 and 1095 °C [26]. On the other hand, products such as clouds, volcanic ashes, and plumes can act as a shield to radiation and partially or completely absorb the radiance emitted to space [27]. Both mechanisms affect the MIR radiance measured by the sensor, therefore, its trend has the potential to discriminate between eruptive styles. Herein, we aim to correlate the radiance growing rate at the beginning of the eruption (ramp) to the specific type of volcanic activity. By exploiting the SEVIRI RSM resolution a time-window of radiance values has been considered and analyzed in terms of signal processing. Changes in "signal" are detected at a certain time (t_0) using a moving window of fixed width, which is incremented at every SEVIRI acquisition.

Signal analysis is performed on SEVIRI MIR data in order to obtain optimal smoothing of radiance time-series. The smoothing removes noise introduced by radiometric calibration and atmospheric effects caused by passing of transitional clouds in the investigated scene. The literature on time-series smoothing and denoising techniques is vast and encompasses different methods. Among the most renown and utilized methods are the Savitzky-Golay and Hodrick-Prescott filters, the Hilbert-Huang transform (HHT), the ample class of kernel filters, as well as wavelet analysis. Herein, we propose a method that uses the wavelet transform [28] to analyze the SEVIRI time-series.

3.1. *Wavelet Transform: the 'à trous' Algorithm*

The *à trous* algorithm of discrete wavelet transform is a powerful tool for multiscale (multiresolution) analysis of images [28]. The *à trous* algorithm allows performing a hierarchical decomposition of an image into a series of *scale layers*, also known as *wavelet planes* [29]. Each layer contains only structures within a given range of characteristic *dimensional scales* in the space of a *scaling function*. The decomposition is done throughout a number of *detail layers* defined at growing characteristic scales, plus a final *residual layer* or *smoothing layer*, which contains the rest of unresolved structures. By isolating significant image structures within specific detail layers, detail enhancement can be carried out with high accuracy. Similarly, if noise occurs at some specific dimensional scales in the image, as is usual in most cases, by isolating it into appropriate detail layers we can reduce or remove it without affecting significant structures. A complete description of the implementation of this algorithm can be found in Bijaoui and Giudicelli [28]. Herein, the *à trous* algorithm has been adapted to remove the effect of background noise and to isolate changes due to volcanic activity in the moving time-window.

Plate 2a of Figure 2 shows a time-series of MIR radiance obtained for a short lava fountain event occurred at Mt. Etna from 13 to 15 December 2013. The detail layers and the residual layer from wavelet transform are shown in Plates 2b–e. Both detail and residual layers have potentials to be used for the assessment of eruptive phenomena. High-frequency changes occur at the earliest decomposition steps. Therefore, detail coefficients of the first scale layer are ideal to trigger the initial phase of the eruption (Plate 2b). One property of the wavelet transform is to have a sampling step proportional to the scale [28]. This implies that significant structures (e.g., local maxima) in lower-detailed scale layers match structures in higher-detailed scale layers (Plates 2b, c, d). This property allows "following" the eruptive signal in subsequent transformations where the higher frequency content of the signal is progressively eliminated.

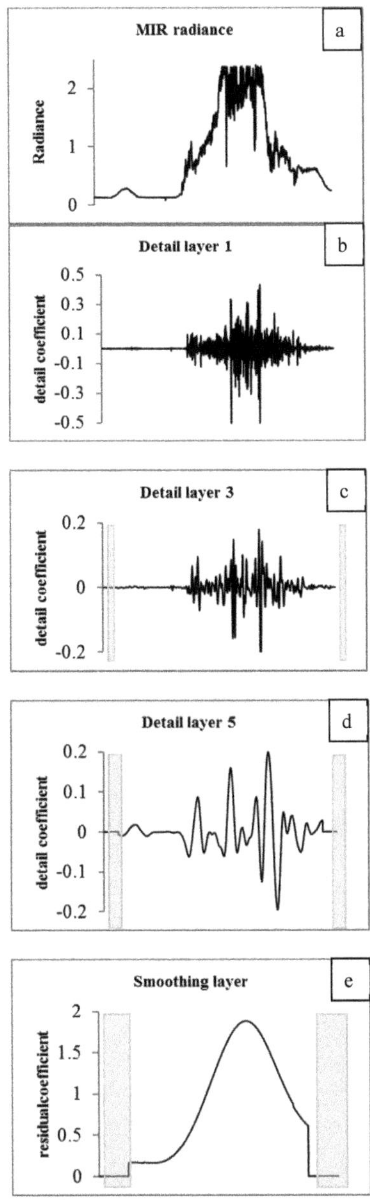

Figure 2. An example of multiscale decomposition with the *à trous* discrete wavelet transform algorithm. Seven detail layers have been generated with dyadic scaling sequence (only three are shown for clarity, **Plates 2b**, **2c**, and **2d**). Detail layers are generated for a growing scaling sequence of powers of two. The layers are generated for scales of 1, 2, 4, 8, . . . samples. For example, the fourth layer contains structures with characteristic scales between five and eight samples. Wavelet decomposition results in a border distortion which increases with the power of two at each decomposition scale (indicated in figures with grey areas). The original SEVIRI MIR radiance time-series is shown in **Plate 2a**. Detail layers at increasing characteristic scales contain larger signal structures. At the end of the sequence is the residual layer (**Plate 2e**), which contains all of the remaining unresolved structures.

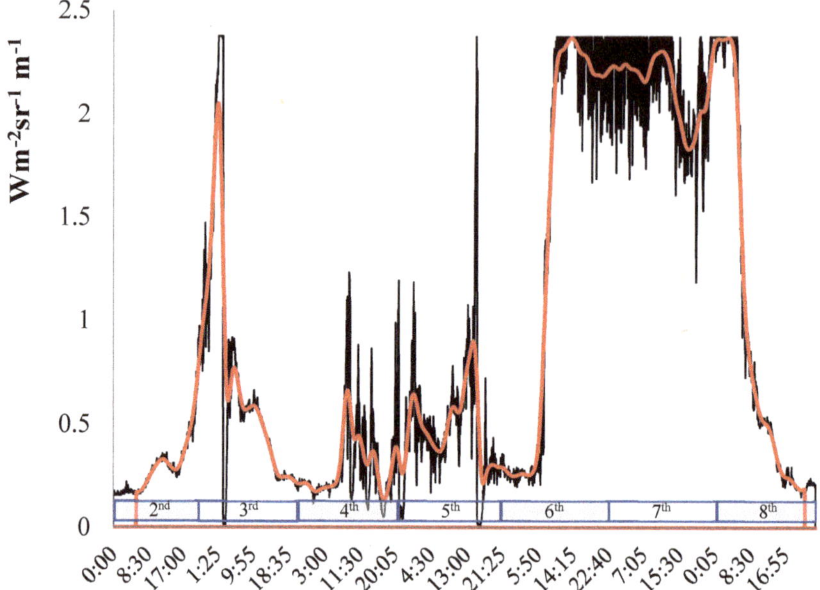

Figure 3. SEVIRI MIR radiance acquired during December 2015 Mt. Etna eruption. Red line is the smoothed layer derived from the *à trous* discrete wavelet transform. The smoothed layer allows for analysis of the overall trend without high-frequency noise due to instrumental noise, transient weather effects, and acquisition failures.

3.2. Statistical Analysis

The *smoothing layer* of the discrete wavelet transform allows capturing important patterns in the radiance data, while leaving out noise or other fine-scale structures/rapid phenomena. Herein, we propose a statistical approach to characterize the variability of the dataset in the *smoothing layer*.

Useful statistics for measuring the variability of a data set are skewness, kurtosis, and gradient. Skewness and kurtosis statistics can help us to assess certain kinds of deviations from normality of our data-generating process. Skewness is a measure of symmetry, or more precisely, the lack of symmetry. A distribution, or data set, is symmetric if it looks the same to the left and right of the center point. Skewness determines whether a distribution is symmetric about its maximum.

Kurtosis is defined as the degree to which a statistical frequency curve is peaked. Kurtosis is a measure of whether the data are heavy-tailed or light-tailed relative to a normal distribution. That is, data sets with high kurtosis tend to have heavy tails or outliers. Data sets with low kurtosis tend to have light tails or lack outliers. Finally, the gradient is defined as the slope of the line that interpolates the dataset of the moving window. The gradient measures the rate of change of radiance with time.

Time series analysis of SEVIRI data from 2011 to 2017 showed that kurtosis and gradient are most appropriate to characterize the relationship between radiance variations and eruptive styles.

4. Results

The sensitivities of the results were tested against different trends. Figure 4 shows the kurtosis (black line) and gradient (blue filled curve) values derived from *smoothing layer* (red line) at every SEVIRI acquisition (every 5 min) during the 2–8 December 2015 Etna eruption.

The kurtosis coefficient is often regarded as a measure of the tail heaviness of a distribution relative to that of the normal distribution. However, it also measures how much "peaked" a distribution is. When interpreting kurtosis, the normal distribution is used as a reference. A positive kurtosis implies a distribution with more extreme possible data values (outliers) than a normal distribution, thus fatter tails (*Leptokurtic distributions*). A negative kurtosis implies a distribution with less extreme possible data values than a normal distribution, thus thinner tails (*Platykurtic distributions*). Finally, distributions with zero kurtosis have roughly the same outlier character as a normal distribution (*Mesokurtic distributions*).

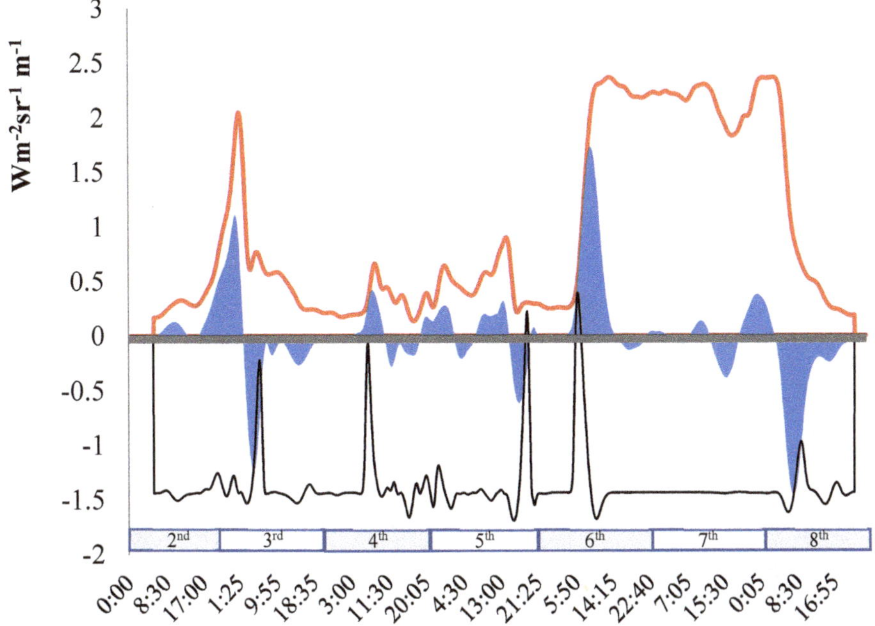

Figure 4. MIR radiance acquired during the December 2015 Etna eruption. The kurtosis (**black line**) and the gradient (**blue filled curve**) trends were derived from the *smoothing layer* (**red line**) at every Meteosat Second Generation (MSG) acquisition (every 5 min). Highest values of kurtosis (> 0) and gradient (> 1.5) occur at the beginning of the lava flow episode (6 December).

The ratio of kurtosis to its standard error can be used as a test of normality (that is, you can reject normality if the ratio is less than −2 or greater than +2). A large positive value for kurtosis indicates that the tails of the distribution are longer than those of a normal distribution. A negative value for kurtosis indicates shorter tails (becoming like those of a box-shaped uniform distribution).

Normalized kurtosis shows that the spread of the observations was slightly less than what would have been expected under the normal distribution. Kurtosis is less than −1.3 during non-eruptive phases and when the eruption reaches the highest point of intensity measurable with SEVIRI data. The normalized kurtosis trend indicates that outliers mainly occur at the beginning and the end of the eruptive episodes. Therefore, outliers in the kurtosis distribution can be used for hot-spot detection to identify the initial and final phases of the eruption. The magnitude of the outliers appears to be different depending on the eruptive episode. The magnitudes of the initial outliers are −1.4, −0.1, and 0.45 for the lava fountain, the explosive activity, and the lava flow episode, respectively.

The gradient (slope) of the line that best fits the *smoothing layer* distribution indicates the steepness of the *moving window* distribution at every SEVIRI acquisition. Positive slope indicates increments in radiance and, therefore, possible increases in volcanic activity. Again, the maximum slope at the onset of each eruptive episode appears to vary depending on the type of volcanic activity. A slope magnitude of 1.0, 0.4, and 1.7 is obtained at the onset peak for the three episodes.

Figure 5 shows the *smoothing layer* (red line), the normalized kurtosis (black line) and slope (blue filled curve) for the 28 February and the 15 March eruptive events. The two events, both characterized by the presence of lava flows, show a relatively high value of gradient and kurtosis at the onset of the eruption. The lava flow generated by the 28 February eruption displays a gradient of 2.1 and normalized kurtosis of 2.0. The lava flow of 15 March, much stronger and longer than the previous episode, shows maximum gradient and kurtosis values of 2.5 and 4.4, respectively.

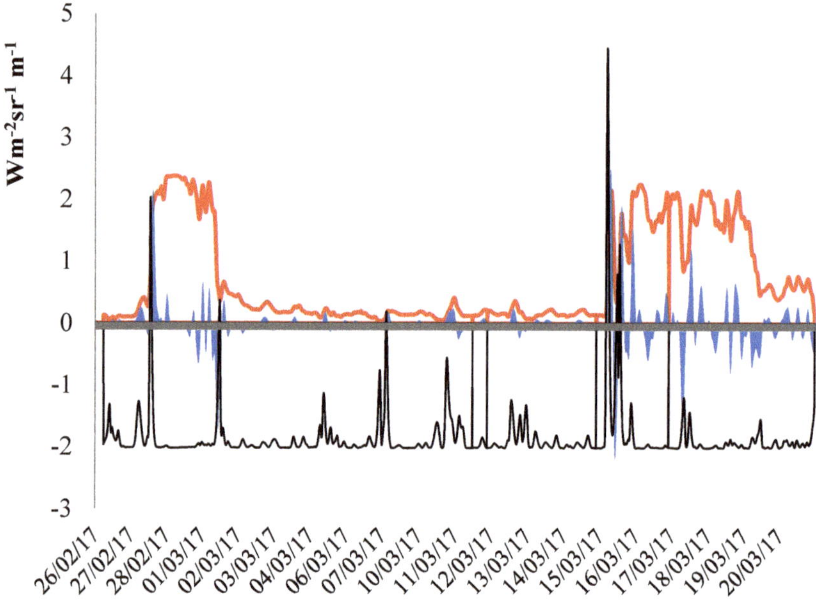

Figure 5. The *smoothing layer* (**red line**), the normalized kurtosis (**black line**) and slope (**blue filled curve**) for the 28 February and 15 March eruptive events. The lava flow generated by the 28 February eruption displays a gradient of 2.1 and normalized kurtosis of 2.0 at the onset of the eruption. The lava flow of 16 March, much stronger and longer than the previous episode, shows maximum gradient and kurtosis values of 2.5 and 4.4 respectively.

Threshold values were calculated using an eight-hour (100 samples) temporal window. The window width was experimentally chosen to optimize the signal-to-noise ratio of the radiance wavelet transform. A larger window implies that signal is smoothed over a longer period which results in losing details. A narrow window is more sensitive to high-frequency noise and, therefore, susceptible to false positives [30]. Figure 6 shows the *smoothing layer* (red line), the normalized kurtosis (black line) and slope (blue filled curve) for the 14 and 15 December 2013 Etna lava fountain. Vigorous Strombolian activity with associated ash plumes was observed at the New Southeast Crater (NSEC) (Volcano Discovery, Etna volcano activity updates: December 2013). The activity generated a small lava flow (probably mainly fed by rapid accumulation of liquid spatter) moving towards the SE flank with the presence of exploding magma bubbles towards the end of the eruption [31]. Under a statistical point of view, this paroxysmal event is very similar to the 3–5 December paroxysm (Ep 1) with kurtosis outlier of 0.05 and a maximum gradient of 0.45. Both episodes were characterized by Strombolian activity, intense ash emissions, and small effusive activity. Many of the paroxysms that took place during the investigation period fall into this type of scenario.

The transitional phases from Strombolian activity to lava fountaining can be observed and detected. On the evening of 16 March 2013, NSEC produced an intense episode of lava fountaining. This event, one of the most violent of the 2013 series of paroxysms, was preceded by a long "prelude" of Strombolian activity [32] that started on the afternoon of 15 March and was followed by weak, discontinuous activity at the Voragine [31]. The transition from the strombolian to the lava fountaining episodes is recognizable in the slope (blue filled curve) trends of Figure 7. It is clear from the plots that we have an increase in the slope magnitude from 0.5 to 1.5 when the lava fountaining begins and the lava starts to overflow through the deep breach in the southeastern rim of the NSEC [33].

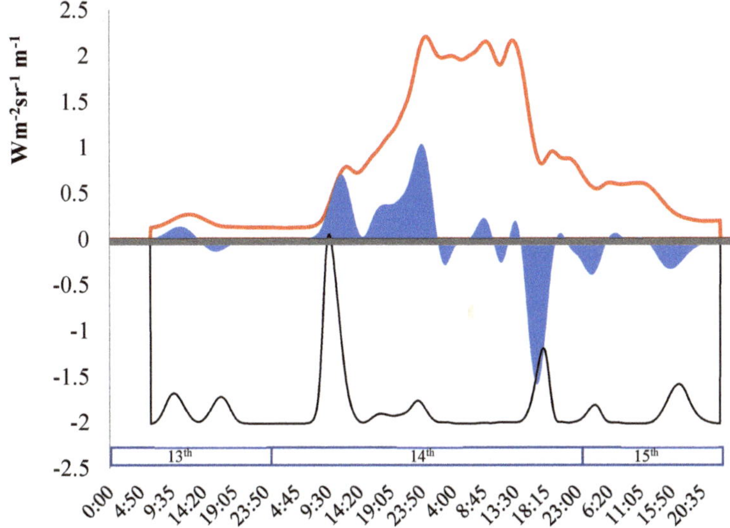

Figure 6. Statistics for the 14–15 December 2013 Etna lava fountain. Relatively low values of kurtosis (<0) and gradient (<0.5) characterize this kind of activity at the onset of the eruption.

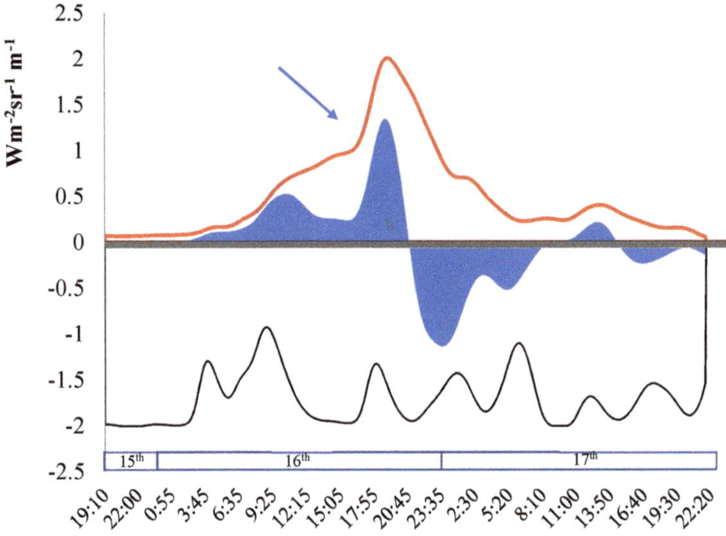

Figure 7. The analysis of statistical indexes allows highlighting transitions between different eruptive phases within the same activity. The transition from the Strombolian phase to the lava fountaining is well recognizable when we look at the peak (**blue arrow**) in gradient values (**blue filled curve**) during the 16 March 2013 Etna eruption.

5. Discussion

Our results suggest a relationship between the rate of increment in radiance (and thus temperature) and the nature of the volcanic process (i.e., eruptive styles as Strombolian activity, lava fountains and lava flow) causing that increment. Statistical analysis of SEVIRI MIR radiance trends highlighted the involvement of at least two different heating mechanisms for eruptions at Mt. Etna.

The first mechanism involves rapid increases in radiance and appears to be associated with effusive activity. Statistic kurtosis and gradient provide a measure of the steepness of the radiance trend. Relatively high values of kurtosis and gradient are mainly associated with the presence of active lava flows. In this case, Kurtosis and gradient magnitudes vary in the range of 0.0–5.0 and 0.5–3.0 respectively. We observed that statistics for small flows, as those generated by lava fountaining lava spatter or intra-crater lava ponding, show lower values of kurtosis (−0.5–0.5) and gradient (0.5–1) compared with values associated with long-term lava flows (see 2013 March case study of Figure 5).

The second mechanism is characterized by slope values usually less than 0.5 and refers to Strombolian events, often associated with paroxysmal activity and ash plumes. Moreover, the kurtosis outlier values remain below zero.

To understand and interpret the differences in radiance rates at the beginning of an eruption is not an easy task. However, we believe that such differences are the results of differences in eruptive and depositional processes of the different hot volcanic products (i.e., lava, ash, volcanic bombs). Eruption style controls both surface heating, the function of the output rate of lava and/or pyroclasts at a different temperature, and surface cooling, the function of the nature, temperature, and dispersal of the products. These heating and cooling mechanisms may be acting at the same time and thus explain the observed radiance behavior.

Temperatures of erupting magmas at Mount Etna are shown to be determined within a few °C in the range 1000–1200 °C [26]. The magmatic temperatures and petrological characters appear closely related to the volcanic activity. The alkaline basic magma of Mount Etna extrudes as lava at a nearly constant temperature of 1080 °C, but during explosive paroxysmal eruptions, the magma temperature is higher (1125 °C and possibly more). However, the small fragments of ash, lapilli, and bombs erupted during paroxysmal eruption fly through the air and cool quickly along their path through the atmosphere. Therefore, pyroclasts ejected in the atmosphere during explosive activity cool much faster than lava flows originated at effusive vents. In addition, gases and ash particles thrown into the atmosphere during volcanic eruptions influence the measured radiance. The particles spewed from volcanoes have the ability to absorb the radiation emitted by erupted materials shading the radiance signal detected at the sensor.

An important role is also played by the volume of the volcanic deposits generated during the eruption. The fraction of the pixel occupied by "hot products" depends on the type of volcanic eruptions and their spatial distribution. For example, explosive eruptions at Mt. Etna are mainly concentrated at the summit craters while lava flows tend to expand and occupy an increasingly larger area as the eruption continues. Therefore, the radiance detected at the sensor is a function of the temperature of the erupted materials and the fractional area occupied by those materials.

Validation was conducted by comparing our results with literature data using a time-series of SEVIRI data from 2011–2017 (4). We used threshold values of 0.1 and 0.5 for the kurtosis and the slope, respectively, to distinguish between effusive and explosive eruptions in the first 8, 16, and 24 h from the onset of the eruption. We found that, 79% of the eruptions were correctly interpreted in the first eight hours, 85% in the first 16 h, and 93% after a day from the beginning of the eruption.

However, several issues need to be taken into account. First, information on eruptions derived from monitoring bulletins is mainly qualitative information. Therefore, it is difficult to accurately determine the intensity of the eruptions and their duration from available literature data. Second, literature data are mainly referring to thermal camera data, remote sensing and field observations. These observations are all weather dependent. Third, our time-series suffers from data loss issues due to acquisition failure. Data loss and noisy images are suitable to generate artifacts that lead to the presence of a false outlier. Finally, events occurring between 2011 and 2017 are mainly short-term paroxysmal eruptions, while long-term effusive eruptions are scarce.

The transition from Strombolian to fountaining activity is well documented [19,32]. Since 1999, Mount Etna has been characterized by episodic lava fountaining, each episode identified by initial Strombolian activity followed by transition to sustained fountaining to feed high effusion rate lava flow [32]. Such transitional behavior introduces further uncertainties into the identification and classification of the erupting style.

Mt. Etna, with its dense, multiparametric array of monitoring networks, provided an ideal test case for our analysis. Future integration of the proposed method will complement the existing networks and provide an additional monitoring tool independent from ground support. Finally, our results represent a promising step toward similar application to less monitored volcanoes, also in remote areas. Such application will require the threshold values empirically calculated for Mt. Etna to be adapted to specific cases or, alternatively, to be derived by ad-hoc models of eruption-dependent thermal radiation.

6. Conclusions

Eruption monitoring by satellite remote sensing still requires validation and refined, ground-tested methodologies. In this study, we explored the capability of high temporal resolution satellite data to discriminate and, to a limited extent, predict erupting styles at Etna volcano. A new statistical approach based on the wavelet transform of time-series of SEVIRI radiance data acquired at 5–15 min acquisition rate in the MIR spectral band (3.9 µm) is presented. Statistical analysis of processed data show potential in predicting eruptive styles. Statistic kurtosis and gradient derived from the analyzed time series allowed to recognize between two different eruptive styles: 1) paroxysmal events, characterized by Strombolian explosive activity, often associated with ash emissions, and lava fountaining (15–17/03/2013, 13–17/12/2013) and 2) lava flows (2–8/12/2015 and 26/02/2017–21/03/2017). It must be stressed that our methodology applies to data acquired at the onset of the eruption in order to give a quick response in terms of alert. The results obtained by this method have a considerable degree of trustworthiness in the first eight hours from the beginning of the eruption.

By considering the reliability of the proposed approach, the authors believe that it can be also applied worldwide, in particular for volcanoes lying in remote areas.

Author Contributions: V.L. conceived the idea, analyzed data, implemented wavelet transform, and wrote the manuscript with contributions from all authors. M.M. and M.S. processed satellite data and exported time-series for wavelet transform. S.C. and J.T. carried out the study of paroxysmal events with particular emphasis on the effects of gaseous volcanic emissions, ash plumes and explosive episodes. All authors contributed to the ideas, writing, and discussion.

Funding: This research received no external funding.

Acknowledgments: We thank the anonymous reviewers for their careful reading of our manuscript and their many insightful comments and suggestions.

Conflicts of Interest: The authors declare no conflict of interest.

References

1. Ganci, G.; Vicari, A.; Fortuna, L.; Del Negro, C. The HOTSAT volcano monitoring system based on combined use of SEVIRI and MODIS multispectral data. *Ann. Geophys.* **2011**, *54*. [CrossRef]
2. Ganci, G.; Harris, A.J.L.; Del Negro, C.; Guehenneux, Y.; Cappello, A.; Labazuy, P.; Calvari, S.; Gouhier, M. A year of lava fountaining at Etna: Volumes from SEVIRI. *Geophys. Res. Lett.* **2012**, *39*, L06305. [CrossRef]
3. Marchese, F.; Filizzola, C.; Genzano, N.; Mazzeo, G.; Pergola, N.; Tramutoli, V. Assessment and improvement of a robust satellite technique (RST) for thermal monitoring of volcanoes. *Remote Sens. Environ.* **2011**, *115*, 1556–1563. [CrossRef]

4. Corradini, S.; Guerrieri, L.; Lombardo, V.; Merucci, L.; Musacchio, M.; Prestifilippo, M.; Scollo, S.; Silvestri, M.; Spata, G.; Stelitano, D. Proximal Monitoring of the 2011–2015 Etna Lava Fountains Using MSG-SEVIRI Data. *Geosciences* **2018**, *8*, 140. [CrossRef]
5. Harris, A.J.L.; Pilger, E.; Flynn, L.P.; Garbeil, H.; Mouginis-Mark, P.J.; Kauahikaua, J.; Thornber, C. Automated, high temporal resolution, thermal analysis of Kilauea volcano, Hawaii, using GOES satellite data. *Int. J. Remote Sens.* **2001**, *22*, 945–967. [CrossRef]
6. Wright, R.; Flynn, L.; Garbeil, H.; Harris, A.; Pilger, E. MODVOLC: Near-real-time thermal monitoring of global volcanism. *J. Volcanol. Geotherm. Res.* **2004**, *135*, 29–49. [CrossRef]
7. Colin, O.; Rubio, M.; Landart, P.; Mathot, E. VoMIR: Over 300 Volcanoes Monitored in Near Real-Time by AATSR. In Proceedings of the Envisat Symposium 2007, Montreux, Switzerland, 23–27 April 2007.
8. Dehn, J.; Dean, K.; Angle, K. Thermal monitoring of North Pacific volcanoes from space. *Geology* **2000**, *28*, 755–758. [CrossRef]
9. Dehn, J.; Dean, K.; Angle, K.; Izbekov, P. Thermal precurs in satellite images of the 1999 eruption Shishaldin Volcano. *Bull. Volcanol.* **2002**, *64*, 525–534. [CrossRef]
10. Patrick, M.; Dehn, J.; Dean, K. Numerical modeling of lava flow cooling applied to the 1997 Okmok eruption, II: Comparison with AVHRR thermal imagery. *J. Geophys. Res.* **2005**, *110*, B02210. [CrossRef]
11. Webley, P.W.; Wooster, M.J.; Strauch, W.; Saballos, J.A.; Dill, K.; Stephenson, P.; Stephenson, J.; Escobar Wolf, R.; Matias, O. Experiences from near-real-time satellite-based volcano monitoring in Central America: Case studies at Fuego, Guatemala. *Int. J. Remote Sens.* **2008**, *29*, 6621–6646. [CrossRef]
12. Pergola, N.; Marchese, F.; Tramutoli, V. Automated detection of thermal features of active volcanoes by means of Infrared AVHRR records. *Remote Sens. Environ.* **2004**, *93*, 311–327. [CrossRef]
13. Harris, A.J.L.; Butterworth, A.L.; Carlton, R.W.; Downey, I.; Miller, P.; Navarro, P.; Rothery, D.A. Low cost volcano surveillance from space: Case studies from Etna, Krafla, Cerro Negro, Fogo, Lascar and Erebus. *Bull. Volcanol.* **1997**, *59*, 49–64. [CrossRef]
14. Harris, A.J.L.; Blake, S.; Rothery, D.A.; Stevens, N.F. A chronology of the 1991 to 1993 Mount Etna eruption using advanced very high resolution radiometer data: Implications for real-time thermal volcano monitoring. *J. Geophys. Res.* **1997**, *102*, 7985–8003. [CrossRef]
15. Wooster, M.J.; Rothery, D.A. Time series analysis of effusive volcanic activity using the ERS along track scanning radiometer: The 1995 eruption of Fernandina volcano, Galapagos Island. *Remote Sens. Environ.* **1997**, *69*, 109–117. [CrossRef]
16. Corradini, S.; Montopoli, M.; Guerrieri, L.; Ricci, M.; Scollo, S.; Merucci, L.; Marzano, F.S.; Pugnaghi, S.; Prestifilippo, M.; Ventress, L.J.; et al. Multi-Sensor Approach for Volcanic Ash Cloud Retrieval and Eruption Characterization: The 23 November 2013 Etna Lava Fountain. *Remote. Sens.* **2016**, *8*, 58. [CrossRef]
17. Musacchio, M.; Silvestri, M.; Buongiorno, M.F. Use of radiance value from MSG SEVIRI and MTSAT data: Application for the monitoring on volcanic area. In Proceedings of the 2011 EUMETSAT Meteorological Satellite Conference, Oslo, Norway, 5–9 September 2011.
18. Musacchio, M.; Silvestri, M.; Buongiorno, M.F. RT Monitoring of active volcanoes: MT Etna. In Proceedings of the SeaSpace International Remote Sensing Conference, San Diego, CA, USA, 21–24 October 2012.
19. Ripepe, M.; Marchetti, E.; Delle Donne, D.; Genco, R.; Innocenti, L.; Lacanna, G.; Valade, S. Infrasonic early warning system for explosive eruptions. *J. Geophys. Res. Solid Earth* **2018**, *123*. [CrossRef]
20. Behncke, B.; Branca, S.; Corsaro, R.A.; De Beni, E.; Miraglia, L.; Proietti, C. The 2011–2012 summit activity of Mount Etna: Birth, growth and products of the new SE crater. *J. Volcanol. Geotherm. Res.* **2014**, *270*, 10–21. [CrossRef]
21. De Beni, E.; Behncke, B.; Branca, S.; Nicolosi, I.; Carluccio, R.; D'Ajello Caracciolo, F.; Chiappini, M. The continuing story of Etna's New Southeast Crater (2012–2014): Evolution and volume calculations based on field surveys and aerophotogrammetry. *J. Volcanol. Geotherm. Res.* **2015**, *303*, 175–186. [CrossRef]
22. Guerrieri, L.; Merucci, L.; Corradini, S.; Pugnaghi, S. Evolution of the 2011 Mt. Etna ash and SO2 lava fountain episodes using SEVIRI data and VPR retrieval approach. *J. Volcanol. Geotherm. Res.* **2015**, *291*, 63–71. [CrossRef]
23. Vulpiani, G.; Ripepe, M.; Valade, S. Mass discharge rate retrieval combining weather radar and thermal camera observations. *J. Geophys. Res. Solid Earth* **2016**, *121*, 5679–5695. [CrossRef]

24. Corsaro, R.A.; Andronico, D.; Behncke, B.; Branca, S.; Caltabiano, T.; Ciancitto, F.; Miraglia, L. Monitoring the December 2015 summit eruptions of Mt. Etna (Italy): Implications on eruptive dynamics. *J. Volcanol. Geotherm. Res.* **2017**, *341*, 53–69. [CrossRef]
25. Volcano Descovery, Etna Volcano—Eruption Update Archive—Part 11. Available online: https://www.volcanodiscovery.com/etna/current-activity-part-11.html (accessed on 1 January 2019).
26. Pompilio, M.; Trigila, R.; Zanon, V. Melting experiment on Mr. Etna lavas: I-The calibration of an empirical geothermometer to estimate the eruptive temperature. *Acta Vulcanol.* **1998**, *10*, 67–76.
27. Harris, A.J.L. *Thermal Remote Sensing of Active Volcanoes: A User's Manual*; Cambridge Press: Cambridge, UK, 2013.
28. Bijaoui, A.; Giudicelli, M. The optimal image addition using the wavelet transform. *Exp. Astron.* **1991**, *1*, 347–362. [CrossRef]
29. Mallat, S.G. A theory for multiresolution signal decomposition: The wavelet representation. *IEEE Trans. Pattern Anal. Mach. Intell.* **1989**, *11*, 674–693. [CrossRef]
30. Lombardo, V. AVHotRR: Near-real time routine for volcano monitoring using IR satellite data. *Geol. Soc. Lond. Spec. Publ.* **2016**, *426*, 73–92. [CrossRef]
31. Volcano Descovery, Etna Volcano Activity Updates: December 2013. Available online: https://www.volcanodiscovery.com/etna/eruption-updates/december2013.html (accessed on 1 January 2019).
32. Bombrun, M.; Spampinato, L.; Harris, A.; Barra, V.; Caltabiano, T. On the transition from strombolian to fountaining activity: A thermal energy-based driver. *Bull. Volcanol.* **2016**, *78*, 15. [CrossRef]
33. INGV Etnean Observatory, the 16 March 2013 Paroxysm of Etna. Available online: http://www.ct.ingv.it/en/11-notizie/news/861-the-16-march-2013-paroxysm-of-etna.html (accessed on 1 January 2019).

Article

Imaging Thermal Anomalies in Hot Dry Rock Geothermal Systems from Near-Surface Geophysical Modelling

David Gomez-Ortiz [1,*], Isabel Blanco-Montenegro [2,3], Jose Arnoso [3,4], Tomas Martin-Crespo [1], Mercedes Solla [5,6], Fuensanta G. Montesinos [3,7], Emilio Vélez [3,4] and Nieves Sánchez [8]

1 Dpt. Biología y Geología, Física y Química Inorgánica, ESCET, Universidad Rey Juan Carlos, C/Tulipán s/n, 28933 Móstoles, Madrid, Spain; tomas.martin@urjc.es
2 Departamento de Física, Escuela Politécnica Superior, Universidad de Burgos, Avda. de Cantabria s/n, 09006 Burgos, Spain; iblanco@ubu.es
3 Research Group 'Geodesia', Universidad Complutense de Madrid, 28040 Madrid, Spain
4 Instituto de Geociencias (IGEO, CSIC-UCM), C/ Doctor Severo Ochoa, 7. Facultad de Medicina (Edificio Entrepabellones 7 y 8), 28040 Madrid, Spain; jose.arnoso@csic.es (J.A.); emilio.velez@csic.es (E.V.)
5 Defense University Center, Spanish Naval Academy, Plaza de España 2, 36920 Marín, Pontevedra, Spain; merchisolla@cud.uvigo.es
6 Research Group 'Geotecnologías Aplicadas', Universidad de Vigo, 36310 Vigo, Spain
7 Facultad de CC. Matemáticas, Universidad Complutense de Madrid, Plaza de Ciencias, 3, 28040 Madrid, Spain; fuensanta.gonzalez@mat.ucm.es
8 Spanish Geological Survey, Unit of Canary Islands, Alonso Alvarado, 43, 2°A, 35003 Las Palmas de Gran Canaria, Spain; n.sanchez@igme.es
* Correspondence: david.gomez@urjc.es; Tel.: +34-914887092

Received: 25 January 2019; Accepted: 19 March 2019; Published: 21 March 2019

Abstract: Convective hydrothermal systems have been extensively studied using electrical and electromagnetic methods given the strong correlation between low conductivity anomalies associated with hydrothermal brines and high temperature areas. However, studies addressing the application of similar geophysical methods to hot dry rock geothermal systems are very limited in the literature. The Timanfaya volcanic area, located on Lanzarote Island (Canary Islands), comprises one of these hot dry rock systems, where ground temperatures ranging from 250 to 605 °C have been recorded in pyroclastic deposits at shallow (<70 m) depths. With the aim of characterizing the geophysical signature of the high ground temperature areas, three different geophysical techniques (ground penetrating radar, electromagnetic induction and magnetic prospecting) were applied in a well-known geothermal area located inside Timanfaya National Park. The area with the highest ground temperatures was correlated with the location that exhibited strong ground penetrating radar reflections, high resistivity values and low magnetic anomalies. Moreover, the high ground temperature imaging results depicted a shallow, bowl-shaped body that narrowed and deepened vertically to a depth greater than 45 m. The ground penetrating radar survey was repeated three years later and exhibited subtle variations of the signal reflection patterns, or signatures, suggesting a certain temporal variation of the ground temperature. By identifying similar areas with the same geophysical signature, up to four additional geothermal areas were revealed. We conclude that the combined use of ground penetrating radar, electromagnetic induction and magnetic methods constitutes a valuable tool to locate and study both the geometry at depth and seasonal variability of geothermal areas associated with hot dry rock systems.

Keywords: Timanfaya volcanic area; HDR geothermal systems; GPR; EMI; magnetic anomalies; seasonality

1. Introduction

One of the most remarkable features of active volcanic areas is the presence of high-enthalpy geothermal systems that exhibit high temperature zones (>150–200 °C) at ground level (e.g., [1]). The most common are convective hydrothermal systems that consist of a magmatic body acting as a heat source, a groundwater system to transport the heat towards the surface and a confining, shallow, impermeable structure (e.g., [2]). Less common are hot dry rock (HDR) geothermal systems that consist of subsurface zones with very low fluid content (e.g., [3]). Most of the literature on the exploration of geothermal resources is related to hydrothermal systems and is primarily based on electrical and electromagnetic methods due to the strong correlation between the low conductivity anomalies associated with hydrothermal brines and the high temperature areas. Among the geophysical techniques more commonly used are magnetotellurics, transient electromagnetics, electrical resistivity, magnetics, self-potential and seismic, with investigation depths ranging from few to hundreds of meters (e.g., [1] and references herein). However, the use of such techniques for studying HDR systems is very limited because HDR systems are not as common as convective hydrothermal ones, and because the presence of vapour as dominant phase instead of hot water makes these kinds of systems more difficult to interpret using electromagnetic methods.

The Timanfaya volcanic area (Figure 1) is located on Lanzarote Island (Canary Archipelago, Spain). The Canary Islands constitute an intraplate oceanic volcanic archipelago consisting of seven major islands and four islets located in the eastern Central Atlantic, approximately 125 km from northwest Africa. The islands of Lanzarote and Fuerteventura developed in the earliest stage of the Canary magmatic activity and constitute the emergent part of the Eastern Canary Ridge, which is located northeast of the archipelago [4]. In Lanzarote, thermal anomalies associated with a HDR geothermal system are still present in different areas as a consequence of the historical eruptive activity that occurred between 1730 and 1736 (the Timanfaya eruption).

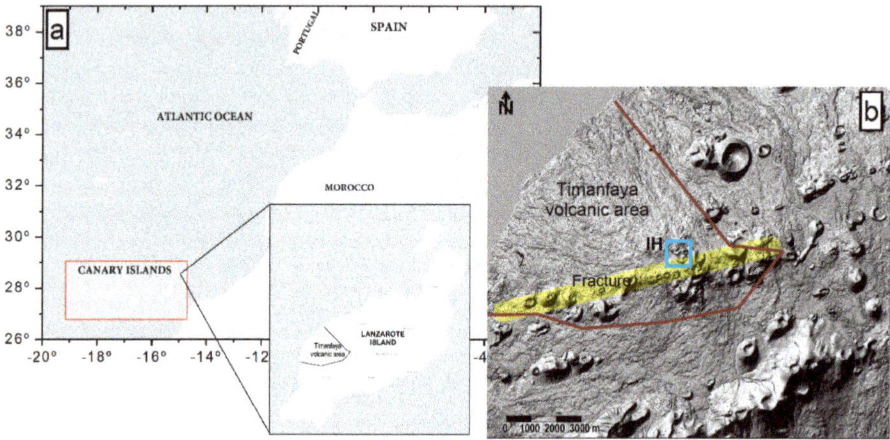

Figure 1. (**a**) location map of Lanzarote Island in the Canary archipelago (Spain); (**b**) shaded relief of the volcanic area of Timanfaya on Lanzarote Island, showing the National Park boundary (red line) and the eruptive fissure (thick yellow line) of the 1730–1736 volcanic eruption. The blue rectangle marks the area of thermal anomalies located at Islote de Hilario (IH).

Over 30 volcanic vents formed during that eruption, mostly characterized by fissural eruptions and low-explosive basaltic magmas. The last eruption on Lanzarote Island occurred in 1824, with the formation of three new volcanoes. Typically, all of those vents are aligned along a system of fractures oriented to N70°E, of approximately 14 km in length, following the path of a central structural rift-type zone [5,6]. As a result of the 1730–1736 eruption, hot magma exists at approximately 4 km depth [7].

Although the magma body undergoes a cooling process, it still has a very high temperature, which explains the presence of superficial thermal anomalies in the area. The anomalies are isolated and limited to some fracture-related alignments and along craters' rims [7–9].

The studied geothermal field is located on Islote de Hilario, within Timanfaya National Park, northwest of the Timanfaya volcanic system (see Figures 1 and 2). The shallow geological structure of Islote de Hilario, determined from the previously available information [10] (surface geological units and several monitoring wells are used for temperature and gas composition determination), is very simple, consisting of a thick (>70 m) unit of pyroclastic materials that originated during the historical eruption of Timanfaya (1730–1736) overlying older basaltic lava flows. The measured temperature in this area is approximately 605 °C inside a 13-m deep well [10]. Other temperature measurements made in shallower wells range from 300 to 380 °C, 450 °C in a natural furnace and approximately 250 °C on the surface [11]. This geothermal field has been defined as a HDR system [8,11,12] that is connected to the base of the Timanfaya volcano through very shallow inclined fractures as revealed by previous surveys [13]. This work presents a study on the geothermal anomalies located at the Timanfaya volcanic area via an analysis and joint interpretation of ground penetrating radar (GPR), magnetic prospecting and electromagnetic induction (EMI) data obtained for areas not previously surveyed. First, we combined three shallow geophysical methods applied on a well-known geothermal anomaly (Figure 2) to fully characterize the geophysical signature and the relationship between the resistivity, magnetic and electromagnetic anomalies with temperature. Then, we have looked for the same geophysical signature allowing us to locate similar unknown geothermal anomalies in the surrounding area. The results help assess the reliability of the combination of the proposed methods to locate the geothermal zones at the surface and to obtain information regarding their distribution and geometry at depth.

2. Geophysical Techniques

2.1. Ground Penetrating Radar (GPR)

GPR is a geophysical method based on the propagation of electromagnetic pulses (1–20 ns) in the frequency band of 10 MHz–2.5 GHz. A transmitting antenna emits an electromagnetic signal into the ground, which is partly reflected when it encounters media with different dielectric properties and partly transmitted into deeper layers. Then, the reflections produced are recorded from the receiving antenna, which is either in a separate antenna box or in the same antenna box as the transmitter. The strength (amplitude) of the reflected fields is proportional to the magnitude of the dielectric constant change, which is characterized by the reflection coefficient (RC). The RC increases as the dielectric contrast increases. As the antenna is moved along the ground surface, a two-dimensional image (radargram) is obtained, which is an XZ graphic representation of the detected reflections. The x-axis represents the antenna displacement along the survey line, and the z-axis represents the two-way travel time of the pulse emitted (in terms of nanoseconds). If the time required to propagate to a reflector and back is measured, and the velocity of this pulse in medium is known, then the depth of the reflector can be determined [14–16].

GPR Data Acquisition and Processing

The same GPR transect was recorded in a time span of three years during two different field campaigns, May 2012 and April 2015 (Figure 2b). This was aimed to investigate possible variations on the subsurface temperature along a well-known geothermal anomaly (TA1) developed over pyroclastic deposits.

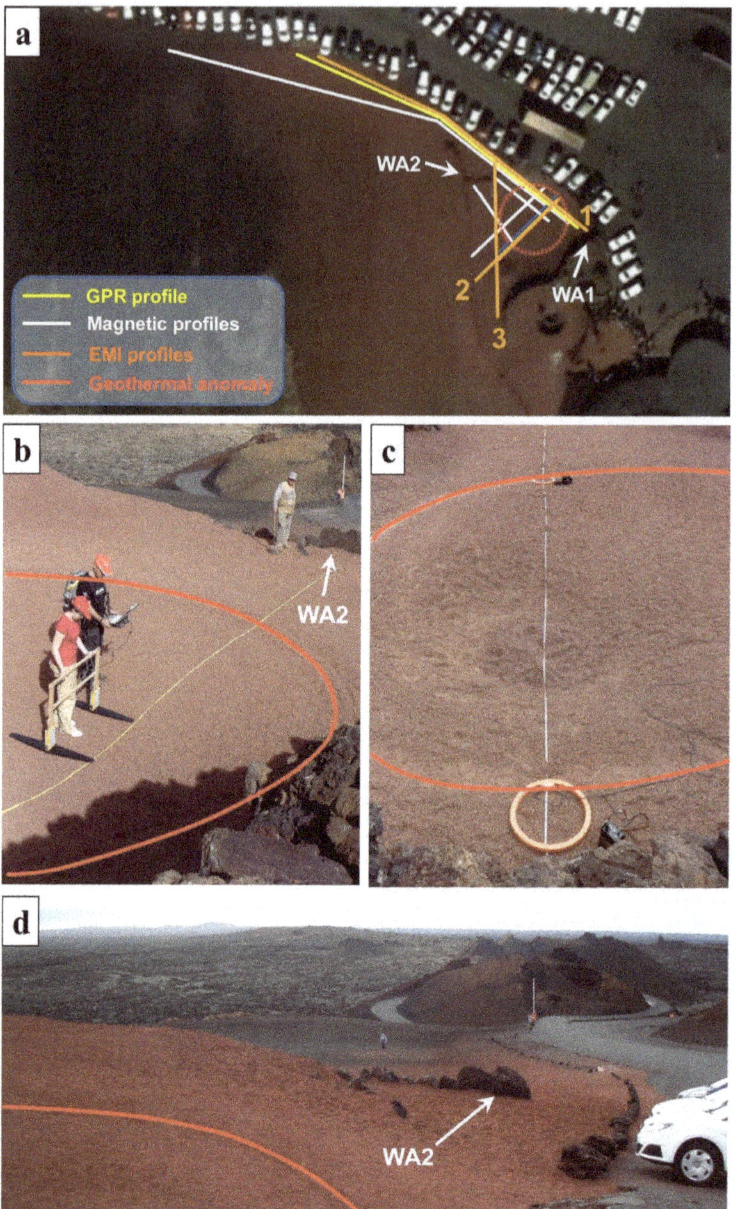

Figure 2. (**a**) location of the different surveys performed at the Islote de Hilario area (GPR, EMI and magnetic survey) and the geothermal anomaly; (**b**) GPR survey data acquisition. One of the man-made walls (WA1) can be seen at the beginning of the profile; (**c**) EMI survey data acquisition. The coils are placed on the boundaries of the circular geothermal anomaly area (red line); and (**d**) a person carrying the proton magnetometer during the magnetic survey data acquisition. A second man-made wall (WA2) can be seen at the middle of the profile.

The GPR surveys were conducted using a RAMAC/GPR system from MALÅ Geosciences (Mala, Sweden). First, a common mid-point mode (CMP) test was performed to determine the velocity of propagation of the GPR waves in the subsurface medium, which is composed of lapilli. In the CMP configuration, the receiver (Rx) and transmitter (Tx) are moved one from the other at a constant step along the survey line, maintaining a common fixed centre point. The antenna spacing is varied at a fixed location and the change in the two-way travel time of the electromagnetic wave from the reflectors is measured [14]. The CMP data acquisition was carried out with a 200-MHz unshielded antenna. The geometry consisted of an 11-m antenna separation with a step size of 0.1 m (0.2 m antenna separation per trace), with reflections collected using a time window of 234 ns, defined by 512 samples per trace. A root-mean-squared velocity of 0.08 m/ns was obtained for the pyroclastic deposit (Figure 3).

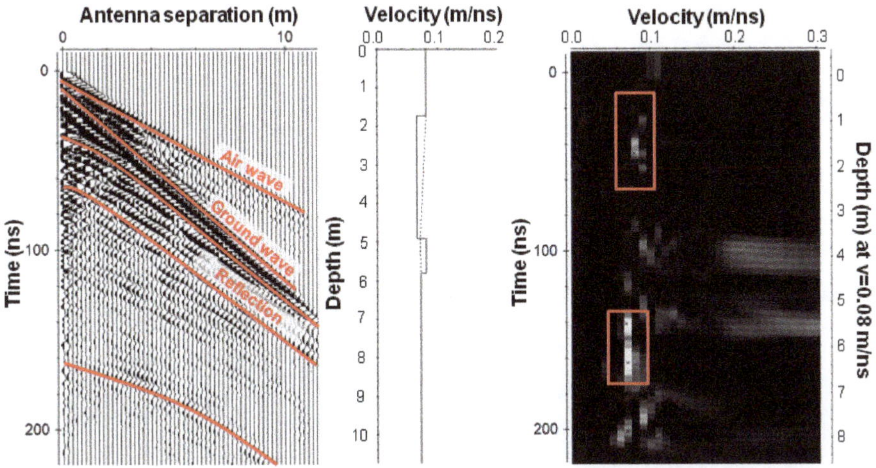

Figure 3. CMP-velocity analysis performed and the corresponding velocity spectrum (velocity semblance panel).

Next, both survey seasons were conducted using a common-offset (CO) mode with a 100-MHz unshielded antenna. In the CO configuration, the receiver and transmitter are moved together along the radar line with a constant distance between them. In this work, point-to-point data were acquired with an offset between transmitter and receiver of 1 m, with reflections collected within a time window of 467 ns defined by 512 samples per trace and trace-distance intervals of 50 and 20 cm (in 2012 and 2015 campaigns, respectively). To measure the profile lengths, and to control the distance between traces, a tape measure was used as a guide (Figure 2b). The 100-MHz unshielded antenna was selected since this frequency provides sufficient penetration (up to 18 m depth) [14]. Moreover, unshielded antennas produce more powerful electromagnetic energy into the ground reaching higher depth penetration. However, they are very sensitive to background noise (e.g., electrical wires and metal) and should be used in cleaned spaces. For our time window, and considering a velocity of propagation of 0.08 m/ns, the maximum depth of penetration is about 18 m.

The reflection profiles recorded during the two different field campaigns were identically filtered using the processing sequence described in Table 1 with the ReflexW software (version 8.5.7, Sandmeier geophysical research, Karlsruhe, Germany) [17]. The velocity of propagation obtained from the CMP (using the algorithm semblance analysis [18]) was used for topographic correction as well as to convert the time window of the radargrams into depth values (Figure 4).

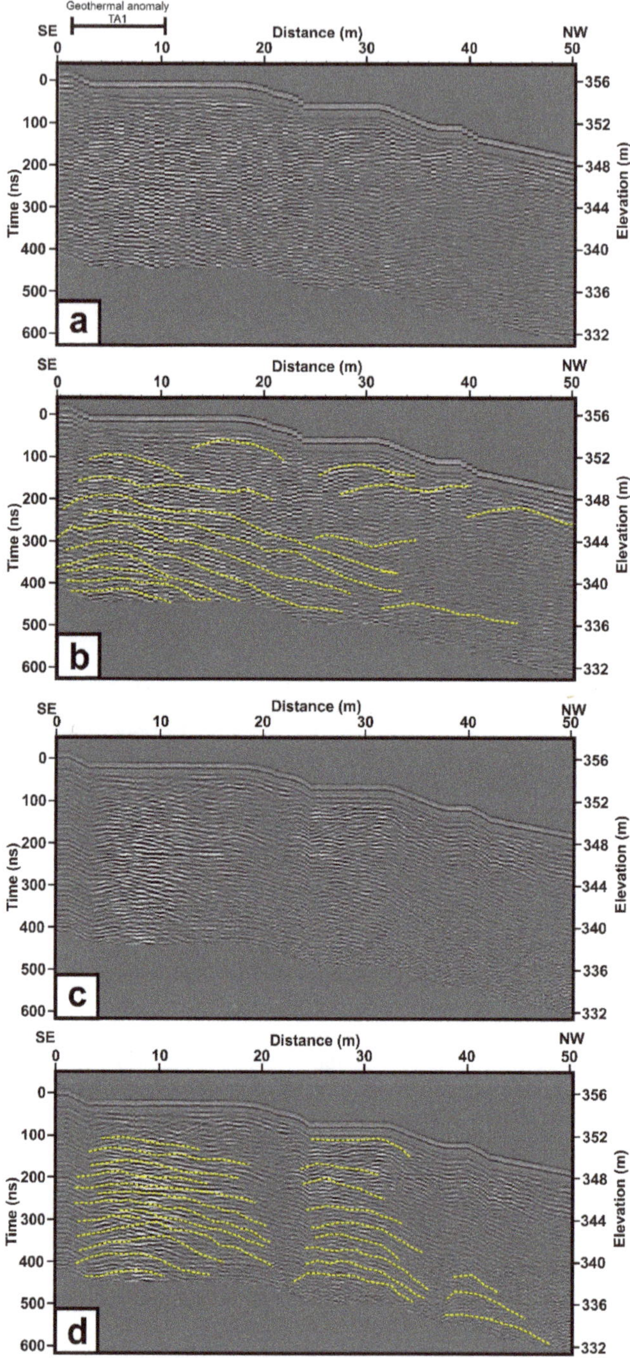

Figure 4. GPR data for the profile across the geothermal anomaly area: (**a**) observed radargram (May 2012); (**b**) interpreted radargram (May 2012); (**c**) observed radargram (April 2015); (**d**) interpreted radargram (April 2015).

Table 1. Data processing applied to the 100-MHz GPR data.

Step	Filter	Parameters
1	Time-zero correction	—
2	Subtract-mean (dewow)	Time window: 10 ns
3	Energy decay	Scaling value: 0.5
4	Subtracting average	Average traces: 100
5	Band-pass (butterworth)	Low pass: 40 MHz; high pass: 150 MHz
6	Topographic correction	Velocity: 0.08 m/ns

2.2. Electromagnetic Survey

Electromagnetic induction (EMI) is a shallow geophysical method that obtains information about the apparent conductivity (or, its inverse, apparent resistivity) distribution at depth without requiring the direct contact of electrodes with the ground, as in the electrical resistivity method. EMI has been widely used to detect arsenic concentrations in groundwater [19], map saline plumes [20], locate faults [21], depict impact structures [22] and identify sinkholes [23], among other uses. Because temperature modifies the electromagnetic properties of the materials, the anomalies mapped by the EMI method might be related to variations in temperature, represented as differences in resistivity contrast.

Typical EMI equipment consists of two coils (a transmitter and receiver) separated by a fixed distance. A time-varying electric current at a fixed frequency generates a primary magnetic field at the transmitting coil. The primary field creates eddy currents in the ground that induce a secondary magnetic field. Both fields are recorded by the receiving coil [24]. From these data, the apparent resistivity of the subsoil can be obtained as it is linearly related to the quadrature component (i.e., the ratio of the secondary to the primary magnetic field) [24].

The depth of investigation depends on both the orientation of the coils (vertical or horizontal coplanar dipoles) and the intercoil spacing. We used a Geonics EM34-3 conductivity meter with three different intercoil spacings of 10 m, 20 m and 40 m, corresponding to transmitter frequencies of 6.4 kHz, 1.6 kHz and 0.4 kHz, respectively [24]. For each intercoil distance, we used two different coil configurations, vertical and horizontal coplanar dipoles (Figure 2c), selecting a station spacing of 1 m to attain a suitable horizontal resolution of the profiles. When the length of the profile is ≥40 m, a total of six measurements at each station can be obtained. The intercoil distances of 10 m, 20 m and 40 m correspond to investigation depths of 15 m, 30 m and 60 m, respectively, in the vertical dipole configuration and 7.5 m, 15 m and 30 m in the horizontal dipole configuration. Considering the availability of space within the study area, three profiles of different lengths and orientation were investigated (Figures 2a and 5) in May 2013. The longest profile (EMI1), with a length of 49 m, runs NW-SE, laterally crossing a known geothermal anomaly area at the SE end (TA1) where the National Park's visitor exhibitions are located (Figure 2c). The readings totalled 160 (80 for each dipole configuration). A second profile (EMI2) trending NE-SW, transverse to EMI1, also crosses the geothermal anomaly area (TA1) at its central part but is only 21 m long because it is limited by a parking area and other man-made walls. Because of space restrictions, an intercoil spacing of only 10 m was used with 26 readings. A third profile (EMI3) running N-S was obtained to image the resistivity values outside but close to the geothermal anomaly area (TA1). The length of this profile is 30 m, allowing the use of both 10 m and 20 m intercoil spacings, with readings totalling 64.

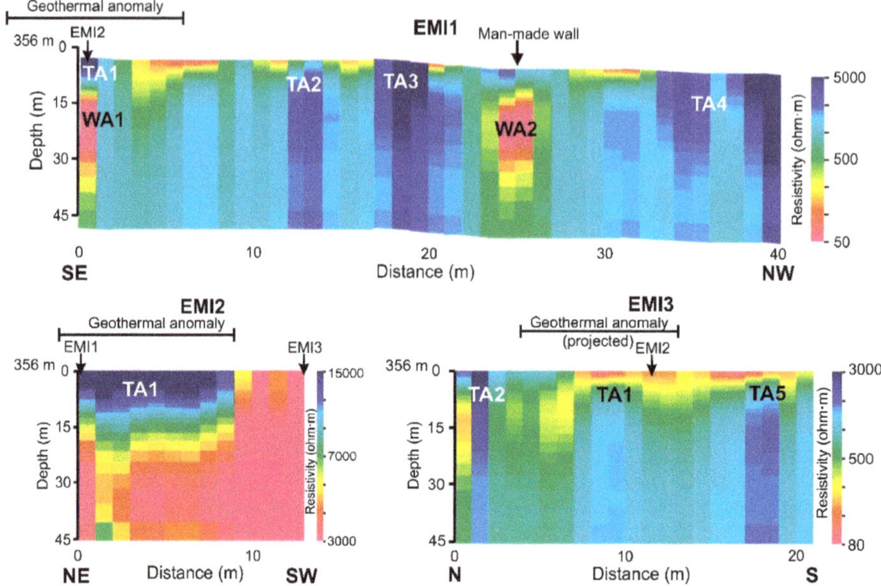

Figure 5. Resistivity models obtained from the 2D inversion of EMI data for the three profiles of Figure 2a. Five areas of high-resistive values have been marked and interpreted as "temperature anomaly" (TA). Two areas of low-resistivity values have been marked and interpreted as "wall anomaly" (WA).

The field data were inverted using EM1DFM [25] software to produce a true resistivity depth image. EM1DFM is a 2D inversion programme that fits a modelled response to the data by minimizing an objective function [26]. The data required are the quadrature response of the secondary magnetic field, the orientation and intercoil separations, transmitter frequencies, height of the coils, and data error. The quadrature response data were inverted to a model consisting of 25 layers with thickness ranging from 0.5 to 2 m, increasing at depth. An L-curve inversion type, with a maximum number of iterations of 15 and a tolerance value of 0.4, was determined to be optimal after several trial inversions. The resulting inverted model presents the resistivity variations in both lateral and vertical directions along the profile. Because the data measured correspond to the midpoint of the different intercoil spacings, the lengths of the inverted models are less than the lengths of the corresponding profiles. Thus, the EMI1 model is 40 m long, the EMI2 model is 13 m long and the EMI3 model is 21 m long.

2.3. Magnetic Survey

Magnetic anomalies represent the contribution to the Earth's magnetic field originating from geological structures due to the presence of small amounts of magnetic minerals in crustal rocks. These anomalies can be studied at very different scales, from a lithospheric perspective when they are measured from satellites to high-resolution local studies based on land surveys aimed at imaging the shallow subsurface structure. In particular, volcanic areas typically display intense magnetic anomalies due to the high concentration of magnetic minerals of volcanic rocks. Most surveys are conducted using scalar magnetometers that measure the magnitude of the Earth's magnetic field vector. Under these conditions, the magnetic anomalies created by the subsurface structures are dipolar in shape, each of them consisting of a high and low (unless the anomalies are measured at the magnetic poles, where the Earth's magnetic field is vertical).

Magnetization (J), which is the physical property responsible for magnetic anomalies, is a vector quantity resulting from the sum of a remanent component acquired in the past (J_{rem}) and an induced component (J_{ind}), which is proportional to the magnetic susceptibility (k) and is parallel to the Earth's present magnetic field (H). Thus, the following relationship may be written: $J = J_{rem} + J_{ind} = J_{rem} + k H$. In volcanic rocks, the remanent magnetization is primarily acquired when the rocks are cooled in the presence of the Earth's magnetic field (thermoremanence) and usually exceeds the induced magnetization. Therefore, in volcanic environments, the ratio J_{rem}/J_{ind}, known as the Köenigsberger ratio (Q), is greater than 1 in most cases. Magnetic anomalies can be related to magnetization contrasts among subsurface structures, which are temperature-dependent, since magnetic minerals lose their properties at temperatures above their Curie point. The primary mineral responsible for magnetic anomalies in volcanic environments is titanomagnetite ($Fe_{3-x}Ti_xO_4$, where $0 \leq x \leq 1$ and x represents the Ti content), which has a Curie temperature that increases as the Ti content decreases, reaching 575 °C for pure magnetite ($x = 0$) [27]. In active volcanic areas with associated thermal anomalies and geothermal systems, magnetic studies revealed that some of the detected anomalies are linked to the decrease of the rocks' magnetization, which can be explained not only by the loss of magnetic properties at high temperatures but also by the chemical alteration affecting magnetic minerals and the circulation of hydrothermal fluids within the rocks [28–32].

In November 2012, a land magnetic survey was performed in the Timanfaya volcanic area. In the Islote de Hilario zone, the total field magnetic data were acquired using an Overhauser magnetometer (GSM-19W, GEM Systems Inc., Toronto, ON, Canada with 0.01 nT resolution and 0.2 nT absolute accuracy), following a 75 m-long profile that coincided in the first 50 m with the GPR profile (Figure 2a,d). Three additional profiles (two of them parallel and one orthogonal to the first) completed the survey. Data sampling along the profiles was performed every 80–100 cm. The magnetic sensor was located approximately 2 m above the ground.

Data in Islote de Hilario area were measured on November 26 (2012) between 4:41 p.m. and 5:12 p.m. local time (GMT-1). Magnetic data acquired at the Güímar magnetic observatory in Tenerife on that date show that the variation in the total field intensity (F) within the same time interval was less than 2 nT. This variation range was so minimal compared with the variations of the crustal field (several hundred nT) that correction of the temporal magnetic variations of external origin was unnecessary.

Magnetic Data Processing and Rock Magnetic Properties

Processing the magnetic data consisted of: (a) subtraction of the main Earth magnetic field modelled using the International Geomagnetic Reference Field [33]; (b) data interpolation into a regular grid with a cell size of 50 cm (Figure 6a); and (c) reduction to the pole assuming a magnetization of induced origin (parallel to the Earth's magnetic field, with declination of –4.7° and inclination of 38.2°) (Figure 6b). This transformation removes the dipolar character of magnetic anomalies and makes their interpretation easier because the maximum of the reduced-to-the-pole anomaly is located roughly over the center of the causative source [34]. In a reduced-to-the-pole map, magnetic lows are associated with negative magnetization/susceptibility contrasts, whereas magnetic highs are related to positive magnetization/susceptibility contrasts in the subsurface beneath the anomaly.

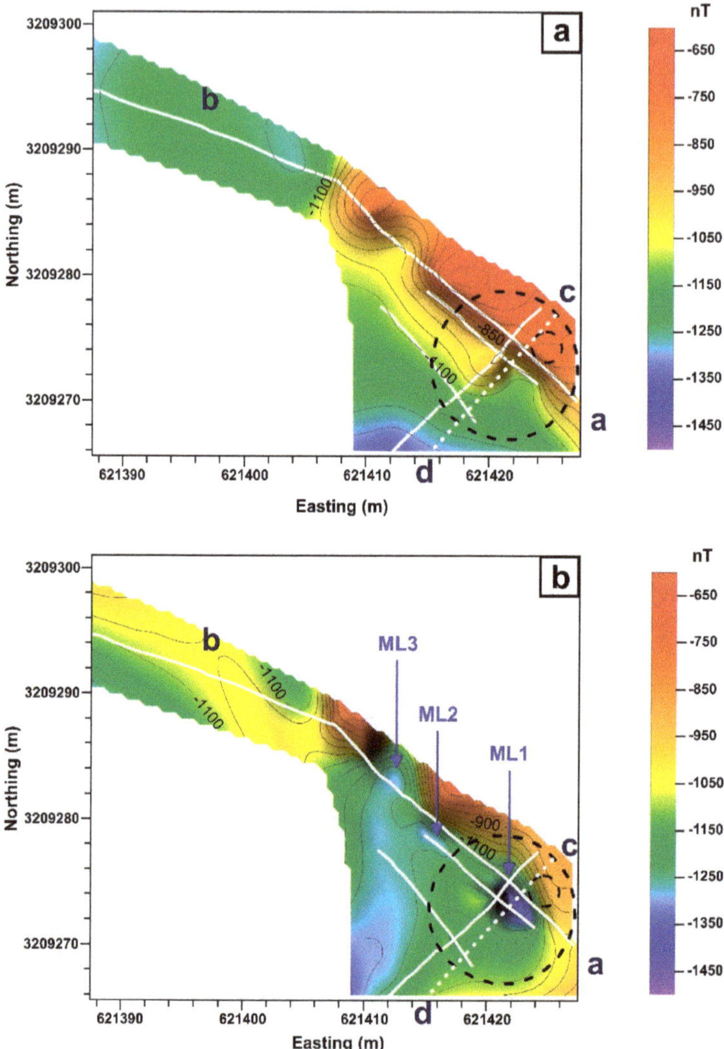

Figure 6. (**a**) magnetic anomaly map obtained through the interpolation of the survey profiles (shown in white); (**b**) reduced-to-the-pole magnetic anomaly map of the studied area. Three magnetic lows can be identified in the map, suggesting high temperatures in the subsoil. a and b mark the limits of the survey line coinciding with the GPR profile and are used for one of the forward models, whereas c and d identify the limits of an orthogonal profile (white dotted line), which was extracted from the reduced-to-the-pole grid and also used for modelling. The dashed black circle marks the surface thermal anomaly. Coordinates correspond to the Universal Transverse Mercator projection (zone 28N).

A paleomagnetic study was conducted on Lanzarote in 2013 [35]. In the Timanfaya volcanic area, three representative basaltic lava flows were sampled: one pertaining to the 1824 Chinero eruption (TM1) and three corresponding to the 1730–1736 eruptions (TM2 to TM4). The magnetic properties relevant to the interpretation of magnetic anomalies are summarized in Table 2 (personal communication). Laboratory experiments revealed that most of the magnetic signal is carried by titanomagnetite with low to medium Ti content and Curie temperatures ranging between approximately 420 °C and 575 °C [35].

Table 2. Summary of the rock magnetic properties of the lava flows in the Timanfaya area.

Site and Lithology	Magnetic Susceptibility (SI)	Induced Magnetization (A/m)	Remanent Magnetization, NRM				Q
			Intensity (A/m)	Declination (°)	Inclination (°)	α_{95}	
TM1 (lava flow 1)	0.011 ± 0.007	0.35 ± 0.22	22.5 ± 14.8	331.8	63.4	24.3	64
TM2 (lava flow 2)	0.019 ± 0.018	0.58 ± 0.57	10.4 ± 4.1	346.0	52.9	11.9	18
TM3 (lava flow 3)	0.020 ± 0.014	0.62 ± 0.42	15.2 ± 9.3	2.8	59.2	8.9	24
TM4 (lava flow 4)	0.025 ± 0.013	0.78 ± 0.40	12.9 ± 1.5	321.4	62.3	22.4	17

Lava flows in the Timanfaya volcanic area reveal very intense natural remanent magnetizations (NRMs). Induced magnetization was calculated as the product of the magnetic susceptibility and magnetic field in Lanzarote (estimated as 31 A/m), and resulted in being much weaker. Both magnetizations present large variability within the same lava flow (see the high standard deviations). The Köenigsberger ratio (Q) reveals that the NRM is the most relevant component of the total magnetization vector in lava.

In the Islote de Hilario area, pyroclastic fall deposits have the same composition as the Timanfaya lava flows (basalts). Therefore, the expected differences between lava and pyroclastic deposits with regard to their magnetic response are: (1) the magnetization of the pyroclastic layer is only of an induced origin because the remanent component (acquired during the rapid cooling of pyroclasts while falling) is chaotically distributed within the layer, so the summation of the remanent magnetization vectors is null; and (2) the induced magnetization of the pyroclastic layer may exceed the induced magnetization of the sampled lava flows, which were quite vesicular and consequently less dense than the pyroclastic deposits.

3. Results

3.1. GPR Results

Figure 4 presents the reflection profiles obtained in May 2012 (A) and April 2015 (B). The depth axis in radargrams was obtained using the velocity of 0.08 m/ns provided by the CMP-velocity analysis (Figure 3). A closer look at the radargram obtained in 2012 shows that no continuous subhorizontal reflections are displayed up to approximately 8 m depth, and the signatures of the signal reflections patterns vary both laterally and vertically. The areas showing maximum signal reflections (or scattering) in depth agree with the location of the higher temperatures at the surface. Moreover, the reflections of high amplitude display a convex-upwards geometry that is clearly seen, especially below the well-known geothermal anomaly located at the beginning of the profile. The strength of the GPR signal reflections diminishes upwards. Thus, a subsurface area with higher temperatures can be outlined at the beginning of the radargram (from 0 to 35 m), indicating that the heat source is shallower at the beginning of the profile and extends laterally and progressively deeper towards the end. With respect to the radargram obtained in 2015, a similar signature is observed and the subsurface area with reflections of higher strength extends from 4 to 35 m, although the greatest signal strength is concentrated at 4 to 15 m. These subtle variations of the GPR signal reflection patterns suggest a certain temporal variation of the ground temperature, which will be discussed later.

3.2. EMI Results

The EMI1 profile (Figure 5, top panel) shows four high resistivity areas at the surface, which are located at the beginning of the profile (0 to 3 m, TA1), at the middle part (11 to 15 m, TA2, and 17 to 22 m, TA3) and at the end (30 to 40 m, TA4). These high resistivity anomaly zones continue vertically at a depth of, at least, 40 m. The first high-resistive zone (TA1) is located at a well-known circular geothermal anomaly area, whereas the last zone (TA4) is located close to some old monitoring wells used to measure ground temperature, reporting temperature values as high as 388 °C at 7.5 m depth [10]. Considering this, the intermediate high-resistive areas must also be related to geothermal anomalies (TA2 and TA3). Thus, the four high-resistive anomaly areas are associated with high ground temperature zones at the surface that extend vertically downwards through the pyroclastic deposits. The lowest resistivity anomaly values (WA1 and WA2 in Figure 5), placed at both the beginning of the profile and at 28–30 m, are located ~15 m below the surface, but they do not continue at depth. These low-resistive anomaly values agree with the location of man-made walls (Figure 2b,d) built using local basaltic lava rocks as bricks. Considering all this, we suggest that the occurrence of fractures in the underlying volcanic rocks act as conduits for the emission of gases at high temperature, and the surface geothermal anomaly areas are located immediately above these fractures. The high temperature values would modify the electromagnetic properties of the pyroclastic rocks, resulting in an increase in their resistivity values. The significance of the localized low resistivity zones WA1 and WA2 is that they are impacted by the effect the man-made walls have on the coils when they are very close to the walls; therefore, in this case, they can be considered an artefact of anthropogenic origin. Moreover, the intermediate resistivity values located at the surface between TA1-TA2 and TA3-TA4 represent pyroclastic deposits with a ground temperature that is lower than the geothermal anomaly areas.

The EMI2 profile (Figure 5, lower left panel) has a clearer image of the geothermal anomaly area TA1. A well-defined high-resistive area extends from the beginning up to 9 m, which is in good agreement with the dimensions of the geothermal anomaly area at the surface (Figure 2a–c). The high-resistive area that corresponds to TA1 in EMI1 is bowl-shaped and extends downwards to ~15 m depth. Close to its central part (2–3 m from the beginning of the profile), a medium resistivity area can be seen extending vertically up to, at least, 45 m depth. A wide area of lower resistivity values surrounds the high-resistive area both horizontally and vertically. As for the EMI1 profile, good correlation exists between the zone of high-resistive values and the location of the geothermal anomaly area (TA1), confirming the relationship between a high ground temperature and the increased resistivity. The narrow vertical area of medium resistivity values corresponds to the location of the vertically ascending hot gases column emanating from a fault located below the pyroclastic deposits.

The EMI3 profile (Figure 5, lower right panel) runs tangentially to the geothermal anomaly area. For reference, the location of the nearby geothermal anomaly zone has been projected. Low to medium resistivity values dominate most of the shallower parts of the profile (up to ~5–7 m depth); however, for profiles EMI1 and EMI2, high-resistive areas are imaged at greater depths extending downwards to ~45 m depth. The first area is located 1–2 m from the beginning of the profile and corresponds to TA2 in the EMI1 profile; the second area at 8–10 m is immediately beneath the projection of the centre of the geothermal anomaly (TA1 in the EMI1 and EMI2 profiles). The third area (TA5) is located at 17–19 m. Considering the previous profiles, the high-resistive areas correspond to the location of hot gases emanating from the faults buried by pyroclastic deposits, which would also extend laterally close to the surface.

3.3. Magnetic Modelling Results

As previously mentioned, the shallow subsurface structure of the studied zone consists of a pyroclastic deposit that is several tens of metres (>70 m) thick overlying the older basaltic lava flows. From a magnetic point of view, it is expected that the chaotic distribution of lapilli fragments cancels the effect of the thermoremanent magnetization acquired during cooling. Consequently, we can assume the magnetic signal in this survey (caused by the shallowest deposit) is only due to the induced magnetization of the pyroclastic blanket, which is proportional to the magnetic susceptibility and parallel to Earth's present magnetic field. Within this context, magnetic anomalies can be associated with variations in the magnetic susceptibility produced by high temperatures and the eventual alteration caused by the circulation of high temperature gases within the rocks. In fact, the reduced-to-the-pole anomaly map displays some magnetic lows (Figure 6b), which we interpret as the effect of high temperatures at shallow depths causing a partial or total demagnetization of the subsurface rocks.

Based on this hypothesis, we performed a 2.75D forward modelling (Figure 7) with Geosoft GM-SYS software (version 8.1, Geosoft Inc., Toronto, Canada) along the two orthogonal profiles shown in Figure 6: profile ab, which coincides with both the GPR and EMI1 profiles; and profile cd, which runs parallel to the electromagnetic profile EMI2 (see location in Figure 2a). Profile cd was selected for modelling considering that it cuts the "magnetic low 1" symmetrically. Both profile data (ab and cd) were obtained by "slicing" the reduced-to-the-pole magnetic grid (Figure 6b).

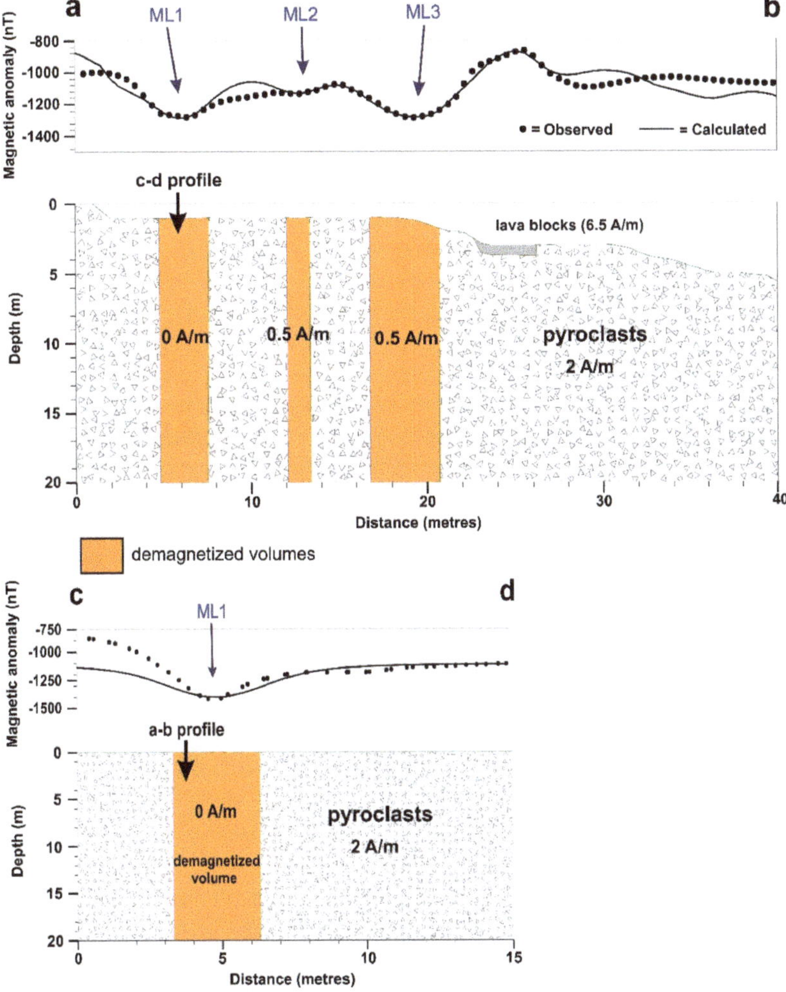

Figure 7. Forward magnetic models along the ab and cd profiles shown in Figure 6; the intersection between both profiles is marked in each of them. Magnetic lows were modelled by means of totally/partially demagnetized vertical bodies as a consequence of high temperatures in the subsoil. The quantities expressed in A/m represent the induced magnetization of each body. The sizes of the demagnetized bodies in the direction perpendicular to the profile are: in profile ab, left body (0.5 m to the North, 2.5 m to the South), middle body (0.5 m to the North, 5 m to the South) and right body (5 m to the North, 5 m to the South); in profile cd (1 m to the East, 2 m to the West).

We assumed that the pyroclastic layer was characterized by a uniform induced magnetization of 2 A/m. The three magnetic lows identified in the reduced-to-the-pole magnetic anomaly map (Figure 6b) were modelled as totally (source of low 1) or partially (sources of lows 2 and 3) demagnetized vertical bodies (lower magnetization than the surrounding pyroclastic deposits) located beneath each magnetic low. We hypothesized that the source of magnetic low 1, which is the most intense and is located inside the circular ground thermal anomaly area, is completely demagnetized. The vertical extension of these demagnetized volumes cannot be established due to the size of the survey (magnetization variations below a depth greater than approximately 20 m are not detected), but we can assume that they have a deep source. The main magnetic high was modelled, including the presence of massive surface lava blocks used to build a wall located very close to the profile (WA2). The magnetic highs located at the beginning of both profiles can also be related to the presence of another man-made wall (WA1) (Figure 2b).

4. Discussion

4.1. Relation between GPR Signal and Ground Temperature

The response of a material to the propagation of electromagnetic waves, and consequently the signal reflection and its strength depends on three factors (e.g., [16,36]): the relative permittivity (ε) of the materials through which the electromagnetic waves pass, the conductivity (σ) and the magnetic permeability (μ). The relative permittivity depends on several factors, primarily the nature of the material and the water content. For most of the GPR studies, the greater the change of material and/or water content, the stronger the reflections obtained. However, there is also a thermal effect resulting in a dependence of the relative permittivity upon the temperature [15,16]. In general, ε tends to increase with decreasing temperature. Although for most practical applications of GPR, this temperature dependence is negligible; it can be important when strong variations of temperature are present in the materials surveyed, and this is the case for the geothermal areas in this study. Similarly, σ primarily depends on the nature of the material and the water content (essentially the charges of dissolved anions and cations), whereas μ depends on the ferromagnetic minerals present in the materials (primarily magnetite, maghemite and hematite). Most of the materials do not contain ferromagnetic minerals; thus, their magnetic effect has little effect on the propagation of the GPR signal [37], which is not the case for volcanic materials because their ferromagnetic mineral content is relevant and has a considerable effect on the GPR wave velocity and signal attenuation [16]. As previously mentioned, the magnetic properties of volcanic minerals depend on temperature, exhibiting a decrease in magnetization at high temperatures. This temperature dependence on the value of μ is very significant for the GPR signal response in geothermal areas.

Because the material composition of Islote de Hilario zone is uniform up to at least 70 m depth, as demonstrated by the loose pyroclastic rocks with similar characteristics found from the drilling of several wells, the differences in the strength of the GPR reflections can only be due to differences in water content and/or temperature. However, the water content is extremely low, both at the surface (the landscape is typical of an arid environment, with a mean annual accumulated rainfall of ~150 mm (source: AEMET, Spanish Meteorological Agency)) and at great depths (0.0003–0.002% at the bottom of the monitoring wells [38]). Therefore, temperature is the only factor that can explain the differences in the strength of the signal reflections observed in the radargrams.

By considering the temperature dependence of both ε and μ, the strongest GPR reflections will correspond to the areas where a strong difference in ground temperature exists. This relationship is confirmed at the beginning of the radargram (Figure 4a,b), with the strongest reflections located immediately below the well-known geothermal anomaly area. The convex-upwards geometry of the reflections is interpreted as the geometry of the overheated gases column emanating from the underlying fractured lava flows and propagating towards the surface. Strong GPR reflections, although of lower strength, can also be seen at the central part of the radargram, indicating that the presence of gases at high temperatures is also present below the surface but with a lower contrast of temperature. The geometry of the reflections, decreasing slightly and shallower towards the end of the profile, seems to be the consequence of a combination of an upwards and lateral movement as the gases extend from the emission point, with a consequent decrease in temperature due to the adiabatic expansion. Thus, in geothermal areas of homogeneous materials and low water content, GPR results are a fast and accurate method for locating the distribution at depth of the higher temperature areas inside the geothermal site.

4.2. Seasonal Temperature Variations from the GPR Signal

Once it is determined that variations in the GPR signatures are related to ground temperature changes at the Islote de Hilario zone, with the strongest reflections associated with the greater temperature contrasts, it becomes possible to study if seasonal variations of temperature exist in the area by repeating the GPR survey at different epochs. Ref. [39] has linked the variations in ground temperature and deformation for the Timanfaya volcanic area with changes in surface thermal anomalies produced by the movement of hydrothermal fluids, and changes due to fluctuations in the water table, which is altered by rainfall periods. Thus, it would be interesting to determine if GPR can detect such seasonal variations of ground temperature. Consequently, the GPR survey performed on May 2012 (Figure 4a,b) at the Islote de Hilario zone was repeated in April 2015 (Figure 4c,d). The same processing sequence was selected to ensure that differences in the GPR signatures (if any) were due to changes on the material properties and not artefacts derived from a different data processing. Both radargrams look very similar at a first glance, although some differences are present: the zone showing greater GPR signal reflections, or scattering, is wider for the May 2012 profile, extending along the first 35 m of the profile and from ~2 m depth to the bottom of the profile, becoming deeper towards the end of the radargram (from ~10 m to the bottom). Looking at the April 2015 profile, instead of a single wide zone of high amplitude signatures, three narrower ones can be seen: one at the beginning of the profile and extending from ~2–4 m depth to the bottom; a second one at the middle of the radargram (25–30 m) and a third one near the end of the profile at ~12 m depth. The three areas can be interpreted as three different zones of overheated gas emissions that could potentially be located over underlying fractures affecting the lava flows located at a depth greater than 70 m. A greater volume of overheated gas emissions (and, consequently, a higher ground temperature) would permit the connection of the three temperature anomaly areas laterally converging in a single wider zone of high GPR signal strength, as observed in May 2012. If the volume of gas emissions decreases, the three areas would laterally disconnect, as seen in the April 2015 profile.

Therefore, a certain seasonal variation of the ground temperature can be confirmed from the GPR surveys, which constitutes a novel result considering that the application of GPR to active geothermal areas is very limited in the literature, primarily researching the location of vents, fractures and sinter deposits [40–42] but not to the study of the ground temperature distribution and its temporal variations.

4.3. Relation between Resistivity Anomalies and Ground Temperature

Electromagnetic methods have been extensively used in the exploration of geothermal systems ([1,43–45] among others). Low resistivity anomalies associated with ascending hydrothermal brines can easily be detected with these methods, and a clear relationship between low resistivity anomalies and high temperature zones can be established ([44], and references herein) which is the case for different hydrothermal areas that have been extensively studied, such as Iceland [43,46], New Zealand [47–49], Guadeloupe Island [50], and Italy [51]. In the Canary Islands, a geophysical and geochemical study conducted at Tenerife Island revealed a prominent low-resistivity structure, interpreted as a clay alteration cap related to the mechanism of the upward motion of deep-seated gases from a volcano-geothermal system [52]. However, the application of electromagnetic systems to HDR geothermal systems (such as the Timanfaya volcanic zone) is very limited in the literature.

Our study has revealed a positive correlation between high-resistive anomaly areas determined from EMI measurements and zones of high ground temperatures. This correlation is very clear at the beginning of profiles EMI1 and EMI2 (Figure 5), where bowl-shaped high (>5000 ohm·m) resistivity anomaly areas are located on a well-known geothermal anomaly of the Islote de Hilario zone, extending downwards to, at least, 45 m depth. This correlation is the opposite of that observed in classical geothermal systems associated with hydrothermal convection. That is, the upper part of a HDR geothermal system is characterized by very low liquid water content, with vapour being the dominant phase. As previously mentioned, earlier studies on the composition of emanating gases in the Islote de Hilario zone have revealed that the composition is primarily N2 (95–98%) with very low values of CO2 (0.5–1%) and water vapour (0.0003–0.0020%). Whereas the presence of liquid water greatly decreases the resistivity value of rocks, the occurrence of overheated gases (the temperature values obtained for the Islote de Hilario zone range from 250 to 605 °C) in the absence of a liquid phase dramatically increases the resistivity, and thus a positive correlation between resistivity and ground temperature is found. This relationship has been described in previous works. For example, Ref. [51] in a geophysical study of the geothermal area of Travale (Italy) states that 'depending on the predominant source of the fluids (liquid or vapour), the fluids produce low- or high-resistivity anomalies.' Similarly, Ref. [53], in a study conducted at the geothermal site of The Geysers (California), concludes that, if the temperature is above boiling in a borehole, a local zone of high-resistive values could indicate the presence of a steam-filled fracture.

Considering this, the high-resistive anomalies displayed in the EMI profiles correspond to high temperature areas due to the emission of overheated gases from underlying fractures. Along the longest profile, EMI1, up to four different zones of gas emission can be obtained (TA1 to TA4) corresponding to, at least, the location of four fractures affecting the underlying lava flows located below the maximum depth of investigation. Profile EMI2 provides a more detailed image of the TA1 anomaly with a narrow (~2 m) high-resistive subvertical zone that corresponds to the vertical ascent of the overheated gases through the pyroclastic deposit. Additionally, a shallower but wide (~10 m) high-resistive area, which is in good agreement with the extension of the geothermal anomaly observed at the surface, constitutes the lateral expansion of the gases reaching the surface. It must be considered that the resolution of the resistivity model obtained from the inversion of the EMI data decreases with depth, and thus the high temperature areas are much better defined near the ground surface. The two low-resistive anomalies WA1 and WA2 observed in the profile EMI1 correspond to the effect of two man-made walls built from basaltic rock, and these have very different resistivity and magnetic properties than those of the pyroclastic deposits. Accordingly, these anomalies are not related to the temperature variations of the geothermal anomaly areas, which constitutes one limitation of the EMI method because, when anthropogenic structures are very close to the transmitter or receiver coil, the effect can be remarkable, generating resistivity anomalies that are not related to any geological feature. Moreover, due to the specific data acquisition and inversion procedures of the method, these anomaly values horizontally displace half the intercoil spacing and at a depth much deeper than its actual position, which is evident for anomalies WA1 and WA2 that are placed 5 m (i.e., half the minimum intercoil spacing of 10 m) to the right of their actual position and at a mean depth of 15 m (the depth related to the inversion method to the EMI data obtained for both horizontal and vertical coplanar dipole modes for the 10 m intercoil spacing).

All of the above suggest that the EMI method constitutes a useful tool to obtain reliable information about the location of fractures where the emission of overheated gases occurs, as well as the lateral and vertical extension of the rocks affected by the diffusion of the gases and the associated high temperature areas. Although it is very difficult to find examples that apply this method for the exploration of HDR geothermal systems, we consider it a very valuable geophysical prospection technique, as it provides clear images of the temperature distribution up to several tens of metres depth.

4.4. Relation between Magnetic Anomalies and Ground Temperature

The results from the magnetic modelling (Figure 7) have allowed the locations of some narrow areas of demagnetization associated with the presence of magnetic lows to be identified. Up to three different totally or partially demagnetized volumes, ranging in width from 1.5 to 4 m, are located at 6 m (source of the "magnetic low 1"), 13 m (source of the "magnetic low 2") and 19 m (source of the "magnetic low 3") of the horizontal distance in profile ab. The source of the "magnetic low 1" was also modelled in the orthogonal direction along profile cd, which can be compared with the TA1 anomaly in the resistivity profile EMI2 (Figure 5). The first demagnetized area (located at 6 m) agrees well with the location of the geothermal anomaly TA1 at the surface. Because magnetic susceptibility is dependent on temperature, it is straightforward to interpret the demagnetized volumes as being discrete high temperature geothermal anomalies in the subsoil. Temperatures measured in the different wells of the Islote de Hilario zone ranged from 250 to 605 °C [10,11]. Considering that Curie temperatures of the Timanfaya basalts vary from approximately 420 to 575°C [35], it becomes evident that the magnetic lows may be associated with the presence of volumes affected by the demagnetization of thermal origin within the pyroclastic deposits. This interpretation is found in previous works where magnetic anomalies were used as exploration tools in geothermal areas i.e., [28–32,54,55]. For example, Ref. [54] linked the negative magnetic anomalies in New Zealand's high temperature geothermal systems with the demagnetization of the host volcanic rocks. In a similar manner but using aeromagnetic data, Ref. [55] interpreted a negative magnetic anomaly located at the Ngatamariki geothermal field (New Zealand) as demagnetized rocks extending to >1 km depth. A similar interpretation of the negative magnetic anomalies can be found in a more regional study of the high temperature reservoirs of the Taupo volcanic zone (New Zealand), where the authors [30] linked the demagnetization of the volcanic rocks with the alteration of magnetite, caused by geothermal systems expressing both liquid and vapour dominant phases.

The positive correlation between high temperature areas and rocks' demagnetization constitutes a valuable and simple tool to locate discrete geothermal anomaly areas because of the fast acquisition of the magnetic data. However, in volcanic areas with a more complex geology, the interpretation of the magnetic lows is not so simple and additional information coming from independent geophysical techniques and/or geological data would be needed to provide reliable results.

4.5. Joint Interpretation of GPR, EMI and Magnetic Data for the Detection of Shallow Geothermal Anomaly Areas

Our results have provided the geophysical signature of shallow, high temperature areas inside a HDR geothermal area using three independent geophysical methods. A positive correlation with the temperature exits between the strength of the GPR signal and the high-resistive areas, whereas, for the magnetic anomalies, the correlation is negative (Figure 8). Up to five geothermal anomaly areas have been outlined, and three of them (TA1, TA2 and TA3) have been detected using the three methods. Only TA4, which is associated with a high GPR signal strength and high-resistive values, cannot be correlated to a magnetic low due to the absence of magnetic data at that zone. The magnetic profile deviates by a few metres from the GPR and EMI profiles, precisely where TA4 is located (see Figure 2a). It must also be noted that the data acquisition of the three geophysical methods was conducted at different times (2012, 2013 and 2015). As a seasonal variation of the subsurface, ground temperature has been revealed by means of GPR data, and differences between the locations of the geothermal anomalies revealed by the three techniques may exist.

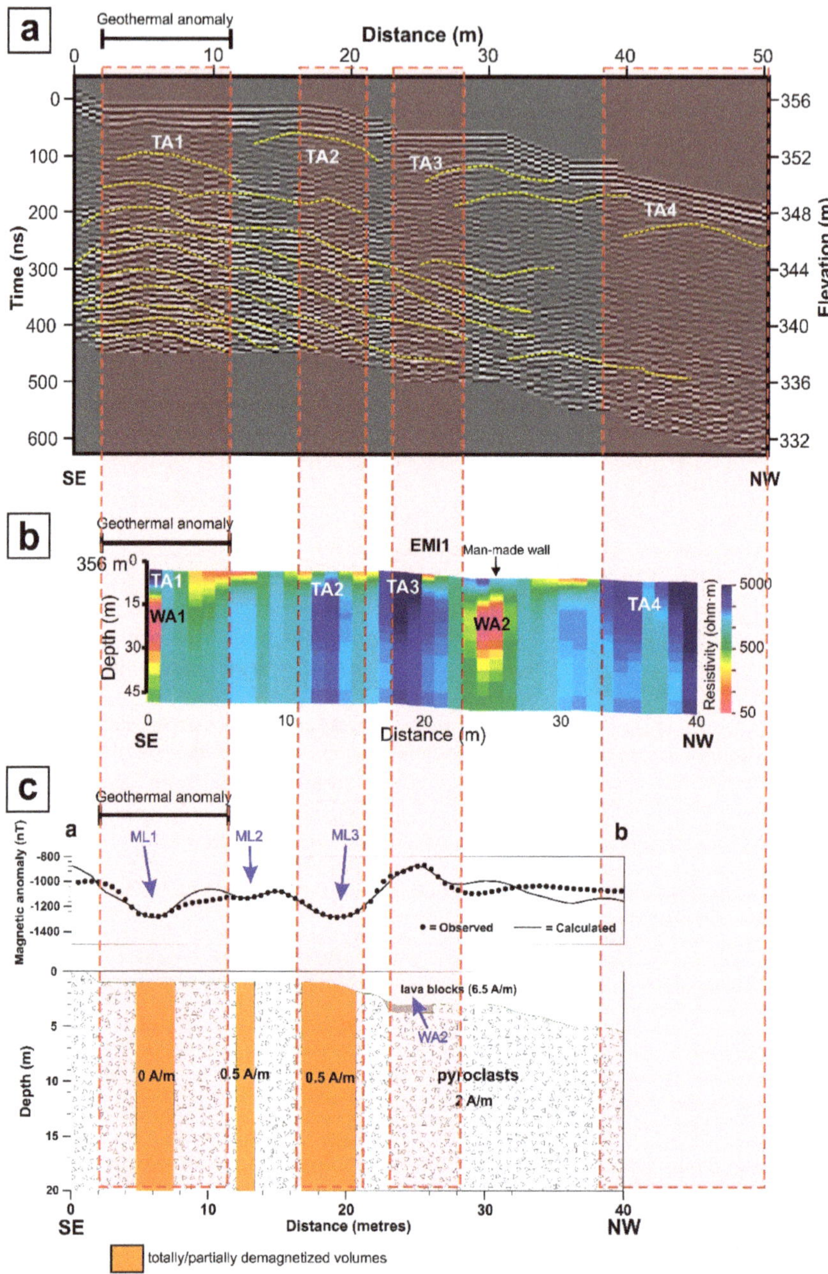

Figure 8. Comparison of the results obtained from the different geophysical techniques along a common profile: (**a**) GPR data, (**b**) resistivity model obtained from EMI data, and (**c**) magnetic model. All the profiles are displayed on the same scale with the same horizontal position to compare the location of the anomalies revealed by the three different methods. For a description of the compared results, see the text. Note that the vertical scales are different.

Due to the ambiguity inherent to geophysical methods, by using only one single technique, some alternative interpretations might be possible. Special care must be taken in areas of complex geology because the different electromagnetic, resistivity and magnetic anomalies might be associated with causes other than temperature variations. For example, Ref. [56], in an area of the Timanfaya volcanic zone very close to the Islote de Hilario zone, has detected hidden lava tubes by using similar GPR, EMI and gravity anomalies that consequently are not related to changes in ground temperature. Moreover, some anomalies have an anthropogenic origin, such as those corresponding to man-made walls that are present in the EMI and magnetic profiles of the present study and that could be misinterpreted as areas of lower ground temperature if additional information is not available. Thus, the best way to reduce any ambiguity is to combine as many independent geophysical methods as possible. Accordingly, the most accurate interpretation when looking for discrete areas of high temperature inside HDR geothermal systems is by referencing the combined occurrence of a high strength GPR signal, a high resistivity anomaly and magnetic low.

Using a combination of different geophysical methods for the exploration of volcanic areas is frequent in the literature. For example, the hydrothermal system of Solfatara (Italy) has been extensively surveyed using resistivity and self-potential methods [57] as well as resistivity and gravity [58]. A combination of magnetotelluric-time domain electromagnetics surveys with resistivity, gravity and magnetics has been used in the geothermal fields of New Zealand [59]. Recently, Ref. [60] surveyed the Piton de la Fournaise volcano (La Réunion Island, France) using self-potential, resistivity, GPR and microgravimetric methods to unravel the air convection and associated thermal field inside the volcano. These works are only a sample among many others that can be found in the literature that primarily focus on the characterization of geothermal and volcanic systems up to several hundreds or even thousands of metres depth. However, we propose here a novel combination of shallow, high-resolution geophysical methods that provide reliable and accurate information for the shallowest part of the geothermal areas. The combination of GPR, EMI and magnetics allows a good compromise between resolution and depth for the investigation of the first several tens of metres below the surface.

5. Conclusions

The shallower part of the HDR geothermal system of the Islote de Hilario, inside the Timanfaya volcanic zone, has been studied using a combination of three independent geophysical techniques: GPR, EMI and magnetic prospecting. The survey of a discrete, well-known geothermal anomaly in the area has been used to depict the geophysical signature of the high-temperature zones, which was then used to infer the location of similar unknown zones distributed around the area. The results from the GPR revealed that, when the material is homogeneous, the signature of the reflections is more intense in the areas with high temperature values, due to the temperature dependence of the physical parameters that determine the response of the rocks to the electromagnetic waves. Similarly, a variation in the subsurface distribution of the thermal anomaly has been detected for the first time through the analysis of GPR at different periods. The resistivity models obtained from the inversion of EMI data demonstrated that high-resistive areas are associated with high temperature zones, due to the increase in resistivity experienced by the volcanic materials when overheated gases propagate through them towards the surface. Regarding the magnetic method, the zones with high temperature values are associated with magnetic lows due to the demagnetization of the volcanic materials when they are heated to temperatures close to or higher than the Curie point of the involved magnetic minerals.

We propose here that a combination of the GPR, EMI and magnetic prospecting methods provides the best results for identifying the location of high ground temperature areas in HDR geothermal systems. Because data acquisition is fast and the methods are non-destructive, they constitute a very useful tool in areas that require special protection, as in the case of Timanfaya National Park. If these techniques can be performed at the same place but at different periods of the year, they can also provide valuable information regarding the seasonality of temperature variations, which can be related to changes in the volume of the gas emissions that cause geothermal anomalies.

Author Contributions: Investigation, D.G.-O., I.B.-M., J.A., T.M.-C., M.S., F.G.M., E.V. and N.S.; processing and analysis of geophysical data, D.G.-O., I.B.-M. and M.S.; funding acquisition, J.A.; writing—original draft preparation, D.G.-O., I.B.-M., J.A. and M.S.; writing—review and editing, D.G.-O., I.B.-M., J.A., T.M., M.S., F.G.M., E.V. and N.S.; creation of models: D.G.-O., I.B.-M. and M.S.

Funding: This research was funded the Spanish Ministry of Science, Innovation and Universities [CGL2015-63799-P] and the National Parks Autonomous Agency [320/2011]. The acquisition of the Overhauser magnetometer (GSM-19W) used in this study has been co-funded by FEDER through the Scientific-Technological Infrastructure Projects Subprogram [IGME10-3E-1261].

Acknowledgments: We wish to acknowledge Manuel Calvo for providing us with unpublished paleomagnetic data and Henrique Lorenzo for his collaboration in this research project. Special appreciation is expressed to the staff of the National Park of Timanfaya, and Jaime Arranz and Orlando Hernández from Casa de los Volcanes-Cabildo Insular of Lanzarote for their assistance during all field work.

Conflicts of Interest: The authors declare no conflict of interest.

References

1. Muñoz, G. Exploring for Geothermal Resources with Electromagnetic Methods. *Surv. Geophys.* **2014**, *35*, 101–122. [CrossRef]
2. DiPippo, R.; Renner, J.L. Geothermal Energy. In *Future Energy*, 2nd ed.; Letcher, T.M., Ed.; Elsevier: Waltham, MA, USA, 2014; pp. 211–223.
3. Brown, D.W.; Duchane, D.V.; Heiken, G.; Hriscu, V.T. *Mining the Earth's Heat: Hot Dry Rock Geothermal Energy*; Springer: Berlin/Heidelberg, Germany, 2012; p. 658.
4. Coello, J.; Cantagrel, J.M.; Hernán, F.; Fúster, J.M.; Ibarrola, B.; Ancochea, E.; Casquet, C.; Jamond, C.; Díaz de Terán, J.R.; Cendrero, A. Evolution of the eastern volcanic ridge of the Canary Islands based on new K-Ar data. *J. Volcanol. Geoth. Res.* **1992**, *53*, 251–274. [CrossRef]
5. Romero, C. *La Erupción de Timanfaya (Lanzarote-1730–1736). Análisis Documental y Estudio Geomorfológico*; Secretariado de Publicaciones de la Universidad de La Laguna, La Laguna: Tenerife, Spain, 1991; p. 136.
6. Carracedo, J.C.; Rodríguez Badiola, E.R.; Soler, V. The 1730–1736 eruption of Lanzarote: An unusually long, high magnitude fissural basaltic eruption in the recent volcanism of the Canary Islands. *J. Volcanol. Geoth. Res.* **1992**, *53*, 239–250. [CrossRef]
7. Ortiz, R.; Araña, V.; Astiz, M.; García, A. Magnetotelluric study of the Teide (Tenerife) and Timanfaya (Lanzarote) volcanic areas. *J. Volcanol. Geoth. Res.* **1986**, *30*, 357–377. [CrossRef]
8. Díez Gil, J.L.; Araña, V.; Ortiz, R.; Yuguero, J. Stationary convection model for heat transport by means of geothermal fluids in post eruptive systems. *Geothermics* **1987**, *15*, 77–87. [CrossRef]
9. Carracedo, J.C.; Troll, V. *The Geology of Canary Islands*; Elsevier: Amsterdam, The Netherlands, 2016; p. 621.
10. Albert, J.F.; Diez Gil, J.L.; Valentin, M.A.; Torres, F. Evaluación de las anomalías geotérmicas de Lanzarote. In *Serie Casa de los Volcanes*; Garcia, A., Felpeto, A., Eds.; Servicio de publicaciones del Excmo, Cabildo Insular de Lanzarote: Lanzarote, Spain, 1994; Volume 3, pp. 41–60.
11. Araña, V.; Díez Gil, J.L.; Ortiz, R.; Yuguero, J. Convection of geothermal fluids in the Timanfaya volcanic area (Lanzarote, Canary Islands). *Bull. Volcanol.* **1984**, *47*, 667–677. [CrossRef]
12. García, A. Estudio magnetotelúrico del área volcánica de Timanfaya (Lanzarote). *Anal. Fis. Ser. B* **1986**, 35–43.
13. García, A. Modelos Corticales a Partir de Sondeos Magnetotelúricos. Aplicación a Zonas Volcánicas Activas. Master's Thesis, University Complutense of Madrid, Madrid, Spain, 1983.
14. Annan, P. *GPR Principles, Procedures & Applications*; Sensors and Software Inc.: Mississauga, ON, Canada, 2003; p. 278.
15. Daniels, D.J. *Ground Penetrating Radar*; The institution of Electrical Engineers: London, UK, 2004; p. 726.
16. Jol, H.M. *Ground Penetrating Radar: Theory and Applications*; Elsevier Science: Amsterdam, The Netherlands, 2009; p. 544.
17. Sandmeier, K.J. *ReflexW Version 8.5. Windows™ XP/7/8/10-program for the processing of seismic, acoustic or electromagnetic reflection, refraction and transmission data*; Sandmeier: Karsruhe, Germany, 2018; p. 327.
18. Tosti, F.; Slob, E. Determination, by using GPR, of the volumetric water content in structures, substructures, foundations and soils. In *Civil Engineering Applications of Ground Penetrating Radar*; Benedetto, A., Pajewski, L., Eds.; Springer: New Delhi, India, 2015; pp. 163–194.

19. Aziz, Z.; van Geen, A.; Stute, M.; Versteeg, R.; Horneman, A.; Zheng, Y.; Goodbred, S.; Steckler, M.; Weinman, B.; Gavrieli, I.; et al. Impact of local recharge on arsenic concentrations in shallow aquifers inferred from the electromagnetic conductivity of soils in Araihazar, Bangladesh. *Water Resour. Res.* **2008**, *44*, 2007WR006000. [CrossRef]
20. Paine, J.G. Determining salinization extent, identifying salinity sources, and estimating chloride mass using surface, borehole, and airborne electromagnetic induction methods. *Water Resour. Res.* **2003**, *39*, 1–10. [CrossRef]
21. Everett, M.E.; Meju, M.A. Near-surface controlled-source electromagnetic induction: Background and recent advances. In *Hydrogeophysics*; Rubin, Y., Hubbard, S.S., Eds.; Springer: New York, NY, USA, 2005; pp. 157–183.
22. Bobee, C.; Schmutz, M.; Camerlynck, C.; Robain, H. An integrated geophysical study of the western part of the Rochechouart-Chassenon impact structure, Charente, France. *Near Surf. Geophys.* **2010**, *4*, 259–270. [CrossRef]
23. Valois, R.; Camerlynck, C.; Dhemaied, A.; Guerin, R.; Hovhannissian, G.; Plagnes, V.; Rejiba, F.; Robain, H. Assessment of doline geometry using geophysics on the Quercy plateau karst (South France). *Earth Surf. Process.* **2011**, *36*, 1183–1192. [CrossRef]
24. McNeill, J.D. *Electromagnetic Terrain Conductivity Measurement at Low Induction Numbers*; Technical Note TN-6; Geonics Ltd.: Mississauga, ON, Canada, 1980; p. 15.
25. UBC-GIF, 2000. EM1DFM: A Program Library for Forward Modelling and Inversion of Frequency Domain Electromagnetic Data over 1D Structures. 2000. Available online: https://em1dfm.readthedocs.io/en/latest/ (accessed on 22 January 2019).
26. Farquharson, C.G.; Oldenburg, D.W.; Rough, P.S. Simultaneous 1D inversion of loop-loop electromagnetic data for magnetic susceptibility and electrical conductivity. *Geophysics* **2003**, *68*, 1857–1869. [CrossRef]
27. Hunt, C.P.; Moskowitz, B.M.; Banerjee, S.K. Magnetic properties of rocks and minerals. In *Rock Physics and Phase Relations, A Handbook of Physical Constants*; Ahrens, T.J., Ed.; American Geophysical Union: Washington, DC, USA, 1995; Volume 3, pp. 189–204.
28. Hildenbrand, T.G.; Rosenbaum, J.G.; Kauahikaua, J.P. Aeromagnetic study of the island of Hawaii. *J. Geophys. Res.* **1993**, *98*, 4099–4119. [CrossRef]
29. Blanco, I.; García, A.; Torta, J.M. Magnetic study of the Furnas caldera (Azores). *Ann. Geophys.* **1997**, *40*, 341–359.
30. Hoechstein, M.P.; Soengkono, S. Magnetic anomalies associated with high temperature reservoirs in the Taupo Volcanic Zone (New Zealand). *Geothermics* **1997**, *26*, 1–24. [CrossRef]
31. Caratori Tontini, F.; de Ronde, C.E.J.; Scott, B.J.; Soengkono, S.; Stagpoole, V.; Timm, C.; Tivey, M. Interpretation of gravity and magnetic anomalies at Lake Rotomahana: Geological and hydrothermal implications. *J. Volcanol. Geoth. Res.* **2016**, *314*, 84–94. [CrossRef]
32. Paoletti, V.; Passaro, S.; Fedi, M.; Marino, C.; Tamburrino, S.; Ventura, G. Sub-circular conduits and dikes offshore the Somma-Vesuvius volcano revealed by magnetic and seismic data. *Geophys. Res. Lett.* **2016**, *43*, 9544–9551. [CrossRef]
33. Finlay, C.C.; Maus, S.; Beggan, C.D.; Bondar, T.N.; Chambodut, A.; Chernova, T.A.; Chulliat, A.; Golovkov, V.P.; Hamilton, B.; Hamoudi, M.; et al. International Geomagnetic Reference Field: The eleventh generation. *Geophys. J. Int.* **2010**, *183*, 1216–1230.
34. Baranov, V. A New Method for Interpretation of Aeromagnetic Maps, Pseudo-Gravimetric Anomalies. *Geophysics* **1957**, *22*, 359–363. [CrossRef]
35. Calvo-Rathert, M.; Morales-Contreras, J.; Carrancho, A.; Gogichaishvili, A. A comparison of Thellier-type and multispecimen paleointensity determinations on Pleistocene and historical flows from Lanzarote (Canary Islands, Spain). *Geochem. Geophys. Geosyst.* **2016**, *17*, 3638–3654. [CrossRef]
36. Conyers, L.B. *Ground-Penetrating Radar for Archaeology*; AltaMira Press: Walnut Creek, CA, USA, 2004; p. 203.
37. Olhoeft, G.R. Electrical, magnetic, and geometric properties that determine ground penetrating radar performance. In Proceedings of the 7th International Conference on Ground Penetrating Radar (GPR'98), Lawrence, KS, USA, 27–30 May 1998; pp. 477–483.
38. Valentín, A. La Fase Gaseosa Relacionada con la Actividad Volcánica y Tectónica. Master's Thesis, University of Barcelona, Barcelona, Spain, 1988.

39. Riccardi, U.; Arnoso, J.; Benavent, M.; Vélez, E.; Tammaroe, U.; Montesinos, F.G. Exploring deformation scenarios in Timanfaya volcanic area (Lanzarote, Canary Islands) from GNSS and ground based geodetic observations. *J. Volcanol. Geoth. Res.* **2018**, *357*, 14–24. [CrossRef]
40. Pettinelli, E.; Beaubien, S.E.; Lombardi, S.; Annan, A.P. GPR, TDR, and geochemistry measurements above an active gas vent to study near surface gas-migration pathways. *Geophysics* **2008**, *73*, A11–A15. [CrossRef]
41. Dougherty, A.J.; Lynne, B.Y. A novel geophysical approach to imagine sinter deposits and other subsurface geothermal features utilizing ground penetrating radar. *Geoth. Res. Trans.* **2010**, *34*, 857–863.
42. Dougherty, A.J.; Lynne, B.Y. Utilizing Ground Penetrating Radar and Infrared Thermography to Image Vents and Fractures in Geothermal Environments. *Geoth. Res. Trans.* **2011**, *35*, 743–750.
43. Hersir, G.P.; Björnsson, A. *Geophysical Exploration for Geothermal Resources, Principles and Application*; UNU-GTP, report 15: Reykjavik, Iceland, 1991; p. 94.
44. Spichak, V.; Manzella, A. Electromagnetic Sounding of Geothermal Zones. *J. Appl. Geophys.* **2009**, *68*, 459–478. [CrossRef]
45. Börner, J.H.; Bär, M.; Spitzer, K. Electromagnetic methods for exploration and monitoring of enhanced geothermal systems—A virtual experiment. *Geothermics* **2015**, *55*, 78–87. [CrossRef]
46. Spichak, V.; Zakharova, O.; Goidina, A. A new conceptual model of the Icelandic crust in the Hengill geothermal area based on the indirect electromagnetic geothermometry. *J. Volcanol. Geoth. Res.* **2013**, *257*, 99–112. [CrossRef]
47. Lumb, J.T.; Macdonald, W.J.P. Near-surface resistivity surveys of geothermal areas using the electromagnetic method. *Geothermics* **1970**, *2*, 311–317. [CrossRef]
48. Risk, G.F.; Caldwell, T.G.; Bibby, H.M. Tensor time domain electromagnetic resistivity measurements at Ngatamariki geothermal field, New Zealand. *J. Volcanol. Geoth. Res.* **2003**, *127*, 33–54. [CrossRef]
49. Simpson, M.P.; Bignall, G. Undeveloped high-enthalpy geothermal fields of the Taupo Volcanic Zone, New Zealand. *Geothermics* **2016**, *59*, 325–346. [CrossRef]
50. Gailler, L.S.; Bouchot, V.; Martelet, G.; Thinon, I.; Coppo, N.; Baltassat, J.M.; Bourgeois, B. Contribution of multi-method geophysics to the understanding of a high-temperature geothermal province: The Bouillante area (Guadeloupe, Lesser Antilles). *J. Volcanol. Geoth. Res.* **2014**, *275*, 34–50. [CrossRef]
51. Spichak, V.; Zakharova, O. Constructing the Deep Temperature Section of the Travale Geothermal Area in Italy, with the use of an Electromagnetic Geothermometer. *Izv. Phys. Solid Earth* **2015**, *51*, 87–94. [CrossRef]
52. Rodríguez, F.; Pérez, N.M.; Padrón, E.; Melián, G.; Piña-Varas, P.; Dionis, S.; Barrancos, J.; Padilla, G.D.; Hernández, P.A.; Marrero, R.; et al. Surface geochemical and geophysical studies for geothermal exploration at the southern volcanic rift zone of Tenerife, Canary Islands, Spain. *Geothermics* **2015**, *55*, 195–206. [CrossRef]
53. Roberts, J.J.; Duba, A.G.; Bonner, B.P.; Kasameyer, P.W. The effects of capillarity on electrical resistivity during boiling in metashale from scientific corehole SB-15-D, The Geysers, California, USA. *Geothermics* **2001**, *30*, 235–254. [CrossRef]
54. Hochstein, M.P.; Hunt, T.M. Seismic, gravity and magnetic studies, Broadlands geothermal field, New Zealand. *Geothermics* **1970**, *2*, 333–346. [CrossRef]
55. Soengkono, S. Magnetic anomalies over the Ngatamariki Geothermal Field, Taupo Volcanic Zone, New Zealand. In Proceedings of the 14th New Zealand Geothermal Workshop, University of Auckland, New Zealand, 4–6 November 1992; pp. 241–246.
56. Gómez-Ortiz, D.; Montesinos, F.; Martín-Crespo, T.; Solla, M.; Arnoso, J.; Vélez, E. Combination of geophysical prospecting techniques into areas of high protection value: Identification of shallow volcanic structures. *J. Appl. Geophys.* **2014**, *109*, 15–26. [CrossRef]
57. Byrdina, S.; Vandemeulebrouck, J.; Cardellini, C.; Legaz, A.; Camerlynckc, C.; Chiodini, G.; Lebourg, T.; Gresse, M.; Bascoua, P.; Motos, G.; et al. Relations between electrical resistivity, carbon dioxide flux, and self-potential in the shallow hydrothermal system of Solfatara (Phlegrean Fields, Italy). *J. Volcanol. Geoth. Res.* **2014**, *283*, 172–182. [CrossRef]
58. Rinaldi, A.P.; Todesco, M.; Vandemeulebrouck, J.; Revil, A.; Bonafede, M. Electrical conductivity, ground displacement, gravity changes, and gas flow at Solfatara crater (Campi Flegrei caldera, Italy): Results from numerical modeling. *J. Volcanol. Geoth. Res.* **2011**, *207*, 93–105. [CrossRef]

59. Chambefort, I.; Buscarlet, E.; Wallis, I.C.; Sewell, S.; Wilmarth, M. Ngatamariki Geothermal Field, New Zealand: Geology, geophysics, chemistry and conceptual model. *Geothermics* **2016**, *59*, 266–280. [CrossRef]
60. Antoine, A.; Finizola, A.; Lopez, T.; Baratoux, D.; Rabinowicz, M.; Delcher, E.; Fontaine, F.R.; Fontaine, F.J.; Saracco, G.; Bachèlery, P.; et al. Electric potential anomaly induced by humid air convection within Piton de La Fournaise volcano, La Réunion Island. *Geothermics* **2017**, *65*, 81–98. [CrossRef]

Article

Determination of Primary and Secondary Lahar Flow Paths of the Fuego Volcano (Guatemala) Using Morphometric Parameters

Marcelo Cando-Jácome and Antonio Martínez-Graña *

Geology Department, External Geodynamics Area, Faculty of Sciences, University of Salamanca, Plaza Merced s/n, 37008 Salamanca, Spain; id00709713@usal.es

* Correspondence: amgranna@usal.es; Tel.: +34-923-294496; Fax: +34-923-923294514

Received: 2 February 2019; Accepted: 25 March 2019; Published: 26 March 2019

Abstract: On 3 June 2018, a strong eruption of the Fuego volcano in Guatemala produced a dense cloud of 10-km-high volcanic ash and destructive pyroclastic flows that caused nearly 200 deaths and huge economic losses in the region. Subsequently, due to heavy rains, destructive secondary lahars were produced, which were not plotted on the hazard maps using the LAHAR Z software. In this work we propose to complement the mapping of this type of lahars using remote-sensing (Differential Interferometry, DINSAR) in Sentinel images 1A and 2A, to locate areas of deformation of the relief on the flanks of the volcano, areas that are possibly origin of these lahars. To determine the trajectory of the lahars, parameters and morphological indices were analyzed with the software System for Automated Geoscientific Analysis (SAGA). The parameters and morphological indices used were the accumulation of flow (FCC), the topographic wetness index (TWI), the length-magnitude factor of the slope (LS). Finally, a slope stability analysis was performed using the Shallow Landslide Susceptibility software (SHALSTAB) based on the Mohr–Coulomb theory and its parameters: internal soil saturation degree and effective precipitation, parameters required to destabilize a hillside. In this case, the application of this complementary methodology provided a more accurate response of the areas destroyed by primary and secondary lahars in the vicinity of the volcano.

Keywords: volcano deformation; lahars hazard; magma accumulation; pyroclastic flows; ash plumes

---

1. Introduction

Primary lahars are flows which are formed as a direct result of a volcanic eruption. They tend to be bulky (107–109 m^3) and record high speeds (>20 m/s). Their maximum flows are commonly between 103–105 m^3/s. These features provide the ability to flow long distances, even hundreds of kilometers downstream. They occur primarily when during an eruptive event incandescent material causes the fast melting of large volumes of ice and snow of the glaciers that cover some volcanic edifices and flows by descent gullies. Secondary lahars mainly include lahars caused by the rains. The unbound, by previous eruptions, pyroclastic material can be easily removed by the rains. In general, these are at lower speed, volume and they travel shorter distances as compared to primary lahars, however, they are most frequent during periods of rain [1].

Guatemala is the only country in the Central American region that has trained local observers to generate reports on volcanic activity, who then transmit the information via radio and/or telephone three times a day. The information is transmitted to the National Institute of Seismology, Volcanology, Meteorology and Hydrology (INSIVUMEH) plant when the following characteristics are present: changes in the release of energy, increase in the number of explosions, increase in the expulsion of ash, increase in seismic activity, rumblings, shock waves, and the manifestation of block avalanches that descend through the ravines in the volcanic perimeter.

On 3 June 2018, a strong eruption of the Fuego volcano in Guatemala produced a dense volcanic ash cloud rising to 10 km high. Following the collapse of the volcanic column, pyroclastic flows and descending lahars caused significant damage and a high number of fatalities around the mountain [2]. After the eruption, INSIVUMEH, with support from the Volcano Disaster Assistance Program (VDAP) of the United States Geological Survey (USGS), the University of Edinburgh, and Michigan Technological University, elaborated 2 scenarios of lahars for medium and heavy rains based on the numerical models obtained from the program LAHAR Z (Figure 1). These scenarios can be seen on the MapAction page [3].

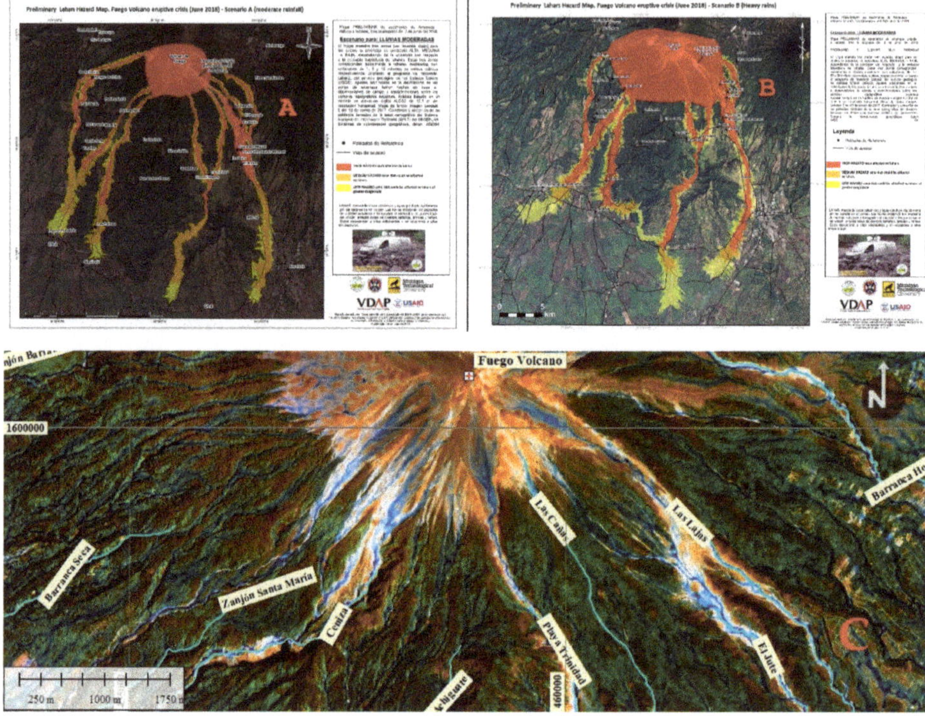

Figure 1. Web Map prepared by MapAction as a collaborator on the HazMap project and in support of National Institute of Seismology, Volcanology, Meteorology and Hydrology (INSIVUMEH). Data from preliminary maps of the hazard of pyroclastic flows and lahars for scenario A (moderate rainfall) and scenario B (very heavy rainfall). Prepared in June 2018 by INSIVUMEH, Volcano Disaster Assistance Program (VDAP), United States Geological Survey (USGS), the University of Edinburgh, and Michigan Technological University. In C, location of Las Lajas, El Jute, Ceniza and Seca ravines.

On 19 November and for the fifth time during 2018, the Fuego volcano erupted, with a 1200-m long lava flow that descended into the Ceniza canyon. This volcanic flow with fragments of rock producws by erosion on the slopes of the volcano moved downhill, incorporating enough water, so that they formed mud flows and volcanic debris or lahars that descended the slopes of the volcano, affecting sectors such as Las Lajas, El Jute, besides the Ceniza and Seca ravines, according to INSIVUMEH in its last special volcanological bulletins (Nos. 157-2018 and 158-2018) [4,5].

The theory of LAHAR Z's operation will not be studied in this article, since it is not the objective of this study. Its operation is reviewed briefly. In this article, reference is made to the LAHAR Z program and its results as a basis for superimposing the results obtained by the Differential Interferometry

Synthetic Aperture Radar Differential (DINSAR) analysis, the morphometric indices and the unstable areas near the volcano.

LAHAR Z was written to delimit areas of potential lahar inundation from one or more user-specified lahar volumes. LAHAR Z is a code that is executed within a geographic information system (GIS), which was created by the USGS [4] and is based on a semi-empirical model that delimits flood risk zones by lahar (that is, areas drawn to represent the probable floods during a lahar event) in a digital elevation model (DEM).

The program uses two semi-empirical equations calibrated by statistical analysis in the cross section of an area flooded by a lahar (A) and the flooded planimetric area (B) measured in 27 lahars deposits of 9 volcanoes in the United States of America, Mexico, Colombia, Canada and Philippines [6] (Figure 2).

The equations are:

$$A = \alpha_1 V^{2/3} \quad (1)$$

$$B = \alpha_2 V^{2/3} \quad (2)$$

where A is the maximum section area flooded, B is the total area flooded and V is the volume of the lahar, $\alpha_1 = 0.05$ and $\alpha_2 = 200$, are constant values.

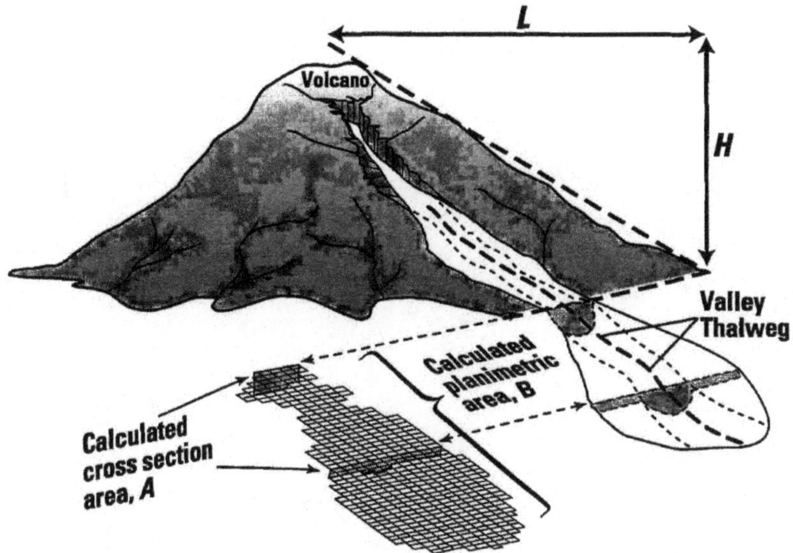

Figure 2. Diagram of association between dimensions of an idealized lahar and cross-sectional (A) and planimetric (B) areas calculated by LAHAR Z for a hypothetical volcano. The ratio of vertical drop (H) to horizontal runout distance (L) describes the extent of proximal volcano hazard. LAHAR Z begins calculations of lahar-inundation hazard zones where a user-specified stream and the proximal-hazard zone boundary intersect [6].

The LAHAR Z model works under assumptions such as:

1. Floods from past lahars can provide the information basis to predict possible future floods.
2. Distant hazard zones are confined to the valleys and are directed to the slopes of the volcano.
3. The volume of the proximal lahar controls the extent of the distal flood forming downstream.
4. Very voluminous lahars occur less frequently.
5. No one can predict the size of the next lahar that will descend by a river.

6. The volume selection for the lahar and the origin site thereof control the flow range.

As a complementary mechanism to the results obtained by DINSAR and Lahar-Z, the morphometric model has been used to determine the trajectories of the destructive lahars with greater precision by combining morphometric parameters obtained from the digital elevation model of 12 m of spatial resolution such as the flow accumulation (FCC), the topographic wetness index (TWI), the length-magnitude factor of slope (LS), the degree of internal saturation of the soil (h/z), where z is soil depth, h is water level above the failure plane and the effective precipitation (q/T) required to destabilize a slope, where h is the height of the water table and z is the thickness of the colluvium that slides above the failure plane. All these parameters are obtained from the shallow landslide slope stability model (SHALSTAB) to complement the mapping of this type of lahars using remote sensing (differential interferometry, DINSAR) in Sentinel images 1A and 2A, to locate areas of deformation of the relief on the flanks of the volcano, areas that are possibly origin of these lahars. To determine the trajectory of the lahars, parameters and morphological indices were analyzed with the software System for Automated Geoscientific Analysis (SAGA). The parameters and morphological indices used were the accumulation of flow (FCC), the TWI, the length-magnitude factor of the slope (LS). Finally, a slope stability analysis was performed using the Shallow Landslide Susceptibility software (SHALSTAB) based on the Mohr–Coulomb theory and its parameters: internal soil saturation degree (h/z) and effective precipitation (q/T), parameters required to destabilize a hillside. In this case, the application of this complementary methodology provided a more accurate response of the areas destroyed by primary and secondary lahars in the vicinity of the volcano [6] (Figure 3). These parameters are described in more detail in the Materials and Methods chapter.

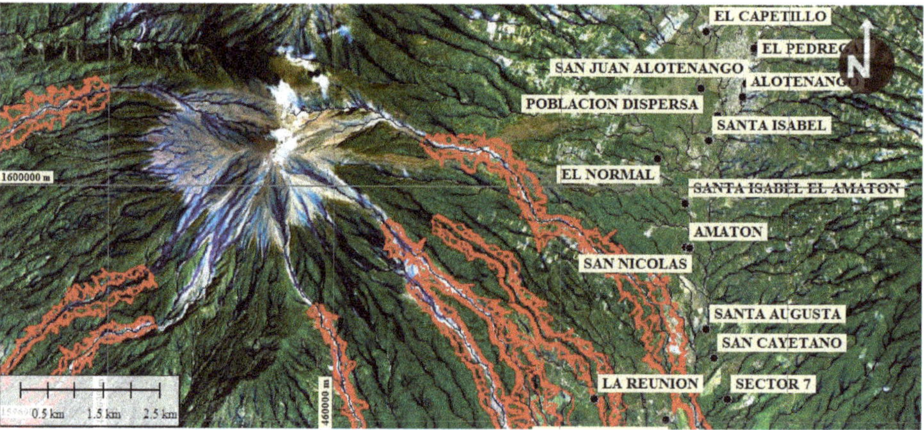

Figure 3. LAHAR Z Model (red lines) and the poor tracing of possible secondary lahars and analysis of their distal trajectories originating from proximal sources. The morphometric model combines parameters such as the flow accumulation (FCC), the topographical humidity index (TWI), and slope-length factor of the slope to more accurately determine the trajectories of the destructive lahars.

Although the limits of the areas delimited by the numerical modeling of lahars must be taken with caution and are considered as references and not as absolutes, the comparison between the limits of hazards by flows of lahars obtained from modeling with LAHAR Z and the morphometric model is mainly differentiated by their theoretical conceptualization.

Based on the damage caused by the pyroclastic flows, it was found that the volcanic hazard maps and physical vulnerability contained errors for two main reasons: a lack of field observation data to adjust the limits of the hazard map simulated with LAHAR Z, and the poor tracing of possible secondary lahars and the analysis of their distal trajectories originating in proximal sources [5].

2. Materials and Methods

The Sentinel 1 and 2 images obtained from the Alaska Satellite Facility [7], were used to determine the deformations of the terrain using the DINSAR technique [8] and those from Sentinel 2 were used to determine the changes in the morphology of the relief, which in this case was the delimitation of primary and secondary lahars, and compare them with the results obtained from the morphological model. The digital elevation model of 12 m of spatial resolution was obtained from the Copernico Project [9]. The planimetric information such as the results of the LAHAR Z software application was provided by the National Coordinator for Disaster Reduction (CONRED).

To complement the results obtained with the DINSAR analysis and the morphological method, in this study, the spatial distribution of potential sources of primary and secondary lahars during eruptions and torrential rains was analyzed using the slope stability model of shallow landslides. SHALSTAB is a plugin of the SAGA GIS.

The general flow of processes used in this study can be seen in Figure 4.

The processes A, B and C are described below:

Process (A): DINSAR analysis. The Sentinel 1 and 2 images were obtained from the Alaska Satellite Facility [9]. Sentinel 1 images were used to determine the deformations of the terrain using the differential SAR interferometry (DINSAR) technique [10] and those from Sentinel 2 were used to determine the changes in the morphology of the relief, which in this case was the delimitation of primary and secondary lahars and compare them with the results obtained from the morphological model. Process B: the planimetric information such as the results of the LAHAR Z software application was provided by CONRED. Process C: digital elevation model of 12 m of spatial resolution was obtained from the Copernico Project. The field observations carried out by CONRED technicians and sent in the format of points in shp format, helped to calibrate and verify the results obtained from the presented methodology, as well as to understand which of the three morphological parameters used had greater weight according to the scenario in the one that developed the lahars.

For the study of the primary and secondary trajectories of lahars using the DINSAR technique, the morphometric method and the stability analysis, the following parameters were analyzed:

The relief deformation analysis by the Sentinel Application Platform (SNAP) [11] followed the workflow indicated in Figure 4 from the Sentinel 1B images taken on May 21 and June 14, 2018.

This analysis was carried out to determine the possible zones of deformation of the ground where the possible secondary lahars start.

Brief synthesis of the processes to obtain the Interferogram:

Base images: Sentinel 1B

S1B_IW_SLC__1SDV_20180521T001317_20180521T001345_011012_0142C5_A797
S1B_IW_SLC__1SDV_20180614T001319_20180614T001347_011362_014DD5_D06F;

Mission: SENTINEL-1B
Pass: ASCENDING
Polarization: VV

Process diagram in ESA-SNAP [10] (Figure 5).

Step 1 Read-Topsar-Split-Orbit file: Open the products (Figure 5A).

S1B_IW_SLC__1SDV_20180521T001317_20180521T001345_011012_0142C5_A797
S1B_IW_SLC__1SDV_20180614T001319_20180614T001347_011362_014DD5_D06F;

View the products. (Figure 5B): In the Sentinel-1 IW SLC products, there are 3 subsets IW1, IW2, IW3. Each subset is for an adjacent strip of satellite data collection by TOPS mode. The Fuego volcano can be seen in IW3 (Figure 5C).

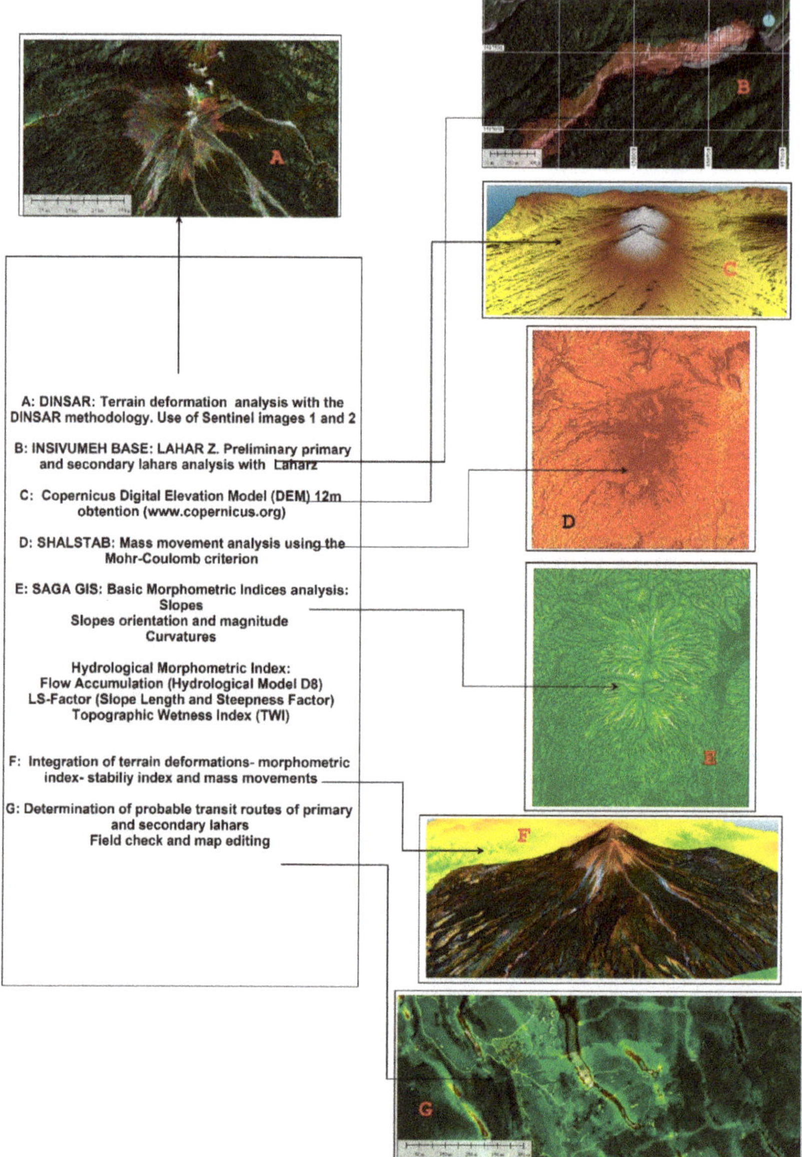

Figure 4. Flow methodological diagram for the determination of primary and secondary lahars flow paths from the following processes: (**A**) Analysis of terrain deformation with differential synthetic aperture radar interferometry DINSAR, (**B**) Use of base information of lahars routes obtained from LAHAR Z -National Institute of Seismology, Volcanology, Meteorology and Hydrology—INSIVUMEHIN. (**C**) Obtaining the digital elevation model of 12m of spatial resolution of Copernicus.org, (**D**) Analysis of Instability of the Relief with Shallow Landslide Susceptibility software (SHALSTAB), (**E**) Determination of Morphometric Indexes of the Relief with System for Automated Geoscientific Analyzes (SAGA), (**F**) Integration of information and determination of possible flow paths of primary and secondary lahars (**G**).

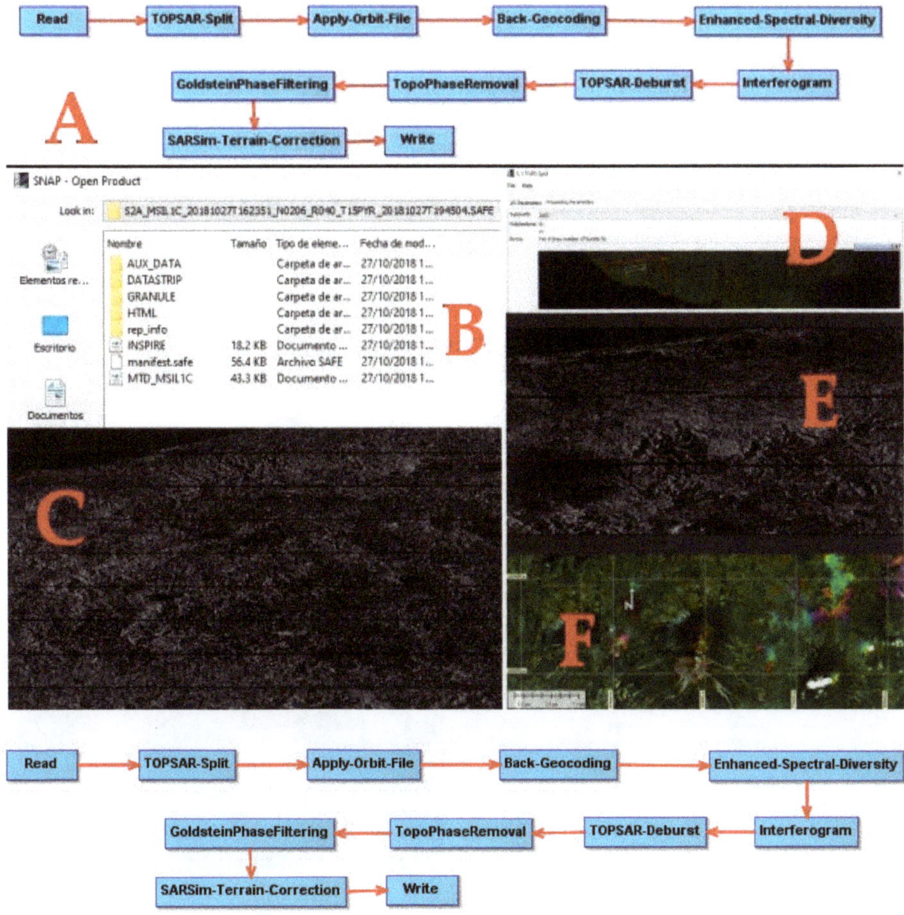

Figure 5. (**A**) The Sentinel Application Platform (SNAP) process used to obtain the deformations of the relief by DINSAR analysis. Steps (**B–F**). Sentinel Application Platform (SNAP) passages of each portion of the routine. See text for further explanation.

Step 2 View a band geocoding-enhanced spectral diversity. To view the data, Intensity_IW3_VV band.

Coregister the images into a stack: For interferometric processing, two or more images must be coregistered into a stack. One image is selected as the master and the other images are the slaves. The pixels in slave images will be moved to align with the master image to sub-pixel accuracy (Figure 5D).

Step 3 Form the Interferogram-Topsar Deburst-TopoPhase Removal-Filtering-Sarsim Terrain Corection-Write. (Figure 5E): Interferogram formation from the InSAR products menu.

The current location of the primary and secondary tabs after the eruption of 3 June 2018, were obtained with a Sentinel 2 image of 10 m of spatial resolution where bands 4, 3 and 2 were combined to obtain the current position of the lahars (Figure 6). A brief synthesis of the processes to obtain the real position of lahars is presented below:

Base Image: Sentinel-2B: 2018-10-22T16: 23: 19.024Z
Pass: DESCENDING

Bands: 4-3-2

Process diagram in ESA-SNAP

Step 1 Read-Open the products. Granule-img-bands (Figure 6A).

View the products. (Figure 6B): In the Sentine-2B-Granule-Img products, there are 13 spectral bands in the visible and near-infrared (VNIR) and short-wavelength infrared (SWIR) spectrum, as show in the below table (Figure 6C).

Step 2 Combine bands 4-3-2 RGB natural colors. Location of primary and secondary lahars (Figure 6D).

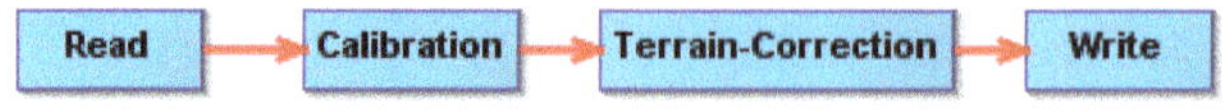

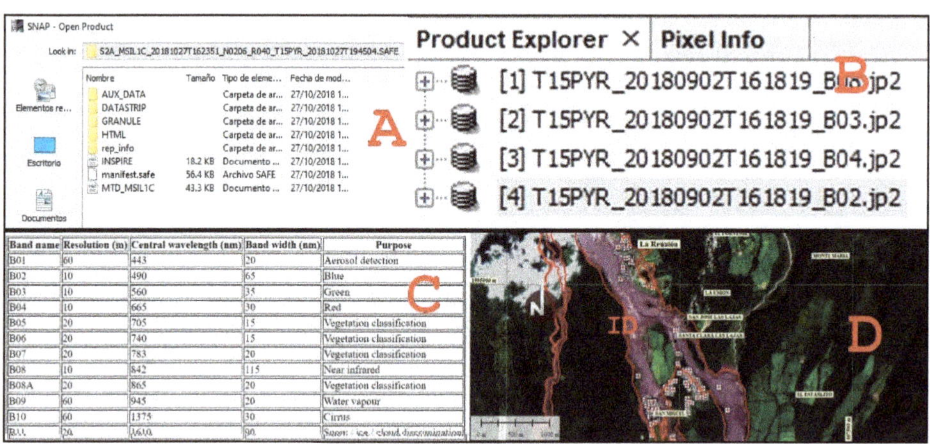

Band name	Resolution (m)	Central wavelength (nm)	Band width (nm)	Purpose
B01	60	443	20	Aerosol detection
B02	10	490	65	Blue
B03	10	560	35	Green
B04	10	665	30	Red
B05	20	705	15	Vegetation classification
B06	20	740	15	Vegetation classification
B07	20	783	20	Vegetation classification
B08	10	842	115	Near infrared
B08A	20	865	20	Vegetation classification
B09	60	945	20	Water vapour
B10	60	1375	30	Cirrus
B11	20	1610	90	Snow / ice / cloud discrimination

Figure 6. Process flow in SNAP. Spatial location of primary and secondary lahars from a Sentinel image 2. Combination of bands 4, 3 and 2. (**A–C**) SNAP passages of each portion of the routine in Sentinel 2A images. Location of primary and secondary lahars, (**D**) Synthesis of final product: limit of the model of lahar in red color obtained from LAHAR Z. In lead color the location to that date of the real lahar product of the eruption of June 3 and nearby places.

Process (D): mass movements analysis. To evaluate the susceptibility for landslides and define potential areas of sediment production through mass removal processes, the SHALSTAB program was used, which is a model to map potential shallow landslides. SHALSTAB combines the characteristics of the stationary subsurface flow (water flow that moves at shallow depths below the surface of the land with a constant velocity), considering the morphology-hydrology relationship [12]. SHALSTAB uses two types of numerical models: the slope stability model and the hydrologic model.

The slope stability model is based on an infinite slope form of the Mohr–Coulomb failure law in which the downslope component of the weight of the soil just at failure, τ, is equal to the strength of resistance caused by cohesion (soil cohesion and/or root strength), C, and by frictional resistance due to the effective normal stress on the failure plane Equation (3).

$$\tau = C + (\sigma - \mu) \tan \varphi \quad (3)$$

where σ is the normal stress, μ is the pore pressure opposing the normal load and tan φ is the angle of internal friction of the soil mass at the failure plane. This model assumes, therefore, that the resistances to movement along the sides and ends of the landslide are not significant. As SHALSTAB's goal is the mapping of landslide hazards, then establishing zero cohesion maximizes the degree of instability possible at the study site. By eliminating cohesion, Equation (4) can be written as:

$$\rho_s gz \cos\theta \sin\theta = \left[\rho_s gzcos^2\theta - \rho_w ghcos^2\theta\right] \tan\varphi \tag{4}$$

where z is soil depth, h is water level above the failure plane, cρs and ρ_w is the soil and water bulk density, respectively; g is gravitational acceleration and θ is slope angle (Figure 7A). This Equation (5) can then be solved for h/z which is the proportion of the soil column that is saturated at instability:

$$\frac{h}{z} = \frac{\rho_s}{\rho_w}\left[1 - \frac{\tan\theta}{\tan\varphi}\right] \tag{5}$$

h/z could vary from zero (when the slope is as steep as the friction angle) to $\frac{\rho_s}{\rho_w}$ when the slope is flat (tan θ = 0). An important assumption, however, will be used below which sets a limit on what h/z can be. It is assumed that the failure plane and the shallow subsurface flow is parallel to hillslope, in which case h/z can only be less than or equal to 1.0 and any site requiring h/z greater than 1 is unconditionally stable—no extreme rain can cause it to fail. Figure 7B illustrates the relationship between h/z and $\tan\theta$ for an angle of internal friction φ of 45° and a bulk density ratio of 1.6. There may be 4 stability states: any slopes equal to or greater than the friction angle will cause the right-hand side of (3) go to zero, hence the site is unstable even if the site is dry (h/z = 0). This scenario is "unconditionally unstable", commonly corresponds to sites of bedrock outcrop. Because h/z cannot exceed 1.0 in this model, if $\tan\theta$ is less than or equal to tanφ(1-(ρs-ρw) then the slope is "unconditionally stable". The two other stability states are "stable" and "unstable", with the former corresponding to the condition in which h/z is greater than or equal to that needed to cause instability (given by the right hand side of Equation (5)) and the latter corresponding to the case in which h/z is less than that needed to cause instability.

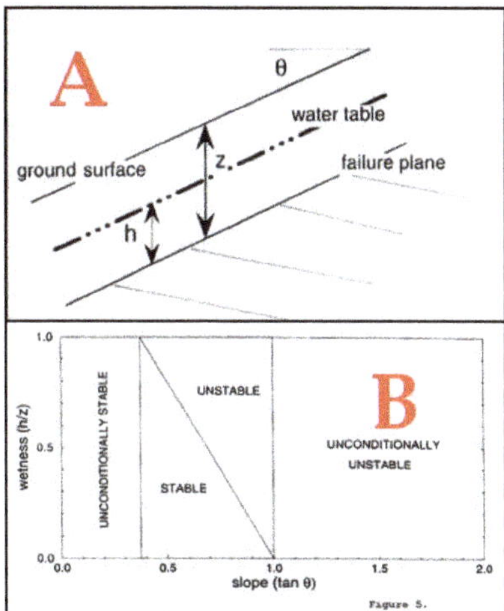

Figure 7. (**A**) The one-dimensional approximation used in the SHALSTAB slope stability model in which the failure plane, water table, and ground surface are assumed parallel. The slope is θ, the height of the water table is h, and the thickness of the colluvium that slides above the failure plane is z. Typically, the failure plane is at the colluvium-weathered bedrock or saprolite boundary. (**B**) Definition of stability fields. For this example, the angle of internal friction is 45 degrees, and the bulk density ratio is 1.6 [13].

Hydrologic model

To model the hydrologic controls on h/z. SHALSTAB use a steady state shallow subsurface flow based on the work by [14,15]. Assume that the steady state hydrologic response model mimics what the relative spatial pattern of wetness (h/z) would be during an intense natural storm which is not in steady state. This assumption would break down if precipitation events are sufficiently intense that thin soils on non-convergent sites can quickly reach destabilizing values of h/z before shallow subsurface flow can converge on unchanneled valleys.

Figure 8 illustrates the geometry and routing of water off the landscape used in the SHALSTAB hydrologic model. If we assume that there is no overland flow, no significant deep drainage, and no significant flow in the bedrock, then q, the effective precipitation (rainfall minus evapotranspiration) times the upslope drainage area, a, must be the amount of runoff that occurs through a grid cell of width b under steady state conditions. Using Darcy's law, we can write that (6):

$$qa = k_s h \cos\theta \sin\theta b \tag{6}$$

where $\sin\theta b$ is the head gradient, k_s is saturated hydraulic conductivity. At saturation the shallow subsurface flow will equal the transmissivity, T, (the vertical integral of the saturated conductivity) times the head gradient, sinq and the width of the outflow boundary, b and this we can approximate as follows (7):

$$\frac{h}{z} = \frac{q}{T}\, WI; WI = \frac{a}{b\ sen\ \theta} \tag{7}$$

where h represents the height of the water table on the sliding surface (m), z the depth of the soil (m), q effective precipitation [mm], T soil transmissivity [m^2·day-1], WI (Wetness Index) factor describes the effect of morphology on the subsurface flow (this is calculated in SAGA as the topogrraphic wetness index (TWI)).

In order to delimit areas of danger, new studies have applied a methodology with a geographical information system (GIS) base, identifying zones where flows start along the hydrographic network [12].

Morphologically, the triggering of a detritic or lahar flow depends on two factors: the slope of the channel and the critical flow. The speed of an elementary gravity wave is called the critical velocity. When the velocity of the surface water in a channel equals the critical velocity, the elementary gravity waves with upstream motion remain at the point where they were generated and the flow in the channel is called critical.

When the superficial velocity of the water in the channel is equal to the critical velocity, the elementary waves of gravity with movement upstream remain at the point where they were generated and the flow in the channel is called critical. Both factors can be obtained from the morphological characteristics of the basin. The first one is evaluated directly from the digital terrain model, while the liquid flow is evaluated indirectly through the drained area [12].

Studies conducted in the Swiss Alps found a relationship for the critical slope (S) that triggered detrital flows and the drained area of the basin (A) (Equation (8)).

$$S = c\ A^{-n} \tag{8}$$

The coefficients c and n have values of 0.32 and 0.20, respectively. The inverse relationship between S and A indicates that a greater slope of the channel will require a smaller liquid flow to trigger a debris flow (a smaller amount of liquid within the flow area).

Other authors have proposed classifying lahar flows into four categories: triggering, propagation, deceleration, and deposit [13].

Triggering occurs on slopes greater than the threshold proposed by other authors; 25° is known as a slope of minimum probable slope for the initiation of debris flow [13–15], but with a lower angle than the internal friction of the ground and a drained area smaller than 10 km^2. The maximum slope corresponding to the internal friction angle of the deposit is such that, in areas with steeper slopes, the amount of detritus is generally modest or negligible, while for drained areas exceeding 10 km^2, the type of transport becomes fluvial in place of detrital. The model calculates the degree of the internal saturation of the soil with the relationships of the height h/soil depth of the groundwater level z; (h/z) (Equation (9)) is the necessary factors to destabilize the slopes.

$$\frac{h}{z} = \frac{\rho s}{\rho w}(1 - \frac{tan\theta}{tan\varnothing}) \quad (9)$$

This formula allows two limit states to be defined. When h/z is negative, the slope is unstable for any degree of saturation and is called "unconditionally unstable".

On the other hand, when h/z is greater than one, it is called "unconditionally stable" because even in cases of saturation, the slope is stable. To define the intermediate situations, it is necessary to add the hydrological model.

The hydrological model considers a subsurface flow in a permanent regime.

Applying the law of Darcy, on the characteristics of the movement of water through a porous medium, it is possible to arrive at the following expression that links the degree of saturation (h/z) with effective precipitation (q), which is the rainwater that enters the soil and is assimilated by the plants (Equation (10)).

$$\frac{h}{z} = \frac{q}{T}\,WI; WI = \frac{a}{b\ sen\ \theta} \quad (10)$$

With

q = effective precipitation [mm];
T = soil transmissivity [$m^2 \cdot day^{-1}$];
θ = slope [°];
a = contribution area [m^2]; and
b = is the length of the contour through which the flow transits [m].

where T is the transmissivity of the soil; a is the drained area; and b is the dimension of the raster cell. The relation q/T represents the magnitude of the precipitation relative to the capacity of the soil to conduct the water. The WI (wetness index) factor describes the effect of ground morphology on the subsurface flow.

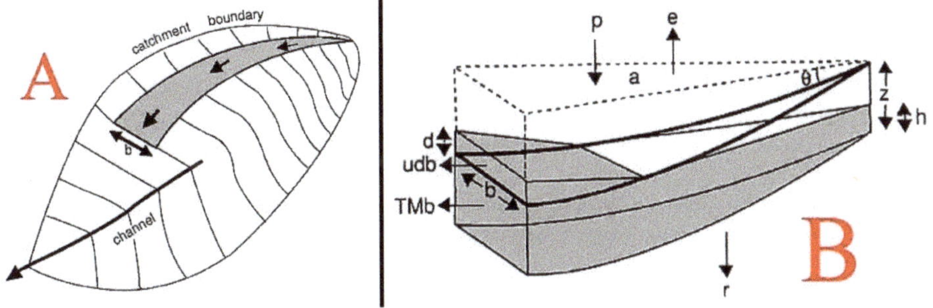

Figure 8. Sketch of hydrological model: (**A**) plan view and (**B**) cross section of a draining area a across a contour of length b. (**B**) The grey volume is the shallow sub-surface flow TMb and the saturation overland flow udb (not calculated by the model). p: precipitation; e: evapotranspiration; r: deep drainage; q: effective rainfall (q = p − e − r); h: vertical depth of the saturated sub-surface flow; z: vertical depth of the failure plane; T: transmissivity; M: sinθ; θ: slope angle. Mettma Ridge study site [13].

The analysis of the stability on the slopes of the volcano with SHALSTAB through the hydrological relationship q/T, the magnitude of precipitation, q, in relation to the capacity of the subsoil to transmit water along the T (transmissivity) slopes determined that the greater q was in relation to T, the more likely the soil would be saturated, and the greater the number of sites with unstable slopes. This relationship used the classification of pixels with stable positive values to quasi stable and chronically unstable values with negative values according to the stability classes (Table 1).

Table 1. Stability classes defined by SHALSTAB and modified from Montgomery and Dietrich [15].

Classes SHALSTAB	Interpretation of Class
Chronic instability	Unconditionally unstable and unsaturated
Log q·T -1 < -3,1	Unconditionally unstable and unsaturated
3,1 < Log q·T -1 < -2,8	Unstable and saturated
-2,8< Log q·T-1 < -2,5	Unstable and unsaturated
-2,5< Log q·T-1 < -2,2	Stable and unsaturated
Log q·T -1 > -2,2	Unconditionally stable and unsaturated
Stable	Unconditionally stable and saturated

Process E: The morphometric method with SAGA. Once the areas of surface deformation of the relief were found with DINSAR (Sentinel 1 images), the current location of the lahars (Sentinel 2 images), the superficial flow accumulators of the runoff and instability zones (SHALSTAB) were determined, the morphological method was applied to complement the location of trajectories of lahar flows with the analysis of the following morphological index.

2.1. LS-Factor (Slope Length and Steepness Factor)

This is a combination of the slope and the length of the slope as when combined, it is a useful attribute to predict erosion potential (Equation (11)). The length factor of the slope, L, calculates the effect of the length of the slope on the erosion, and the slope factor of the slope, S, calculates the effect of the slope on the erosion [16,17]. The higher values of brown color represent a greater susceptibility to erosion (Figure 9).

$$LS = L\hat{}(1/2)/100*(1.36 + 0.97 + 0.1385*S\hat{}2) \quad (11)$$

where L is the length of the slope in meters, and S is the gradient of the slope in %.

2.2. Topographic Wetness Index (TWI)

The topographic wetness index models the dynamics of surface and subsurface flows based on the topographic control of runoff, which offers a better perspective in relation to the prediction of sites where the saturation and high concentration of runoff can act as initial flow paths to processes greater than flood (Figure 10).

The TWI combines the contribution to runoff from a local area drained and the pending of it, and it is commonly used to quantify topographic control on hydrological processes and is defined as [6–10,18–20] (Equation (12):

$$TWI = Ln\ (af/\tan\beta) \quad (12)$$

where af is the local area drained for a calculation point, and tan β is the directional slope of the cell of interest (and of the eight neighbors in the case of using an algorithm D8. Ln is natural logarithm.

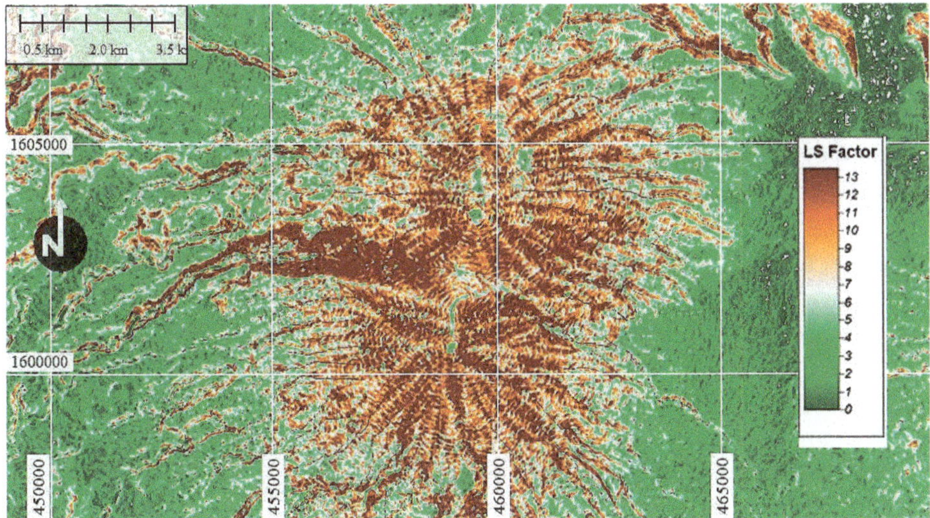

Figure 9. LS-factor (slope length and steepness factor) calculated in SAGA geographic information system (GIS). The low susceptibility to erosion is represented with green colors and the high with brown color.

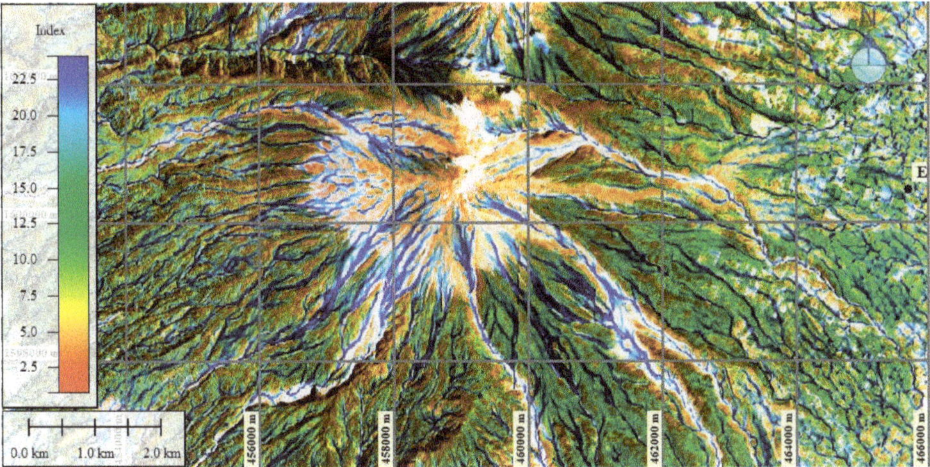

Figure 10. Topographic wetness index (TWI) calculated in SAGA GIS. Higher values of TWI (tendency to blue color) represent drainage depressions and lower values (tendency to red color) represent crests and ridges.

2.3. Flow Accumulation (FCC)

Flow accumulation is calculated as the upslope contributing (catchment) area using the D8 model or multiple flow direction approach [17–19]. The result is an accumulated flow raster for each cell, determined by the accumulation of the weight of all of the cells that flow into each cell of the descending slope. The accumulated flow is based on the total number of cells, or a fraction of them, flowing to each cell in the output raster.

The output cells with a high flow accumulation are areas of concentrated flow and can be used to identify watercourse channels (Figure 11).

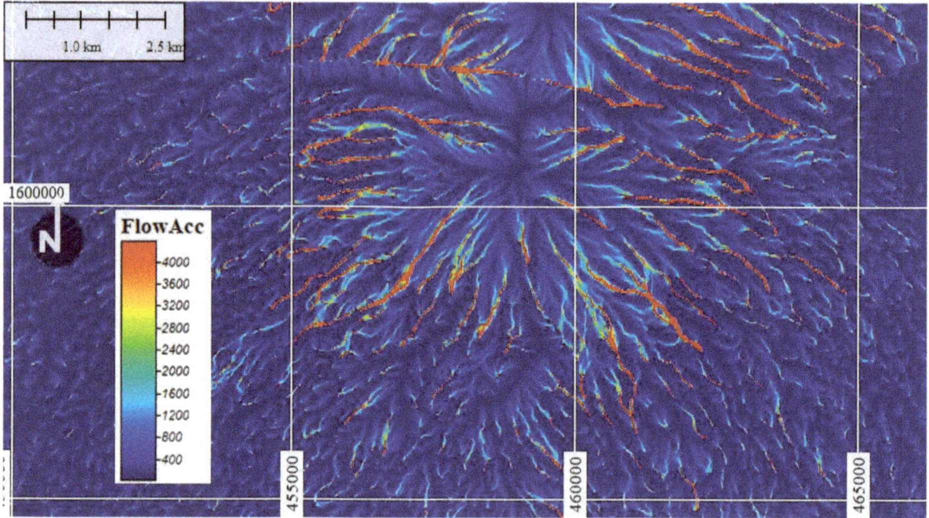

Figure 11. Cumulative flow calculated in SAGA GIS. The higher values (tendency to red color) represent greater accumulations of flow and areas of concentrated flow that can be used to identify watercourse channels.

3. Results

The differential synthetic aperture radar interferometry (DINSAR) analysis, based on the Sentinel 1 images, showed the lack of magmatic deformation before the eruption, but there were surface subsidences on the associated slopes due to the passage and deposits of lahars (Figure 12). This lack of magmatic deformation corroborates the studies undertaken by several researchers in the volcanoes of the area [20–22], which attribute the absence of deformation to differences in the storage of magma in relation to other continental arcs. The absence of magmatic deformation before the lahars can mean that the intrusion of magma is not the trigger of the lahars. Most likely, the rain is the cause.

The analysis of stability on the slopes of the volcano with SHALSTAB determined that the zones near the volcano were the most unstable with a qualification of quasi stable to chronic instability through the erosion of the channels where the lahars had flowed and where there had been greater destruction as is the case of El Rodeo, San Miguel, Monte María, and sectors near the Guacalate River (Figure 12B,C).

The current locations of the lahars produced by the eruption of 3 June 2018 and later were delimited with the Sentinel 2 images and their destructive power was verified with field data mainly from La Réunion, San José Las Lajas, Santa Clara Las Lajas, San Miguel, and El Rodeo, amongst other towns (Figure 12C). Figure 13A shows the sectors before the eruption. The implementation of lead-colored lahars on that date. In line of red color, the limits of the model of lahars obtained with LAHAR Z. In blue are the ravines and canals where the primary and secondary lahars flowed that flooded the sectors, and in green-yellow-ocher color the areas of ridges and crests. The field control points of the destroyed sites are represented by red crosses. In (B) Sectors close to La Réunion, in (C) sectors near San Miguel and in (D) close to Monte María.

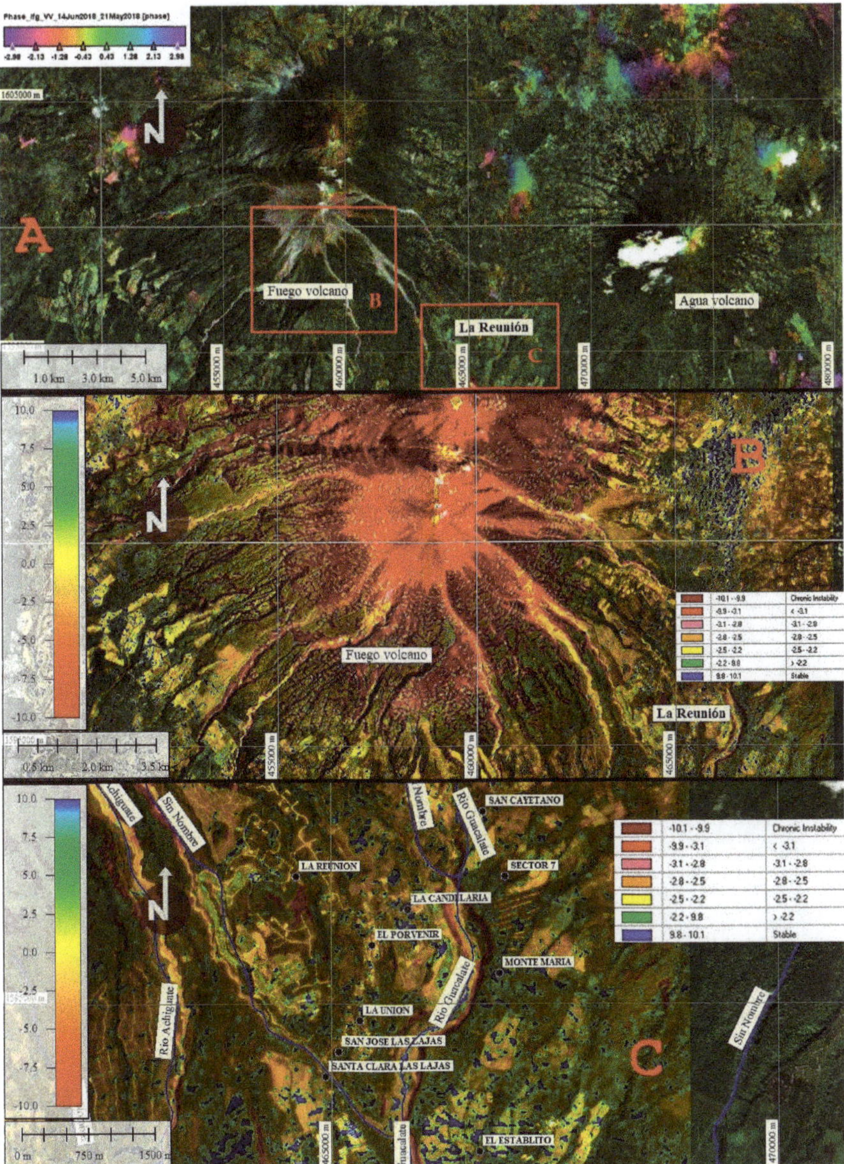

Figure 12. DINSAR analysis and SHALSTAB unstable and erosion areas. (**A**) Iridescent colors indicate the tendency of deformation in the region in cm. The positive values represent raised areas and the negative subsidence zones with minimal deformation in the volcano on the SE flank. (**B**) SHALSTAB results: the relief on the flanks of the volcano is very irregular with raised and sunken areas possibly due to the permanent morphological change resulting from the eruptions that have occurred. (**C**) Detail of the areas near the volcano that are the most unstable, with a rating of almost stable to chronic instability due to erosion of the channels that the lahars have flowed in, have overflowed and caused flooding with the destruction of the civil infrastructure mainly in El Rodeo, San Miguel, Monte María, and sectors near the Guacalate river. Negative red colors and trend towards –10 are unstable areas and stable positive zones with colors towards the blue and values with trend toward 10. The value is dimensionless.

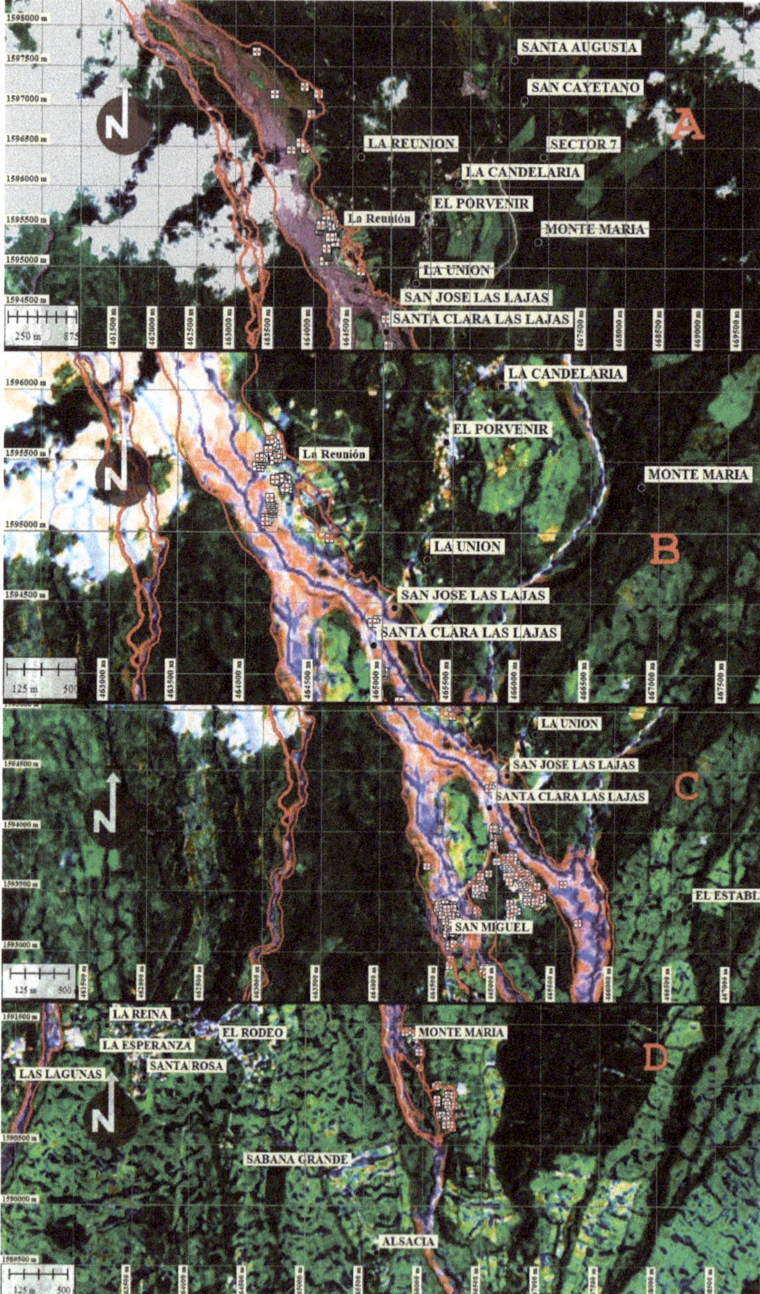

Figure 13. Sequence of images from the upper parts of the volcano of the areas destroyed by the lahars, erosion of channels and floods obtained from the combination of the response of the Sentinel 2A, morphometric index LS, TWI, FlowAcc(acronym of Flow Accumulation), and stabilized in the sectors the day of the eruption of Fuego volcano. (**A**) The implementation of lead colored lahars on that date. In line of red color, the limits of the model of lahars obtained with LAHAR Z. In blue the ravines and canals where the primary and secondary lahars flowed that flooded the sectors, and in green-yellow-ocher color the areas of ridges and crests. The field control points of the destroyed sites are represented by red crosses. In (**B**) Sectors close La Reunion, in (**C**) sectors near San Miguel and in (**D**) close to Monte María.

From the analysis of the morphometric parameters LS, TWI, and FlowAcc, the areas with greater susceptibility to erosion were obtained as well as those with the highest concentration and accumulation of superficial flow of rainwater that were directly related to the primary and secondary lahars (Figure 14). These areas of blue color were observed to have been directed by the lahars that destroyed the populated areas. These three morphometric indices [23–26] are valid parameters to determine the flow paths of primary and secondary lahars. The areas with high values of these indices are areas of moisture concentration in the slopes, flow paths and concentration of sediments that can move from the high parts by intense rains and gravity generating secondary lahars, generally on the slopes.

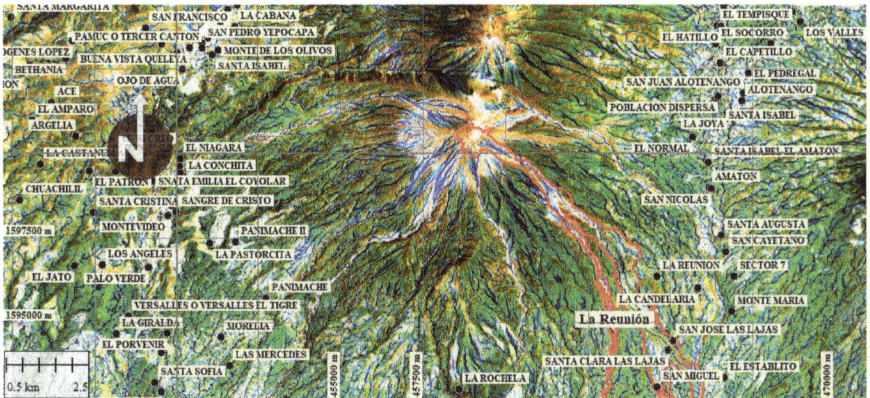

Figure 14. The flow paths, calculated by using the three morphometric indices TWI, LS, and FlowAcc in SAGA, determined the areas that were flooded by primary and secondary lahars.

The field control was carried out by technicians of CONRED.

4. Discussion

Unlike the LAHARZ model, the presented model uses morphometry as a value that changes the trajectory of the primary and secondary lahars, which allowed us to obtain the real flow paths of these phenomena and their effects. A beneficial coincidence of the application of both models is that they kept a central axis path of the main lahars, due to the fact that they were drains or main rivers. However, LAHARZ does not separate secondary lahars, due to this action; the morphological method can determine the flow paths of the mentioned phenomenon.

The lahar danger map, created for the Fuego volcano, based on the LAHARZ program, published in 2018 by CONRED, established high hazards zones (marked in red color) that can be affected by lahar flow paths. Torrential rains have generated secondary lahars that have affected the surrounding areas of the volcano, even those ones of low hazard. The medium hazard zones were affected by lahar flow paths from the La Reunión site to the Alsacia site, as can be evidenced in Figure 13.

In the current paper, with the proposed methodology, a study of the lahars flow paths has been carried out, being the main study areas the valleys of these Barranca Seca, Ceniza, Guacalate, Playa Trinidad, Las Cañas, El Jute and Las Lajas rivers. Lahars originated due to the erosion of the cause ways of detritus-ash-water deposited during the volcanic eruption and torrential rains. The axis of the primary lahar paths studied with the proposed methodology are like the ones obtained to delimit the high hazards zones in the mentioned hazard map.

On the other side, is has been proved that the modeled secondary lahar deposits with the proposed methodology destroyed civilian structures in La Reunión, San José de Las Lajas, Santa Clara de Las Lajas, San Miguel and other areas had location limits in accordance with the data obtained with LAHARZ and the location obtained with Sentinel 2A images, but with an added element that was not obtained with LAHARZ which is the outline of the secondary lahars flow paths, some as contributors to primary lahars and others as independent elements with no connection to the main ones.

The range of the secondary lahars and its respective modeled central axis is similar and bigger to the June and November lahars of 2018, observed in the Sentinel images. This is probably due to the resolution of the digital elevation model and the data given by the analyzed morphometric index.

Finally, a comparison of the affected areas has been carried out and its reach on transit zones, obtained with the proposed methodology and LAHARZ, observing that the entire area affected by lahars is within the high danger zone. The lahars corresponding to the Las Lajas y Guacalate rivers, take the biggest amount of area of the affected areas, with most of the civilian infrastructure destroyed as a result, turning the mentioned areas to maximum danger zones.

The surface terrain deformations, obtained from Sentinel 1B, located in the high mass erosion and removal areas can be part of secondary lahars which were generated from high ground. This is a contribution when the DINSAR analysis is used.

In addition to this work, further research studies can improve the location of the lahars flow paths using the deformed areas of the relief obtained with the DINSAR analysis, the stability criterion of slopes and analysis of morphometric parameters in digital models of high elevation spatial resolution obtained from Lidar images and integrating them into specialized programs in lahars and floods analysis such as LAHAR Z, Titan 2D, Ash3D, Flo-2D, IRIC, Iber, among others. The final objective of these investigations is to put in the hands of government authorities and the community, vulnerability maps to this type of threats to reduce them.

5. Conclusions

As a contribution of the application of this methodology, based on a flow modeler such as LAHAR Z or another program, to know the Flow Paths of the primary and secondary lahars that have caused the destruction in the vicinity of the volcano, this methodology deepened the study of these trajectories based on the analysis of the morphometric parameters, which allowed us to have a better knowledge of the possible paths before a new volcanic eruption. This methodology is currently being applied in the Fuego volcano by INSIVUMEH technicians with the objective of putting it in the hands of the government authorities, as well as the institutions linked to the subject, which allows these conditions to be ascertained in advance thereby implementing structural measures and non-structural measures to reduce this danger.

To complement the results obtained with the DINSAR analysis and the morphological method, in this study, the spatial distribution of the sources and potential routes of primary and secondary lahars during the eruptions and torrential rains that can erode and produce mass movements in areas of critical instability was obtained from the stability model of shallow landslide slopes with SHALSTAB.

The integration of the DINSAR methodologies to determine the surface deformations of the ground, where the SHALSTAB model determines the stability-erosion of the slopes of the volcano, and the morphometric method to analyze the primary and secondary lahar currents, is a procedure that can help to prevent the destructive effects that they cause during these events as demonstrated in the same destroyed localities that were not mapped in the maps of existing hazards by CONRED.

The field observations carried out by CONRED technicians and sent in the format of points in shp format, helped to calibrate and verify the results obtained from the presented methodology, as well as in understanding which of the three morphological parameters used had greater weight according to the scenario in the one that developed the lahars. This parameter corresponds to the TWI, the same one that determined that the lower areas of the streams are the ones that get wet and fill quickly overflowing and flooding nearby areas.

The morphological model could obtain the primary and secondary lahar paths in all sectors near the volcano in a very short time, unlike the numerical models such as the LAHAR Z, which are applied in a single axis of a current or drainage. In this case, the paths of the lahars obtained are sufficient to produce maps of the vulnerability reduction of civil structures and socio-economic systems including the preservation of human lives.

This combination of procedures can be applied in any volcanic landscape based on digital models of high-resolution elevation, obtaining very good agreement with the paths of the real flows.

Author Contributions: The research work presented in this paper is a collaborative development by both authors. Conceptualization, M.C. and A.M.; Methodology, M.C. and A.M.; Software, M.C.; Validation, M.C. and A.M.; Formal Analysis, M.C.; Investigation, M.C. and A.M.; Resources, M.C.; Data Curation, M.C.; Writing-Original Draft Preparation, M.C. and A.M.; Writing-Review and Editing, A.M.; Visualization, Supervision: A.M.; Project Administration, A.M.; Funding Acquisition, A.M.

Funding: This research received no external funding.

Acknowledgments: This research was funded by projects Junta Castilla y León SA044G18 and projects from the Ministry of Economy and Competitiveness CGL2015-67169-P and CGL2015-69919-R.

Conflicts of Interest: The authors declare no conflict of interest.

References

1. Almeida, S.; Daniel, S.; Daniel, A. Definition, Primary and Secondary Lahars, Flow Types, Behavior, afectation And Hazard Monitoring. Geophysical Institute National Polytechnic School, 2017. Available online: https://www.igepn.edu.ec/publicaciones-para-la-comunidad/comunidad-eng/19268-triptych-lahars-mud-flows/file (accessed on 21 February 2019).
2. National Coordinator for Disaster Reduction CONRED. Situation Report No. 1 (as of 04 June 2018). Internal Report. Available online: https://reliefweb.int/sites/reliefweb.int/files/resources/2018-06-04%20GT%20SITREP%20-%20Volcanic%20Activity%20%28ENG%29.pdf (accessed on 21 February 2019).
3. Department of Vulcanology, INSIVUMEH. Preliminary Map of Hazard Scenarios by Pyroclastic Flows, after the Eruption of June 3, 2018. Available online: http://www.insivumeh.gob.gt/?page_id=70 (accessed on 21 February 2019).
4. National Coordinator for Disaster Reduction CONRED. Informative Bulletin No. 3162018—The Drop of Lahares Is Monitored on The Fuego Volcano. 2018. Available online: https://conred.gob.gt/site/Boletin-Informativo-3162018andhttps://conred.gob.gt/www/index.php?option=com_content&view=article&id=7076&catid=37&Itemid=1010 (accessed on 21 February 2019).
5. León, X. Vulnerability Assessment Associated with the Hazard of the Fuego Volcano. Internal Study. Universidad de San Carlos de Guatemala, 2012. Available online: http://biblioteca.usac.edu.gt/tesis/02/02_3433.pdf (accessed on 24 February 2019).
6. Schilling, S.P. TI-LAHARZ: GIS Programs for Automated Mapping of Lahar-Inundation Hazard Zones. USGS Publications Warehouse, 1998. Available online: http://pubs.er.usgs.gov/publication/ofr98638 (accessed on 25 February 2019). [CrossRef]
7. Dietrich, W.E.; Bellugi, D.; Real de Asua, R. Validation of the Shallow Landslide Model, SHALSTAB, for forest management. Land Use and Watersheds: Human Influence on Hydrology and Geomorphology in Urban and Forest Areas. *Water Sci. Appl.* **2001**, 2, 195–227. Available online: https://www.researchgate.net/publication/284902139_Validation_of_the_Shallow_Landslide_Model_SHALSTAB_for_forest_management (accessed on 25 February 2019). [CrossRef]
8. Montoya, S. Tutorial to Download Sentinel 2 Images (10 m Resolution) and Process Them in QGIS. GIDAHATARI, 2017. Available online: http://gidahatari.com/ih-es/tutorial-para-descargar-imgenes-sentinel-2-resolucin-10-m-y-procesarlas-en-qgis (accessed on 25 February 2019).
9. Zeni, G.; Pepe, A.; Zhao, Q.; Bonano, M.; Gao, W.; Li, X.; Ding, X. A differential SAR interferometry (DInSAR) investigation of the deformation affecting the coastal reclaimed areas of the Shangai megacity. In Proceedings of the International Geoscience and Remote Sensing Symposium (IGARSS), Quebec City, QC, Canada, 13–18 July 2014; pp. 482–485. [CrossRef]
10. The Copernicus Open Access Hub. The Open Access Hub Provides Complete, Free and Open Access to Sentinel-1, Sentinel-2, Sentinel-3 and Sentinel-5P User Products. Available online: https://scihub.copernicus.eu/userguide/ (accessed on 21 February 2019).

11. Foumelis, M.; Delgado Blasco, J.M.; Desnos, Y.-L.; Engdahl, M.; Fernández, D.; Veci, L.; Lu, J.; Wong, C. ESA SNAP—StaMPS Integrated processing for Sentinel-1 Persistent Scatterer Interferometry. 2018. Available online: https://www.researchgate.net/publication/327253039_ESA_SNAP_-_StaMPS_Integrated_processing_for_Sentinel-1_Persistent_Scatterer_Interferometry (accessed on 23 February 2019).
12. Martins, T.; Vieira, B.; Fernandes, N.; Oka-Fiori, C.; Montgomery, D.R. Application of the SHALSTAB model for the identification of areas susceptible to landslides: Brazilian case studies. *Revista de Geomorfologie* **2017**, *19*, 136–144.
13. Panagos, P.; Borrelli, P.; Meusburger, K. A New European Slope Length and Steepness Factor (LS-Factor) for Modeling Soil Erosion by Water. *Geosciences* **2015**, *5*, 117–126. [CrossRef]
14. Zimmermann, M.; Mani, P.; Romang, H. Magnitude-frequency aspects of alpine debris flows. *Eclogae Geol. Helv.* **1997**, *90*, 415–420. Available online: https://www.e-periodica.ch/cntmng?pid=egh-001:1997:90::738 (accessed on 21 February 2019).
15. Carmona-Álvarez, J.E.; Ruge-Cárdenas, J.C. Análisis de las correlaciones existentes del ángulo de fricción efectivo para suelos del piedemonte oriental de Bogotá usando ensayos in situ. *Tecno Lógicas* **2015**, *18*, 93–104. Available online: http://www.scielo.org.co/pdf/teclo/v18n35/v18n35a09.pdf (accessed on 23 February 2019). [CrossRef]
16. Dietrich, W.E.; Montgomery, D.R. A Digital Terrain Model for Mapping Shallow Landslide Potential. University of California-University of Washington, 1998. Published as a Technical Report by NCASI. Available online: http://calm.geo.berkeley.edu/geomorph/shalstab/index.htm (accessed on 23 February 2019).
17. O'Loughlin, E.M. Prediction of surface saturation zones in natural catchments by topographic analysis. *Water Resour. Res.* **1986**, *22*, 794–804. Available online: https://www.scirp.org/(S(i43dyn45teexjx455qlt3d2q))/reference/ReferencesPapers.aspx?ReferenceID=729751 (accessed on 21 February 2019). [CrossRef]
18. Sebastiano, T.; Cavalli, M. Topography-based flow-directional roughness: potential and challenges. *Earth Surf. Dyn.* **2016**, *4*, 343–358. Available online: https://www.earth-surf-dynam.net/4/343/2016/esurf-4-343-2016.pdf (accessed on 26 February 2019).
19. Beven, K.J.; Kirkby, M.J. A physically based variable contributing area model of basin hydrology. *Hydrol. Sci. Bull.* **1979**, *24*, 43–69. Available online: https://www.tandfonline.com/doi/abs/10.1080/02626667909491834 (accessed on 26 February 2019). [CrossRef]
20. Cando, M.; Martínez-Graña, A. Numerical modeling of flow patterns applied to the analysis of the susceptibility to movements of the ground. *Geosciences* **2018**, *8*, 340. Available online: https://www.mdpi.com/2076-3263/8/9/340 (accessed on 26 February 2019). [CrossRef]
21. Hojati, M.; Mokarram, M. Determination of a Topographic Wetness Index using high resolution Digital Elevation Models. *Eur. J. Geogr.* **2016**, *7*, 41–52. Available online: http://www.eurogeographyjournal.eu/articles/4.%20DETERMINATION%20OF%20A%20TOPOGRAPHIC%20WETNESS%20INDEX%20USING%20HIGH%20RESOLUTION%20DIGITAL%20ELEVATION%20MODELS.pdf (accessed on 25 February 2019).
22. Ebmeier, S.; Biggs, J.; Mather, T.; Amelung, F. Applicability of InSAR to tropical volcanoes: Insights from Central America. *Geol. Soc. Lond. Spec. Publ.* **2013**, *380*, 15–37. Available online: https://www.researchgate.net/publication/259130455_Applicability_of_InSAR_to_tropical_volcanoes_Insights_from_Central_America (accessed on 21 February 2019). [CrossRef]
23. Baumann, V.; Bonadonna, C.; Cuomo, S.; Moscariello, M.; Manzella, I. Slope stability models for rainfall-induced lahars during long-lasting eruptions. *J. Volcanol. Geotherm. Res.* **2018**. Available online: https://www.researchgate.net/publication/326139979_Slope_stability_models_for_rainfall-induced_lahars_during_long-lasting_eruptions (accessed on 25 February 2019). [CrossRef]
24. Vidal Montes, R.; Martínez-Graña, A.M.; Martínez Catalán, J.R.; Ayarza, P.; Sánchez San Román, F.J. Vulnerability to groundwater contamination, SW Salamanca, Spain. *J. Maps* **2016**, *12*, 147–155. [CrossRef]
25. Martínez-Graña, A.M.; Goy, J.L.; Zazo, C. Geomorphological applications for susceptibility mapping of landslides in natural parks. *Environ. Eng. Manag. J.* **2016**, *15*, 1–12. [CrossRef]
26. Santos-Francés, F.; Martínez-Graña, A.; Rojo, P.A.; Sánchez, A.G. Geochemical background and baseline values determination and spatial distribution of heavy metal pollution in soils of the Andes mountain range (Cajamarca-Huancavelica, Peru). *Int. J. Environ. Res. Public Health* **2017**, *14*, 859. [CrossRef] [PubMed]

Letter

Volcanic Cloud Top Height Estimation Using the Plume Elevation Model Procedure Applied to Orthorectified Landsat 8 Data. Test Case: 26 October 2013 Mt. Etna Eruption

Marcello de Michele [1,*], Daniel Raucoules [1], Stefano Corradini [2], Luca Merucci [2], Giuseppe Salerno [3], Pasquale Sellitto [4] and Elisa Carboni [5]

1 Bureau de Recherches Géologiques et Minières, 3 av. C. Guillemin, 45000 Orléans, France; d.raucoules@brgm.fr
2 Istituto Nazionale di Geofisica e Vulcanologia, Sez. Roma Via di Vigna Murata, 00100 Rome, Italy; stefano.corradini@ingv.it (S.C.); luca.merucci@ingv.it (L.M.)
3 Istituto Nazionale di Geofisica e Vulcanologia, Sez Catania, Piazza Roma, 95100 Catania, Italy; giuseppe.salerno@ingv.it
4 Laboratoire Interuniversitaire des Systèmes Atmosphériques, Université Paris-Est, 94000 Créteil, France; pasquale.sellitto@u-pec.fr
5 COMET, Atmospheric, Oceanic and Planetary Physics, University of Oxford, Clarendon Laboratory, Parks Road, Oxford OX1 3PU, UK; elisa.carboni@physics.ox.ac.uk
* Correspondence: m.demichele@brgm.fr

Received: 22 January 2019; Accepted: 28 March 2019; Published: 2 April 2019

Abstract: In this study, we present a method for extracting the volcanic cloud top height (VCTH) as a plume elevation model (PEM) from orthorectified Landsat 8 data (Level 1). A similar methodology was previously applied to raw Landsat-8 data (Level 0). But level 0 data are not the standard product provided by the National Aeronautics and Space Administration (NASA)/United States Geological Survey (USGS). Level 0 data are available only on demand and consist on 14 data stripes multiplied by the number of multispectral bands. The standard product for Landsat 8 is the ortho image, available free of charge for end-users. Therefore, there is the need to adapt our previous methodology to Level 1 Landsat data. The advantages of using the standard Landsat products instead of raw data mainly include the fast -ready to use- availability of the data and free access to registered users, which is of major importance during volcanic crises. In this study, we adapt the PEM methodology to the standard Landsat-8 products, with the aim of simplifying the procedure for routine monitoring, offering an opportunity to produce PEM maps. In this study, we present the method. Our approach is applied to the 26 October 2013 Mt. Etna episodes comparing results independent VCTH measures from the spinning enhanced visible and infrared imager (SEVIRI) and the moderate resolution imaging spectroradiometer (MODIS).

Keywords: volcanic cloud; Landsat 8; elevation model

1. Introduction

In volcanology, the volcanic cloud-top height (VCTH) is one of the most critical parameters to retrieve. It affects the quantitative estimation of volcanic cloud ash and gases parameters [1–3], the mass eruption rate needed for the transport and deposition models [4–6] and the definition of the most dangerous zone for air traffic. Exploiting their global coverage (in time and space), satellite sensors offer the unique possibility for an effective monitoring of VCTH. In recent years, many techniques have been developed exploiting the dark pixel brightness temperature [2], the CO_2 [7,8] and O_2 [9,10]

absorption bands, the radio occultation [11] and backward trajectory modelling [12]. Among the different techniques (for a complete review of advantages and drawbacks please refer to [13]) and satellite active systems as CALIPSO [14], several algorithms have been developed exploiting the parallax between remote sensing measurements collected by different views of the same object by using one or more instruments. The use of the parallax was introduced by Prata and Turner [15] using the dual view of the along track scanning radiometer (ATSR) and developed further by Mims et al. [16], Nelson et al. [17], and Flower and Kahn [18] using dedicated multiangle imager spectro radiometer (MISR) measurements. Zakšek et al. [19], Corradini et al. [20], and Merucci et al., [13] developed algorithms based on the combined use of polar-geostationary, ground based-geostationary and geostationary-geostationary measurements respectively.

De Michele et al. [21] generalized the method based on small parallax for virtually all push broom sensor data. The extraction of the VCTH in the form of a plume elevation model from the high-resolution push broom operational land imager (OLI) sensor on board a Landsat-8 satellite, has been demonstrated starting from raw data. The main idea expressed in [21] is that the physical distance between the panchromatic sensor (PAN) and the multi-spectral sensors (MS), both on Landsat-like satellites, yields a baseline and a time lag between the PAN and MS image acquisitions during a single passage of the satellite. This information can be used to extract a spatially detailed map of VCTH from virtually any multi spectral push broom system, called a plume elevation model (PEM).

The main difficulty of the data processing comes from the fact that one Landsat image is composed of 14 focal plane modules (FPMs) arranged in the so called 'staggered' geometry, which makes the joint retrieval of plume velocities and heights challenging. De Michele et al. [21] addressed this problem by reconstructing a new OLI image starting from the raw OLI data stripes (courtesy of National Aeronautics and Space Administration (NASA), pers. comm.). However, the raw Landsat-8 data are not the standard Landsat-8 products provided by NASA/ United States Geological Survey (USGS). The raw data are available only on demand and consist of 14 data stripes (one data stripe for each FPM) multiplied by the number of multispectral bands. The standard product for Landsat 8 is the ortho image, available at no cost for the end-user. For PEM extractions, the advantages of using the standard Landsat products instead of the raw data mainly include the fast -ready to use- availability of the data, free to registered users, which is of major importance during volcanic crises. In this study, we adapt the methodology described in [21] to standard Landsat-8 products, with the aim of simplifying the procedure for routine use, thus widening the usability of this method for producing PEM maps. In this study, the procedure will be applied to the standard Landsat 8 data collected during the 26 October 2013 Mt. Etna eruptive episode and the results compared with those obtained using different satellites systems.

The paper is organized as follows: Section 1 outlines the 26 October 2013 Etna eruption and Section 2 describes the PEM procedure applied to the standard Landsat-8 products. In Section 3 the results obtained are compared with the VCTH retrieved from the PEM procedure applied to the estimations realized using the spinning enhanced visible and infrared imager (SEVIRI) and the moderate resolution imaging spectroradiometer (MODIS). In Sections 4 and 5 the discussion and the conclusions are presented.

The 26 October 2013 Mt. Etna Eruption

Mt. Etna activity in 2013 was characterized by a sequence of 16 episodes of intense eruptive activity at the summit of the volcano, fed by the New Southeast Crater [22,23]. The 26 October 2013 episodes stand as the 14th of the year and the 39th paroxysm episode of the sequence started earlier in 2011 [24–26]. The eruption occurred after a few months of quiescence and started in the early morning on 25 October, displaying mild intra-crater Strombolian activity. On 26 October, the eruptive activity gradually increased in magnitude and frequency of explosions, and lava started pouring from the crater slowly expanding towards the Valle del Bove. In the early morning of 26 October, the explosion intensity increased markedly, and between 2:00 and 10:00 UTC the activity climaxed into a lava fountain. Over the paroxysm, the height of the lava fountain steadily reached ~200 m above the crater rim. A significant emission of gas, ash, and lapilli formed an eruptive column that rose convectively several kilometres above the summit of the volcano. The eruption ceased progressively in the late evening, and marked ash fall was reported to be dispersed by the wind southwest of the volcano proximally and distally down to the Ionian Mediterranean Sea [27]. Figure 1 shows the Orthorectified Landsat 8 image collected the 26 October 2013 at 09:37 UTC. The volcanic cloud is clearly visible.

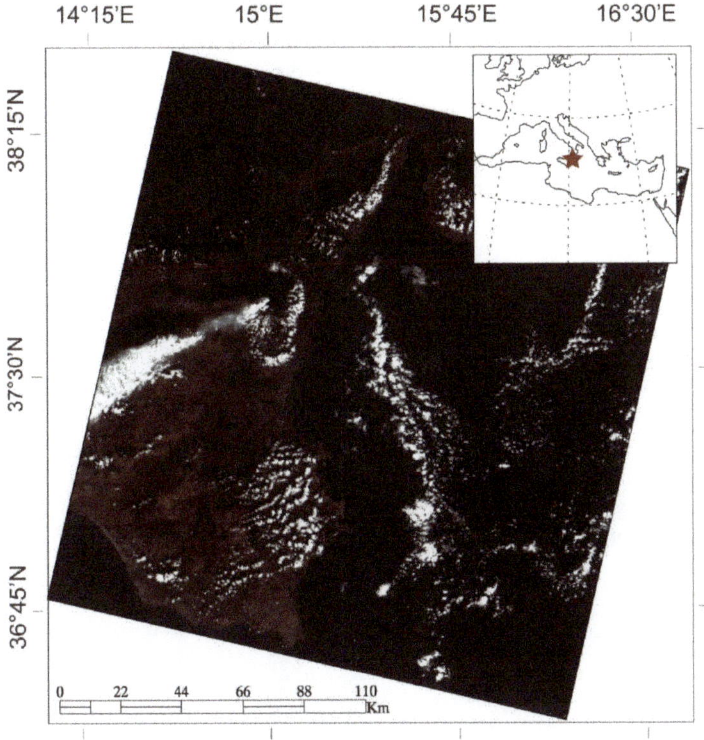

Figure 1. Orthorectified Landsat 8 data acquired the 26 October 2013 at 09:37 UTC on Mt. Etna volcano (image courtesy of National Aeronautics and Space Administration (NASA)/United States Geological Survey (USGS)).

2. Materials and Methods

The Landsat 8 OLI is a push-broom (linear array) imaging system that collects visible, near infra-red, and short-wave infra-red spectral band imagery at 30 m multi-spectral and 15 m panchromatic ground sample distances. It collects 190 km wide image swaths from ~705 km orbital altitude [28].

The OLI focal plane layout is very well described in [28] and [29]. The OLI detectors are distributed across 14 separate FPMs, each of which covers a portion of the 15° OLI cross-track field of view. Adjacent FPMs are offset in the along-track direction to allow for FPM-to-FPM overlap, avoiding any gaps in the cross-track coverage. The internal layout of all 14 FPMs is the same, with alternate FPMs being rotated by 180° to keep the active detector areas as close together as possible. This feature has the effect of inverting the along-track order of the spectral bands in adjacent FPMs. Consequently, this has the effect of inverting the signs of the cross-correlation measurements when calculating pixels offsets between PAN and MS bands, related to the volcanic cloud velocity and parallax.

The general concept of PEM methodology is that the PAN and the MS sensors on board a satellite platform cannot occupy the same position in the focal plane of the push-broom instrument. There is a physical separation between them. This separation yields a baseline and a time lag between the PAN and MS image acquisitions. Two directions are considered: The epipolar direction (EP), i.e., the azimuth direction of the satellite or the flight direction, and the perpendicular (P2E) to the EP direction. The pixel offset between PAN and MS in the EP direction is proportional to the height of the plume plus the pixel offset contribution induced by the motion of the plume itself in between the two acquisitions. The pixel offsets in the P2E direction, also controlled by the time lag, are proportional to the plume motion only, as there is no parallax in the P2E direction by definition. In principle, the offset in the P2E direction is proportional to movements of every feature in the imaged scene (e.g., meteorological clouds, lahars, rivers flow, ocean waves, vehicles). In our case study, we are interested in the volcanic cloud motion only. We use this latter information to compensate for the apparent parallax recorded in the EP offset.

If the data are downloaded in a staggered and orthorectified geometry, the processing is not straightforward, since FPMs are rotated 180°, and the offset analysis by cross-correlation would yield opposite signs at adjacent image stripes. This hampers the correct deployment of the method described in [21]. To avoid this inconvenience, we propose the following 5 step procedure:

I. The dataset is rotated, so that the columns of the image matrix are aligned to the nominal azimuth direction of the satellite reported in the ancillary data files.

II. A correlator to perform pixel (or sub-pixel) offset measurements is used (e.g., [30]). If one considers the OLI image as a matrix made of lines and columns, offsets among lines are the EP offsets (O_e), while offsets among columns are the P2E offsets (O_{p2e}). In the offset results, there could exist a ramp resulting from band mis-registration; the ramp, if found, is removed.

III. The direction of the plume is measured with respect to the azimuth direction, with the convention depicted in Figure 2.

IV. The absolute value of the pixel offsets due to the VCTH is calculated as O_h. Generally, O_h from staggered sensors should be calculated as follows:

$$|O_h| = |O_e| - \left|O_{p2e}\right| |tan\ \theta| \tag{1}$$

if theta is between zero and 180, or

$$|O_h| = |O_e| + \left|O_{p2e}\right| |tan\ \theta| \tag{2}$$

if theta is between 180 and 360. O_h is then converted into a VCTH using the formula provided in [20]:

$$h = |O_h|.\frac{s.H}{V.t} \tag{3}$$

for every pixel, which makes it a PEM. h is the plume height (m), s is the pixel size (m), V is the platform velocity (m/s), t is the temporal lag between the two Landsat 8 bands (s) and H is the platform height (m). Peculiar cases are: θ = 0° and θ = 180°. In these cases, the system is no longer sensitive to plume velocity. Therefore,

$$|O_h| = |O_e| \tag{4}$$

V. Finally, the results are re-rotated to their original position. Then, one has to choose a known reference altitude value on land and attribute it to the corresponding pixel. In our case study, we choose to set to zero the coastline close to the city of Catania.

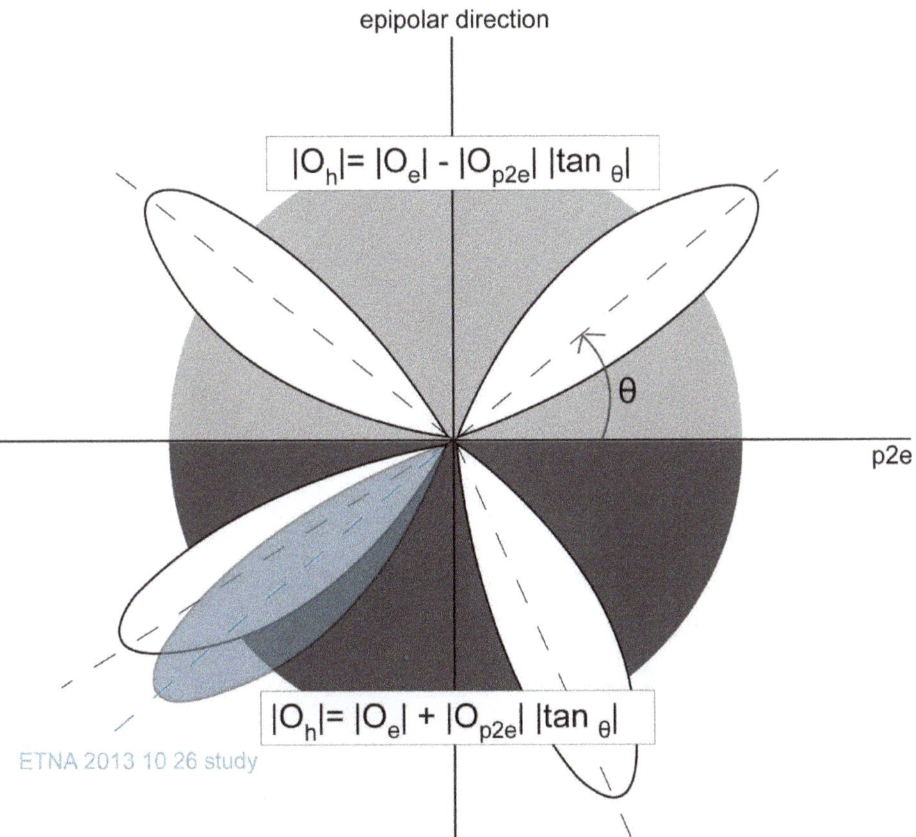

Figure 2. Angles and geometrical conventions, imagining this is a rotated Landsat 8 scene with a volcano at its centre. The epipolar (flight motion) direction and the perpendicular (p2e) to the epipolar direction (EP) direction are indicated. Dark gray and light gray colours indicate quadrants where the offsets are either summed up, either subtracted respectively. Ovals represent possible ash clouds directions. The light blue colour indicate the ash cloud direction of the case study presented here.

Figure 3 shows the offsets results. We show the raw results of the correlator on the left sides and the corrected results on the right side. The vertical stripes on the left sides are due to volcanic (and non-volcanic) cloud velocities and parallax: As the FPMs are inverted 180°, the correlator yields velocities with sign opposition (as explained in the introduction). The correlation results are corrected by using $|O_e|$, $|O_{p2e}|$ as described in the above paragraphs. The pixel offsets are expressed in meters. It is interesting to note that the P2E offset correspond to cloud (volcanic cloud and non-volcanic cloud) velocities, which values are comparable to the wind speed represented in Figure 6.

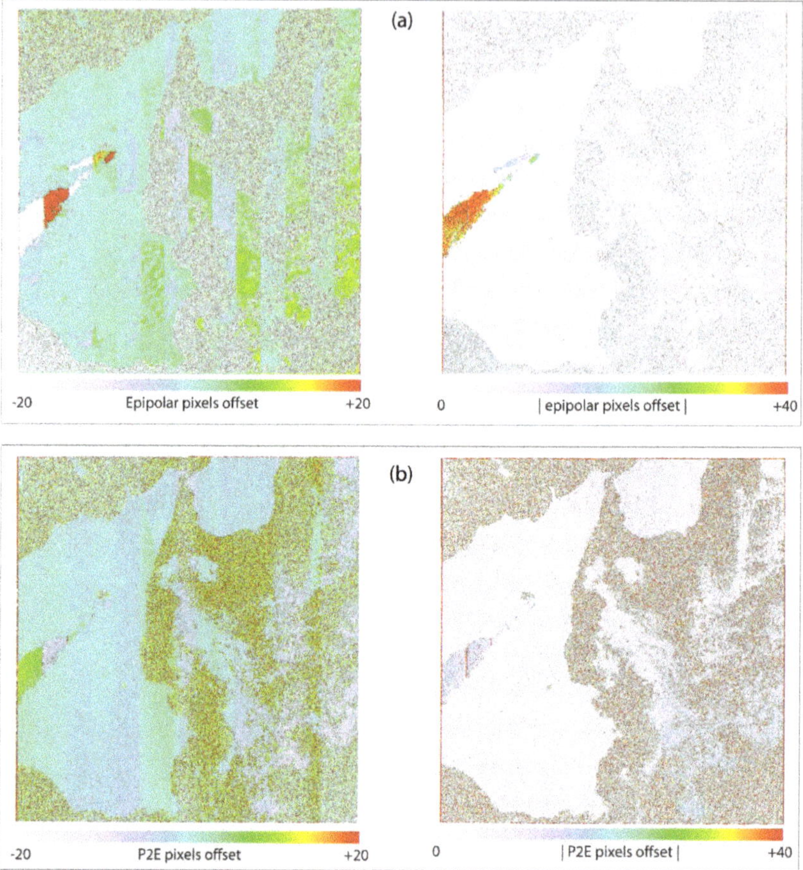

Figure 3. (**a**): Results from the correlator show vertical stripes due to volcanic (and non-volcanic) cloud velocities. As focal plane modules (FPMs) are inverted 180°, the correlator yields velocities with sign opposition (see text for more details). (**b**): Correlation results are corrected by using $|O_e|$, $|O_{p2e}|$ as described in this study. EP offsets (O_e), P2E offsets (O_{p2e}). The pixel offset is expressed in meters. It is interesting to note that the P2E offset correspond to cloud velocities.

3. Results and Cross-Comparisons

Figure 4 shows the VCTH map obtained from the PEM procedure applied to the Landsat orthorectified data. The volcanic cloud height vary from about 6 up to 9.5 km above the sea level (a.s.l.) with the higher values that lie in the central region of the cloud.

The estimated VCTH has been compared the VCTH extracted by using different procedure applied to other satellite sensors. Note that the cross-comparison here is not used as a validation. It allows us to assess the consistency of the results (i.e., our results are in the same order/scale as independent measurements). A thorough validation is not possible since acquisition times (repeat cycle) of different sensors are not the same as Landsat. Since the VCTH evolves with time, we prefer to call it « cross-comparisons » rather than « validation ». In addition, different sensors acquire data from different positions, introducing a bias in the eventual validation campaign.

Here the VCTH of the Etna 26 October 2013 eruption, used for the cross-comparison with Landsat results, are estimated by using geostationary (SEVIRI) and polar (MODIS) satellite sensors.

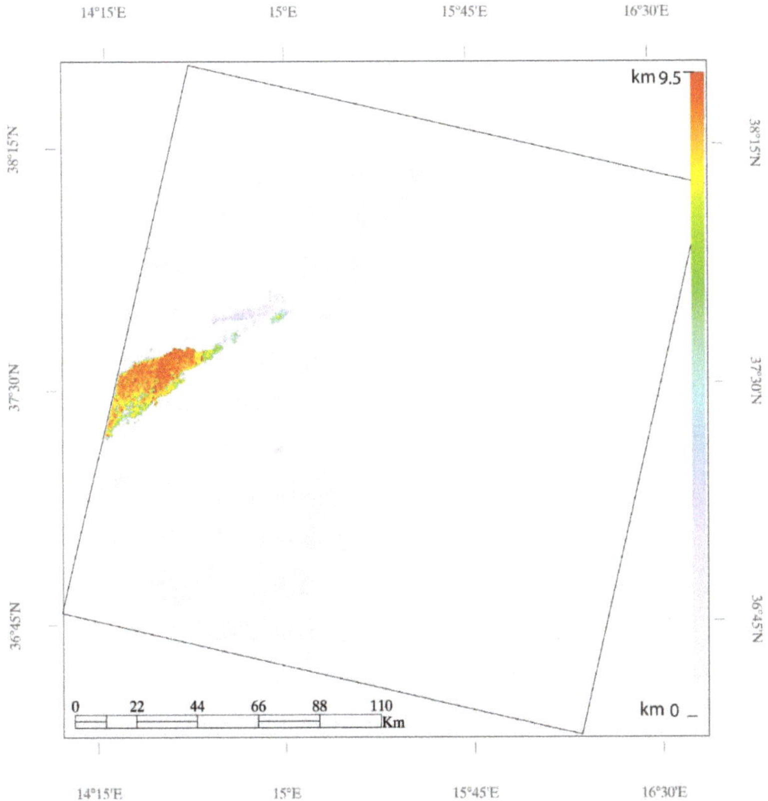

Figure 4. The volcanic cloud elevation (km), extracted from Landsat 8 orthorectified image.

3.1. MODIS VCTH Estimation

MODIS is a multispectral radiometer on board the NASA Terra and Aqua polar satellites. It has 36 spectral channels from visible (VIS) to thermal infrared (TIR), with a spatial resolution at sub-satellite of 1 km in the TIR and repetition cycle of 1–2 days. The VCTH is computed by exploiting the well known "dark pixels" procedure, based on the comparison between the brightness temperature at 11 μm of the coldest volcanic cloud pixel ($T_{b,11}$), with the atmospheric temperature profile of the same region at same time [2]. The temperature profile has been obtained from the National Centers for Environmental Prediction (NCEP)/National Center for Atmospheric Research (NCAR) [31], considering a box with 2.5° × 2.5° centred on Etna at 12:00 UTC. The left panel of Figure 5 shows the $T_{b,11}$ for the MODIS-Terra image collected the 26 October 2013 at 09:00 UTC. The volcanic cloud is clearly visible as dark signature on the right side of the image, while the cyan region indicates no data. The right panel of Figure 5 shows the NCEP temperature profile (grey line) and the $T_{b,11}$ of the dark pixel (red vertical line). Being $T_{b,11} = -37.4$ °C, VCTH result 8.9 km with an uncertainty of +/− 500 m, computed considering $T_{b,11}$ +/− 2 °C [2,32].

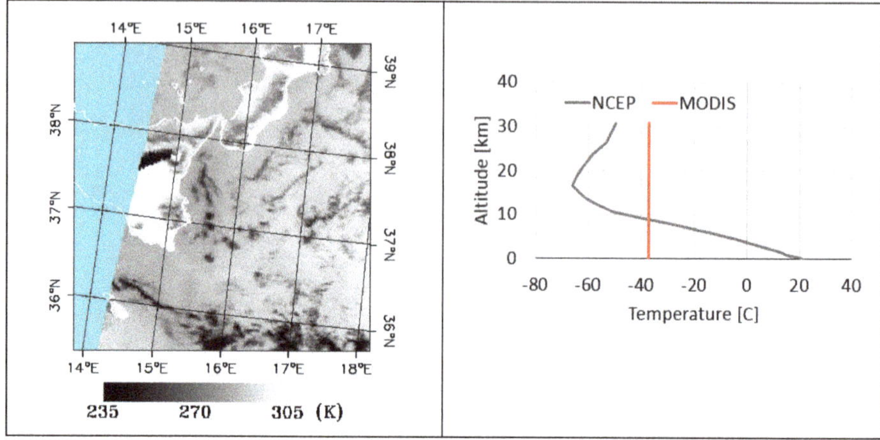

Figure 5. Left panel: brightness temperature at 11 µm for the moderate resolution imaging spectroradiometer (MODIS)-Terra image collected the 26 October 2013 at 09:00 UTC. **Right panel**: National Centers for Environmental Prediction (NCEP) temperature profile (grey line) and dark pixel brightness temperature (red vertical line).

3.2. SEVIRI VCTH Estimation

The SEVIRI instrument on board METEOSAT Second Generation (MSG) geostationary satellites is a 12 channel VIS-TIR multi-channel imager, which operates from 15 min repeat cycle on entire hemisphere (Full Disk) to 5 min over Europe (Rapid Scan). The sub-satellite point spatial resolution is 3×3 km^2, and the pixel dimension in the Etnean area is about 4.3×3.3 km. Exploiting the high data frequency of the SEVIRI images, the volcanic cloud speed can be retrieved by following the volcanic cloud centre of mass. By making a basic assumption that the estimated centre of mass speed is the whole volcanic cloud speed, VCTH can be obtained by comparing this value with the wind speed profile collected in the same time and position [26]. Also in this case the wind profile derive from NCEP/NCAR considering a box with 2.5° x 2.5° centred on Etna at 12:00 UTC. The upper panels of Figure 6 show the volcanic cloud ash mass maps obtained from the SEVIRI images collected at 09:00 UTC (left panel) and at 10:00 UTC (right panel). In these two images the different position of the volcanic cloud centre of mass is clearly identifiable. The distance of the centre of mass from the vents and the time of acquisition of the SEVIRI images, allows the computation of the volcanic cloud speed (see lower-left panel). This retrieved value is then compared with the wind speed NCEP/NCAR profile (see lower-right panel). In this case the wind speed of the volcanic cloud centre of mass is 18.0 m/s that yield to a VCTH of 10.5 km. An uncertainty of +/- 500 m is associated to take into account the uncertainty in the centre of mass identification. It is interesting to note that the P2E offset in Figure 3 corresponding to volcanic cloud velocities, is comparable to the wind speed measured by NCEP/NCAR (~18 m/s) shown in Figure 6.

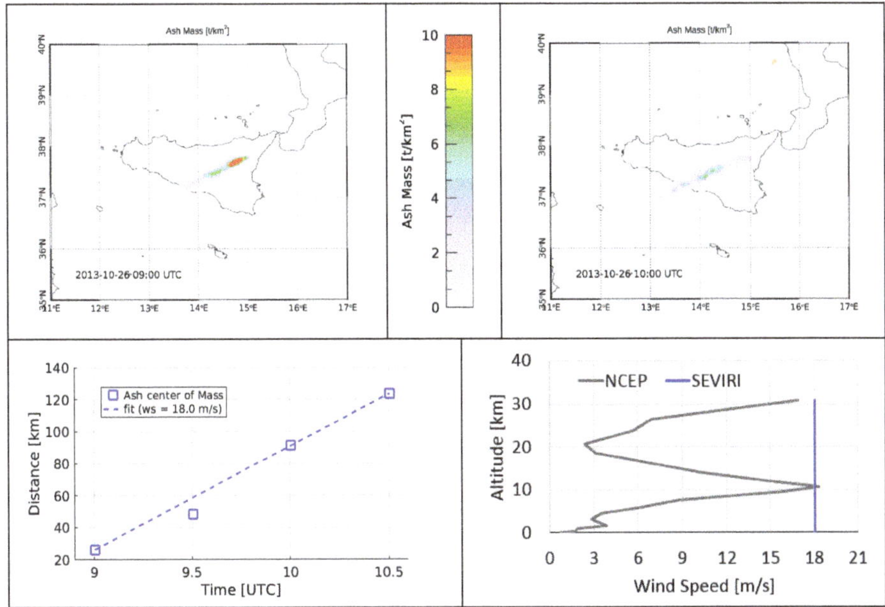

Figure 6. Upper panels: volcanic cloud ash mass maps obtained from the spinning enhanced visible and infrared imager (SEVIRI) images collected at 09:00 UTC (**left panel**) and at 10:00 UTC (**right panel**). Lower panels: Computation of the volcanic cloud speed from SEVIRI images (**left panel**) and comparison between the value obtained and the wind speed NCEP/ National Center for Atmospheric Research (NCAR) profile (**right panel**).

4. Discussion

The simple cross comparison presented here indicates that VCTH retrievals from Landsat 8, SEVIRI, and MODIS are in good agreement, taking into account the different methodologies and the different times of images acquisition (9:37 UTC for Landsat, 9:00 UTC for MODIS and from 9 to 10:30 UTC for SEVIRI). The volcanic plume is a complex medium made of different materials and chemical species. Different materials (such as ash or ice) and different chemicals (such as SO_2, H_2O, and CO_2, and halogens) are dispersed at different altitudes. As an example, sensors that capture or model SO_2 dispersion in the atmosphere (such as the Infrared Atmospheric Sounding Interferometer, Iasi) will, therefore, measure a different plume altitude compared to sensors that measure visible plume particles (such as Landsat 8 and SEVIRI). This important distinction has to be taken into account when comparing results from different sensors. The PEM method works best for volcanic clouds made of ash particles and optically thick material.

The accuracy of the PEM extracted from Landsat 8 has been already assessed against ground cameras in de Michele et al. (2016) for the Holuraun fissural eruption (Iceland). They reported an accuracy of 300 m. Landsat 8 acquires data at a high spatial resolution (15/30 m grid) at the cost of a medium revisit time (every 14 days). Instead, SEVIRI has the advantage of acquiring data every half an hour everywhere in the globe. This is done at the cost of medium spatial resolution (1 km grid). Statistics performed over VCTH retrieved by Landsat 8 (PEM), SEVIRI and MODIS data on a common area show very good agreement with a maximum height value of 8 (with 1 km precision). The strength in the results show the complementarity between these systems. Assimilation models could benefit from complementary constraints brought by these sensors.

Some questions remain open to future studies. How large must the optical depth or concentration of the particles be to enable VCTH-assessment? How far from the source is the retrieval of VCTH typically possible with Landsat 8 ? We believe that this information would be essential for designing a new, dedicated mission. In general, in our approach, the plume height is not measureable if the plume is not visible in the image (Table 1).

Table 1. Summary of the comparison between the different volcanic cloud-top height (VCTH) retrievals derived from Landsat 8, MODIS and SEVIRI.

VCTH [km]	VCTH	Uncertainty
LANDSAT	9.93	0.3
MODIS	8.9	0.5
SEVIRI	10.5	0.5

5. Conclusions

Landsat 8 satellites offer a unique opportunity for monitoring volcanic plumes at a high spatial resolution together with the opportunity to study volcanic eruptions back in time by exploiting the data archives. De Michele et al. [21] presented a method to extract the PEM (i.e., the digital elevation model of a volcanic plume) from raw Landsat 8 data. Nevertheless, raw Landsat 8 data are not available to the general public, limiting de facto the straight-forward application of the method. Moreover, the Landsat 8 OLI sensor presents a staggered geometry, which needs tuned processing. In this study, we push the methodology forward: Here we present a generalized, simplified methodology to extract the VCTH as a PEM from the common push broom staggered sensors, such as the standard Landsat 8 products. The advantages are manifold. First, it takes advantage of the freely available Landsat 8 data archives. Second, the method exploits the already-orthorectified Landsat 8 dataset, which is a standard Landsat 8 product, readily available on Landsat 8 archives. Third, the method could be adapted to the Copernicus Sentinel 2 data, a staggered sensor for which the products available to general public are orthorectified (similar to Landsat 8). The synergetic use of multiple sensors could improve the revisit time over a given volcanic eruption. It represents a step towards the routine monitoring of volcanic plumes height from space, at high spatial resolution.

The VCTH retrievals obtained from the geostationary SEVIRI and the polar MODIS satellite instruments indicate a good agreement with the Landsat VCTH product.

Funding: This study benefited from funding of the European Union and BRGM within the APHORISM (Advanced Procedures for Volcanic and Seismic Monitoring) project, 7th framework programme of the European Union. We are thankful to NASA/USGS Landsat 8 program for the Landsat 8 OLI data.

Acknowledgments: We are thankful to NASA/USGS for the Landsat8 data.

Conflicts of Interest: The authors declare no conflict of interest.

References

1. Wen, S.; Rose, W.I. Retrieval of sizes and total masses of particles in volcanic clouds using AVHRR bands 4 and 5. *J. Geophys. Res. Atmos.* **1994**, *99*, 5421–5431. [CrossRef]
2. Prata, A.J.; Grant, I.F. Retrieval of microphysical and morphological properties of volcanic ash plumes from satellite data: Application to Mt Ruapehu, New Zealand. *Q. J. R. Meteorol. Soc.* **2001**, *127*, 2153–2179. [CrossRef]
3. Corradini, S.; Merucci, L.; Prata, A.J. Retrieval of SO2 from thermal infrared satellite measurements: Correction procedures for the effects of volcanic ash. *Atmos. Meas. Tech.* **2009**, *2*, 177–191. [CrossRef]
4. Mastin, L.G.; Guffanti, M.; Servranckx, R.; Webley, P.; Barsotti, S.; Dean, K.; Durant, A.; Ewert, J.W.; Neri, A.; Rose, W.I.; et al. A multidisciplinary effort to assign realistic source parameters to models of volcanic ash-cloud transport and dispersion during eruptions. *J. Volcanol. Geotherm. Res.* **2009**, *186*, 10–21. [CrossRef]
5. Stohl, A.; Prata, A.J.; Eckhardt, S.; Clarisse, L.; Durant, A.; Henne, S.; Kristiansen, N.I.; Minikin, A.; Schumann, U.; Seibert, P.; et al. Determination of time- and height-resolved volcanic ash emissions and their use for quantitative ash dispersion modeling: The 2010 Eyjafjallajökull eruption. *Atmos. Chem. Phys.* **2011**, *11*, 4333–4351.

6. Poret, M.; Corradini, S.; Merucci, L.; Costa, A.; Andronico, D.; Montopoli, M.; Vulpiani, G.; Freret-Lorgeril, V. Reconstructing volcanic plume evolution integrating satellite and ground-based data: Application to the 23rd November 2013 Etna eruption. *Atmos. Chem. Phys. Discuss.* **2018**. [CrossRef]
7. Richards, M.S. Volcanic Ash Cloud Heights Using the MODIS CO2-Slicing Algorithm. Master's Thesis, University of Wisconsin, Madison, WI, USA, 6 January 2006.
8. Chang, F.-L.; Minnis, P.; Lin, B.; Khaiyer, M.M.; Palikonda, R.; Spangenberg, D.A. A modified method for inferring upper troposphere cloud top height using the GOES 12 imager 10.7 and 13.3 μm data. *J. Geophys. Res.* **2010**, *115*, D06208. [CrossRef]
9. Corradini, S.; Cervino, M. Aerosol extinction coefficient profile retrieval in the Oxygen A-band considering multiple scattering atmosphere. Test case: SCIAMACHY nadir simulated measurements. *J. Quant. Spectrosc. Radiat. Transf.* **2006**, *97*, 354–380. [CrossRef]
10. Dubuisson, P.; Frouin, R.; Dessailly, D.; Duforêt, L.; Léon, J.-F.; Voss, K.; Antoine, D. Estimating the altitude of aerosol plumes over the ocean from reflectance ratio measurements in the O_2 A-band. *Remote Sens. Environ.* **2009**, *113*, 1899–1911. [CrossRef]
11. Biondi, R.; Steiner, A.K.; Kirchengast, G.; Brenot, H.; Rieckh, T. Supporting the detection and monitoring of volcanic clouds: A promising new application of Global Navigation Satellite System radio occultation. *Adv. Space Res.* **2017**, *60*, 2707–2722. [CrossRef]
12. Pardini, F.; Burton, M.; de' Michieli Vitturi, M.; Corradini, S.; Salerno, G.; Merucci, L.; Di Grazia, G. Retrieval and intercomparison of volcanic SO_2 injection height and eruption time from satellite maps and ground-based observations. *J. Volcanol. Geotherm. Res.* **2017**. [CrossRef]
13. Merucci, L.; Zakšek, K.; Carboni, E.; Corradini, S. Steeoscopic estimation of volcanic cloud-top height from two geostationary satellites. *Remote Sens.* **2016**, *8*, 206. [CrossRef]
14. Winker, D.M.; Liu, Z.; Omar, A.; Tackett, J.; Fairlie, D. CALIOP observations of the transport of ash from the Eyjafjallajökull volcano in April 2010. *J. Geophys. Res.* **2012**, *117*, D00U15. [CrossRef]
15. Prata, A.J.; Turner, P.J. Cloud top height determination from the ATSR. *Remote Sens. Environ.* **1997**, *59*, 1. [CrossRef]
16. Mims, S.R.; Kahn, R.A.; Moroney, C.M.; Gaitley, B.J.; Nelson, D.L.; Garay, M.J. MISR stereo heights of grassland fire smoke plumes in Australia. *IEEE Trans. Geosci. Remote Sens.* **2010**, *48*, 25–35. [CrossRef]
17. Nelson, D.L.; Garay, M.J.; Kahn, R.A.; Dunst, B.A. Stereoscopic Height and Wind Retrievals for Aerosol Plumes with the MISR INteractive eXplorer (MINX). *Remote Sens.* **2013**, *5*, 4593–4628. [CrossRef]
18. Flower, V.J.; Kahn, R.A. Assessing the altitude and dispersion of volcanic plumes using MISR multi-angle imaging from space: Sixteen years of volcanic activity in the Kamchatka Peninsula, Russia. *J. Volcanol. Geotherm. Res.* **2017**, *337*, 1–15. [CrossRef]
19. Zakšek, K.; Hort, M.; Zaletelj, J.; Langmann, B. Monitoring volcanic ash cloud top height through simultaneous retrieval of optical data from polar orbiting and geostationary satellites. *Atmos. Chem. Phys.* **2013**, *13*, 2589–2606. [CrossRef]
20. Corradini, S.; Montopoli, M.; Guerrieri, L.; Ricci, M.; Scollo, S.; Merucci, L.; Marzano, F.S.; Pugnaghi, S.; Prestifilippo, M.; Ventress, L.; et al. A multi-sensor approach for the volcanic ash cloud retrievals and eruption characterization. *Remote Sens.* **2016**, *8*, 58. [CrossRef]
21. De Michele, M.; Raucoules, D.; Arason, Þ. Volcanic plume elevation model and its velocity derived from Landsat 8. *Remote Sens. Environ.* **2016**, *176*, 219–224. [CrossRef]
22. Ferlito, C.; Bruno, V.; Salerno, G.; Caltabiano, T.; Scandura, D.; Mattia, M.; Coltorti, M. Dome-like behavior at Mt. Etna: The case of the 28 December 2014 South East Crater paroxysm. *Sci. Rep.* **2017**, *7*, 5361. [CrossRef]
23. Patanè, D.; Aiuppa, A.; Aloisi, M.; Behncke, B.; Cannata, A.; Coltelli, M.; Di Grazia, G.; Gambino, S.; Gurrieri, S.; Mattia, M.; et al. Insights into magma and fluid transfer at Mount Etna by a multi parametric approach: A model of the events leading to the 2011 eruptive cycle. *J. Geophys. Res.* **2013**, *118*, 1–21. [CrossRef]
24. Behncke, B.; Branca, S.; Corsaro, R.A.; De Beni, E.; Miraglia, L.; Proietti, C. The 2011–2012 summit activity of Mount Etna: Birth, growth and products of the new SE crater. *J. Volcanol. Geotherm. Res.* **2014**, *270*, 10–21. [CrossRef]
25. Spampinato, L.; Sciotto, M.; Cannata, A.; Cannavò, F.; La Spina, A.; Palano, M.; Salerno, G.G.; Privitera, E.; Caltabiano, T. Multiparametric study of the February–April 2013 paroxysmal phase of Mt. Etna New South-East crater. *Geochem. Geophys. Geosyst.* **2015**, *16*, 1932–1949. [CrossRef]

26. Corradini, S.; Guerrieri, L.; Lombardo, V.; Merucci, L.; Musacchio, M.; Prestifilippo, M.; Scollo, S.; Silvestri, M.; Spata, G.; Stelitano, D. Proximal monitoring of the 2011-2015 Etna lava fountains using MSG-SEVIRI data. *Geosciences* **2018**, *8*, 140. [CrossRef]
27. Sellitto, P.; di Sarra, A.; Corradini, S.; Boichu, M.; Herbin, H.; Dubuisson, P.; Sèze, G.; Meloni, D.; Monteleone, F.; Merucci, L.; et al. Synergistic use of Lagrangian dispersion and radiative transfer modelling with satellite and surface remote sensing measurements for the investigation of volcanic plumes: The Mount Etna eruption of 25–27 October 2013. *Atmos. Chem. Phys.* **2016**, *16*, 6841–6861. [CrossRef]
28. Storey, J.; Choate, M.; Lee, K. Landsat-8 operational land imager on-orbit geometric calibration and performance. *Remote Sens.* **2014**, *6*, 11127–11152. [CrossRef]
29. Knight, E.; Kvaran, G. Landsat-8 operational land imager design, characterization, and performance. *Remote Sens.* **2014**, *6*, 10286–10305. [CrossRef]
30. Leprince, S.; Barbot, S.; Ayoub, F.; Avouac, J.P. Automatic and precise orthorectification, coregistration, and subpixel correlation of satellite images, application to ground deformation measurements. *IEEE Trans. Geosci. Remote Sens.* **2007**, *45*, 6. [CrossRef]
31. The NCEP/NCAR Reanalysis Project at the NOAA/ESRL Physical Sciences Division. Available online: http://www.esrl.noaa.gov/psd/data/reanalysis/reanalysis.shtml (accessed on 1 December 2017).
32. Corradini, S.; Merucci, L.; Prata, A.J.; Piscini, A. Volcanic ash and SO_2 in the 2008 Kasatochi eruption: Retrievals comparison from different IR satellite sensors. *JGR* **2010**, *115*, D00L21. [CrossRef]

Article

Chronology of the 2014–2016 Eruptive Phase of Volcán de Colima and Volume Estimation of Associated Lava Flows and Pyroclastic Flows Based on Optical Multi-Sensors

Norma Dávila [1], Lucia Capra [2,*], Dolors Ferrés [2], Juan Carlos Gavilanes-Ruiz [3] and Pablo Flores [4]

1 Laboratorio de Ciencia y Tecnología de Información Geográfica, Facultad de Geografía, Universidad Autónoma del Estado de México, Toluca 50100, Estado de México, Mexico; nadavilah@uaemex.mx

2 Centro de Geociencias, Universidad Nacional Autónoma de México, Juriquilla, Querétaro 76230, Mexico; dferres@igeofisica.unam.mx

3 Facultad de Ciencias, Universidad de Colima, Colima 28045, Mexico; gavilan@ucol.mx

4 Estación Recepción México (ERMEX), 22a Zona Militar SEDENA, Sta Ma. Rayón 52360, Estado de México, Mexico; pflores@agentetecnico.com

* Correspondence: lcapra@geociencias.unam.mx; Tel.: +52-55-5623-4104

Received: 7 March 2019; Accepted: 17 April 2019; Published: 16 May 2019

Abstract: The eruption at Volcán de Colima (México) on 10–11 July 2015 represents the most violent eruption that has occurred at this volcano since the 1913 Plinian eruption. The extraordinary runout of the associated pyroclastic flows was never observed during the past dome collapse events in 1991 or 2004–2005. Based on *Satellite Pour l'Observation de la Terre* (SPOT) and *Earth Observing-1* (EO-1) ALI (Advanced Land Imager), the chronology of the different eruptive phases from September 2014 to September 2016 is reconstructed here. A digital image segmentation procedure allowed for the mapping of the trajectory of the lava flows emplaced on the main cone as well as the pyroclastic flow deposits that inundated the Montegrande ravine on the southern flank of the volcano. Digital surface models (DSMs) obtained from SPOT/6 dual-stereoscopic and tri-stereopair images were used to estimate the volumes of some lava flows and the main pyroclastic flow deposits. We estimated that the total volume of the magma that erupted during the 2014–2016 event was approximately 40×10^7 m^3, which is one order of magnitude lower than that of the 1913 Plinian eruption. These data are fundamental for improving hazard assessment because the July 2015 eruption represents a unique scenario that has never before been observed at Volcán de Colima. Volume estimation provides complementary data to better understand eruptive processes, and detailed maps of the distributions of lava flows and pyroclastic flows represent fundamental tools for calibrating numerical modeling for hazard assessment. The stereo capabilities of the SPOT6/7 satellites for the detection of topographic changes and the and the availability of EO-1 ALI imagery are useful tools for reconstructing multitemporal eruptive events, even in areas that are not accessible due to ongoing eruptive activity.

Keywords: Volcán de Colima; lava flow volume estimation; pyroclastic flows; SPOT; EO-1 ALI

1. Introduction

Volcán de Colima is one of the most active volcanoes in Mexico (Figure 1a) [1–3]. Before 2015, the last major volcanic crisis occurred in 2004–2005 and was characterized by several episodes of dome growth and collapse accompanied by the emplacement of block-and-ash flow (BAF) deposits that reached up to 7 km from the volcano's summit [4,5]. The July 2015 eruption represented an extraordinary episode; a fast-growing dome collapsed and generated two major BAFs on the 10th and

11th of July and a maximum runout of 10.5 km [6,7]. No similar runout for BAFs has even been observed in previous dome collapse events or in the stratigraphic record. The only BAFs that have reached similar distances correspond to the deposits associated with the Soufriére-type eruptive phase that preceded the 1913 Plinian eruption [8]. The unexpected 2015 scenario provides evidence supporting the need to revise hazard assessments associated with pyroclastic flows from dome collapses (i.e., [5,9]). Rigorous estimations of the volumes and inundated areas of the lava flows and pyroclastic flow deposits are thus fundamental, especially for numerical model calibration.

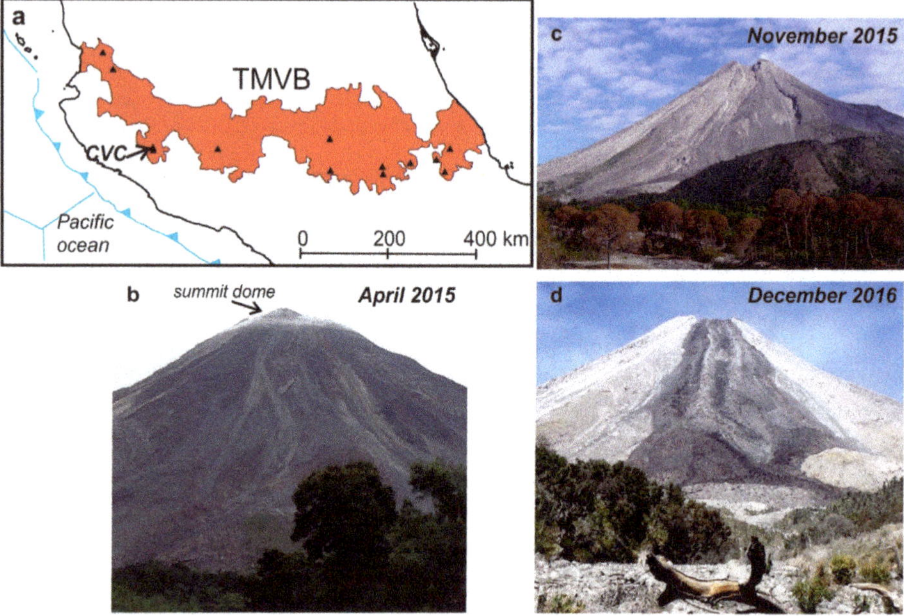

Figure 1. (**a**) Sketch map of the Trans-Mexican Volcanic Belt (TMVB) showing the location of the Colima Volcanic Complex (CVC). (**b**) Picture of the cone taken in April 2015 in which the summit dome is clearly visible. (**c**) Panoramic view of the cone after the 10–11 July 2015 eruption. Note the channel eroded by the pyroclastic flow that was subsequently filled by a lava flow. (**d**) Picture of the southern flank of the volcano in which the lava flows emplaced in 2016 are clearly visible.

Satellite remote sensing data, such as those collected from synthetic aperture radar (SAR) and optical, multi-, hyperspectral and thermal images, have been largely used for near real-time data acquisition regarding ongoing eruptive activity to define the locations of eruptive events, lava flow inundated areas, volumes and discharge rates [10–12], which are key factors for hazard assessment and numerical simulations [13,14]. Examples can be found from Hawaii, the Etna volcano, the lava fields in Iceland [15,16] and, more recently, the Sinabung volcano (Sumatra), for which the dome growth and pyroclastic flow deposit distributions and their magnitudes have been defined for the 2013–2015 eruptive crises [17]. Thermal remote sensing represents an optimal tool for lava flow discharge-rate estimation relative to optical products for which cloud coverage frequently impedes object observation, and is generally the same for volcanoes in tropical environments. Additionally, the lava discharge rates retrieved using thermal images agree well with those measured with ground-based methods, as recently demonstrated for the 2014–2015 eruption at Holuhraun, Iceland [18]. Finally, in recent years, SAR interferometry radar techniques (InSAR) and time series analysis have been used to identify deformation areas and surficial changes linked to lava flow deposits after major eruptions to determine their distributions, extrusion rates and thicknesses [19–22].

A broad variety of remote sensing data are freely available, including Advanced Spaceborne Thermal Emission and Reflection Radiometer (ASTER), Earth Observing-1 (EO-1) Advanced Land Imager (ALI), Landsat and Copernicus constellation (i.e., Sentinel 1 A-B and Sentinel 2) data. All those data are provided at a medium spatial resolution (between approximately 10 and 30 m) including in the thermal bands (i.e., the ASTER-TIR channel, 90 m ground resolution). In Mexico, Satellite Pour l'Observation de la Terre (SPOT) constellation data have been freely distributed since 2004 under an agreement between academic institutions and the Station of Telemetry Reception in Mexico (Estación de Recepción México-ERMEX), which is under the command of the Office for National Defense. SPOT data utilizes stereo and tri-stereoscopic capabilities to generate digital surface models (DSMs) with very high resolutions that allow for the identification of topographic changes after a major volcanic eruption. Thus, accessing the SPOT data represents an invaluable opportunity for performing multitemporal remote sensing analyses of natural phenomena at more detailed scales in Mexico.

In this paper, we assess the capabilities of stereo and tri-stereo SPOT6/7 images as well as EO-1 (ALI) images for the mapping of lava flows and pyroclastic flow deposits emplaced during the 2014–2016 eruptive phase of Volcán de Colima, including their volume estimations. We also assess the usefulness of the data for complementing the interpretation of the eruptive mechanism and the volume of the magma involved and for improving the calibration of numerical modeling of lava flows and granular flows. This work represents the first effort to apply remote sensing to the reconstruction of an eruption of an active Mexican volcano.

2. Chronology of the Eruptive Phases Prior to, During and After the 10–11 July 2015 Climatic Event

After a period of a relative calm since the last dome destruction period in early 2013, in May 2014, renewed effusive activity formed a lava flow on the W flank (WLF) of the Volcán de Colima during the extrusion of a new dome that, in late September, overspilled the crater with the emplacement of a lava flow on the SW slope of the volcano (SWLF) that advanced 2.2 km from the summit. On November 21, 2014, an ash plume rose ~7 km above the crater, and from this plume, a 3-km long pyroclastic flow descended along the San Antonio ravine [23]. In January 2015, several explosions caused the partial destruction of the dome and generated a 3-km ash plume that dispersed ash towards the NE [24]. In February 2015, the volcano produced several gas-and-ash plumes per day that rose to altitudes of 5.5–7.3 km (above sea level) [25]. Similar activity continued through the following months. In May, a thermal anomaly was detected inside the crater [26] and suggested a new dome extrusion phase that, by the beginning of June, was overflowing the crater (Figure 1b). In early July, two main lava flows were rapidly advancing (a few hundred m); one was on the northern flank, and the other (the larger of the two) was on the southern slope. During the days of July 7–9, gradual increases in the frequency of ash plumes, rock falls and small BAFs were observed [27]. By July 10, a lava flow had reached ~700 m down the southern flank (Figure 2a).

On 10 July, without any detected precursory activity, the summit dome collapsed, which generated a pyroclastic flow that emplaced along the Montegrande ravine up to approximately 8 km from the volcano summit. On the morning of 11 July, a second event, which was probably associated with the collapse of a new, fast-growing dome, emplaced a pyroclastic flow that traveled up to 10.5 km from the volcano along the same ravine and caused important damage to the vegetation alongside the channel (Figure 2b) [6,7,28,29]. This second pyroclastic flow promoted the partial failure and erosion of the upper portion of the crater leaving a v-shaped scarp (Figure 1c). The same day, a new lava flow began to flow out from this scarp, and by August 6th, it reached a distance of 2.5 km from the summit [6]. During the next months, the activity was characterized by small explosions and the emplacement of pyroclastic flows with a maximum runout of 2.5 km. In September 2016, a new lava flow filled the v-shaped scarp that was formed during the dome collapse activity and flowed along the southern flank to reach a distance of 2.2 km in late October (Figure 1d) [30,31]. Since then, the volcano has decreased its activity, which has included only sporadic explosions. Based on the above reconstruction, the last eruptive phase at Volcán de Colima began at the end of 2014, culminated in July 2015 and closed after the emplacement of the September 2016 lava flow.

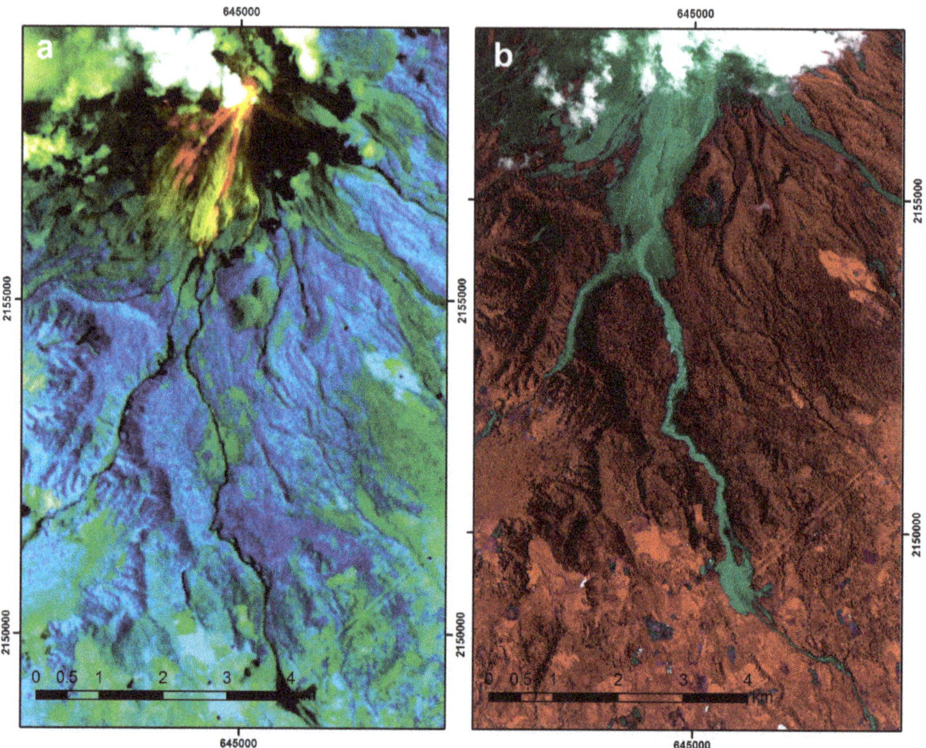

Figure 2. (**a**) Earth Observing-1 (EO-1) Advanced Land Imager (ALI) RGB composite (bands: 7, 4, 2, spatial resolution: 30 m) image taken on 10 July 2015 a few hours before the dome collapse occurred at 20:17 LT (Local Time). Incandescence is clearly visible on the southern sector of the cone (yellow-reddish colored area). (**b**) 2015-Satellite Pour l'Observation de la Terre (SPOT) composite image (bands: 4, 3, 2, spatial resolution 6 m) acquired on 25 July 2015 in which the area affected by the emplacement of the pyroclastic flows is clearly visible (gray area). (Map Projection: Universal Transverse Mercator—UTM).

3. Data

3.1. SPOT 6/7 Constellation

The SPOT constellation is one of the few satellite sensor systems with the stereoscopic and tri-stereoscopic capabilities to generate DSMs with very high pixel resolution and coverage over large areas. Specifically, SPOT-6 and SPOT-7 are two twin satellites orbiting at an altitude of 694 km that were launched September 2012 and February 2014, respectively, and have a swath coverage of 60 km × 60 km at nadir. SPOT-6 satellite and SPOT-7 represent the service continuity of SPOT- 4 and SPOT-5 satellites (launched 1998 and 2002), and they provide a revisit on ground each 26 days with a total coverage of 6 million km^2 per day.

As mention before, the SPOT data were acquired from the Station of Telemetry Reception in Mexico (Estación de Recepción México—ERMEX) through an inter-institutional agreement for delivering SPOT6 and 7 panchromatic (resolution 1.5 m) and multispectral (resolution 10 m) modes: blue: 0.450–0.520 µm, green: 0.530–0.590 µm, red: 0.625–0.695 µm and near infrared: 0.760–0.890 µm. Table 1 provides details of the images used here for stereoscopic processes and their acquisition dates.

Table 1. SPOT6 data used to generate the digital surface models (DSMs).

Satellite	Date	Mode	Orientation Angle	Incidence Angle	Image Resolution	DSM Resolution
SPOT6	28/11/2014	Panchromatic with tri-stereo capacity *	+169.66°	+19.08°	1.5 m	10 m
SPOT6	28/11/2014	Panchromatic with tri-stereo capacity	+74.31°	+8.22°	1.5 m	
SPOT6	28/11/2014	Panchromatic with tri-stereo capacity	+42.95°	+14.05°	1.5 m	
SPOT6	27/07/2015	Panchromatic with stereo capacity-duo stereo		+16.4°	1.5 m	21 m
SPOT6	27/07/2015	Panchromatic with stereo capacity-duo stereo		+16.8°	1.5 m	
SPOT6	11/04/2017	Panchromatic with stereo capacity-duo stereo	+306.71°	+19.22°	1.5 m	7.5 m
SPOT6	29/04/2017	Panchromatic with stereo capacity-duo stereo	+250.10°	+20.92°	1.5 m	

* This image was used for the eruptive temporal sequence (see Section 3.2).

The SPOT stereoscopy capability consists of illuminating the same location from different points of view to create a stereoscopic view. The images are captured along the same orbit with customizable positions and orientations of the target (Figure 3).

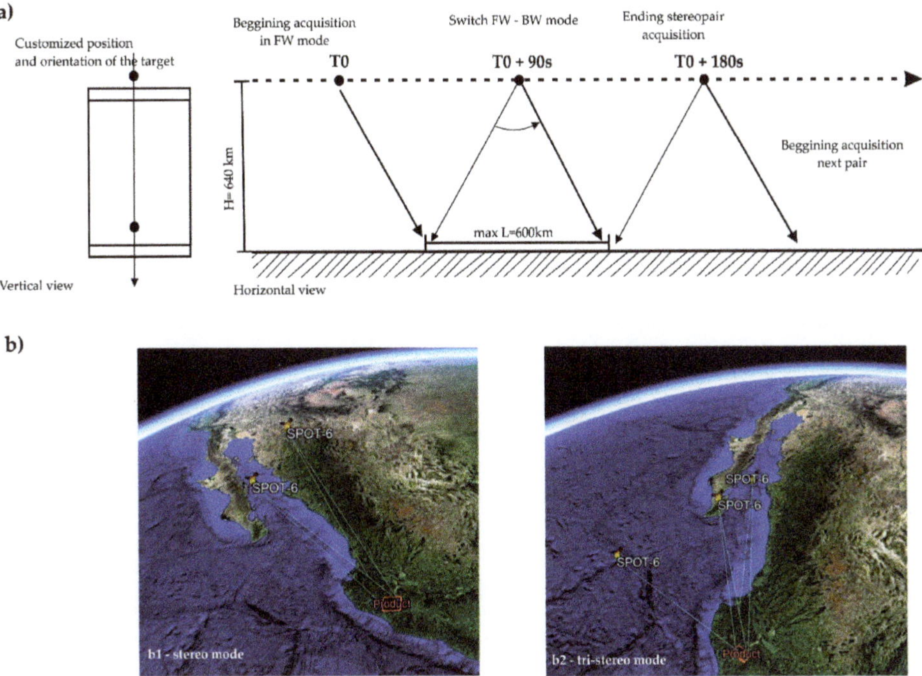

Figure 3. (**a**) Stereoscopic and tri-stereo SPOT configuration capabilities. (**b**) Google Earth preview of the SPOT satellite configurations in stereo and tri-stereo modes. FW, forward acquisition; BW backward acquisition.

Here, we used the Level 1A data. All images were acquired with a cloud cover until 25% (according with the technical acquisition procedure), and these clouds were located in some areas over the volcano. Therefore, after the stereoscopy process, a mask was applied to remove the cloud cover because, although the software attempts to triangulate distal data to compensate for areas without height information, the final height measurements could have been wrong.

SPOT6/7 works over a simple stereoscopy configuration, termed "agile view", which implies a customized position and orientation of the observed target. The sensor begins with forward acquisition (FW), switches to backward acquisition (BW) after 90 s and repeats this process once again and ends the stereopair acquisition after 180 s. Therefore, each line of the image is acquired twice with a 90 s delay. To acquire tri-stereo full coverage, an additional nadir acquisition is performed with 30° as the maximum incidence angle [32] (Figure 3).

3.2. Remote Sensing Data (EO-1 (ALI) and SPOT 6/7) Used for the Interpretation of the Eruptive Temporal Sequence

SPOT6/7 and EO-1 (ALI) images were used to support the chronology of the eruptive phase during the period of 2014–2016 (Table 2). The technical characteristics of the SPOT 6/7 satellites were mentioned above. ALI is one of the instruments onboard the EO-1 satellite. This is a multispectral sensor of 10 bands: R, G, B, VNIR, SWIR (30 m of resolution) and Panchromatic (10 m resolution), the satellite provides a standard swath around 37 km and length of 42 km (and it is possible to get increased length of 185 km); and a revisit on ground each 26 days. The EO-1 was part of NASA's project "Volcano Sensor Web" [33], focused on monitoring volcanic activity in selected areas including the Kilauea volcano, Iceland's Eyjafjallajökull and Anak Krakatau Volcano; EO-1 was used principally for tracking lava flows, discharge-rate estimation relative and thermal remote sensing for hotspot detection [33–36].

Table 2. Remote Sensing and SPOT data used for the eruptive temporal sequence.

Sensor	Date	Spectral Mode	Pixel Resolution (m)	Level Acquisition
EO-1 (ALI)	15/07/2015	Multispectral	10	1B
SPOT/6	25/07/2015	Panchromatic/Multispectral	1.5 10	1A
SPOT/7	13/04/2017	Multispectral	6	1A

All SPOT images were acquired in L1-A level, EO-1 (ALI) image in L1-B format. Radiometric corrections were applied using the "Apply Gain and Offset" algorithm and FLAASH atmospheric correction module of the ENVI 5.3 software. These corrections consist of removing atmospheric distortions due to the presence of water vapor and aerosols based on estimations of the correct wavelength positions of the bands. To achieve an accurate coregistration of each DSM, a 5-m DSM based on light detection and ranging data (LIDAR) was used as a spatial reference. The LIDAR data were acquired in 2012 (2012-DSM) and obtained from the Mexican National Institute of Statistics and Geography [37]. Moreover, to improve the visual interpretation of the images, spectral enhancement was applied to maximize the contrast of the data by applying a non-linear contrast stretch-enhanced RGB composite image [38]. Specifically, from the 2015 SPOT image, a 2.5 m resolution image was obtained with a fusion process between the 10 m resolution multispectral image and the panchromatic data (Figure 2b), which, in combination with the E0-1 (ALI) data, was used to better define the spatial distribution of the lava flows prior to and after the climatic activity of July 2015 (Figure 2).

4. Methods

The work flow applied for the volume estimation of the lava and pyroclastic flows is shown in Figure 4, which is described in the sub-sections below.

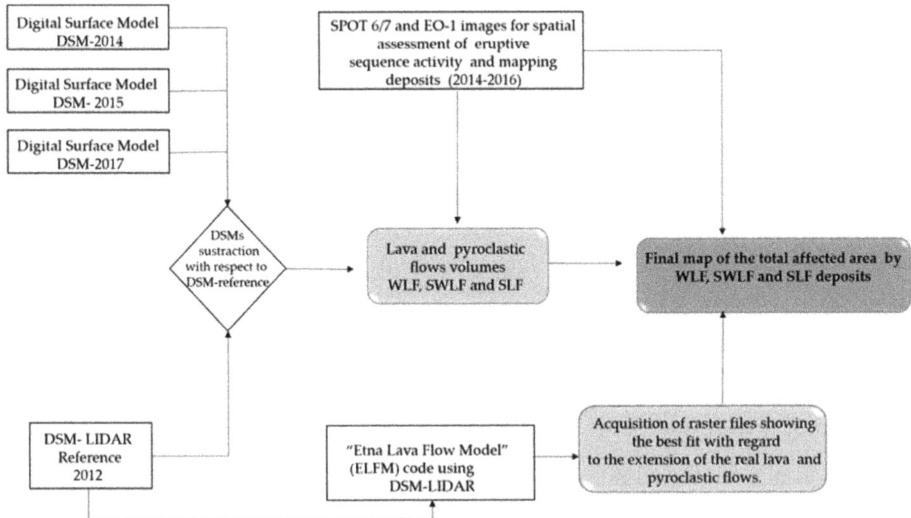

Figure 4. Work flow applied for volume estimations of the lava and pyroclastic flows linked to the eruptive phase (2014–2016). DSM-LIDAR, digital surface model light detection and ranging data; WLF, lava flow on the W flank; SWLF, lava flow on the SW slope; SLF, south lava flow.

4.1. Digital Surface Models (DSM) Processing

To perform the digital stereoscopy processing, we used the IMAGINE Photogrammetry 2015 software. This is a specialized module of ERDAS IMAGINE that focuses on high-accuracy full photogrammetry processing, such as digital terrain model generation, full analytical triangulation and orthophoto-mosaicking production. We used an intuitive interface with a linear workflow that was based on the model sensors that were previously defined for each aircraft and satellite sensor. In accordance with the SPOT image, the rational polynomial coefficients (RPC) file contains all of the orbital ephemeris coefficients required to resolve the software model and thus acquire the point cloud needed to generate the final DSM. To improve the absolute horizontal (x, y) position, an ortho-rectified photo mosaic with a pixel resolution of 2 m was used as a reference. This mosaic was acquired from the National Institute of Geography and Statistics (INEGI) and was, therefore, the highest quality georeference in Mexico. Approximately 25 and 30 ground control points (GCPs) were settled for each generated DSM. Twenty tie points were also selected to increase the spatial correlation between the stereo and tri-stereo images. Finally, three DSMs were extracted using the WGS 84 datum and the Universal Transversal of Mercator (UTM) projection with a horizontal resolution between 7 m and 21 m (Table 1). From this point on, the three DSMs will be abbreviated as follows: 2014-DSM, 2015-DSM, and 2017-DSM.

To evaluate the DSM accuracies, the root mean squared error (RMSE) was calculated for each DSM. The highest National Geodesic Network accuracy was used as a spatial reference; this network is a collection constituted by more than 100,000 geodetic control points distributed over all territory since 1978 (Table 3). The 2015- and 2017-DSMs have RMSEs below 1 m; in contrast, the 2014-DSM extracted from the tri-stereo 2014 data has a RMSE of 1.29 m, which represents the lowest accuracy among the three DSMs extracted with the stereoscopic techniques. We would have expected a higher vertical accuracy of the DSMs extracted from the tri-stereo images due to the tri-stereoscopic capacity, which would reduce the number of voids caused by the high topography in volcanic areas. However, this result can be explained by the small angle formed between the triplet image acquisitions (+19.08°/+8.22°/+14.05°) with respect to an optimal tri-stereo configuration of −15°–20°/±0°/+15°+20°.

Table 3. Root mean squared error (RMSE) estimation based on the National Geodesic Network of Mexico.

DSM	RMS (m)	S2	S	Number of Geodesic Stations (NGRS) Considered
Stereo-pair, 2017	0.281	0.07370655	0.27148951	63
Stereo-pair, 2015	0.803	0.2833703	0.532325379	42
Tri-stereo, 2014	1.2991	0.44845054	0.669664499	53

4.2. Volume Estimation of the Associated Lava Flows and Pyroclastic Flows

As mention before, a 5 m DSM from the LIDAR data acquired in 2012 (2012-DSM) was used as a pre-eruptive DSM topographic reference. For the estimation of the lava flow volumes, the 2012-DSM and 2017-DSM were used as pre- and post-topography references to obtain the volumes of the main lava flows that were emplaced on the cone during the 2014–1016 period.

The estimation was performed with raster operations using @ArcGis 10.2 (Table 4, Figure 5). The 2014-DSM was used only to evaluate the partial advance of the SW lava flow that, at the time of the 2014 SPOT image, was still advancing downwards (Figure 6a,b). Specifically, for the two lava flows that were emplaced in May and September 2014, the 2012-DSM was appropriated as the pre-event surface because, between 2012 and 2014, no major events affected the W or SW sectors of the cone besides the emplacement of small pyroclastic flows and lavas (Figure 5, black arrow). In contrast, for the lava flow that was emplaced in September 2016 on the S flank, the 2015-DSM could not be used as a pre-event surface because the cone was covered by clouds. Thus, the 2012-DSM was used, but this was still inappropriate because several morphological changes affected the southern slope of the cone during the activity in 2015 (Figure 1b,c). Other than this approximation, the raster operation performed between the 2012-DSM and 2017-DSM defined the lava flow extensions well (Figure 5), and the calculated volume for the south lava flow (SLF) is probably overestimated because, with respect to the 2012 topography, at least one lava flow lies below the 2016 lava flows. To calculate the error in the volume estimation (z-axis), the altitudes of control points on known morphological features that were not affected by the eruption were compared in both the 2012 and 2017 DSMs. The differences in the heights were on the order of +/−1.5 m. This value was multiplied by the area covered by each lava flow to obtain an estimate of the error of the volume calculation (Table 4). The volume of the 10–11 July 2015 BAF deposits was previously calculated [6,28], and this result is taken here to estimate the total magma volume involved in the 2014–2016 eruptive phase (Table 5).

Table 4. Estimated lava flow parameters.

Lava Flow	Date	Travel Distance (km)	Volume (m^3)	Area (m^2)	Error in Volume (m^3)	Slope
LAVA W	May 2014	1.2	7.5×10^6	2.3×10^5	$\pm 3.45 \times 10^5$	31°
LAVA SW	September 2014	2.2	4.7×10^6 (by Nov. 2014) 12.2×10^6 final	5.6×10^5	$\pm 7.8 \times 10^5$	27.8°
LAVA S	September 2016	2.2	16×10^6	8.1×10^5	$\pm 1.1 \times 10^6$	29.5°

Table 5. Estimated parameters for the 10–11 July 2015 pyroclastic flows emplaced along the Montegrande ravine.

Facies	Area (m^2)	Volume (m^3)	Error in Volume (m^3)	References
channel	8.6×10^5	4.2×10^6	$\pm 6.75 \times 10^5$	[6]
channel	9.4×10^5	5.8×10^6	$\pm 1 \times 10^6$	[28]

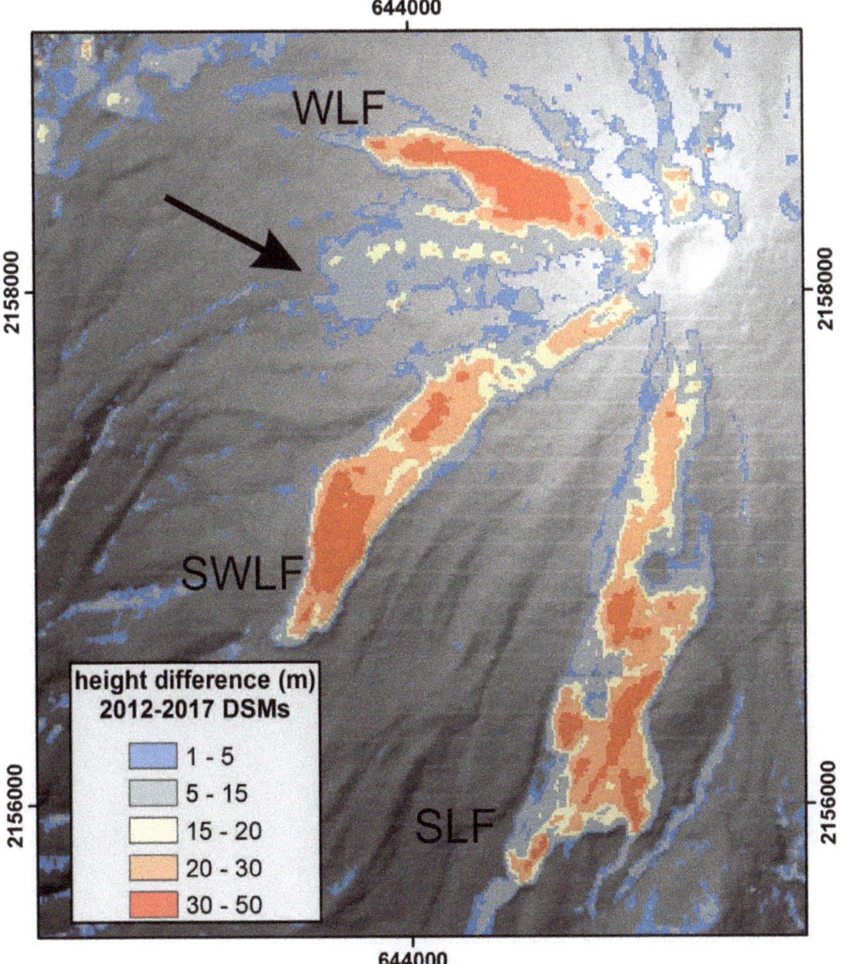

Figure 5. Raster (7.5 m resolution) resulting from the subtraction between the 2012 and 2017 DSMs showing the main changes related to lava flow emplacement during the 2014–2016 eruptive phase. The raster was classified for values > 1 m considering the z-error of −/+ 1.5 m. The black arrow points to pyroclastic and lava products that were emplaced in 2013. (Map Projection: Universal Transverse Mercator—UTM).

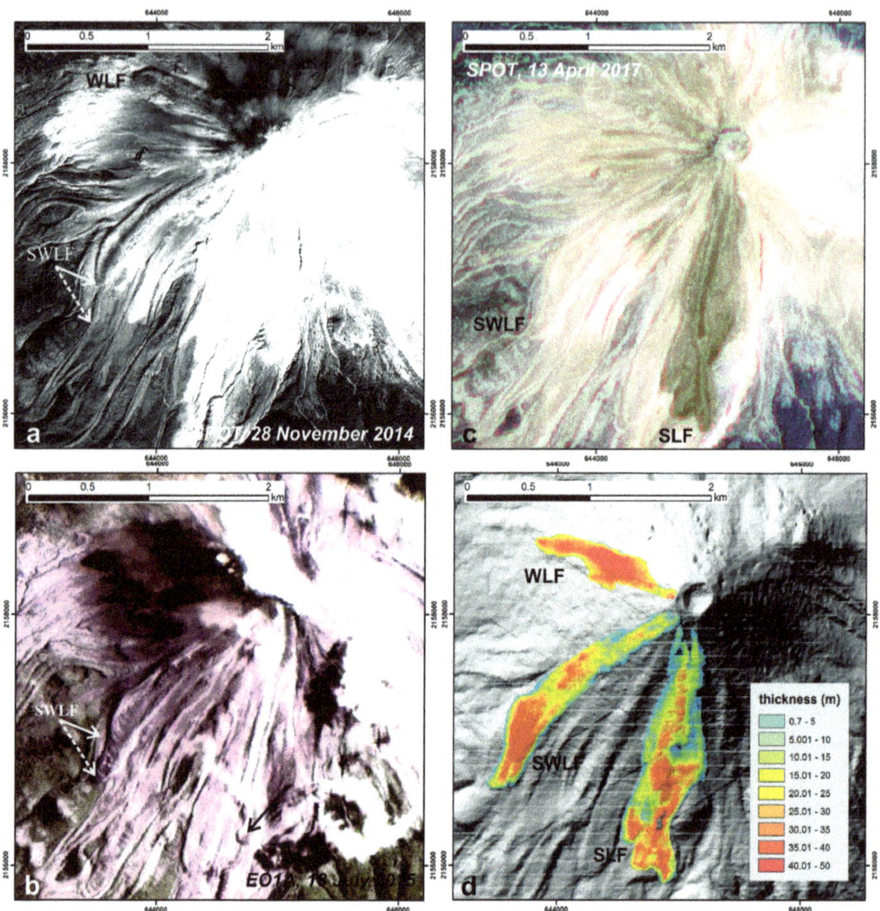

Figure 6. (**a**) 28 November 2014 SPOT image in which the 2014 lava flow areas are clearly visible. By comparing this image with the EO-1 (ALI) (bands 3, 2, 1) of 18 July 2015 (**b**), it is possible to observe that, by November 2015, the SW lava flow was still advancing (white arrows, the dotted line points to the final extent). In this image, the lava flow emplaced just after the July eruption is also visible (black arrow). This flow was subsequently covered by the S lava flow in 2016 as evident in the 13 April SPOT image (bands: 3, 4. 2) in (**c**). (**d**) Raster showing the lava flow thicknesses as obtained by the raster operation between the 2012 and 2017 DSMs (Map Projection: Universal Transverse Mercator- UTM).

4.3. Using the Etna Lava Flow Model Simulation Code

The previously described pre- and post-eruption DSMs constitute an excellent opportunity to calibrate lava flow simulation software for the construction of hazard maps of this volcanic phenomenon.

The calibration stage is essential for the simulation of different scenarios of lava flows (or for any type of volcanic process in general) and consists of reproducing the extensions and runouts of known lava flows, which requires knowledge of the topographies prior to emplacement.

Here, we used the Etna Lava Flow Model (ELFM) code [39], which has previously been applied to other Mexican volcanoes, including Popocatépetl and Ceboruco [40,41]. The ELFM is a probabilistic model that is similar to other types of lava flow models [42,43] and simulates different paths that lava might follow when emitted from a volcanic vent [39,44]. In addition to the digital elevation model of the volcano (DEM) and the coordinates of the vent, the input parameters for the ELFM are the following:

the maximum thickness of the flow (m), the possible maximum length that the lava can cross on the DEM (in the number of steps or pixels), and how the thickness of the flow varies during emplacement until its movement stops [39]. The software is based on Felpeto's algorithm [45] but incorporates improvements to solve the flow loop problem with a dynamic DEM, which makes it possible to establish the length and the height of each lava flow simulation. Moreover, these improvements allow for changes to the lava flow height over the flow path based on the distance via the use of different functions (i.e., constant, linear or logarithmic). This last parameter correlates with the viscosity and yield strength of the flow during its emplacement [46], which enables the simulation of more evolved lava flows compared with those of Etna (which are basaltic in composition) for which the software was originally created. Another variable parameter is the number of iterations in each run of the code. A value of 1000 iterations is commonly considered sufficient (i.e., no significant changes occur with greater numbers of iterations) as verified in previous studies [39,44,45,47].

In the above-mentioned research, it was necessary to reconstruct the paleo-topographies of historical lava flows because digital elevation models were not available prior to the emplacement of the flows.

This process, which is primarily performed manually, incorporates considerable error into the calibration and simulation processes because the results of simulations of lavas are highly dependent on the resolution of the DEM used [37], as repeatedly noted in simulations of other volcanic and geological flow phenomena [48–51]. For the case of Volcán de Colima, the September 2014 and September 2016 lava flows were reproduced on the existing DSM obtained from the 2012 INEGI LIDAR dataset [37], which was resampled to 10-m horizontal resolution.

The results of the ELFM were ASCII (see glossary) files that were processed in the ArcGIS 10.2® software according to the following steps: (a) transformation the ASCII files to raster files; (b) reclassification of each raster file into eight groups of values, excluding the pixels with values between 1 and 5 (which corresponded to 0.5% of the total iterations and were considered to be overestimations in the simulations [44]); and (c) transformation of the raster files into shape files (polygons) to calculate the inundated areas in each of the calibration runs and compare them with the real areas of the considered lava flows.

5. Results

5.1. Lava Flow Inundation Area and Volume Estimation

Three main lava flows were identified from the analysis of the SPOT images, and they are named here based on the directions of their emplacements as follows: May 2014 West Lava Flow (WLF), September 2014 South-West Lava Flow (SWLF), and September 2016 South Lava Flow (SLF) (Figure 6 and Table 4). Based on the raster operations, their volumes were estimated and ranged from a minimum of 7.5×10^6 m^3 for the WLF to a maximum of 16×10^6 m^3 for the SLF (Table 4). As previously mentioned, the volume estimation for the most recent SLF is probably overestimated because the 2012-DSM does not reflect all of the morphological modifications of the southern slope of the cone during and after the 10–11 July 2015 eruptive activity. Lava flows were emplaced over slopes of 29.5° for the WLF, 27.8° for the SWLF and 31° for the SLF and reached maximum distances of 1.2, 2.2 and 2.2 km, respectively (Table 4). Interestingly, in the SPOT images from 28 November 2014, the SWLF is still advancing because its extent does not correspond with the maximum runout visible in the 18 July 2015 EO-1 (ALI) and 2017 SPOT images (Figure 6a–c). Therefore, by 28 November 2014, the lava traveled 1.9 km (for a volume of 4.7×10^6 m^3 based on the 2012- and 2014-DSM raster operations; Table 4). Its final runout is 2.2 km for a total volume of 12.2×10^6 m^3 (Figure 6a,b). For this lava flow, the exact dates of the initial and final extrusion times are unknown. If its initial stage of extrusion was detected in September 2014, by 28 November (~75 days), an extrusion rate of ~0.7 m^3/s could be estimated, which is in the range of the effusion rate calculated by [52] for the same lava flow. If this estimation is correct, the lava flow probably stopped by March 2015, but no images were found to confirm this assumption.

In contrast, a more precise estimation could be performed for the SLF. A thermal anomaly related to the SLF was first reported on 27 September 2016, and the lava flow maximum runout was reported on October 19th [23,24]. Based on this information, an extrusion rate of ~8 m^3/s was estimated here, which is much higher than that of the SWLF but very similar to the extrusion rates calculated for lava flows in 2004 [53]. From the raster values of the lava flow bodies and based on the topographic profiles, the lava flows exhibit a maximum thickness of approximately 50 m towards their distal fronts (Figure 6d). The WLF shows a general convex profile with a clear blocky superficial texture, which contrasts with the SWLF, which shows a concave profile with a lateral levee (Figure 6; Figure 7). The SLV consists of three main lava bodies with concave profiles similar to the SWLF.

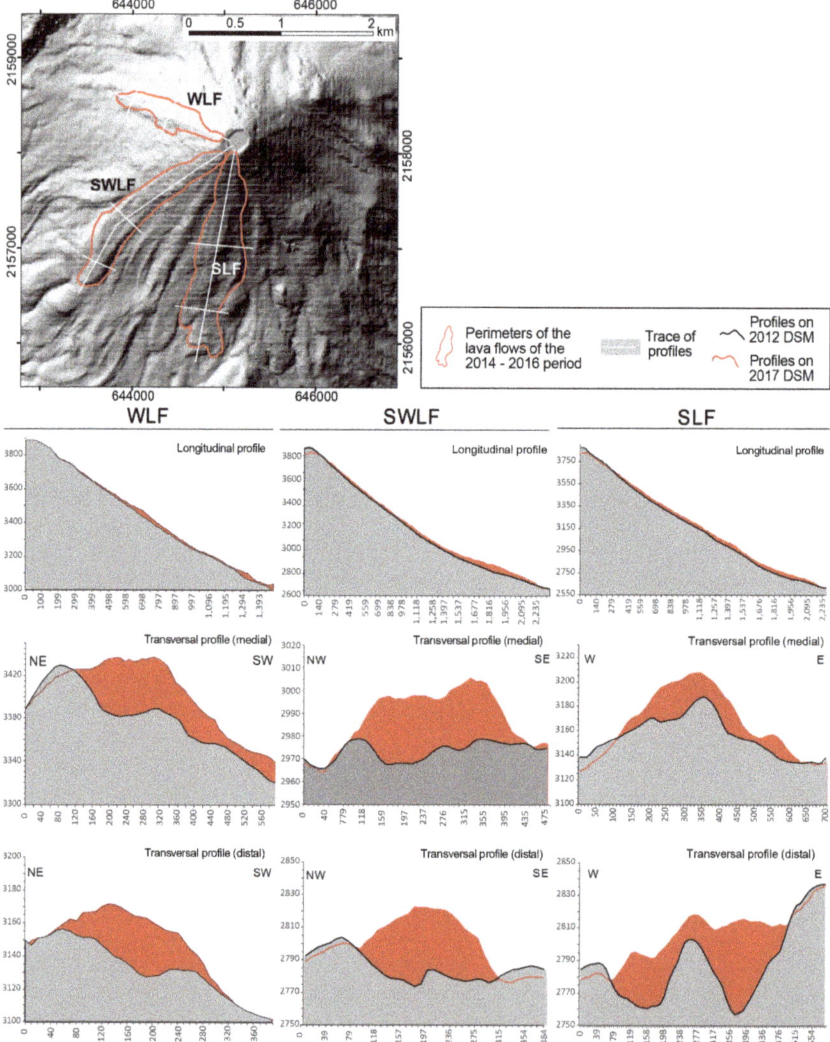

Figure 7. Longitudinal and transverse profiles of the WLF, SWLF and SLF described in the text. The transverse profiles were delineated in the middle part and at the distal end of the path of each lava flow. (Map Projection: Universal Transverse Mercator—UTM).

5.2. Block-and-Ash Flow Deposits: Inundation Limits, Volume and Thickness Variation

The 10–11 July 2015 collapse of the summit dome produced pyroclastic flows that inundated the Montegrande ravine (Figure 2). The 2015 SPOT product was segmented to define the affected area [6] (Figure 2b) and distinguish between the valley-confined facies that consist of massive, matrix-supported BAF deposits that are up to a 20 m thick with cm to m blocks imbedded in an ashy matrix, and the overbank facies that correspond to 1 to 3 m thick massive BAF deposits that are covered by centimetric layers of massive ash. These fine, homogenous ash layers on top of the overbank unit comprise the feature that allowed for its discrimination from the valley-confined facies during image segmentation. A volume of 4.2×10^6 m^3 was estimated for the valley-confined BAF based on the difference between the 2012-DSM and the 2015-DSM (Table 5). The dense rock equivalent (DRE) [54] was then estimated based on an andesitic lava density of 2.4 g/cm^3 and a BAF deposit density of 2 g/cm^3, yielding a DRE (magma) volume of approximately 3.5×10^6 m^3. [28] analyzed two 50-cm resolution stereopairs of Pleiade-1A satellite images acquired in April 2013 and January 2016. Based on photogrammetry of these two stereo pairs, these authors obtained two 1 m spatial resolution DEMs based on which a volume of 5.8×10^6 m^3 for the same valley-confined facies was estimated. Considering the errors (Table 5), both volume estimations are comparable.

5.3. Calibration of the ELFM at the Colima Volcano

Software calibration is a fundamental step prior to lava flow simulations of different eruptive scenarios for hazard evaluation. It consists of several test-error simulations and modification of the input parameters until the best fit of the area covered by the real lava flow is found. The 2014 SWLF and the 2016 SLF were reproduced on the 2012 DSM [37] after totals of 42 and 25 runs, respectively, using 1000 iterations in each run as a fixed parameter.

A fundamental aspect of the calibration was the selection of the exact possible vent of each of the lava flows. Tests were performed on eight different pairs of coordinates for the 2014 SWLF and three different vent position for the 2016 SLF, all of which were located in the proximal area of the lava flow polygons obtained by the raster subtraction (Figures 5 and 6d). Notably, the positions of the three emission centers used for each flow are separated by an average of 30 m.

For both lava flows and each of the vents selected, tests with linear and logarithmic variations of the thickness of the flow were performed and found that a linear increase in the thickness (value 1 in the software) over the path of the flows was the best choice for the lavas emplaced on high slopes; this is also the case for the 2014 and 2016 Colima lava flows (slopes between 28° and 31°, Table 2) in which the maximum thicknesses (35 to 50 m) are reached in the middle and distal portions of the total lengths (Figure 7).

The other input parameters needed for the ELFM were varied in each of the runs to search for the datasets that best fit the extents of the real lava flows. The maximum thickness of the flow was modified between 10 and 40 m, and the possible maximum length (in number of steps or pixels) that the lava could cross was varied between 300 and 550 steps. The best parameters for reproducing the Colima 2014 SWLF were a thickness of 30 m and 400 steps (Figure 8a). For the 2016 SLF, a thickness of 30 m and 370 steps were sufficient to reach the length of the observed lava flow (Figure 8b). A thickness value of 30 m was selected as the average value of the thickness of the lava flows. The maximum value of 50 m, which was obtained in the calculation of the thickness in the raster operations (Figure 4d), was not considered in the calibration simulations because this value was only obtained at specific points of the total area of each lava flow, as observed in the transverse profiles (Figure 7).

The selected simulations revealed a low percentage of underestimated areas (i.e., areas flooded by the real lava flows and not covered by the simulation) and fit with the full spatial extent of the reproduced lavas. However, in both cases, the simulations overestimated the flooded areas—by 40% in the case of the SWLF and 35% in the case of the SLF. This overestimation was probably due to the fact that the 2012 DSM does not account for changes at the summit of the cone that occurred between 2012 and 2014.

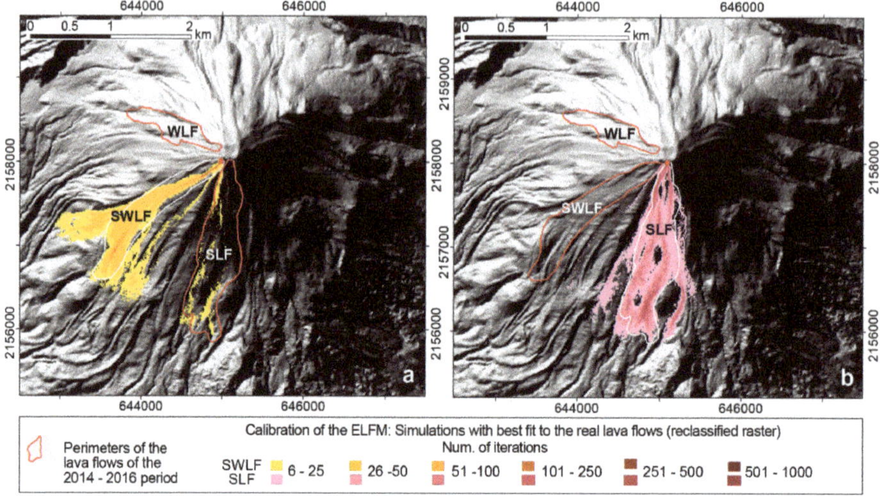

Figure 8. Results of the simulations showing the best fits with regard to the extensions of the real lava flows. The simulations were performed with 1000 iterations. The ASCII (see glossary) files resulting from the Etna Lava Flow Model (ELFM) software runs were transformed into raster files and later reclassified in a discrete number of groups of iterations as shown in the legend. (**a**) Reclassified raster of the SWLF simulation using 400 steps for the flow path, 30 m for the final thickness and a linear increase in the flow thickness. (**b**) Reclassified raster of the SLF simulation using 370 steps for the flow path, 30 m for the final thickness and a linear increase in thickness. (Map Projection: Universal Transverse Mercator- UTM).

Nevertheless, based on this calibration, a lava flow hazard evaluation could be extended to the entire cone, including other vents around the crater rim, for lavas with similar extensions and volumes, and also for other eruptive scenarios that could involve the emission of lava flows of greater length and volume.

6. Discussion

Using SPOT and EO-1 (ALI) data, we were able to map and estimate the volume of the deposits emitted during the 2014–2016 eruptive phase of Volcán de Colima, which is one of the most active volcanoes in Mexico. These data were freely available, and their temporality acquisition was optimal for defining the chronology of the eruption and the maximum extensions of the pyroclastic flows and lava flows, their volumes and their extrusion rates. Although products of higher resolution are available, such as the Pleiades 1A data (Pleiades-HR is composed by two-spacecraft constellation, optical high-resolution panchromatic -0.7 m- and multispectral -2.8 m-), most are prohibitively expensive, especially for institutions working in Latin America. Despite this limitation, the results presented here demonstrate that SPOT images can be a useful tool and provide a spatial resolution that is sufficient to obtain DSMs from which volume estimations can be retrieved. Moreover, these volume estimations agree with those obtained, for example, with DEMs based on Pleiades 1A data [28] for the BAF deposits (Table 5) and by direct observation of lava flows [52]. A map of the total affected area that distinguishes between the lava flows and pyroclastic deposits is presented here (Figure 9). This map is an invaluable tool that can be used to calibrate numerical simulations and upgrade the hazard map of Volcán de Colima, for which a similar scenario has not previously been considered [9]. Specifically, the calibration of the lava flow simulation software presented here benefited from the availability of the real topography prior to the event and the total distribution of the simulated lava flow. Based on the best-fit values obtained in this work, a range of values for each parameter can be established to simulate the emission of andesitic lava flows of a few kilometers in length from eruptive centers on the edges of central crater and to establish other sets of data to reproduce different scenarios of more voluminous lavas.

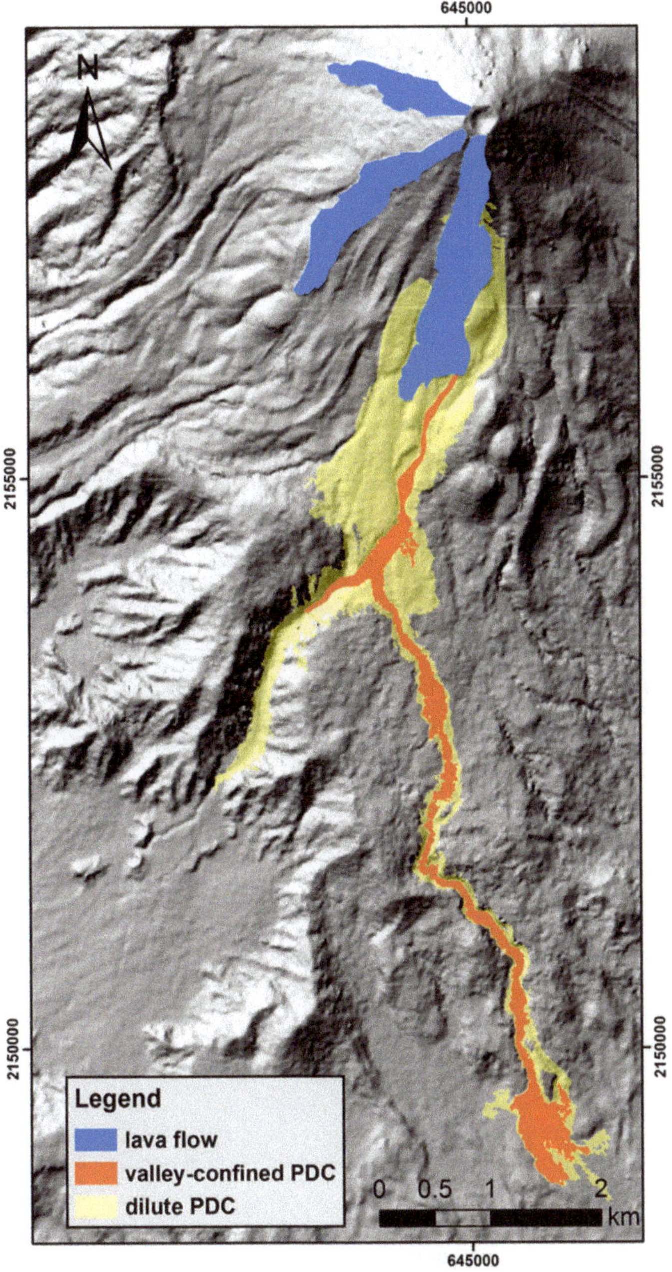

Figure 9. Map showing the distribution of lava flow deposits and PDC (Pyroclastic Density Currents) deposits associated with the 2014–2016 eruptive phase of Volcán de Colima. (Map Projection: Universal Transverse Mercator—UTM).

An additional result of this study relates to the eruptive behavior at Volcán de Colima. Based on the chronology of the effusive phases and the volume of the lava flows, the extrusion rates were estimated. The calculated extrusion rates spanned from a minimum of 0.7 to a maximum of 8 m^3/s, in agreement with the information reported in a previous study (i.e., [52]). The lower rate corresponds to the SWLF that emplaced over a lower slope gradient (27.8°) and with a lower magma volume (12.2×10^6 m^3), and the higher value corresponds to the SLV that emplaced over a 31° slope with a larger volume (16×10^6 m^3). The correlations between the slope gradient, magma volume and extrusion rates fit with the general behavior of lava flow [55], which again confirms the validity of the parameters calculated for the 2014–2016 lava flows.

Finally, based on the estimation of the total volume of the magma involved in the 2014–2016 eruptive phase, which was considered to be an anomalous scenario associated with historical activity (i.e., [6,7]), some considerations about the eruptive style can be presented. Considering the lava flows and pyroclastic flow deposits, we estimated a total magma volume of approximately 40×10^6 m^3. This value is much higher than previous activities associated with dome collapses, which involve BAF volumes on the order of 10^5 m^3 (i.e., [4,5]), but one order of magnitude lower than that of the 1913 Plinian eruption, which had a total volume of 3×10^8 m^3 [8]. Thus, even if the 2015 scenario was unexpected (primarily due to the maximum runout of the BAFs, which have usually extended up to 7 km during previous eruptive episodes), the magma volume in the magma chamber was at the lower limit of the estimation at which a Plinian or even sub-Plinian scenario could be expected based on numerical models [56]. A larger magma volume (10^8 m^3) is needed to generate an overpressure sufficient to trigger a Plinian scenario. This is an important result that helps to better understand the eruptive behavior of Volcán the Colima and supports the theoretical results obtained with numerical models.

7. Conclusions

This analysis focused on defining the chronology of the events, the affected area and the total magma volume involved in the eruption. As demonstrated here, the SPOT data were optimal for generating a first approximation of the magnitude of the eruption and producing a map with the distribution of the deposits. These are invaluable data for understanding a volcano's behavior and simulating possible future scenarios, especially considering the reduced accessibility of the majority of the area around an active volcano. Based on here presented results, the hazard assessments for the block-and-ash flows and lava flows at Volcán de Colima can be revised and improved. Additionally, based on numerical models [56], the estimated magma volume involved in the 2014–2016 eruptive phase was too low to trigger a Plinian scenario similar to the 1913 activity. A larger magma volume stored in the magma chamber would be needed, and, as observed in the precursory activity of the 1913 eruption, dome destruction would be accompanied by explosive events; no such events were observed during the 2015 activity.

Remote sensing is an invaluable tool for the rapid mapping of active volcanoes after major eruptions. As recently observed at Volcán de Fuego in 2018 (Guatemala), a quick mapping to evaluate damage can be realized and used for rescue activities in the few days following a disaster (i.e., Information Technology for Humanitarian Assistance, Cooperation and Action (ITHACA), http://www.ithacaweb.org/; United Nations Institute for Training and Research Operational Satellite Applications Programme (UNITAR-UNOSAT), www.unitar.org/unosat [57]). Subsequently, the same products can be used to define a map of the different deposits (lava flows and pyroclastic flow) that enable volume estimation, identify key features for numerical model calibration, and allow the construction of hazard maps.

Author Contributions: Conceptualization, L.C. and D.F.; Data curation, N.D.; Formal analysis, L.C.; Investigation, J.C.G.-R.; Methodology, N.D., L.C., D.F. and P.F.; Software, P.F.; Supervision, L.C.; Writing—review & editing, N.D., L.C., D.F. and J.C.G.-R.

Funding: This research was funded by the PAPIIT-DGAPA project IN105116 to Lucia Capra. DF was benefited by a postdoctoral fellowship from the Conacyt project PN-360 to LC.

Acknowledgments: We thank to Station of Telemetry Reception in Mexico (Estación de Recepción México-ERMEX) for having provided all stack of SPOT6/7 images. DF thanks Dr. Gianluca Groppelli, of the Istituto per la Dinamica dei Processi Ambientali (Milan, Italy), who provided access to ELFM software and advised the work of modeling lava flows. Finally, we thank to earthexplorer webserver (https://earthexplorer.usgs.gov/) for having provided EO-1 (ALI) image.

Conflicts of Interest: The authors declare no conflict of interest.

Glossary

ASCII	American Standard Code for Information Interchange
ASTER	Advanced Spaceborne Thermal Emission and Reflection Radiometer
BAF	block-and-ash flow
BW	switches to backward acquisition
DEM	Digital Elevation Model
DRE	dense rock equivalent
DSM	Digital Surface Model
ELFM	Etna Lava Flow Model (code)
EO-1 (ALI)	Earth Observing-1 ALI (on) Advanced Land Imager
ERMEX	Estación de Recepción México
FW	forward acquisition
GCP	ground control points
INEGI	National Institute of Geography and Statistics
InSAR	Interferometry Synthetic Aperture Radar
ITHACA	Information Technology for Humanitarian Assistance, Cooperation and Action
LIDAR	Light Detection and Ranging Data
PDC	Pyroclastic Density Currents
RMSE	root mean squared error
RPC	rational polynomial coefficients
SLF	south lava flow
SPOT	Satellite Pour l'Observation de la Terre (SPOT/6)
SWLF	South-West Lava Flow
UNITAR	United Nations Institute for Training and Research
UNOSAT	UNITAR Operational Satellite Applications Programme
UTM	Universal Transversal of Mercator
WLF	West Lava Flow

References

1. Luhr, J.F.; Carmichael, I.S.E. The Colima Volcanic Complex, México. *Contrib. Mineral. Petrol.* **1981**, *76*, 127–147. [CrossRef]
2. Medina, F. Analysis of the eruptive history of the Volcán Colima, México, 1560–1980. *Geof. Int.* **1983**, *22*, 157–178.
3. De la Cruz-Reyna, S. Random patterns of occurrence of explosive eruptions at Colima volcano, México. *J. Volcanol. Geotherm. Res.* **1993**, *55*, 51–68. [CrossRef]
4. Macias, J.L.; Saucedo, R.; Gavilanes-Ruiz, J.C.; Varley, N.; Velasco-Garcia, S.; Bursik, M.I.; Vargas-Gutierres, V.; Cortes, A. Flujos piroclásticos asociados a la actividad explosiva del volcán de Colima y perspectivas futuras. *GEOS* **2006**, *25*, 340–351.
5. Sulpizio, R.; Capra, L.; Sarocchi, D.; Saucedo, R.; Gavilanes, J.C.; Varley, N.R. Predicting the block-and-ash flow inundation areas at Volcán de Colima (Colima, Mexico) based on the present day (February 2010) status. *J. Volcanol. Geotherm. Res.* **2010**, *193*, 49–66. [CrossRef]
6. Capra, L.; Macías, J.L.; Cortés, A.; Dávila, N.; Saucedo, R.; Osorio-Ocampo, S.; Arce, J.L.; Gavilanes-Ruiz, J.C.; Corona-Chávez, P.; García-Sánchez, L.; et al. Preliminary report on the July 10–11, 2015 eruption at Volcán de Colima: Pyroclastic, density currents with exceptional runouts and volume. *J. Volcanol. Geotherm. Res.* **2016**, *310*, 39–49. [CrossRef]

7. Reyes-Dávila, G.; Arámbula-Mendoza, R.; Espinasa-Pereña, R.; Pankhurst, M.J.; Navarro-Ochoa, C.; Savov, I.; Vargas-Bracamontes, D.M.; Cortés-Cortés, A.; Gutiérrez-Martínez, C.; Valdés-González, C.; et al. Volcán de Colima dome collapse of July, 2015 and associated pyroclastic density currents. *J. Volcanol. Geotherm. Res.* **2016**, *320*, 100–106. [CrossRef]
8. Saucedo, R.; Macias, J.L.; Gavilanes, J.C.; Arce, J.L.; Komorowski, J.C.; Gardner, J.E.; Valdez, G. Corrigendum to Eyewitness, stratigraphy, chemistry, and eruptive dynamics of the 1913 Plinian eruption of Volcan de Colima, Mexico. *J. Volcanol. Geotherm. Res.* **2010**, *191*, 149–166. [CrossRef]
9. Capra, L.; Gavilanes-Ruiz, J.C.; Bonasia, R.; Saucedo-Giron, R.; Sulpizio, R. Re-assessing volcanic hazard zonation of Volcán de Colima, México. *Nat. Hazards* **2014**, *76*, 41–51. [CrossRef]
10. Harris, A.J.L.; Butterworth, A.L.; Carlton, R.W.; Downey, I.; Miller, P.; Navarro, P.; Rothery, D.A. Low-cost volcano surveillance from space: Case studies from Etna, Krafla, Cerro Negro, Fogo, Lascar and Erebus. *Bull. Volcanol.* **1997b**, *59*, 49–64. [CrossRef]
11. Harris, A.J.L. *Thermal Remote Sensing of Active Volcanoes, a User's Manual*; Cambridge University Press: Cambridge, UK, 2013; p. 736.
12. Ramsey, M.S.; Harris, A.J.L. Volcanology 2020: How will thermal remote sensing of volcanic surface activity evolve over the next decade? *J. Volcanol. Geotherm. Res.* **2013**, *249*, 217–233. [CrossRef]
13. Harris, A.J.L.; Rowland, S.K. FLOWGO: A kinematic thermo-rheological model for lava flowing in a channel. *Bull. Volcanol.* **2001**, *63*, 20–44. [CrossRef]
14. Wright, R.; Garbeil, H.; Harris, A.J.L. Using infrared satellite data to drive a thermo-rheological/stochastic lava flow emplacement model: A method for near-real-time volcanic hazard assessment. *Geophys. Res. Lett.* **2008**, *35*, L19307. [CrossRef]
15. Oppenheimer, C. Lava Flow Cooling Estimated from Landsat Thematic Mapper Infrared Data: The Lonquimay Eruption (Chile, 1989). *J. Geophys. Res.* **1991**, *96*, B13. [CrossRef]
16. Aufaristama, M.; Hoskuldsson, A.; Orn Ulfarsson, M.; Jonsdottir, I.; Thordarson, T. The 2014–2015 Lava Flow Field at Holuhraun, Iceland: Using Airborne Hyperspectral Remote Sensing for Discriminating the Lava Surface. *Remote Sens.* **2019**, *11*, 476. [CrossRef]
17. Pallister, J.; Wessels, R.; Griswold, J.; McCausland, W.; Kartadinata, N.; Gunawan, H.; Budianto, A.; Primulyana, S. Monitoring, forecasting collapse events, and mapping pyroclastic deposits at Sinabung volcano with satellite imagery. *J. Volcanol. Geotherm. Res.* **2018**. [CrossRef]
18. Bonny, E.; Thordarson, T.; Wright, R.; Höskuldsson, A.; Jónsdóttir, I. The Volume of Lava Erupted During the 2014 to 2015 Eruption at Holuhraun, Iceland: A Comparison Between Satellite- and Ground-Based Measurements. *J. Geophys. Res. Solid Earth* **2018**, *123*, 5412–5426. [CrossRef]
19. Dietterich, H.R.; Poland, M.P.; Schmidt, D.A.; Cashman, K.V.; Sherrod, D.R.; Espinosa, A.T. Tracking lava flow emplacement on the east rift zone of Kīlauea, Hawai'i, with synthetic aperture radar coherence. *Geochem. Geophys. Geosyst.* **2012**, *13*, 5. [CrossRef]
20. Schaefer, L.; Lu, Z.; Oommen, T. Post-Eruption Deformation Processes Measured Using ALOS-1 and UAVSAR InSAR at Pacaya Volcano, Guatemala. *Remote Sens.* **2016**, *8*, 73. [CrossRef]
21. McAlpin, D.B.; Meyer, F.J.; Gong, W.; Beget, J.E.; Webley, P.W. Pyroclastic Flow Deposits and InSAR: Analysis of Long-Term Subsidence at Augustine Volcano, Alaska. *Remote Sens.* **2016**, *9*, 4. [CrossRef]
22. Cando, M.; Martínez, A. Determination of Primary and Secondary Lahar Flow Paths of the Fuego Volcano (Guatemala) Using Morphometric Parameters. *Remote Sens.* **2019**, *11*, 727. [CrossRef]
23. GVP. Report on Colima (Mexico). In *Weekly Volcanic Activity Report, 19 November–25 November 2014*; Smithsonian Institution: Washington, DC, USA; US Geological Survey: Reston, VA, USA, 2014.
24. GVP. Report on Colima (Mexico). In *Weekly Volcanic Activity Report, 7 January–13 January 2015*; Smithsonian Institution: Washington, DC, USA; US Geological Survey: Reston, VA, USA, 2015.
25. GVP. Report on Colima (Mexico). In *Weekly Volcanic Activity Report, 18 February–24 February 2015*; Smithsonian Institution: Washington, DC, USA; US Geological Survey: Reston, VA, USA, 2015.
26. GVP. Report on Colima (Mexico). In *Weekly Volcanic Activity Report, 13 May–19 May 2015*; Smithsonian Institution: Washington, DC, USA; US Geological Survey: Reston, VA, USA, 2015.
27. GVP. Report on Colima (Mexico). In *Weekly Volcanic Activity Report, 8 July–14 July 2015*; Smithsonian Institution: Washington, DC, USA; US Geological Survey: Reston, VA, USA, 2015.

28. Macorps, E.; Charbonnier, S.J.; Varley, N.R.; Capra, L.; Atlas, Z.; Cabré, J. Stratigraphy, sedimentology and inferred flow dynamics from the July 2015 block-and-ash flow deposits at Volcán de Colima, Mexico. *J. Volcanol. Geotherm. Res* **2018**, *349*, 99–116. [CrossRef]
29. Pensa, A.; Capra, L.; Giordano, G.; Corrado, S. Emplacement temperature estimation of the 2015 dome collapse of Volcán de Colima as key proxy for flow dynamics of confined and unconfined pyroclastic density currents. *J. Volcanol. Geotherm. Res.* **2018**, *357*, 321–338. [CrossRef]
30. GVO. Global Volcanism Program, 2016. Report on Colima (Mexico). In *Weekly Volcanic Activity Report, 28 September–4 October 2016*; Smithsonian Institution: Washington, DC, USA; US Geological Survey: Reston, VA, USA, 2016.
31. GVO. Global Volcanism Program, 2016. Report on Colima (Mexico). In *Weekly Volcanic Activity Report, 19 October–25 October 2016*; Smithsonian Institution: Washington, DC, USA; US Geological Survey: Reston, VA, USA, 2016.
32. Astrium and Eads company: The SPOT 6 & SPOT 7 Imagery User Guide. Available online: http://www.intelligence-airbusds.com/en/5280-spot-6-technical-documents (accessed on 30 October 2018).
33. Davies, A.G.; Chien, S.; Baker, V. Monitoring active volcanism with the Autonomous Sciencecraft experiment on EO-1. *Remote Sens. Environ.* **2006**, *101*, 427–446. [CrossRef]
34. NASA Earth Observatory. Available online: https://earthobservatory.nasa.gov/images/51583/volcanic-acitivity-at-krakatau (accessed on 14 April 2019).
35. Davies, A.G.; Chien, S.; Doubleday, J.; Tran, D.; Thordarson, T.; Gudmundsson, M.T.; Höskuldsson, A.; Jakobsdóttir, S.S.; Wright, R.; Mandl, D. Observing Iceland's Eyjafjallajo¨kull 2010 eruptions with the autonomous NASA Volcano Sensor Web. *J. Geophys. Res.* **2013**, *118*, 1–21. [CrossRef]
36. Patrick, M.R.; Kauahikaua, T.; Orr, A. Operational thermal remote sensing and lava flow monitoring at the Hawaiian Volcano Observatory. In *Detecting, Modelling and Responding to Effusive Eruptions*; Geological Society: London, UK, 2016; Volume 426, p. 489.
37. Mexican National Institute of Statistics and Geography-INEGI: Digital Elevation Model. Available online: http://www.beta.inegi.org.mx/app/geo2/elevacionesmex (accessed on September 2018).
38. Lira, J. *Tratamiento Digital de Imágenes*, 3rd ed.; Ciudad de México: Mexico City, Mexico, 2018; pp. 235–318.
39. Damiani, M.L.; Groppelli, G.; Norini, G.; Bertino, E.; Gigliuto, A.; Nucita, A. A lava flow simulation model for the development of volcanic hazard maps for Mount Etna (Italy). *Comput. Geosci.* **2006**, *32*, 512–526. [CrossRef]
40. Martin Del Pozzo, A.L.; Alatorre, M.; Arana, L.; Bonasia, R.; Capra, L.; Cassata, W.; Córdoba, G.; Cortés, J.; Delgado, H.; Ferrés, M.D.; et al. *Memoria Técnica del Mapa de Peligros del Volcán Popocatépetl*, 1st ed.; Monografías del Instituto de Geofísica, Universidad Nacional Autónoma de México: Mexico City, Mexico, 2008; p. 166.
41. Sieron, K.; Ferrés, D.; Siebe, C.; Constantinescu, R.; Capra, L.; Connor, C.; Connor, L.; Gropelli, G.; González-Zuccolotto, K. Ceboruco hazard map: Part II—modeling volcanic phenomena and construction of the general hazard map. *Nat. Hazards* **2019**. [CrossRef]
42. Costa, A.; Macedonio, G. Computational modeling of lava flows: A review. In *Kinematics and Dynamics of Lava Flows*, 1st ed.; Geological Society of America: Boulder, CO, USA, 2005; pp. 209–218.
43. Favalli, M.; Tarquini, S.; Fornaciai, A. DOWNFLOW code and LIDAR technology for lava flow analysis and hazard assessment at Mount Etna. *Ann. Geophys.* **2011**, *54*, 552–566. [CrossRef]
44. Bertino, E.; Damiani, M.L.; Groppelli, G.; Norini, G.; Aldighieri, B.; Borgonovo, S.; Comoglio, F.; Pasquaré, G. Modelling lava flow to assess hazard on Mount Etna (Italy). From geological data to a preliminary hazard map. In *Proceedings of the iEMSs*; International Environmental Modelling and Software Society: Burlington, NJ, USA, 2006; pp. 1–8.
45. Felpeto, A.; Araña, V.; Ortiz, R.; Astiz, M.; García, A. Assessment and modelling of lava flow hazard on Lanzarote (Canary Islands). *Nat. Hazards* **2001**, *23*, 247–257. [CrossRef]
46. Connor, L.; Connor, C.B.; Meliksetian, K.; Savov, I. Probabilistic approach to modeling lava flow inundation: A lava flow hazard assessment for a nuclear facility in Armenia. *J. Appl. Volcanol.* **2012**, *1*, 1–19. [CrossRef]
47. Aldighieri, B.; Groppelli, G.; Norini, G.; Bertino, E.; Borgonovo, S.; Comoglio, F.; Pasquaré, G. Proposta di una metodología per la valutazione della pericolositá vulcanica del Monte Etna. *Rend. Soc. Geol. It.* **2007**, *4*, 23–25.

48. Stevens, N.F.; Manville, V.; Heron, D.W. The sensitivity of a volcanic flow model to digital elevation model accuracy: Experiments with.zed map contours and interferometric SAR at Ruapehu and Taranaki volcanoes, New Zealand. *J. Volcanol. Geotherm. Res.* **2002**, *119*, 89–105. [CrossRef]
49. Claessens, L.; Heuvelink, G.B.M.; Schoorl, J.M.; Veldkamp, A. DEM resolution effects on shallow landslide hazard and soil redistribution modeling. *Earth Surf. Process. Landf.* **2005**, *30*, 461–477. [CrossRef]
50. Capra, L.; Manea, V.C.; Manea, M.; Norini, G. The importance of digital elevation model resolution on granular flow simulations: A test case for Colima volcano using TITAN2D computational routine. *Nat. Hazards* **2011**, *59*, 665–680. [CrossRef]
51. Caballero, L.; Capra, L.; Vázquez, R. Evaluating the performance of FLO2D for simulating past lahar events at the most active Mexican volcanoes: Popocatépetl and Volcán de Colima. In *Natural Hazard Uncertainty Assessment: Modeling and Decision Support*; Geophys Monograph: Washington, DC, USA, 2017; Volume 223, pp. 179–189.
52. Arámbula-Mendoza, R.; Reyes-Dávila, G.; Dulce, M.V.B.; González-Amezcua, M.; Navarro-Ochoa, C.; Martínez-Fierros, A.; Ramírez-Vázquez, A. Seismic monitoring of effusive-explosive activity and large lava dome collapses during 2013–2015 at Volcán de Colima, Mexico. *J. Volcanol. Geotherm. Res* **2018**, *351*, 75–88. [CrossRef]
53. Zobin, V.; Varley, N.; Gonzalez, M.; Orozco, J. Monitoring the 2004 andesitic block-lava extrusion at Volcán de Colima, México from seismic activity and SO2 emission. *J. Volcanol. Geotherm. Res* **2008**, *177*, 367–377. [CrossRef]
54. Pyle, D.M. Chapter 13—Sizes of Volcanic Eruptions. In *The Encyclopedia of Volcanoes*, 2nd ed.; Academic Press: San Francisco, CA, USA, 2015; Volume 1, pp. 257–263.
55. Harris, J.L.; Rowland, S.K. Chapter 17—Lava Flows and Rheology. In *The Encyclopedia of Volcanoes*, 2nd ed.; Academic Press: San Francisco, CA, USA, 2015; Volume 1, pp. 321–342.
56. Massaro, M.; Sulpizio, R.; Costa, A.; Capra, L.; Lucci, F. Understanding eruptive style variations at calc-alkaline volcanoes: The 1913 eruption of Fuego de Colima volcano (Mexico). *Bull. Volcanol.* **2018**, *80*, 62. [CrossRef]
57. United Nations Institute for Training and Research (UNITAR-UNOSAT). Guatemala, Volcán de Fuego. Flujo piroclástico, 4 de Junio de 2018. Available online: https://unitar.org/unosat/node/44/2815?utm_source=unosat-unitar&utm_medium=rss&utm_campaign=maps (accessed on 15 October 2018).

Article

Space- and Ground-Based Geophysical Data Tracking of Magma Migration in Shallow Feeding System of Mount Etna Volcano

Marco Laiolo [1,*], Maurizio Ripepe [2], Corrado Cigolini [1], Diego Coppola [1], Massimo Della Schiava [2], Riccardo Genco [2], Lorenzo Innocenti [2], Giorgio Lacanna [2], Emanuele Marchetti [2], Francesco Massimetti [1,2] and Maria Cristina Silengo [2]

1 Dipartimento di Scienze della Terra, Università di Torino, V. Valperga Caluso 4; 10125 Torino, Italy; corrado.cigolini@unito.it (C.C.); diego.coppola@unito.it (D.C.); francesco.massimetti@unito.it (F.M.)

2 Dipartimento di Scienze della Terra, Università di Firenze, V. G. La Pira 4; 50121 Firenze, Italy; maurizio.ripepe@unifi.it (M.R.); massimo.dellaschiava@unifi.it (M.D.S.); riccardo.genco@unifi.it (R.G.); lorenzo.innocenti@unifi.it (L.I.); giorgio.lacanna@unifi.it (G.L.); emanuele.marchetti@unifi.it (E.M.); mariacristina.silengo@unito.it (M.C.S.)

* Correspondence: marco.laiolo@unito.it

Received: 29 April 2019; Accepted: 16 May 2019; Published: 18 May 2019

Abstract: After a month-long increase in activity at the summit craters, on 24 December 2018, the Etna volcano experienced a short-lived lateral effusive event followed by a rapid resumption of low-level explosive and degassing activity at the summit vents. By combining space (Moderate Resolution Imaging Spectroradiometer; MODIS and SENTINEL-2 images) and ground-based geophysical data, we track, in near real-time, the thermal, seismic and infrasonic changes associated with Etna's activity during the September–December 2018 period. Satellite thermal data reveal that the fissural eruption was preceded by a persistent increase of summit activity, as reflected by overflow episodes in New SouthEast Crater (NSE) sector. This behavior is supported by infrasonic data, which recorded a constant increase both in the occurrence and in the energy of the strombolian activity at the same crater sectors mapped by satellite. The explosive activity trend is poorly constrained by the seismic tremor, which shows instead a sudden increase only since the 08:24 GMT on the 24 December 2018, almost concurrently with the end of the infrasonic detections occurred at 06:00 GMT. The arrays detected the resumption of infrasonic activity at 11:13 GMT of 24 December, when tremors almost reached the maximum amplitude. Infrasound indicates that the explosive activity was shifting from the summit crater along the flank of the Etna volcano, reflecting, with the seismic tremor, the intrusion of a gas-rich magma batch along a ~2.0 km long dyke, which reached the surface generating an intense explosive phase. The dyke propagation lasted for almost 3 h, during which magma migrated from the central conduit system to the lateral vent, at a mean speed of 0.15–0.20 m s^{-1}. Based on MODIS and SENTINEL 2 images, we estimated that the summit outflows erupted a volume of lava of 1.4 Mm^3 (±0.5 Mm^3), and that the lateral effusive episode erupted a minimum volume of 0.85 Mm^3 (±0.3 Mm^3). The results presented here outline the support of satellite data on tracking the evolution of volcanic activity and the importance to integrate satellite with ground-based geophysical data in improving assessments of volcanic hazard during eruptive crises.

Keywords: MODIS data; SENTINEL-2 images; infrasonic activity; open-vent activity; fissural eruption; long- and short-term precursors

1. Introduction

Multidecadal satellite data from different platforms have provided great support for volcano monitoring, and are enlarging our knowledge on the processes that drive the eruptions (e.g., [1–5]).

For example, infrared sensors with low to moderate spatial resolution (i.e., >1 km) make it possible to detect and quantify almost continuously the heat flux sourced from remote volcanoes [6,7], and may now be used to estimate, in near real-time, the lava discharged rates and erupted volumes, during worldwide effusive eruptions [5,8–10].The availability of high spatial resolution data (i.e., 30–90 m) acquired from different platforms constitutes a further improvement of space-based thermal volcano monitoring [11], making it possible to localize with high accuracy the position of an eruptive vents or, more generally, to characterize the thermal contribution of different volcanic sectors inferring magnitude and spatial information (e.g., location and size) of small hot spots (i.e., fumaroles, hot cracks) [12–15]. High-temporal resolution thermal data make it possible to track fast and energetic events (such as lava flows) once they have started [9,10], but they not have sufficient spatial resolution to detect small and long-term precursory thermal anomalies.

On the other hand, ground-based monitoring networks, installed on active volcanoes, acquire, store and process several fundamental parameters (e.g., seismicity, infrasound) that, combined together, make a great contribution in the short-term forecasting of eruptive crises [16].

Open-vent basaltic volcanoes, such as Etna, continuously emit magmatic-related products into the atmosphere [17], and can be sporadically affected by more energetic phenomena, such as paroxysmal explosions or flank eruptions [18,19] that can be preceded by short- or long-term precursors [20]. Hence, the ability to combine multiple dataset and to decipher the processes occurring within a volcanic system at different timescales represents a challenge and a future mandatory prerequisite for hazard assessment [21,22].

In this view, the great impact related to the explosive ash-rich paroxysms experienced by the Etna volcano in the latter years, pushed the scientific community to improve multiparametric monitoring systems by investigating long- and short-term variations in the behavior of the Etna volcano [23–26]. For example, Ripepe and coauthors [22] successfully developed a fully-automated early warning system that is able to track and notify the transition from violent intermittent strombolian activity to LFs- related eruptive column, based on infrasonic and thermal data [22].

In this work, we combine space- and ground-based geophysical data in order to investigate the September-December 2018 eruptive phase of the Etna volcano (Sicily, Italy), with a focus on the long- and short-term precursor signals that preceded the main effusive event of the 24 December 2018, and the successive resuming of the summit explosive activity.

Heat flux data, derived by MODIS sensor, are used to calculate and track the evolution of time averaged lava discharge rates (TADRs) and erupted volumes, while the high-spatial resolution potentiality of SENTINEL-2 images are used to locate the thermal activity at the multiple summit active vents of Etna. Infrasonic arrays and tremor amplitude measures are used to track the intensity, the frequency and the source of the explosive events occurring at summit craters.

The integration of satellite and ground-based data, give us an in-depth ability to record the shifting from open-vent conditions, represented by sustained summit strombolian activity, to the 24–26 December flank effusion promoted by a 2-km long feeder-dyke intrusion. This approach is particularly useful for a complex and multiple-crater volcano as Etna, with composite eruptive pathway and high variability in distribution and impact of future opening vents and effusive fissures.

2. Background

Etna is an open-vent basaltic volcano, characterized by a persistent degassing and frequent explosive activity, ranging from low strombolian explosions to lava fountaining and sub-plinian ash-rich paroxysms [27–29]. This activity occurs at several active vents located at three main sectors of the summit area, named: Bocca Nuova–Voragine (BN/VOR, also cited as Central Craters—CC; [29]), NorthEast Crater (NEC) and Southeast Crater–New Southeast Crater (SEC/NSEC) ([30,31]; see Figure 1).

This open-vent activity is periodically interrupted by effusive eruptions, occurring at the summit or along the flanks of the volcano, that drain portion of the shallow magmatic system of Etna [29,32–35]. While the so-called "summit eruptions" are fed by degassed magma likely stored at a shallow level and rising through an open central conduit [29], the flank and fissural eruption seems to be fed by a deeper, gas-rich magma. These eccentric eruptions take place along three main "rift zones"—NE Rift, S Rift and W Rift—outlining the strict control played by the tectonic structural framework of the volcano on the opening of lateral vents [34,36]. The eccentric flank eruptions may be promoted by magma reaching the surface via new, secondary conduits and/or dike intrusions, successively being erupted from propagating lateral fissures with multiple aligned vents [29,36]. In this view, a main hazard source at the Etna volcano is related to the opening of distal eruptive fissures with lava flows that potentially threatened nearby villages, infrastructures and tourist facilities [37].

During the last decade, the Etna volcano has exhibited a migration in persistence and magnitude of eruptive activity from SEC toward the NSEC crater, due to a preferential NE–SW oriented dilatation in the summit sector promoted by NE volcano flank instability during inflation phases [38]. Since 2011, its activity has been characterized by the occurrence of several (54) Lava Fountain (LF) episodes occurring at NSEC and VOR, generally preceded by an hour- to day-long increase of strombolian activity [22,25,29,39], and recurrently accompanied by short-lived (0.5–9 h), low-volume (0.5–3 Mm^3) lava flows. The 2011–2018 period was also characterized by 5 minor effusive eruptions (January–April 2014, July–August 2014, February 2017, March–April 2017, August 2018) not accompanied by lava fountain activity [40,41]. These events were characterized by duration from 5 to 75 days, and by limited erupted volumes (between 1 and 12 Mm^3; [42–45],) remarking a clear difference from the voluminous eruptions occurred between 2001–2009 (having duration of 20–420 days and volumes of 30–60 Mm^3; [46]).

The last effusive episodes occurred on 24 December 2018, and were characterized by the opening of an eruptive fissure at the base of the New South-East Crater on the western flank of Valle del Bove. The eruption produced gas and ash-rich plumes from the summit vents an intense Strombolian activity along the fissure, feeding several eastward lava flows [47]. Notably, this event was accompanied by an intense seismic swarm (more than 130 earthquakes in 3 h) culminating, on 26 December at 2:19 am (UTC), with a Mw 4.9 earthquake with the epicenter between the village of Lavinaio and Viagrande (CT), in correspondence of the Flandaca Fault. The earthquake was very shallow (1.2 km depth) and affected buildings, facilities and injured a dozen of people close to the epicentral areas along the lower south-eastern flank of Etna [48].

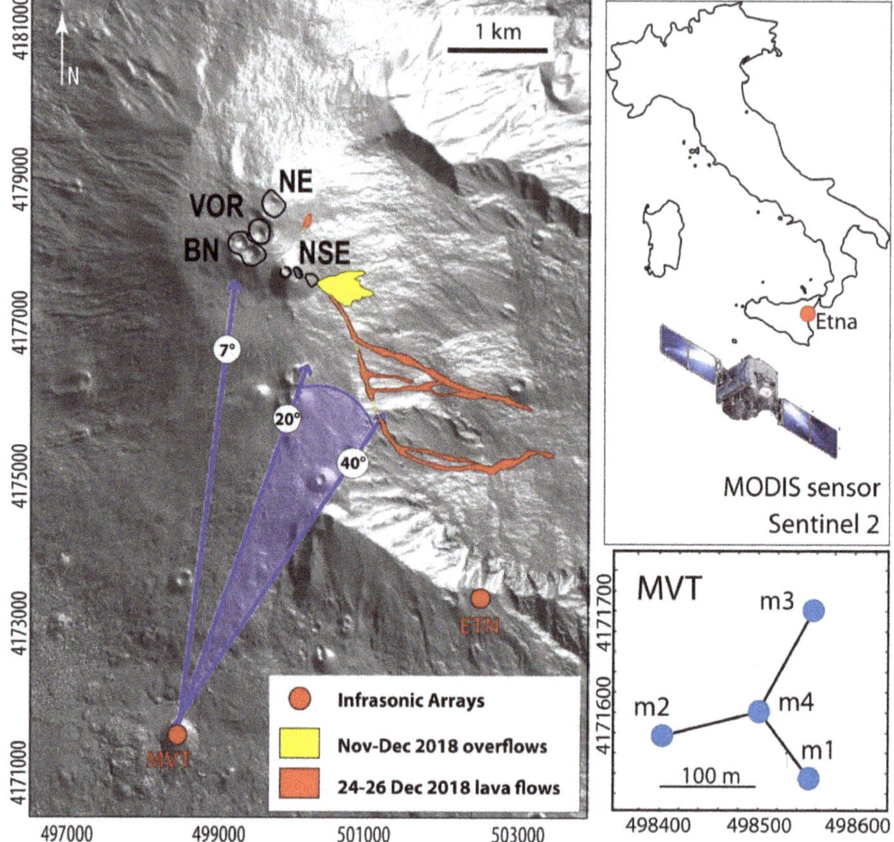

Figure 1. Shaded relief map of Etna summit area, with the indication of the main crater sectors and approximate maps of lava flows emplaced during the September–December 2018 activity. The location and the configuration of the MVT infrasonic array are also represented (see Method section). Blue arrows show the main localized back-azimuth during summit (7°, BN/VOR sector) and lateral activity (20°–40°, eruptive fissure), respectively. The adopted satellite sensors are also presented.

3. Dataset and Methods

Our work is based on the analysis of 2 satellite-based thermal datasets (provided by MODIS and SENTINEL 2 sensors, respectively) and 2 geophysical datasets, constituted by seismic and infrasonic signals recorded at ETN and MVT stations, respectively (Figure 1). Here, we summarize the methodology used to analyze these four sets of data in order to track the activity of the Etna volcano occurred between September and December 2018.

3.1. *Volcanic Radiative Power by Using MODIS-MIROVA Data*

MIROVA is a fully automatic volcano-dedicated hot-spot detection system ([49]) based on the images acquired by the MODIS sensor mounted on TERRA and AQUA NASA's satellites.

The combination of spatial and spectral principles to the middle infrared images acquired by MODIS makes possible the identification of the hot-spot contaminated pixel(s) (1 km resolution) and, consequently, the quantification of high temperature (>600 K) thermal anomalies, in terms of Volcanic Radiative Power (VRP, in Watt) (see [50]). The VRP (in Watt) is calculated through the MIR-method [50]:

$$VRP_{pIX} = A_{pIX} \times 18.7 \times (L_{4alert} - L_{4bk}) \quad (1)$$

where A_{Pix}, L_{4alert} and L_{4bk} are the pixel area (10^6 m^2 for MODIS), and the middle infrared radiance (W m^{-2} sr^{-1} μm^{-1}) recorded by the alerted (L_{4alert}) and background (L_{4bk}) pixels, respectively. According to [50] for hot targets that have an integrated temperature comprised between 600 and 1500 K, the use of the constant of 18.7 sr·μm provides reliable estimates of radiant power with an uncertainty of ±30%. This makes VRP particularly appropriate for calculating the heat flux sourced by only the active portions of lava flows, being almost insensitive to cooling lava surfaces having temperature lower than ~300 °C [49].

Overall, during the September 2018–January 2019 period, MIROVA elaborated more than 657 images (four overpasses per day) of which 184 (ca. 28%) detected a thermal anomaly. The VRP values range from less than 1 MW to a maximum of 2295 MW, this last being recorded on December 24 at 21:15 UTC.

3.2. Time Average Discharge Rate and Erupted Volume via MODIS Data

Satellite thermal data represent a useful parameter to estimate of Time Averaged lava Discharge Rate (TADR) and erupted volumes during effusive eruptions [51]. This thermal approach relies on the observed relationships that characterize the effusion rates, the active flow area, and the thermal flux, as documented by [51–54].

Here, we adopted an empirical approach, proposed by [55], that directly relates the volcanic radiant power (VRP), measured via MODIS, to the Time Average Discharge Rate (TADR) through a unique, best-fit parameter, called radiant density (c_{rad}; in J m^{-3});

$$TADR = \frac{VRP}{c_{rad}} \quad (2)$$

This empirical parameter represents the bulk efficiency of the lava body to radiate heat from its active surface, and is mainly controlled by the rheological, insulation, and topographic conditions at the time of emplacement [49,55]. Consequently, by integrating VRP measurements in time to obtain the total Radiant Energy (VRE) we are able, to estimate the erupted volume by assuming the appropriate $c_{rad,}$ for the observed lava flow.

Based on previous calibration of the radiant density typical of Etnean lava flows [49,55], here we used a c_{rad} between 2 and 3.6 × 10^8 J m^{-3} that makes it possible to constrain the erupted lava volumes between a minimum and maximum value.

It should be noted that the above method relies on the assumption that all the heat recorded by space is produced by the outpoured lava from the surface and the emplacing of lava flow or lava body. However, during open-vent activity, no net lava can be outpoured from the volcano, despite a thermal anomaly is still detected. In these cases the thermal approach can be used to infer the rate at which magma reaches the uppermost levels of the conduit (i.e., the bottom of a crater), before being cycled back in the convective magma column [56–59]. In this view, a variation in magnitude and persistence of heat radiation from summit craters could indicate a different level of magma column [60].

3.3. SENTINEL 2 Images

The Multispectral Instrument (MSI) is carried on-board SENTINEL-2A/2B platforms, two ESA satellites launched on polar, sun-syncronous orbit on June 2015 and March 2017, respectively. MSI provides multispectral data in 13 bands from VNIR to SWIR spectral region, useful for volcano monitoring applications. Indeed, the 20 m/pixel high-spatial resolution in the SWIR bands makes it possible to detail morphometric thermal features of the heat volcanic sources and distinguish active volcanic craters, or sectors, during ongoing eruptions (with a decameter spatial detail; [14,42]).

The SENTINEL-2 data are made available by the ESA Copernicus Service Data Hub (https://cophub.copernicus.eu/dhus/#/home), were downloaded through the cloud storage service of Amazon Web Service S3 (AWS-S3, https://registry.opendata.aws/SENTINEL-2/). Over the Etna geographical area, SENTINEL-2 satellites have a revisit time of about 2–3 days considering the two inspecting platforms 2A and 2B, enabling us to acquire about 10 high-resolution images per month, on average.

Following Massimetti et al. [61] we used the TOA (Top of the Atmosphere) reflectance data with a band combination 12–11–8a in the SWIR spectral region (R: 2190 nm, ρ12; G: 1610 nm, ρ11; B: 865 nm, ρ8a).

To analyze the thermal signature and to enlighten the presence of hot-spots, we applied a simple approach based on spectral ratios analysis of 12–11–8a bands, detecting and locating the number of "hot" pixels for each image in which a thermal volcanic emission is ongoing [62]. The algorithm to detect hot-spot contaminated pixel applied here was already successfully tested on several worldwide volcanic targets in comparison with MODIS-MIROVA heat flux [61]. For each pixel detected as "hot", we calculated a Thermal Index (T.I.) as the sum of the reflectances in the three SWIR bands analyzed (e.g., T.I. = $\rho_{12}+\rho_{11}+\rho_{8A}$) which is here considered as a proxy of the heat source temperature. We use the Thermal Index map to create stacked thermal profiles over the summit crater area of the Etna volcano (Figure 4), enabling us to distinguish the different craters in terms of persistence and to quantify the thermal contribute of each defined sector (see section Results).

To summarize, the SENTINEL-2 data are used to:

- observe qualitatively the ongoing eruptive state, comparing the chronology of Etna activity with the visual inspection of satellite images;
- track the thermal activity of each summit crater sector via the number of "hot" pixels detected during September–December 2018;
- evaluate the area and the length of the lava body produced during the analyzed timespan by image processing.

3.4. Infrasound Arrays and Seismic Tremor

Infrasound monitoring at the Etna volcano is performed with two 4 elements small aperture infrasound arrays [22,63,64], named ETN and MVT and installed, respectively, on September 2007 and in 2015. The ETN array was installed at 2100 m a.s.l., at a distance of 5500 m from the summit craters on the southern rim of Valle del Bove, while the MVT was installed on the southern flank of volcano at 1800 m a.s.l. and at a distance of about 6500 m from the craters. Both arrays are equipped with a FIBRA (www.item-geophysics.it) with four elements deployed following a triangular geometry and with an aperture (maximum distance between two array elements) of ~250 m and ~150 m, respectively ([22]; see Figure 1). Each array element is equipped with a differential pressure transducer with a sensitivity of 25 mV/Pa, maximum pressure range of ± 100 Pa, and flat frequency response between 0.01 and 100 Hz. Infrasound pressure data are converted to digital at each array element, transmitted with fiber optic cable to the central array element, where data is collected and GPS time stamped. The use of fiber optic made it possible to increase the signal-to-noise ratio and reducing the damages related to lightning and electrical discharges, with the ETN array being operational continuously since 2007. At ETN the array is co-located with a Guralp CMG-6T broadband seismometer, with eigen-period of 10 s and sensitivity of 2000 V/m s^{-1}.

Infrasound array data are processed in the time domain by applying a grid search procedure for a source within the crater area. The analysis is applied on 5-s-long time window. The sound speed velocity is fixed within the 330–360 m s^{-1} apparent velocity range, which accounts for a temperature effect on the acoustic propagation velocity and for the source-receiver elevation differences (1300 ± 100 m for the ETN array and 1600 ± 100 m for the MVT array). These two constraints allow the automatic removal of all the sources different from the volcano. The algorithm ([65,66] for details) searches for coherent signals recorded across the array, and we define a detection when the semblance exceeds 0.5. For each detection, the corresponding back-azimuth and the mean acoustic pressure are thus calculated according to the delay times observed among the different channels. Considering the aperture of the two arrays (<250 m) and the typical frequency of 1 Hz, the expected azimuth resolution is ~2° [67], which corresponds to ~190–230 m horizontal resolution at a source-to-receiver distance of 5500–6500 m. This processing is applied to the infrasonic data recorded at both the ETN and MVT arrays.

Here, we analyze the MVT array data because its capability, due to the location (Figure 1), to discriminate between Central Craters (BN/VOR and NE) and SE (NSE and SE) explosive activity.

4. Results

The complete time-series of the parameters recorded during September– December 2018 is shown in Figure 2. Based on our acquired datasets, inspection of satellite images and published reports (www.ingv.it), we subdivide the analyzed period into five main phases:

1. 1 September–4 November 2018; characterized by a monthly-long phase characterized by a low-level explosive activity at summit vents;
2. 5–28 November 2018; characterized by a gradual increase in explosive activity at summit vents;
3. 29 November–23 December 2018; characterized by the concurrent strombolian and overflows activity at BN vents and NSE sector, respectively;
4. 24–26 December 2018; the 24 December short-live lateral effusive episode along an NNW–SSE oriented 2 km-long eruptive fissure;
5. 27 December 2018–15 January 2019; the resumption of a highly-energetic explosive activity, localized at Central Craters sector, overlapping the end of the fissural effusive phase.

4.1. Phase I (1 September–4 November 2018): Low-Level Explosive Activity at Summit Craters

After the short effusive eruption of 23–29 August 2018 a low activity persisted at the Etna volcano from September to mid-November (Phase I; Figure 2). This period was characterized by persistent degassing, and low-level strombolian activity at the Central Craters (www.ct.ingv.it). Thermal and geophysical observations during this phase outlined: (i) sporadic low to moderated MODIS-MIROVA thermal alerts (VRP < 20 MW; Figure 2a); (ii) tremor amplitude at background level (Figure 2b) and (iii) low-pressure infrasonic detections (<1 Pa at MVT array; Figure 2c) localized at the BN crater (back-azimuth about 7°). In particular, SENTINEL-2 images show that the MODIS-detected thermal alerts were sourced by two distinct vents hosted within BN crater and from NSE sector with a sustained degassing from NE and Central Craters (Figure 3a–c).

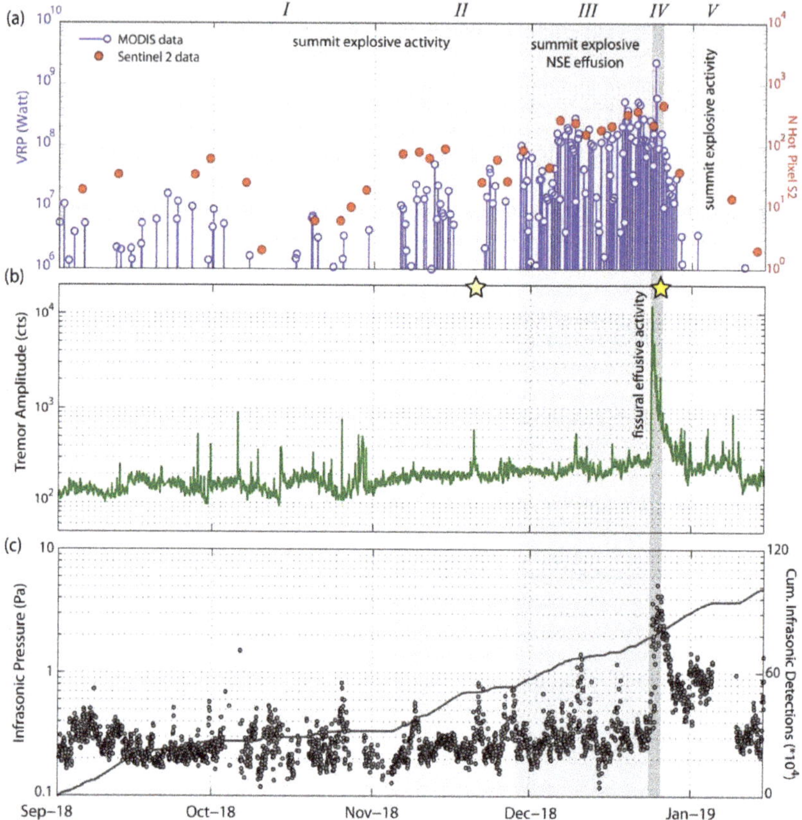

Figure 2. 1 September 2018 –15 January 2019 time-series of thermal, seismic and infrasonic activity at the Etna volcano: (**a**) VRP (blue stem) and Number of Hot Pixel (red circles) as retrieved by MODIS-MIROVA and SENTINEL-2 images, respectively; (**b**) hourly-mean tremor amplitude expressed as RSAM measured at ETN station; (**c**) hourly-mean amplitude pressure of the infrasonic detections recorded at MVT array (see Figure 1) and cumulative number of detections at the same array. See Figure 1 for ETN and MVT location. All datasets are expressed as logarithmic scale, excepted for the Number of Infrasonic detections. Yellow stars represent the occurrence of the 20 November seismic swarm and the 26 December earthquake (Mw = 4.9), see text for details.

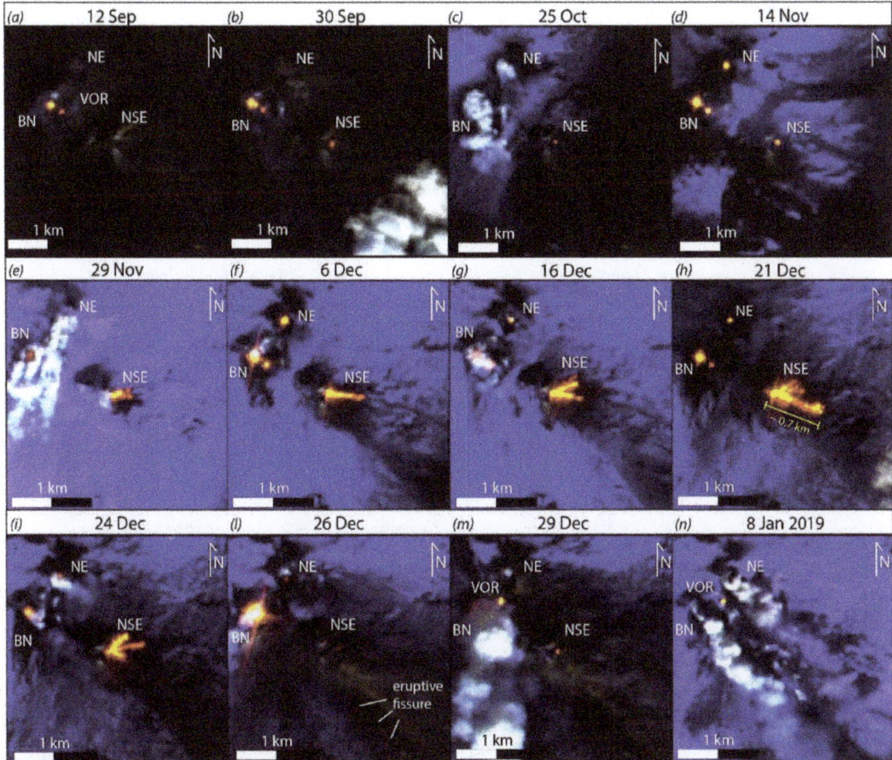

Figure 3. SENTINEL-2 class images acquired during the investigated Etna activity, in with summit craters are marked. The imagery is processed in RGB colors (bands 12–11–8a) to enhance the presence of hot targets (see Method section). Images represents the most representative shots (**a**–**n**) of the activity characterized Etna volcano during September–December 2018 period.

4.2. Phase II (5 November–28 November 2018): Gradual Increase in Explosive Activity at Summit Craters

Since early November, an intensification on summit thermal activity at NSE and NE craters was tracked by high-spatial resolution SENTINEL-2 images (Figure 3d). A clear increase in the number of hot pixels (Figure 2a) was in fact coupled with a VRP that reached a value above 50 MW, for the first time on 12 November, (see Figures 2a and 3d). This thermal increasing trend was confirmed by an increment in clustered medium infrasonic pressure transient (up to 1 Pa at MVT array; Figure 2c) associated to strombolian explosions. On 20 November 2018, a 15 h-long seismic swarm affected the western sector of the volcano [68].

4.3. Phase III (29 November–24 December 2018): Strombolian and Effusive Activity at Summit Craters

A few days after the seismic swarm of 20 November, the Etna volcano showed a further increase of summit activity, with the beginning of a sporadic lava effusion from a vent opened at the eastern base of the NSE crater. SENTINEL-2 images firstly captured this overflow activity on 29 November (see Figure 3e) when emissions become more continuous. Similarly, on 29 November at 00:50 UTC, MODIS data firstly record VRP value above 100 MW (Figure 2a). This marked change in the thermal activity detected by space marks the transition from Phase II to Phase III. Satellite observations tracked the persistence of the lava overflows from the NSE Crater producing short lava flows towards East-South East direction (maximum length of ca. 0.7 km, i.e., 21 December; Figure 3e–h).

Overall, MODIS thermal alerts during December 2018 show a significant augmentation in terms of both frequency (from 15% to 75% of the overpass) and intensity (from 25 to 160 MW, on average). The further increase in VRP measured since 18 December (above the 200 MW; Figure 2a), corresponds to the occurrence of several lava flows outpouring from the vent(s) at the base of the NSE crater, as reported by field observations [69]. This heat flux increment is coherent with a general increase in the number of hot pixels detected by SENTINEL-2 images during the same period (Figure 2a).

In the Phase *III*, the effusive episodes at NSE was accompanied by the concurrent explosive activity at BN and NE craters (Figure 3e–h). This is also outlined by the increased thermal anomalies detected by thermal profiles obtained by SENTINEL-2 Thermal Index map processing (Figure 4a,b). Despite of a sustained activity at the others summit craters, no thermal anomalies were detected inside VOR throughout Phase *I*, *II* and *III* (Figures 3e–h and 4a,b). It is worth noting that the overall heightened eruptive activity at the summit craters was accompanied by only a moderate increase in frequency and amplitude of infrasonic detections (Figure 2b).

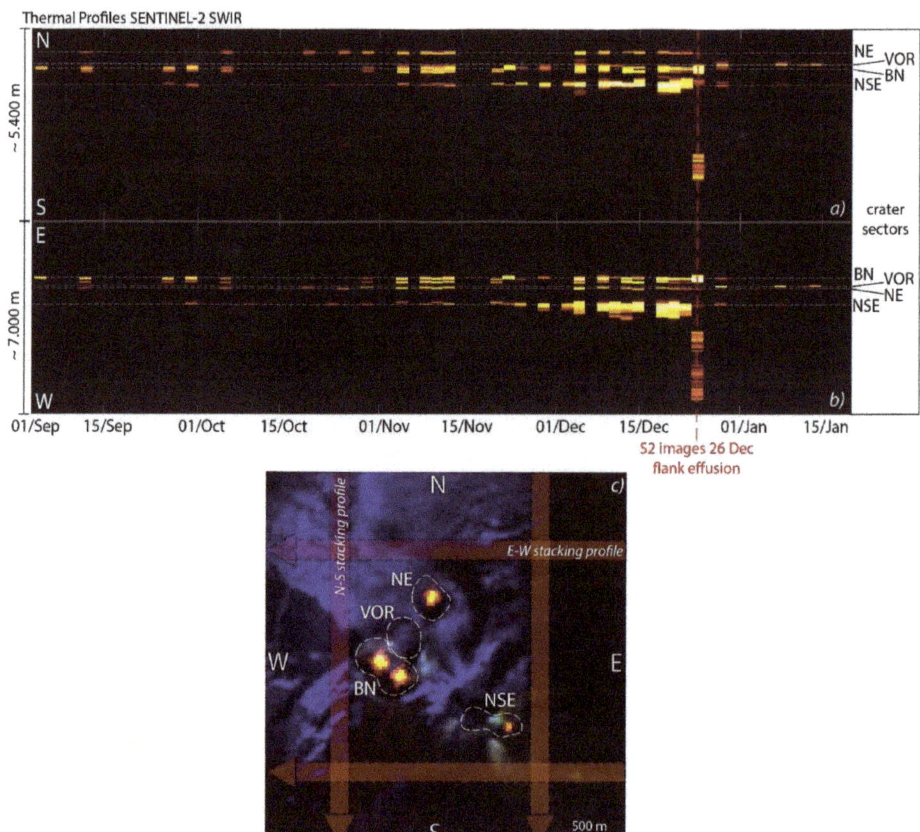

Figure 4. N–S and E–W thermal profile(s) acquired at summit crater sector of the Etna volcano during September 2018– January 2019 (**a**,**b**). The profiles are produced stacking each SENTINEL-2 image, analyzed in the SWIR bands and with at least one hot pixel anomaly, in two direction, from north to south (for each row) and from east to west (for each column; (**c**)). With white dotted lines are marked the Etna craters locations, thus letting to enlighten the presence of thermal activity over time for each vent(s). The hot pixel spread toward south and west in both profiles (red sign) indicate the occurrence of eruptive fissure feeding the lava flow of the 24 December.

4.4. Phase IV (24 December–26 December 2018): Opening of Lateral Fissure and Distal Effusive Activity

The general increase of eruptive activity recorded during the Phases *II* and *III* culminated on 24 December when the geophysical network recorded a drastic change on infrasonic and seismic signals (Figure 5). At 06:00 UTC there was a sudden, almost complete cessation of infrasonic activity at the Central Craters (back-azimuth close to 7°; see Figures 1 and 5b), followed at 08:24 UTC, by a sharp increase in volcanic tremor (Figure 5a,c). Intense grey- and reddish-ash plumes were observed from NE and BN craters [47] and accompanied the rapid rise of the tremor amplitude that reached a peak value at about the 11:00 UTC (Figure 5c). Hence, the volcanic tremor started to decline, while infrasonic activity resumed, but with back-azimuth consistent with a new source, located at SE from the summit craters (back-azimuth spanning from 20° to 40° angle; see Figure 5b). During this period, ground surveys reported the opening of a 2 km-long NNW–SSE eruptive fissure (EF) extending from the south-east base of the NSE crater, (at 3000 m a.s.l.) toward the upper portions of the west rim walls of the Valle del Bove (2400 m a.s.l.) [47]; cf. Figures 1 and 6). This episode was accompanied by violent strombolian explosions and ash emissions from distinct fissure segments, as well as by the emission of several lava flows units descending along the steep walls of the Valle del Bove [47]. The end of fissure propagation was marked by a new, sharp shift of infrasonic detections that, since 13:00 UTC were localized again at BN crater (Figure 5). However, SENTINEL-2 and MODIS images acquired, at 10:00 UTC and 11:50 UTC, respectively, did not show yet any clear increase in terms of intensity and extension of thermal anomaly at the summit craters, still indicating the occurrence of a multiple-branch overflow from NSE sector (Figures 5 and 6a). Interestingly, the SENTINEL-2 image shows a dark-grey plume, apparently sourced at the base of northwards side of Serra Giannicola Grande (Figure 6a), that could be possibly associated to a landslide generated in response to the ongoing fissure propagation and/or related to the intense seismic activity. On the evening of 24 December, at the 21:15 UTC, the MODIS image recorded a peak value of 2294 MW suggesting that an important effusive activity was ongoing (Figure 2a). However, the successive image, acquired just few hours later at 01:25 UTC of 25 December, detected a much lower thermal radiation (612 MW), likely resulting from a rapidly waning phase of lava emission. The decrease of thermal activity was, in turn, accompanied by a marked return of intense explosive activity at the Central Craters, as outlined by the infrasonic pressure measured at MVT array (Figure 5a). In particular, the peaking period of explosive activity recorded at BN sector on 25 December at 12:00 UTC (with infrasonic pressure up to 15 Pa) was coupled by VRP values close to 100 MW, suggesting that the feeding of lateral lava flow was drastically decreased or, at least, ceased.

The end of the lateral eruption was accompanied by the closure of the effusive crack, and by the reactivation of the explosive activity at the summit vents, as clearly testified by the SENTINEL-2 image acquired on 26 December at 10:00 UTC (see Figure 6b). This image captured the cooling of the southernmost lava flow unit (about 2.5 km in length), as well as a clear thermal anomaly at BN that, outline a sudden resumption of explosive activity at the summit craters (Figures 5b and 6b), as supported by infrasound detections.

Notably, during the ongoing eruption, at 02:19 UTC on 26 December, an earthquake of Mw 4.9 localized on the SE flank of the volcano and linked to the reactivation of the Fiandaca Fault, injured 28 people and damaged several buildings in the villages surrounding the fault [70].

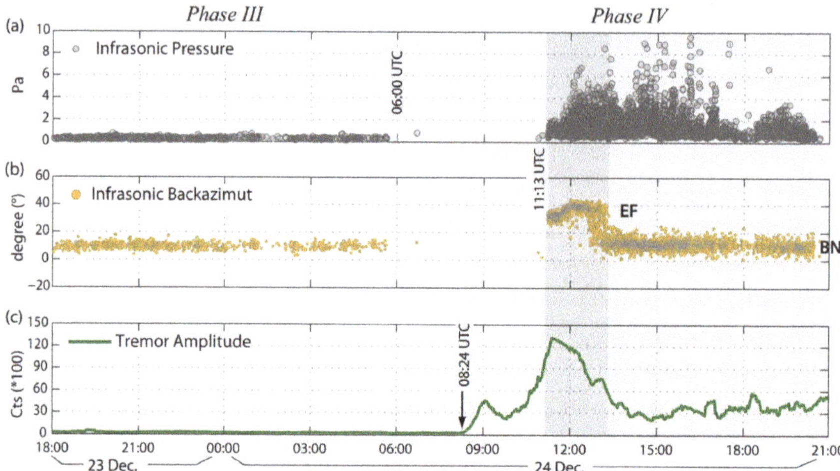

Figure 5. Timeseries of the 23–24 December of: (**a**) amplitude of infrasonic detections; (**b**) direction of the infrasound activity, EF—eruptive fissure; BN—Bocca Nuova; (**c**) tremor amplitude recorded at ETN site. (**a**) and (**b**) are those recorded at MVT, see Figure 1 for location of ETN and MVT sites.

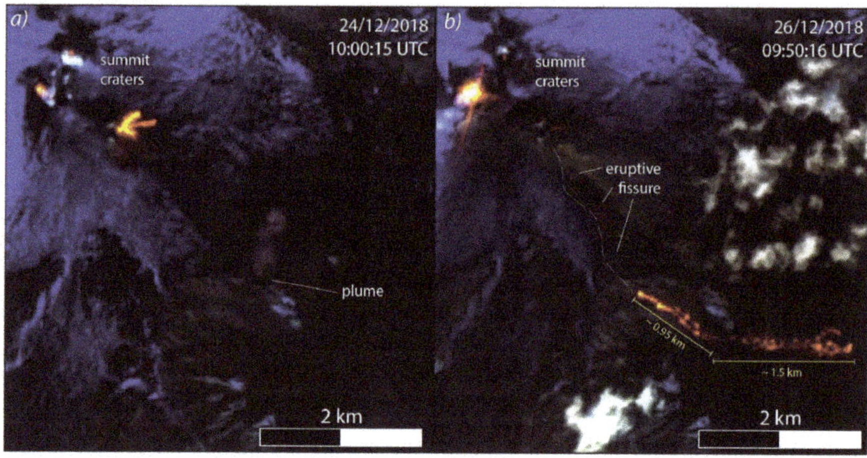

Figure 6. Focus on 24 (**a**) and 26 (**b**) December 2018, Phase *III* of the Etna volcano activity. SENTINEL-2 class images are processed in RGB colors (bands 12–11–8a), and to enhance the presence of hot targets (see Method section). The images resume the eastern summit portion of Etna edifice, covering from craters area to the upper portion of Valle of the Bove.

4.5. Phase V (26 December 2018–15 January 2019): Summit Explosive Activity

The end of the short-lived lateral effusion was followed by an intense and powerful explosive activity mainly localized at BN/VOR sector as recorded by the integrated geophysical measurements and supported by space-based thermal data. The thermal output detected by MODIS-MIROVA persisted at moderate levels (10–100 MW), as typical of high-explosive summit activity, while the SENTINEL-2 images make it possible to confine thermal anomalies within the Central crater sector, showing an extensive and intense anomaly rising from BN sector (Figure 3l,m). Moreover, since 29 December, the VOR crater was "thermally reactivated", as shown in Figure 3l and in the thermal profiles (Figure 4a,b) retrieved by high-resolution images. Since early January 2019 all the monitored parameters suggest a general decrease and a return to moderate explosive and sustained degassing activity, particularly from BN and NE craters (Figures 2 and 3m,n).

5. Discussion

The datasets presented in the previous chapter outline the variations of thermal, infrasonic and seismic activity continuously monitored at the Etna volcano by space- and ground-based sensors. The trends preceding and following the 24 December fissural eruption are well represented by plotting the cumulative curves of the above-mentioned parameters as recorded by SENTINEL 2 images, MVT infrasonic array and ETN seismic station, respectively (Figure 7).

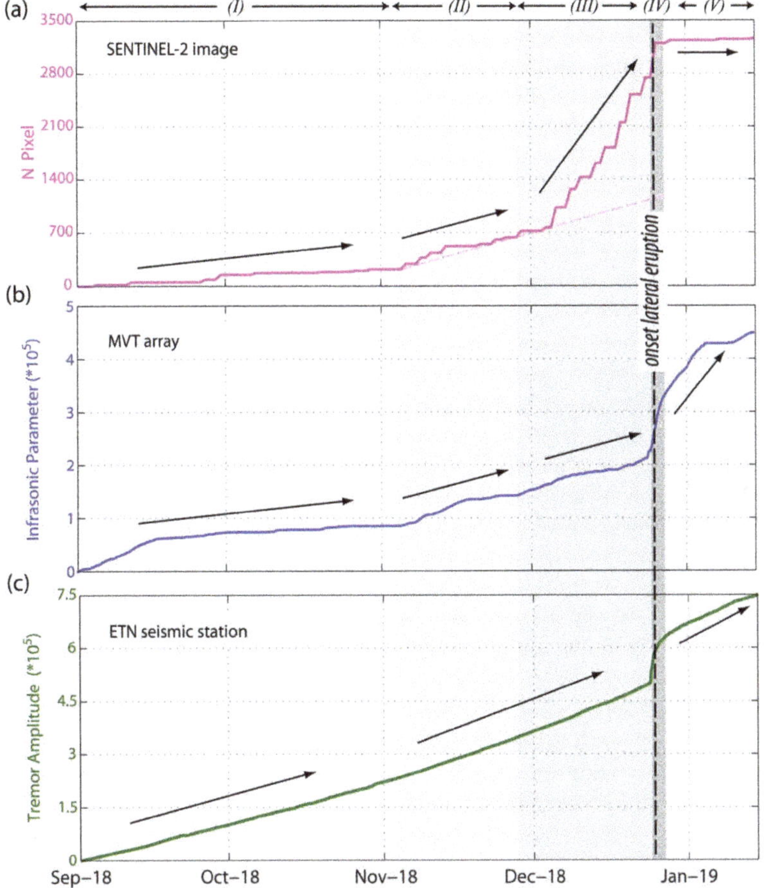

Figure 7. Cumulative trend of: (**a**) number of hot-detected pixels retrieved by SENTINEL-2 images, (**b**) the number of hourly infrasonic detection multiplied per the hourly-average infrasonic pressure of the detected events, and (**c**) the hourly-mean seismic tremor amplitude counts In the Phase II the increasing of explosive activity marked by infrasonic measures was accompanied by a similar trend in thermal behavior track by SENTINEL-2 images. During the Phase III the thermal activity show a clear increase related to the NSE overflow episodes that cause the separation in infrasonic and thermal signals. In (**a**) the grey dotted line highlights the hypothetical contribution of explosive activity on the whole thermal activity following the infrasonic activity. Black dotted vertical line marks the onset of the lateral effusive activity as recorded by the seismic tremor amplitude.

Firstly, it can be noted that the increase in thermal activity at the end of November 2018 (Phase III in Figure 7a), was preceded by a slight increment in both thermal and infrasonic activity started since early November 2018 (Phase II; Figure 7a,b). Notably, SENTINEL 2 images allowed us to recognize the contribution of two distinct thermal sources: (i) the overflows at the eastern base of the NSE crater; (ii) the Strombolian activity localized at BN, NE and NSE vents. In contrast, the cumulate in tremor amplitude highlights that before the sharp increase preceding the 24 December event, seismic tremor stays at background level and showing any significant variations during September-November timespan (Figure 7c).

In our view, the occurrence of strombolian and outflows activity at different summit craters, coupled by sustained and growing infrasonic activity, likely represented the surface expression of a slow but continuous increase in the magma supply within the shallow portion of the magma column.

In fact, by considering Time Averaged Discharge Rates (TADR, retrieved by MODIS heat flux; see Method Section), the Phase III was characterized by an overall value of 0.5–0.9 m^3s^{-1}, for a total volume outpoured during the summit overflow episodes of 0.9–1.7 Mm^3 (Figure 8). On a whole, the ongoing summit effusive regime in relation with the increase in thermal flux, seems to indicate a value of 100 MW (0.3–0.5 m^3 s^{-1}), as a possible threshold for the transition between open-vent conditions to effusive activity, as already suggested by Coppola et al. [57].

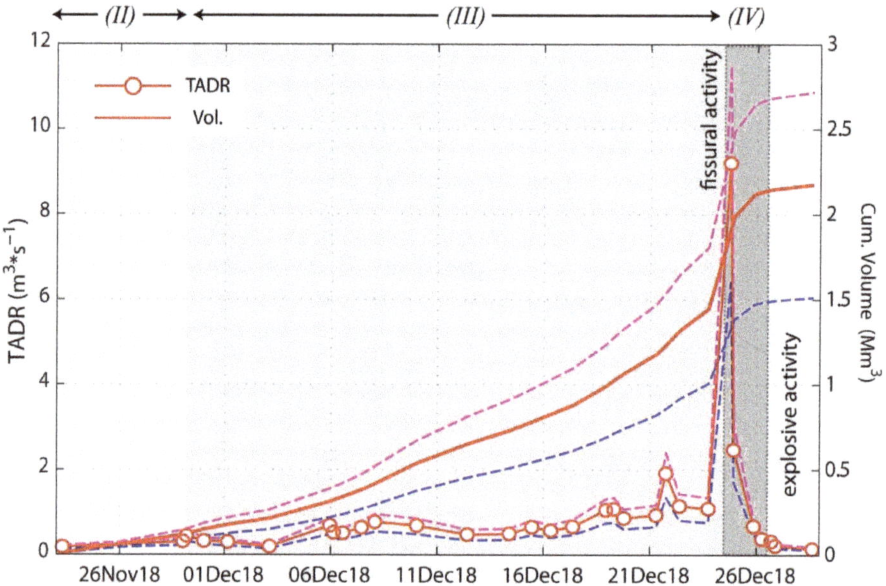

Figure 8. TADR and Volume trend retrieved from MODIS data, measured during the effusive phase experienced by Mt. Etna on November–December 2018 by using a proposed c_{rad} of 2.5 × 108 J m−3. Blue and magenta dotted lines represent ranging estimates of plotted parameters as obtained by c_{rad} equal to 2 × 10^8 J m−3 and 3.6 × 10^8 J m−3, respectively (see Method section).

On 24 December from the 06:00 UTC and the 11:13 UTC we record: (i) the abrupt interruption of infrasonic activity at BN crater; (ii) the beginning of a sudden increment in tremor amplitude that (iii) peaked at about the 11:13 UTC concurrently with the reawakening of a high-energetic infrasonic activity localized along the eruptive fissure (Figure 9). We interpret the three hours-long increase in tremor amplitude as a tracer of the magma- dyke injection and its propagation from the central conduit along the 2 km-long eruptive fissure extending on the western rim of the Valle del Bove. As a result, the timing of the increasing tremor let us suggest a velocity of the intrusion of 0.15–0.2 m s^{-1}. The end of this excited short-term behavior of seismic and infrasonic activity substantially marked the onset of the fissural activity (Figure 9). This effusive episode was mainly characterized by the violent explosions testified by the high infrasonic activity (Figure 9) and by eruptive vents feeding several lava flow units, extending up to about 2.5 km (see Figure 6). Notably, the activity along the eruptive fissure caused the interruption of the overflows from the NSEC, suggesting the drainage of the shallow portion of the conduit, because of the opening of the lateral dyke. Moreover, a similar scenario may also explain the lack of infrasonic summit activity as the result of the lateral magma migration from the central conduit to the feeder dyke.

MODIS thermal data acquired after about 10 h after the beginning of this eruptive episode (24 December at 21:15 UTC) suggests that the lava discharged from the fissure vent(s) with a TADR of 9.2 (±3) $m^3 s^{-1}$ (Figure 8). However, on the early 25 December (01:25 UTC) this rate declined to a value of 2.4 (±0.7) $m^3 s^{-1}$, indicating a rapid decrease in the effusive activity. This decline is supported by the SENTINEL-2 image of the 26 December 10:00 UTC which tracked the cooling of the lava flow, occurring after less than 48 h from the beginning of the effusive activity. As a whole, during this fissural eruption we measured an erupted lava volume of about 0.8 (±0.2) Mm^3, to give a total lava volume erupted during the November–December phases of 2.1 (±0.6) Mm^3 (Figure 8).

We must remind that because of the MODIS revisit time (4 image per day), all the short-lived eruptions (such as the 24 December eruption) may be severely underestimated. In this case, our estimate was mainly based to a unique peak thermal value that, however, has been recorded about 12 h after the onset of the effusive/explosive fissural eruption. Moreover, we know that Etna paroxysmal events such as effusions may be characterized by eruption rate during the first, most energetic phases, largely above the 10 $m^3 s^{-1}$ [44,71,72]. Taking into account that a field report highlighted the high-energetic behavior of the 24 December eruption onset, we stress that the proposed whole erupted volume of 0.8 Mm^3 for the fissure effusive phase may be significantly underestimated.

As previously cited, the activity that followed the onset of the 24 December episode was marked by the resumption of high infrasonic activity at the BN/VOR craters accompanying strombolian explosions. Notably, this quick renewal of summit explosive activity was coupled with a tremor amplitude that remained well above the pre-eruptive level, also after the complete cessation of the effusive activity (Figure 9).

This feature seems to imply that the shallow magma drainage, occurred during the lateral effusion, enhanced a perturbation in the conduit feeding the Central Craters. Possibly, this fast emptying may have induced a decompression of the active magma column (unloading), with a consequent gas exsolution promoting a resumption of the summit infrasonic detections.

Previously, this interpretation has been invoked to explain sudden explosive behaviors ("paroxysms") at effusive basaltic systems promoting the rapid ascent of a deep undegassed magma batch [72–76]. Here the process has many similarities, with a partial drainage of magma stored in the central conduit [73–75] and a consequent downward depressurization created by the flank effusion. However, the small volume involved during the 24 December short-lived event may fails to trigger the unloading of deeper magma portion but rather cause a strong gas separation promoting related infrasonic and seismic activity.

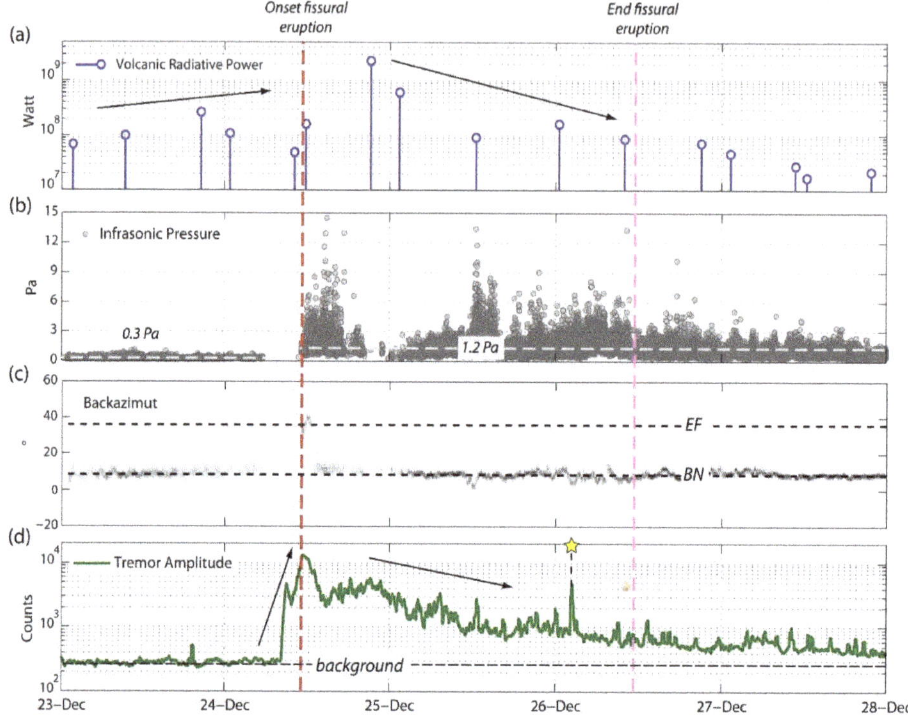

Figure 9. Timeseries of the (**a**) VRP, (**b**) Infrasonic Pressure, (**c**) Infrasonic back-azimuth and (**d**) Tremor Amplitude focused on the behavior pre-during and after the 24 December Etna fissural activity. See text for details.

6. Conclusions

The combined acquisition of ground and space-based data during the September 2018–January 2019 timespan allowed us to track the changes on the Etna volcanic activity and, particularly, the long- and short-term precursor signals preceding the 24–26 December fissural eruptive episode.

Our measurements showed that this short-lived activity has been heralded by monthly-long thermal precursor, tracked by MODIS and SENTINEL-2 satellite data, which clearly detect the overflow episodes at NSE crater sector. We link this activity as the surface expression of an upward migration of the active magma column that, moreover, promoted a concurrent increase in summit explosive activity measured by infrasonic arrays. By using MODIS data, we estimate that during this phase about 1.3 Mm^3 (±0.4 Mm^3) was erupted from summit vents. The continuous increase in the supply rate pushing in the shallow portion of the magma conduit drive the intrusion of a feeder dyke producing a 2-km long NW–SE direction eruptive fissure. Acquired datasets suggest that the propagation and the emplacement of the dyke occurred at a rate of 0.15–0.20 m s^{-1} during the three-hour long sudden increase in seismic tremor amplitude. This event was the prelude to the take place, along the fissure, of an activity characterized by violent strombolian ash-rich explosions and multiple lava flows and the concurrent interruption of overflows at NSE crater. Notably, the shift from summit and lateral activity has been revealed by the interruption of infrasonic detections at Central Craters and by a subsequent reawakening localized along the opened fissure. During the 24–26 fissural eruption, heat flux reach value typical of effusive activity (2294 MW at 24 December 21:15 UTC), such as the 26 December SENTINEL-2 image acquired at 10:00UTC capture the ending phase of this activity. The resumption of a sustained degassing activity at BN/VOR sector, recorded by the infrasonic data, seems the result of a decompression mechanism acting on the central conduit likely due by the magma lateral drainage feeding the fissural activity.

Results highlight that satellite, infrasonic and seismic tremor data represent an invaluable support for the hazard assessment of the wide and complex Etna eruptive scenarios with an impact on the forecasting of its most energetic eruptive episodes. Finally, we stress how the integration of thermal, infrasonic and seismic data make it possible to track the shallow magma movements causing migration of the surface activity and, in general, to detect the main volcanic changes occurred at open-vent volcanoes.

Author Contributions: All of the authors contributed to the manuscript. In detail: M.L. and M.R. conceive the paper. M.L. and F.M. wrote the main sections of the manuscript. D.C., E.M. and F.M. prepare the Method section. M.L., F.M. and D.C. prepared the figures. M.L. and D.C. analyzed the MODIS data. F.M. and D.C. analyze the SENTINEL-2 images. E.M. and G.L. analyzed infrasonic data. E.M., G.L., L.I., M.D.S., M.C.S. and R.G. worked on processing of infrasonic and seismic data. C.C., D.C., M.R. and E.M. supervised the preparation of the manuscript.

Funding: Part of the research leading to these results was performed within the EUROVOLC project (https://eurovolc.eu) and received funding from the European Community's Horizon 2020 program (grant agreement 731070). The ETN array and infrasound based early warning procedure of eruptive activity at the Etna volcano is performed by UNIFI for the Italian Civil Protection and received funding under the DPC-DEVNET project

Acknowledgments: MIROVA is a collaborative project between the Universities of Turin and Florence (Italy). We acknowledge the LANCE-MODIS data system for providing MODIS Near Real Time products. We acknowledge Copernicus Open Access Hub (https://scihub.copernicus.eu/) and the storage service of Amazon Web Service S3 (AWS-S3, https://registry.opendata.aws/SENTINEL-2/) for the availability and accessibility of SENTINEL-2 dataset. The comments and suggestions of two anonymous reviewers help us to improve the quality of the manuscript.

Conflicts of Interest: The authors declare no conflict of interest.

References

1. Bonny, E.; Wright, R. Predicting the end of lava flow-forming eruptions from space. *Bull. Volcanol.* **2017**, *79*, 52. [CrossRef]
2. Carn, S.A.; Clarisse, L.; Prata, A.J. On the detection and monitoring of effusive eruptions using satellite SO_2 measurements. *Geol. Soc. Lond. Spec. Publ.* **2016**, *426*, 277–292. [CrossRef]
3. Coppola, D.; Macedo, O.; Ramos, D.; Finizola, A.; Delle Donne, D.; del Carpio, J.; White, R.; McCausland, W.; Centeno, R.; Rivera, M.; et al. Magma extrusion during the Ubinas 2013–2014 eruptive crisis based on satellite thermal imaging (MIROVA) and ground-based monitoring. *J. Volcanol. Geotherm. Res.* **2015**, *302*, 199–210. [CrossRef]
4. Pyle, D.M.; Mather, T.A.; Biggs, J. Remote sensing of volcanoes and volcanic processes: Integrating observation and modelling—Introduction. *Geol. Soc. Lond. Spec. Publ.* **2013**, *380*, 1–13. [CrossRef]
5. Harris, A.J.L.; Villeneuve, N.; Di Muro, A.; Ferrazzini, V.; Peltier, A.; Coppola, D.; Favalli, M.; Bachèlery, P.; Froger, J.-L.; Gurioli, L.; et al. Effusive crises at Piton de la Fournaise 2014–2015: A review of a multi-national response model. *J. Appl. Volcanol.* **2017**, *6*, 1. [CrossRef]
6. Wright, R.; Glaze, L.; Baloga, S.M. Constraints on determining the composition and eruption style of terrestrial lavas from space. *Geology* **2011**, *39*, 1127–1130. [CrossRef]
7. Wright, R.; Blackett, M.; Hill-Butler, C. Some observations regarding the thermal flux from Earth's erupting volcanoes for the period 2000 to 2014. *Geophys. Res. Lett.* **2014**, *42*, 282–289. [CrossRef]
8. Coppola, D.; Ripepe, M.; Laiolo, M.; Cigolini, C. Modelling satellite-derived magma discharge to explain caldera collapse. *Geology* **2017**, *45*, 523–526. [CrossRef]
9. Ganci, G.; Vicari, A.; Cappello, A.; Del Negro, C. An emergent strategy for volcano hazard assessment: From thermal satellite monitoring to lava flow modeling. *Remote Sens. Environ.* **2012**, *119*, 197–207. [CrossRef]
10. Gouhier, M.; Guéhenneux, Y.; Labazuy, P.; Cacault, P.; Decriem, J.; Rivet, S. HOTVOLC: A web-based monitoring system for volcanic hot spots. *Geol. Soc. Lond. Spec. Publ.* **2018**, *426*, 223–241. [CrossRef]
11. Blackett, M. Early Analysis of Landsat-8 Thermal Infrared Sensor Imagery of Volcanic Activity. *Remote Sens.* **2014**, *6*, 2282–2295. [CrossRef]
12. Aufaristama, M.; Hoskuldsson, A.; Jonsdottir, I.; Ulfarsson, M.O.; Thordarson, T. New Insights for Detecting and Deriving Thermal Properties of Lava Flow Using Infrared Satellite during 2014–2015 Effusive Eruption at Holuhraun, Iceland. *Remote Sens.* **2018**, *10*, 151. [CrossRef]

13. Cigolini, C.; Coppola, D.; Yokoo, A.; Laiolo, M. The thermal signature of Aso Volcano during unrest episodes detected from space and ground-based measurements. *Earth Planets Space* **2018**, *70*, 67. [CrossRef]
14. Marchese, F.; Neri, M.; Falconieri, A.; Lacava, T.; Mazzeo, G.; Pergola, N.; Tramutoli, V. The Contribution of Multi-Sensor Infrared Satellite Observations to Monitor Mt. Etna (Italy) Activity during May to August 2016. *Remote Sens.* **2018**, *10*, 1948. [CrossRef]
15. Mia, M.B.; Fujimitsu, Y.; Nishijima, J. Thermal Activity Monitoring of an Active Volcano Using Landsat 8/OLI-TIRS Sensor Images: A Case Study at the Aso Volcanic Area in Southwest Japan. *Geosciences* **2017**, *7*, 118. [CrossRef]
16. Tilling, R.I. The critical role of volcano monitoring in risk reduction. *Adv. Geosci.* **2008**, *14*, 3–11. [CrossRef]
17. Rose, W.I.; Palma, J.L.; Delgado Granados, H.; Varley, N. Open-vent volcanism and related hazards: Overview. *Geol. Soc. Lond. Spec. Publ.* **2013**. [CrossRef]
18. Andronico, D.; Branca, S.; Calvari, S.; Burton, M.; Caltabiano, T.; Corsaro, R.A.; Del Carlo, P.; Garfì, G.; Lodato, L.; Miraglia, L.; et al. A multi-disciplinary study of the 2002–03 Etna eruption: Insights into a complex plumbing system. *Bull. Volcanol.* **2005**, *67*, 314–330. [CrossRef]
19. Calvari, S.; Spampinato, L.; Lodato, L. The 5 April 2003 vulcanian paroxysmal explosion at Stromboli volcano (Italy) from field observations and thermal data. *J. Volcanol. Geotherm. Res.* **2006**, *149*, 160–175. [CrossRef]
20. Aiuppa, A.; Burton, M.; Allard, P.; Caltabiano, T.; Giudice, G.; Gurrieri, S.; Liuzzo, M.; Salerno, G. First observational evidence for the CO2-driven origin of Stromboli's major explosions. *Solid Earth* **2011**, *2*, 135–142. [CrossRef]
21. Ripepe, M.; Pistolesi, M.; Coppola, D.; Delle Donne, D.; Genco, R.; Lacanna, G.; Laiolo, M.; Marchetti, E.; Ulivieri, G.; Valade, S. Forecasting effusive dynamics and decompression rates by magmastatic model at open-vent volcanoes. *Sci. Rep.* **2017**, *7*, 3885. [CrossRef]
22. Ripepe, M.; Marchetti, E.; Delle Donne, D.; Genco, R.; Innocenti, L.; Lacanna, G.; Valade, S. Infrasonic early warning system for explosive eruptions. *J. Geophys. Res. Solid Earth* **2018**, *123*, 9570–9585. [CrossRef]
23. Alparone, A.; Andronico, D.; Sgroi, T.; Ferrari, F.; Lodato, L.; Reitano, D. Alert system to mitigate tephra fallout hazards at Mt. Etna volcano, Italy. *Nat. Hazards* **2007**, *43*, 333–350. [CrossRef]
24. Bonaccorso, A.; Cannata, A.; Corsaro, R.A.; Di Grazia, G.; Gambino, S.; Greco, F.; Miraglia, L.; Pistorio, A. Multidisciplinary investigation on a lava fountain preceding a flank eruption: The 10 May 2008 Etna case. *Geochem. Geophys. Geosyst.* **2011**, *12*, Q07009. [CrossRef]
25. Calvari, S.; Cannavò, F.; Bonaccorso, A.; Spampinato, L.; Pellegrino, A.G. Paroxysmal Explosions, Lava Fountains and Ash Plumes at Etna Volcano: Eruptive Processes and Hazard Implications. *Front. Earth Sci.* **2018**. [CrossRef]
26. Scollo, S.; Prestifilippo, M.; Pecora, E.; Corradini, S.; Merucci, L.; Spata, G.; Coltelli, M. Eruption column height estimation of the 2011–2013 Etna lava fountains. *Ann. Geophys.* **2014**, *57*, S0214. [CrossRef]
27. Allard, P. Endogenous magma degassing and storage at mount etna. *Geophys. Res. Lett.* **1997**, *24*, 2219–2222. [CrossRef]
28. Allard, P.; Behncke, B.; D'Amico, S.; Neri, M.; Gambino, S. Mount Etna 1993–2005: Anatomy of an evolving eruptive cycle. *Earth Sci. Rev.* **2006**, *78*, 85–114. [CrossRef]
29. Corsaro, R.A.; Andronico, D.; Behncke, B.; Branca, S.; Caltabiano, T.; Ciancitto, F.; Cristaldi, A.; De Beni, E.; La Spina, A.; Lodato, L.; et al. Monitoring the December 2015 summit eruptions of Mt. Etna (Italy): Implications on eruptive dynamics. *J. Volcanol. Geotherm. Res.* **2017**, *341*, 53–69. [CrossRef]
30. Behncke, B.; Branca, S.; Corsaro, R.A.; De Beni, E.; Miraglia, L.; Proietti, C. The 2011–2012 summit activity of Mount Etna: Birth, growth and products of the new SE crater. *J. Volcanol. Geotherm. Res.* **2014**, *270*, 10–21. [CrossRef]
31. Neri, M.; De Maio, M.; Crepaldi, S.; Suozzi, E.; Lavy, M.; Marchionatti, F.; Calvari, S.; Buongiorno, M.F. Topographic maps of Mount Etna's summit craters, updated to December 2015. *J. Maps* **2017**, *13*, 674–683. [CrossRef]
32. Andronico, D.; Lodato, L. Effusive activity at Mount Etna volcano (Italy) during the 20th century: A contribution to volcanic hazard assessment. *Nat. Hazards* **2005**, *36*, 407–443. [CrossRef]
33. Branca, S.; Del Carlo, P. Types of eruptions of Etna Volcano AD 1670–2003: Implications for short-term eruptive behaviour. *Bull. Volcanol.* **2005**, *67*, 732–742. [CrossRef]
34. Neri, M.; Acocella, V.; Behncke, B.; Giammanco, S.; Mazzarini, F.; Rust, D. Structural analysis of the eruptive fissures at Mount Etna (Italy). *Ann. Geophys.* **2011**, *54*, 464–479.

35. Giuffrida, M.; Viccaro, M.; Ottolini, L. Ultrafast syn-eruptive degassing and ascent trigger high-energy basic eruptions. *Sci. Rep.* **2018**, *8*, 147. [CrossRef] [PubMed]
36. Acocella, V.; Neri, M. What makes flank eruptions? The 2001 Etna eruption and the possible triggering mechanisms. *Bull. Volcanol.* **2003**, *65*, 517–529. [CrossRef]
37. Bonaccorso, A.; Calvari, S.; Boschi, E. Hazard mitigation and crisis management during major flank eruptions at Etna volcano: Reporting on real experience. *Geol. Soc. Lond. Spec. Publ.* **2016**, *426*, 447–461. [CrossRef]
38. Acocella, V.; Neri, N.; Behncke, B.; Bonforte, A.; Del Negro, C.; Ganci, G. Why does a mature volcano need new vents? The case of the New Southeast Crater at Etna. *Front. Earth Sci.* **2016**, *4*, 67. [CrossRef]
39. De Beni, E.; Behncke, B.; Branca, S.; Nicolosi, I.; Carluccio, R.; D'Ajello Caracciolo, F.; Chiappini, M. The continuing story of Etna's New Southeast Crater (2012–2014): Evolution and volume calculations based on field surveys and aerophotogrammetry. *J. Volcanol. Geotherm. Res.* **2015**, *303*, 175–186. [CrossRef]
40. Gambino, S.; Cannata, A.; Cannavò, F.; La Spina, A.; Palano, M.; Sciotto, M.; Spampinato, L.; Barberi, G. The unusual 28 December 2014 dike-fed paroxysm at Mount Etna: Timing and mechanism from a multidisciplinary perspective. *J. Geophys. Res.* **2016**, *121*, 2037–2053. [CrossRef]
41. Barberi, G.; Giampiccolo, E.; Musumeci, C.; Scarfì, L.; Bruno, V.; Cocina, O.; Díaz-Moreno, A.; Sicali, S.; Tusa, G.; Tuvè, T.; et al. Seismic and volcanic activity during 2014 in the region involved by TOMO-ETNA seismic active experiment. *Ann. Geophys.* **2016**, *59*, 4. [CrossRef]
42. Cappello, S.; Ganci, G.; Bilotta, G.; Herault, A.; Zago, V.; Del Negro, C. Satellite-driven modeling approach for monitoring lava flow hazards during the 2017 Etna eruption. *Ann. Geophys.* **2018**, *61*, 13. [CrossRef]
43. Ganci, G.; Bilotta, G.; Cappello, A.; Hérault, A.; Del Negro, C. HOTSAT: A multiplatform system for the satellite thermal monitoring of volcanic activity. *Geol. Soc. Lond. Spec. Publ.* **2016**, *426*, 207–221. [CrossRef]
44. Ganci, G.; Cappello, S.; Bilotta, G.; Herault, A.; Zago, V.; Del Negro, C. Mapping Volcanic Deposits of the 2011–2015 Etna Eruptive Events Using Satellite Remote Sensing. *Front. Earth Sci.* **2018**. [CrossRef]
45. Laboratorio di Geofisica Sperimentale. Available online: http://lgs.geo.unifi.it/index.php/reports/etna (accessed on 26 December 2018).
46. Harris, A.J.L.; Steffke, A.; Calvari, S.; Spampinato, L. Thirty years of satellite-derived lava discharge rates at Etna: Implications for steady volumetric output. *J. Geophys. Res.* **2011**, *116*, B08204. [CrossRef]
47. Istituto Nazionale di Geofisica e Vulcanologia. Bollettino Settimanale, Rep. N° 01/2019. Available online: http://www.ct.ingv.it/it/rapporti/multidisciplinari.html (accessed on 26 December 2018).
48. Istituto Nazionale di Geofisica e Vulcanologia. Catalogo Nazionale Terremoti. Available online: http://cnt.rm.ingv.it/event/21285011 (accessed on 26 December 2018).
49. Coppola, D.; Laiolo, M.; Cigolini, C.; Delle Donne, D.; Ripepe, M. Enhanced volcanic hot-spot detection using MODIS IR data: Results from the MIROVA system. *Geol. Soc. Lond. Spec. Publ.* **2016**, *426*, 181–205. [CrossRef]
50. Wooster, M.J.; Zhukov, B.; Oertel, D. Fire radiative energy for quantitative study of biomass burning: Derivation from the BIRD experimental satellite and comparison to MODIS fire products. *Remote Sens. Environ.* **2003**, *86*, 83–107. [CrossRef]
51. Harris, A.J.L.; Baloga, S.M. Lava discharge rates from satellite-measured heat flux. *Geophys. Res. Lett.* **2009**, *36*, L19302. [CrossRef]
52. Pieri, D.; Baloga, S.M. Eruption rate, area, and length relationships for some Hawaiian lava flows. *J. Volcanol. Geotherm. Res.* **1986**, *30*, 29–45. [CrossRef]
53. Wright, R.; Blake, S.; Harris, A.J.L.; Rothery, D.A. A simple explanation for the space-based calculation of lava eruption rates. *Earth Planet. Sci. Lett.* **2001**, *192*, 223–232. [CrossRef]
54. Harris, A.J.L. *Thermal Remote Sensing of Active Volcanoes. A User's Manual*; Cambridge University Press: Cambridge, UK, 2013; p. 736.
55. Coppola, D.; Laiolo, M.; Piscopo, D.; Cigolini, C. Rheological control on the radiant density of active lava flows and domes. *J. Volcanol. Geotherm. Res.* **2018**, *249*, 39–48. [CrossRef]
56. Francis, P.; Oppenheimer, C.; Stevenson, D. Endogenous growth of persistently active volcanoes. *Nature* **1993**, *366*, 554–557. [CrossRef]
57. Harris, A.J.L.; Stevenson, D. Magma budgets and steady-state activity of Vulcano and Stromboli. *Geophys. Res. Lett.* **1997**, *24*, 1043–1046. [CrossRef]

58. Aiuppa, A.; de Moor, J.M.; Arellano, S.; Coppola, D.; Francofonte, V.; Galle, B.; Giudice, G.; Liuzzo, M.; Mendoza, E.; Saballos, A.; et al. Tracking formation of a lava lake from ground and space: Masaya volcano (Nicaragua), 2014–2017. *Geochem. Geophys. Geosyst.* **2018**, *19*, 496–515. [CrossRef]
59. D'Aleo, R.; Bitetto, M.; Delle Donne, D.; Coltelli, M.; Coppola, D.; McCormick Kilbride, B.; Pecora, E.; Ripepe, M.; Salem, L.C.; Tamburello, G.; et al. Understanding the SO2 Degassing Budget of Mt Etna's Paroxysms: First Clues from the December 2015 Sequence. *Front. Earth Sci.* **2018**. [CrossRef]
60. Coppola, D.; Piscopo, D.; Laiolo, M.; Cigolini, C.; Delle Donne, D.; Ripepe, M. Radiative heat power at Stromboli volcano during 2000–2011: Twelve years of MODIS observations. *J. Volcanol. Geotherm. Res.* **2012**, *215–216*, 48–60. [CrossRef]
61. Massimetti, F.; Coppola, D.; Laiolo, M.; Cigolini, C.; Ripepe, M. First comparative results from SENTINEL-2 and MODIS-MIROVA volcanic thermal dataseries. In Proceedings of the CoV10 IAVCEI General Assembly, Naples, Italy, 2–7 September 2018.
62. Murphy, S.W.; de Souza Filho, C.R.; Wright, R.; Sabatino, G.; Pabon, R.C. HOTMAP: Global hot target detection at moderate spatial resolution. *Remote Sens. Environ.* **2016**, *177*, 78–88. [CrossRef]
63. Marchetti, E.; Ripepe, M.; Ulivieri, G.; Caffo, S.; Privitera, E. Infrasonic evidences for branched conduit dynamics at Mt. Etna volcano, Italy. *Geophys. Res. Lett.* **2009**, *36*, L19308. [CrossRef]
64. Ulivieri, G.; Ripepe, M.; Marchetti, E. Infrasound reveals transition to oscillatory discharge regime during lava fountaining: Implication for early warning. *Geophys. Res. Lett.* **2013**, *40*, 3008–3013. [CrossRef]
65. Ripepe, M.; Marchetti, E. Array tracking of infrasonic sources at Stromboli volcano. *Geophys. Res. Lett.* **2002**, *29*, 2076. [CrossRef]
66. Ripepe, M.; Marchetti, E.; Ulivieri, G. Infrasonic monitoring at Stromboli volcano during the 2003 effusive eruption: Insights on the explosive and degassing process of an open conduit system. *J. Geophys. Res.* **2007**, *112*, B09207. [CrossRef]
67. Ulivieri, G.; Marchetti, E.; Ripepe, M.; Chiambretti, I.; De Rosa, G.; Segor, V. Monitoring snow avalanches in Northwestern Italian Alps using an infrasound array. *Cold Reg. Sci. Technol.* **2011**, *69*, 177–183. [CrossRef]
68. Istituto Nazionale di Geofisica e Vulcanologia. Bollettini Settimanali Rep. N° 48/2018. Available online: http://www.ct.ingv.it/it/rapporti/multidisciplinari.html (accessed on 27 December 2018).
69. Istituto Nazionale di Geofisica e Vulcanologia. Bollettini Settimanali Rep. N° 52/2018. Available online: http://www.ct.ingv.it/it/rapporti/multidisciplinari.html (accessed on 27 December 2018).
70. Istituto Nzionale di Geofisica e Vulcanologia. Available online: https://ingvterremoti.wordpress.com/2018/12/26/approfondimento-e-aggiornamento-sullattivita-sismica-e-vulcanica-in-area-etnea/ (accessed on 27 December 2018).
71. Wadge, G. Effusion rate and the shape of aa lava flow-fields on Mount Etna. *Geology* **1978**, *6*, 503–506. [CrossRef]
72. Calvari, S.; Salerno, G.G.; Spampinato, L.; Gouhier, M.; La Spina, A.; Pecora, E.; Harris, A.J.L.; Labazuy, P.; Biale, E.; Boschi, E. An unloading foam model to constrain Etna's 11–13 January 2011 lava fountaining episode. *J. Geophys. Res. Solid Earth* **2011**, *116*, 1–18. [CrossRef]
73. Calvari, S.; Spampinato, L.; Bonaccorso, A.; Oppenheimer, C.; Rivalta, E.; Boschi, E. Lava effusion—A slow fuse for paroxysms at Stromboli volcano? *Earth Planet. Sci. Lett.* **2011**, *301*, 317–323. [CrossRef]
74. Valade, S.; Lacanna, G.; Coppola, D.; Laiolo, M.; Pistolesi, M.; Delle Donne, D.; Genco, R.; Marchetti, E.; Ulivieri, G.; Allocca, C.; et al. Tracking dynamics of magma migration in open-conduit systems. *Bull. Volcanol.* **2016**, *78*, 78. [CrossRef]
75. Ripepe, M.; Delle Donne, D.; Genco, R.; Maggio, G.; Pistolesi, M.; Marchetti, E.; Lacanna, G.; Ulivieri, G.; Poggi, P. Volcano seismicity and ground deformation unveil the gravity-driven magma discharge dynamics of a volcanic eruption. *Nat. Commun.* **2015**, *6*, 6998. [CrossRef]
76. Coppola, D.; Di Muro, A.; Peltier, A.; Villeneuve, N.; Ferrazzini, V.; Favalli, M.; Bachèlery, P.; Gurioli, L.; Harris, A.J.L.; Moune, S.; et al. Shallow system rejuvenation and magma discharge trends at Piton de la Fournaise volcano (La Réunion Island). *Earth Planet. Sci. Lett.* **2017**, *463*, 13–24. [CrossRef]

Article

Changes in SO_2 Flux Regime at Mt. Etna Captured by Automatically Processed Ultraviolet Camera Data

Dario Delle Donne [1,2,*], Alessandro Aiuppa [1], Marcello Bitetto [1], Roberto D'Aleo [3], Mauro Coltelli [4], Diego Coppola [5], Emilio Pecora [4], Maurizio Ripepe [6] and Giancarlo Tamburello [7]

1 Earth and Marine Science Dept. (DiSTeM)—University of Palermo, 90123 Palermo, Italy; alessandro.aiuppa@unipa.it (A.A.); marcello.bitetto@unipa.it (M.B.)
2 INGV—Osservatorio Vesuviano, 80124 Naples, Italy
3 INGV, Sez. Palermo, 90123 Palermo, Italy; robertodaleo@gmail.com
4 INGV–Osservatorio Etneo, 95125 Catania, Italy; mauro.coltelli@ingv.it (M.C.); emilio.pecora@ingv.it (E.P.)
5 Earth Science Dept., University of Turin, 10125 Torino Italy; diego.coppola@unito.it
6 Earth Science Dept., University of Florence, 50121 Firenze, Italy; maurizio.ripepe@unifi.it
7 INGV, Sez. Bologna, 40128 Bologna, Italy; giancarlo.tamburello@ingv.it
* Correspondence: dario.delledonne@ingv.it

Received: 9 April 2019; Accepted: 15 May 2019; Published: 20 May 2019

Abstract: We used a one-year long SO_2 flux record, which was obtained using a novel algorithm for real-time automatic processing of ultraviolet (UV) camera data, to characterize changes in degassing dynamics at the Mt. Etna volcano in 2016. These SO_2 flux records, when combined with independent thermal and seismic evidence, allowed for capturing switches in activity from paroxysmal explosive eruptions to quiescent degassing. We found SO_2 fluxes 1.5–2 times higher than the 2016 average (1588 tons/day) during the Etna's May 16–25 eruptive paroxysmal activity, and mild but detectable SO_2 flux increases more than one month before its onset. The SO_2 flux typically peaked during a lava fountain. Here, the average SO_2 degassing rate was ~158 kg/s, the peak emission was ~260 kg/s, and the total released SO_2 mass was ~1700 tons (in 3 h on 18 May, 2016). Comparison between our data and prior (2014–2015) results revealed systematic SO_2 emission patterns prior to, during, and after an Etna's paroxysmal phases, which allows us to tentatively identify thresholds between pre-eruptive, syn-eruptive, and post-eruptive degassing regimes.

Keywords: SO_2 fluxes; UV Camera; Etna Volcano; explosive basaltic volcanism

1. Introduction

Active volcanoes are monitored by a variety of increasingly sophisticated volcanic gas sensing techniques [1–3], which are contributing to improved understanding and monitoring of volcanic activity. In particular, trends in volcanic gas composition and fluxes help detect subtle changes in the rates of magma ascent and degassing within shallow volcano plumbing systems, and allow to better confinepre-eruptive and syn-eruptive processes (gas in magmas being the main drivers of volcanic processes) [4]. Nevertheless, gas monitoring still lags behind more established seismic and geodetic techniques [5]. This is because the low temporal resolution of gas observations has traditionally hampered real-time analysis of fast-occurring volcanic processes, such as shallow intrusion of magma prior to eruption, and rapid gas ascent and release during explosive eruptions.

Volcanic SO_2 emissions provide key information on the rates of magma ascent in shallow (<3 km) magma plumbing systems, and are extensively monitored worldwide using instrumental networks of scanning ultraviolet spectrometers using the Differential Optical Adsorption Spectroscopy (DOAS) technique [6]. The advantage of this method is that acquisition and processing are relatively easy to automate [7], which yield exceptionally continuous records of volcanic SO_2 fluxes at relatively high temporal resolution (tens of minutes) [8–10]. Biases in the technique include limited spatial resolution (making it impossible to distinguish contemporaneous degassing from different vents), temporal resolution inadequate to resolve individual explosive events, and errors related to poor knowledge of plume speed [11].

The recent advent of UV cameras [12] has paved the way to volcanic SO_2 flux observations of much improved temporal and spatial resolution (see References [13–15] for recent reviews), which contributes to more effective integration between gas and geophysical datasets [16–29].

Even though space-based techniques are becoming increasingly performant in detecting volcanic SO_2 fluxes, ground-based SO_2 flux observations are still important and fundamental in monitoring basaltic volcanoes. While satellites become invaluable during paroxysmal explosive eruptions, when measurements from ground are complicated by volcanic ash within the plume. UV-camera measurements are more effective in monitoring more sluggish quiescent emissions in the low troposphere, and perhaps more useful to capture the early phases of unrest with escalating degassing activity [30].

The first examples of fully automated, permanent UV camera systems [14,28,31,32] are particularly promising, since they are opening the way to routinely monitoring volcanic SO_2 flux at a high rate continuously (daily hours only). For example, Reference [32] has demonstrated the ability of UV cameras to capture a precursory phase of heightened SO_2 flux in the weeks prior to the 2014 effusive eruption at the Stromboli volcano (Italy).

A current limitation of permanent UV camera systems is that, while data acquisition is fully autonomous, data processing is still time-consuming and operator-managed, e.g., data streamed by these systems are archived, and post-processed with ad-hoc codes [33]. To fully exploit the volcano monitoring potentials of UV cameras, automation of UV camera acquisition and processing routines is now timely and important.

The objectives of this study are: (i) to describe a new automatic routine for nearly real-time processing and visualization of UV camera data, (ii) to demonstrate the ability of the automatic routine to capturing temporal changes in SO_2 flux regime at Mt. Etna, and (iii) to contribute to better constraining Etna's degassing and eruptive behavior in 2016. To this aim, we reported on automatically processed data streamed by a permanent UV camera deployed on Etna. In order to fully characterize the Etna's 2016 behavior, we integrated our SO_2 flux results with independent geophysical parameters, traditionally used at Mt. Etna to constrain volcanic activity state and evolution [34–42]. More specifically, we used Moderate Resolution Imaging Spectroradiometer (MODIS) satellite-based thermal data obtained from the MIROVA (Middle InfraRed Observation of Volcanic Activity) system [43], ground-based thermal data streamed by monitoring cameras [44] of the Osservatorio Etneo (INGV-OE), and seismic tremor data [34,36,42].

2. Materials and Methods

2.1. UV Camera System

In the framework of the ERC-funded project BRIDGE (www.bridge.unipa.it), we designed a multi-instrument UV-absorption spectroscopy system for robust SO_2 flux measurements (Figure S1). The system is composed of (i) an instrument module and (ii) an acquisition/processing module. The instrument module is equipped with two JAI CM-140GE-UV cameras sensible to UV-radiation, and one Ocean-Optics USB2000+ Spectrometer coupled to a telescope of rectangular, vertically-oriented Field Of View (FOV $\approx 0.3° \times 14°$), and is spatially filtered to match the $\approx$12° vertical width. Two different band-pass optical filters with Full Width at Half Maximum (FWHM) of 10 nm, and a central wavelength of 310 and 330 nm, respectively, are applied in front of the cameras to enhance differential UV absorption in the SO_2 bandwidth [45,46]. In addition, 520 × 676 pixel images are acquired at 10-bit resolution with a frame rate of 0.5 Hz.

To obtain a quantitative measure of SO_2 column density within the volcanic cloud, we calculate the proportionality ratio between absorbance and SO_2 concentration in a defined region of the image pointed by the Ocean-Optic USB2000+ Spectrometer [47]. The use of the UV spectrometer allows us to quantitatively measure the full UV spectrum, and then fit the theoretical SO_2 absorption cross-section [48] with the differential absorption between two consecutive spectra (acquired every 5 seconds). The Ocean-Optic USB2000+ Spectrometer in use has on-board a Sony ILX511B Linear Silicon CCD Array Detector at 2048 pixels, with a wavelength response of 200–1100 nm, a dynamic range of 8.5 × 108, and a SNR of 250:1 at full signal. Calibrated SO_2 column densities over the entire images are then obtained by integrating images achieved by the UV camera with information achieved by the spectrometer. The instrument module is powered with a 12 V power supply, and requires 15 W in a fully operational mode. A Fujitsu RX100 Workstation, connected to the instrument module, automatically acquires synchronous data from the instrument module, and processes data without the need of the operator. To do this, we designed algorithms to control acquisition and processing parameters, such as the automatic tuning of camera's and spectrometer's exposures, and automatic evaluation of optimal viewing conditions (see Section 2.3.2). The computer internal time drift is controlled by a specifically designed application that reads the time-stamp from an NMEA (National Marine Electronics Association) standard message coming from a GPS antenna. The instrument module communicated with the acquisition/processing module via wired or wireless TCP/IP connection.

This UV camera system was installed at the Montagnola site (on Etna Volcano, Figure 1), and designed to stream real-time SO_2 flux results using a Wi-Fi data link. The objective is to capture SO_2 emissions associated with diverse volcanic processes and dynamics, including quiescent (passive) degassing, explosive eruptions (strombolian activity/lava fountaining), and effusive eruptions [49]. Montagnola is located at ~3 km distance from the active summit vents and grants perfect views of the southern sector of the summit crater area (Figure 1).

2.2. Seismic and Thermal Data

We compared our SO_2 fluxes with other independent geophysical parameters such as tremor and thermal radiance primary to have a benchmark in the calibration of automatic SO_2 flux calculation algorithm. Last but not least, we combined these data to characterize volcanic activity.

We used seismic data recorded by the ETN station [42,50], located at Lapide Malerba, at 5 km from the summit area. ETN is equipped with a broad-band seismometer (Guralp CMG-40T, with a sensitivity of 800 V/(m/s) and eigen period of 30 s). The link between SO_2 flux and volcanic tremor at Mt. Etna [22,51,52] suggests that the tremor is generated by the same degassing dynamics. We then calculated volcanic tremor amplitude from raw traces recorded at ETN station, by averaging within a 1-minute length window the maximum RMS amplitude taken within a 1-s window.

Thermal remote sensing offers a great opportunity to follow volcanic unrest from ground and space to characterize volcanic activity in near-real time [53] and to estimate near vent large pyroclastic products and lava flow discharged during eruptions. We used satellite data from the MIROVA system [43,54] and ground-based thermal cameras [44] to constrain onset, duration, and intensity through the time of eruptive events occurring at Etna during 2016. In particular, MIROVA uses the data provided by the MODIS sensor, which acquires four images per day (two daytime and two nighttime) with a spatial resolution of 1 km in the infrared bands [43]. The heat flux retrieved from MIROVA data, called Volcanic Radiative Power (VRP), is a combined measurement of the area and the integrated temperature of the hot emitters (hot vents, lava flows, etc.) at the time of each image acquisition. These data are provided without correction for cloud cover and satellite view geometry [43]. These factors may introduce noise in the dataset, which has been proven not to affect the general trend associated with the activity of Mt. Etna [43]. A mask 5 × 5 km around the volcano summit is used to filter out thermal anomalies due to wildfires.

2.3. The Algorithm for Automatic Processing of UV Camera Data

Manual processing of UV camera results [33] has the advantage that images are checked and validated manually by an operator. However, manual procedures are time-consuming, especially when dealing with huge data flows from permanent monitoring stations. To overcome this limit, we designed a MATLAB-based algorithm to automatically process images, and thus obtain SO_2 fluxes in near real-time. Using a "low-cost" PC-workstation (with 8 GB RAM and Xeon E3 type CPU), we can successfully process data at ~5× speed (i.e., a 5 minutes-long record is processed in ~1 min), which allows nearly real-time monitoring. The automatic routine has been calibrated using the results of the manual procedure and includes: (1) automatic determination of background absorbance levels, (2) automatic determination of image goodness, using image quality indexes, (3) estimation of gas plume speed and its distribution across the crater area, and (4) calculation of SO_2 flux distribution throughout the summit crater area. These are described in more detail below.

2.3.1. Image Processing and SO_2 Column Densities

Relative UV absorption by volcanic plume SO_2 is quantified by applying the Beer–Lambert law. Sets of synchronous images, taken by the two co-aligned cameras using different filters, are combined to obtain single absorbance images. This method, known as the "double filter method" [12,45], implies the use of two cameras with different filters, with one centered at 310 nm and the other centered on 330 nm, where UV radiation is/is not absorbed, respectively. The use of the two filters method allows compensating for aerosol attenuation/backscattering, to avoid any temporal mismatch associated with filter change while using a single camera, and to maintain the sampling rate at up to 0.5 Hz (two synchronous images every 2 s taken by two cameras [45]).

In our automatic routine, once a new raw image is acquired, a first quality check is made by the system, in order to keep the best exposure times to compensate any subtle changes in sunlight intensity. An automatic real-time tuning of exposure time is also applied to the UV spectrometer data, in order to obtain the best measurement dynamics within the UV bandwidth.

Synchronous images from the two cameras are then real-time corrected for vignetting effects associated with filters and optics, and normalized for relative exposure times. Residual intensities are then combined to obtain an un-calibrated absorbance image using the Lambert-Beer Law equation.

$$A = -\log_{10} \frac{I^{310}}{I^{330}} - A_0 \quad (1)$$

where A is the absorbance, I^{310} and I^{330} are pixel intensities associated with cameras mounting the 310 or 330 nm filter, while A_0 is the absorbance level associated with a clear background sky sub-area of the image (assumed to be unaffected by SO_2 absorption, ${I_0}^{310}$ and ${I_0}^{330}$ of Figure 2) and calculated as $-\log_{10}(I^{310}/I^{330})$, following Kern [55].

In the automatic processing module, this background sky sub-area is automatically selected for each image by monitoring a distal sky horizontal section with respect to the vent position, and selecting the sector with the lowest absorbance intensity.

Residual absorbance is then converted into SO_2 column density integrating data from the co-located ultraviolet spectrometer, which is pointing to a known sub-area within the camera field of view. This procedure yields, in real-time, the proportionality ratio between absorbance and SO_2 column densities using the method described in McGonigle [47].

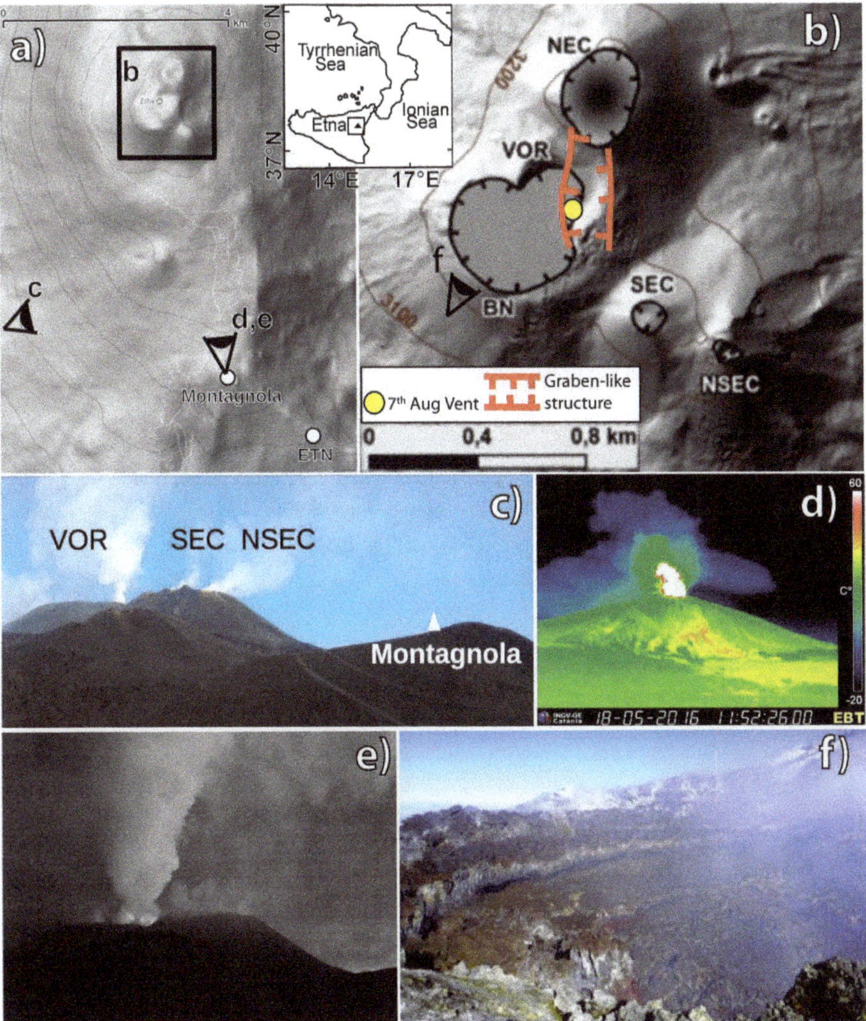

Figure 1. (**a**) The Montagnola site where the UV camera system was installed (position of ETN seismic station is also indicated). Sites from where pictures (**c**), (**d**), and (**e**) have been taken are shown. Black inset identifies the zoomed area of Figure 1b. (**b**) Etna summit area (redrawn from Reference [56]) shows the active summit craters (BN: Bocca Nuova; VOR: Voragine; NEC: North-East Crater; SEC: South-East Crater; NSEC: New South East Crater), the 7 August degassing vent, and the graben-like structure discussed in the text. The site where picture f has been taken is also shown. (**c**) Photo showing the Montagnola site and the summit area. (**d**) Thermal snapshot from INGV-OE monitoring camera capturing the 18 May lava fountaining episode. (**e**) Vigorous degassing from the vent opened on 7 August. (**f**) Picture taken from Reference [56] showing BN crater collapse occurred on 10 October, and marking the end of enhanced degassing activity (see text).

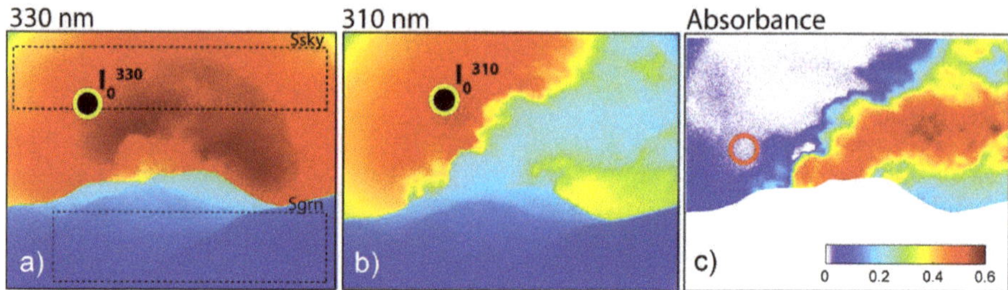

Figure 2. Calculation of the SO_2 column densities using the dual camera method with 330 nm (**a**) and 310 nm (**b**) optical filters. Absorbance image (**c**) is obtained using the Lambert-Beer equation after image normalization with respect to background absorbance intensities (black circles in (**a**) and (**b**), automatically located). Black circles in (**a**) and (**b**) represent the sky areas where absorbance is assumed to be SO_2 free (minimum absorbance level). Black rectangles (*Ssky Sgrn*) in (**a**) represent the areas used for calculation of the visibility index. The red circle in (**c**) shows the FOV of co-located UV-scanning spectrometer used to convert un-calibrated absorbance intensities into SO_2 column densities.

2.3.2. Automatic Determination of the Optimal Viewing Condition

The optimal plume viewing conditions, and the presence of a clear sky, are required for reliable SO_2 density measurements. However, weather conditions are extremely variable on Etna's summit, and often prevent optimal SO_2 observation. To minimize uncertainties due to poor weather conditions, we set-up a sub-routine for real-time calculation of two visibility indexes. The first visibility index (Fog index) is calculated as the unsigned ratio between the mean pixel intensity associated with the camera FOV's portions capturing sky and ground, respectively. Tests we conducted on real and synthetic images show that the higher this ratio is, the better the visibility condition is. SO_2 measurements are then selected by setting a threshold on the visibility index, and discarding measurements below the threshold (e.g., biased by a poor visibility condition).

Detecting a "sky" signal well above the "ground" signal (Figure 2) is a required but not sufficient condition for reliable SO_2 measurements. This is especially true in the presence of a highly condensed plume, where the SO_2 absorbance signal can be masked. Thus, a second automatic procedure was developed and run in real-time, which allows us to select only images with a clear SO_2 signal above atmospheric noise.

This latter procedure is based on the principle of combining absorbance and 310 nm images associated with the plume. A well detected and measurable SO_2 signal requires that lower intensities in the 310 nm image are measured in the plume relative to its surroundings (because SO_2 is absorbing solar radiation), and that higher intensities are consistently obtained in the absorbance image (see the Lambert Beer equation). This condition is only verified if the SO_2 signal is high enough to emerge above the atmospheric noise. To discriminate this condition in real-time, we defined a correlation index (Figure 3) as the correlation coefficient between absorbance and 310-nm pixel intensities over a cross-section intersecting the plume (Figures 3 and 4). The correlation coefficient is defined as $C(i,j)/((C(i,i)*C(j,j))^(1/2)$, where C is the covariance matrix, i and j are pixel intensities over the cross-section of absorbance and 310-nm images, respectively. In such plots, the closer the correlation coefficient is to the value −1, the more absorbance can be related to gas. Images that do not satisfy this condition (e.g., that have a Correlation Index < −0.5) are disregarded by the automatic computation (Figure 4).

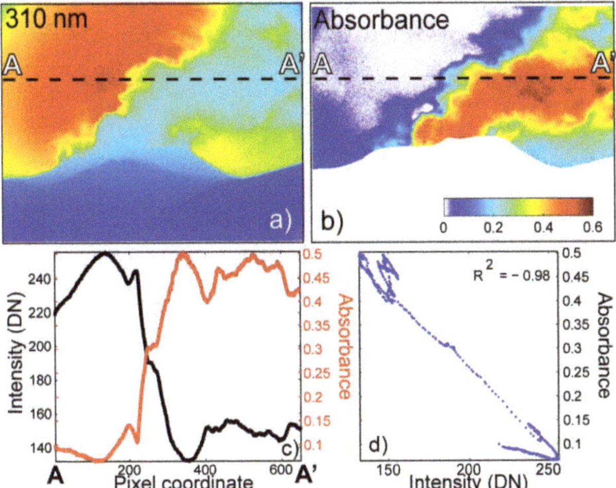

Figure 3. The image quality calculation method using the correlation coefficient between the 310 nm filter image (**a**) and the corresponding absorbance image (**b**). An intensity profile associated with a section (dashed line) crossing the volcanic plume, for both the 310-nm filter and absorbance images, is obtained (**c**). If these profiles are negatively correlated with a high correlation coefficient (**d**), then the gas is visible within the plume.

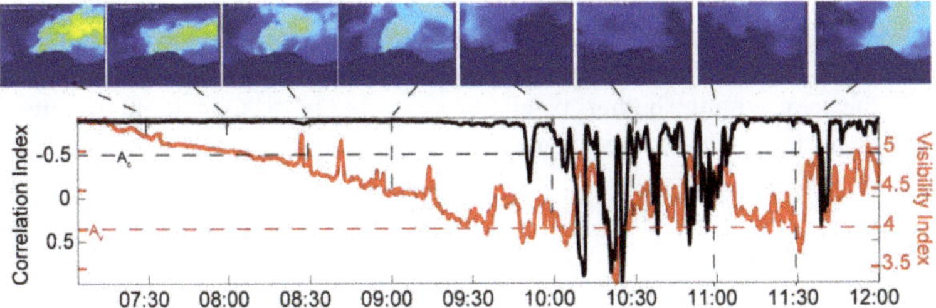

Figure 4. Example of output of the quality indexes sub-routine. The visibility (fog) and correlation indexes fluctuate through time as visibility conditions change (see snapshots on top of the figure). Gas is visible only when the fog index is greater than 4 and the correlation index is less than −0.5.

2.3.3. Plume Velocity Field

A robust plume velocity field is mandatory for reliable SO_2 flux measurements. Errors in plume velocity have shown to contribute to 40% or more of the overall error in the determined fluxes [13,57].

The UV camera approach offers the unique opportunity to track the gas while dispersing right after atmospheric emission, which minimizes errors in plume speed determination of yet more established DOAS and COSPEC methods. These methods indirectly infer plume speed from either on-site measurement of wind velocity or from meteorological models [11,15].

The UV camera approach allows us to derive the velocity profile over the summit craters by applying an optical flow algorithm that tracks gas fronts in consecutive frames [58].

Optical flow consists ofthe apparent motion pattern of image objects between two consecutive frames, caused by the movement of either the object or the camera, and is valid under the assumptions that the pixel intensities of an object do not change significantly between consecutive frames, and that the neighboring pixels have similar motion.

If a pixel, with intensity *I(x,y,t)*, where *(x,y)* are the pixel coordinates and *t* is the time in first frame, moves by distance *(dx,dy)* in the next frame taken after time *dt*, it can be assumed that:

$$I(x, y, t) = I(x + dx, y + dy, t + dt) \tag{2}$$

Then, from the Taylor series approximation of the right-hand side, removing common terms and dividing by dt, one gets the following equations.

$$\frac{\partial I}{\partial x}u + \frac{\partial I}{\partial y}v + \frac{\partial I}{\partial t} = 0 \tag{3}$$

$$u = \frac{dx}{dt}; v = \frac{dy}{dt} \tag{4}$$

where $\frac{\partial I}{\partial x}$ and $\frac{\partial I}{\partial y}$ are the image gradients, $\frac{\partial I}{\partial t}$ is the gradient along time, and u and v are horizontal and vertical velocities that are unknown. Lucas & Kanade [59] provide a method (LK) to derive these unknown velocities, by solving the basic optical flow Equations (3) and (4) for all the pixels using the least squares criterion, and by combining information from nearby pixels. We then applied the LK algorithm, included within the Open-CV toolbox, to our dataset [60]. We tested the performance of this method by applying it to artificial images with known particle velocities. The method has successfully determined velocity field with an error of <5%.

Absorbance images, obtained using the two filters method, contain gas-rich and ash-free portions of the plume with a higher absorbance relative to the background, and/or to ash-rich or particle-rich plume portions. We exploit this feature to track only gas moving fronts in consecutive frames by filtering them from other moving features such as lapilli and ash and by applying the LK method to the absorbance images rather than to the raw images directly acquired by our dual camera system. Velocities are then calculated by selecting the best features to track within the image, and which correspond to the areas with the highest pixel intensities (i.e., high SO_2 column densities) and high spatial coherence in consecutive frames (taken every 2 s).

2.3.4. Image Analysis and SO_2 Flux Calculation

To enhance the contrast between volcanic emissions and atmospheric noise, and to limit dispersion effects and chemical conversions of SO_2 in the atmosphere, image processing was conducted on a restricted image portion capturing an image sub-region above the crater area (Figure 2).

We also aimed at resolving gas contributions from different vents, and therefore, capturing changes in degassing dynamics and location [28]. To do this, we selected a rectangular sub-area over the crater terrace, along which we calculated the distribution of SO_2 column densities and a plume velocity field (Figure 2). We calculated SO_2 column densities and plume speed as close as possible to the vent, which minimizes the effects of wind and air entrainment within the plume that would produce dilution of SO_2 concentrations farther downwind. This allowed us to detect changes in degassing dynamics across the crater terrace that were associated with changes in volcanic activity and regimes.

We then calculated velocity (mean, maximum, and associated standard deviation) and absorbance distribution along an ideal profile positioned in the middle of the sub-area, derived from averaging a series of parallel profiles within the area of analysis. From this, the SO_2 density flux (in $kgm^{-1}s^{-1}$) was calculated by multiplying column densities associated with each pixel of the profile with the corresponding normal velocity component of motion (Figure 5, see also Reference [32]). Velocity profiles (Figure 5) were obtained by averaging the calculated two-dimensional velocity fields, and filtering out velocity points with low coherence. We also derived uncertainty in velocity determination along the profile by calculating the standard deviation associated with the velocity values used to determine the average velocity.

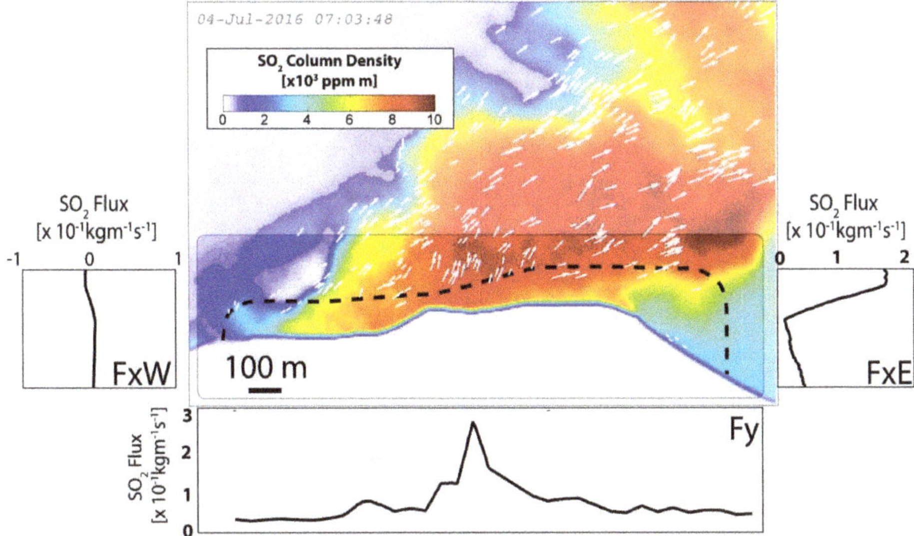

Figure 5. SO_2 density flux calculated in the sub-area encompassing the summit craters. White arrows correspond to gas velocity vectors calculated on high coherence regions of the images. SO_2 density flux distributions along the entire crater area and over the black dashed profile are calculated within the highlighted area, corresponding to vertical (Fy), western (FxW), and eastern (FxE) horizontal components, respectively. The SO_2 total flux, for a given sector, was then calculated by integrating the density flux over the total length of the profile.

The SO_2 flux was obtained by spatial 1D integration over the profiles. Discrimination of degassing contributions from the different craters, specifically the central (VOR+BN) and southeastern (SEC+NSEC) craters, was obtained by vectorial summation of the various SO_2 flux components that border each crater. In doing so, we took into account if gas is moving away or toward the vent, which corresponds to a positive/negative flux contribution respectively.

Given the position of the station relative to the summit crater area (Figure 1), the gas contribution from the North-East crater (NEC) was not resolvable from our images, as hidden behind the SEC crater's ridge.

2.3.5. Validation of the Automatic Method

Validity of the automatically processed SO_2 fluxes was tested for a comparison with SO_2 fluxes manually obtained using the Vulcamera software [33] (Figure 6). In particular, this was conducted for some selected days of acquisition characterized by a good weather condition and by clear and well-visible volcanic plume. The manual Vulcamera procedure involves calibrating images by individuation of a clear sky portion unaffected by degassing. Then, SO_2 fluxes were calculated over an integration profile roughly perpendicular to wind direction, using plume speeds obtained from cross-correlation between consecutive frames along the selected profile [33].

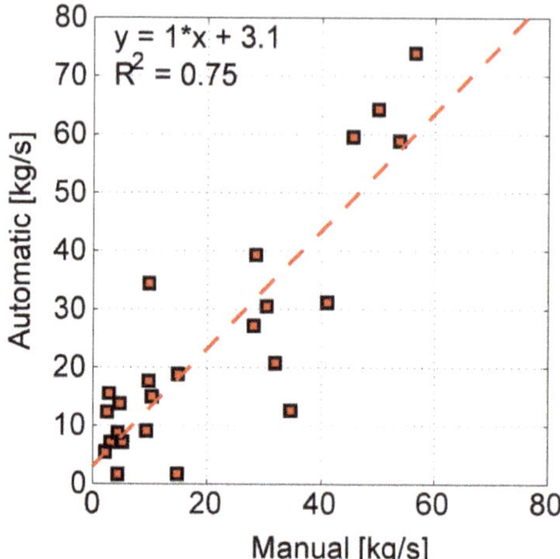

Figure 6. Comparison between daily averaged SO_2 fluxes calculated by a manual operator (using Vulcamera software, [33]) and those obtained by the automatic algorithm. The correlation coefficient between the two datasets is 0.75, with a 1:1 proportionality ratio. This correlation validated the automatic algorithm as an alternative, reliable method for SO_2 flux calculation.

Comparison between manually and automatically calculated fluxes (Figure 6) demonstrated a correlation coefficient (R^2) of ~0.75, with a best-fit regression line showing a ~1 proportionality factor. The main source of errors was associated with the algorithm for the automatic selection of the background sky area within an image, in particular during highly variable cloud conditions and strong winds. In such cases, background sky detection within images might not be optimal and may then result in a main underestimation of the real SO_2 column densities, as shown by the position of outliers (Figure 6). Overall, this comparison validated the use of the automatic SO_2 flux determination procedure, which paves the way to its full exploitation in real-time volcano monitoring, as already started on the Stromboli volcano (Italy) [32].

3. Results: Application of the Automatic Real Time Algorithm: the Etna 2016 Case

Etna is one of the volcanoes worldwide with the longest and most continuous SO_2 flux record. SO_2 flux measurements have become fundamental in volcano monitoring to define the rates of magma ascent and degassing within the shallow (<3 km) plumbing system. SO_2 fluxes have been measured on Etna since the 1970s using the COSPEC (Correlation Spectrometer) [49,61–64] and, more recently, a network of Differential Optical absorption spectrometers (FLAME [65,66]). In view of its recurrent activity and robust past SO_2 flux record, Mt. Etna is an ideal test site for validating our automatic processing method.

We reported below on the SO_2 data automatically acquired and processed during 2016, which is a period characterized by substantial temporal changes in activity styles, including a phase of reduced degassing in the aftermaths of the December 2015 eruption [30,67]. This was followed by gradual activity escalation culminating into the May 2016 eruptive phase, and in a new vent opening episode on the eastern VOR crater rim in August [68–70].

Our aim was to test if different SO_2 degassing regimes, related to such diverse activity styles, could be resolved and characterized in automatic (and in nearly real-time) using our permanent SO_2 camera system.

3.1. Etna's Activity in 2016, SO_2 Flux Records, and Comparison with Seismic, Thermal Dataset, and Field Observations

The SO_2 flux time-series of 2016, which were generated by using the automated processing algorithm proposed in this study, is illustrated in Figure 7. The figure highlights that the significant variability in volcanic activity style in 2016 reflected a highly dynamic SO_2 flux behavior (Figure 7). As illustrated in Figure 7, our 2016 temporal record shows daily averaged SO_2 fluxes ranging between a few hundred to ~6000 tons per day (t/d). The associated standard deviations range from 100 to 4000 t/d. To assist interpretation of SO_2 flux variations, we also reported seismic tremor and thermal radiance time-series (Figure 7), where the latter is expressed as Volcanic Radiative Power (VRP) [71].

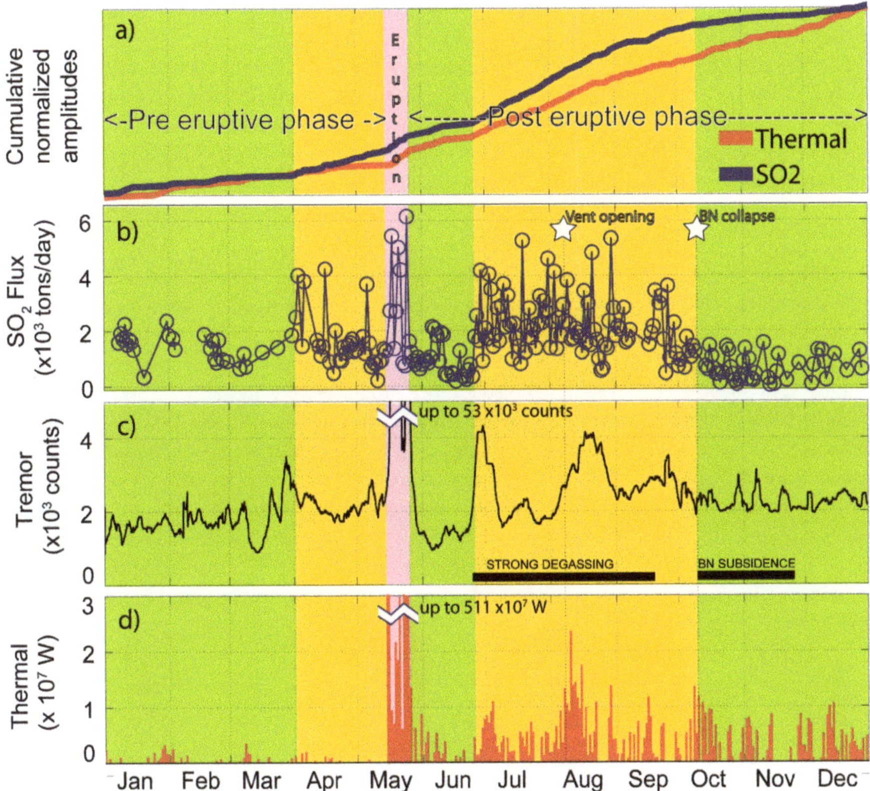

Figure 7. (**a**) Cumulative daily averaged SO_2 fluxes are compared with thermal cumulative activity detected by MIROVA. (**b**) SO_2 flux daily average time-series are compared with seismic tremor amplitude (**c**), and thermal radiance (**d**). Three volcanic degassing levels are distinguished by different colors: green, low average degassing, orange, enhanced degassing, pink: intense degassing. The May 2016 Eruptive phase is preceded by more than one month long period of enhanced degassing activity when no significant increase in tremor and thermal is identified. Enhanced degassing activity is also detected between July and September, according to relatively higher tremor and thermal signals.

Combined analysis of field observations reported in Reference [56] including thermal, seismic, and SO_2 fluxes time-series allowed us to distinguish three main periods of activity: (1) a pre-eruptive period (January to 16 May 2016), in which volcanic activity remained low (January to March) or gradually resuming (April to 16 May), (2) an eruptive period between 16 May and 25 May, characterized by intense strombolian activity, lava flows, and three short-lived lava fountaining episodes that occurred in a brief time lapse between 18 May and 21 May, (3) a post-eruptive period (26 May to 31 December), that included a brief period of reduced activity until the end of June, which is followed by gradual (re)intensification of volcanic activity culminating with opening of new, strongly degassing incandescent vent at VOR on 7 August. This strong degassing declined during the subsidence of the Bocca Nuova (BN) crater's floor that occurred on October 10. These periods are described in more detail in the following sections.

3.1.1. Volcanic Activity from 1 January 2016 to 15 May 2016 (Pre-Eruptive Period)

Volcanic activity from January to March 2016 was mainly characterized by sporadic ash emissions from the NSEC, and by passive degassing mainly from NEC. The daily average of SO_2 fluxes fluctuated at low levels around ~1500 tons/day. The seismic tremor was stable at around low levels for Etna, and thermal activity occurred at reduced levels (<4 MW) (Figure 7).

Starting from the beginning of April, SO_2 emissions gradually intensified and reached daily averaged fluxes of ~2000 t/d, and seismic tremor fluctuated within a subtle increasing trend. No significant thermal anomaly was still observed (Figure 7), and no significant volcanic activity change was reported from field observations, with NEC still passively degassing and NSEC producing a little more frequent ash emissions with occasional blocks ejected.

3.1.2. The May Eruptive Period (16–25 May)

In the early morning of 16 May, strombolian activity resumed at NSEC and NEC, and became very strong at the latter crater on 17 May (also reported by References [68–70]). On 18 May at 10:50 UTC, a lava fountain started at VOR, which had been quiescent since 3 December, 2015. This event marked the beginning of a paroxysmal sequence, lasting until 25 May. The sequence included two additional short-lived lava fountaining episodes at VOR, on 19 May and 21 May, and ended with an intense strombolian and lava flow activity that lasted several hours (Figures 7 and 8). Lava effusion accompanied all the strongest explosive episodes, issuing from both fissures on the summit cone and overflow from its crater rim. In particular, overflowing from the BN crater rim formed a large lava field that extended downslope up to 3 km. The cumulative volume of lava flows and pyroclastic fall deposits was preliminary estimated at 7 to 10 Mm^3 [56], which is similar to what erupted during the December 2015 paroxysmal sequence [30]. Lastly, the summit crater's area was affected by intense deformation with fracturing, subsidence, and a formation of a ~1 km-long and nearly NS oriented graben-like structure (Figure 1b).

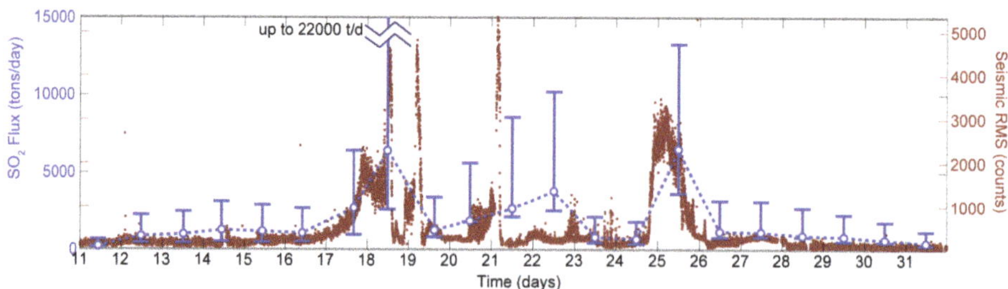

Figure 8. SO_2 fluxes measured from 11 May 2016 to 31 May 2016 encompassed the 16–25 May eruptive period (blue bars show the maximum and minimum values while white dot indicates the daily average), which show a good agreement with seismic tremor amplitude, with relative maxima fitting the phases of enhanced seismic activity linked with paroxysmal events during the eruptive period.

Thermal radiance and the seismic tremor showed coherent increases of three and two order of magnitude, respectively (Figures 7 and 8). The three lava fountains were clearly marked by short-lived peaks in seismic tremor amplitude on 18, 19, and 21 May (Figure 8). The 24–25 May episode of intense strombolian activity/lava fountaining was associated with a wider, longer-lived phase of seismic tremor increase (Figure 8).

The daily averaged SO_2 fluxes increased up to 6000 tons/day (Figure 7), pointing to heightened degassing, nearly tripled with respect to the pre-eruptive phase. A detail of the SO_2 fluxes recorded during the May 2016 paroxysmal sequence is illustrated in Figure 8, where alternation of eruptive and repose periods were evident in the degassing record, with peaks in daily averaged SO_2 flux in three (18, 21 and 25 May) out of four days of paroxysmal activity.

It is also worth noting that only the first (18 May) lava fountaining episode occurred during the UV-camera acquisition temporal interval (see Section 3.2). The high daily averaged SO_2 fluxes obtained on this specific day (Figure 8), thus, reflected the "explosive" SO_2 contribution during the paroxysmal event. In contrast, the SO_2 flux peaks on 22 and 25 May could not be explained by syn-explosive SO_2 release (the lava fountains occurred outside the camera acquisition hours), but rather reflected heightened passive degassing, and/or milder (strombolian) explosive activity, prior to/after the paroxysmal episode itself.

3.1.3. Volcanic Activity after the Eruptive Phase (26 May–31 Dec)

Reduced degassing emissions (<2000 t/d) were measured after the eruptive phase, from May 26th until the end of June (~1 month), which they could interpret as the aftermath of voluminous gas/magma release during the previous December 2015 [30,39,67,72] and May 2016 eruptive sequences. Field observations indicated that the summit craters were weakly fuming and occluded by lavas and pyroclasts. A progressive subsidence occurred on the VOR crater's floor, where lunar cracks formed on the crust of the spatter deposits that had filled the crater. Since early July, intensification of SO_2 degassing (average daily fluxes increased from ~900 to ~3000 tons/day, Figure 7) was accompanied by crack widening near the eastern VOR's crater rim, along the graben-like structure (Figure 1), which culminated, on 7 August, by opening a new 20-m large pit vent, characterized by vigorous high temperature degassing and glowing at night (Figure 1c) [69]. Consistently, thermal activity, as detected by MIROVA, increased in early July from <5 to ~10 MW, which was also reported by Reference [69], and a notable increase was also observed on the seismic tremor (Figure 7). However, repeated inspections on the summit area did not reveal any evidence of explosive activity such as fallout material deposited around the pit vent (as reported in Reference [56]).

From 10 October, after a small explosion, a large subsidence of the BN north-western inner floor was observed. This affected lavas and pyroclastic materials that had filled the central craters during the 2015–2016 paroxysmal sequences. This inner crater subsidence, characterized by episodic collapses, was nicely paralleled by declining SO_2 fluxes down to low values (~1000 tons/day on average). RMS seismic amplitude and thermal radiance slightly decreased as well.

3.2. *Syn-Explosive SO_2 Emissions during the Lava Fountaining Event*

The 18 May lava fountaining event, entirely captured by the SO_2 camera (Figure 9), allowed us to explore to what extent SO_2 cameras could resolve the degassing dynamics associated with a lava fountaining event. Lava fountains are of special concern at Etna since the volcanic ash they inject into the atmosphere is a potential threat to aviation and population living in the surroundings [73]. These events, while very well monitored and understood [34,36–42,50,74–77], are poorly characterized in terms of their associated gas emission rates and volumes. Our 18 May results (Figure 9), therefore, represent one of the first syn-explosive gas records on the volcano [28,30].

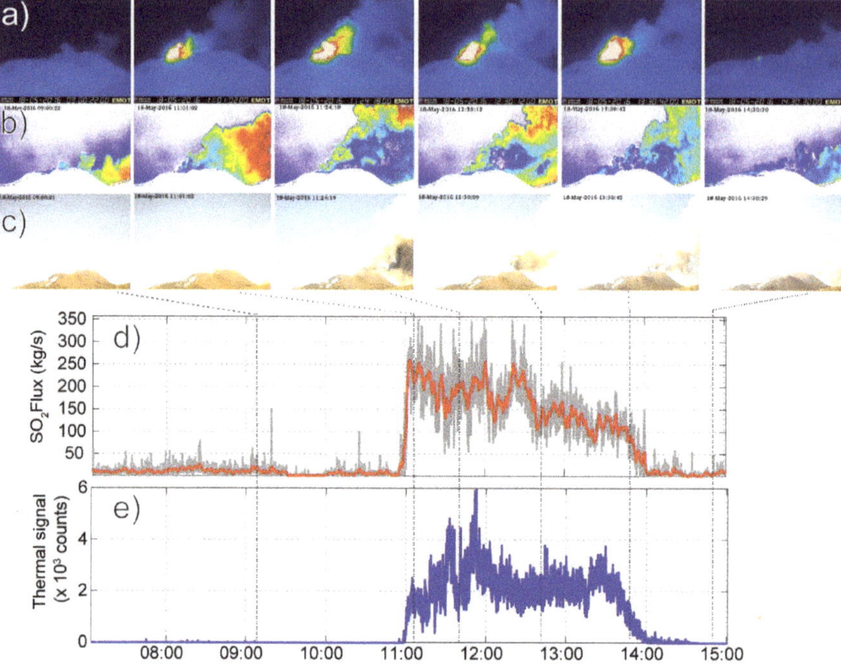

Figure 9. SO_2 camera record during the 18 May lava fountain compared with the thermal signal recorded by the co-located INGV-OE TIR camera. The upper panels show thermal (**a**), absorbance (**b**), and visible (**c**) images of the summit area in specific time intervals. SO_2 fluxes are shown in (**d**) as raw measurements (gray line) with the associated 10 second-window moving average (red line), while the thermal signal is shown in (**e**).

Onset of the lava fountain at 10:57 UTC, as constrained by co-acquired images of the INGV-OE thermal and visual cameras (Figure 9), was clearly detected as a visible SO_2 flux increase up to ~260 kg/s (22,000 t/d), relative to a pre-eruptive level of ~12 kg/s (~1000 t/d). In the following hour, while activity escalated to peak at ~ 11.30–12.00 (see thermal records), a fluctuating and irregular SO_2 flux trend is registered (Figure 9). Co-acquired thermal and visual images (see panels in Figure 9), clearly indicated that negative peaks in the SO_2 flux time-series were systematically associated with the presence of ash. The latter severely impacts SO_2 detection via UV cameras [21], particularly in near-vent measurements where plumes can be very ash-rich and, thus, optically opaque. Thermal and visual observations showed that ash caused high-frequency fluctuations (short-lived negative peaks) in the SO_2 flux record particularly during the paroxysm climax (~11.30–12.00), but also prior to the lava fountain onset, e.g., after 09.00 when the visual camera captured the first ash emission with no thermal anomaly yet detected (Figure 9).

4. Discussion

Our results demonstrated the ability of the novel automatic processing routine to capture fluctuations in the SO_2 regime, which responded to changes in Etna's activity style (Figure 7). The automatically processed SO_2 data were consistent with those manually obtained (Figure 6). However, requiring no operator time and being obtained/delivered in nearly real-time, they represented a clear advantage for monitoring purposes.

Our 2016 UV camera-based dataset also provided novel insights into the relationship between rates/modes of SO_2 release and eruptive/degassing styles. These latter were quite diverse in 2016 since they ranged from non-eruptive quiescent degassing to intense paroxysmal activity in May. Nicely, our automatically processed SO_2 fluxes peaked during heightened activity (Figures 7 and 8) during the May 2016 eruptive sequence, and during the July-August degassing unrest that led to opening of the VOR incandescent vent on 7 August.

One important observation is the mild but significant SO_2 flux increase in April 2016, as demonstrated by a clear change in the slope of the cumulative SO_2 mass (see orange-coloured area in Figure 7a). The SO_2 flux increase started more than one month before onset of the May eruptive sequence, which is a period when the seismic tremor was at average levels and no significant thermal anomaly was detected (Figure 7). We interpreted these escalating SO_2 fluxes as reflecting the slow but systematic increase in the magma supply rate to the shallow (<3 km [78]) Etna's magma feeding system that triggered the May 2016 paroxysmal sequence [68]. However, while the relatively subtle changes in SO_2 passive degassing in April 2016 may have represented a precursory sign for the imminent (May 2016) eruption, we still noticed that a similar SO_2 increase was observed from July to September 2016 (orange-coloured sub-interval in Figure 7a). This did not culminate into an eruption yet (if not for the opening of the 7 August summit vent). Thus, the volcano's feedback to increased shallow magma emplacement (as indicated by increasing SO_2 fluxes) may be different from time to time, perhaps in response to distinct stress regimes prevailing on the upper part of the edifice, and/or temporally varying feedbacks between magma ascent and rates of gravitational spreading of the mobile eastern flank [79,80].

We also characterized SO_2 emissions associated with a lava fountaining event at a high spatial and temporal resolution [28,30]. Even though ash is a serious issue for ground-based SO_2 remote sensing during explosive eruptions, we still noted that the entirety of 18 May eruptive episode was well marked by elevated SO_2 fluxes well above background emissions in the pre-event and post-event phases. In fact, SO_2 emissions manifestly dropped down at only 14:00 UTC, when declining thermal emissions consistently marked lava fountaining termination. From such, we found it useful to tentatively estimate the cumulative SO_2 mass released by the 18 May lava-fountain, by integrating the signal over the eruption duration. We obtained a total SO_2 mass released in the event of ~1700 tons, and an average flux of 158 kg/s (13,600 t/d) for a total duration of ~3 h. We caution this inferred mass corresponds to a lower range estimate, due to the presence of volcanic ash that severely depressed the measured SO_2 signal during the eruption climax.

SO_2 Fluxes during the May 2016 Eruptive Sequence: Comparison with 2014–2015 Results

It is well established that SO_2 fluxes are directly linked to the rate of magma ascent and degassing [78]. Thus, temporal variations in SO_2 fluxes do reflect changes in magma feeding to the volcano's shallow plumbing system, and, as such, may help track transition in activity style, from quiet passive degassing to eruptive periods [81]. On Etna, we now have 3 years of UV camera observations available ([28,30], this study), during which transition from quiescence to eruption has frequently been observed. We, thus, examined our dataset in the attempt to tentatively identify any possible systematic SO_2 flux threshold/trend corresponding to such an activity switch.

For this purpose, in Figure 10, we compared the cumulative SO_2 flux trends (time-normalized) for three different periods encompassing three eruptive paroxysmal episodes, which occurred in August 2014, December 2015, and May 2016, respectively. For each of three events, we calculated the averaged SO_2 fluxes (corresponding to the slopes of the cumulative curve) in the periods before, during, and after eruption, and we found significant similarities between the three events. In each of the three 2014–2016 events, the pre-eruptive fluxes fell in a relatively narrow range, between 1900 and 2500 t/d. The syn-eruptive (during the paroxysmal sequence) fluxes were typically higher, and spanned between 3000 and 5200 t/d (Figure 10), and were the highest during the December 2015 paroxysmal sequence that was consistently the most energetic in the past few years [67]. Lastly, each of the three post-paroxysmal phases was characterized by reduced SO_2 emissions, ranging between 600 and 900 t/d, which impliesa reduced magma supply and degassing of a volatile depleted (residual) magma after each eruptive episode. These ranges were fully in agreement with those indicated by Reference [49] and by the analysis of a 13-year-long dataset.

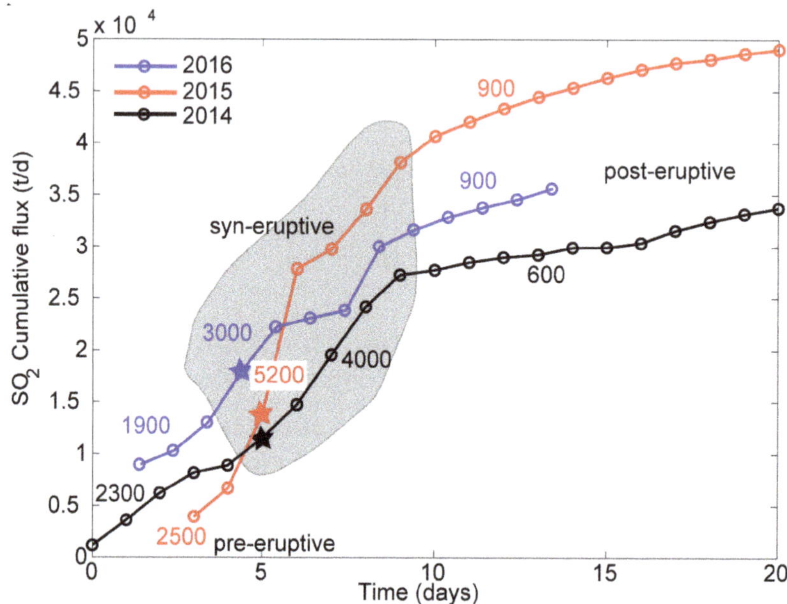

Figure 10. Comparison between cumulative SO_2 flux trends associated with three lava fountaining paroxysmal sequences that occurred at NSEC (black line) in August 2014, and at VOR in December 2015 and May 2016 (red line and blue line). Stars indicate onsets of the paroxysmal sequence. Pre-eruptive, syn-eruptive, and post-eruptive phases show similar SO_2 fluxes for all these three events. In each of these three events, pre-eruptive fluxes range between 1900 and 2500 t/d, syn-eruptive fluxes are the highest (3000–5200 t/d), while post-eruptive fluxes are systematically the lowest (<900 t/d).

Our preliminary results are suggestive of the existence of a systematic pattern in SO_2 emissions that, if confirmed, would imply a recurrent degassing process/mechanism prior to, during, and after the Etna's eruptive periods. Clearly, additional data are required to corroborate this initial hypothesis.

5. Conclusions

SO_2 imaging at Mt Etna during 2016 revealed different styles of gas emissions, which reflected changes in volcanic activity, from quietly passive degassing to eruptive activity (lava fountaining, intense strombolian activity, and lava flowing).

To real-time characterize and monitor this very dynamic activity period, we designed a novel routine to automatically calculate SO_2 fluxes including computer-based detection of image quality, and calculation of plume speed time-series using computer vision libraries. This automatic processing routine allowed us to obtain real-time information on volcano degassing dynamics at high spatial and temporal resolution, which results in a further step in instrumental volcanic gas monitoring. We validated the methodology through a comparison with manually processed results and integration with independent thermal and seismic observations. All these independent datasets showed coherent temporal variations that validated the use of UV cameras for detecting subtle changes in volcanic and degassing activity. Our novel method, thus, promises a step ahead in instrumental volcanic gas monitoring.

Our automatically derived SO_2 flux time-series were used to constrain degassing regimes on Mt. Etna in 2016. We have shown that our automatically processed SO_2 fluxes peaked during heightened activity, such as during the May 2016 eruptive sequence, and during a phase of elevated degassing in July–August 2016, which culminated with the opening of a new summit incandescent vent. The May 2016 sequence was preceded by a circa one-month-long phase of a mild but detectable SO_2 flux increase, and was followed by an abrupt drop in degassing. Comparison between our SO_2 flux time-series in May 2016 and those associated with two other eruptive paroxysmal sequences (occurred in 2014 and 2015) highlighted strong similarities in SO_2 flux dynamics, which implies the possible existence of thresholds that distinguish between degassing regimes, prior to, during, and after eruptions. Pre-paroxysm SO_2 fluxes were found to have consistent values (of ~2000 t/d) during the three episodes. Similarly, the highest SO_2 fluxes (from 3000 t/d up to 5200 t/d on a daily average basis) were identified during the three eruptive sequences, while post-eruptive were systematically characterized by reduced degassing (<1000 t/d). This result, if confirmed by future observations, may bring implications for identifying switches in volcanic activity regime. We believe this methodology can be successfully exported on other open-vent active volcanoes, after a relatively brief period of calibration.

We also tested the ability of UV camera records to characterizing SO_2 emissions during a lava fountain episode. This is, to the best of our knowledge, one of the first SO_2 camera record at high spatial and temporal resolution of an ongoing lava fountaining [28,30]. We have reported on high levels of degassing during the lava fountain (of up to 260 kg/s), from which we assessed a lower limit (due to the ash presence within the plume) for the cumulative SO_2 mass of ~1700 tons emitted during a lava fountain episode.

Supplementary Materials: The following are available online at http://www.mdpi.com/2072-4292/11/10/1201/s1.

Author Contributions: D.D.D., A.A., and M.R. conceived the manuscript. D.D.D. developed the automatic UV camera processing algorithm. M.B. designed the UV camera system hardware. R.D. and G.T. validated the automatic algorithm. D.C. processed thermal remote sensing data. E.P. collected and processed thermal camera data. M.C. supervised the work and contributed to write the manuscript.

Funding: The European Community's Seventh Framework Program under grant agreement 305377 (BRIDGE Project) funded this research.

Acknowledgments: We wish to thank Salvo Caffo (Ente Parco dell'Etna) for administrative support. Nino Giuffrida (Funivia dell'Etna) and Filippo Greco (INGV-OE) are acknowledged for support in the field. Three anonymous reviewers are acknowledged for their constructive comments, which improved the manuscript.

Conflicts of Interest: The authors declare no conflict of interest. The funders had no role in the design of the study, in the collection, analyses, or interpretation of data, in the writing of the manuscript, or in the decision to publish the results.

References

1. Oppenheimer, C.; Fischer, T.P.; Scaillet, B. Volcanic Degassing: Process and Impact. In *Treatise on Geochemistry, The Crust*; Holland, H.D., Turekian, K.K., Eds.; Elsevier: Amsterdam, The Netherlands, 2014; Volume 4, pp. 111–179.
2. Fischer, T.P.; Chiodini, G. Volcanic, Magmatic and Hydrothermal Gas Discharges. In *Encyclopaedia of Volcanoes*, 2nd ed.; Academic Press: Cambridge, MA, USA, 2015; pp. 779–797. [CrossRef]
3. Aiuppa, A. Volcanic Gas Monitoring. In *Volcanism and Global Environmental Change*; Schmidt, A., Fristad, K.E., Elkins-Tanton, L.T., Eds.; Cambridge University Press: Cambridge, UK, 2015; pp. 81–96.
4. Wallace, P.J.; Kamenetsky, V.S.; Cervantes, P. Melt inclusion CO_2 contents, pressures of olivine crystallization, and the problem of shrinkage bubbles. *Am. Mineral.* **2015**, *100*, 787–794. [CrossRef]
5. Saccorotti, G.; Iguchi, M.; Aiuppa, A. In situ Volcano Monitoring: Present and Future. In *Volcanic Hazards, Risks, and Disasters*; Shroder, J.F., Papale, P., Eds.; Elsevier: Amsterdam, The Netherlands, 2015; pp. 169–202.
6. Galle, B.; Oppenheimer, C.; Geyer, A.; McGonigle, A.J.S.; Edmonds, M.; Horrocks, L.A. A miniaturized ultraviolet spectrometer for remote sensing of SO_2 fluxes: A new tool for volcano surveillance. *J. Volcanol. Geotherm. Res.* **2003**, *119*, 241–254. [CrossRef]

7. Galle, B.; Johansson, M.; Rivera, C.; Zhang, Y.; Kihlman, M.; Kern, C.; Lehmann, T.; Platt, U.; Arellano, S.; Hidalgo, S. Network for Observation of Volcanic and Atmospheric Change (NOVAC): A global network for volcanic gas monitoring: Network layout and instrument description. *J. Geophys. Res.* **2010**, *115*, D05304. [CrossRef]
8. Christopher, T.; Edmonds, M.; Humphreys, M.C.S.; Herd, R.A. Volcanic gas emissions from Soufrière Hills Volcano, Montserrat 1995–2009, with implications for mafic magma supply and degassing. *Geophys. Res. Lett.* **2010**, *37*, L00E04. [CrossRef]
9. Mori, T.; Shinohara, H.; Kazahaya, K.; Hirabayashi, K.; Matsushima, T.; Mori, T.; Ohwada, M.; Odai, M.; Iino, H.; Miyashita, M. Time-averaged SO_2 fluxes of subduction-zone volcanoes: Example of a 32-year exhaustive survey for Japanese volcanoes. *J. Geophys. Res. Atmos.* **2013**, *118*, 8662–8674. [CrossRef]
10. Hidalgo, S.; Battaglia, J.; Arellano, S.; Steele, A.; Bernard, B.; Bourquin, J.; Galle, B.; Arrais, S.; Vásconez, F. SO_2 degassing at Tungurahua volcano (Ecuador) between 2007 and 2013: Transition from continuous to episodic activity. *J. Volcanol. Geothem. Res.* **2015**, *298*, 1–14. [CrossRef]
11. McGonigle, A.J.S.; Hilton, D.R.; Fischer, T.P.; Oppenheimer, C. Plume velocity determination for volcanic SO_2 flux measurements. *Geophys. Res. Lett.* **2005**, *32*. [CrossRef]
12. Mori, T.; Burton, M.R. The SO_2 camera: A simple, fast and cheap method for ground-based imaging of SO_2 in volcanic plumes. *Geophys. Res. Lett.* **2006**, *33*, L24804. [CrossRef]
13. Burton, M.R.; Prata, F.; Platt, U. Volcanological applications of SO_2 cameras. *J. Volcanol. Geothem. Res.* **2015**, *300*, 2–6. [CrossRef]
14. Burton, M.R.; Salerno, G.; D'Auria, L.; Caltabiano, T.; Murè, F.; Maugeri, R. SO_2 flux monitoring at Stromboli with the new permanent INGV SO_2 camera system: A comparison with the FLAME network and seismological data. *J. Volcanol. Geothem. Res.* **2015**, *300*, 95–102. [CrossRef]
15. McGonigle, A.J.S.; Pering, T.D.; Wilkes, T.C.; Tamburello, G.; D'Aleo, R.; Bitetto, M.; Aiuppa, A.; Willmott, J.R. Ultraviolet Imaging of Volcanic Plumes: A New Paradigm in Volcanology. *Geosciences* **2017**, *7*, 68. [CrossRef]
16. Dalton, M.P.; Waite, G.P.; Watson, I.M.; Nadeau, P.A. Multiparameter quantification of gas release during weak Strombolian eruptions at Pacaya Volcano, Guatemala. *Geophys. Res. Lett.* **2010**, *37*. [CrossRef]
17. Holland, P.A.S.; Watson, M.I.; Phillips, J.C.; Caricchi, L.; Dalton, M.P. Degassing processes during lava dome growth: Insights from Santiaguito lava dome, Guatemala. *J. Volcanol. Geotherm. Res.* **2011**, *202*, 153–166. [CrossRef]
18. Kazahaya, R.; Mori, T.; Takeo, M.; Ohminato, T.; Urabe, T.; Maeda, Y. Relation between single very long period pulses and volcanic gas emissions at Mt. Asama, Japan. *Geophys. Res. Lett.* **2011**, *38*. [CrossRef]
19. Kazahaya, R.; Shinohara, H.; Mori, T.; Iguchi, M.; Yokoo, A. Pre-eruptive inflation caused by gas accumulation: Insight from detailed gas flux variation at Sakurajima volcano, Japan. *Geophys. Res. Lett.* **2016**, *43*, 21. [CrossRef]
20. Nadeau, P.A.; Jose, L.P.; Waite, G.P. Linking volcanic tremor, degassing, and eruption dynamics via SO_2 imaging. *Geophys. Res. Lett.* **2011**, *38*, L01304. [CrossRef]
21. Tamburello, G.; Aiuppa, A.; Kantzas, E.P.; McGonigle, A.J.S.; Ripepe, M. Passive vs. active degassing modes at an open-vent volcano (Stromboli, Italy). *Earth Planet. Sci. Lett.* **2012**, *359–360*, 106–116. [CrossRef]
22. Tamburello, G.; Aiuppa, A.; McGonigle, A.J.S.; Allard, P.; Cannata, A.; Giudice, G.; Kantzas, E.P.; Pering, T.D. Periodic volcanic degassing behavior: The Mount Etna example. *Geophys. Res. Lett.* **2013**, *40*, 4818–4822. [CrossRef]
23. Waite, G.P.; Nadeau, P.A.; Lyons, J.J. Variability in eruption style and associated very long period events at Fuego volcano, Guatemala. *J. Geophys. Res. Solid Earth* **2013**, *118*, 1526–1533. [CrossRef]
24. Pering, T.D.; Tamburello, G.; Mc Gonigle, A.J.S.; Aiuppa, A.; Cannata, A.; Giudice, G.; Patanè, D. High time resolution fluctuations in volcanic carbon dioxide degassing from Mount Etna. *J. Volcanol. Geotherm. Res.* **2014**, *270*, 115–121. [CrossRef]
25. Pering, T.D.; Tamburello, G.; McGonigle, A.J.S.; Aiuppa, A.; James, M.R.; Lane, S.J.; Sciotto, M.; Cannata, A.; Patanè, D. Dynamics of mild strombolian activity on Mt. Etna. *J. Volcanol. Geothem. Res.* **2015**, *300*, 103–111. [CrossRef]
26. Pering, T.D.; McGonigle, A.J.S.; James, M.R.; Tamburello, G.; Aiuppa, A.; Delle Donne, D.; Ripepe, M. Conduit dynamics and post explosion degassing on Stromboli: A combined UV camera and numerical modeling treatment. *Geophys. Res. Lett.* **2016**, *43*, 5009–5016. [CrossRef] [PubMed]

27. Nadeau, P.A.; Werner, C.A.; Waite, G.P.; Carn, S.A.; Brewer, I.D.; Elias, T.; Sutton, A.J.; Kern, C. Using SO_2 camera imagery and seismicity to examine degassing and gas accumulation at Kīlauea Volcano, May 2010. *J. Volcanol. Geotherm. Res.* **2015**, *300*, 70–80. [CrossRef]
28. D'Aleo, R.; Bitetto, M.; Delle Donne, D.; Tamburello, G.; Battaglia, A.; Coltelli, M.; Patanè, D.; Prestifilippo, M.; Sciotto, M.; Aiuppa, A. Spatially resolved SO_2 flux emissions from Mt Etna. *Geophys. Res. Lett.* **2016**, *43*. [CrossRef] [PubMed]
29. Wilkes, T.C.; Pering, T.D.; McGonigle, A.J.S.; Tamburello, G.; Willmott, J.R. A Low-Cost Smartphone Sensor-Based UV Camera for Volcanic SO_2 Emission Measurements. *Remote Sens.* **2017**, *9*, 27. [CrossRef]
30. D'Aleo, R.; Bitetto, M.; Delle Donne, D.; Coltelli, M.; Coppola, D.; Mc Cormick Kilbride, B.; Pecora, E.; Ripepe, M.; Salem, L.; Tamburello, G.; et al. Understanding the SO_2 Degassing Budget of Mt Etna's Paroxysms: First Clues from the December 2015 Sequence. *Front. Earth Sci.* **2019**, *6*, 239. [CrossRef]
31. Kern, C.; Sutton, J.; Elias, T.; Lee, L.; Kamibayashi, K.; Antolik, L.; Werner, C. An automated SO_2 camera system for continuous, real-time monitoring of gas emissions from Kīlauea Volcano's summit Overlook Crater. *J. Volcanol. Geothem. Res.* **2015**, *300*, 81–94. [CrossRef]
32. Delle Donne, D.; Tamburello, G.; Aiuppa, A.; Bitetto, M.; Lacanna, G.; D'Aleo, R.; Ripepe, M. Exploring the explosive-effusive transition using permanent ultraviolet cameras. *J. Geophys. Res. Solid Earth* **2017**, 4377–4394. [CrossRef]
33. Tamburello, G.; Kantzas, E.P.; McGonigle, A.J.S.; Aiuppa, A. Vulcamera: A program for measuring volcanic SO_2 using UV cameras. *Ann. Geophys.* **2011**, *54*, 2.
34. Alparone, S.; Andronico, D.; Lodato, L.; Sgroi, T. Relationship between tremor and volcanic activity during the Southeast Crater eruption on Mount Etna in early 2000. *J. Geophys. Res.* **2003**, *108*, 2241. [CrossRef]
35. Vergniolle, S.; Ripepe, M. From Strombolian explosions to fire fountains at Etna Volcano (Italy): What do we learn from acoustic measurements? *Geol. Soc. Lond. Spec. Publ.* **2008**, *307*, 103–124. [CrossRef]
36. Patanè, D.; Aiuppa, A.; Aloisi, M.; Behncke, B.; Cannata, A.; Coltelli, M.; Di Grazia, G.; Gambino, S.; Gurrieri, S.; Mattia, M. Insights into magma and fluid transfer at Mount Etna by a multiparametric approach: A model of the events leading to the 2011 eruptive cycle. *J. Geophys. Res. Solid Earth* **2013**, *118*, 3519–3539. [CrossRef]
37. Bonaccorso, A.; Calvari, S.; Currenti, G. From source to surface: Dynamics of Etna's lava fountains investigated by continuous strain, magnetic, ground and satellite thermal data. *Bull. Volcanol.* **2013**, *75*. [CrossRef]
38. Bonaccorso, A.; Calvari, S.; Linde, A.; Sacks, S. Eruptive processes leading to the most explosive lava fountain at Etna volcano: The 23 November 2013 episode. *Geophys. Res. Lett.* **2014**, *41*, 4912–4919. [CrossRef]
39. Bonaccorso, A.; Calvari, S. A new approach to investigate an eruptive paroxysmal sequence using camera and strainmeter networks: Lessons from the 3–5 December 2015 activity at Etna volcano. *Earth Planet. Sci. Lett.* **2017**, *475*, 231–241. [CrossRef]
40. Spampinato, L.; Sciotto, M.; Cannata, A.; Cannavò, F.; La Spina, A.; Palano, M.; Salerno, G.; Privitera, E.; Caltabiano, T. Multiparametric study of the February–April 2013 paroxysmal phase of Mt. Etna New South-East crater. *Geochem. Geophys. Geosyst.* **2015**, *16*, 1932–1949. [CrossRef]
41. Gambino, S.; Cannata, A.; Cannavò, F.; La Spina, A.; Palano, M.; Sciotto, M.; Spampinato, L.; Barberi, G. The unusual 28 December 2014 dike-fed paroxysm at Mount Etna: Timing and mechanism from a multidisciplinary perspective. *J. Geophys. Res. Solid Earth* **2016**, *121*, 2037–2053. [CrossRef]
42. Ripepe, M.; Marchetti, E.; Delle Donne, D.; Genco, R.; Innocenti, L.; Lacanna, G.; Valade, S. Infrasonic Early Warning System for Explosive Eruptions. *J. Geophys. Res. Solid Earth* **2018**, *123*, 9570–9585. [CrossRef]
43. Coppola, D.; Laiolo, M.; Cigolini, C.; Delle Donne, D.; Ripepe, M. Enhanced volcanic hot-spot detection using MODIS IR data: Results from the MIROVA system. *Geol. Soc. Lond. Spec. Publ.* **2015**, *426*, SP426. [CrossRef]
44. Andò, B.; Pecora, E. An advanced video-based system for monitoring active volcanoes. *Comput. Geosci.* **2006**, *32*, 85–91. [CrossRef]
45. Kantzas, E.P.; McGonigle, A.J.S.; Tamburello, G.; Aiuppa, A.; Bryant, R.G. Protocols for UV camera volcanic SO_2 measurements. *J. Volcanol. Geothem. Res.* **2010**, *194*, 55–60. [CrossRef]
46. Kern, C.; Deutschmann, T.; Vogel, L.; Wöhrbach, M.; Wagner, T.; Platt, U. Radiative transfer corrections for accurate spectroscopic measurements of volcanic gas emissions. *Bull. Volcanol.* **2010**, *72*, 233–247. [CrossRef]
47. McGonigle, A.J.S. Measurement of volcanic SO_2 fluxes with differential optical absorption spectroscopy. *J. Volcanol. Geoth. Res.* **2007**, *162*, 111–122. [CrossRef]

48. Vandaele, A.C.; Simon, P.C.; Guilmot, J.M.; Carleer, M.; Colin, R. SO_2 absorption cross section measurement in the UV using a Fourier transform spectrometer. *J. Geophys. Res.* **1994**, *99*, 25–599. [CrossRef]
49. Caltabiano, T.; Burton, M.; Giammanco, S.; Allard, P.; Bruno, N.; Murè, F.; Romano, R. Volcanic Gas Emissions from the Summit Craters and Flanks of Mt. Etna, 1987–2000. In *Mt. Etna: Volcano Laboratory*; Bonaccorso, A., Calvari, S., Coltelli, M., Del Negro, C., Falsaperla, S., Eds.; Geophysical Monograph Series; American Geophysical Union: Washington, DC, USA, 2004; Volume 143, pp. 111–128.
50. Ulivieri, G.; Ripepe, M.; Marchetti, E. Infrasound reveals transition to oscillatory discharge regime during lava fountaining: Implication for early warning. *Geophys. Res. Lett.* **2013**, *40*, 3008–3013. [CrossRef]
51. Zuccarello, L.; Burton, M.R.; Saccorotti, G.; Bean, C.J.; Patanè, D. The coupling between very long period seismic events, volcanic tremor, and degassing rates at Mount Etna volcano. *J. Geophys. Res. Solid Earth* **2013**, *118*, 4910–4921. [CrossRef]
52. Salerno, G.G.; Burton, M.R.; Di Grazia, G.; Caltabiano, T.; Oppenheimer, C. Coupling Between Magmatic Degassing and Volcanic Tremor in Basaltic Volcanism. *Front. Earth Sci.* **2018**, *6*, 157. [CrossRef]
53. Harris, A.J.L. *Thermal Remote Sensing of Active Volcanoes: A User's Manual*; Cambridge University Press: Cambridge, UK, 2013; p. 716.
54. Aiuppa, A.; De Moor, J.M.; Arellano, S.; Coppola, D.; Francofonte, V.; Galle, B. Tracking formation of a lava lake from ground and space: Masaya volcano (Nicaragua), 2014–2017. *Geochem. Geophys. Geosyst.* **2018**, *19*, 496–515. [CrossRef]
55. Kern, C. Spectroscopic Measurements of Volcanic Gas Emissions in the Ultra-Violet Wavelength Region. Ph.D. Thesis, University of Heidelberg, Heidelberg, Germany, 2009.
56. INGV—Osservatorio Etneo. 2016. Available online: http://www.ct.ingv.it/en/rapporti/multidisciplinari.html (accessed on 26 February 2019).
57. Bluth, G.S.J.; Shannon, J.M.; Watson, I.M.; Prata, A.J.; Realmuto, V.J. Development of an ultra-violet digital camera for volcanic SO_2 imaging. *J. Volcanol. Geothem. Res.* **2007**, *161*, 47–56. [CrossRef]
58. Thomas, H.E.; Prata, A.J. Computer vision for improved estimates of SO_2 emission rates and plume dynamics. *Int. J. Remote Sens.* **2018**, *39*, 1285–1305. [CrossRef]
59. Lucas, B.D.; Kanade, T. An iterative image registration technique with an application to stereo vision. *IJCAI* **1981**, *81*, 674–679.
60. Bradski, G.; Kaehler, A. *Learning OpenCV: Computer Vision with the OpenCV Library*; O'Reilly Media: Newton, MA, USA, 2008.
61. Moffat, A.J.; Millan, M.M. The applications of optical correlation techniques to the remote sensing of SO_2 plumes using sky light. *Atmos. Environ.* **1971**, *5*, 677–690. [CrossRef]
62. Stoiber, R.E.; Malinconico, L.L., Jr.; Williams, S.N. Use of the Correlation Spectrometer at Volcanoes. In *Forecasting Volcanic Events*; Tazieff, H., Sabroux, J.C., Eds.; Elsevier: New York, NY, USA, 1983; pp. 425–444.
63. Caltabiano, T.; Romano, R.; Budetta, G. SO_2 flux measurements at Mount Etna (Sicily). *J. Geophys. Res.* **1994**, *99*, 12809–12819. [CrossRef]
64. Bruno, N.; Caltabiano, T.; Romano, R. SO_2 emissions at Mt.Etna with particular reference to the period 1993–1995. *Bull. Volcanol.* **1999**, *60*, 405–411. [CrossRef]
65. Burton, M.R.; Caltabiano, T.; Salerno, G.; Murè, F.; Condarelli, D. Automatic measurements of SO_2 flux on Stromboli using a network of scanning ultraviolet spectrometers. *Geophys. Res. Abstr.* **2004**, *6*, 03970.
66. Salerno, G.G.; Burton, M.R.; Oppenheimer, C.; Caltabiano, T.; Tsanev, V.; Bruno, N. Novel retrieval of volcanic SO_2 abundance from ultraviolet spectra. *J. Volcanol. Geotherm. Res.* **2009**, *181*, 141–153. [CrossRef]
67. Pompilio, M.; Bertagnini, A.; Del Carlo, P.; Di Roberto, A. Magma dynamics within a basaltic conduit revealed by textural and compositional features of erupted ash: the December 2015 Mt. Etna paroxysms. *Sci. Rep.* **2017**, *7*, 4805. [CrossRef] [PubMed]
68. Cannata, A.; Di Grazia, G.; Giuffrida, M.; Gresta, S. Space Time Evolution of Magma Storage and Transfer at Mt. Etna Volcano (Italy): The 2015–2016 Reawakening of Voragine Crater. *Geochem. Geophys. Geosyst.* **2018**, *19*, 471–495. [CrossRef]
69. Marchese, F.; Neri, M.; Falconieri, A.; Lacava, T.; Mazzeo, G.; Pergola, N.; Tramutoli, V. The Contribution of Multi-Sensor Infrared Satellite Observations to Monitor Mt. Etna (Italy) Activity during May to August 2016. *Remote Sens.* **2018**, *10*, 1948. [CrossRef]

70. Edwards, M.J.; Pioli, L.; Andronico, D.; Scollo, S.; Ferrari, F.; Cristaldi, A. Shallow factors controlling the explosivity of basaltic magmas: The 17–25 May 2016 eruption of Etna Volcano (Italy). *J. Volcanol. Geothem. Res.* **2018**, *357*, 425–436. [CrossRef]
71. Coppola, D.; Piscopo, D.; Laiolo, M.; Cigolini, C.; Delle Donne, D.; Ripepe, M. Radiative heat power at Stromboli volcano during 2000–2011: Twelve years of MODIS observations. *J. Volcanol.Geoth. Res.* **2012**, *215*, 48–60. [CrossRef]
72. Vulpiani, G.; Ripepe, M.; Valade, S. Mass discharge rate retrieval combining weather radar and thermal camera observations. *J. Geophys. Res. Solid Earth* **2016**, *121*, 5679–5695. [CrossRef]
73. Scollo, S.; Boselli, A.; Coltelli, M. Monitoring Etna volcanic plumes using a scanning LiDAR. *Bull. Volcanol.* **2012**, *74*, 2383–2395. [CrossRef]
74. Harris, A.J.L.; Neri, M. Volumetric observations during paroxysmal eruptions at Mount Etna: Pressurized drainage of a shallow chamber or pulsed supply? *J. Volcanol. Geothem. Res.* **2002**, *116*, 79–95. [CrossRef]
75. Dubosclard, G.; Donnadieu, F.; Allard, P.; Cordesses, R.; Hervier, C.; Coltelli, M.; Privitera, E.; Kornprobst, J. Doppler radar sounding of volcanic eruption dynamics at Mount Etna. *Bull. Volcanol.* **2004**, *66*, 443–456. [CrossRef]
76. Andronico, D.; Corsaro, R.A. Lava fountains during the episodic eruption of South–East Crater (Mt. Etna), 2000: Insights into magma-gas dynamics within the shallow volcano plumbing system. *Bull. Volcanol.* **2011**, *73*, 1165. [CrossRef]
77. Calvari, S.; Salerno, G.G.; Spampinato, L.; Gouhier, M.; La Spina, A.; Pecora, E.; Harris, A.J.L.; Labazuy, P.; Biale, E.; Boschi, E. An unloading foam model to constrain Etna's 11–13 January 2011 lava fountaining episode. *J. Geophys. Res.* **2011**, *116*, B11207. [CrossRef]
78. Allard, P. Endogenous magma degassing and storage at Mount Etna. *Geophys. Res. Lett.* **1997**, *24*, 2219–2222. [CrossRef]
79. Bonforte, A.; Guglielmino, F.; Puglisi, G. Interaction between magma intrusion and flank dynamics at Mt. Etna in 2008, imaged by integrated dense GPS and DInSAR data. *Geochem. Geophys. Geosyst.* **2013**, *14*, 2818–2835. [CrossRef]
80. Bonaccorso, A.; Bonforte, A.; Gambino, S. Twenty-five years of continuous borehole tilt and vertical displacement data at Mount Etna: Insights on long-term volcanic dynamics. *Geophys. Res. Lett.* **2015**, *42*, 10222–10229. [CrossRef]
81. Oppenheimer, C.; Scaillet, B.; Martin, R.S. Sulfur degassing from volcanoes: Source conditions, surveillance, plume chemistry and earth system impacts. *Rev. Mineral. Geochem.* **2011**, *73*, 363–421. [CrossRef]

Article

Deformations and Morphology Changes Associated with the 2016–2017 Eruption Sequence at Bezymianny Volcano, Kamchatka

René Mania [1,*], Thomas R. Walter [1], Marina Belousova [2], Alexander Belousov [2] and Sergey L. Senyukov [3]

1 Department of Geophysics, GFZ German Research Centre for Geosciences, Telegrafenberg, 14473 Potsdam, Germany; twalter@gfz-potsdam.de

2 Institute of Volcanology and Seismology, Far East Branch, Russian Academy of Sciences, Piipa boulevard, 9, 683006 Petropavlovsk-Kamchatskii, Russia; chikurachki1@gmail.com (M.B.); belousov@mail.ru (A.B.)

3 Kamchatkan Branch of Geophysical Survey, Russian Academy of Sciences, Piipa boulevard, 9, 683006 Petropavlovsk-Kamchatskii, Russia; ssl@emsd.ru

* Correspondence: rene.mania@gfz-potsdam.de; Tel.: +49-331-288-28665

Received: 29 April 2019; Accepted: 24 May 2019; Published: 29 May 2019

Abstract: Lava domes grow by extrusions and intrusions of viscous magma often initiating from a central volcanic vent, and they are frequently defining the source region of hazardous explosive eruptions and pyroclastic density currents. Thus, close monitoring of dome building processes is crucial, but often limited to low data resolution, hazardous access, and poor visibility. Here, we investigated the 2016–2017 eruptive sequence of the dome building Bezymianny volcano, Kamchatka, with spot-mode TerraSAR-X acquisitions, and complement the analysis with webcam imagery and seismic data. Our results reveal clear morphometric changes preceding eruptions that are associated with intrusions and extrusions. Pixel offset measurements show >7 months of precursory plug extrusion, being locally defined and exceeding 30 m of deformation, chiefly without detected seismicity. After a short explosion, three months of lava dome evolution were characterised by extrusions and intrusion. Our data suggest that the growth mechanisms were significantly governed by magma supply rate and shallow upper conduit solidification that deflected magmatic intrusions into the uppermost parts of the dome. The integrated approach contributes significantly to a better understanding of precursory activity and complex growth interactions at dome building volcanoes, and shows that intrusive and extrusive growth is acting in chorus at Bezymianny volcano.

Keywords: Bezymianny; volcano deformation; monitoring; lava dome; inflation; SAR imaging; radar pixel offsets

1. Introduction

Many active volcanoes, about 200 worldwide [1], generate lava domes that are often characterised by hazardous explosive eruptions that involve flank instability [2]. As domes grow, the outer flanks oversteepen until they collapse and perilous pyroclastic flows are produced that purge down the slopes, affecting regions at kilometres distance to the dome [3]. Lava domes are thought to grow by interactions between magma injections into the dome (i.e., endogenous dome growth) and the addition of extrusion layers on the top of the carapace (i.e., exogenous dome growth) [4–7]. So far, these two styles of growth are considered as endmembers, with few examples showing higher complexity that could be instrumentally recorded in nature, such as at the dome building volcanoes Mount St. Helens, Unzen, or Soufrière Hills [5–7]. Geophysical sensors often observe short-term precursors, such as seismicity, enhanced rockfall intensity, alternating volcanic gas emissions, or localised deformation when magma reaches shallow depths prior to imminent eruptions [8]. Interferometric Synthetic Aperture Radar (InSAR), for example, provides an estimation of precursory deformation on the mm to cm scale over short [9] and long [10] periods of time, yet the technique is affected by atmosphere and it is less effective when volcanoes are covered in snow or when ground motion exceeds the maximum detectable deformation gradient [11]. Moreover, determining deformation that is associated with dome building volcanoes, and therefore identifying the particularities of lava dome growth, is challenging due to the small dimensions and the hazardous access of most domes. Successful approaches barely include in-situ and, more often, remote sensing approaches, such as ground-fixed cameras [12,13] or satellite radar amplitude images [14–16]. A noteworthy case of the strength of camera monitoring for tracking deformation is that of Mount St. Helens during the 2004–2008 dome growth episode, which allowed for the spatial and temporal quantification of endogenous and exogenous growth [17]. The value of satellite radar observation, on the other hand, was underlined during the 2010 cataclysmic eruption at Merapi, where Synthetic Aperture Radar (SAR) amplitude scenes provided vital support in the early detection of dome growth and the associated hazard assessment [14]. The weaknesses of SAR, in turn, come from the poor revisit period (several days), and geometric distortions that limit interpretations due to the regions of shadow, foreshortening, and layover effects.

Here, we integrate the strengths of these techniques to better understand the current dome growth mechanisms acting during the January 2016–June 2017 eruption sequence at Bezymianny. We use camera monitoring to roughly identify topographic changes at Bezymianny's flank, and we employ a pixel offset tracking algorithm on high-resolution TerraSAR-X amplitude images to quantify ground motion in range and azimuth direction. We show the details of plug extrusion that were identified at least seven months before the first documented effusive eruption, and that exogenous growth at Bezymianny was likely preceded by intrusions into the northern part of the composite dome. The complexity of the observed cascade suggests that this finding may also provide a basis for dome growth observation at other dome building volcanoes, ultimately promoting the understanding of dome growth and related hazard assessment.

2. Bezymianny

2.1. Volcanological Background

Bezymianny is an andesitic, dome building volcano (~3000 m a.s.l.) that is located within the Klyuchevskoy Group of Volcanoes (KGV) in Kamchatka, Russia (Figure 1a). It is thought that Bezymianny, Klyuchevskoy, but also the further south located Tolbachik, derive their fluids from a common deep parental magma chamber at 30 km depth [18,19]. During ascent beneath Bezymianny, the volatile-rich magma arrests at different levels, which are likely associated with magma chambers, at approximately 15 km and 5 km, but also possibly at 1.5 km depth [18,20–22].

Bezymianny is a relative young volcano (5.5 ka) whose geologic history was characterised by major eruptive activity between 2400 and 1700 and 1350–1000 before present [23]. In 1955–1956, Bezymianny re-emerged with a phase that culminated in a cataclysmic sector collapse and directed lateral blast eruption, which left behind a horseshoe-shaped crater moat (Figure 1b) [24–26]. Eruption characteristics showed strong similarities to the catastrophic eruption at Mount St. Helens in 1980 [27–29]. After the 1956 eruption at Bezymianny, near-continuous, mostly endogenous dome growth started to fill the horseshoe shaped crater floor until 1965 [24,25,30,31]. Since 1977, on average, 1–2 explosive eruptions occurred, which showed a characteristic cyclic behaviour: initially, days to weeks long-lasting summit plug extrusions were followed by Vulcanian explosions and pyroclastic flows; eruptions then eventually ceased with lava flow emplacements and degassing until the volcano became quiet again [30,31]. Only few eruptions during the 1980's and 1990's were solely characterised by effusive activity, or lava flow emplacements prior to explosions [30]. By 2004, Bezymianny's dome was completely covered with lava flows, and multiple explosions on its top between 2005 and 2012 a new relatively stable summit crater [19,32,33]. After four years of quiescence, activity initiated in 2016 and was followed by long effusive activity (5 December 2016–9 March 2017) and two strong explosive eruptions on 9 March and 16 June 2017 [31]. Today, the morphology of Bezymianny is characterized by the remnants of the 1956 sector collapse amphitheatre (the "somma") and the presence of an approximately 500–600 m high composite dome in the centre (Figure 1b).

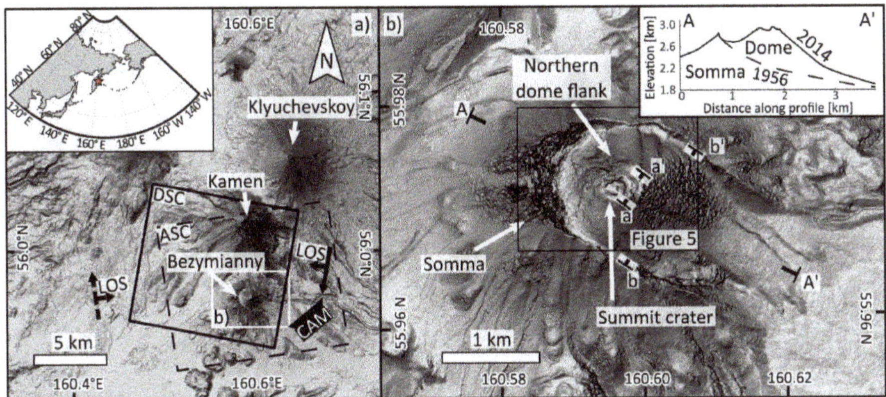

Figure 1. (**a**) Shaded relief map (TanDEM digital elevation model from 2014) of Bezymianny and its closest neighbouring volcanoes in Kamchatka, Far East Russia (star in inset map). Location of the time-lapse camera and footprint of the TerraSAR-X (TSX) satellite are indicated by CAM and black box, respectively. Orthogonal arrows show flight and line-of-sight (LOS) directions of the descending (DSC) and ascending (ASC) TSX satellites, respectively. ASC is shown in dashed lines, as this paper focuses on the more regular DSC data. White box denotes area shown in (**b**). (**b**) Close-up shaded relief map of Bezymianny showing the 1956 collapse scar (somma), the subsequently evolved central composite dome, and its recent summit crater. Profile A-A' indicates approximate 1956 collapse plane (dashed line). Small letter profiles a–a' and b–b' show landmarks that are used for scale approximation of camera images.

2.2. Monitoring Activities at Bezymianny

Bezymianny is one of the most active volcanoes in Kamchatka that poses a risk to air traffic between North America and Asia. However, multiparametric and long-term monitoring is challenging due to the remoteness of the volcano. During the past two decades, seismic monitoring was realized by the Kamchatkan Branch of Geophysical Survey [34], allowing for eruption precursor identification days to weeks before eruptions [35]. Enhanced frequency of rockfalls from the central dome is easily identified and indicative of renewed activity at Bezymianny [35,36]. Besides characteristic tremors and high frequency seismic signatures, long-period (LP) seismicity may also identify heralding eruptions [19], but earlier studies suggest that only one out of four eruptions were preceded by LP events [36]. Eruptions at Bezymianny are sometimes concurrent with activity at Klyuchevskoy, which may strongly obscure the records of Bezymianny's seismic activity [35,36].

Besides the routine seismic monitoring, increasing importance has been ascribed to remote sensing data analysis. Remote sensing comes with two main motivations: first, general monitoring of the volcanic activities exploiting cost-free data and web portals; second, experimental and scientific in-depth analysis of selected volcanic crisis. For instance, the instruments of the Advanced Spaceborne Thermal Emission and Reflectance Radiometer (ASTER) have been episodically used to study heat radiation during eruptions of the last few decades [37–39]. Overall, these studies have identified enhanced ground temperature anomalies as a common precursor for Bezymianny's eruptions, although two eruptions were reported without a preceding change in the thermal level [39].

Yet, existing real-time monitoring methods, as well as event-based observations, could not assess detailed dome growth processes of Bezymianny. Here, we investigate webcam images and high-resolution satellite radar data that cover the December 2016–June 2017 eruption sequence at Bezymianny. The data catalogue enabled the observation of the volcano with unprecedented precision of precursory activity, as well as exogenous and endogenous dome growth.

3. Data and Methods

This study concentrates on camera monitoring and satellite radar data acquired in 2016 and 2017. The eruption had a precursory phase, as identified by rockfalls and seismicity, then an effusive eruption between 5 December 2016 and 28 February 2017, followed by (i) the effusive and explosive eruption on 9 March, and then (ii) the explosive 16 June 2017 eruption. Details of these three stages (precursor–effusive–explosive) were identified in the data. We compare our camera and SAR results to the seismic records of Bezymianny [40].

3.1. Camera Monitoring and Mimatsu-Diagrams

Previous studies have demonstrated the strength of time-lapse camera analysis for the determination of morphology changes at volcanoes, which substantially contributed to the spatial deformation monitoring. Mimatsu (1962) already highlighted the significance of optical volcano monitoring by constantly recording the volcano's changing shape on his office paper window during the dome growth episode at Showa Shinzan volcano, Japan. Johnson et al. [41] used video-derived imagery to track dome uplift at Santiaguito volcano, Guatemala, and successfully correlated the results with long-period seismic signals. A fixed camera network that was installed around Mount St. Helens, USA, permitted the precise estimation of growth and strain rates during the 2004–2008 spine extrusion [13,17]. Based on displacements between fixed time-lapse photographs, Walter et al. [42] showed that dome deformation at Merapi volcano, Indonesia, is strongly governed by the local topography. In this context, we used time-lapse imagery of Bezymianny during the 2016–2017 eruption series to record the changes at the summit following Mimatsu's approach.

The employed day and night capable network camera (Axis P1346) that is operated by the Kamchatkan Branch of Geophysical Survey [43] has a focal length of 4 mm and it produces one image (2048 × 1536 pixels) per second (Figure 2a). The time-lapse camera is located 7 km to the southeast (160.696E, 55.94N; Figure 1a) and it captures Bezymianny as well as the neighbouring Kamen and Klyuchevskoy volcanoes. The investigated period from May 2016 to August 2017 encompasses 579,180 time lapse images. The images were weeded out for night, no-operation, and cloudy records, but also records where the camera lens was covered with snow. The remaining dataset was then visually checked for clear view and high contrast images that were taken at approximately the same daytime, of which 12 representative images were manually selected and cropped to the area of Bezymianny (Figure 2a,e and Figure S1). From the image stack, we follow the silhouette of the volcano image-by-image, which is referred to as a Mimatsu diagram. Image offsets due to strong winds, recurrent snow cover on top of the camera, and/or temperature changes of the installed camera gear are corrected by manual alignment (translation in x and y, rotation around the image centre) of all the images with respect to the master scene (7 May 2016). We favoured manual over automatic alignment as Bezymianny's dome was recurrently covered with snow, limiting automatic algorithm performance.

To quantify topographic changes in the field-of-view (FOV), two scales were derived by measuring pixel distances between conspicuous landmarks on the master image (summit crater, 1956 collapse scar) that correspond to landmarks on the digital elevation model (Figures 1b and 2b). Thus, the scale varies between ~4.2–5.3 m/pixel, which is related to the reduction of the three-dimensional topography into the camera's two-dimensional FOV. As the pixels of the images are approximately squared, the scale is assumed to be valid for horizontal and vertical changes. Since most of the images unveiled insufficient contrast conditions and strongly varying colours (e.g., recurrent snow cover), the outlines of Bezymianny's composite dome were manually mapped and stacked in the resulting Mimatsu diagram.

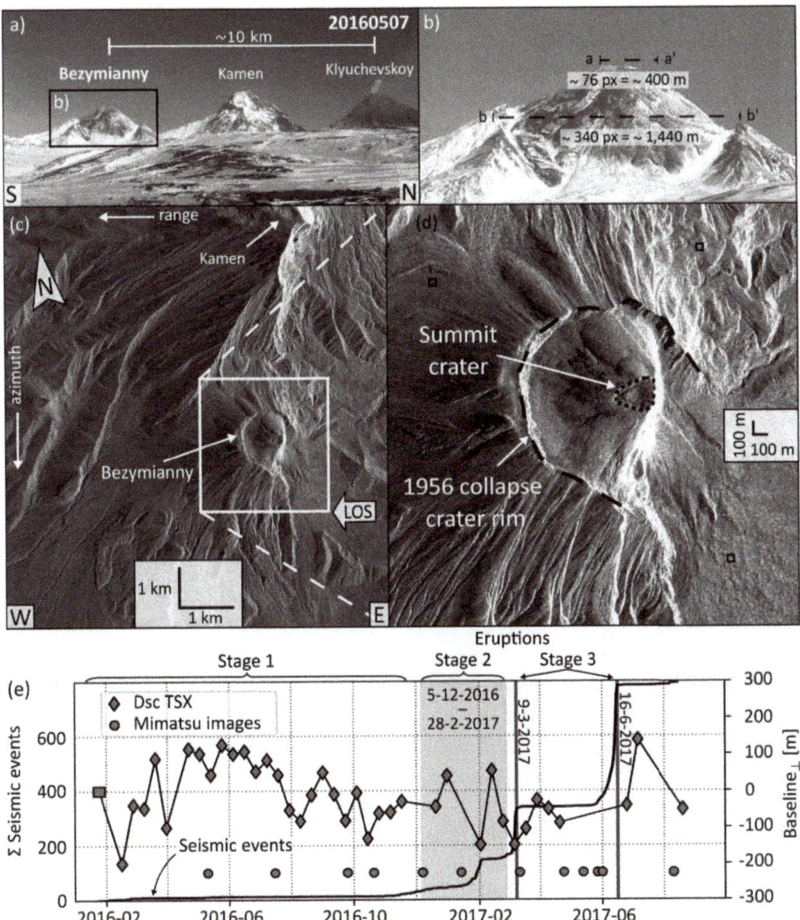

Figure 2. (**a**) Monitoring camera image of Bezymianny and its neighbouring volcanoes on 7 May 2016. View is to the west. Black box denotes image details shown in: (**b**) Close-up view of Bezymianny. Pixel and metric distances between a–a′ and b–b′ are derived from a terrain model (cf. Figure 1b). (**c**) Descending non-geocoded spotlight-mode TSX amplitude image from 23 September 2016 (cf. Figure 1a). Flight direction (azimuth) and line-of-sight (LOS) or range direction of the satellite are indicated. White box shows area used for pixel offset tracking displayed in: (**d**) Close-up of Bezymianny. Small black boxes mark assumed stable areas (i.e., no deformation) referred to later in displacement analysis. (**e**) Cumulative number of seismic events in a 6 km radius to the volcano. Available TSX acquisitions with their perpendicular baselines (Baseline$_{\perp}$) to adjacent acquisitions.

3.2. Synthetic Aperture Radar (SAR)

3.2.1. SAR Data Set and Amplitude Images

SAR systems emit electromagnetic pulses to the Earth's surface. Based on the backscattered intensity (amplitude) and the time delay, the amplitude images are obtained from the illuminated surface independent of daytime and weather conditions [11]. Earlier applications have used SAR amplitude imagery to monitor and comprehend the evolution of dome building [14–16,44] and other volcanic processes [45–47]. At Bezymianny, we employ 39 descending (track 11) spotlight [48] TerraSAR-X satellite (TSX, wavelength = 31 mm) amplitude images that have been acquired between

January 2016 and August 2017, with a recurrence time of mostly 11 days (Figure 2a and Figure S2, Table S1). The images were recorded with an incidence angle of 38.7°, and have a resolution cell (pixel) spacing of 0.9 × 1.25 m in slant-range and azimuth direction, respectively. We also studied eight ascending (track 64) (Figures S9, S10 and Table S2) spotlight-mode TSX images (incidence angle = 49°; 0.9 × 1.20 m in slant-range x azimuth, respectively). The backscattered radar signal is confined by the acquisition geometry, as well as the roughness and dielectric properties of the illuminated surface [11]. Thus, rougher and smoother surfaces correspond to brighter and darker pixels in the amplitude image (Figure 2d,e), respectively. To visualise the reflectivity changes between amplitude images, we created change difference maps [15] that show regions of unchanged, decreased, and increased reflectivity values with yellow, magenta, and green colours, respectively. We note that the amplitude information is strongly influenced by steep topography and the oblique radar acquisition geometry, which causes distortions, such as the foreshortening of Bezymianny's eastern flank or shadowing at the summit crater floor (both track 11; Figure 2e). We mainly focus on track 11, as the viewing geometry of track 64 creates pronounced foreshortening and shadowing of the western and eastern flanks. However, although distortions in track 11 prohibit comprehensive observations of Bezymianny, the TSX amplitude data set provides unique information to determine exogenous and endogenous dome growth processes during the 2016–2017 eruptive sequence.

3.2.2. SAR Co-registration and Pixel Offset Measurements

Tracking pixel offsets of the co-registered SAR amplitude images may provide unambiguous range and azimuth quantification of surface displacements, where InSAR measurements become decorrelated [11,49]. Here, we co-registered all of the descending scenes with respect to the reference image (master) from 25 January 2016 with the Gamma remote sensing software (Gamma) [50] (Figure S3). We used a Pléiades digital elevation model (DEM) with a grid size of 2 m for real to SAR coordinate conversion. Look-up tables were calculated for the first scene (sub-master) of individual amplitude pairs, which subtracts topographic effects in the sub-master scene from range and azimuth pixel offsets. Moreover, to retain the deformation signal, orbital related offsets were subtracted via application of a cross-correlation based offset estimation, which determines a linear fit all over the offset tracking image pairs. Lastly, the TSX spotlight SAR scenes were deramped to remove azimuth ramps in the Doppler frequency. Subsequently, we employed an iterative image offset tracking algorithm with Gamma on the amplitude data with maximum resolution (i.e., no multilooking). Initially, we used a large tracking patch of 256 × 256 pixels (step-size = 4 pixels) to estimate the large pixel offsets. Then, we refined the offset estimation in a subsequent step with smaller patches that varied from 160 × 160 to 32 × 32 pixels to identify smaller displacements. To avoid aliasing of the spectrum, we oversampled the data by a factor of two. Appendix A details the error estimation.

4. Results

4.1. *Precursory Ground Movement*

4.1.1. Precursory TSX Observations

Analysis of 26 TSX amplitude images that were acquired between January and November 2016 provides detailed evolution of surface motion at Bezymianny before the first documented effusive eruption in December 2016. For this episode, we calculated range offset maps based on three cross-correlation patches (32, 64, 96 pixels), and derived mean range offset rates under consideration of the TSX recurrence time (Table S1) for a selected region within the summit crater (Figure 3h).

Initial ground motion (0–0.08 m d^{-1}) is observed between January–April 2016, while the amplitude images do not show clear changes in reflectivity (Figure 3a,e). Simultaneously, only few seismic events occurred between February–March 2016. In May 2016, a new radar shadow area appears at the western portion of the crater floor, which gradually increases in size until August 2016 (Figure 3b). During this time, the crater floor moves at increased, but near constant, rates of 0.07–0.13 m d^{-1} towards the satellite (Figure 3i). By the beginning of September 2016, the shadow area considerably increased, but it started to diverge with brighter pixels in-between at its western portion (Figure 3c). Moreover, new radar shadows appeared at the north-eastern summit rim. Concurrent range pixel offsets within the summit crater show a stepwise increase from 0.16 to 0.23–0.25 m d^{-1} during October (Figure 3g,i). At the same time, and after five months of quiescence, seismic events were recorded again, and seismicity continued throughout November 2016. Simultaneously, the summit floor radar shadow considerably increased eastwards, and the ground motion is marked by the most significant stepwise increase from 0.43 m d^{-1} to 0.63 m d^{-1} (Figure 3d,i). Eventually, the total detected ground displacement at the summit crater floor amounts to approximately 39 m towards the satellite (Figure 3h).

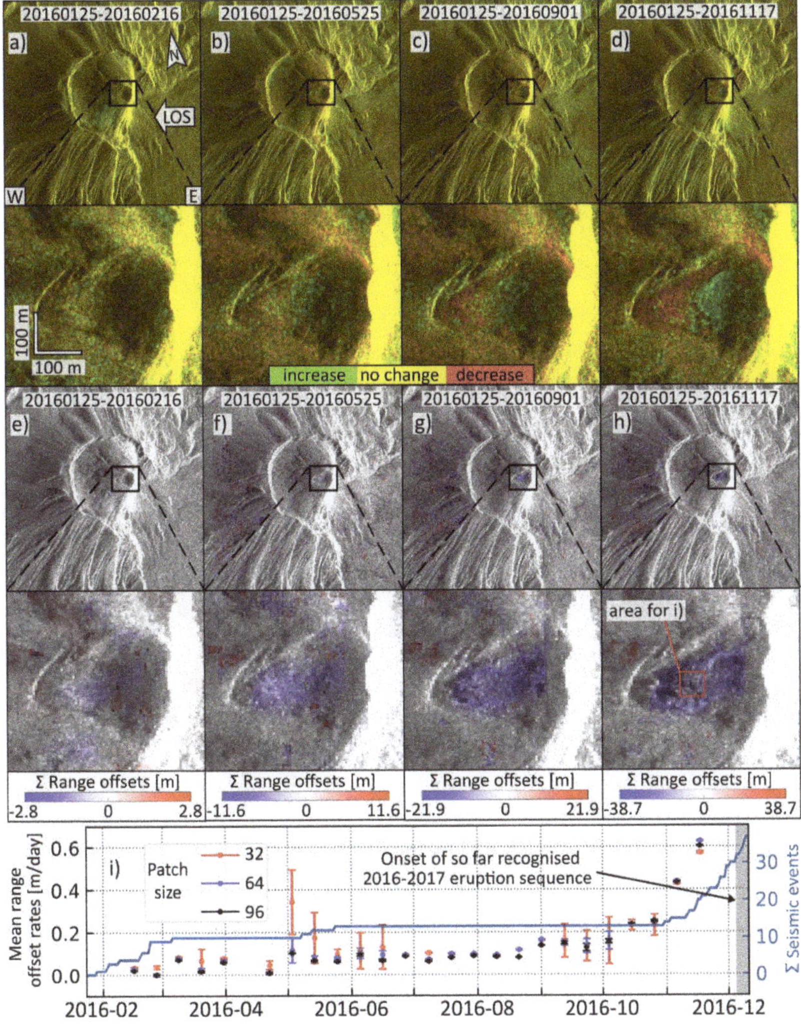

Figure 3. Eruption precursory deformation at Bezymianny between January and November 2017 determined in radar amplitude imagery change difference maps (**a**–**d**). Close-ups of these maps show the gradual emergence of a rigid body that produces a successively larger radar shadow at the summit crater floor displayed by magenta colours.

Contemporaneously, new shadows appear at the northern summit rim. Displayed cumulative range offset maps of consecutive amplitude images (**e–h**) are calculated with a cross-correlation patch of 64 pixels. Red and blue pixel displacements reflect motion away or towards the satellite, respectively. (**i**) The lower row shows the temporal evolution of ground movement calculated for a 10 × 10 pixels area located within the crater (red box in (**h**)). Error bars correspond to offset deviations in selected stable regions (cf. Figure S4). See text for details.

4.1.2. Precursory Webcam Observations

To visually confirm the overall detected precursory TSX ground motion, we created a Mimatsu-diagram based on five clear webcam images acquired between May and December 2016 (Figure 4a,b). The camera images reveal slight growth of the summit between May–September 2016, whereas other images of the same period show intermittent translucent degassing and white steaming (Figure S7). First clearly distinguishable topographic changes are discernable in October 2016, where the eastern summit uplifts by 9–22 m, and elevations at the southern summit changes by 15 m in FOV. This topographic growth occurs at approximately the same time as the first significant rise of TSX range offsets (see above) and the onset of seismic activity in October 2016. Between September and beginning of December 2016, summit degassing significantly enhanced (Figure S7), and the 7 December 2016 Mimatsu image reveals a striking topographic uplift of 9–37 m of the eastern summit. The latter observations concurrently occurred with the gradual increase of detected seismic events as well as with the most significant increase of radar-derived ground motion detected in November 2016 (Figure 3).

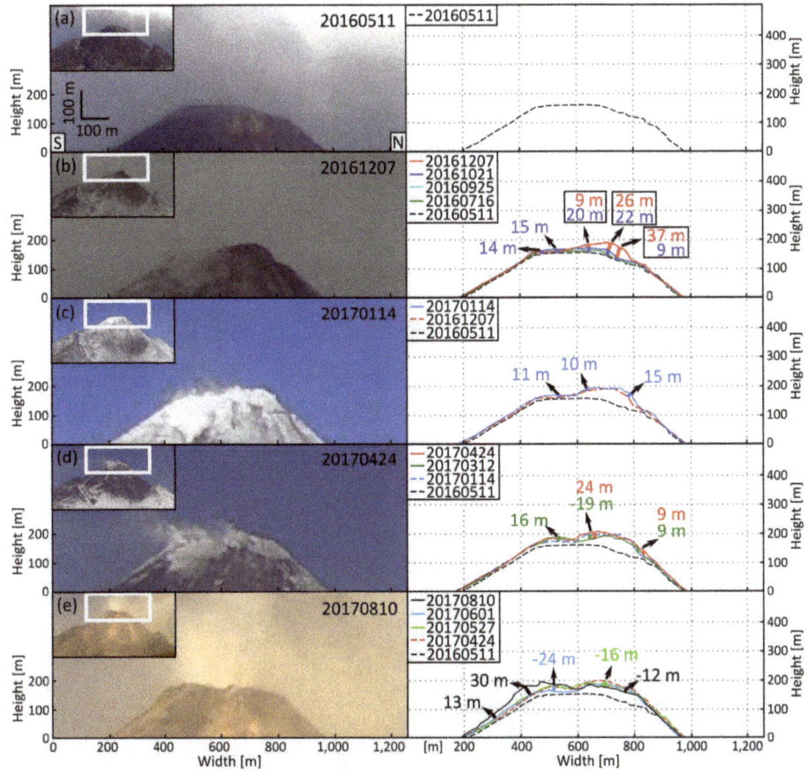

Figure 4. Mimatsu diagrams depicting Bezymianny's summit: (**a**) before onset of the eruption sequence, (**b**,**c**) during the 5 December 2016 and 28 March 2017 eruption, (**d**,**e**) after the first (9 March 2017) and second (16 June 2017) explosive eruptions, respectively. Coloured bold lines correspond to the elevation change of the summit with respect to the previous camera image.

4.2. Co-eruptive Ground Movement Observations

4.2.1. Co-eruptive TSX Observations

Co-eruptive TSX amplitude images reveal a tongue-like reflectivity change at the western dome, which depicts the emplacement of an extensive lava flow (flow 1) (Figure 5a2 and Figure S5). Its lower parts are marked by radar shadow casting crevasses, as well as radially and flow parallel oriented shadow casting ridges that are bisected by a significantly larger and irregularly oriented, but also flow perpendicular, shadow-casting ridge (Figure 6a). Between 17 November and 20 December 2016, the lava effusion is accompanied by northward-directed rigid (inelastic) bulging of the northern composite dome flank, and the detected azimuth offset rates increase with elevation from approximately 0.05 to 0.1 m d^{-1} (Figure 5a3 and Figure S5). During the end of December 2016, no motion of the northern carapace was determined (Figure S6), whereas between 31 December 2016 and 2 February 2017 the same motion direction and elevation-related distribution is observed, but at significantly higher rates of approximately 0.1–0.5 m d^{-1} (Figure 5b3 and Figure S5). A substantial increase in seismic activity is observed during the second half of January 2017, which rapidly decreases by the end of January 2017 (Figure 2e).

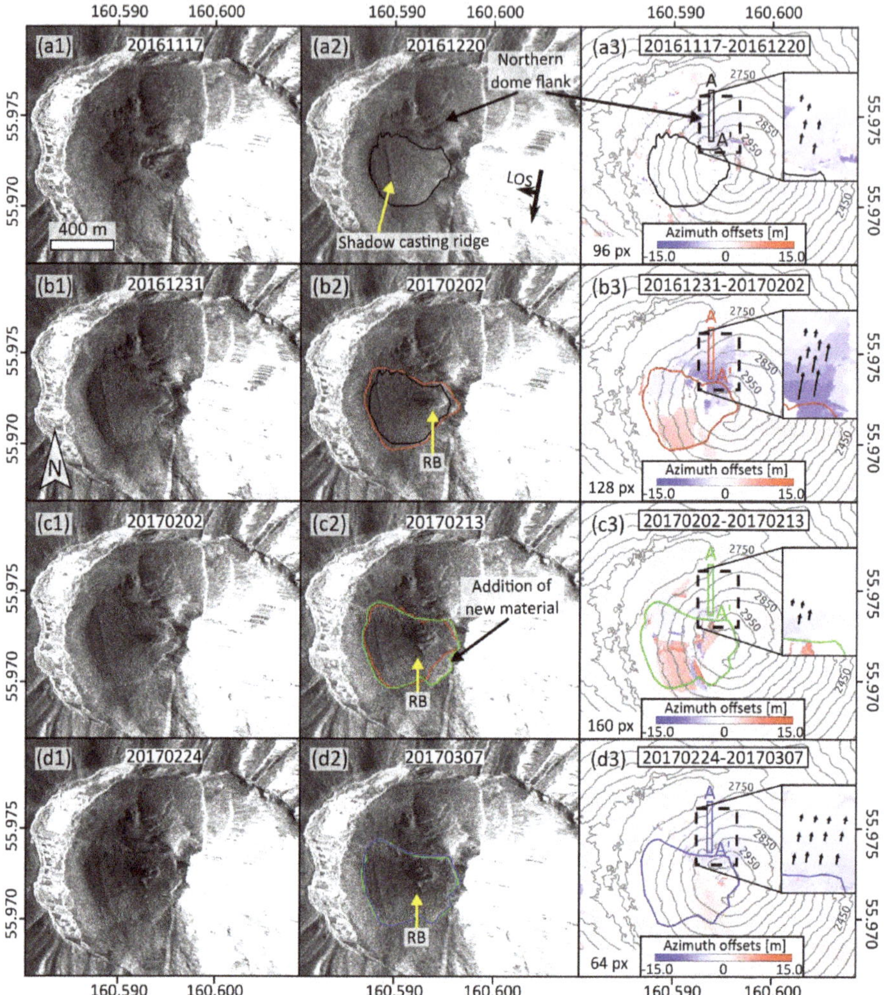

Figure 5. Ground motion detected in TSX amplitude images during the 5 December 2016–28 February 2017 effusive eruption. Extent is indicated in Figure 1b. (**a1,2**), (**b1,2**), (**c1,2**), (**d1,2**) Geocoded TSX amplitude images with indicated changes. Left column are master images, central column are slave images. RB = rigid body. The colour coded outlines show extent of lava flow 1 compared with previous extent. (**a3–d3**) Azimuth offset maps and 50 m contour lines of Bezymianny. Red and blue colours reflect movement along (approximately southward) or against (approximately northward) the TSX flight direction (LOS). Employed cross-correlation patches are indicated. Arrows in close-up views indicate relative increase of azimuth displacements from the foot to the summit of the dome. Displacement increases towards the summit. Compare Figure S5 for swath profiles A–A′.

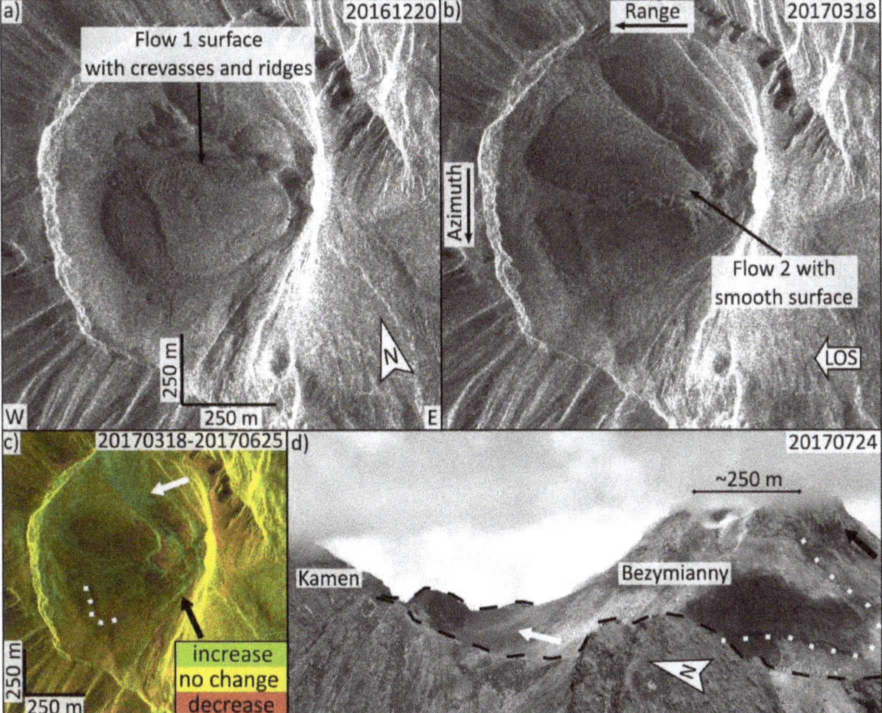

Figure 6. Topographic changes at Bezymianny between December 2016 and June 2017. (**a**,**b**) show amplitude images of Bezymianny after the 5 December 2016 and 9 March 2017 eruptions, both marked by large lava flows. (**c**) Amplitude change difference map and (**d**) aerial image after the 16 June 2017 eruption. White and black arrows (**c**,**d**) indicate deposition of pyroclastic deposits and an extrusive body, respectively. Both, the pyroclastic deposits and the extrusive body are indicated by reflectivity increases in (**c**) (green colouring), whereby the extrusive body is barely perceptible.

By beginning of February 2017, the surface fractures of flow 1 are widely distributed, and a new contrasting radar shadow appears at the summit that indicates uplift of a new rigid body (Figure 5b2). The subsequent scene from 13 February 2017 shows that the reflectivity pattern of flow 1 at the summit area significantly extended, and the shadow-producing rigid body moved westwards (Figure 5c2). Simultaneously, the northern flank bulges again northwards with rates between 0.1–0.3 m d^{-1} that increase with elevation (Figure 5c3 and Figure S5). In addition, azimuth offset maps between December 2016 and February 2017 reveal differential motion of flow 1, where its northern and southern segments move approximately north- and southwards, respectively. The following amplitude pair does not reveal significant azimuth ground motion anymore (Figure S6).

The azimuth offset map of the last descending amplitude image pair (24 February–7 March 2017) again reveals northward directed motion of the northern composite dome flank, where the rates repeatedly increase with increasing elevation from 0.1 to 0.4 m d^{-1} (Figure 5d3 and Figure S5). In addition, minor southward directed motion indicates the bulging of the southern flank. Lastly, the only ascending amplitude image pair (17–28 February 2017) of this episode shows southward directed motion of the southern composite dome flank (Figure S11), which implies that the southward and northward directed bulging detected in the last descending pair occurred at different times.

The beginning of March 2017 is characterised by a significant increase in seismic activity that culminated on 9 March 2017, when the first explosive eruption occurred (Figure 2e). The descending amplitude image from 18 March 2017 shows another tongue-like reflectivity change at the north-western composite dome flank, which reveals the outpouring of the second lava flow (flow 2) (Figure 6b). Subsequent scenes do not unveil significant changes neither of flow 2 nor of the dome flanks (Figure S2). The surface reflectivity of flow 2 is more homogeneously distributed than that of flow 1.

By end of May 2017, seismicity picked-up and significantly increased until 16 June 2017, when the second explosive eruption occurred. The descending change difference map between the 18 March and 25 June 2017 amplitude images shows strongly lifted reflectivity within the western and northern 1956 crater moat (Figure 6c). Most of the surface of the two previously emplaced lava flows is now covered by new material, which corresponds to the pyroclastic deposits that were observed in aerial photographs from July 2017 (Figure 6d).

4.2.2. Co-eruptive Webcam Observations

Only one clear image could be selected to determine morphologic changes during the co-eruptive episode because of strong fumarolic activity at Bezymianny and predominant poor weather conditions between 8 December 2016 and 9 March 2017 (cf. Figure 2e). The corresponding Mimatsu diagram between 7 December 2016 and 14 January 2017 reveals growth of the northern and southern summit with approximately 15 m and 11 m, respectively (Figure 4c). The timing of summit growth correlates with the onset of enhanced seismicity, and may also correspond to the beginning of the uplift of the rigid body that produce the significant radar shadow detected on 2 February 2017 (Figures 2e and 5b2).

The 12 March 2017 Mimatsu image demonstrates partial destruction of Bezymianny's summit after the explosive 9 March 2017 eruption, yet most of the previously developed morphology remained (Figure 4d). The latter agrees with radar shadows that are discernible in both the 7 and 18 March 2017 descending amplitude images (Figures 5d2 and 6a). However, the subsequent 24 April 2017 Mimatsu image shows growth of the summit by 24 m, while during May and June 2017, a repeated partial destruction of the summit is observed (Figure 4e).

Eventually, the 10 August 2017 Mimatsu image reveals that significant parts of the previously determined summit accumulation are destroyed, whereas the southern summit portion is characterised by extrusion of new material (Figures 4e and 6c,d). This last and most significant optically detected summit morphology change corresponds well with the compelling reflectivity change of the summit determined in the 25 June 2017 TSX amplitude image, whereby the southern summit growth is only weakly represented in this amplitude image.

4.3. *Three Stage activity*

Overall, the TSX amplitude data provided the most detailed (spatially and temporally) observations of precursory and co-eruptive ground motion during the 2016–2017 eruption sequence. Near constant precursory ground motion is observed between January and October 2016, which then rapidly increased two months (stage 1) prior the 5 December 2016–28 February 2017 eruption. The latter observation agrees well with Mimatsu-derived topographic growth of Bezymianny's eastern summit that was detected in December 2016. During the effusive December 2016–February 2017 eruption (stage 2), the radar data unveil recurrent flank motion at different rates that always increase with increasing elevation. Moreover, SAR data show differential lava flow motion as well as the uplift of a rigid body during January–February 2017 that subsequently moved westwards as the summit reflectivity significantly changed. The timing agrees well with considerably enhanced seismicity and optically derived growth of the eastern summit. The surface texture of the two lava flows 1 and 2 (stage 3, 9 March 2017) differ markedly, as flow 1 is characterised by shadow casting crevasses that are absent on flow 2. Eventually, the 16 June 2017 eruption (stage 3) produced pyroclastic deposits that cover most of the two lava flows and fill the northern 1956 crater moat.

5. Discussion

Our data set captured seven to nine months of precursory ground motion as a rigid body extruded at the summit prior to the first documented effusive December 2016–February 2017 eruption. We interpret the rigid body as extruded, solidified conduit material that we refer to as a plug. Subsequent determined differential lava flow motion was accompanied by a second plug extrusion that rafted westwards as new lava was emplaced near the summit. Besides exogenous growth, the SAR amplitude images also unveiled distinct, recurrent endogenous growth stages as Bezymianny's dome bulged northwards multiple times. Hereinafter, we first shed light on the limitation of employed techniques, and we will then discuss our observations of the different dome growth stages at Bezymianny.

5.1. Limitations

Seismic activity beneath Bezymianny is monitored by a widespread array of seismometers that covers activity of all volcanoes within the Klyuchevskoy Group of Volcanoes (see locations for stations in Shapiro et al. [19]). The data used here is from a local catalogue and reflects seismic events detected beneath the volcano within a radius of 6 km. However, accurate allocation of events is impeded when other nearby volcanoes are active. In fact, Klyuchevskoy was very active in 2016 (cf. Figure S7), which caused the detection of only few events that are directly associated with Bezymianny. This attracts the attention to other methods to determine activity at Bezymianny, such as SAR and optical observations. However, these techniques are also not immune to shortcomings that have to be considered for interpretation.

Visual observations of volcanic unrest and eruptions are important at volcano observatories to examine topographic changes, levels of gas emissions, and other processes. Time-lapse cameras are increasingly used for documentation and observation as they require low budget and maintenance [13,51]. However, the number of cameras, their location, installation, and weather conditions have strong effects on the resolution to retrieve quantitative information. Multiple terrestrial cameras enable to break down the three dimensional deformation over time, as was demonstrated for dome growth at, for instance, Mount St. Helens during the 2004–2008 eruption [17]. Single cameras, in turn, may have the disadvantage of reducing the three-dimensional displacement into its two-dimensional FOV. Therefore, determined and quantified topographic changes at Bezymianny may over- or underestimate the total amount of deformation, as the absolute displacement may encompass deformation further away or closer towards the camera's FOV. Moreover, the low spatial resolution is confined to the large distance (7 km) of the camera, which is focused on Bezymianny and its neighbouring volcanoes. This significantly lowers the image contrast and it causes blurry edges at Bezymianny's dome (Figure S8), which, together with fumarolic activity and background clouds, may have strong impact on the outline mapping quality. While most of the error contributions cannot be further quantified, the mapping error may be equal to the calculated pixel-size-range (i.e., 4.2–5.3 m/pixel), as the choice of the pixels along the cone outline depends on subjective and biased decisions during the mapping. In addition, the employed metric pixel conversion strongly depends upon topography, distance, and image distortion, which was not corrected for in the images. Thus, the mapping error may be even higher. Yet, the webcam imagery provided valuable qualitative information that supports and complements deductions from seismic and TSX observations, such as deformation in foreshortening areas of the radar data. Thus, time-lapse camera observations constitute an indispensable tool to monitor Bezymianny.

Tracking changes at dome building volcanoes is vital for hazard assessment because of the close link to their explosive potential, dome collapse, and associated pyroclastic density currents generation [5,14]. Yet, lava domes are often tied to the volcanic summit, which is often obscured by frequent cloud cover. SAR systems, in turn, penetrate this cover, and hence may significantly aid in identifying dome growth processes. Here, we analysed the amplitude information of TSX data and employed a pixel offset tracking technique to estimate deformation at Bezymianny during the 2016–2017 eruption series. However, specific steps within the processing chain may have substantial influence on the distribution of pixel offsets. Speckle, for instance, causes random noise in the amplitude information that may result in the occurrence of randomly distributed offsets. On the other hand, speckle on surfaces also increases the tracking quality of these features. Multilooking (down-sampling), in turn, may significantly reduce amplitude noise. Yet, it also decreases the spatial resolution and may lead to the omission of small scale pixel offsets, such as the episodically detected distension of the dome or smaller differential offsets during the precursory episode. Additionally, strong scatterers within the cross-correlation window may obtain a high displacement weighting that dominates the whole patch. This causes the appearance of patch-like offsets, where the strong reflector related offset propagates throughout several overlapping and adjacent windows [52]. Patch like offsets occur, for instance, outside the 1956 crater rim (Figure 3e–h), or in the azimuth offset map of the northern dome between images 31 December 2016–2 February 2017 as offsets decrease stepwise downslope (Figure 5b3 and Figure S5). The latter may be caused by bright scatterers at the edges of older lava flows located along the northern dome flank. In addition, offset tracking between May and September 2016 showed that larger cross-correlation windows (64 and 96 pixels) detected near continuous range offset rates with minor errors, whereas the smallest window (32 pixels) revealed varying offsets with much larger errors (Figure 3i and Figure S4). Also, calculated SNRs do not sufficiently aid in the identification for erroneous offsets as both low and high SNRs were calculated for offsets close to zero in the stable areas. The latter is most prominent during the summer months where SNRs in stable areas are temporarily significantly higher, while SNRs in the summit region do not reveal a considerable change (Figure S4). However, other potential offset tracking error sources may result from changing amplitudes that are related to slope processes (e.g., gravity driven toppling rocks as moving bright scatterers), downslope block addition onto lava flows after its emplacement increasing the surface roughness, changing amplitude values due to intermittent snow cover [46], or layover effects as observed at Cleveland volcano [16]. Layover effects may be observable at the 1956 collapse scar rim, but they could not be observed at the summit. Other limitations in pixel offset tracking occur when pixels disappear due to strong motion as observed at the front of flow 1, or when surfaces are shifted into foreshortening areas as observed at the summit crater rim during the precursory plug extrusion episode. Despite the error sources, the method enabled the detailed quantification and analysis of exogenous and endogenous growth stages at Bezymianny.

5.2. *Implications and Interpretations of Eruptive Events*

The observations from the TSX data allowed us to identify different stages of ground motion activity and to distinguish processes from plug extrusion over endogenous dome deformation to lava flow emplacement. This enables us to derive a conceptual model of volcanic growth at Bezymianny.

5.2.1. Precursory Deformation

After approximately four years of quiescence, our amplitude data reveal persistent range motion at Bezymianny's summit seven to nine months prior to the first documented effusive December 2016–March 2017 eruption. We associate this motion with the extrusion of cold crystalline upper conduit material (plug) along a pre-existing, reactivated fracture network (Figure 7a, Table 1). Initially, the plug extrusion was characterised by intermittent offsets and few detected seismic events, which may constrain the extrusion onset to January–April 2016. The subsequently derived range offset rates remained near constant until August 2016, whereas the seismicity of Bezymianny could not be differentiated from that of the active Klyuchevskoy volcano. Moreover, the plug extrusion was accompanied by observed intermittent degassing (cf. Figures S1 and S7), which, in contrast to the observed continuous range offset rates, might indicate a discontinuous precursory plug-extrusion behaviour. Yet, alternating weather conditions, such as daily changing wind directions and atmospheric pressures, may have had major impact on the irregular degassing pattern [53]. Previous seismic and petrographic studies at Bezymianny, in turn, showed that the rising magma is being stored at different depths prior to eruptions, which may also reflect alternating emission patterns [18,20,22,54] during the 2016 precursory stage.

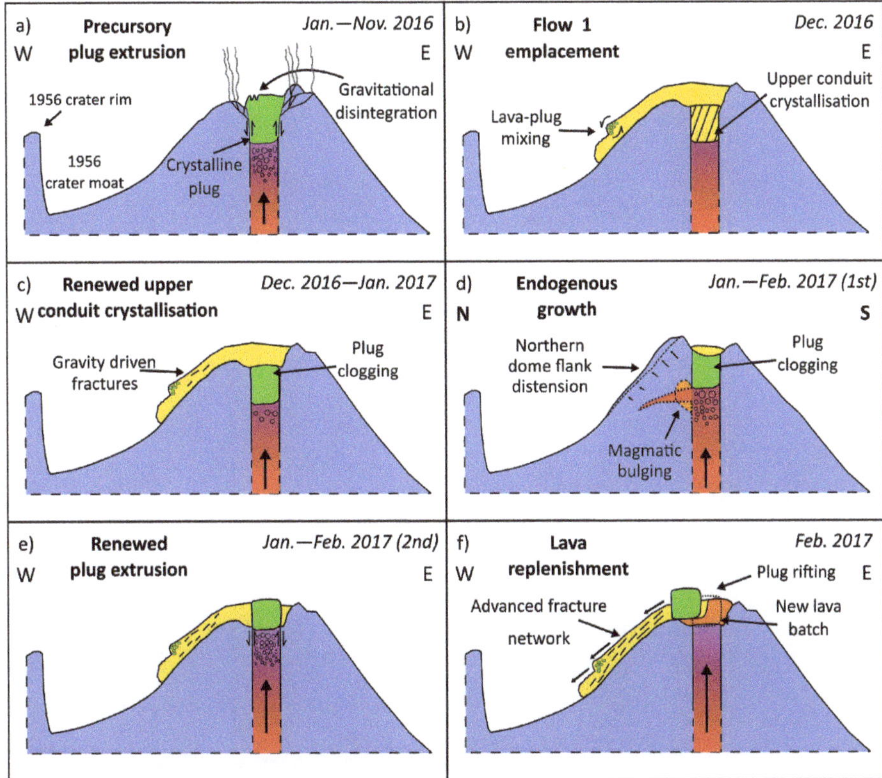

Figure 7. Schematic sketch of endogenous and exogenous growth episodes at Bezymianny. (**a**) Plug extrusion accompanied with degassing through a pre-existing, reactivated fracture network. Circles indicate gas pressurisation, ellipses show shearing at the conduit walls. (**b**) December 2016 lava flow emplacement and mixing of flow with precursory plug material causing significant compressional folding. The remaining magma batch in the upper conduit starts to solidify. (**c**) The new plug clogged the vent and (**d**) deflected the rising magma into distinct parts of the carapace. These mechanisms may account for all detected distension episodes. (**e**) Gas pressurisation exceeds yield strength of flow 1, and the previously formed new plug extruded. (**f**) Lava replenishment pushed the new plug and flow 1 westwards, which caused enhanced fracturing of the lava flow.

However, between September–October 2016, we observed a gradual change from slower to faster plug extrusion rates, which may have been related to the gradual ascent of magmatic fluids into shallower reservoirs. The simultaneously observed Mimatsu-derived summit growth, as well as new radar shadows at the rim, may have formed due to the presumably related increased gas pressurisation, which eventually pushed volcanic material over the rim, and stress-parallel oriented shadow-casting tensile fractures were formed at the rim. During November 2016, the observed seismicity and extrusion rates in range direction increased simultaneously, and the plug-related radar shadow was bisected by brighter surface reflectivity. This may indicate plug disintegration due to exhumation and concurrent loss of the previously existing circumferential pressure that was induced by the surrounding dense composite dome crust. As the largest Mimatsu displacement was determined between October–December 2016, and because the observed seismicity substantially increased during end of November 2016 and the beginning of December 2016, we assume that the determined cumulative range motion (~40 m) constitutes the minimum of the total amount of plug extrusion.

Table 1. Chronology of volcanic processes at Bezymianny between January 2016 and August 2018. The ra- and az-rates correspond to range and azimuth offset rates, respectively. FOV = field of view.

Observation period	Description
January–April 2016	Discontinuous plug extrusion (ra-rates: 0–0.08 m d^{-1}) Weak seismicity between (January and March 2016)
May–August 2016	Near constant plug extrusion (ra-rate: 0.07–0.13 m d^{-1}) Brief seismicity in May 2016 Intermittent (apparent) translucent degassing at Bezymianny Klyuchevskoy active (steaming) July 2016: onset of tensile crack formation at eastern summit rim
September 2016	Faster plug extrusion (ra-rate: 0.15 m d^{-1}) Klyuchevskoy active (steaming) Onset of plug disintegration at the western portion
October 2016	Onset of eastern summit uplift (9–22 m in Mimatsu diagram) Increased plug extrusion rates (ra-rate: 0.24 m d^{-1}) Further widening of tensile cracks
November–beginning of December 2016	Accelerated plug extrusion (ra-rate: 0.43–0.63 m d^{-1}) Onset of continuous seismicity Substantial widening of summit rim tensile cracks Strong disintegration of precursory plug Significant summit uplift (9–37 m in Mimatsu diagram)
December 2016	Persistently increasing seismicity Inelastic bulging of northern composite dome (17 November–20 December 2016; az-rate: 0.05–0.1 m d^{-1}) related to magmatic fluid intrusion Az-rates of 1st flank bulging increased towards the summit Bulging occurred likely prior to emplacement of flow 1 (SAR-scene: 20 December 2016) Flow perpendicular shadow-casting ridge interpreted as compressional fold due to lava-plug mixing
End of December 2016–beginning of February 2017	Minor destruction of previously determined summit uplift in Mimatsu diagram Inferred upper conduit solidification (formation of plug 2) that clogged the vent Inelastic bulging of northern carapace (31 December 2016–2 February 2017; az-rate: ~0.1–0.6 m d^{-1}) related to magmatic fluid intrusion Az-rates of 2nd flank bulging increased towards the summit Substantial increase of seismicity during second half of January 2017 Extrusion of plug 2 Intrusion and plug 2 extrusion possibly related to enhanced seismicity
Mid of February–beginning of March 2017	Repeated inelastic bulging of northern carapace (2–13 February 2017; az-rate: ~0.1–0.3 m d^{-1}) related to magmatic fluid intrusion Lava replenishment at summit that pushed plug 2 westwards Repeated inelastic bulging of northern carapace (24 February–7 March 2017; az-rate: ~0.1–0.4 m d^{-1}) related to magmatic fluid intrusion Az-rates of the 3rd and 4th bulging events increased towards the summit Strongly enhanced seismic activity Inferred upper conduit crystallisation by end of February and beginning of March 2017
March 2017	Significant increase in seismic activity Explosive eruption on 9 March 2017 Emplacement of flow 2 (SAR-scene: 18 March 2017) Smooth reflectivity characteristics of flow 2 due to thorough degassing in reservoir
June 2017	Strongest explosive eruption on 16 June 2017 depicted by solely deposition of pyroclastic deposits Summit bulge significantly destroyed, appearance of new extrusive body at southern summit Southern summit crater extrusion (up to 30 m in Mimatsu diagram) not clearly resolved in SAR data

In general, plug extrusions at Bezymianny were commonly observed days to weeks prior to explosive eruptions [30,31]. Similarly, satellite thermal observations showed enhanced anomalies that are associated with exogenous growth 15 to 20 days prior to >20 eruptions between 1993 and 2008 [39]. Our data set, in turn, documented seven to nine months of precursory ground motion related to plug extrusion prior to the documented effusive 5 December 2016–28 February 2017 eruption. The absence of explosive activity of Bezymianny during the effusive eruption may be caused by insufficient gas pressurisation during magma ascent likely as a consequence of persistent degassing. Similar observations were made prior to a non-explosive eruption at MSH during the dome building phase in 1981 [55]. Lastly, pixel offsets that are derived from the precursory plug extrusion episode may be used in future to refine models and investigate the corresponding source of deformation with, for example, a discrete element method, as described in [56,57].

5.2.2. Effusive 5 December 2016–7 March 2017 Eruption

Our data set has shown that the so far first registered effusive eruption initiated with emplacement of flow 1, its surface characterised by few crevasses and a dominant, radar-shadow producing, flow perpendicular ridge. This ridge does not exhibit the typically observed radial orientation of surface folds of many silicic lava flows, which form as the flows stretch and rotate in the flow direction [58]. Therefore, the ridge on flow 1 could be interpreted as a result of a pronounced step in the paleotopography, but our terrain model from 2014 does not reveal any significant elevation changes on the western composite dome flank (see inset in Figure 2b). Instead, it may reflect the compressional folding [59] of a mixture of the lower viscous flow 1 with the highly viscous precursory plug material (Figure 7b and Table 1). The partial preservation of the plug would emphasize the weak explosiveness of the effusive 5 December 2016–28 February 2017 eruption.

Between the end of December 2016 and the beginning of February 2017, we observed significant unidirectional bulging of the northern dome flank, which were not reversed again and therefore can be considered inelastic. This was either accompanied or preceded by the emergence of a new radar shadow at the summit as well as enhanced seismicity during the second half of January 2017. The expansion may be explained by near vertical magmatic intrusions into the carapace without an existing plug. Yet, deformation experiments of conical shaped volcanoes have shown that, under these conditions, the summit always concurrently subsides [60].

Since summit subsidence was not observed at Bezymianny, we assume that a second plug formed between December 2016 and beginning of January 2017 that caused the unidirectional bulging (Figure 7b,c). This plug may have formed in response to low effusion rates, shallow degassing (see degassing in Figure 4c), and microlite crystallisation [7]. A process also inferred for spine formations during dome growth episodes at Unzen and Soufrière Hills volcanos [6,7]. The second plug at Bezymianny possibly clogged the upper conduit, thereby causing unidirectional bulging of magmatic fluids in the upper conduit, or the deflection of magmatic fluids into structurally weaker parts of the composite dome (Figure 7d). Layer boundaries, interlayered unconsolidated pyroclastic deposits, or pre-existing fractures related to explosive events that formed older summit craters might depict the latter. Eventually, the new summit radar shadow indicates that the second plug was extruded by the beginning of February 2017. By mid February 2017, we observed a general amplitude change at the summit, which we associate with a second pulse of lava emplaced at the summit. Simultaneously observed enhanced crevasse density on flow 1 and the location change of the plug related summit shadow indicate that the new lava flow pushed both flow 1 and the previously extruded plug westwards (Figure 7f).

As inflations were recurrently observed at the northern composite dome flank, a clogging plug may also explain the other observed distinct lateral flank movements. Lava flow emplacements followed the bulging events during February and March 2017 (including flow 2), which reveals that exogenous growth succeeded endogenous growth. Therefore, it may be possible that flank bulging between November and December 2016 also preceded flow 1. This would agree with precursory endogenous growth of Bezymianny that was derived from remotely detected enhanced thermal activity prior to most eruptions during 1993–2008 [39]. Since each bulging event of the recent eruption series increased in magnitude towards the summit (Figure 5 and Figure S5), the nucleus of unidirectional intrusions was likely located in the uppermost few hundred meters (~100–400 m) of the volcanic conduit, thus above the base of Bezymianny's composite dome (cf. Figure 1b(inset) and Figure 7). This agrees with the deformation observations at Colima in Mexico, where an inferred shallowly located (~200–300 m) clogging plug may have also governed the pathway of rising magmatic fluids prior to the volcano's 2013 eruption series [9]. Evidence of shallow conduit pressurisation that is derived from ground deformation and seismicity was also identified at other dome building volcanoes, such as Soufrière Hills on Montserrat [61], Unzen in Japan [6], or Lascar in Chile [62].

Lastly, we cannot differentiate this motion from potential endogenous growth motion, since the size of flow 1 continuously changed between adjacent amplitude scenes. Thus, we did not consider employing an elastic modelling approach to identify the source of deformation at the northern flank. In fact, the detected southward motion of the southern flank in the descending and ascending data (Figure 5b3 and Figure S11) indicates that magma may have also been deflected into the southern carapace, but this was observed only once and it occurred very localised near the summit.

5.2.3. Exogenous and Endogenous Dome Growth—Comparison with other Volcanoes

Volcanic activity at Bezymianny identified by TSX radar and optical data indicate precursory plug extrusion, as well as explosive eruptions, both being followed by lava flow emplacements and renewed summit excavation. All of these activities originated from the central summit. At the dome building Colima volcano, Mexico, activity between 1998 and 2010 also originated from a central summit, yet in this case the recurrent growth of blocky domes, from which short lava flows emanated, dominated the summit [63]. Soon after their formation, the summit domes were destroyed by recurrent Vulcanian eruptions that excavated new summit craters similar to the observed reshaped summit craters of Bezymianny. Yet, the repeated emergence of blocky domes at Colima contrasts with the recurrently observed plug extrusions at Bezymianny that depict the stiffened upper conduit. However, minor summit deformation at Colima's summit (2013) also suggested the existence of a shallow plug, which caused the deflection of the rising magmatic fluids, but was not extruded after all [9]. Yet, deformation at Colima only occurs days or hours prior to new eruptions [9]. Moreover, our data has shown that Bezymianny produced multiple eruptions within one year that significantly increased in explosiveness, whereas Colima produced eruptions with rather similar explosive character [63].

In addition, we showed that the recurrent bulging events of Bezymianny's northern dome flank occurred at strikingly different azimuth rates near the summit (0.1–0.6 m d^{-1}). These were likely related to intermittent conduit plug formation, which clogged the vent as the magmatic fluids were deflected into the northern carapace. The variance of observed endogenous growth rates at Bezymianny's mature composite dome strongly contrast with observations of linear endogenous growth rates during the formation of the relatively young domes of Mount St. Helens (1980–1986) and Unzen (1990–1995), which emphasizes the very distinct character of dome growth behaviour at different volcanoes.

Overall, we have shown that Bezymianny evolved during 2016–2017 from precursory plug extrusion over mostly effusive and unidirectional endogenous growth to successively stronger explosions, which produced large amounts of pyroclastic deposits that cover most of the two major lava flows. This may point out that Bezymianny's dome evolution is on the verge to a stratocone volcano, as was described prior to the 1956 eruption. In fact, all of the observed eruptive activity during the 2016–2017 eruption sequence at Bezymianny was confined to the central summit crater, which is a common feature for many stratocone volcanoes [64].

6. Conclusions

Here, we studied endogenous and exogenous dome growth before and during the 2016–2017 eruption sequence at Bezymianny. Multitemporal TSX amplitude imagery uncovered seven to nine months lasting precursory plug extrusion prior to the known onset of the eruption series. Deformation analysis of the ensuing effusive December 2016–March 2017 eruption revealed repeated exogenous lava flow emplacements that were accompanied and/or preceded by intermittent unidirectional bulging of the northern carapace. These events are likely related to the intermittent rapid formation of upper conduit plugs that deflected the rising magmatic fluids into the uppermost regions of the composite dome. The corresponding endogenous growth rates of Bezymianny's relatively mature dome significantly varied. Thus, dome growth at Bezymianny may have reached an advanced stage in its evolution close to the formation of a stratocone. Although endogenous growth could not be resolved by the webcam imagery, the images unveiled exogenous growth near the summit undetected by radar data. Yet, the images' poor resolution only contributed qualitatively to inferences that are drawn from seismic and SAR observations.

In this study, we have demonstrated the strengths of high-resolution SAR amplitude images as an effective observation tool to derive information regarding the detailed course of precursor activity as well as for differentiation of distinct lava flow surface and dome growth processes. However, as these processes rapidly changed, it becomes apparent that more frequent SAR acquisitions in different acquisition geometries would make it even more useful for real-time observations. This becomes obvious for the acquisition gaps prior the June 2017 eruption, which concealed possible precursor ground motion. In contrast, continuous seismic observations revealed a clear picture of magmatic activity and/or associated rockfalls prior to the eruption. Therefore, integration and analysis of different geophysical data sets is a vital base for the monitoring of remote volcanoes, such as Bezymianny, who poses a permanent threat to intercontinental aviation.

Supplementary Materials: The following are available online at http://www.mdpi.com/2072-4292/11/11/1278/s1, Table S1: Descending spotlight TSX amplitude image pairs of Bezymianny. Table S2: Ascending spotlight TSX amplitude image of Bezymianny. Figure S1: Selected clear view images that cover the 2016–2017 eruptive sequence of Bezymianny. Figure S2: Selected co-registered descending TSX amplitude images. Figure S3: Co-registration processing chain of the TSX amplitude data time-series. Figure S4: Summit range pixel offsets compared with stable areas (i.e., no deformation). Figure S5: Average azimuth pixel offset rates during December 2016–March 2017. Figure S6: Azimuth offset maps for descending acquisition intervals without northward dome bulging. Figure S7: Webcam images showing episodic degassing (cf. Figure S1) at Bezymianny during May–November 2016. Figure S8: Illustration demonstrating constraints on the Mimatsu mapping quality because of blurry edges of Bezymianny's dome. Figure S9: Ascending non-geocoded spotlight-mode TSX amplitude image acquired on 17 February 2017. Figure S10: Co-registered ascending TSX amplitude images. Figure S11: Azimuth offset map between the ascending TSX acquisitions from 17–28 February 2017.

Author Contributions: Conceptualization, R.M. and T.R.W.; methodology, R.M., T.R.W, S.B., and M.B.; software, R.M.; field work, R.M., T.R.W., M.B. and A.B.; formal analysis, R.M.; validation, R.M. and T.R.W.; formal analysis, R.M.; data curation, R.M.; writing–original draft preparation, R.M.; writing–review and editing, T.R.W., A.B., and M.B.; visualization, R.M.; supervision, T.R.W.; project administration, R.M.; funding acquisition, T.R.W. and S.L.S.

Funding: This is a contribution to VOLCAPSE, a research project funded by the European Research Council under the European Union's H2020 Programme/ERC consolidator grant ERC-CoG Q7 646858. We thank the DLR for support; the acquisition of the spot-mode TerraSAR-X data was realized through proposal GEO1505. DV-B is grateful to CONACYT-PDCAPN project 2579. This study was also supported by Scientific Research Work of Russian Academy of Science: "Complex geophysical studies of the volcanoes of Kamchatka and of the northern Kuril Islands in order to detect the signs of the future eruption, as well as forecast its dynamics with an assessment of the ash hazard to aviation", # AAAA-A19-119031590060-3.

Acknowledgments: We would like to thank Jacqueline Salzer and Henriette Sudhaus for the numerous discussions on the SAR processing and Bodo Bookhagen for his thoughts on the error analysis of the SAR data. Finally we thank Ilyas Abkadyrov for supporting us during field work.

Conflicts of Interest: The authors declare no conflict of interest.

Appendix A

Error estimation for offset measurements during the precursory plug extrusion

To differentiate significant from erroneous pixel offsets during the precursory plug extrusion stage, we applied an analytical approach that is usually used to remove DEM errors for change analysis of river beds and slope failures. This approach uses stable areas for calculation of errors [65] that follow a normal distribution around 0 (i.e., no changes), and are assumed to be independent [66,67]. Following the approach of Lane et al. [68], we assume that the error of the point-to-point distance between DEMs is equal to the offset uncertainty, which may be expressed as:

$$\sigma_{offset} = \sqrt{\sigma^2 + \sigma^2}, \tag{A1}$$

where σ is the standard deviation of offsets in a stable area with a size of 100 × 100 pixels. The statistic t-score is then calculated by the following equation of Bennet et al. [69]:

$$t_{score} = \frac{\Delta_{px}}{\sigma_{offset}}, \tag{A2}$$

where Δpx is the absolute pixel offset within the stable area. To determine whether the offsets of individual pixels in the stable areas are significant, a simple one-sided t-Test with a confidence interval of 80% (tc > 0.845) was applied to the area of real surface motion. Thus, only pixel offsets in the deforming area larger than the median of Δpx > tc were considered as real displacements. Moreover, since motion in the stage prior the first recognised eruption was directed towards the satellite, pixel offsets away from the satellite were omitted in the deformation area. Finally, pixel offsets for the same stage were compared with the Signal-to-Noise Ratio (SNR) to estimate the degree of error estimation:

$$SNR = \frac{ccp}{std}, \tag{A3}$$

where ccp depicts the cross-correlation peak and std the corresponding standard deviation.

References

1. Sheldrake, T.E.; Sparks, R.S.J.; Cashman, K.V.; Wadge, G.; Aspinall, W.P. Similarities and differences in the historical records of lava dome-building volcanoes: Implications for understanding magmatic processes and eruption forecasting. *Earth Sci. Rev.* **2016**, *160*, 240–263. [CrossRef]
2. Ogburn, S.E.; Loughlin, S.C.; Calder, E.S. The association of lava dome growth with major explosive activity (VEI ≥ 4): DomeHaz, a global dataset. *Bull. Volcanol.* **2015**, *77*, 40. [CrossRef]
3. Voight, B.; Elsworth, D. Instability and collapse of hazardous gas-pressurized lava domes. *Geophys. Res. Lett.* **2000**, *27*, 1–4. [CrossRef]
4. Huppert, H.E.; Shepherd, J.B.; Sigurdsson, H.; Sparks, R.S.J. On lava dome growth, with application to the 1979 lava extrusion of the Soufrière of St. Vincent. *J. Volcanol. Geotherm. Res.* **1982**, *14*, 199–222. [CrossRef]
5. Fink, J.H.; Malin, M.C.; Anderson, S.W. Intrusive and extrusive growth of the Mount St. Helens lava dome. *Nature* **1990**, *348*, 435–437. [CrossRef]
6. Nakada, S.; Shimizu, H.; Ohta, K. Overview of the 1990–1995 eruption at Unzen Volcano. *J. Volcanol. Geotherm. Res.* **1999**, *89*, 1–22. [CrossRef]
7. Watts, R.B.; Herd, R.A.; Sparks, R.S.J.; Young, S.R. Growth patterns and emplacement of the andesitic lava dome at Soufrière Hills Volcano, Montserrat. *Geol. Soc. Lond. Mem.* **2002**, *21*, 115–152. [CrossRef]
8. Sparks, R.S.J. Forecasting volcanic eruptions. *Earth Planet. Sci. Lett.* **2003**, *210*, 1–15. [CrossRef]
9. Salzer, J.T.; Nikkhoo, M.; Walter, T.R.; Sudhaus, H.; Reyes-Dávila, G.; Bretón, M.; Arámbula, R. Satellite radar data reveal short-term pre-explosive displacements and a complex conduit system at Volcán de Colima, Mexico. *Front. Earth Sci.* **2014**, *2*, 12. [CrossRef]
10. Lu, Z.; Dzurisin, D.; Biggs, J.; Wicks, C.; McNutt, S. Ground surface deformation patterns, magma supply, and magma storage at Okmok volcano, Alaska, from InSAR analysis: 1. Intereruption deformation, 1997–2008. *J. Geophys. Res. Solid Earth* **2010**, *115*, B00B02. [CrossRef]
11. Massonnet, D.; Feigl, K.L. Radar interferometry and its application to changes in the Earth's surface. *Rev. Geophys.* **1998**, *36*, 441–500. [CrossRef]
12. James, M.R.; Pinkerton, H.; Robson, S. Image-based measurement of flux variation in distal regions of active lava flows. *Geochem. Geophys. Geosy.* **2007**, *8*, 3. [CrossRef]
13. Walter, T.R. Low cost volcano deformation monitoring: Optical strain measurement and application to Mount St. Helens data. *Geophys. J. Int.* **2011**, *186*, 699–705. [CrossRef]
14. Pallister, J.S.; Schneider, D.J.; Griswold, J.P.; Keeler, R.H.; Burton, W.C.; Noyles, C.; Newhall, C.G.; Ratdomopurbo, A. Merapi 2010 eruption-chronology and extrusion rates monitored with satellite radar and used in eruption forecasting. *J. Volcanol. Geotherm. Res.* **2013**, *261*, 144–152. [CrossRef]

15. Wadge, G.; Cole, P.; Stinton, A.; Komorowski, J.-C.C.; Stewart, R.; Toombs, A.C.; Legendre, Y. Rapid topographic change measured by high-resolution satellite radar at Soufrière Hills Volcano, Montserrat, 2008–2010. *J. Volcanol. Geotherm. Res.* **2011**, *199*, 142–152. [CrossRef]
16. Wang, T.; Poland, M.P.; Lu, Z. Dome growth at Mount Cleveland, Aleutian Arc, quantified by time series TerraSAR-X imagery. *Geophys. Res. Lett.* **2015**, *42*, 10. [CrossRef]
17. Major, J.J.; Dzurisin, D.; Schilling, S.P.; Poland, M.P. Monitoring lava-dome growth during the 2004–2008 Mount St. Helens, Washington, eruption using oblique terrestrial photography. *Earth Planet. Sci. Lett.* **2009**, *286*, 243–254. [CrossRef]
18. Koulakov, I.; Abkadyrov, I.; Al Arifi, N.; Deev, E.; Droznina, S.; Gordeev, E.I. Three different types of plumbing system beneath the neighboring active volcanoes of Tolbachik, Bezymianny, and Klyuchevskoy in Kamchatka. *J. Geophys. Res. Solid Earth* **2017**, *122*, 3852–3874. [CrossRef]
19. Shapiro, N.M.; Droznin, D.V.; Droznina, S.Y.; Senyukov, S.L.; Gusev, A.A.; Gordeev, E.I. Deep and shallow long-period volcanic seismicity linked by fluid-pressure transfer. *Nat. Geosci.* **2017**, *10*, 442–445. [CrossRef]
20. Thelen, W.; West, M.; Senyukov, S. Seismic characterization of the fall 2007 eruptive sequence at Bezymianny Volcano, Russia. *J. Volcanol. Geotherm. Res.* **2010**, *194*, 201–213. [CrossRef]
21. Shcherbakov, V.D.; Plechov, P.Y.; Izbekov, P.E.; Shipman, J.S. Plagioclase zoning as an indicator of magma processes at Bezymianny Volcano, Kamchatka. *Contrib. Mineral. Petrol.* **2011**, *162*, 83–99. [CrossRef]
22. Turner, S.J.; Izbekov, P.; Langmuir, C. The magma plumbing system of Bezymianny Volcano: Insights from a 54 year time series of trace element whole-rock geochemistry and amphibole compositions. *J. Volcanol. Geotherm. Res.* **2013**, *263*, 108–121. [CrossRef]
23. Braitseva, O.A.; Melekestsev, I.V.; Bogoyavlenskaya, G.E.; Maksimov, A.P. Bezymyannyi: Eruptive history and dynamics. *Volcanol. Seismol.* **1991**, *12*, 165–194.
24. Gorshkov, G.S. Gigantic eruption of the volcano Bezymianny. *Bull. Volcanol.* **1959**, *20*, 77–109. [CrossRef]
25. Bogoyavlenskaya, G.Y.; Braitseva, O.A.; Melekestsev, I.V.; Maksimov, A.P.; Ivanov, B.V. Bezymianny volcano. In *Active Volcanoes of Kamchatka*; Fedotov, S.A., Masurenkov, Y.P., Eds.; Moscow Nauka Publishers: Moscow, Russia, 1991; Volume 1, pp. 195–197.
26. Belousov, A. Deposits of the 30 March 1956 directed blast at Bezymianny volcano, Kamchatka, Russia. *Bull. Volcanol.* **1996**, *57*, 649–662. [CrossRef]
27. Lipman, P.W.; Moore, J.G.; Swanson, D.A. Bulging of the north flank before the May 18 eruption-geodetic data. In *The 1980 Eruptions of Mount St. Helens, Washington*; Lipman, P.W., Mullineaux, D.R., Eds.; United States Geological Survey: Reston, VA, USA, 1981; pp. 143–155.
28. Voight, B.; Glicken, H.; Janda, R.J.; Douglass, P.M. Catastrophic rockslide avalanche of May 18. In *The 1980 Eruptions of Mount St. Helens, Washington*; Lipman, P.W., Mullineaux, D.R., Eds.; United States Geological Survey: Reston, VA, USA, 1981; pp. 347–377.
29. Belousov, A.; Voight, B.; Belousova, M. Directed blasts and blast-generated pyroclastic density currents: A comparison of the Bezymianny 1956, Mount St Helens 1980, and Soufrière Hills, Montserrat 1997 eruptions and deposits. *Bull. Volcanol.* **2007**, *69*, 701–740. [CrossRef]
30. Belousov, A.; Voight, B.; Belousova, M.; Petukhin, A. Pyroclastic surges and flows from the 8–10 May 1997 explosive eruption of Bezymianny volcano, Kamchatka, Russia. *Bull. Volcanol.* **2002**, *64*, 455–471. [CrossRef]
31. Girina, O.A. Chronology of Bezymianny Volcano activity, 1956–2010. *J. Volcanol. Geotherm. Res.* **2013**, *263*, 22–41. [CrossRef]
32. Carter, A.J.; Ramsey, M.S.; Belousov, A.B. Detection of a new summit crater on Bezymianny Volcano lava dome: Satellite and field-based thermal data. *Bull. Volcanol.* **2007**, *69*, 811–815. [CrossRef]
33. Dvigalo, V.N.; Svirid, I.Y.; Shevchenko, A.V.; Sokorenko, A.V.; Demyanchuk, Y.V. Active volcanoes of Northern Kamchatka as seen from aerophotogrammetric data in 2010. In *Proceedings of regional conference "Volcanism and associated processes"*; Institute of Volcanology and Seismology FEB RAS: Petropavlovsk-Kamchatsky: Petropavlovsk-Kamchatsky, Russia, 2011; pp. 26–36. (In Russian)
34. Chebrov, V.N.; Droznin, D.V.; Kugaenko, Y.A.; Levina, V.I.; Senyukov, S.L.; Sergeev, V.A. The system of detailed seismological observations in Kamchatka in 2011. *J. Volcanol. Seismol.* **2013**, *7*, 16–36. [CrossRef]
35. Senyukov, S.L. Monitoring and prediction of volcanic activity in Kamchatka from seismological data: 2000–2010. *J. Volcanol. Seismol.* **2013**, *7*, 86–97. [CrossRef]
36. West, M.E. Recent eruptions at Bezymianny volcano-A seismological comparison. *J. Volcanol. Geotherm. Res.* **2013**, *263*, 42–57. [CrossRef]

37. Ramsey, M.; Dehn, J. Spaceborne observations of the 2000 Bezymianny, Kamchatka eruption: The integration of high-resolution ASTER data into near real-time monitoring using AVHRR. *J. Volcanol. Geotherm. Res.* **2004**, *135*, 127–146. [CrossRef]
38. Carter, A.J.; Girina, O.; Ramsey, M.S.; Demyanchuk, Y.V. ASTER and field observations of the 24 December 2006 eruption of Bezymianny Volcano, Russia. *Remote Sens. Environ.* **2008**, *112*, 2569–2577. [CrossRef]
39. Van Manen, S.M.; Dehn, J.; Blake, S. Satellite thermal observations of the Bezymianny lava dome 1993–2008: Precursory activity, large explosions, and dome growth. *J. Geophys. Res. Solid Earth* **2010**, *115*, 1–20. [CrossRef]
40. Monitoring of Volcanic Activity in Kamchatka. Available online: http://www.emsd.ru/~{}ssl/monitoring/main.htm (accessed on 30 September 2017).
41. Johnson, J.B.; Lees, J.M.; Gerst, A.; Sahagian, D.; Varley, N. Long-period earthquakes and co-eruptive dome inflation seen with particle image velocimetry. *Nature* **2008**, *456*, 377–381. [CrossRef]
42. Walter, T.R.; Ratdomopurbo, A.; Subandriyo; Aisyah, N.; Brotopuspito, K.S.; Salzer, J.; Lühr, B. Dome growth and coulée spreading controlled by surface morphology, as determined by pixel offsets in photographs of the 2006 Merapi eruption. *J. Volcanol. Geotherm. Res.* **2013**, *261*, 121–129. [CrossRef]
43. Chebrov, V.N.; Droznin, D.V.; Zakharchenko, N.Z.; Melnikov, D.V.; Mishatkin, V.N.; Nuzhdina, I.N. The development of the system of integrated instrumental monitoring of volcanoes in the Far East region. *Seism. Instrum.* **2013**, *49*, 254–264. [CrossRef]
44. Chaussard, E. A low-cost method applicable worldwide for remotely mapping lava dome growth. *J. Volcanol. Geotherm. Res.* **2017**, *341*, 33–41. [CrossRef]
45. Di Traglia, F.; Nolesini, T.; Ciampalini, A.; Solari, L.; Frodella, W.; Bellotti, F. Tracking morphological changes and slope instability using spaceborne and ground-based SAR data. *Geomorphology* **2018**, *300*, 95–112. [CrossRef]
46. Arnold, D.W.D.; Biggs, J.; Wadge, G.; Mothes, P. Remote sensing of environment using satellite radar amplitude imaging for monitoring syn-eruptive changes in surface morphology at an ice-capped stratovolcano. *Remote Sens. Environ.* **2018**, *209*, 480–488. [CrossRef]
47. Arnold, D.W.D.; Biggs, J.; Dietterich, H.R.; Vargas, S.V.; Wadge, G.; Mothes, P. Lava flow morphology at an erupting andesitic stratovolcano: A satellite perspective on El Reventador, Ecuador. *J. Volcanol. Geotherm. Res.* **2019**, *372*, 34–47. [CrossRef]
48. Mittermayer, J.; Wollstadt, S.; Prats-Iraola, P.; Scheiber, R. The TerraSAR-X staring spotlight mode concept. *IEEE Trans. Geosci. Remote Sens.* **2014**, *52*, 3695–3706. [CrossRef]
49. Michel, R.; Avouac, J.-P.; Taboury, J. Measuring ground displacements from SAR amplitude images: Application to the Landers earthquake. *Geophys. Res. Lett.* **1999**, *26*, 875–878. [CrossRef]
50. Werner, C.; Wegmüller, U.; Strozzi, T.; Wiesmann, A. GAMMA SAR and interferometric processing software. In Proceedings of the ERS-ENVISAT Symposium, Gothenburg, Sweden, 16–20 October 2000.
51. Poland, M.P.; Dzurisin, D.; LaHusen, R.G.; Major, J.J.; Lapcewich, D.; Endo, E.T. Remote camera observations of lava dome growth at Mount St. Helens, Washington, October 2004 to February 2006. *US Geol. Surv. Prof. Pap.* **2008**, *1750*, 225–236.
52. Wang, T.; Jonsson, S. Improved SAR amplitude image offset measurements for deriving three-dimensional coseismic displacements. *Sel. Top. Appl. Earth Obs. Remote Sens.* **2015**, *8*, 3271–3278. [CrossRef]
53. Zimmer, M.; Walter, T.R.; Kujawa, C.; Gaete, A.; Franco-Marin, L. Thermal and gas dynamic investigations at Lastarria volcano, Northern Chile. The influence of precipitation and atmospheric pressure on the fumarole temperature and the gas velocity. *J. Volcanol. Geotherm. Res.* **2017**, *346*, 134–140. [CrossRef]
54. López, T.; Ushakov, S.; Izbekov, P.; Tassi, F.; Cahill, C.; Neill, O.; Werner, C. Constraints on magma processes, subsurface conditions, and total volatile flux at Bezymianny Volcano in 2007–2010 from direct and remote volcanic gas measurements. *J. Volcanol. Geotherm. Res.* **2013**, *263*, 92–107. [CrossRef]
55. Casadevall, T.J.; Rose, W.; Gerlach, T.; Greenland, L.P.; Ewert, J.; Wunderman, R.; Symonds, R. Gas emissions and the eruptions of Mount St. Helens through 1982. *Science* **1983**, *221*, 1383–1385. [CrossRef]
56. Husain, T.; Elsworth, D.; Voight, B.; Mattioli, G.; Jansma, P. Influence of extrusion rate and magma rheology on the growth of lava domes: Insights from particle-dynamics modeling. *J. Volcanol. Geotherm. Res.* **2014**, *285*, 100–117. [CrossRef]
57. Harnett, C.E.; Thomas, M.E.; Purvance, M.D.; Neuberg, J. Using a discrete element approach to model lava dome emplacement and collapse. *J. Volcanol. Geotherm. Res.* **2018**, *359*, 68–77. [CrossRef]

58. Farrell, J.; Karson, J.; Soldati, A. Multiple-generation folding and non-coaxial strain of lava crusts. *Bull. Volcanol.* **2018**, *80*, 84. [CrossRef]
59. Anderson, S.W.; Stofan, E.R.; Plaut, J.J.; Crown, D.A. Block size distributions on silicic lava flow surfaces: Implications for emplacement conditions. *Geol. Soc. Am. Bull.* **1998**, *110*, 1258–1267. [CrossRef]
60. Rincón, M.; Márquez, A.; Herrera, R. Contrasting catastrophic eruptions predicted by different intrusion and collapse scenarios. *Sci. Rep.* **2018**, *8*, 6178. [CrossRef] [PubMed]
61. Melnik, O.; Sparks, R.S.J. Nonlinear dynamics of lava dome extrusion. *Nature* **1999**, *402*, 37–41. [CrossRef]
62. Matthews, S.J.; Gardeweg, M.C.; Sparks, R.S.J. The 1984 to 1996 cyclic activity of Lascar Volcano, northern Chile: Cycles of dome growth, dome subsidence, degassing and explosive eruptions. *Bull. Volcanol.* **1997**, *59*, 72–82. [CrossRef]
63. Varley, N.R.; Arámbula-Mendoza, R.; Reyes-Dávila, G.; Stevenson, J.; Harwood, R. Long-period seismicity during magma movement at Volcán de Colima. *Bull. Volcanol.* **2010**, *72*, 1093–1107. [CrossRef]
64. Pinel, V.; Jaupart, C. Magma chamber behavior beneath a volcanic edifice. *J. Geophys. Res.* **2003**, *108*, 1–17. [CrossRef]
65. Westaway, R.M.; Lane, S.N.; Hicks, D.M. The development of an automated correction procedure for digital photogrammetry for the study of wide, shallow, gravel-bed rivers. *Earth Surf. Process. Landf.* **2000**, *25*, 209–226. [CrossRef]
66. Durand, V.; Mangeney, A.; Haas, F.; Jia, X.; Bonilla, F.; Peltier, A. On the link between external forcings and slope instabilities in the Piton de la Fournaise Summit Crater, Reunion Island. *J. Geophys. Res. Earth Surf.* **2018**, *123*, 2422–2442. [CrossRef]
67. Brasington, J.; Rumsby, B.T.; McVey, R.A. Monitoring and modelling morphological change in a braided gravel-bed river using high resolution GPS-based survey. *Earth Surf. Process. Landf.* **2000**, *25*, 973–990. [CrossRef]
68. Lane, S.N.; Westaway, R.M.; Hicks, D.M. Estimation of erosion and deposition volumes in a large, gravel-bed, braided river using synoptic remote sensing. *Earth Surf. Process. Landf.* **2003**, *28*, 249–271. [CrossRef]
69. Bennett, G.L.; Molnar, P.; Eisenbeiss, H.; Mcardell, B.W. Erosional power in the Swiss Alps: Characterization of slope failure in the Illgraben. *Earth Surf. Process. Landf.* **2012**, *37*, 1627–1640. [CrossRef]

Article

Recent Developments and Applications of Acoustic Infrasound to Monitor Volcanic Emissions

Silvio De Angelis [1,*], Alejandro Diaz-Moreno [1] and Luciano Zuccarello [2,3]

1 School of Environmental Sciences, University of Liverpool, Liverpool L69 3GP, UK; aledm@liverpool.ac.uk
2 Departamento de Fisica Teorica y del Cosmos, University of Granada, 18071 Granada, Spain; lzuk@ugr.es
3 Istituto Nazionale di Geofisica e Vulcanologia, 56126 Pisa, Italy
* Correspondence: s.de-angelis@liverpool.ac.uk or silvioda@liverpool.ac.uk; Tel.: +44-(0)151-794-5161

Received: 25 April 2019; Accepted: 15 May 2019; Published: 31 May 2019

Abstract: Volcanic ash is a well-known hazard to population, infrastructure, and commercial and civil aviation. Early assessment of the parameters that control the development and evolution of volcanic plumes is crucial to effective risk mitigation. Acoustic infrasound is a ground-based remote sensing technique—increasingly popular in the past two decades—that allows rapid estimates of eruption source parameters, including fluid flow velocities and volume flow rates of erupted material. The rate at which material is ejected from volcanic vents during eruptions, is one of the main inputs into models of atmospheric ash transport used to dispatch aviation warnings during eruptive crises. During explosive activity at volcanoes, the injection of hot gas-laden pyroclasts into the atmosphere generates acoustic waves that are recorded at local, regional and global scale. Within the framework of linear acoustic theory, infrasound sources can be modelled as multipole series, and acoustic pressure waveforms can be inverted to obtain the time history of volume flow at the vent. Here, we review near-field (<10 km from the vent) linear acoustic wave theory and its applications to the assessment of eruption source parameters. We evaluate recent advances in volcano infrasound modelling and inversion, and comment on the advantages and current limitations of these methods. We review published case studies from different volcanoes and show applications to new data that provide a benchmark for future acoustic infrasound studies.

Keywords: acoustic infrasound; volcanic emissions; ground-based remote sensing

1. Introduction

Approximately 10% of the Earth's population live under the direct threat of one of 1508 active volcanoes [1,2]. Understanding the nature and impact of volcanic hazards and developing monitoring systems to detect and track eruptions in (near) real time, is central to effective early warning and successful risk mitigation. The type and severity of hazards posed by volcanic activity depend on the style of eruption. Effusive eruptions often affect settlements and infrastructure such as during the recent 2018 rift eruption of Kilauea (Hawaii, USA) [3]. At the other end of the spectrum of volcanism, explosive and sustained activity can result in loss of life, such as during the 2018 eruption of Volcan de Fuego (Guatemala), and have a severe impact on local communities [4]. In general, long-lived eruptions can have significant, negative, repercussions on the economy at the national and international scale. During volcanic explosions fragments of pulverised rock, referred to as volcanic ash, are ejected from one or more active vents at high speed, driven into the atmosphere by hot gasses. It is well-known that airborne volcanic ash presents a direct threat to aviation; it can cause disruption to flight operations such as during the 2010 eruption of Eyafjallajokull (Iceland) and damage to infrastructure, with consequent loss of revenue [5].

During volcanic crises nine international Volcanic Ash Advisory Centers (VAAC), part of the International Civil Aviation Organization's (ICAO) Air Navigation Commission, are deputed to dispatch Volcanic Ash Advisories (i.e., alerts on the presence of airborne volcanic ash and estimates of its movement) to aviation authorities [6]. Numerical models are routinely used by VAACs to forecast atmospheric dispersal of ash clouds in order to evaluate the potential impacts of eruptions on aviation. The volume of erupted material, the rate at which mass is ejected from volcanic vents and the initial height of eruption plumes, collectively referred to as eruption source parameters [5], are key inputs into these models. Assessment of these parameters relies on empirical laws, which were derived from measurements of well-documented eruptions [7–10]. These empirical relationships are integrated in routine protocols to estimate mass eruption rates (MER) from measurements of ash plume height [6], despite large uncertainties when results are compared with values obtained from direct measurements of eruption deposits and durations [5]. Mastin et al. [5] discussed whether MER could be alternatively estimated by means of numerical modelling: based on the results of 1D numerical simulations, they argued that empirical equations still perform better than numerical models in reproducing field observations.

In recent times, acoustic infrasound has emerged as an increasingly popular tool for volcano remote sensing, and its potential to assess volcanic emissions in (near) real time has been extensively investigated. The term infrasound identifies atmospheric acoustic waves with frequencies typically <20 Hz, below the audible range of humans. Volcanoes are prolific radiators of infrasound, generated by large-scale eruptive processes, which inject gas and pyroclasts into the atmosphere causing its rapid acceleration [11]; these low-frequency acoustic waves can travel distances of up to several thousands of kilometres [1] lending themselves to volcano monitoring at different scales, from local to global. The use of acoustic infrasound in regional and local eruption monitoring, with applications in volcano early warning has become increasingly popular [12–16]. A substantial body of literature demonstrate the use of infrasound to detect, locate and track explosive volcanic eruptions, and its potential to provide estimates of eruption source parameters to inform volcanic plume rise and ash dispersal modelling [17–23].

In this manuscript, we provide a review of past work in the field of volcano infrasound, and describe recent developments of its use to assess eruption source parameters. Here, we focus on applications of acoustic infrasound at local distances (i.e., within 10–15 km from eruptive vents) and their potential to provide robust estimates of the amount of material ejected into the atmosphere during eruptions. We present a summary of the theory of linear acoustics and discuss equivalent models for the representation of volcano-acoustic sources. Methods derived from this theory that exploit volcano-acoustic time series to retrieve eruptive jet velocities, and invert infrasound waveforms to obtain estimates of mass and volume flow rates (of the eruptive mixture of gas and particles) are reviewed. We discuss the recent integration of numerical modelling of acoustic wavefield propagation within these workflows. We show examples from the published literature, as well as applications to new case studies. Finally, we offer a perspective on the potential for use of acoustic infrasound in real-time volcano monitoring.

2. The Acoustic Fingerprint of Volcanoes

Volcanic activity encompasses a broad spectrum of eruptive styles ranging from effusive discharge of lava onto the Earth's surface, to energetic explosions that inject large amounts of gas-laden pyroclasts in the atmosphere. Acoustic waves are produced when the atmosphere is perturbed from its background state by a moving source causing unsteady fluid motion, and pressure disturbances propagate through the air. Multiple processes occur in volcanic environments making them ideal sources of acoustic waves; these include gas-driven oscillations of the surface of lava lakes or magma columns [24], surface mass movements and impacts (rockfalls, pyroclastic flows and lahars [25]) explosive fragmentation of magma at different scales (from Strombolian to Plinian [17,20]), and vigorous ash-and-gas jetting during sustained eruptions [18]. Volcanoes produce acoustic waves that deliver energy over a broad spectrum of frequencies; however, because of the comparatively large spatial scale of source mechanisms, the majority of sound is radiated in the infrasonic band, approximately between 0.01 and 20 Hz [11].

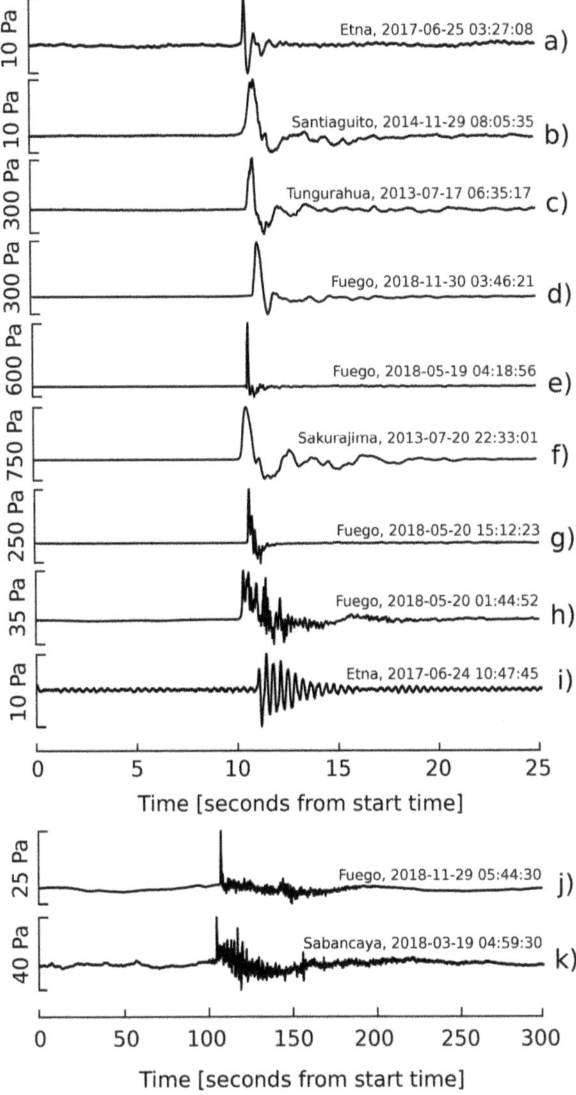

Figure 1. Infrasound waveforms associated with explosions at volcanoes that exhibit regular activity. Excess pressures are reduced to a distance of 1 km from the vent for comparison across volcanoes: (**a**) small degassing explosions recorded at Etna, Italy; (**b**) small degassing explosions recorded at Santiaguito, Guatemala; (**c**,**d**) ash-rich impulsive explosions from Tunghurahua, Ecuador, and Fuego, Guatemala; (**e**,**f**) large amplitude signals exhibiting blast wave characteristics recorded at Fuego, Guatemala, and Sakurajima, Japan; (**g**,**h**) short-duration signals featuring multiple pulses from Fuego, Guatemala; (**i**) monochromatic signal recorded at Mt. Etna associated with degassing activity from one of the summit vents; abd (**j**,**k**) explosion signals with impulsive onsets followed by an extended coda from Fuego, Guatemala, and Sabancaya, Peru, associated with observations of sustained ash plumes.

Volcano infrasound waveforms range from impulsive transients (Figure 1) to tremor-like sustained signals, with a number of other intermediate types [26]. Waveforms with short duration (5–15 s) and sharp onsets (Figure 1a–f) are frequently generated by impulsive explosion-like sources. These signals may feature compression onsets followed by rarefaction with a nearly symmetrical shape (Figure 1a); this symmetry is indicative of a flow rate source time function symmetrically distributed, in time, around a peak value [27]. In many instances, waveforms are less symmetrical, exhibiting rapid compression onsets (e.g., Figure 1e) followed by rarefaction phases with reduced amplitudes (Figure 1b–f); waveform asymmetry may represent either a non-symmetrical flow rate source function, or reflect a shock-type source mechanism similar to blast waves produced by chemical explosions [27,28]. Typically, blast waves can be separated from other explosion mechanisms due to their characteristic appearance and much larger peak amplitudes, of the order of several hundreds of Pa within a few hundred metres from the source. Diversity in the characteristics of infrasound signals reflects variety in source mechanisms, and may additionally provide clues on whether explosions are gas- or tephra-rich [29]. Observational evidence suggests that waveforms featuring impulsive onsets followed by several additional pulses (Figure 1g–h), or by a prolonged coda (Figure 1j–k), are frequently associated to tephra-rich plumes with durations up to a few minutes. Open-vent volcanic systems, where active degassing is persistent, often produce nearly monochromatic waveforms [24] due to resonance taking place within the shallow conduit and crater system (Figure 1i).

Longer duration (several minutes to hours) infrasound is typically associated with sustained open-vent degassing, or continuous eruptive activity. Figure 2 illustrates three examples of extended duration waveforms associated with persistent pulse-like degassing, a sequence of intermittent small Strombolian explosions, and an episode of sustained lava fountaining. A wider variety of tremor-like signals have been reported in the literature; a comprehensive account of these signals is, however, beyond the scope of this manuscript. For additional details, the reader is referred to a number of insightful reviews published in recent years on the nature and characteristics of volcano infrasound [11,26,30,31].

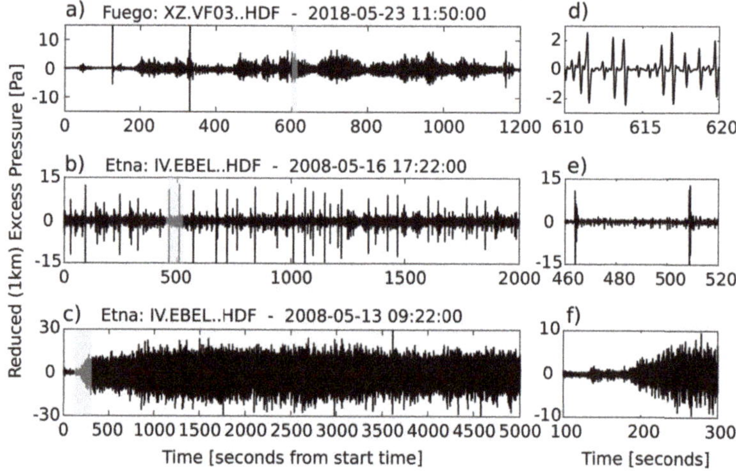

Figure 2. Extended duration infrasound waveforms: (**a**) Twenty minutes of very regular, pulse-like, degassing recorded at Fuego, Guatemala. This type of activity is often referred to as "chugging" and accompanied by audible degassing with a noise similar to that of a steam train. (**b**) Sequence of small Strombolian explosions recorded at Mt. Etna. This activity recorded in the summer of 2008, persisted for days. (**c**) Infrasound signature of a lava fountain recorded at Etna, Italy in May 2008. This episode lasted for several hours with relatively constant infrasonic amplitudes. (**d**–**f**) Enlargements of selected portions (shaded gray) of signals shown in (**a**–**c**).

3. Linear Acoustic Theory and Multipole Acoustic Sources

Fluid flow associated with volcanic explosions causes acceleration of the atmosphere making them efficient radiators of acoustic infrasound. The propagation of density (and thus, pressure) perturbations in a fluid medium at rest can be investigated with the Navier–Stokes equations, by solving an associated wave equation [32]. Early, seminal, work by Lighthill [33] and Woulff and McGetchin [32] set out an elegant formulation for the solution of the acoustic wave equation in free- and half-space, to describe the radiation of sound from monopole, dipole, and quadrupole sources. A monopole is the simplest source of acoustic waves, linked to unsteady fluid flow (i.e., injection of mass) such as during a volcanic explosion, and it is the most efficient form of acoustic radiation. Dipole radiation results from the application of a force field at the source location, and it is obtained as the linear superposition of two adjacent monopoles oscillating out of phase by 180°. Dipole sources, unlike monopoles, do not introduce net mass at the source; they impart momentum to the fluid causing its movement, and thus, to radiate sound. Woulff and McGetchin [32] suggested that dipole radiation may result from the interaction of fluid flow from volcanic vents with the crater walls. Quadrupole radiation represents turbulence within the gas jet itself, and it results from the combination of two opposite dipoles. Woulff and McGetchin [32], building on previous work by Lighthill [33], and using dimensional analysis, presented a suite of scaling laws that link the acoustic power radiated by monopole (Π_m), dipole (Π_d) and quadrupole (Π_q) sources to gas velocity during fluid flow from volcanic vents:

$$\Pi_m = K_m \frac{\rho_p A_v}{c} v_e^4 \tag{1}$$

$$\Pi_d = K_d \frac{\rho_p A_v}{c^3} v_e^6 \tag{2}$$

$$\Pi_q = K_q \frac{\rho_p A_v}{c^5} v_e^8 \tag{3}$$

where K_m, K_d and K_q are dimensionless constants, ρ_p is the density of the volcanic jet, c is the speed of sound in the atmosphere, v_e is the gas exit velocity at the vent, and A_v is the cross-sectional area of the vent.

Recently, Kim et al. [34] presented a detailed review of the derivation of the formal integral solution to the wave equation in a half-space for monopole and dipole sources, using the Green's Function (GF) solution to the inhomogeneous Helmholtz wave equation [35,36]. Under the assumptions of compact acoustic sources (i.e., their characteristic dimensions are much smaller than the acoustic wavelength) and that far-field volcano infrasound propagates in half-space (i.e., an atmosphere bounded by a flat solid topography), the transient solutions to the wave equation for monopole, and horizontal and vertical dipole radiation at distance r from the source can be written as:

$$p_m(\mathbf{r}, t) = \frac{1}{2\pi r} \dot{S}\left(t - \frac{r}{c}\right) \tag{4}$$

$$p_h(\mathbf{r}, t) = \frac{1}{2\pi r c^2} \left[\frac{x}{r} \ddot{D}_x(t - r/c) + \frac{y}{r} \ddot{D}_y(t - r/c)\right] \tag{5}$$

$$p_v(\mathbf{r}, t) = \frac{1}{2\pi r c^2} \left[\frac{z}{r} \frac{z_0 \partial^3 D_z(t - r/c)}{\partial t^3} \cos\theta\right] \tag{6}$$

where p_m, p_h, and p_v are monopole, horizontal and vertical dipole pressure fields, respectively; $\dot{S}$ is the time history of mass acceleration at the source; (x, y, z) are coordinates in a Cartesian system centred at the source location; $D_{x,y,z}$ are dipole momenta (in units of kg m s^{-1}) in the (x, y, z) directions; θ is the angle between the axis of the dipole and the vertical direction; z_0 is source elevation; and c is the sound speed. Any infrasound time series can be written as the radiation from a multipole source, that is the linear superposition of a monopole, with horizontal and vertical dipoles (and terms of higher order):

$$p(\mathbf{r}, t) = p_m(\mathbf{r}, t) + p_h(\mathbf{r}, t) + p_v(\mathbf{r}, t) \tag{7}$$

The formulation of Equations (4)–(7) represents the foundation for acoustic waveform inversion that is discussed below.

4. Eruption Source Parameters and Their Potential Use in Ash Plume Modelling

In this section, we discuss the use of infrasound for the evaluation of eruption source parameters, and their integration into models of ash plume rise. We review some case studies and expand on past work to include recent developments in numerical modelling of volcano acoustic wavefields, and their integration into infrasound waveform inversion schemes for the assessment of volume and mass flow rates during explosive eruptions.

4.1. Fluid Flow Velocity from Acoustic Infrasound

Early work from Woulff and McGetchin [32] demonstrated the application of Equations (1)–(3) to assess volcanic jet dynamics—in particular fluid flow velocity—from records of the sound emitted by explosions at Stromboli (Italy) and fumaroles at Acatenango (Guatemala). An experiment at Acatenango where acoustic power, the cross-sectional area of the vent, and flow velocity could be directly measured in the field, allowed calibration of the dimensionless constants K_m, K_d, and K_d in Equations (1)–(3), and to infer that the sound radiated in either case had, predominantly, dipolar nature. Several authors [17–20] built on the original work of Woulff and McGetchin [32] on translating time series of acoustic pressure into gas flow velocities. Acoustic power (Π) can be estimated directly from the recorded infrasound [20] as:

$$\Pi = \frac{\pi r^2}{\rho_{atm} c \tau} \int_0^{\tau} \Delta p^2(t) dt \tag{8}$$

where $\Delta p(t)$ is a time series of atmospheric pressure, r is the source-receiver distance, ρ_{atm} is atmospheric air density, c is the atmospheric sound speed and τ is the source duration. Once acoustic power is calculated by solving the definite integral in Equation (8), the flow velocity at the source can be easily retrieved by inversion of Equations (1)–(3). Gas velocity is an important eruption parameter easily translated into volume flow rate (for a known cross-sectional area of the vent), which is used as input into one of many empirical models to evaluate the initial height of eruption plumes. The most used of models is the one from Sparks et al. [8]:

$$Q_{magma} = \left[\frac{H_{plume}}{1.67}\right]^{\frac{1}{0.259}} \tag{9}$$

where Q_{magma} is the magma volume flow rate in units of m^3/s and H_{plume} is the plume height in units of kilometres. Vergniolle and Caplan-Auerbach [20] used acoustic data recorded during the sub-Plinian phase of the 1999 eruption of Shishaldin volcano (USA) to infer gas velocity and total gas volume erupted; they assumed dipole radiation during the sub-Plinian phase of the eruption, and confirmed that the gas velocities obtained were realistic using independent plume height measurements and Equation (9). They noted that the assumption of dipole radiation would not be appropriate outside the period of sub-Plinian activity; for instance, discrete explosions are more effectively described as monopole sources. A dipole source is often the preferred choice for explosive events lasting up to few tens of seconds associated with plumes rising to their highest elevation within a few minutes. In these situations, the force field exerted by the conduit and crater walls on the volcanic flow, and the interactions of the gas phase with solid pyroclasts are well described by a dipole. During sustained eruptions, lasting up to several hours, infrasound is generated by internal turbulence within the plume, a source best represented as a quadrupole [17].

Caplan-Auerbach et al. [17] also favoured a dipole source to describe the infrasound recorded during the 2006 eruption of Augustine volcano (USA); they analysed nine eruptive events with durations of the order of several minutes that generated largely buoyant plumes reaching elevations of

8–15 km above the vent. Here (Figure 3), we have reproduced their results following the workflow presented in the original manuscript for two of the explosions (Events 1 and 4, in the original publication). Estimates of flow velocity at the vent were obtained, in the original manuscript and, here, by using Equation (8) to calculate acoustic power and then inverting Equation (2). They calculated volume flow rates for all nine explosions assuming a vent with circular cross-section, and used these values to predict plume heights. Modelling the plumes as buoyant thermals rather than continuously fed by eruption (an implicit assumption of Equation (9)), they showed that gas fluxes retrieved from analysis of acoustic data are a good indicator for the rise height of volcanic plumes. Ideally, the choice of a specific source mechanism should be cross-validated using other geophysical measurements, when available. For example, high-resolution visual and thermal infrared imagery can be used to assess eruption dynamics and to calculate the exit velocity of volcanic jets in the near-vent region, two crucial factors in controlling sound generation. Ripepe et al. [18] employed multi-parameter data to investigate the 2011 eruption of Eyjafjallajokull (Iceland); plume heights were retrieved from infrasound-derived fluid flow velocities assuming a dipole acoustic source, and cross-validating these measurements with continuous thermal IR measurements of the plume in the near-vent region.

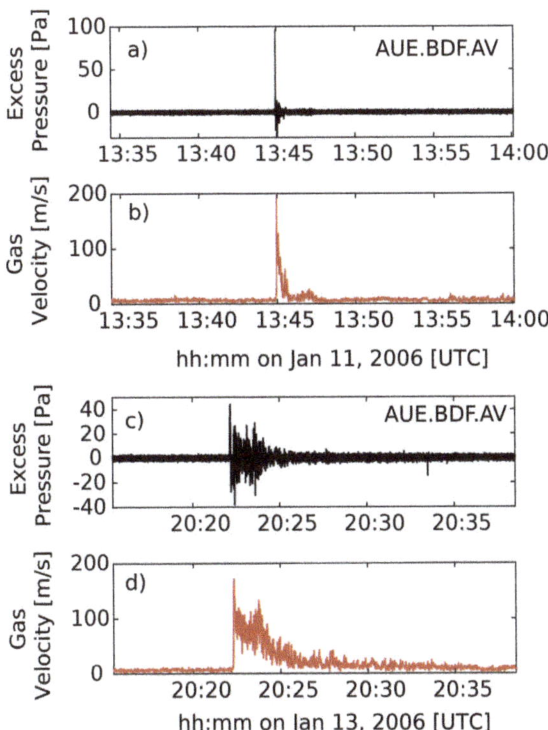

Figure 3. Infrasound-derived time series of gas velocity for two explosions during the 2006 eruption of Augustine volcano, USA: (**a**,**c**) Raw infrasound of explosive events on 11 January 2006 and 13 January 2006, respectively. Signals recorded by a microphone at 3.2 km from the vent. (**b**,**d**) Five-second moving-window gas velocity time series obtained by inverting Equation (2), assuming a dipole acoustic source. The figure reproduces results originally presented in Caplan-Auerbach et al. [17].

Careful consideration, when using infrasound-derived flow velocities for the assessment of plume rise, should be given to the nature of the plume itself, and whether its initial development is affected by local atmospheric conditions. Empirical relationships such as Equation (9) are usually derived for strong plumes where internal drag is negligible and entrainment of atmospheric air is approximately proportional to the vertical velocity of the flow [37]. For weak plumes, where internal shear is near normal to the plume axis, turbulence may be important and some of the empirical relationships found in the literature may not be appropriate [38]. Furthermore, plumes that develop from low to intermediate velocity volcanic jets are more likely to be affected by local wind conditions in the lower troposphere, causing plumes to bend. These effects are included in more complex numerical models of ash plume rise. Woodhouse et al. [39] developed an integral model of volcanic plume rising in a windy atmosphere and investigated how wind restricted the rise height of volcanic plumes at Eyjafjallajokull. Lamb et al. [19] used the same plume rise model to investigate two events during the 2009 eruption of Mt. Redoubt, USA. Fluid flow velocities were estimated from acoustic infrasound for monopole, dipole and quadrupole sources, and used to produce models of plume rise, including local wind conditions (Figure 4). Comparison of plume modelling [39] with syn-eruptive Doppler radar measurements of the plumes [40] showed good agreement with the results obtained for infrasound dipole radiation.

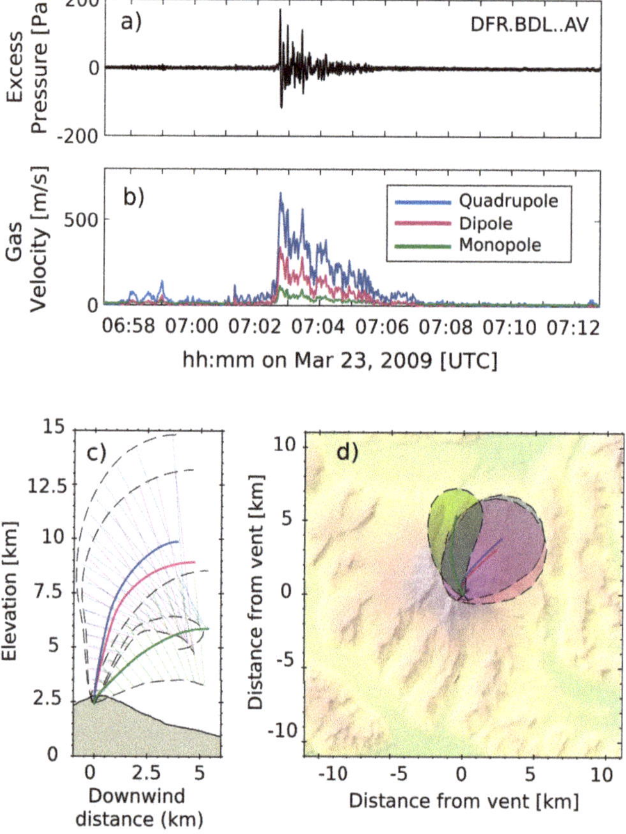

Figure 4. Infrasound-derived gas flow velocities, and plume models for an explosion at Mt. Redoubt, USA, on 23 March 2009: (**a**) raw infrasound signal recorded at 12 km from the vent; (**b**) five-second moving window gas velocity time series calculated for monopole, dipole, and quadrupole infrasound sources by inverting Equations (1)–(3); and (**c**,**d**) cross-section (in the downwind direction) and map view of modelled plumes using measurements of local atmospheric conditions at the time of eruption. Initial flow velocities in the plume models corresponds to peak gas velocities from infrasound for monopole, dipole and quadrupole sources (colour coding consistent across (**b**–**d**)). The figure reproduces and expands results originally presented in Lamb et al. [19].

4.2. Acoustic Monopoles: Volume and Mass Flow Rate

One of the advantages in the use of acoustic infrasound for assessment of volcanic emissions is that its source is directly linked to the unsteady injection of mass into the atmosphere and its acceleration, and to other forces acting in the near-source region. Equations (4)–(6), thus, provide a framework for retrieval of volume or mass flow rates of volcanic material ejected during eruptions. For a pure monopole, the relationship between volume or mass flow rate and the resulting acoustic pressure time series is given by Equation (4), as illustrated in Figure 5; for a source radiating as a monopole in a homogeneous and isotropic atmosphere, Equation (4) can be inverted to retrieve the rate at which atmospheric mass is displaced during an explosion by direct integration over time of the recorded pressure time series. Johnson [41], Johnson and Lees [42], and Johnson and Miller [43] demonstrated the application of the monopole source model to explosions recorded at Erebus (Antarctica), Santiaguito (Guatemala), and Sakurajima (Japan).

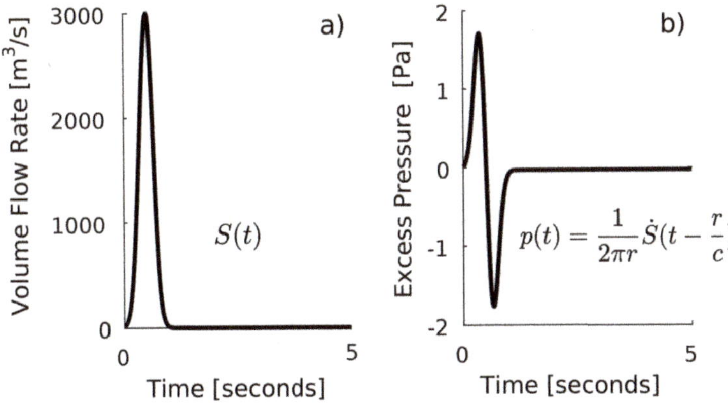

Figure 5. Modelling according to linear acoustic theory for a monopole source: (**a**) time-history of fluid injection into a uniform and homogeneous atmosphere (Gaussian); and (**b**) excess pressure time series at 1 km from the source, obtained as the time derivative of the source time function in (**a**).

Direct waveform integration according to Equation (4) rests on the assumption that infrasound amplitudes attenuate proportionally to the inverse of distance from the source ($1/r$, or geometrical spreading), and neglects the effect of topography on the acoustic wavefield. It is, however, well-known that acoustic wave amplitudes often do not attenuate proportionally to $1/r$ (e.g., Lacanna and Ripepe [44]), thus limiting the application of the monopole model to single-station recordings. Johnson and Miller [43] at Sakurajima proposed its use with data recorded at sites at proximal distance from the source and with unobstructed line-of-sight to the vent, conditions that provide a reasonable approximation to geometrical spreading and allow discarding major effects of topography. In Figure 6, we show an application of the monopole model to data collected by a temporary infrasound network at Fuego (Guatemala) during a moderate-size explosion in May 2018. This example illustrates well the effects of non-$1/r$ attenuation and topography at different locations around the source. Acoustic waveform shape (Figure 6a–e) varies across the network, evidence of diffraction and scattering from topography; the source time functions of volume flow rate (Figure 6f,j) show variable shapes and inconsistent peak values, suggesting that Equation (4) cannot account for the true attenuation pattern, and for realistic scattering of the acoustic wavefield.

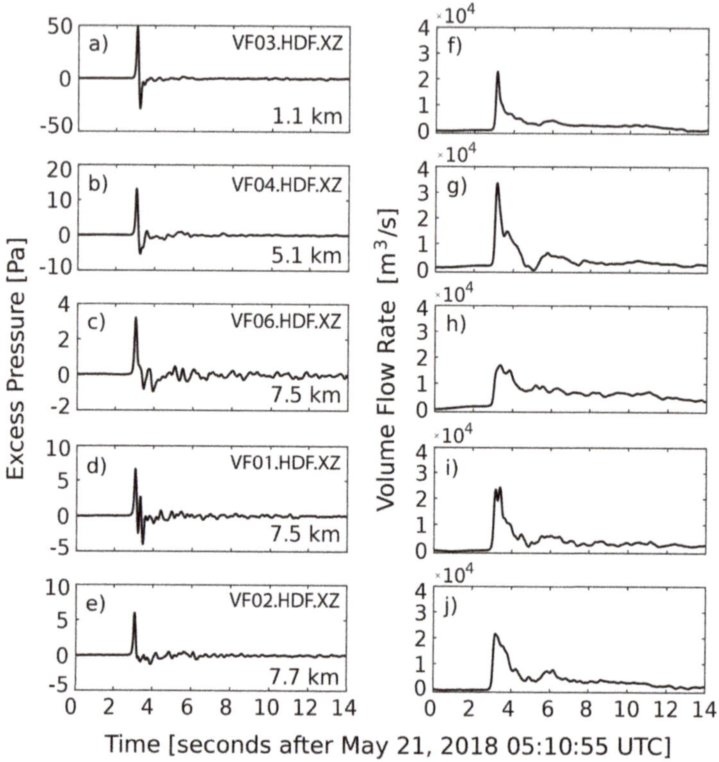

Figure 6. Volume flow source time functions derived from integration of infrasound waveforms at Fuego volcano, Guatemala: (**a–e**) explosion infrasound recorded at a range of distances from the source; and (**f–j**) time series of volume flow rate calculated from (**a–e**) according to Equation (10). The inconsistency of peak volume flow rate and variability in the shape of the volume flow rate source time functions suggest that: (i) the attenuation does not follow simple geometrical spreading corrects; and (ii) the effects of topography on acoustic wave propagation are present.

4.3. Acoustic Multipole Sources and Infrasound Waveform Inversion

The increasing availability of infrasound data over the last decade has allowed important advances in volcano acoustics, and helped overcoming some of the limitations of past studies based on single-station infrasound records. While true amplitude attenuation cannot fully be accounted for without resorting to numerical methods [44,45], using signals from multiple sensors across a network can help mitigating the bias introduced by station location. If available, multiple signals can be simultaneously inverted to constrain volume or mass flow rate source time functions. Johnson et al. [46] performed multi-station inversion solving separately for pure monopole, and pure 3D dipole sources using data from a three-station deployment at Mt. Erebus. They suggested that a more realistic acoustic radiation would be modelled as the combination of the sound produced by the combination a volumetric source (monopole) and a directional force field (dipole); however, inversion for such a source would require data from at least four stations, which were not available. More recently, infrasound waveform inversion was performed considering multipole sources. A second-order multipole source (i.e., combined monopole and dipole) is well-justified when short-duration acoustic signals associated with non-turbulent plumes are analysed. Higher-order source terms (i.e., quadrupoles) should be considered when infrasound is recorded during sustained eruptions, which may generate continuous jet noise due to internal turbulence within the fluid flow. Studies of the sound radiated by sustained volcanic eruptions, until now, have focused on the characterisation of their spectral signatures in analogy to similarity spectra of aeroacoustic jet-noise, or on investigation of the asymmetrical distribution of the recorded amplitudes [1]. We are not aware of any studies, to date, that performed waveform inversions to constrain third-order acoustic multipole sources, thus including a quadrupole term.

A pressure time series, p, can be written as the linear superposition of three contributions (Equation (7)), i.e., a monopole, a horizontal dipole and a vertical dipole. The vertical dipole term is often neglected in infrasound studies; vertical dipoles cannot be reliably constrained by data gathered exclusively on the ground, and their inclusion would introduce errors in source estimates [34]. Kim et al. [34] and De Angelis et al. [47] neglected the vertical dipole component in infrasound waveform inversions at Tunghuraua volcano (Ecuador) and Santiaguito (Guatemala); they combined Equation (7) with Equations (4) and (5) to obtain:

$$p(\mathbf{r},t) = \frac{1}{2\pi r}\left[\dot{S}\left(t-\frac{r}{c}\right) + \frac{x}{cr}\ddot{D}_x\left(t-\frac{r}{c}\right) + \frac{y}{cr}\ddot{D}_y\left(t-\frac{r}{c}\right)\right] \tag{10}$$

which represents the acoustic pressure radiated in a half-space (a homogeneous and isotropic atmosphere with constant sound speed, bounded by a flat topography) by a combined monopole-horizontal dipole, and recorded at distance r from the source. A system of linear equations can be obtained by discretising Equation (10):

$$p_i^k = \frac{1}{2\pi r_i}\left[m_1^k + \frac{x_i}{cr_i}m_2^k + \frac{y_i}{cr_i}m_3^k\right] \tag{11}$$

where p_i^k is the kth value of pressure recorded at the ith station with coordinates (x_i, y_i), at distance r_i from the source; $m_i^k = [m_1^k, m_2^k, m_3^k] = [\dot{S}, \ddot{D}_x, \ddot{D}_y]$ is a vector of parameters that represent monopole and dipole strengths. Equation (11) can be written in matrix form:

$$\mathbf{P}^k = \mathbf{G}\mathbf{m}^k \tag{12}$$

where $\mathbf{P}^k$ is a vector of n values of pressure measured at n stations. Equation (12) can be iteratively solved over all samples in each of the n time-series, for the model vector $\mathbf{m}^k$:

$$\mathbf{m}^k = \mathbf{G}^{-1}\mathbf{P}^k \tag{13}$$

Equations (10)–(13) provide a framework for inversion of acoustic infrasound waveforms for a multipole source. In Figure 7, we show an example of multipole source inversion for an explosion recorded at Fuego (Guatemala) in May 2018. The time history of monopole (i.e., atmospheric volume flow rate) and dipole (i.e., the magnitude of the dipole force vector) strengths are shown in Figure 7a; it should be noted that the vector m_1^k retrieved by iteratively solving Equation (13), provides a time history of the rate of atmospheric mass displacement, which is then translated into a volume flow rate using the atmospheric density at vent elevation. This formulation assumes that the volume of displaced atmosphere is equivalent to the volume of material ejected from the vent. In Figure 7b, we show the dominant dipole direction over the source time function. Figure 7c shows the fit between the inversion model and observations. The quality of fit is measured, following De Angelis et al. [47], using variance reduction:

$$VR = \frac{1}{N}\sum_{i=1}^{N}\left[1 - \frac{\int(s_i - s_i)^2}{\int(d_i)^2}\right] \tag{14}$$

where s_i and d_i are the synthetic and data at each of the N stations, and integrals are performed over the duration of the signals. For comparison, in Figure 7d,e, we show the results for a monopole-only source, obtained by excluding any dipole terms in Equation (11) from the inversion. Comparison of volume flow rates obtained from multipole (black) and monopole-only (yellow) inversions (Figure 7d,e) shows that the inclusion of dipole terms does not produce significant differences in volume flow rate, although it improves waveform fit between model and data.

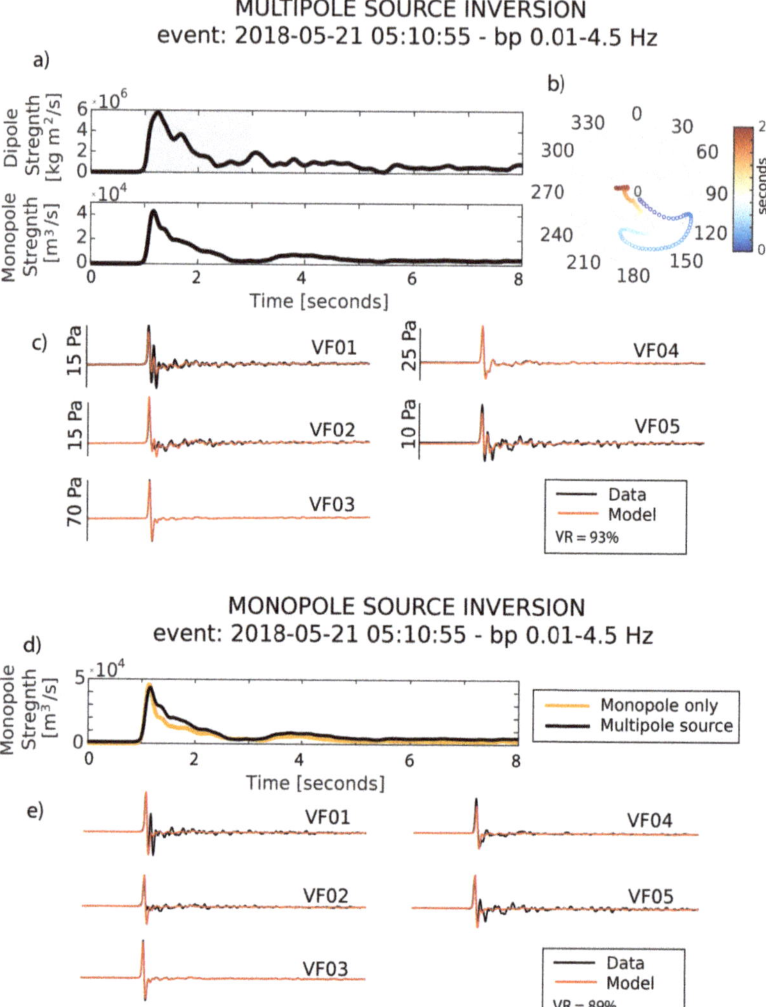

Figure 7. Comparison of acoustic multipole and monopole-only source inversions for an explosion recorded in May 2018 at Fuego, Guatemala: (**a**) monopole and (horizontal) dipole source time functions obtained from multipole inversion; (**b**) dominant direction of the horizontal dipole over the source time function (grey shaded segment in (**a**)); (**c**) waveform fit between data and model; (**d**) comparison of volume flow rates obtained from monopole-only (yellow) and multipole (black) inversions; and (**e**) waveform fit between data and model for monopole-only inversion. Note that these inversions are based on a representation of the acoustic wavefield as in Equation (10). VR is variance reduction (please, see Equation (14) in main text).

All methods discussed thus far to evaluate eruption source parameters from infrasound data do not consider the effects of topography and local atmospheric conditions on the propagation of the acoustic wavefield. Waveform inversion as described in Equations (10)–(13) assumes half-space sound propagation, and relies on the simplest possible 1D formulation of the atmosphere's GF, that is one that only depends on the distance between source and receiver. Kim and Lees [48] and Lacanna and Ripepe [44] first demonstrated by means of numerical modelling the effects of both crater morphology and local volcano topography on infrasound radiation. More recently, Kim et al. [45] introduced an acoustic waveform inversion method that exploits a numerical approximation to the 3D atmospheric GFs computed by a Finite Difference Time Domain (FTDT) scheme. A GF represents the impulse response of the atmosphere, which is the acoustic pressure generated by an impulsive source, and recorded at a given distance from it. GFs are calculated by solving a set of first-order, velocity-pressure coupled differential equations [49], accounting for the propagation of the acoustic wavefield over realistic topography. In these models at local scale (less than about 10 km source-receiver distance), the atmosphere is considered homogeneous and non-moving, with constant sound speed [45]. Figure 8 illustrates results of FDTD modelling of infrasound propagation at Fuego (Guatemala); both the sound pressure level (i.e., acoustic wave intensity measured in decibels; SPL in Figure 8a), and the GFs computed at different locations on the volcano (Figure 8b) demonstrate complex scattering and attenuation effects. Numerical modelling, thus, reveals that infrasound waveform complexity may, at least partially, reflect topography rather than source effects. Kim et al. [45] and Fee et al. [50] included 3D GFs into a new acoustic source inversion workflow, akin to ordinary seismic moment tensor inversion in seismology. A time series of acoustic pressures, p, produced by a monopole and recorded at given location from the source can be written as a convolutional model:

$$p(\mathbf{r},t) = G(\mathbf{r},t;\mathbf{r}_0,t) * S(t) \tag{15}$$

where $S(t)$ is the source mass flow rate (i.e., monopole strength) and $G(\mathbf{r},t;\mathbf{r}_0,t)$ is the Green's function excited by a source at position $\mathbf{r}_0$ and evaluated at position $\mathbf{r}$. Equation (15) can be written in matrix form:

$$\mathbf{d} = \mathbf{Gm} \tag{16}$$

where $\mathbf{d}$ is a vector of recorded pressures at different stations (locations), $\mathbf{G}$ is a matrix of GFs at each of these locations, and $\mathbf{m}$ is a vector of the unknown monopole source strength. Equation (16) can be solved for $\mathbf{m}$ using an iterative least square solver [51]. In Figure 9a,b, we show an application of monopole inversion using 3D, numerical, GFs at Fuego volcano (Guatemala), and compare the results with those from traditional monopole inversion using 1D GFs. The main observation is that volume flow rates calculated without considering topography are overestimated by about 130% when compared to inversion using 3D GFs. Similar results were reported by Kim et al. [45] at Sakurajima volcano, Japan, where overestimates of volume flow rate were up to 155%.

Numerical simulations of acoustic infrasound over realistic topography are rapidly becoming commonplace due to rapid advances in computational methods, and increasing availability of High Performance Computing resources at comparatively low cost. Several ongoing studies are investigating multipole source inversion using 3D GFs (e.g., [52,53]), and successful attempts to include a vertical dipole component have been made [52]. Including dipole terms in the inversion appears to marginally reduce estimates of fluid flow rates, and has variable influence on the quality of waveform fit between data and the final inversion model. The influence of station density and azimuthal coverage is a factor likely to play an important role on the stability of multipole source inversions although the still relatively small number of experiments with high enough quality datasets means that debate on this subject remains open. For example, due its intrinsic directivity, dipole radiation can only be detected at certain locations, making network azimuthal coverage vital to correctly resolve its orientation and strength. The extent to which the contribution of dipole terms can be confidently resolved during inversion, also depends on the quality and resolution of the digital topography used in GFs numerical modelling, in particular in the near-vent region where more pronounced terrain discontinuities and gradients are encountered. Frequent cloud cover and rapid changes in crater morphology make the availability of up-to-date, high-resolution, digital terrain models difficult, posing challenges to the implementation of acoustic source inversion in (near) real time.

Preliminary results from both Iezzi et al. [52] and Diaz-Moreno et al. [53], however, suggest that for impulsive and short-duration explosions, if the effects of topography are correctly accounted for, monopole-only inversions are stable and may be a reasonable approximation to the infrasound source mechanism. Finally, until now, numerical models of acoustic wave propagation have only considered 1D homogeneous representations of the atmosphere with no wind or density changes. This is a reasonably well-justified assumption in the very near-field, although its validity may break down as source-receiver distances approach several kilometres.

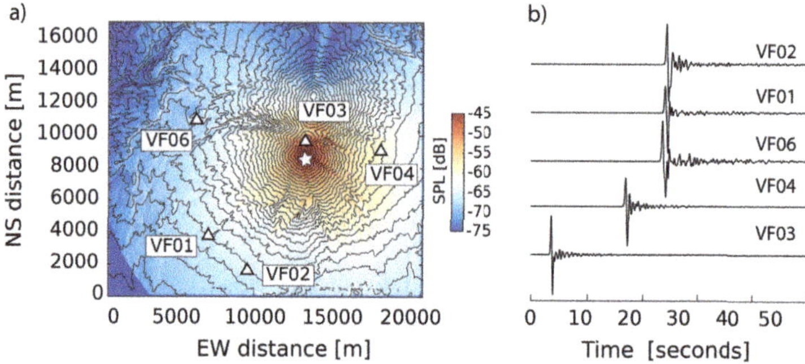

Figure 8. Numerical simulation of acoustic wavefield propagation at Fuego volcano (Guatemala): (**a**) distribution of Sound Pressure Level (i.e., acoustic wavefield intensity measured in decibels relative to background atmospheric pressure) around the source (white star); and (**b**) 3D Green's Functions excited by a monopole located at the source position (white star) and recorded at different stations across a local infrasound network (station positions on the volcano indicated in (**a**)).

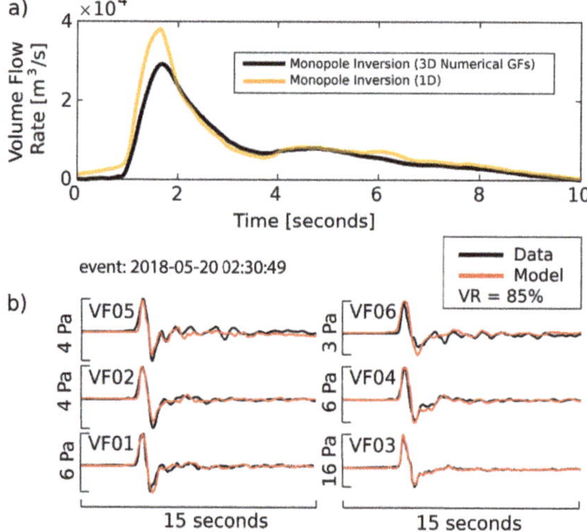

Figure 9. Monopole source inversion using 3D numerical Green's functions for an explosion recorded at Fuego volcano (Guatemala): (**a**) comparison of volume flow rates retrieved by monopole source inversion using 3D and 1D Green's functions (black and yellow, respectively); and (**b**) plot of waveform fit between recorded infrasound and best model from 3D monopole inversion. VR is variance reduction (please, see Equation (14) in main text).

We recommend that future infrasound studies should be supported by data from high-density networks, and exploit better digital terrain models to: (i) resolve the relative influence of monopole and dipole terms in evaluating volcanic emissions; and (ii) assess the importance of considering the

3D nature of dipoles on the results and stability of inversions. We suggest that the influence of variable atmospheric conditions should be further explored, as well. The theory and methods of acoustic inversion need to be extended beyond case studies of short-duration transients, to include the more complex infrasound associated with sustained eruptions for which internal plume turbulence may have significant influence on the radiated acoustic wavefield. Finally, we speculate that, despite several caveats, real-time implementation of acoustic source inversion in an operational sense is within close reach. Similar to seismic moment tensor inversion, libraries of GFs could be implemented and stored for use in (near) real-time inversions, and periodically updated if required. We stress the temporal resolution offered by acoustic infrasound in retrieving eruption source parameters in order to inform ash plume rise and transport models remains unmatched, and can play a crucial role for rapid assessment of airborne eruption hazards.

5. Conclusions

We have presented a comprehensive overview of the most recent developments in the use of acoustic infrasound to assess volcanic emissions, and discussed the theoretical framework of linear acoustic wave theory and its application to volcanic explosions. A wealth of studies over the past two decades have used acoustic infrasound data ranging from single-station pressure time-series to high-density, multi-station datasets to decipher the complexity of the processes that radiate sound at volcanoes. Fluid flow velocities and volume and mass flow rates can be retrieved by means of inversion based on equivalent representations of the acoustic source mechanisms as multipole series. We have discussed the recent introduction of numerical modelling to approximate the atmosphere's impulse response in the presence of realistic topography, and how it has been integrated into inversion workflows. A large body of literature demonstrates the potential of acoustic infrasound to provide rapid estimates of eruption source parameters and their use as input into models of ash plume rise. The still unmatched temporal resolution offered by infrasound makes its use in (near) real time attractive for rapid assessment of airborne hazards during volcanic crises.

Author Contributions: All authors contributed equally to the ideas and data analyses presented in the manuscript. S.D.A. initially wrote the manuscript; and all authors worked on final revisions and copy-editing.

Funding: Silvio De Angelis and Alejandro Diaz-Moreno are funded by NERC grant number NE/P00105X/1. Luciano Zuccarello has received funding from the European Union's Horizon 2020 research and innovation programme under the Marie Sklodowska-Curie grant agreement No. 798480.

Acknowledgments: The authors thank D. Fee and A. Iezzi (Geophysical Institute at the University of Alaska, USA), M. Ripepe and G. Lacanna (Dipartimento di Scienze della Terra, University of Firenze), and J.B. Johnson (Boise State University) for many insightful discussions during the preparation of this manuscript. This manuscript was partly written during a period of research leave spent by S. De Angelis at the University of Alaska Fairbanks; S. De Angelis is indebted to UAF for their generous hospitality and support during his stay. All data used in this manuscript are publicly available through the facilities of the IRIS Data Management Center (http://ds.iris.edu/ds/nodes/dmc/data/#requests) with the exception of those in: (1) Figure 2b,c,e,f available via direct request to the Istituto Nazionale di Geofisica e Vulcanologia, INGV, Sezione di Catania; and (2) Figure 1b, available via direct request to the authors, or publicly from the National Geoscience Data Centre, UK (https://www.bgs.ac.uk/services/ngdc/). The IRIS DMC is funded through the Seismological Facilities for the Advancement of Geoscience and EarthScope (SAGE), Proposal of the National Science Foundation under Cooperative Agreement EAR-1261681.

Conflicts of Interest: The authors declare no conflict of interest.

References

1. Matoza, R.S.; Fee, D.; Green, D.N.; Le Pichon, A.; Vergoz, J.; Haney, M.M.; Mikesell, T.D.; Franco, L.; Valderrama, O.A.; Kelley, M.R.; et al. Local, Regional, and Remote Seismo-acoustic Observations of the April 2015 VEI 4 Eruption of Calbuco Volcano, Chile. *J. Geophys. Res. Solid Earth* **2018**, *123*, 3814–3827. [CrossRef]
2. Brown, S.K.; Jenkins, S.F.; Sparks, R.S.J.; Odbert, H.; Auker, M.R. Volcanic fatalities database: Analysis of volcanic threat with distance and victim classification. *J. Appl. Volcanol.* **2017**, *6*, 15. [CrossRef]

3. Neal, C.A.; Brantley, S.R.; Antolik, L.; Babb, J.L.; Burgess, M.; Calles, K.; Cappos, M.; Chang, J.C.; Conway, S.; Desmither, L.; et al. The 2018 rift eruption and summit collapse of Kilauea Volcano. *Science* **2019**, *363*, 367–374. [CrossRef] [PubMed]
4. Report on Fuego (Guatemala). In *Weekly Volcanic Activity Report*; Sennert, S.K., Ed.; 2018. Available online: https://volcano.si.edu/volcano.cfm?vn=342090#June2018 (accessed on 30 May 2019).
5. Mastin, L.G.; Guffanti, M.; Servranckx, R.; Webley, P.; Barsotti, S.; Dean, K.; Durant, A.; Ewert, J.W.; Neri, A.; Rose, W.I.; et al. A multidisciplinary effort to assign realistic source parameters to models of volcanic ash-cloud transport and dispersion during eruptions. *J. Volcanol. Geotherm. Res.* **2009**, *186*, 10–21. [CrossRef]
6. VAAC Operational Dispersion Model Configuration Snap Shot. Available online: https://www.wmo.int/aemp/sites/default/files/VAAC_Modelling_OperationalModelConfiguration-March2016_v3.pdf (accessed on 15 March 2019).
7. Dürig, T.; Gudmundsson, M.T.; Dioguardi, F.; Woodhouse, M.; Björnsson, H.; Barsotti, S.; Witt, T.; Walter, T.R. REFIR- A multi-parameter system for near real-time estimates of plume-height and mass eruption rate during explosive eruptions. *J. Volcanol. Geotherm. Res.* **2018**, *360*, 61–83. [CrossRef]
8. Sparks, R.S.J.; Bursik, M.I.; Carey, S.N.; Gilbert, J.S.; Glaze, L.; Sigurdsson, H.; Woods, A.W. *Volcanic Plumes*; John Wiley & Sons, Inc.: Hoboken, NJ, USA, 1997.
9. Settle, M. Volcanic eruption clouds and the thermal power output of explosive eruptions. *J. Volcanol. Geotherm. Res.* **1978**, *3*, 309–324. [CrossRef]
10. Wilson, L.; Sparks, R.S.J.; Huang, T.C.; Watkins, N.D. The control of volcanic column heights by eruption energetics and dynamics. *J. Geophys. Res. Solid Earth* **1978**, *83*, 1829–1836. [CrossRef]
11. Garces, M.; Fee, D.; Matoza, R. Volcano Acoustics. *Modeling Volcanic Processes: The Physics and Mathematics of Volcanism*; Cambridge University Press: Cambridge, UK, 2013.
12. De Angelis, S.; Fee, D.; Haney, M.; Schneider, D. Detecting hidden volcanic explosions from Mt. Cleveland Volcano, Alaska with infrasound and ground-coupled airwaves. *Geophys. Res. Lett.* **2012**, *39*, 21312. [CrossRef]
13. Garcés, M.; Fee, D.; Steffke, A.; McCormack, D.; Servranckx, R.; Bass, H.; Hetzer, C.; Hedlin, M.; Matoza, R.; Yepes, H.; Ramon, P. Capturing the Acoustic Fingerprint of Stratospheric Ash Injection. *EOS Trans. Am. Geophys. Union* **2008**, *89*, 377–378. [CrossRef]
14. Fee, D.; Steffke, A.; Garces, M. Characterization of the 2008 Kasatochi and Okmok eruptions using remote infrasound arrays. *J. Geophys. Res.* **2010**, *115*, D00L10. [CrossRef]
15. Kamo, K.; Ishihara K.; Tahira, M. Infrasonic and seismic detection of explosive eruptions at Sakurajima volcano, Japan, and the PEGASAS-VE early-warning system. In *First International Symposium on Volcanic Ash and Aviation Safety*; Casadevall, T.J., Ed.; U.S. Geological Survey Bulletin: Seattle, WA, USA, 1994; Volume 2047, pp. 357–365.
16. Ripepe, M.; Marchetti, E.; Delle Donne, D.; Genco, R.; Innocenti, L.; Lacanna, G.; Valade, S. Infrasonic Early Warning System for Explosive Eruptions. *J. Geophys. Res. Solid Earth* **2018**, *123*, 9570–9585. [CrossRef]
17. Caplan-Auerbach, J.; Bellesiles, A.; Fernandes, J.K. Estimates of eruption velocity and plume height from infrasonic recordings of the 2006 eruption of Augustine Volcano, Alaska. *J. Volcanol. Geotherm. Res.* **2010**, *189*, 12–18. [CrossRef]
18. Ripepe, M.; Bonadonna, C.; Folch, A.; Donne, D.D.; Lacanna, G.; Marchetti, E. Ash-plume dynamics and eruption source parameters by infrasound and thermal imagery: The 2010 Eyjafjallaj okull eruption. *Earth Planet. Sci. Lett.* **2013**, *366*, 112–121. [CrossRef]
19. Lamb, O.D.; De Angelis, S.; Lavallée, Y. Using infrasound to constrain ash plume rise. *J. Appl. Volcanol.* **2015**, *4*, 20. [CrossRef]
20. Vergniolle, S.; Caplan-Auerbach, J. Basaltic thermals and Subplinian plumes: Constraints from acoustic measurements at Shishaldin volcano, Alaska. *Bull. Volcanol.* **2006**, *68*, 611–630. [CrossRef]
21. Caudron, C.; Taisne, B.; Garcés, M.; Alexis, L.P.; Mialle, P. On the use of remote infrasound and seismic stations to constrain the eruptive sequence and intensity for the 2014 Kelud eruption. *Geophys. Res. Lett.* **2015**, *42*, 6614–6621. [CrossRef]
22. Matoza, R.S.; Vergoz, J.; Le Pichon, A.; Ceranna, L.; Green, D.N.; Evers, L.G.; Ripepe, M.; Campus, P.; Liszka, L.; Kvaerna, T.; Kjartansson, E.; Höskuldsson, Á. Long-range acoustic observations of the Eyjafjallajökull eruption, Iceland, April–May 2010. *Geophys. Res. Lett.* **2011**, *38*. [CrossRef]

23. Green, D.N.; Evers, L.G.; Fee, D.; Matoza, R.S.; Snellen, M.; Smets, P.; Simons, D. Hydroacoustic, infrasonic and seismic monitoring of the submarine eruptive activity and sub-aerial plume generation at South Sarigan, May 2010. *J. Volcanol. Geotherm. Res.* **2013**, *257*, 31–43. [CrossRef]
24. Johnson, J.B.; Watson, L.M.; Palma, J.L.; Dunham, E.M.; Anderson, J.F. Forecasting the Eruption of an Open-Vent Volcano Using Resonant Infrasound Tones. *Geophys. Res. Lett.* **2018**, *45*, 2213–2220. [CrossRef]
25. Johnson, J.B.; Palma, J.L. Lahar infrasound associated with Volcán Villarrica's 3 March 2015 eruption. *Geophys. Res. Lett.* **2015**, *42*, 6324–6331. [CrossRef]
26. McNutt, S.R.; Thompson, G.; Johnson, J.; De Angelis, S.; Fee, D. *Seismic and Infrasonic Monitoring*, 2nd ed.; Elsevier Inc.: Amsterdam, The Netherlands, 2015; pp. 1071–1099.
27. Brogi, F.; Ripepe, M.; Bonadonna, C. Lattice Boltzmann modeling to explain volcano acoustic source. *Sci. Rep.* **2018**, *8*, 9537. [CrossRef] [PubMed]
28. Marchetti, E.; Ripepe, M.; Delle Donne, D.; Genco, R.; Finizola, A.; Garaebiti, E. Blast waves from violent explosive activity at Yasur Volcano, Vanuatu. *Geophys. Res. Lett.* **2013**, *40*, 5838–5843. [CrossRef]
29. Matoza, R.S.; Fee, D.; Lopez, T.M. Acoustic Characterization of Explosion Complexity at Sakurajima, Karymsky, and Tungurahua Volcanoes. *Seismol. Res. Lett.* **2014**, *85*, 1187–1199. [CrossRef]
30. Johnson, J.B.; Ripepe, M. Volcano infrasound: A review. *J. Volcanol. Geotherm. Res.* **2011**, *206*, 61–69. [CrossRef]
31. Fee, D.; Matoza, R.S. An overview of volcano infrasound: From hawaiian to plinian, local to global. *J. Volcanol. Geotherm. Res.* **2013**, *249*, 123–139. [CrossRef]
32. Woulff, G.; McGetchin, T.R. Acoustic Noise from Volcanoes: Theory and Experiment. *Geophys. J. Int.* **1976**, *45*, 601–616. [CrossRef]
33. Lighthill, M.J. On sound generated aerodynamically I. General theory. *Proc. R. Soc. Lond. Ser. A Math. Phys. Sci.* **1952**, *211*, 564–587.
34. Kim, K.; Lees, J.M.; Ruiz, M. Acoustic multipole source model for volcanic explosions and inversion for source parameters. *Geophys. J. Int.* **2012**, *191*, 1192–1204. [CrossRef]
35. Morse, P.M.C.; Ingard, K.U. *Theoretical Acoustics*; International Series in Pure And Applied Physics; McGraw-Hill: New York, NY, USA, 1968.
36. Lighthill, M.J. *Waves in Fluids*; Cambridge University Press: Cambridge, UK, 1978; p. 504.
37. Turner, J.S. *Buoyancy Effects in Fluids*; Cambridge Monographs on Mechanics, Cambridge University Press: Cambridge, UK, 1973.
38. Briggs, G.A. Optimum Formulas for Buoyant Plume Rise. *Philos. Trans. R. Soc. Lond. Ser. A Math. Phys. Sci.* **1969**, *265*, 197–203. [CrossRef]
39. Woodhouse, M.J.; Hogg, A.J.; Phillips, J.C.; Sparks, R.S.J. Interaction between volcanic plumes and wind during the 2010 Eyjafjallajökull eruption, Iceland. *J. Geophys. Res. Solid Earth* **2013**, *118*, 92–109. [CrossRef]
40. Schneider, D.J.; Hoblitt, R.P. Doppler weather radar observations of the 2009 eruption of Redoubt Volcano, Alaska. *J. Volcanol. Geotherm. Res.* **2013**, *259*, 133–144. [CrossRef]
41. Johnson, J.B. Volcanic eruptions observed with infrasound. *Geophys. Res. Lett.* **2004**, *31*, L14604. [CrossRef]
42. Johnson, J.B.; Lees, J.M. Sound produced by the rapidly inflating Santiaguito lava dome, Guatemala. *Geophys. Res. Lett.* **2010**, *37*, 1–6. [CrossRef]
43. Johnson, J.B.; Miller, A.J.C. Application of the Monopole Source to Quantify Explosive Flux during Vulcanian Explosions at Sakurajima Volcano (Japan). *Seismol. Res. Lett.* **2014**, *85*, 1163–1176. [CrossRef]
44. Lacanna, G.; Ripepe, M. Influence of near-source volcano topography on the acoustic wavefield and implication for source modeling. *J. Volcanol. Geotherm. Res.* **2013**, *250*, 9–18. [CrossRef]
45. Kim, K.; Fee, D.; Yokoo, A.; Lees, J.M. Acoustic source inversion to estimate volume flux from volcanic explosions. *Geophys. Res. Lett.* **2015**, *42*, 5243–5249. [CrossRef]
46. Johnson, J.; Aster, R.; Jones, K.R.; Kyle, P.; McIntosh, B. Acoustic source characterization of impulsive Strombolian eruptions from the Mount Erebus lava lake. *J. Volcanol. Geotherm. Res.* **2008**, *177*, 673–686. [CrossRef]
47. De Angelis, S.; Lamb, O.D.; Lamur, A.; Hornby, A.J.; von Aulock, F.W.; Chigna, G.; Lavallée, Y.; Rietbrock, A. Characterization of moderate ash-and-gas explosions at Santiaguito volcano, Guatemala, from infrasound waveform inversion and thermal infrared measurements. *Geophys. Res. Lett.* **2016**, *43*, 6220–6227. [CrossRef]
48. Kim, K.; Lees, J.M. Finite-difference time-domain modeling of transient infrasonic wavefields excited by volcanic explosions. *Geophys. Res. Lett.* **2011**, *38*, 2–6. [CrossRef]

49. Ostashev, V.E.; Wilson, D.K.; Liu, L.; Aldridge, D.F.; Symons, N.P.; Marlin, D. Equations for finite-difference, time-domain simulation of sound propagation in moving inhomogeneous media and numerical implementation. *J. Acoust. Soc. Am.* **2005**, *117*, 503–517. [CrossRef]
50. Fee, D.; Izbekov, P.; Kim, K.; Yokoo, A.; Lopez, T.; Prata, F.; Kazahaya, R.; Nakamichi, H.; Iguchi, M. Eruption mass estimation using infrasound waveform inversion and ash and gas measurements: Evaluation at Sakurajima Volcano, Japan. *Earth Planet. Sci. Lett.* **2017**, *480*, 42–52. [CrossRef]
51. Saunders, M.A. Solution of sparse rectangular systems using LSQR and CRAIG. *BIT Numer. Math.* **1995**, *35*, 588–604. [CrossRef]
52. Iezzi, A.M.; Fee, D.; Matoza, R.S.; Jolly, A.D.; Kim, K.; Christenson, B.W.; Johnson, R.; Kilgour, G.; Garaebiti, E.; Austin, A.; et al. 3-D acoustic waveform simulation and inversion supplemented by infrasound sensors on a tethered weather balloon at Yasur Volcano, Vanuatu. In *AGU Fall Meeting Abstracts, Proceedings of the American Geophysical Union, Fall Meeting 2017, New Orleans, LO, USA, 11–15 December 2017*; American Geophysical Union: Washington, DC, USA, 2017.
53. Diaz-Moreno, A.; Iezzi, A.; Lamb, O.; Zuccarello, L.; Fee, D.; De Angelis, S. Assessment of eruption intensity using infrasound waveform inversion at Mt. Etna, Italy. In *AGU Fall Meeting Abstracts, Proceedings of the American Geophysical Union, Fall Meeting 2017, New Orleans, LO, USA, 11–15 December 2017*; American Geophysical Union: Washington, DC, USA, 2017.

remote sensing

Article

Towards Global Volcano Monitoring Using Multisensor Sentinel Missions and Artificial Intelligence: The MOUNTS Monitoring System

Sébastien Valade [1,2,*], Andreas Ley [1], Francesco Massimetti [3,4], Olivier D'Hondt [1], Marco Laiolo [3], Diego Coppola [3], David Loibl ' Olaf Hellwich [1] and Thomas R. Walter [2]

1. Dep. Computer Vision & Remote Sensing, Technische Universität Berlin, 10587 Berlin, Germany
2. GFZ German Research Centre for Geosciences, Telegrafenberg, 14473 Potsdam, Germany
3. Dipartimento di Scienze della Terra, University of Torino, Via Valperga Caluso 35, 10125 Torino, Italy
4. Dipartimento di Scienze della Terra, University of Firenze, Via La Pira 4, 50121 Firenze, Italy
5. Geography Department, Humboldt University Berlin, Unter den Linden 6, 10099 Berlin, Germany

* Correspondence: sebastien.valade@tu-berlin.de

Received: 30 April 2019; Accepted: 18 June 2019; Published: 27 June 2019

Abstract: Most of the world's 1500 active volcanoes are not instrumentally monitored, resulting in deadly eruptions which can occur without observation of precursory activity. The new Sentinel missions are now providing freely available imagery with unprecedented spatial and temporal resolutions, with payloads allowing for a comprehensive monitoring of volcanic hazards. We here present the volcano monitoring platform MOUNTS (Monitoring Unrest from Space), which aims for global monitoring, using multisensor satellite-based imagery (Sentinel-1 Synthetic Aperture Radar SAR, Sentinel-2 Short-Wave InfraRed SWIR, Sentinel-5P TROPOMI), ground-based seismic data (GEOFON and USGS global earthquake catalogues), and artificial intelligence (AI) to assist monitoring tasks. It provides near-real-time access to surface deformation, heat anomalies, SO_2 gas emissions, and local seismicity at a number of volcanoes around the globe, providing support to both scientific and operational communities for volcanic risk assessment. Results are visualized on an open-access website where both geocoded images and time series of relevant parameters are provided, allowing for a comprehensive understanding of the temporal evolution of volcanic activity and eruptive products. We further demonstrate that AI can play a key role in such monitoring frameworks. Here we design and train a Convolutional Neural Network (CNN) on synthetically generated interferograms, to operationally detect strong deformation (e.g., related to dyke intrusions), in the real interferograms produced by MOUNTS. The utility of this interdisciplinary approach is illustrated through a number of recent eruptions (Erta Ale 2017, Fuego 2018, Kilauea 2018, Anak Krakatau 2018, Ambrym 2018, and Piton de la Fournaise 2018–2019). We show how exploiting multiple sensors allows for assessment of a variety of volcanic processes in various climatic settings, ranging from subsurface magma intrusion, to surface eruptive deposit emplacement, pre/syn-eruptive morphological changes, and gas propagation into the atmosphere. The data processed by MOUNTS is providing insights into eruptive precursors and eruptive dynamics of these volcanoes, and is sharpening our understanding of how the integration of multiparametric datasets can help better monitor volcanic hazards.

Keywords: volcano monitoring; Sentinel missions; Convolutional Neural Network (CNN); Synthetic Aperture Radar (SAR) imaging; InSAR processing; infrared remote sensing; SO_2 gas emission

---

1. Introduction

About 1500 volcanoes are considered active worldwide [1], with about 50–85 erupting volcanoes each year [2]. Unfortunately, due to the cost and difficulty to maintain instrumentation in volcanic environments, less than half of the potentially active volcanoes are monitored with ground-based sensors, and even less are considered well-monitored [3]. Numerous volcanic crises have sadly illustrated that this lack of monitoring capabilities can have dramatic consequences [4], as it was the case during the recent 2018 eruptions at Fuego (Guatemala) and Anak Krakatau (Indonesia), which were both responsible for over 430 dead and missing persons according to authorities [5,6]. Moreover, volcanoes considered dormant (no recent eruption) or extinct (no eruption for >10,000 years) are commonly not instrumentally monitored, but may experience large and unexpected eruptions, as it was the case for Chaiten (Chile) in 2008 which erupted after 8000 years of inactivity [7]. In this regard, satellite remote sensing can provide crucial observations when ground-based monitoring is limited or lacking, due to remote environments and/or limited resources. In turn, continuous long-term observations (i.e., monitoring) from space is key to identify background levels of activity, that will allow to better recognize signs of unrest when the activity is deviating from these baselines. Provided that precursor signals can be recorded, pre-eruptive monitoring can lead to better eruption early warnings, and syn-eruptive monitoring can provide critical information for hazard mitigation [8].

Indeed, eruptions are often (but not always) preceded by precursory signals which may last a few hours to a few years, indicating a state of unrest. These signals include changes in seismicity, ground deformation, gas emissions, and/or thermal anomalies [9–11]. Apart from seismicity, all of these can be monitored from space by exploiting various wavelengths across the electromagnetic spectrum: Synthetic Aperture Radar (SAR) is widely used for quantification of surface deformation [12,13], infrared (IR) for quantification of heat radiation [14], and ultraviolet (UV) for quantification of SO_2 degassing [15,16]. During an eruption on the other hand, these space-borne sensors can provide additional support to track eruptive products. In particular, IR spectroscopy is used to estimate lava flow discharge rates [17,18], UV and IR spectroscopy to detect volcanic gas and ash clouds [19,20], whereas SAR processing allows for detection of areas covered by lava flows [21,22] and pyroclastic deposits [23,24].

Efforts to analyze these multi-sensor datasets on global and multi-decadal scales have provided insights into volcanic eruption dynamics and precursors [25–29], and are contributing to the development of strategies for global volcano monitoring. Operational implementations of such multi-sensor approaches into monitoring platforms, aiming at providing near-real-time (NRT) access to gas, thermal, and deformation data from satellites have been initiated by a number of transnational projects. Among the most prominent are the GlobVolcano project (2007–2010, [30]), the European Volcano Observatory Space Services (EVOSS, 2010–2013, [31]), and more recently the Disaster Risk Management (DRM) volcano pilot project led by the Committee on Earth Observation Satellites CEOS (2014–2017, [28]). In parallel, data and tools are shared on research infrastructures such as the European Plate Observing System (EPOS) or the World Organization of Volcano Observatories (WOVO, [32]) to stimulate communication and cooperation. Nevertheless, operational monitoring systems providing integrated multi-sensor analysis for volcano surveillance on global scale is still lacking.

Today, a growing number of new Earth Observation (EO) satellites is providing freely available imagery, with global coverage at unprecedented spatial and temporal resolutions, which is a game changer for volcano monitoring. In particular, the Copernicus Sentinel missions, launched by the European Space Agency (ESA), operate a range of instruments [33] which provide the potential for a comprehensive monitoring of volcanic unrest and eruptive dynamics, opening pathways to global, multisensor volcano monitoring. In parallel, due to the sharply-increasing satellite data volume, novel ways of data reduction and analysis by artificial intelligence (AI) are gaining relevance [34–38].

We here present the first operational volcano monitoring platform, incorporating multisensor satellite-based information (Sentinel-1 SAR, Sentinel-2 SWIR, Sentinel-5P TROPOMI), ground-based earthquakes information (Global Earthquake Catalogues GEOFON and USGS), and AI-assisted monitoring. The system currently monitors 17 volcanoes, located in various climatic and geologic settings across the globe: subduction zones (Colima, Popocatépetl, Fuego, Pacaya, Sangay, Sabancaya, Bezymianny, Klyuchevskoy, Udina, Krakatau, Ambrym, Etna), oceanic hotspots (Piton de la Fournaise, Kilauea), and continental rift zones (Nyiragongo, Nyamulagira, Erta Ale). The system is however designed to easily incorporate new targets to monitor, thanks in particular to the global-coverage and free-access of the data. We use freely-available processing toolboxes to perform routine data processing, and allow the user to investigate the multi-parametric results related to a given volcano through an open-access website (www.mounts-project.com). We demonstrate the utility of such an interdisciplinary approach through a number of recent eruptions, describing a range of volcanic processes which can be tracked, from subsurface magma migration, to surface eruptive deposit emplacement, pre/syn-eruptive morphological changes, and aerosol propagation into the atmosphere. In addition, we show how artificial intelligence can play a key role in monitoring tasks. We show how a pre-trained Convolutional Neural Network (CNN) can be incorporated into the processing pipeline to detect large deformation in InSAR interferograms (e.g., related to dyke intrusions), in an automated, timely, and robust fashion.

2. Background: Existing Space-Based Monitoring Systems

The majority of existing volcano monitoring systems are set up and operated by national monitoring agencies, and consequently focus on volcanoes located within their own political boundaries or limited regions of interest. The U.S. Geological Survey (USGS) for example, provides earthquake information on a global scale, but their Volcano Hazards Program monitors volcanoes on U.S. territory only (https://volcanoes.usgs.gov). Each monitored volcano has a dedicated webpage, where a standardized red-yellow-green Volcano Alert Level System (VALS) informs on the state of volcanic activity [39]. Similar systems are operated by several other countries in charge of monitoring active volcanoes on their territory. However, many volcanoes are located on political boundaries between neighboring countries and states, and therefore leave uncertainties concerning the responsibilities during eruptive crisis. In this regard, only a few satellite-based platforms have the scope to monitor volcanoes on a global scale. Such platforms are often operated by scientists at research institutions and universities, and have proven to be particularly important for volcano observatories which do not have the resources to operate a monitoring network. These however, usually focus on the analysis of one type of data: spectroradiometric imagery (IR and UV), or SAR. We hereafter briefly describe a selection of relevant operating satellite-based platforms which provide open data access on a global scale (Table 1), and expose afterwards the rationale behind MOUNTS.

2.1. Spectroradiometry-Based Systems (IR, UV)

The development of thermal remote sensing volcanic activity has seen a long evolution since the 1960s, with a significant expansion and development of autonomous monitoring systems after the beginning of the new millennium [14].

MODVOLC, developed in 2002 by the Hawai'i Institute of Geophysics and Planetology [40–42], is a widely used volcano thermal monitoring system which can be considered as the archetype of open-access satellite monitoring services. It is based on the autonomous analysis of MODIS (Moderate Resolution Imaging Spectroradiometer) IR data, which has 1 km spatial resolution. The system provides the location of high-temperature thermal anomalies, spectral radiance data, and heat flux estimates in NRT over all volcanoes around the globe, through an open interactive Web-GIS interface (http://modis.higp.hawaii.edu/).

MIROVA (Middle InfraRed Observation of Volcanic Activity, www.mirovaweb.it) is an enhanced NRT volcanic hot-spot detection system, based on the analysis of the MODIS Middle InfraRed (MIR) bands, developed by the University of Torino and the University of Firenze [43]. It currently monitors 215 volcanoes, and displays time series of various measurements, including the Volcanic Radiative Power (VRP) which measures the heat radiated by the volcanic activity, and higher level post-processed products such as lava effusion rates [44,45]. MOUNTS is strongly inspired by MIROVA, whose website template was used.

HOTVOLC is a monitoring system designed to achieve NRT monitoring of volcanic activity using geostationary satellite data (e.g., MSG-0 SEVIRI). It uses both IR and UV spectroradiometry analysis to monitor volcanic thermal activity, lava effusion rate [46], SO_2 gas plume [47], and volcanic ash plume [48]. It is developed at the Observatoire de Physique du Globe de Clermont-Ferrand (France), and monitors ~50 volcanoes worldwide. HOTVOLC provides an interactive Web-GIS interface on which geocoded raster images and time series of the above mentioned parameters are visualized (http://hotvolc.opgc.fr).

The Global Sulfur Dioxide Monitoring Home Page, developed and maintained by NASA's Atmospheric Chemistry and Dynamics Laboratory (www. https://so2.gsfc.nasa.gov/), is a reference portal that brings together volcanic and anthropogenic SO_2 emission data from various monitoring systems. The results are based on the analysis of IR and UV data from different sensors that allow the development of the first inventory of volcanic SO_2 emission at global scale [49]. The website provides an archive of daily SO_2 emissions detected over different regions of volcanic interest that can be freely downloaded. NRT maps are also available through the various systems linked to the Global Sulfur Dioxide Monitoring Home Page.

2.2. SAR-Based Systems

SARVIEWS, developed by the University of Alaska Fairbanks (http://sarviews-hazards.alaska.edu), is a monitoring service based on the analysis of SAR imagery. The processing flow is triggered by an external alert system informing on an ongoing hazardous event, including volcanic eruptions, earthquakes, floods, or fires. By default it provides wrapped interferograms (6-day and 12-day) and Radiometric Terrain Correction (RTC) images computed over large spatial scales (~250 km wide). A Hybrid Pluggable Processing Pipeline (HyP3) allows on-demand higher level analysis, such as change detection algorithms. The derived products, although not available in open-access, have proven to be useful for operational volcano monitoring [50,51].

LiCS (Looking inside the Continents from Space) is developed by COMET (Centre for Observation and Modelling of Earthquakes, Volcanoes and Tectonics). The InSAR portal (https://comet.nerc.ac.uk/COMET-LiCS-portal/) is an interactive map which provides links to download interferograms and coherence maps computed from Sentinel-1 images on large spatial scales. Although the initial focus was the Alpine–Himalayan tectonic belt, the system has recently expanded to process volcanic areas globally. A volcano deformation database provides access to interferograms and coherence maps at several hundreds of volcanoes (https://comet.nerc.ac.uk/volcanoes/), with the intention to be used for monitoring purposes in the future [26].

The University of Miami Geodesy Laboratory developed a system which provides InSAR displacement time-series at a smaller number of selected volcano (https://insarmaps.miami.edu/). The system relies on automatic InSAR time series analysis of Sentinel-1 products, which allows visualization of the temporal evolution of deformation on a large number of points across the volcano. Such approaches can be useful to detect pre-eruptive inflation, which is often observed in closed volcanic systems [52].

Table 1. Selected operational space-based volcano monitoring systems having a global scope. Although both MOUNTS (Monitoring Unrest from Space) and the Insar Viewer (Univ. Miami) incorporate ground-based open-access data in their platform, we here only describe satellite-based data. We refer the reader to the websites and publications for more in-depth detail on the various systems. Revisit and repeat frequencies are indicative, as these vary depending on the latitude and acquisition plan of the space agencies. The revisit frequency indicates the rate at which a volcano is imaged without considering repetitiveness of relative orbits, whereas the repeat frequency considers coverage ensured from the same repetitive relative orbit (i.e., necessary for InSAR analysis in particular). Acronyms used in the table: MIR = Middle InfraRed, TIR = Thermal InfraRed, TADR = Time-Averaged Discharge Rate (m^3/s), NTI = Normalized Thermal Index, VRP = Volcanic Radiative Power (Watt), TSR = Total Spectral Radiance, VFR = Volume Flow Rate (m^3/s), BTD = Brightness Temperature Difference. Webpage links: (1) http://modis.higp.hawaii.edu/, (2) www.mirovaweb.it, (3) http://hotvolc.opgc.fr, (4) https://so2.gsfc.nasa.gov/, (5) http://sarviews-hazards.alaska.edu, (6) https://comet.nerc.ac.uk/COMET-LiCS-portal/, https://comet.nerc.ac.uk/volcanoes/ (7) https://insarmaps.miami.edu/, (8) www.mounts-project.com.

System	Satellite/Sensor Characteristics Used by the System				System Characteristics				
	Satellite (Sensor)	Wave-Length	Orbit, Revisit/Repeat	Resolution	Algorithm	Key Products	Availability after Sensing	N. Volcanoes	Web link
MODVOLC	Aqua/Terra (MODIS)	MIR-TIR	polar-orbiting ~4 img/day revisit	1 km	spectral algorithm [40]	Thermal monitoring: . N hot spot (pixels) . radiant flux (timeseries)	<12 h	>1000	1
MIROVA	Aqua/Terra (MODIS)	MIR-TIR	polar-orbiting ~4 img/day revisit	1 km	contextual algorithm [43]	Thermal monitoring: . VRP (timeseries) . NTI (map) Lava monitoring: . TADR (timeseries, offline)	<4 h	215	2
HOTVOLC	MSG-0 (SEVIRI)	MIR-TIR	geostationary 1 img/15 min	3 km (nadir)	contextual algorithm [46]	Thermal monitoring: . spectral radiance (map, timeseries) Lava monitoring: . TSR (timeseries) . N hot spots (timeseries) . VFR (timeseries)	<30 min	~50	3
					[48]	Ash monitoring: . BTD 5-band, altitude, mass concentration, plume contour (map) . area, altitude, mass (timeseries)			
					[47]	SO_2 monitoring: . BTD tropospheric (map, timeseries) . area (timeseries)			

Table 1. *Cont.*

System	Satellite/Sensor Characteristics Used by the System					System Characteristics			
	Satellite (Sensor)	**Wave-Length**	**Orbit, Revisit/Repeat**	**Resolution**	**Algorithm**	**Key Products**	**Availability after Sensing**	**N. Volcanoes**	**Web link**
NASA Global SO_2 Monitoring	Aura (OMI, OMPS) Sentinel-5P (TROPOMI)	UV	polar-orbiting ~1 day revisit	13 × 24 km (OMI) 50 × 50 km (OMPS) 3.5 × 7 km (TROPOMI)	[53]	SO_2 monitoring: · concentration (map) · mass (timeseries)	<2 d (OMI-OMPS) >7 d (TROPOMI)	>1000	4
SARVIEWS	Sentinel-1 A/B (SAR)	C-band	polar-orbiting ~6–12 days repeat	3.1 × 14.1 m (SLC, IW)	DInSAR (Gamma software) Change Detection [50,51]	Deformation monitoring: · wrapped interferogram (80 × 80 m px) Reflectivity monitoring: · RTC image · change map	≤24 h	variable	5
LICS	Sentinel-1 A/B (SAR)	C-band	polar-orbiting ~6–12 days repeat	3.1 × 14.1 m (SLC, IW)	DInSAR (Gamma software) [54]	Deformation monitoring: · wrapped interferogram + coherence map · unwrapped interferogram	<2 weeks	>1000	6
Univ. Miami InSAR Viewer	Sentinel-1 A/B (SAR)	C-band	polar-orbiting ~6–12 days repeat	3.1 × 14.1 m (SLC, IW)	INSAR timeseries (ISCE + MindPy)	Deformation monitoring: · ground displacement timeseries map	2–3 days	~10	7
MOUNTS	Sentinel-1 A/B (SAR)	C-band	polar-orbiting ~6–12 days repeat	3.1 × 14.1 m (SLC, IW)	DInSAR (SNAP software) Artificial Intelligence (CNN)	Deformation/reflectivity monitoring: · wrapped interferogram + coherence map (14 × 14 m px) · unwrapped interferogram · SAR intensity image · deformation and decorrelation timeseries	≤24 h	~20	8
	Sentinel-2 A/B (MSI)	VIS-SWIR	polar-orbiting ~5 days revisit	10 m (VIS) 20 m (SWIR)	contextual algorithm [55]	Thermal monitoring: · SWIR image B12-B11-B8A (20 × 20 m px) · N hot pixels (timeseries)	≤12 h		
	Sentinel-5P (TROPOMI)	UV	polar-orbiting ~1 day revisit	3.5 × 7 km	[16,56]	SO_2 monitoring: · concentration map in PBL (3.5 × 7 km px) · mass (timeseries)	≤6 h		

2.3. Commercial Platforms

In recent years, commercial platforms have emerged, offering free visualization of open-access satellite imagery. Even though these are not specifically dedicated to volcano monitoring applications, they can be useful to volcano observatories which do not have the capability to download and process large volumes of data.

EO-Browser and Sentinel-hub Playground in particular (https://apps.sentinel-hub.com/sentinel-playground/ and https://apps.sentinel-hub.com/eo-browser/, respectively), make it possible to browse within a Web-GIS the complete archive of ESA Sentinel missions (Sentinel-1, Sentinel-2, Sentinel-3, Sentinel-5P), NASA Landsat missions (Landsat-5, 7 and 8), NASA MODIS imagery, and more. It gives the possibility to export image products in various formats, create time-lapse animation clips, and provides simple statistical analysis tools. Additional features are offered in commercial packages.

Google Earth Engine (GEE) gives access to several catalogs of open-access satellite imagery and geospatial datasets, and provides a platform to develop and run algorithms on Google's cloud infrastructure at no cost. While the GEE Explorer provides a simple web interface to visualize satellite imagery, the GEE Code Editor (https://code.earthengine.google.com/) provides a web-based IDE (Integrated Development Environment) for the GEE JavaScript API (Application User Interface). This allows users to develop algorithms and run them on the cloud without having to download the data. Although useful for processing large amounts of data, such service is not at the moment an immediate resource for volcano monitoring.

3. Materials and Methods

3.1. Rationale of the MOUNTS System

Global volcano monitoring has the aim to assemble and provide timely information of physical and geochemical parameters at a large number of (or all) volcanoes worldwide. The information can in turn be useful to both the scientific and operational communities, to understand volcanic processes and mitigate volcanic hazards. The rationale behind MOUNTS is based on the following three requisites:

(1) To analyze the multi-sensor dataset offered by the Sentinel missions to provide the most complete and comprehensive understanding of volcanic processes (Figure 1). Specifically, the datasets considered are: (i) SAR imagery from Sentinel-1 (S1) to monitor surface deformation, morphological changes, and reflectivity changes, (ii) SWIR imagery from Sentinel-2 (S2) to monitor surface thermal anomalies, and (iii) TROPOMI SO_2 data from Sentinel-5P (S5P) to monitor volcanic SO_2 gas plumes. Ground-based data is also incorporated, by querying global earthquake catalogues (GEOFON and USGS) for earthquakes recorded in the vicinity of the volcano. The volcanic processes which can be apprehended from the integration of these datasets is summarized in Figure 1, and those flagged with an asterisk are illustrated in the paper.

(2) To disseminate the results of the analysis as: (i) geocoded images, and (ii) time series of key parameters extracted from these, so as to have a comprehensive understanding of their evolution through time. The results are published on an open-access website (www.mounts-project.com), with a design similar to MIROVA, whereby each monitored volcano is assigned a webpage following a standard template. Dissemination of these results is achieved in NRT, typically within <1–6 h following the availability of the source product on ESA's Sentinel Data Hub. The availability of such products after sensing varies depending on the data type and processing level: typically <24 h from sensing for Sentinel-1 SLC (Single Look Complex) products, between 2 and 12 h from sensing for Sentinel-2 L1C (Level-1 C) products, and <3 h from sensing for Sentinel-5P NRTI (Near Real Time) SO_2 products.

(3) To implement a modular architecture, allowing easy incorporation of (i) new data types to analyze, (ii) new targets to monitor (i.e., new volcanoes, or new regions of interests for a given volcano depending on the activity, e.g., Kilauea 10 × 10 km extent during regular volcanic activity, and/or

Kilauea East Rift Zone 40 × 60 km extent during flank eruption), and (iii) new algorithms to process the data (i.e., new processing chains, or trained neural networks to solve specific tasks).

MOUNTS is the first system to integrate all these components on a unique platform, with open-access, and global coverage capability.

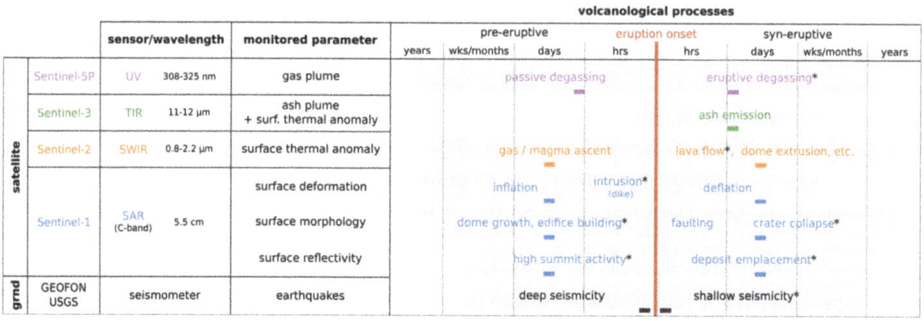

		sensor/wavelength		monitored parameter	volcanological processes: pre-eruptive (years, wks/months, days, hrs)	eruption onset	volcanological processes: syn-eruptive (hrs, days, wks/months, years)
satellite	Sentinel-5P	UV	308-325 nm	gas plume	passive degassing		eruptive degassing*
satellite	Sentinel-3	TIR	11-12 μm	ash plume + surf. thermal anomaly			ash emission
satellite	Sentinel-2	SWIR	0.8-2.2 μm	surface thermal anomaly	gas / magma ascent		lava flow*, dome extrusion, etc.
satellite	Sentinel-1	SAR (C-band)	5.5 cm	surface deformation	inflation; intrusion* (dike)		deflation
				surface morphology	dome growth, edifice building*		faulting; crater collapse*
				surface reflectivity	high summit activity*		deposit emplacement*
grnd	GEOFON USGS	seismometer		earthquakes	deep seismicity		shallow seismicity*

Figure 1. Utility of different sensor types, both spaceborne and ground-based, used by MOUNTS to monitor various volcanic processes. The occurrence of these may vary strongly from one volcano to another [11,26], and therefore the timings reported are given in an indicative manner. Processes flagged with an asterisk are those illustrated in this paper. The approximate revisit frequency of each satellite product used is indicated with a colored-marker (see Table 1 for details). Apart from Sentinel-3, all data are available on MOUNTS website. USGS: U.S. Geological Survey; SAR: Synthetic Aperture Radar; SWIR: short-wave infrared (SWIR).

3.2. *Workflow of the MOUNTS System*

MOUNTS is based on freely available remote sensing data, open source toolboxes, and a modular architecture. The workflow is outlined in Figure 2, and the basic processing information is described hereafter. We refer the reader to the Supplementary Material S1 for more in-depth technical details on various aspects of the processing chain. The entire framework is managed with Python, and is composed of the following elements:

(1) Data query: the ESA Sentinel Data Hub is queried for new products at regular time intervals (typically every hour) using the Open Search API. Query options include the product type (i.e., platform name, file type), the product sensing time, and the footprint of the area of interest (AOI). Products types used by MOUNTS are the following: Sentinel-1 Level 1 (L1) Single Look Complex (SLC) Interferometric Wide (IW) products, Sentinel-2 Level 1C (L1C) products, and Sentinel-5P Level 2 (L2) Near Real Time (NRTI) SO_2 products. The standard AOI for a specific volcano is defined as a 10 × 10 km mask centered around the volcano summit, but can be adapted based on specific requirements (e.g., Kilauea Rift Zone 40 × 60 km to monitor flank eruptions). Volcano name, coordinates, and identification number are taken from the Global Volcanism Program (GVP) database [1], maintained by the Smithsonian Institution. The volcanoes monitored by MOUNTS and the associated relevant information for data query and processing are stored in a local database.

(2) Data download: when a new product is found on the Sentinel data hub, the data is automatically downloaded, and its metadata is stored in a local database.

(3) Data processing: as soon as a product is downloaded it goes through the dedicated processing chain, which intends to generate both image files (geocoded) and extract key parameters which help visualize volcanic activity. Products are opened and processed using ESA's open-source SNAP software (Sentinel Application Platform) through its Python API. The processing techniques applied to Sentinel-1, Sentinel-2, and Sentinel-5P products are described in Section 3.3.

(4) Data dissemination: the outputs of the processing chain are displayed on an open-access website (www.mounts-project.com). The web-design is based upon that of MIROVA, whereby all monitored volcanoes have a dedicated webpage displaying the same set of information. A range of

tools have been incorporated to conveniently visualize the multi-parametric data, and thereby evaluate the volcanic activity (Figure 3).

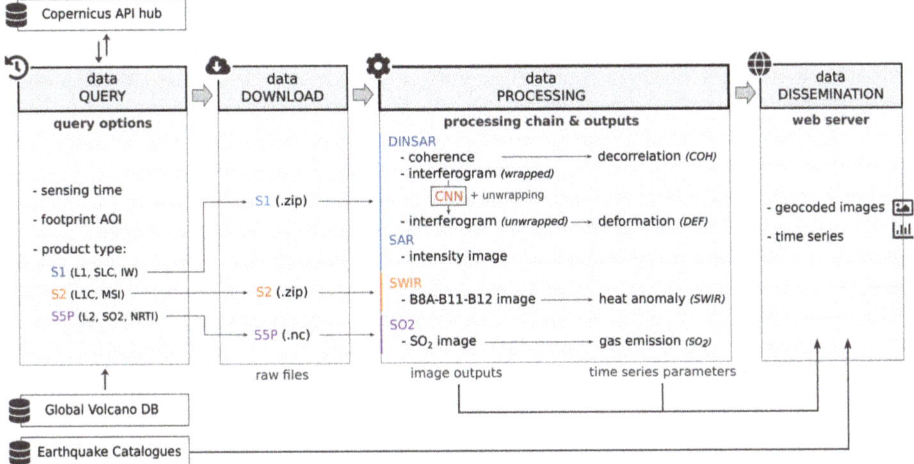

Figure 2. Workflow of MOUNTS monitoring system, managing automated data query, download, processing, and dissemination of satellite-based (Sentinel-1 (S1), -2 (S2) and -5P (SP5)) and ground-based (global earthquake catalogues USGS and GEOFON) data at a number of volcanoes worldwide.

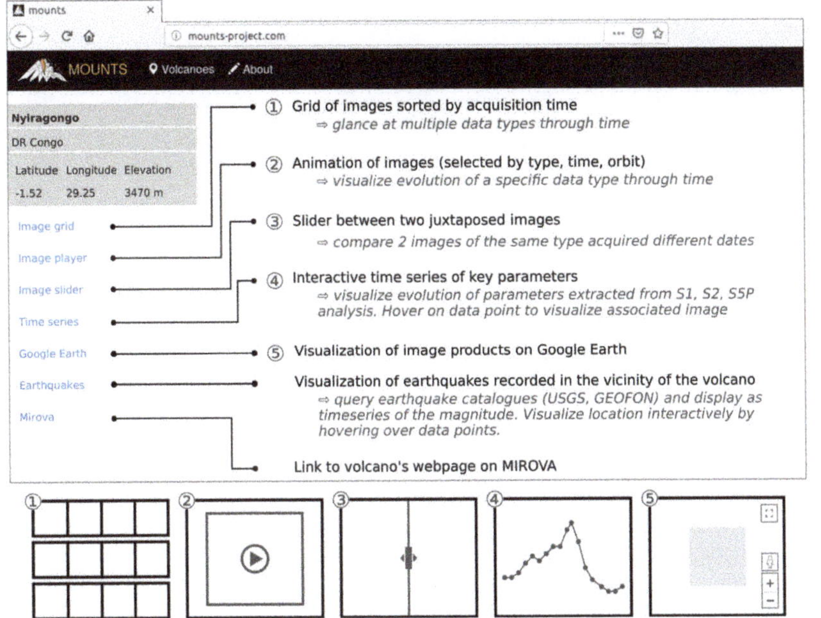

Figure 3. Snapshot of the standard webpage identical for each monitored volcano, and description of the menu content.

3.3. Processing Techniques

3.3.1. Sentinel-1

The Sentinel-1 mission comprises two polar-orbiting satellites (Sentinel-1A and Sentinel-1B, launched in April 2014 and April 2016, respectively), each carrying a SAR operating at C-band [57]. As it is an active remote sensing instrument sending its own electromagnetic radiation, and as radar wavelengths are mostly unaffected by weather clouds, it provides data in all-weather, day and night conditions, making it particularly advantageous for volcano monitoring. The two-satellite constellation offers a 6 day exact repeat cycle at the equator. However, since the orbit track spacing varies with latitude, the revisit rate is significantly greater at higher latitudes than at the equator. MOUNTS uses Interferometric Wide (IW), Single Look Complex (SLC) products, where each pixel is encoded as a complex number containing information on both the intensity and the phase of the signal backscattered to the radar sensor. Both information can be exploited to retrieve parameters relevant to volcano monitoring [12,26]. In turn, MOUNTS operates two processing pipelines: a SAR pipeline to monitor reflectivity changes due to morphological changes (incoherent change detection), and a DInSAR (Differential SAR Interferometry) pipeline to monitor surface deformation and emplacement of eruptive deposits (coherent change detection).

SAR Processing

The amplitude component of a SAR image measures the energy backscattered to the radar antenna after the electromagnetic wave hits the surface. It depends on the surface slope (i.e., incidence angle of the wave front onto the surface), the surface roughness (i.e., specular or diffuse reflector), and the surface material (i.e., the dielectric constant, particularly affected by moisture content). For instance, a terrain with a slope facing towards the radar sensor will backscatterer a lot of energy to the radar (high amplitude signal), which will appear as bright pixels on the SAR image. Conversely, a slope facing away from the radar will appear as dark pixels (low amplitude). Volcanic ash will tend to smooth the terrain on which it is deposited, thereby acting like a specular reflector, which will tend to backscatter more energy if the radar viewing angle is orthogonal to the slope, but less energy if looking at an oblique angle [23].

As soon as an S1 product is downloaded, it goes through the SAR processing chain which creates a geocoded SAR intensity image (see Supplementary Material S1 for details). Visual inspection of this image can be useful for volcano monitoring even for non-experts, especially when optical images are not available because of cloud coverage. In particular, large surface topography changes and eruptive deposits are usually easily identifiable (see Section 4.3).

Retrieving parameters which can help characterize volcanic activity over time in a robust and automated fashion is however a difficult task. Various approaches have been used to detect and visualize changes in SAR imagery. "Low-cost" approaches usually involve "incoherent change detection", which identifies surface intensity changes in successive SAR images. These changes can be visualized in a number of ways: (i) as difference maps, whereby the arithmetic difference, logarithmic ratio, or normalized change index (NCI) between two images is plotted [58–60], or (ii) as composite RGB image, where two bands are assigned to the first and second SAR image, and the third band to one of the difference map described above (e.g., [23,24]). More advanced techniques track features in the radar image prior to geocoding (i.e., radar coordinates in range and azimuth) to quantify morphological changes. Tracking radar shadows, for example, has proven useful to detect crater deepening [23], infilling of valleys by pyroclastic flows [24], and lava lake level variations [61]. Tracking the position of specific reflectors on the other hand, has proven capable of quantifying volcanic dome growths [62]. Such techniques unfortunately require a lot of manual tuning to adapt to each volcano, and are consequently a difficult approach for global monitoring. In this paper we instead favor the logarithmic ratio approach, in which we color code areas where the intensity has increased and decreased to red and blue respectively (see Section 4.2.2).

DInSAR Processing

SAR interferometry exploits the phase difference between two SAR images having similar viewing geometries, in order to measure both surface topography and displacements. More specifically, InSAR processing is commonly used to generate digital elevation maps, whereas DInSAR processing (Differential InSAR) successively subtracts the terrain elevation to retrieve surface deformation between the two acquisitions [63–66]. While these techniques typically exploit only two images, more advanced techniques known as "time series processing" exploit multiple images to retrieve more accurate deformation time series (i.e., Permanent Scatters PS and Small Baseline SBAS, see [67] for a review). The accuracy difference between DInSAR and time series approaches is about one order of magnitude, i.e., from centimeter to millimeter accuracy [68]. However, time series techniques require a lot of computing power and are therefore not performed by MOUNTS, which instead focuses on the two-image DInSAR approach.

Each time a new SAR image (S1 IW SLC product) is downloaded by MOUNTS, it goes through a DInSAR processing chain with the preceding acquisition (shortest temporal baseline, typically 6–12 days) from the same relative orbit. Processing of the image pair is performed with SNAP, freely distributed by ESA, and managed through a Python API. Back-geocoding and terrain correction are by default achieved with the SRTM-3 DEM. No multilooking is performed, so that the ground pixel resolution of the output products is 14 × 14 m. We refer the reader to the Supplementary Material S1 for more details on the processing chain. The outputs are typically an interferogram and a coherence map. The interferogram depicts the deformation of the ground along the sensor's line-of-sight (LOS) direction, and is displayed with values ranging between 0 and 2π which appear as a series of fringes (i.e., "wrapped" interferogram). The coherence is a measure of the interferometric quality, as it is related to the level of noise affecting the phase difference between the two SAR images. It ranges between 0 and 1, and reflects the degree of surface change, where values towards 0 indicate a loss of InSAR coherence (i.e., decorrelation). The main contributions to decorrelation are phase noise due to (i) the temporal change of the scatterers (e.g., vegetation, water, sand-covered areas will appear highly incoherent as the scatterers change continuously), (ii) geometric decorrelation (i.e., images have slightly different look-angles), and (iii) variation in atmospheric water vapor content. Although decorrelation usually renders interferograms useless for measuring ground deformation, it can be used to detect changes in the ground properties (e.g., emplacement of lava flows and eruptive deposits, delineation of eruptive fissures, etc.).

Both the interferogram and coherence image are successively analyzed to derive quantitative parameters, which are meant to be plotted as time series and consequently inform on the evolution of relevant volcanic processes through time. The interferogram analysis is intended to detect strong ground deformation occurring in short time intervals (i.e., between two successive S1 acquisitions, usually 6 to 12 days apart), which typically imprint as multiple narrow fringes (one fringes in Sentinel-1 interferograms represents 2.8 cm displacement in the radar line-of-sight). Processes such as magma intrusion (dykes), caldera collapse, or sudden flank movements are typically detected. The deformation detection is performed using artificial intelligence, which recovers filtered phase gradients in wrapped geocoded interferograms. Unwrapping of the phase gradients is successively performed to recover a deformation map, from which a deformation score (DEF) informing on the amount of ground displacement is computed. The approach is described in detail in Section 3.4 and in Supplementary Material S2. The coherence map, on the other hand, is analyzed to derive a decorrelation score (COH), which counts the number of pixels below a given coherence threshold. Although this parameter can fluctuate due to a number of non-volcanic processes (see phase noise sources described above), it can be useful to track eruptive deposits (e.g., fresh lava flows, pyroclastic deposits, etc.).

Hence, for each new SAR acquisition, an interferogram and coherence map are generated (shortest temporal baseline possible, typically 6 or 12 days), and disseminated on the website. From these, a deformation and decorrelation score are respectively computed, which are plotted as time series on the website.

3.3.2. Sentinel-2

The Sentinel-2 mission comprises two polar-orbiting satellites, Sentinel-2A and Sentinel-2B (launched in June 2015 and March 2017, respectively), placed at 180° from each other in the same sun-synchronous orbit. They carry a Multispectral Instrument (MSI) providing multispectral data in 13 bands spanning from visible (VIS) to short-wave infrared (SWIR) wavelengths. The SWIR channels in particular, allow for detection of thermal emission produced by hot bodies. Considering a revisit frequency of five days under cloud-free conditions (reduced to 2–3 days at mid-latitudes), and a spatial resolution of 20 m/pixel in the SWIR bands, Sentinel-2 provide important measurements for volcano thermal monitoring. Specifically, analysis of thermal and morphometric features of heat sources facilitates tracking of a series of volcanic processes, including lava flow advancements, lava lake pulses, extrusion phases of lava domes, fumarolic activity, thermal activity at multiple active craters, and rise of magma column in open-vent volcanic systems (e.g., [69,70]).

MOUNTS downloads and analyzes Sentinel-2 Level 1C products. We use an approach presented in [55], which is an enhancement of the HOTMAP detection algorithm developed by [71]. It is based on fixed ratios between SWIR bands, and on a contextual threshold derived from a statistical distribution of the thermal anomaly clusters. Images are analyzed considering the TOA (Top of the Atmosphere) reflectance of the 12-11-8A bands (R: 2190 nm, $\varrho 12$; G: 1610 nm, $\varrho 11$; B: 865 nm, $\varrho 8a$). The algorithm detects the number of "hot" pixels in each downloaded image, and stores this parameter (SWIR) which can be displayed in time series. This thermal algorithm was already successfully verified on different volcanic cases worldwide, and compared with MIROVA data, showing a strong correlation between the detected number of "hot" pixels in Sentinel-2, and the VRP [55].

3.3.3. Sentinel-5P

SO_2 emissions are recovered from the imaging spectrometer TROPOMI on-board the Sentinel-5P satellite (launched in 2017, data available since December 2018). It offers data with a revisit frequency of 1 day, at exceptionally high spatial resolution (7×3.5 km^2) and detection limit (respectively 13 and 4 times better than the heritage Ozone Monitoring Instrument (OMI)), thereby opening the possibility to detect volcanic SO_2 with an unprecedented precision [16]. SO_2 slant column densities are retrieved by TROPOMI using differential optical absorption spectroscopy (DOAS) in the UV spectra (312–326 nm). Final gas densities are provided as vertical column densities (VCDs), for 1 km thick boxes at three different altitudes: 0–1 km (planet boundary layer, PBL), 6.5–7.5 km (mid-troposphere), and 14.5–15.5 km (upper troposphere). ESA disseminates the data as near-real time (NRTI) products, available during 1 month after acquisition, after which they are replaced by Offline (OFFL) products. OFFL products are based upon consolidated calibration and auxiliary data and thus provide better data quality data compared to NRTI products.

MOUNTS downloads Level 2 NRTI products, and processes them to retrieve SO_2 mass found in a 500×500 km box centered around the volcano. Vertical column SO_2 densities are first converted from mol·m^{-2} to the more commonly used Dobson Unit (DU), using a multiplication factor of 2241,15. SO_2 mass is then calculated following [72]:

$$\mathrm{MSO_2} = 0.0285 \sum\nolimits_{i=0}^{n} \mathrm{A_i SO2_i}, \tag{1}$$

where MSO_2 is the mass of SO_2 (in tons), and A_i and $SO2_i$ are respectively the area (7×3.5 km^2) and the density (in DU) at each pixel i of the 0–1 km VCD (PBL). Pixels contaminated with SO_2 are successively isolated by creating a mask where DU > 1, to which is successively applied a morphological filter (i.e., erosion + dilatation operators, structuring element of 5×5 pixels), which allows for removal of noisy pixels. The remaining pixel clusters are assumed to be of volcanic origin, and the resulting SO_2 mass is assigned to the volcano for the time series display.

The 500×500 km box size is chosen arbitrarily to capture large SO_2 plumes. However, several monitored volcanoes may be located within this area, potentially resulting in redundant SO_2 anomalies.

Although visual inspection of the image helps identify the emitting volcano, future developments will focus on using wind models to discriminate the emission source, and therefore recover more accurate SO_2 emissions for each volcano.

3.4. Machine Learning in Support of Deformation Detection

Machine learning algorithms can help solve specific tasks without being explicitly programmed, relying instead on patterns learned from a training dataset. We here show how trained Convolutional Neural Networks (CNN) can be plugged into the workflow to automatically detect large ground deformation in interferograms (Figure 4).

Synthetic data was first generated to train the network. The goal was to produce synthetic interferograms, wrapped and orthorectified, depicting realistic deformation fringes and phase decorrelation related to various artifacts (i.e., atmospheric phase delays, geometric distortions due to orthorectification, phase decorrelation due to vegetated areas, etc.). We synthesized the data from a combination of procedural noise and empirical rules, which although not physically motivated, can express a wide range of patterns similar to those found in real data (at least locally). The network was designed as a fully convolutional auto encoder built out of residual learning blocks (inspired by ResNet, [73]). Details on the synthetic data generation and network architecture can be found in Supplementary Material S2, and the trained CNN can be accessed on a Github repository (https://github.com/Andreas-Ley/SAR-InterfPhaseFilter).

The network was then trained to recover from the synthetic interferograms (provided as input), the associated phase gradients and phase decorrelation mask (expected outputs), Figure 4a. Once trained, the network is plugged into the processing pipeline (Figure 4b): newly generated interferograms are fed through this pre-trained network, which yields both the phase gradients and phase decorrelation (i.e., noise mask). Unwrapping of these phase gradients is successively performed to recover approximate ground displacements (Figure 4c), where robustness against artifacts that might trigger false alarms is preferred over precise phase differences. The unwrapping simply solves for a displacement map whose gradients match the gradients estimated by the CNN. Since the CNN performs the filtering of the gradients and detection of noise, treatment of decorrellated areas in the unwrapping is rather simple. From this displacement map a deformation score is ultimately computed (DEF), as the standard deviation of the displacements expressed in meters in the radar line-of-sight (LOS). This data is then disseminated (Figure 4d) and can be used to plot time series that allow for identification of strong deformation events. In addition, automated email alerts are sent to interested users when a threshold is exceeded. Based on empiric evaluation of all 17 volcanoes currently monitored by MOUNTS, representing a total of >1360 interferograms, the threshold was set to 0.001 m. This value can however be tuned for each volcano, to adapt to specific volcanic activity or monitoring requirements.

A similar approach involving CNN has been presented by [34] to detect large deformation signals in short-duration, wrapped and geocoded interferograms. The main differences to our approach are the following: (i) training is performed on real interferograms, combined with a data augmentation approach to increase the number of interferograms where deformation is recorded; we instead create synthetic training data, allowing the generation of an unlimited number of interferograms, and avoiding the time-consuming task of labeling interferograms where deformation is identified; (ii) edge detection is applied to the interferograms (during both training and prediction processes), to identify phase jumps between fringes, which are subsequently considered as noise-free regions where deformation can be detected; our network instead is trained to output a decorrelation mask informing on the degree of phase noise, successively used to filter noisy regions in the interferogram where erroneous gradients could be detected, (iii) the network output is a deformation probability; our network outputs clean phase gradients (and decorrelation mask for additional cleaning), which can be successively unwrapped to compute ground deformation maps, from which a physically meaningful score can be derived (i.e., standard deviation expressed in meters), (iv) the authors fine-tuned a popular network

(AlexNet, [74]), whereas the architecture of our network was designed from scratch and trained on our synthetic dataset, thereby allowing more flexibility in the desired outputs.

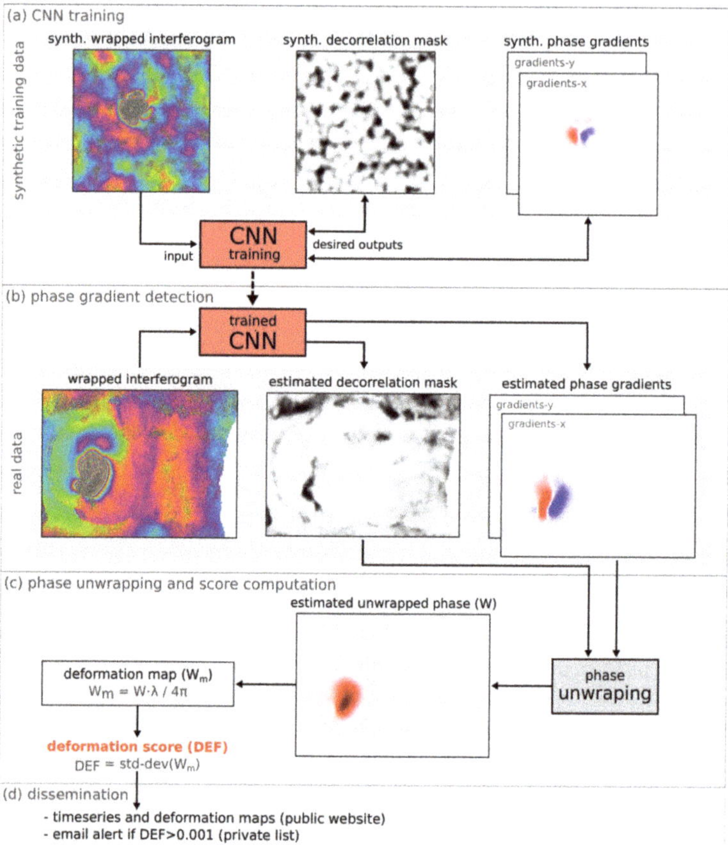

Figure 4. Workflow to detect ground deformation using a pre-trained convolutional network: (**a**) Convolutional Neural Network (CNN) training using synthetic data (see Supplementary Material S2), (**b**) operational usage of the pre-trained CNN to detect to detect phase gradients and decorrelation mask in real interferograms (Piton de la Fournaise in this example), (**c**) phase unwrapping to recover ground displacements, and derive a deformation score defined as the standard deviation of the displacement map, (**d**) data dissemination, as images and time series on a public website (www.mounts-project.com), and email alerts to interested users when a threshold is overcome.

We compare the performances of the two approaches using the same evaluation metrics used in their study, which are derived from the confusion matrix: accuracy (ratio between correctly predicted samples and all testing samples), true positive rate (TPR, ratio between correctly identified positive samples and all positive samples), and true negative rate (TNR, ratio between correctly identified negative samples and all negative samples). Table 2 shows the performances of both networks, tested on the same volcanoes (Erta Ale, Ethiopia, and Etna, Italy), and highlights the good performances of our CNN.

Table 2. Evaluation of the performances of the proposed CNN, and comparison with the CNN presented by [34] (AlexNet). The metrics are derived from the confusion matrix, where P is the number of real positives in the dataset (i.e., number of interferograms where deformation is recorded), N is the number of real negatives, TPR is the true positive rate, and TNR is the true negative rate. For each volcano we evaluate on the entire interferogram dataset (since training is performed on an independent synthetic dataset), whereas [34] report the average over multiple random splits into training and testing groups of equal size. Interferograms where deformation is expected were marked manually for ground truth (i.e., real positives P), and during evaluation deformation detection threshold was DEF = 0.001, the same threshold as for the email notifications.

Volcano	Method	N Interferograms (P + N)	Time Span	Accuracy	TPR	TNR	Exec. Time (Interf. Size)
Erta Ale	proposed CNN	134 (2 + 132)	3 February 2016–19 May 2019	1.000	1.000	1.000	0.18 s (875 × 817)
	AlexNet [1]	205 (12 + 193)	11 December 2016–6 December 2017	0.994	1.000	0.988	1.50 s (500 × 500)
Etna	proposed CNN	126 (2 + 124)	8 May 2018–21 May 2019	0.993	1.000	0.993	-
	AlexNet [1]	189 (2 + 187)	3 September 2016–9 November 2017	0.871	0.747	0.981	-

[1] From [34].

Synthetic data can only seldomly cover the full range of variations and all the corner cases that can be observed in the real world, and is thus usually inferior to real data. However, there are also benefits that make it an attractive choice: with synthetic data the ground truth values of what is to be predicted are known and can be used for the training. Also, since arbitrary amounts of data can be generated, overfitting due to insufficient data is not an issue, and the generalization capability of the final model depends primarily on the realism of the synthetic data. While the present work was under review, a study using synthetic interferograms to train neural networks was published [35]. Unlike our study, the authors generate deformation patterns based on analytic models simulating realistic deformation sources in volcanic settings (i.e., Mogi and Okada sources in particular), and model stratified atmospheric effects from weather models (i.e., GACOS). This approach incorporates expert knowledge into the network via the training data, and opens pathways to potentially recover deformation sources properties from the interferogram in an automatic fashion.

4. Results

We hereafter show how the products derived from S1, S2, and S5P analysis can help monitor various aspects of volcanic activity. The usefulness of each of these parameters varies from one volcano to another, due to both varying volcanic activity (i.e., effusive vs. explosive, open-vent vs. close-vent) and climatic setting (i.e., desert vs. tropical environment). For this reason, we here take various eruptive case examples, where the utility of each of these parameters is best illustrated. Below we successively describe processes related to magma migration towards the surface, effusive and explosive eruptive deposit emplacement, as well as SO_2 gas plume propagation in the atmosphere.

4.1. Detection of Surface Deformation (DInSAR, AI)

A number of processes can cause ground deformation at volcanoes: dyke intrusions, reservoir pressurization, caldera subsidence, cooling of eruptive deposits (lava/pyroclastic flows), landslides, etc. These can to some extent be distinguished based on distinctive patterns of ground displacement imaged in the interferograms [75]. As described previously, MOUNTS focuses on the automatic detection of processes generating large ground deformations in short time intervals (i.e., between two successive S1 acquisitions, usually 6 to 12 days apart), which typically imprint on interferograms as many colored fringes. In particular, dyke intrusions (whereby magma intrudes into a fissure and pushes the surrounding rock aside), are commonly characterized by two lobes with opposite displacements directions, that imprint as a "butterfly" shape. If the dyke reaches the surface, it will result in eruptive fissures/eruptive vents from which lava flows are emplaced.

Figure 5 illustrates the efficiency of this automated deformation detection system, taking the case of Piton de la Fournaise (Réunion Island) as an example. The volcano experienced 5 intrusive episodes in the past year (April 2018–April 2019), all of which reached the surface and resulted in lava flows that lasted between 0.7–47 days (episode 1 and 4 respectively). The system successfully detected the interferograms where ground deformation is recorded (Figure 5a–c), illustrating its robustness against various artifacts in the geocoded interferogram (related to strong topography gradients, strong atmospheric phase noise, and decorrelation in vegetated areas). Furthermore, thermal anomalies detected by S2 SWIR analysis (i.e., number of hot pixels) are compared with the Volcanic Radiative Power (VRP) provided by the MIROVA system (Figure 5d) to testify the good correspondence between the active flow area and the heat radiated by the flow surface.

The systematic analysis of both ground deformation (amplitude, pattern, orientation) related to the intrusion of magma in the shallow portions of the edifice, and lava effusion rate once the magma breaches the surface, should help better understand the mechanisms controlling effusive eruptions in closed-vent volcano systems such as Piton de la Fournaise.

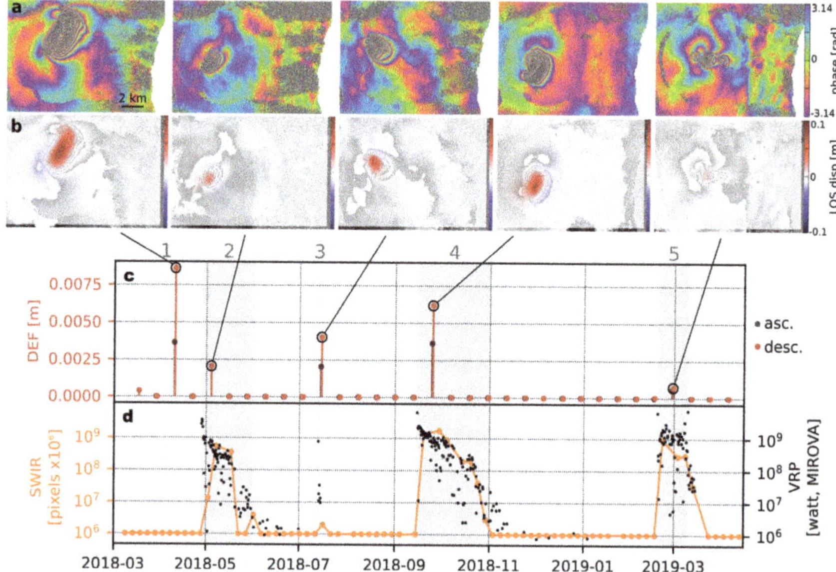

Figure 5. Example of surface deformation detection related to magma intrusions at Piton de la Fournaise (Réunion Island). (**a**) Wrapped interferograms and (**b**) deformation maps in radar line-of-sight (LOS), obtained from interferogram unwrapping (see Figure 4). (**c**) Deformation score DEF computed from the deformation map. (**d**) Number of hot pixels detected in the S2 SWIR image (orange curve, computed by MOUNTS), and Volcanic Radiative Power (VRP) recovered from MODIS data (black markers, computed by MIROVA). The eruptive episodes are numbered from 1 to 5, and highlighted by gray areas based on the eruption timing provided by the OVPF (Observatoire Volcanologique du Piton de la Fournaise): 0.7, 34.6, 0.8, 47, and 19.9 days respectively, publically available at http://www.ipgp.fr/fr/ovpf/activite-recente-piton-de-fournaise. Eruptive episode 5 is shown in detail in Figure 9, in which SO_2 data is also displayed.

4.2. Detection of Eruptive Deposits (SWIR, DInSAR, SAR)

Volcanic eruptions can generate various types of eruptive deposits, including lava flows, pyroclastic flows, mud flows, pyroclastic deposits, ash fall, etc. The best-suited parameter to detect these from space will depend on both the volcanic product type and the surface on which it is emplaced (i.e., arid surface, vegetated surface). We here describe two extreme eruptive case scenarios: the Erta Ale eruption (Ethiopia, ongoing since January 2017), characterized by an effusive activity where lava flows are emplaced in a desert environment, and the Fuego eruption (Guatemala, June 2018), characterized by a violent explosive activity which generated pyroclastic flows (i.e., avalanche of hot blocks and ash which flowed down the surrounding valleys in a densely vegetated environment).

4.2.1. Using DInSAR Coherence and SWIR

Detecting lava flows in non-vegetated regions is particularly efficient using the interferometric coherence [76]. Active lava flows will appear as highly incoherent areas (coherence values close to zero), contrasting with the coherent surface on which they are emplaced (coherence values typically >0.5). Moreover, because the summits of active volcanoes are usually vegetation-free, the coherence can also be utilized to monitor summital volcanic activity, as eruptive deposits will appear incoherent.

Both applications are illustrated in Figure 6, which depicts the ongoing Erta Ale eruption and its precursory activity. Indeed, in the months preceding the eruption onset, intense summit activity is indicated from the coherence map analyzed in a 2 × 2 km box around the summit (Figure 6a black curve). In particular, from June 2016 onwards the northwestern lake becomes active (in addition to the permanently active southeastern lake), and in the following months the activity at both lakes progressively intensifies, leading to multiple lava overflows (e.g., small overflow on 20 August 2016 visible on coherence image in Figure 6(i), and large overflow on 19 January 2017 visible on Sentinel-2 image). On 28 January 2017, the S1 interferogram shows a strong deformation signal with a distinctive pattern characteristic of a dyke intrusion (Figures 6c and 6(ii)), which marks the onset of a >2-year eruption which is still ongoing today [77]. The eruptive fissure opened multiple new vents located SE of the summit lava lakes (clearly identifiable from S1 coherence and amplitude images), which progressively focused on a single new eruptive vent from which lava flowed during the following months. During this time, the activity at the summit lava lakes significantly decreases (i.e., coherence analysis on 2 × 2 km extent, Figure 6a black curve), suggesting a possible drainage of the summit lava lakes. Both the coherence and the SWIR analyses computed on a large >20 km spatial scale (Figure 6a blue curve and Figure 6b respectively), depict the lava flow emplacement. The good agreement between the two highlights the fact that coherence can be used to track the active flow front, even when S2 images may not be usable due to cloud coverage. From this basic analysis, more elaborate parameters can be extracted offline, such as the lava flow front position through time, or the lava effusion rate, both key for hazard mitigation issues.

The coherence threshold can be adapted to each volcano, in order to account for specific environments (i.e., how incoherent is the surrounding land surface). More elaborate strategies can also be implemented to enhance the decorrelation sensibility to volcano-related processes, such as using NDVI masks (computed from S2 images) to exclude vegetated areas from the analysis. Nevertheless, in certain climatic settings coherence-based detection is simply not the best suited, and intensity-based detection can prove more useful.

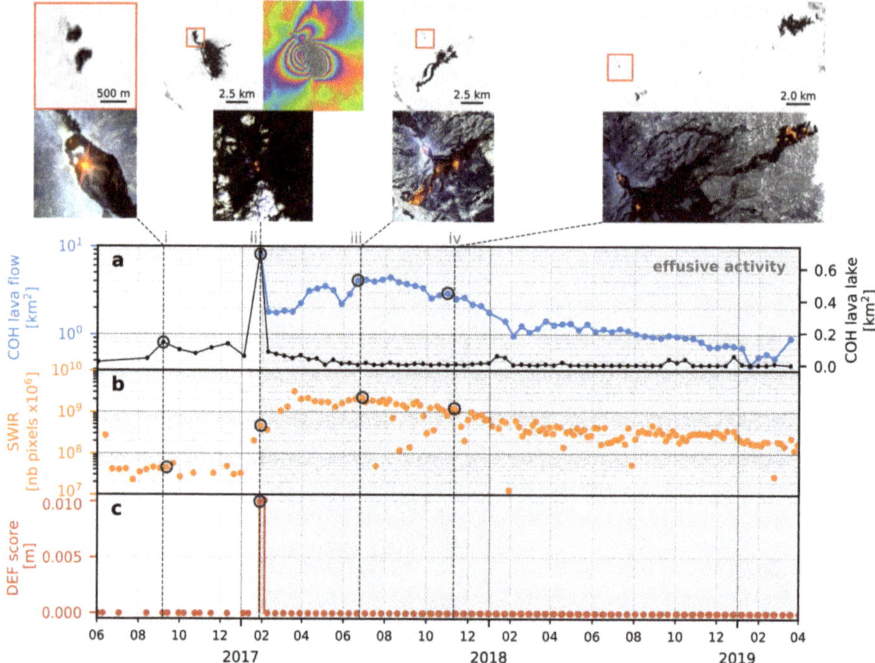

Figure 6. Example of lava flow emplacement detection at Erta Ale (Ethiopia) using both S1 interferometric coherence and S2 SWIR analysis. (**a**) Decorrelation area related to volcanic activity, calculated as the number of pixels where coherence <0.5, multiplied by the pixel area (14 × 14 m). Black curve: decorrelation on 2 × 2 km mask centered at the summit to image the lava lake activity. Blue curve: decorrelation on 50 × 50 km mask to image the lava flow emplacement. Because on such a large scale the coherence map also includes sandy areas where coherence is <0.5, these regions where manually excluded for this specific figure so that the curve only depicts the area of the active lava flow. (**b**) Number of hot pixels (×10^6) detected in the S2 SWIR image (50 × 50 km mask). (**c**) Deformation score DEF computed from the deformation map (unwrapped interferogram). Coherence maps (top image row) and SWIR images (bottom image row) are displayed at selected dates, with spatial scales varying according to the eruptive phase: (i) pre-eruptive phase with intense activity at the summit lava lakes, 2 × 2 km mask; (ii) eruptive onset, manifested by surface deformation and aperture of a new eruptive vent, ~12 × 12 km mask; (iii) early stages of the effusive activity (highlighted by a gray box), showing the emplacement of lava flows on both NE and SW flanks of the volcano, ~12 × 12 km mask; (iv) advanced stages of the effusive activity, when the lava flow front reaches its maximum distance from the vent, 11.5 × 19.5 km mask. The red box displayed in the coherence images refer to the 2 × 2 km mask centered on the active lava lakes.

4.2.2. Using SAR Intensity

The interferometric coherence is hardly exploitable in regions with very dense vegetation, or where the surface is likely to change rapidly due to various environmental factors (e.g., snow or sand). In such context, changes in the intensity of SAR images can help identify eruptive deposits. Figure 7 illustrates this, taking as example the 2018 eruption of Fuego (Guatemala) which killed over 200 persons due to pyroclastic flows [5]. The interferometric coherence image (Figure 7c) computed between SAR image 1 and 2, respectively acquired before and after the eruption, is entirely incoherent and therefore unusable. However, computing the log ratio between the two intensity images (Figure 7a) reveals substantial changes: areas in blue are those where the intensity has decreased, whereas areas in red are those where the intensity has increased. The latter are mainly confined in the valleys, and likely correspond to the rougher block-and-ash deposit [24]. Conversely, areas where intensity decreases are concentrated around the summit vent, which could be associated to the deposition of ash.

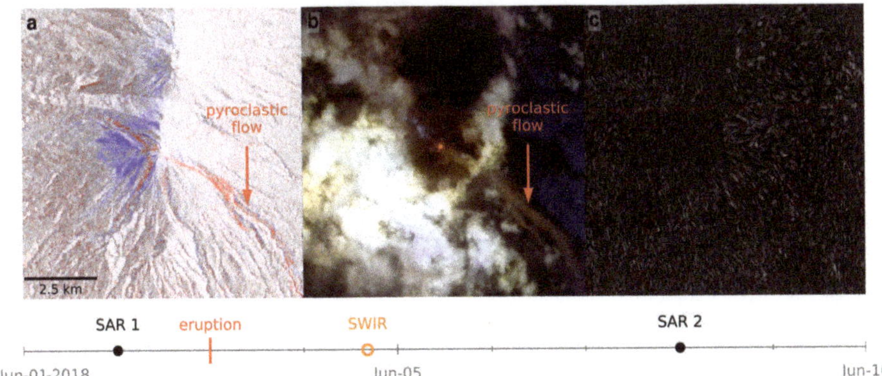

Figure 7. Example of pyroclastic flow detection after the June 2018 Fuego (Guatemala) eruption. (**a**) SAR intensity log-ratio (VV polarization) computed between image SAR 1 acquired before eruption (2 June 2018 00:13), and image SAR 2 acquired after (8 June 2018 00:14). Blue and red colors indicate respectively decrease and increase in backscattered intensity. (**b**) S2 SWIR image acquired on 4 June 2018 16:18 UTC. (**c**) Interferometric coherence between images SAR 1 and SAR 2.

4.3. Detection of Morphological Changes (SAR)

The intensity of SAR images is strongly dependent on the terrain slope, and is therefore useful to monitor morphological changes affecting the volcano. Because radar wavelengths penetrate through clouds, SAR intensity images provide crucial insights into the volcanic activity when optical imagery is obstructed by atmospheric and/or volcanic gas clouds. We here give two examples taken from very different volcanological settings: the summit crater collapse of Kīlauea (Hawai'i) during the 2018 effusive eruption (Figure 8a), and the Anak Krakatau (Indonesia) island growth and destruction during the 2018 explosive eruption (Figure 8b).

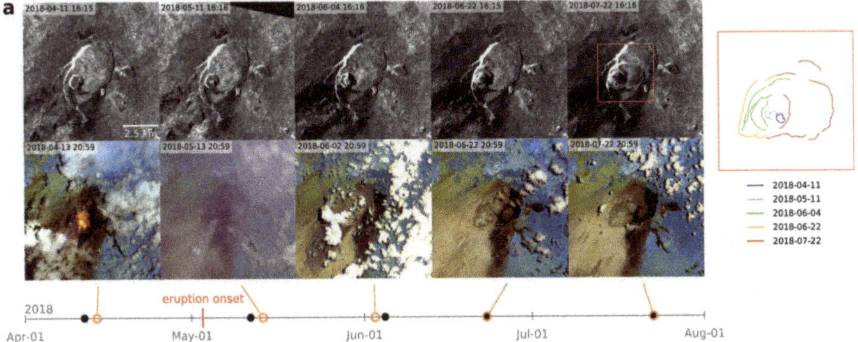

Figure 8. *Cont.*

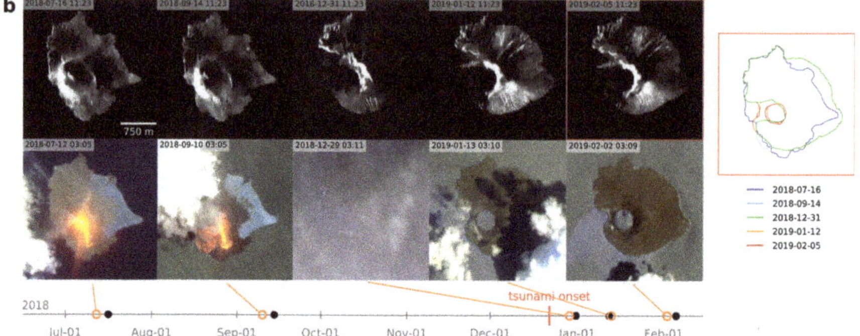

Figure 8. Two examples of morphological changes detected from SAR intensity (images on top rows, black markers on timeline), and closest S2 SWIR acquisition (images in bottom rows, orange markers on timeline). The temporal evolution of the most prominent morphological changes visible in SAR images are sketched on the right. (**a**) Caldera collapse of Kilauea (Hawai'i) during the first months of the 2018 flank eruption. (**b**) Anak Krakatau (Indonesia) island growth during the months preceding the 2018 tsunamigenic landslide, and horseshoe-shaped caldera after the landslide. In this figure speckle is removed from SAR images using a non-local means filter [78] (NDSAR, https://github.com/odhondt/ndsar), and tone mapping of SWIR images is fixed so that colors and contrasts are the same in each image. See Supplementary Material S3 and S4 for video animations spanning several months.

Kilauea is well known for its persistent active lava lake. In 2018, it experienced its largest flank eruption and caldera collapse in the last 200 years [79,80]. During spring 2018, the lava lake activity was high, which was clearly detected as a hotspot in the S2 SWIR images (Figure 8a, 13 April 2018). On 30 April 2018, seismicity indicated the intrusion of a dyke along the East Rift Zone, which generated a ~38 km long deformation zone, and multiple eruptive fissures with lava flows rapidly reaching the sea (see Supplementary Material S5 for analysis of S1 and S2 over the entire rift zone). During this time the summit underwent significant changes: lava lake withdrawal (i.e., hotspot disappears in SWIR images, Figure 8a), accompanied by summit subsidence (i.e., deflation detected in interferogram), progressively evolving in a ~3 km wide caldera collapse (see LIDAR digital elevation model in [80]), as the shallow magma reservoir was being drained. The progression of the caldera collapse is clearly imaged with SAR intensity images, which reveal the progressive formation of fractures and the profound summit morphological changes accompanying the flank eruption (Figure 8a and video in Supplementary Material S3). Ash deposits following the eruption onset is also captured, identified by a decrease in the SAR backscattered intensity on the SE flank of the volcano, also visible in the SWIR images. This decrease can be explained by the fact that fresh ash is less reflective than bare rock owing to its loose structure and high porosity, and that ash deposits smooth the surface, resulting in a more specular reflector which backscatters less energy towards when the slope is facing away from the sensor [23].

Krakatau is well-known for its volcano-induced tsunamis. Just over 135 years after the famous 1883 event, the volcano triggered on 22 December 2018 another deadly wave. Analysis of the SAR intensity images clearly shows the progressive island growth in the months preceding the tsunami, due to multiple lava flows reaching the sea and extending the island's coast line (Figure 8b and Supplementary Material video S4). This likely increased the instability of the volcano's flank, which on 22 December 2018 collapsed, generating a tsunami wave [81]. Post flank sector collapse images first reveal an amphitheater-shaped scar opened to the sea (Figure 8b, 31 December 2018 image), which was closed shortly after by an explosion tuff ring, resulting in a ~400 m wide water filled crater (Figure 8b, 12 January and 15 February 2019 images).

4.4. Detection of SO_2 Gas Flux (UV)

On 18 February 2019 a new eruption of the Piton de la Fournaise began. According to OVPF Reports, 14 ± 5 Mm^3 of lava were erupted during the 18 days of activity, fed by several eruptive fissures located on the upper east flank of the volcanic cone. Following a phase of gradual increase in volcanic tremor and the intensification of surface activity, the eruption ended abruptly on March 10 (OVPF Reports).

Figure 9 shows the time series of SO_2 mass burden recovered from S5P (Figure 9b), complemented by thermal anomalies recorded from S2 SWIR data (Figure 9c). The SO_2 mass obtained from MOUNTS is compared with the SO_2 mass computed by NASA as a measure of correlation between the two datasets. In addition, thermal anomalies detected by S2 are overlaid with the VRP data provided by MIROVA (Figure 9c) to show correspondence between the active flow area and the heat radiated by the flow surface. Once suitably calibrated, the combination of SO_2 and thermal data provides a synoptic view of the gas and magma fluxes during the course of the February–March 2019 eruption which can be used to track eruptive trends and patterns in real time [82]. Notably for this specific case, the two datasets show consistent trends, indicating a gradual intensification of the effusive and degassing activity during the final phases of the eruption. Thermal anomalies recorded after 10 March and in the absence of gas emission are attributed to the cooling of the lava field. The last SO_2 detection (10 March 2019 09:38 UTC) is attributed to the gas plume, no longer fed from the eruptive vent but still inside the 500 × 500 km AOI as it slowly drifts away from the island.

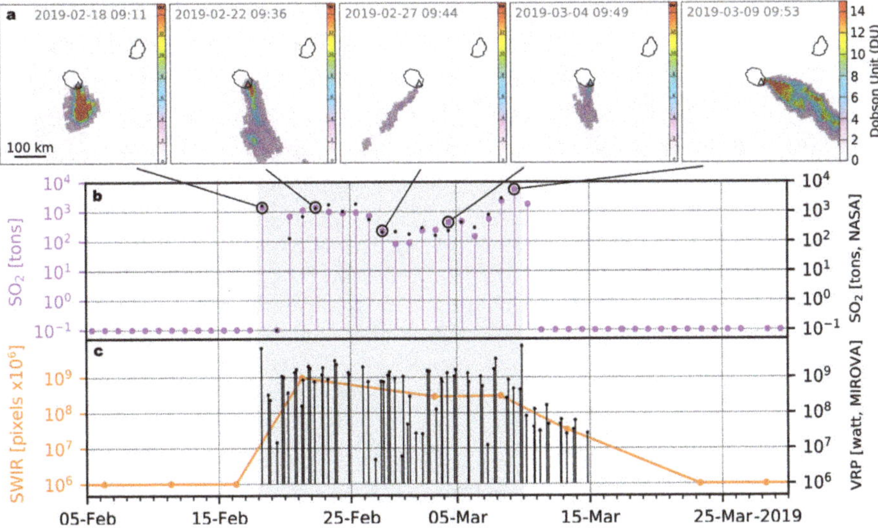

Figure 9. Example of SO_2 emission detection during the February 2019 eruptive crisis at Piton de la Fournaise (Reunion Island). (**a**) SO_2 images from Sentinel-5P at selected dates (planet boundary layer PBL, 500 × 500 km mask). The detected pixels contaminated with volcanic SO_2 are overlaid with a semi-transparent gray mask. (**b**) SO_2 mass recovered by MOUNTS (purple markers) and by NASA (black markers, data available at https://so2.gsfc.nasa.gov/pix/daily/0319/reunion_0319tr.html, computed on 1000 × 1000 km mask). (**c**) Number of hot pixels (×10^6) detected in the S2 SWIR image (orange curve, computed by MOUNTS), and Volcanic Radiative Power (VRP) recovered from MODIS data (black markers, computed by MIROVA).

4.5. Combining Ground-Based and Space-Based Sensors

Magma migration within the crust generates stresses, which can result in earthquakes as the surrounding rocks are displaced or fractured. This seismicity, commonly known as volcano-tectonic

(VT) seismicity, is often recorded both prior and during volcanic eruptions, within and around the volcanic edifice [83]. High magnitude VT earthquakes can be recorded and located by global seismological networks, even when the nearest seismic stations are installed several hundreds of kilometers away. In turn, their timing, location, magnitude, and sometimes focal mechanism, are stored in open access global earthquake databases, particularly GEOFON and USGS catalogues. MOUNTS facilitates the interrogation of such catalogs, recovering potential earthquakes recorded in a region centered around the monitored volcano. This data can support the analysis of the volcanic phenomena, especially when the volcano is not equipped with ground-based monitoring instrumentation.

Figure 10 shows the recent eruption of Ambrym (Vanuatu), and illustrates how combining ground-based and space-based sensors helps understand the eruptive dynamics of this volcano located in a very remote and cloud-prone region. On 15 December 2018 and in the days that followed, a swarm of volcano-tectonic earthquakes were recorded in the vicinity of the volcano, with magnitudes ranging between ~4.5 and 5.5 (Figure 10d). The volcano was known until then for its persistent activity characterized by two active volcanic lakes (Figure 10(b.1)), responsible for high heat and gas fluxes (Figure 10b,c respectively), [84,85]. Analysis of the SAR intensity images immediately before and after this swarm reveal profound morphological changes (Figure 10(d.1,d.2)), in particular the collapse and enlargement of the summit crater. DInSAR analyses indicate very strong ground deformation during this period (Figure 10a red curve, Figure 10(a.1)), related to dyke intrusion and caldera subsidence [86]. Simultaneously, the decorrelation in the coherence map increases (Figure 10a blue curve), due to both the ground deformation and perhaps also pyroclastic deposits. Once stabilized, the coherence map reveals the presence of a new eruptive vent (Figure 10(a.2)), from which lava was most likely emitted, as suggested by the SWIR image acquired on 15 December 2018 (Figure 10(b.2)). Interestingly, following this event the volcano completely changed dynamics: the summit lava lakes were most likely drained, as suggested by the absence of thermal anomalies and the cessation of SO_2 gas emissions.

5. Discussion

The key to detecting volcano unrest and understanding the underlying mechanisms is to be able to recognize when a volcano is deviating from its background level of activity. Once the eruption starts on the other hand, the key to decipher the eruptive dynamics and to mitigate the related hazards, is to integrate multiparametric dataset streaming from both space- and ground-based sensors, in order to provide the most comprehensive view of the eruptive phenomena. Both require "monitoring", i.e., observing the volcanic activity over long periods of time, during both quiescent and eruptive phases. As such, the aim of monitoring platforms such as MOUNTS is twofold: (1) a scientific one, aiming at deepening our understanding of volcanic processes and patterns at stake at active volcanoes, by processing in a systematic way large amounts of data in an effort to construct global databases, and (2) a societal one, aiming at producing more successful eruption forecasts, and providing additional information to the operational community (e.g., local volcano observatories and civil protection) in order to mitigate the risks related to volcanic hazards.

The results presented in this paper intend to demonstrate how the monitoring platform MOUNTS can contribute to both scientific and operational aims. We here discuss the benefits, limitations, and future developments of the system.

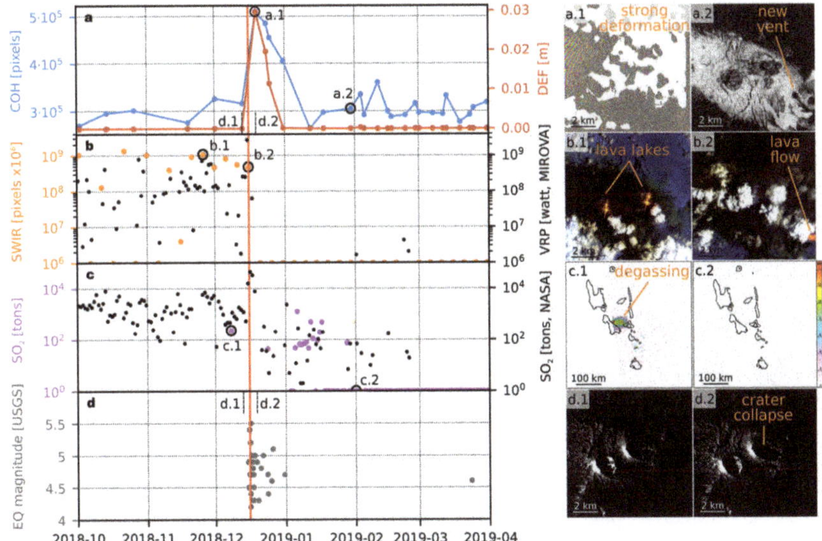

Figure 10. Multiparametric dataset combining space-based and ground-based sensors, which help decipher the recent eruptive dynamics of Ambrym (Vanuatu). The eruption onset on 15 December 2018 is indicated by a red vertical line across the plots. Time series of: (**a**) decorrelation score COH (pixels with coherence <0.5) in blue indicating surface reflectivity change, and deformation score DEF in red; (**b**) thermal anomalies from S2 SWIR analysis in orange (N hot pixels $\times 10^6$) and MODIS MIR analysis in black (VRP), processed by MOUNTS and MIROVA, respectively; (**c**) SO_2 gas mass in the atmosphere from Sentinel-5P analysis in purple and OMPS analysis in black, processed by MOUNTS and NASA respectively; (**d**) magnitudes of the earthquake recorded in the vicinity of the volcano, recovered from the USGS global earthquake catalog. Images: (a.1) wrapped interferogram computed between SAR images d.1 and d.2, revealing very strong ground deformation (areas with strong phase decorrelation are masked to show only where deformation fringes are visible); (a.2) interferometric coherence, on which a new eruptive vent is identifiable; (b.1) S2 SWIR image prior to eruption onset, showing the persistent lava lakes activity; (b.2) S2 SWIR image after eruption onset, showing the lava flow emplacement from the new eruptive vent; (c.1) S5P image showing SO_2 emissions prior to eruption onset; (c.2) S5P image showing strong decrease of SO_2 emissions after eruption onset; (d.1,2) S1 intensity images immediately before and after the eruption onset, revealing profound morphological changes. Spatial extent of images a.1,2, b.1,2 and d.1,2 = 10 × 10 km, extent of images c.1,2 = 500 × 500 km.

5.1. Benefits of MOUNTS

The benefits of the developed system are the following:

- Automated processing of free multisensor dataset which provide key parameters for volcano monitoring: surface deformation and reflectivity changes (Sentinel-1), heat anomalies (Sentinel-2), SO_2 gas emission (Sentinel-5P), and seismic activity (USGS and GEOFON earthquake catalogues). This interdisciplinary approach allows for assessment of a variety of volcanic phenomena in various volcanological contexts. Moreover, exploiting multiple sensors spanning across the electromagnetic spectrum reduces the dependency to sensing conditions (e.g., night, clouds).
- Flexible design allowing fast implementation of new targets to monitor, with freedom regarding the size and shape of the region of interest. This allows to rapidly respond to new eruptive crisis, and adapt to the specific scenarios (e.g., monitoring of summit activity on small ~1–2 km spatial extent, and/or monitoring of effusive activity on large >50 km extent). The system currently monitors 17 volcanoes in various volcanological and climatic settings across the globe, many of which recently experienced large eruptive crisis.

- Visualization through an open-access website (www.mounts-project.com) of both geocoded images (i.e., DInSAR interferograms wrapped/unwrapped, DInSAR coherence map, SAR VV intensity image, SWIR B12-B11-B8A image, and SO_2 PBL concentration), and time series of parameters extracted from each image type (i.e., deformation score, number of decorrelated pixels, number of hot pixels, and SO_2 mass, respectively). This allows to apprehend the evolution through time of the volcanic activity and eruptive products. Download of full resolution images and time series graphs (PNG format) is readily possible from the website; more specific data download based on user-defined queries is planned, but for the moment possible on-demand only.
- Based on the free SNAP toolboxes, providing a unique framework to manipulate data from various satellites, with state-of-the-art processing algorithms (e.g., DInSAR). MOUNTS is open-source, with a Github repository (https://github.com/sebastienValade/mounts) storing both the source code and a changelog informing on all the notable changes made to the system and website.
- Modular architecture, allowing the implementation of new processing algorithms to extract relevant volcanological parameters, or solve specific tasks. As a matter of example, a pre-trained CNN was plugged to detect strong deformation in the interferograms generated by the system.
- Automated email alert messaging to dedicated users when specific thresholds are overcome. Interaction with other monitoring systems such as MIROVA is achieved by facilitating access to volcano-dedicated webpages (Figure 3). Strengthening the interactivity between the systems is planned, in particular by sharing database access in order to confront datasets more easily.

The operational community such as volcano observatories can use MOUNTS and contribute to its development in a number of ways. The IGEPN (Instituto Geofísico de la Escuela Politécnica Nacional) for example, responsible for volcano monitoring in Ecuador, suggested to add Sangay to the list of monitored volcanoes in order to contribute to the surveillance of this remote edifice. The data available on the platform was used freely, and a collaborative exchange was initiated upon request to provide more specific data processing. The resulting material was further analyzed by IGEPN staff according to their needs, and was used in the activity reports describing the ongoing crisis for public information [87]. (Disclaimers on the data usage and appropriate acknowledgements can be found on the website). Scientific collaborations to investigate specific volcanic processes, or to develop specific methods (based on either the dataset available on the website, or on datasets resulting from more complex analysis) are also welcomed.

5.2. Limitations of MOUNTS and Future Developments

The quantity of data available for volcano monitoring is increasing exponentially, but so is the difficulty to transform it into knowledge. Indeed a number of limitations arise, related to the extraction of meaningful parameters (are we looking at the right variables?), resolution issues (is the sampling in time and space accurate enough?), and data handling issues (how do we deal with the growing mass of data?). MOUNTS is at this stage still a proof-of-concept, which has large potential for improvement. We hereafter discuss the main limitations and development directions.

1. Improve MOUNTS' capability to recover parameters informing on the state of volcanic activity, eruptive precursors in particular. While IR and UV spectroradiometry is able to provide rather straightforward parameters (i.e., heat and gas flux respectively), recovering parameters from SAR in a robust and automated fashion is more challenging. In this paper, we show how trained neural networks can achieve complex tasks in a timely and reliable manner, and can be easily implemented in operational processing chains. In particular, strong deformation typically imprint on interferograms as many colored fringes, which are successfully detected. Further development however is needed to detect slow deformation mechanisms, which do not generate deformation patterns with numerous fringes. Future developments should also focus on designing and training neural networks to recover from SAR data other relevant parameters that can inform on volcanic activity. For example, efficient change detection able to exclude changes non related to volcanic activity (e.g., snow fall, vegetation growths, etc.) would prove extremely useful during both pre-eruptive and syn-eruptive phases.

2. Incorporate additional data types in the processing chain to provide further insights into the volcanic activity. A priority is to analyze Sentinel-3 (S3) TIR data routinely, crucial to monitor ground thermal anomalies at high sampling rate (sensor characteristics similar to MODIS), but also to detect ash plumes in the atmosphere. Ash detection is commonly achieved using the brightness temperature difference (BTD) procedure [88], applied to two channels centered around 11 and 12 μm. This approach is easily applicable, but also prone to generate false alarms. A number of methods have been developed to overcome this issue [20], including 3-band algorithms [48], the BTD algorithm with water vapor correction (BTD-WVC), the Robust Satellite Technique (RST) specifically configured for volcanic ash, and shallow neural networks [89,90]. Future developments should therefore implement automated S3 processing to monitor volcanic ash propagation in the atmosphere, which poses a major threat for air traffic in particular.
3. Incorporate modeling tools to predict the propagation of volcanic eruptive products, using the recovered multiparametric data as input source terms for the models. Such strategies are not new, and are now, thanks to increasing computing power and data availability, becoming achievable in NRT to forecast the propagation of lava flows [91] or the dispersion of ash plumes into the atmosphere (e.g., [92,93]), as well as to predict the geometry and depth of magma bodies responsible for volcano deformation (e.g., [94–96]). Monitoring platforms such as MOUNTS should not necessarily include such modeling routines, but should at least strive to provide parameters than can be fed to such models.
4. Migrate processing tasks on cloud platforms where data is archived. The development of MOUNTS was done by automatizing data download and processing, using free and open-source data and software. In doing so, we were able to keep the costs of this proof-of-concept platform very low (i.e., a single desktop computer manages data download, processing, and hosting of the web server). Nevertheless, this architecture limits the ability to process larger amounts of data, which would require the analysis of the entire available Sentinel dataset over hundreds of volcanoes worldwide. To achieve this, cloud computing strategies are preferable, whereby algorithms would run on a platform where the data is hosted, thereby preventing data download, and at the same time offering higher computing power. Commercial platforms offering such services exist: the Copernicus DIAS platforms (Data and Information Access Services, which include Sobloo, Onda, Creodias, Mundi, and Wekeo), the Amazon Web Services (AWS), or the Google Cloud Platform.
5. Analyze the recovered multiparametric volcanic time-trends. Monitoring efforts such as the one presented here, allow the development of consistent multiparametric databases on a variety of volcanic settings (i.e., various volcano types, tectonic settings and magma compositions). Such databases are crucial to decipher eruptive patterns, and potentially better estimate future activity [25–27,29]. As a matter of example, decadal heat and gas emission time-trends help decrypt slow mechanisms of magma/gas accumulation and release at active volcanoes [97]. Clustering of these trends in categorically similar patterns, together with time-series analysis and probabilistic approaches (e.g., [98,99]), should be investigated to help decision making and potentially lead to better eruption forecasting. Moreover, incorporating standardized volcano alert level classifications (whereby color-codes help flag the activity of volcanoes [39,100]), will help better communicate the level of volcanic unrest and eruption likelihood to local populations and governmental authorities [101,102], and potentially lead to better early warning systems.

6. Conclusions

We present an operational volcano monitoring system, based on the automated download and processing of multisensor satellite-based data (Sentinel-1 SAR, Sentinel-2 SWIR, Sentinel-5P TROPOMI). The recovered data aim at providing key parameters able to inform on the state of volcanic activity, namely: surface deformation, surface reflectivity changes, surface heat anomalies, and SO_2 gas emissions. The results are disseminated in NRT on a public website (www.mounts-project.com), where both geocoded images and multi-parametric time series help understand the activity. Moreover, we demonstrate how artificial intelligence can be used in such monitoring system to solve complex tasks. In particular, we designed and trained a convolutional neural network to detect large deformation signals in wrapped interferograms with no atmospheric corrections. The training was done on synthetically generated interferograms, and evaluated on >1360 real interferograms produced by MOUNTS. Due to the very good performances of the network, it is now incorporated into the operational processing chain, which delivers automatic email alerts to dedicated users when strong deformation is recorded at the monitored volcanoes.

In addition to the set of parameters recovered from spaceborne sensors, we incorporate information available from global earthquake catalogues (GEOFON and USGS) to inform on the seismicity located in the vicinity of the volcano. The utility of integrating both satellite-based parameters (deformation, heat and gas) and ground-based parameters (seismicity) are demonstrated through a number of recent eruptions: Erta Ale 2017, Piton de la Fournaise 2018–2019, Fuego 2018, Kilauea 2018, Anak Krakatau 2018, and Ambrym 2018. We show how this interdisciplinary approach allows for assessment of a variety of volcanic phenomena, ranging from subsurface magma migration, to surface eruptive deposit emplacement, pre/syn-eruptive morphological changes, and SO_2 gas emission into the atmosphere. The data processed by MOUNTS is providing insights into the eruptive dynamics of these volcanoes, and is sharpening our understanding of how the integration of such multiparametric datasets can help better monitor volcanic hazards.

Supplementary Materials: The following are available online at http://www.mdpi.com/2072-4292/11/13/1528/s1, S1: Details on DInSAR, SAR, and SWIR processing chains; S2: Details on neural network architecture, training, and synthetic data generation; S3: Video of the summit crater morphological changes detected from SAR imagery during the 2018 eruptive episodes at Kilauea (Hawai'i); S4: Video of the island growth and destruction before/after the December 2018 eruption of Krakatau (Indonesia); S5: Kilauea (Hawai'i) East Rift Zone flank eruption monitored by MOUNTS on large spatial scale.

Author Contributions: Conceptualization: (S.V., T.R.W., A.L.), inspired by MIROVA (D.C., M.L.); Methodology: system architecture and management (S.V., A.L.), website development (S.V.), Convolutional Neural Network (A.L.), SWIR hot pixel detection (F.M.), SAR speckle filtering and change detection, (O.D.), SO_2 emission analysis (S.V., M.L.); Writing: original draft preparation (S.V.), review and editing by (A.L., D.C., D.L., F.M., O.D., M.L., T.R.W., O.H.).

Funding: This publication was financially supported by Geo.X, the Research Network for Geosciences in Berlin and Potsdam (Project Number: SO_087_GeoX).

Acknowledgments: We thank four anonymous reviewers for their comments which helped improve the manuscript. We also wish to acknowledge N. Anantrasirichai for providing details regarding the training dataset used in their study [34], as well as M. Gouhier, F. Amelung and F.J. Meyer for sharing details on the monitoring systems compiled in Table 1. Sentinel data are made freely available online by Copernicus Open Access Hub (https://scihub.copernicus.eu/), and are partially processed with the free SNAP toolboxes (https://step.esa.int/main/toolboxes/snap/). Earthquake catalogs are provided by GEOFON (GFZ Potsdam) and USGS, and interrogated using the Pyrocko Toolbox (https://pyrocko.org/, [103]).

Conflicts of Interest: The authors declare no conflict of interest.

References

1. Siebert, L.; Simkin, T.; Kimberly, P. *Volcanoes of the World*, 3rd ed.; University of California Press: Berkley, CA, USA, 2011.
2. Loughlin, S.C.; Vye-Brown, C.; Sparks, R.S.J.; Brown, S.K.; Barclay, J.; Calder, E.; Cottrell, E.; Jolly, G.; Komorowski, J.-C.; Mandeville, C.; et al. An introduction to global volcanic hazard and risk. In *Global Volcanic Hazards and Risk*; Cambridge University Press: Cambridge, UK, 2015; pp. 1–80. ISBN 9781316276273.
3. Brown, S.K.; Loughlin, S.C.; Sparks, R.S.J.; Vye-Brown, C.; Barclay, J.; Calder, E.; Cottrell, E.; Jolly, G.; Komorowski, J.-C.; Mandeville, C.; et al. Global volcanic hazard and risk. In *Global Volcanic Hazards and Risk*; Cambridge University Press: Cambridge, UK, 2015; pp. 81–172. ISBN 9781316276273.
4. Auker, M.R.; Sparks, R.S.J.; Siebert, L.; Crosweller, H.S.; Ewert, J. A statistical analysis of the global historical volcanic fatalities record. *J. Appl. Volcanol.* **2013**, *2*, 1–24. [CrossRef]
5. Agencia Guatemalteca de Noticias. Available online: https://agn.com.gt/gobierno-de-guatemala-erogo-3609-millones-de-quetzales-para-atender-a-victimas-de-erupcion-del-volcan-de-fuego/ (accessed on 5 June 2019).

6. ReliefWeb—Indonesia: Earthquakes and Tsunami—Sunda Straits Tsunami—MDRID013 EPoA update n° 15; (n° 2 for Sunda Straits Tsunami Operation). Available online: https://reliefweb.int/report/indonesia/indonesia-earthquakes-and-tsunami-sunda-straits-tsunami-mdrid013-epoa-update-n-15-n (accessed on 5 June 2019).
7. Major, J.J.; Lara, L.E. Overview of Chaitén Volcano, Chile, and its 2008–2009 eruption. *Andean Geol.* **2013**, *40*, 196–215. [CrossRef]
8. Poland, M. Volcano monitoring from space. In *Global Volcanic Hazards and Risk*; Loughlin, S.C., Sparks, R.S.J., Brown, S.K., Jenkins, S.F., Vye-Brown, C., Eds.; Cambridge University Press: Cambridge, UK, 2015; pp. 311–316.
9. Pallister, J.; McNutt, S.R. Synthesis of Volcano Monitoring. In *The Encyclopedia of Volcanoes*; Elsevier: Amsterdam, The Netherlands, 2015; pp. 1151–1171. ISBN 9780123859389.
10. Sparks, R.S.J.; Biggs, J.; Neuberg, J.W. Geophysics. Monitoring volcanoes. *Science* **2012**, *335*, 1310–1311. [CrossRef]
11. Phillipson, G.; Sobradelo, R.; Gottsmann, J. Global volcanic unrest in the 21st century: An analysis of the first decade. *J. Volcanol. Geotherm. Res.* **2013**, *264*, 183–196. [CrossRef]
12. Pinel, V.; Poland, M.P.; Hooper, A. Volcanology: Lessons learned from Synthetic Aperture Radar imagery. *J. Volcanol. Geotherm. Res.* **2014**, *289*, 81–113. [CrossRef]
13. Dzurisin, D. A comprehensive approach to monitoring volcano deformation as a window on the eruption cycle. *Rev. Geophys.* **2003**, *41*, 1–29.
14. Harris, A. *Thermal Remote Sensing of Active Volcanoes*; Cambridge University Press: Cambridge, UK, 2013.
15. Carn, S.A.; Fioletov, V.E.; Mclinden, C.A.; Li, C.; Krotkov, N.A. A decade of global volcanic SO2 emissions measured from space. *Sci. Rep.* **2017**, *7*, 1–12. [CrossRef]
16. Theys, N.; Hedelt, P.; De Smedt, I.; Lerot, C.; Yu, H.; Vlietinck, J.; Pedergnana, M.; Arellano, S.; Galle, B.; Fernandez, D.; et al. Global monitoring of volcanic SO2 degassing with unprecedented resolution from TROPOMI onboard Sentinel-5 Precursor. *Sci. Rep.* **2019**, *9*, 1–10. [CrossRef]
17. Harris, A.J.L.; Dehn, J.; Calvari, S. Lava effusion rate definition and measurement: A review. *Bull. Volcanol.* **2007**, *70*, 1–22. [CrossRef]
18. Harris, A.J.L.; De Groeve, T.; Garel, F.; Carn, S.A. *Detecting, Modelling and Responding to Effusive Eruptions*; Geological Society of London: London, UK, 2016.
19. Mackie, S.; Cashman, K.; Ricketts, H.; Rust, A.; Watson, M. *Volcanic Ash: Hazard Observation*; Elsevier: Amsterdam, The Netherlands, 2016.
20. Zehner, C. Monitoring Volcanic Ash from Space. In Proceedings of the ESA-EUMETSAT Workshop on the 14 April to 23 May 2010 Eruption at the Eyjafjöll Volcano, South Iceland, Frascati, Italy, 26–27 May 2010; Zehner, C., Ed.; ESA: Frascati, Italy, 2012.
21. Zebker, H.A.; Rosen, P.; Hensley, S.; Mouginis-Mark, P.J. Analysis of active lava flows on Kilauea volcano, Hawaii, using SIR-C radar correlation measurements. *Geology* **1996**, *24*, 495–498. [CrossRef]
22. Lu, Z.; Fielding, E.; Patrick, M.R.; Trautwein, C.M. Estimating lava volume by precision combination of multiple baseline spaceborne and airborne interferometric synthetic aperture radar: The 1997 eruption of okmok volcano, alaska. *IEEE Trans. Geosci. Remote Sens.* **2003**, *41*, 1428–1436.
23. Arnold, D.W.D.; Biggs, J.; Wadge, G.; Mothes, P. Using satellite radar amplitude imaging for monitoring syn-eruptive changes in surface morphology at an ice-capped stratovolcano. *Remote Sens. Environ.* **2018**, *209*, 480–488. [CrossRef]
24. Wadge, G.; Cole, P.; Stinton, A.; Komorowski, J.C.; Stewart, R.; Toombs, A.C.; Legendre, Y. Rapid topographic change measured by high-resolution satellite radar at Soufriere Hills Volcano, Montserrat, 2008–2010. *J. Volcanol. Geotherm. Res.* **2011**, *199*, 142–152. [CrossRef]
25. Biggs, J.; Ebmeier, S.K.; Aspinall, W.P.; Lu, Z.; Pritchard, M.E.; Sparks, R.S.J.; Mather, T. A Global link between deformation and volcanic eruption quantified by satellite imagery. *Nat. Commun.* **2014**, *5*, 3471. [CrossRef]
26. Ebmeier, S.K.; Andrews, B.J.; Araya, M.C.; Arnold, D.W.D.; Biggs, J.; Cooper, C.; Cottrell, E.; Furtney, M.; Hickey, J.; Jay, J.; et al. Synthesis of global satellite observations of magmatic and volcanic deformation: Implications for volcano monitoring & the lateral extent of magmatic domains. *J. Appl. Volcanol.* **2018**, *7*, 1–26.
27. Furtney, M.A.; Pritchard, M.E.; Biggs, J.; Carn, S.A.; Ebmeier, S.K.; Jay, J.A.; McCormick Kilbride, B.T.; Reath, K.A. Synthesizing multi-sensor, multi-satellite, multi-decadal datasets for global volcano monitoring. *J. Volcanol. Geotherm. Res.* **2018**, *365*, 38–56. [CrossRef]

28. Pritchard, M.E.; Biggs, J.; Wauthier, C.; Sansosti, E.; Arnold, D.W.D.; Delgado, F.; Ebmeier, S.K.; Henderson, S.T.; Stephens, K.; Cooper, C.; et al. Towards coordinated regional multi-satellite InSAR volcano observations: Results from the Latin America pilot project. *J. Appl. Volcanol.* **2018**, *7*, 5. [CrossRef]
29. Reath, K.; Pritchard, M.; Poland, M.; Delgado, F.; Carn, S.; Coppola, D.; Andrews, B.; Ebmeier, S.K.; Rumpf, E.; Henderson, S.; et al. Thermal, Deformation, and Degassing Remote Sensing Time Series (CE 2000–2017) at the 47 most Active Volcanoes in Latin America: Implications for Volcanic Systems. *J. Geophys. Res. (Solid Earth)* **2019**, *124*, 195–218. [CrossRef]
30. Borgström, S.; Bianchi, M.; Bronson, W.; Tampellini, M.L.; Ratti, R.; Seifert, F.M.; Komorowski, J.C.; Kaminski, E.; Peltier, A.; Van der Voet, P. Globvolcano: Earth Observation Services for Global Monitoring of Active Volcanoes. In Proceedings of the Fringes 2009 Workshop, Frascati, Italy, 30 November–4 December 2009; ESA: Frascati, Italy, 2010.
31. Tait, S.; Ferrucci, F. A real-time, space borne volcano observatory to support decision making during eruptive crises: European volcano observatory space services. In Proceedings of the 2013 UK Sim 15th International Conference on Computer Modelling and Simulation, Cambridge, UK, 10–12 April 2013; pp. 283–289.
32. Newhall, C.G.; Costa, F.; Ratdomopurbo, A.; Venezky, D.Y.; Widiwijayanti, C.; Win, N.T.Z.; Tan, K.; Fajiculay, E. WOVOdat—An online, growing library of worldwide volcanic unrest. *J. Volcanol. Geotherm. Res.* **2017**, *345*, 184–199. [CrossRef]
33. Berger, M.; Moreno, J.; Johannessen, J.A.; Levelt, P.F.; Hanssen, R.F. ESA's sentinel missions in support of Earth system science. *Remote Sens. Environ.* **2012**, *120*, 84–90. [CrossRef]
34. Anantrasirichai, N.; Biggs, J.; Albino, F.; Hill, P.; Bull, D. Application of Machine Learning to Classification of Volcanic Deformation in Routinely Generated InSAR Data. *J. Geophys. Res. Solid Earth* **2018**, *123*, 6592–6606. [CrossRef]
35. Anantrasirichai, N.; Biggs, J.; Albino, F.; Bull, D. A deep learning approach to detecting volcano deformation from satellite imagery using synthetic datasets. *Remote Sens. Environ.* **2019**, *230*, 111–179. [CrossRef]
36. Ebmeier, S.K. Application of independent component analysis to multitemporal InSAR data with volcanic case studies. *J. Geophys. Res. Solid Earth* **2016**, *121*, 8970–8986. [CrossRef]
37. Gaddes, M.E.; Hooper, A.; Bagnardi, M.; Inman, H.; Albino, F. Blind Signal Separation Methods for InSAR: The Potential to Automatically Detect and Monitor Signals of Volcanic Deformation. *J. Geophys. Res. Solid Earth* **2018**, *123*, 10,226–10,251. [CrossRef]
38. Witze, A. How AI and satellites could help predict volcanic eruptions. *Nature* **2019**, *567*, 156–157. [CrossRef] [PubMed]
39. Fearnley, C.J.; McGuire, W.J.; Davies, G.; Twigg, J. Standardisation of the USGS Volcano Alert Level System (VALS): Analysis and ramifications. *Bull. Volcanol.* **2012**, *74*, 2023–2036. [CrossRef]
40. Wright, R.; Flynn, L.; Garbeil, H.; Harris, A.; Pilger, E. Automated volcanic eruption detection using MODIS. *Remote Sens. Environ.* **2002**, *82*, 135–155. [CrossRef]
41. Wright, R.; Flynn, L.P.; Garbeil, H.; Harris, A.J.L.; Pilger, E. MODVOLC: Near-real-time thermal monitoring of global volcanism. *J. Volcanol. Geotherm. Res.* **2004**, *135*, 29–49. [CrossRef]
42. Wright, R. MODVOLC: 14 years of autonomous observations of effusive volcanism from space. *Geol. Soc. Lond. Spec. Publ.* **2016**, *426*, 23–53. [CrossRef]
43. Coppola, D.; Laiolo, M.; Cigolini, C.; Delle Donne, D.; Ripepe, M. Enhanced volcanic hot-spot detection using MODIS IR data: Results from the MIROVA system. *Geol. Soc. Lond. Spec. Publ.* **2016**, *426*, 181–205. [CrossRef]
44. Coppola, D.; Laiolo, M.; Delle Donne, D.; Ripepe, M.; Cigolini, C. Hot-spot detection and characterization of strombolian activity from MODIS infrared data. *Int. J. Remote Sens.* **2014**, *35*, 3403–3426. [CrossRef]
45. Coppola, D.; Barsotti, S.; Cigolini, C.; Laiolo, M.; Pfeffer, M.; Ripepe, M. Monitoring the time-averaged discharge rates, volumes and emplacement style of large lava flows by using MIROVA system: The case of the 2014-2015 eruption at Holuhraun (Iceland). *Ann. Geophys.* **2019**, *61*, 52. [CrossRef]
46. Gouhier, M.; Guéhenneux, Y.; Labazuy, P.; Cacault, P.; Decriem, J.; Rivet, S. HOTVOLC: A web-based monitoring system for volcanic hot spots. *Geol. Soc. Lond. Spec. Publ.* **2016**, *426*, 223–241. [CrossRef]
47. Gauthier, P.-J.; Sigmarsson, O.; Gouhier, M.; Haddadi, B.; Moune, S. Elevated gas flux and trace metal degassing from the 2014–2015 fissure eruption at the Bárðarbunga volcanic system, Iceland. *J. Geophys. Res. Solid Earth* **2016**, *121*, 1610–1630. [CrossRef]
48. Guéhenneux, Y.; Gouhier, M.; Labazuy, P. Improved space borne detection of volcanic ash for real-time monitoring using 3-Band method. *J. Volcanol. Geotherm. Res.* **2015**, *293*, 25–45. [CrossRef]

49. Carn, S.A.; Clarisse, L.; Prata, A.J. Multi-decadal satellite measurements of global volcanic degassing. *J. Volcanol. Geotherm. Res.* **2016**, *311*, 99–134. [CrossRef]
50. Meyer, F.J.; McAlpin, D.B.; Gong, W.; Ajadi, O.; Arko, S.; Webley, P.W.; Dehn, J. Integrating SAR and derived products into operational volcano monitoring and decision support systems. *ISPRS J. Photogramm. Remote Sens.* **2015**, *100*, 106–117. [CrossRef]
51. Ajadi, O.A.; Meyer, F.J.; Webley, P.W. Change detection in synthetic aperture radar images using a multiscale-driven approach. *Remote Sens.* **2016**, *8*, 482. [CrossRef]
52. Chaussard, E.; Amelung, F.; Aoki, Y. Characterization of open and closed volcanic systems in Indonesia and Mexico using InSAR time series. *J. Geophys. Res. Solid Earth* **2013**, *118*, 3957–3969. [CrossRef]
53. Li, C.; Krotkov, N.A.; Carn, S.; Zhang, Y.; Spurr, R.J.D.; Joiner, J. New-generation NASA Aura Ozone Monitoring Instrument (OMI) volcanic SO2 dataset: Algorithm description, initial results, and continuation with the Suomi-NPP Ozone Mapping and Profiler Suite (OMPS). *Atmos. Meas. Tech.* **2017**, *10*, 445–458. [CrossRef]
54. González, P.J.; Walters, R.J.; Hatton, E.L.; Spaans, K.; Hooper, A.J.; Wright, T.J. LiCSAR: Tools for automated generation of Sentinel-1 frame interferograms. In Proceedings of the AGU Fall Meeting, San Francisco, CA, USA, 12–16 December 2016.
55. Massimetti, F.; Coppola, D.; Laiolo, M.; Cigolini, C.; Ripepe, M. First comparative results from SENTINEL-2 and MODIS-MIROVA volcanic thermal dataseries. In Proceedings of the CoV10 IAVCEI General Assembly, Naples, Italy, 2–7 September 2018.
56. Theys, N.; De Smedt, I.; Yu, H.; Danckaert, T.; Van Gent, J.; Hörmann, C.; Wagner, T.; Hedelt, P.; Bauer, H.; Romahn, F.; et al. Sulfur dioxide retrievals from TROPOMI onboard Sentinel-5 Precursor: Algorithm theoretical basis. *Atmos. Meas. Tech.* **2017**, *10*, 119–153. [CrossRef]
57. Torres, R.; Snoeij, P.; Geudtner, D.; Bibby, D.; Davidson, M.; Attema, E.; Potin, P.; Rommen, B.Ö.; Floury, N.; Brown, M.; et al. GMES Sentinel-1 mission. *Remote Sens. Environ.* **2012**, *120*, 9–24. [CrossRef]
58. Chaussard, E. A low-cost method applicable worldwide for remotely mapping lava dome growth. *J. Volcanol. Geotherm. Res.* **2017**, *341*, 33–41. [CrossRef]
59. Wadge, G.; Scheuchl, B.; Stevens, N.F. Spaceborne radar measurements of the eruption of Soufrière Hills Volcano, Montserrat. *Geol. Soc. Lond. Mem.* **2002**, *21*, 583–594. [CrossRef]
60. Bernhard, E.-M.; Stein, E.; Twele, A.; Gähler, M. Synergistic Use of Optical and Radar Data for Rapid Mapping of Forest Fires in the European Mediterranean. *ISPRS—Int. Arch. Photogramm. Remote Sens. Spat. Inf. Sci.* **2012**, *XXXVIII-4*, 27–32. [CrossRef]
61. Barrière, J.; d'Oreye, N.; Oth, A.; Geirsson, H.; Mashagiro, N.; Johnson, J.B.; Smets, B.; Samsonov, S.; Kervyn, F. Single-Station Seismo-Acoustic Monitoring of Nyiragongo's Lava Lake Activity (D.R. Congo). *Front. Earth Sci.* **2018**, *6*, 1–17. [CrossRef]
62. Wang, T.; Poland, M.P.; Lu, Z. Dome growth at Mount Cleveland, Aleutian Arc, quantified by time series TerraSAR-X imagery. *Geophys. Res. Lett.* **2015**, *42*, 10614–10621. [CrossRef]
63. Rosen, P.A.; Hensley, S.; Joughin, I.R.; Li, F.K.; Madsen, S.N.; Rodriguez, E.; Goldstein, R.M. Synthetic aperture radar interferometry Synthetic aperture radar interferometry. *Inverse Probl.* **1998**, *14*, 55.
64. Gabriel, A.K.; Goldstein, R.M.; Zebker, H.A. Mapping small elevation changes over large areas: Differential radar interferometry. *J. Geophys. Res.* **1989**, *94*, 9183. [CrossRef]
65. Simons, M.; Rosen, P.A. Interferometric Synthetic Aperture Radar Geodesy. In *Treatise Geophys*, 2nd ed.; Elsevier: Oxford, UK, 2015; Volume 3, pp. 339–385.
66. Lu, Z. InSAR Imaging of Volcanic Deformation over Cloud-prone Areas—Aleutian Islands. *Photogramm. Eng. Remote Sens.* **2013**, *73*, 245–257. [CrossRef]
67. Hooper, A.; Bekaert, D.; Spaans, K.; Arikan, M. Recent advances in SAR interferometry time series analysis for measuring crustal deformation. *Tectonophysics* **2012**, *514–517*, 1–13. [CrossRef]
68. Pepe, A.; Calò, F. A Review of Interferometric Synthetic Aperture RADAR (InSAR) Multi-Track Approaches for the Retrieval of Earth's Surface Displacements. *Appl. Sci.* **2017**, *7*, 1264. [CrossRef]
69. Marchese, F.; Neri, M.; Falconieri, A.; Lacava, T.; Mazzeo, G.; Pergola, N.; Tramutoli, V. The Contribution of Multi-Sensor Infrared Satellite Observations to Monitor Mt. Etna (Italy) Activity during May to August 2016. *Remote Sens.* **2018**, *10*, 1948. [CrossRef]
70. Laiolo, M.; Ripepe, M.; Cigolini, C.; Coppola, D.; Della Schiava, M.; Genco, R.; Innocenti, L.; Lacanna, G.; Marchetti, E.; Massimetti, F.; et al. Space-and Ground-Based Geophysical Data Tracking of Magma Migration in Shallow Feeding System of Mount Etna Volcano. *Remote Sens.* **2019**, *11*, 1182. [CrossRef]

71. Murphy, S.W.; de Souza Filho, C.R.; Wright, R.; Sabatino, G.; Correa Pabon, R. HOTMAP: Global hot target detection at moderate spatial resolution. *Remote Sens. Environ.* **2016**, *177*, 78–88. [CrossRef]
72. Krueger, A.J.; Walter, L.S.; Bhartia, P.K.; Schnetzler, C.C.; Krotkov, N.A.; Sprod, I.; Bluth, G.J.S. Volcanic sulfur dioxide measurements from the total ozone mapping spectrometer instruments. *J. Geophys. Res.* **1995**, *100*, 14057. [CrossRef]
73. He, K.; Zhang, X.; Ren, S.; Sun, J. Deep residual learning for image recognition. In Proceedings of the IEEE Conference on Computer Vision and Pattern Recognition, Las Vegas, NV, USA, 26 June–1 July 2016; pp. 770–778.
74. Krizhevsky, A.; Sutskever, I.; Hinton, G.E. ImageNet classification with deep convolutional neural networks. In Proceedings of the 26th Annual Conference on Neural Information Processing Systems 2012, Lake Tahoe, NV, USA, 3–6 December 2012; pp. 1106–1114.
75. Biggs, J.; Pritchard, M.E. Global volcano monitoring: What does it mean when volcanoes deform? *Elements* **2017**, *13*, 17–22. [CrossRef]
76. Dietterich, H.R.; Poland, M.P.; Schmidt, D.A.; Cashman, K.V.; Sherrod, D.R.; Espinosa, A.T. Tracking lava flow emplacement on the east rift zone of Kīlauea, Hawai'i, with synthetic aperture radar coherence. *Geochem. Geophys. Geosyst.* **2012**, *13*, 1–17. [CrossRef]
77. Xu, W.; Rivalta, E.; Li, X. Magmatic architecture within a rift segment: Articulate axial magma storage at Erta Ale volcano, Ethiopia. *Earth Planet. Sci. Lett.* **2017**, *476*, 79–86. [CrossRef]
78. D'Hondt, O.; López-Martínez, C.; Guillaso, S.; Hellwich, O. Nonlocal filtering applied to 3-D reconstruction of tomographic SAR data. *IEEE Trans. Geosci. Remote Sens.* **2018**, *56*, 272–285. [CrossRef]
79. Patrick, M.R.; Anderson, K.R.; Poland, M.P.; Orr, T.R.; Swanson, D.A. Lava lake level as a gauge of magma reservoir pressure and eruptive hazard. *Geology* **2015**, *43*, 831–834. [CrossRef]
80. Neal, C.A.; Brantley, S.R.; Antolik, L.; Babb, J.L.; Burgess, M.; Calles, K.; Cappos, M.; Chang, J.C.; Conway, S.; Desmither, L.; et al. The 2018 rift eruption and summit collapse of Kīlauea Volcano. *Science* **2019**, *363*, 367–374. [CrossRef]
81. Williams, R.; Rowley, P.; Garthwaite, M.C. Small flank failure of Anak Krakatau Volcano caused catastrophic December 2018 Indonesian tsunami. *EarthArXiv* **2019**. [CrossRef]
82. Coppola, D.; Di Muro, A.; Peltier, A.; Villeneuve, N.; Ferrazzini, V.; Favalli, M.; Bachèlery, P.; Gurioli, L.; Harris, A.J.L.; Moune, S.; et al. Shallow system rejuvenation and magma discharge trends at Piton de la Fournaise volcano (La Réunion Island). *Earth Planet. Sci. Lett.* **2017**, *463*, 13–24. [CrossRef]
83. Zobin, V.M. Origin of Volcano-Tectonic Earthquakes. In *Introduction to Volcanic Seismology, Volume 6*, 3rd ed.; Elsevier: Amsterdam, The Netherlands, 2017; pp. 31–45.
84. Coppola, D.; Laiolo, M.; Cigolini, C. Fifteen years of thermal activity at Vanuatu's volcanoes (2000–2015) revealed by MIROVA. *J. Volcanol. Geotherm. Res.* **2016**, *322*, 6–19. [CrossRef]
85. Allard, P.; Aiuppa, A.; Bani, P.; Métrich, N.; Bertagnini, A.; Gauthier, P.J.; Shinohara, H.; Sawyer, G.; Parello, F.; Bagnato, E.; et al. Prodigious emission rates and magma degassing budget of major, trace and radioactive volatile species from Ambrym basaltic volcano, Vanuatu island Arc. *J. Volcanol. Geotherm. Res.* **2016**, *322*, 119–143. [CrossRef]
86. Hamling, I.J.; Cevuard, S.; Garaebiti, E. Large-Scale Drainage of a Complex Magmatic System: Observations From the 2018 Eruption of Ambrym Volcano, Vanuatu. *Geophys. Res. Lett.* **2019**, 4609–4617. [CrossRef]
87. IGEPN: Informe Especial del Volcán Sangay N° 3. 2019. Available online: https://www.igepn.edu.ec/servicios/noticias/1733-informe-especial-del-volcan-sangay-n-3-2019 (accessed on 6 June 2019).
88. Prata, A.J. Infrared radiative transfer calculations for volcanic ash clouds. *Geophys. Res. Lett.* **1989**, *16*, 1293–1296. [CrossRef]
89. Piscini, A.; Corradini, S.; Marchese, F.; Merucci, L.; Pergola, N.; Tramutoli, V. Volcanic ash cloud detection from space: A comparison between the RST ASH technique and the water vapour corrected BTD procedure. *Geomat. Nat. Hazards Risk* **2011**, *2*, 263–277. [CrossRef]
90. Piscini, A.; Picchiani, M.; Chini, M.; Corradini, S.; Merucci, L.; Del Frate, F.; Stramondo, S. A neural network approach for the simultaneous retrieval of volcanic ash parameters and SO2 using MODIS data. *Atmos. Meas. Tech.* **2014**, *7*, 4023–4047. [CrossRef]
91. Harris, A.; Chevrel, M.; Coppola, D.; Ramsey, M.; Hrysiewicz, A.; Thivet, S.; Villeneuve, N.; Favalli, M.; Peltier, A.; Kowalski, P.; et al. Validation of an integrated satellite-data-driven response to an effusive crisis: The April–May 2018 eruption of Piton de la Fournaise. *Ann. Geophys.* **2019**, *61*. [CrossRef]

92. Fu, G.; Prata, F.; Lin, H.X.; Heemink, A.; Segers, A.; Lu, S. Data assimilation for volcanic ash plumes using a satellite observational operator: A case study on the 2010 Eyjafjallajökull volcanic eruption. *Atmos. Chem. Phys.* **2017**, *17*, 1187–1205. [CrossRef]
93. Scollo, S.; Prestifilippo, M.; Spata, G.; D'Agostino, M.; Coltelli, M. Monitoring and forecasting Etna volcanic plumes. *Nat. Hazards Earth Syst. Sci.* **2009**, *9*, 1573–1585. [CrossRef]
94. Bagnardi, M.; Hooper, A. Inversion of Surface Deformation Data for Rapid Estimates of Source Parameters and Uncertainties: A Bayesian Approach. *Geochem. Geophys. Geosyst.* **2018**, *19*, 2194–2211. [CrossRef]
95. Nikkhoo, M.; Walter, T.R. Triangular dislocation: An analytical, artefact-free solution. *Geophys. J. Int.* **2015**, *201*, 1119–1141. [CrossRef]
96. Nikkhoo, M.; Walter, T.R.; Lundgren, P.R.; Prats-Iraola, P. Compound dislocation models (CDMs) for volcano deformation analyses. *Geophys. J. Int.* **2017**, *208*, 877–894. [CrossRef]
97. Laiolo, M.; Massimetti, F.; Cigolini, C.; Ripepe, M.; Coppola, D. Long-term eruptive trends from space-based thermal and SO2 emissions: A comparative analysis of Stromboli, Batu Tara and Tinakula volcanoes. *Bull. Volcanol.* **2018**, *80*, 68. [CrossRef]
98. Young, P.C. New approaches to volcanic time-series analysis. In *Statistics in Volcanology*; IAVCEI Special Publications; Geological Society of London: London, UK, 2006; pp. 143–160.
99. Ho, C.H. Volcanic time-trend analysis. *J. Volcanol. Geotherm. Res.* **1996**, *74*, 171–177. [CrossRef]
100. Guffanti, M.; Miller, T.P. A volcanic activity alert-level system for aviation: Review of its development and application in Alaska. *Nat. Hazards* **2013**, *69*, 1519–1533. [CrossRef]
101. Winson, A.E.G.; Costa, F.; Newhall, C.G.; Woo, G. An analysis of the issuance of volcanic alert levels during volcanic crises. *J. Appl. Volcanol.* **2014**, *3*, 1–12. [CrossRef]
102. Papale, P. Rational volcanic hazard forecasts and the use of volcanic alert levels. *J. Appl. Volcanol.* **2017**, *6*, 13. [CrossRef]
103. Heimann, S.; Kriegerowski, M.; Isken, M.; Cesca, S.; Daout, S.; Grigoli, F.; Juretzek, C.; Megies, T.; Nooshiri, N.; Steinberg, A.; et al. Pyrocko—An open-source seismology toolbox and library. V. 0.3. *GFZ Data Serv.* **2017**. [CrossRef]

Article

Volcano Monitoring from Space Using High-Cadence Planet CubeSat Images Applied to Fuego Volcano, Guatemala

Anna Aldeghi [1,2], Simon Carn [1], Rudiger Escobar-Wolf [1] and Gianluca Groppelli [3,*]

1 Department of Geological and Mining and Engineering Sciences, Michigan Technological University, 1400 Townsend Drive, Houghton, MI 49931, USA
2 Department of Earth and Environmental Sciences, University of Milano-Bicocca, 20126 Milano, Italy
3 Istituto di Geologia Ambientale e Geoingegneria-sez. di Milano, C.N.R, 20131 Milano, Italy
* Correspondence: gianluca.groppelli@cnr.it; Tel.: +39-388-9999-492

Received: 23 June 2019; Accepted: 6 September 2019; Published: 16 September 2019

Abstract: Fuego volcano (Guatemala) is one of the most active and hazardous volcanoes in the world. Its persistent activity generates lava flows, pyroclastic density currents (PDCs), and lahars that threaten the surrounding areas and produce frequent morphological change. Fuego's eruption deposits are often rapidly eroded or remobilized by heavy rains and its constant activity and inaccessible terrain makes ground-based assessment of recent eruptive deposits very challenging. Earth-orbiting satellites can provide unique observations of volcanoes during eruptive activity, when ground-based techniques may be too hazardous, and also during inter-eruptive phases, but have typically been hindered by relatively low spatial and temporal resolution. Here, we use a new source of Earth observation data for volcano monitoring: high resolution (~3 m pixel size) images acquired from a constellation of over 150 CubeSats ('Doves') operated by Planet Labs Inc. The Planet Labs constellation provides high spatial resolution at high cadence (<1–72 h), permitting space-based tracking of volcanic activity with unprecedented detail. We show how PlanetScope images collected before, during, and after an eruption can be applied for mapping ash clouds, PDCs, lava flows, or the analysis of morphological change. We assess the utility of the PlanetScope data as a tool for volcano monitoring and rapid deposit mapping that could assist volcanic hazard mitigation efforts in Guatemala and other active volcanic regions.

Keywords: satellite remote sensing; volcano monitoring; ash fall; lava flows; pyroclastic density currents; mapping; volcanic hazard

---

1. Introduction

Persistently active volcanoes are an ever-present threat to populations and infrastructure exposed to primary and secondary volcanic hazards on their flanks. Continuous, ground-based monitoring by local or regional volcano observatories plays an essential role in hazard mitigation in such locations. When ground-based techniques are impractical or too hazardous, Earth-orbiting satellites can be a valuable tool for volcano monitoring. However, satellite remote sensing techniques can be limited by temporal and/or spatial resolution, particularly in rapidly evolving situations.

Satellite remote sensing (or Earth observation, EO) has played an increasingly important role in volcano surveillance over the past few decades, concurrent with technological advances in satellite sensors [1–5]. The principle applications of EO data in volcanology include volcanic gas and ash monitoring [6–8], monitoring of heat fluxes [9–13], optical measurements [14–16], mapping of ground deformation [17,18], geological mapping [19,20] and geological hazard assessment [21–23]. The main factors limiting the efficacy of EO data for volcano monitoring are the temporal and spatial resolution of the measurements, which are seldom optimized simultaneously in satellite instruments. Temporal resolution is arguably the most critical, since volcanic activity remains largely unpredictable and can produce dynamic phenomena that can be difficult to track and analyze in real-time during rapidly evolving crisis situations. Several volcanic eruptions in 2018 demonstrated the need for rapid response, including: Kilauea (Hawaii) in May 2018; Agung and Anak Krakatau (Indonesia) in November–December 2018; Mayon (Philippines); Piton de la Fournaise (Reuniòn Island); and Mount Etna (Italy) [24].

Commonly-used pushbroom visible-infrared satellite sensors, such as those deployed on the USGS Landsat or European Sentinel-2 satellites, have intermediate-high spatial (~10–20 m), but low temporal (days to weeks) resolution (Table 1). Since 2008, with the open access of Landsat archives, "time series" images have been used to analyze natural processes. High acquisition frequency allows for a more complete understanding of geological dynamics [25]. Here, we explore some volcanological applications of a new EO paradigm that can provide observations with both high spatial and temporal resolution: constellations of small satellites or 'CubeSats'. The use of a constellation of CubeSats provides a higher cadence than is possible with the use of a single satellite, whilst also providing high spatial resolution by minimizing swath width. One of the pioneers in this field is Planet Labs Inc. [26], which has operated an expanding constellation of over 100 CubeSats (informally called 'Doves') since 2014. The Planet constellation provides three or four band (visible to near-infrared) PlanetScope (PS) imagery of the entire land surface of the Earth with ~3 m spatial resolution and a temporal cadence of ~1–72 h (Table 1). Although the temporal resolution of current CubeSat constellations is lower than that provided by geostationary satellites, their spatial resolution is several orders of magnitude higher, with a pixel size of ~3 m compared with pixel sizes of 0.25–2 km for geostationary imagers. In some cases, consecutive data acquisition by Planet Doves in the same 'flock' (or overlapping flocks) can provide multiple images of the same region within a few minutes. These characteristics, along with an academic open data access policy, make PS data appealing for monitoring of volcanic activity or other dynamic geophysical phenomena resulting in surface change (e.g., landslides, earthquakes).

Table 1. Characteristics of optical satellite instruments currently used for volcano monitoring.

Criteria	Planet Labs	Sentinel 2	Landsat 8	MODIS	ASTER	VIIRS	Digital Globe
Bands (wave length)	4–5 (440–900 nm)	13 (497–2190 nm)	8 (0.45–12.5 μm)	36 (0.405–14.385 μm)	14 (0.52–11.62 μm)	22 (0.412–12.01 μm)	4–16 (450–2373 nm)
Night-time imagery	No	No	Yes	Yes	Yes	No	No
TIR/SWIR	No	Yes	Yes	Yes	Yes	Yes	Yes
Spatial resolution	0.8–5 m	10–60 m	15–30 m	250–1000 m	15–90 m	500–1000 m	0.5–1.85 m
Cadence	<1–72 h	5–10 d	16 d	6–12 h	Highly variable (days to weeks)	24 h	Highly variable (days to months)
Product level	1B	1C	2	1A–B	1C	1	/

We use Fuego volcano (Guatemala; Figure 1) as the target of this initial study. Fuego is a persistently active volcano. It has produced several paroxysmal eruptive events in the last three years, providing a rich opportunity to analyze PS imagery. Paroxysmal eruptions at Fuego follow extended periods of lower-level activity, and typically begin with continuous lava fountains and lava flows, followed by tephra fallout and pyroclastic density currents (PDCs), which present the greatest threat to the surrounding areas. Due to its persistent activity and subsequent remobilization of deposits by rainfall, the morphology of Fuego and the barrancas (valleys) that radiate from its summit is in a constant state of flux, both in the near-vent region and on the lower flanks.

We focus here on activity at Fuego in February 2018. We assess the utility of optical PS data for monitoring activity within Fuego's active vent region and for mapping eruption deposits using the Normalized Difference Vegetation Index (NDVI) [27,28] and visual analysis. Of particular interest is whether PS data can assist with 'rapid response' mapping of eruption deposits that could be eroded or removed by rainfall soon after emplacement, since the volume and extent of such deposits can provide critical information on eruption magnitude and on the threat of secondary volcanic hazards, such as lahars. Monitoring and mapping of such ephemeral deposits is also important for probabilistic mapping of volcanic hazards around Fuego.

Figure 1. Location of the Fuego-Acatenango Complex. (**a**) Location of Fuego volcano in the Central American Volcanic Arc. (**b**) Detail of the main barrancas channels surrounding the Fuego edifice.

2. Background

Fuego (alt. 3763 m; 14.4° N, 90.8° W) is the southernmost and youngest of five vents comprising the Fuego-Acatenango massif, located on the Central American volcanic arc (CAVA) in Guatemala (Figure 1). The CAVA is a chain of Quaternary stratovolcanoes linked to subduction of the Cocos plate under the Caribbean plate [29,30]. Fuego is one of the most active volcanoes in Central America, with over 60 recorded eruptions since 1524 [31]. Most of Fuego's historic eruptions have been classified with a Volcanic Explosivity Index (VEI) of 2 or 3, interspersed with several larger VEI 4 events, the most recent of which occurred in 1974 [31–34]. The region surrounding Fuego is highly populated, with ~9000 people living within the high hazard zone potentially threatened by PDCs, and more than one million within 30 km of the volcano. Larger eruptions of Fuego also potentially threaten the national capital (Guatemala City) and main international airport, which are only ~40 km away. During the 3 June 2018 eruption of Fuego the ash emissions forced the closure of the La Aurora International Airport in Guatemala City.

Fuego undergoes long periods of relatively low background activity, interrupted by periods of intense, or 'paroxysmal', activity that consists of lava flows, lava fountains and PDCs [33,34]. The latter are typically channelled down one or more of the barrancas (valleys) on the flanks of the volcano (Figure 1b), and can be a significant hazard, either by directly impacting populated areas or by providing material for lahars during the rainy season [32,35]. This broad range of eruptive styles makes Fuego's activity both scientifically interesting and challenging to forecast. Furthermore, the persistent activity results in constant morphological variation in the vent region and on the flanks (barrancas), which is difficult to track using ground-based observations.

Fuego was very active in 2018, with 3 paroxysmal eruptions in February, June, and November. We focus here on the 31 January–2 February 2018 eruption, a Strombolian event of VEI 2 (Figure 2) that was well-captured in PS data. The June 2018 eruption, which resulted in extensive loss of life and damage to infrastructure, involved a more complex sequence of events, and will be the subject of a separate paper. The 31 January–2 February 2018 eruption started with continuous lava fountains and two lava flows, directed NW from the summit. Subsequently, PDCs formed on the east flank and deposited material in two of the seven barrancas which surround the edifice. A significant amount of ash was also emitted and dispersed predominantly to the west by prevailing winds. The activity was also monitored by webcams, providing a visible record of the sequence of events. PS images collected before, during and after the eruption provide a novel, detailed perspective on the volcanic activity for monitoring and mapping the distribution of the volcanic deposits.

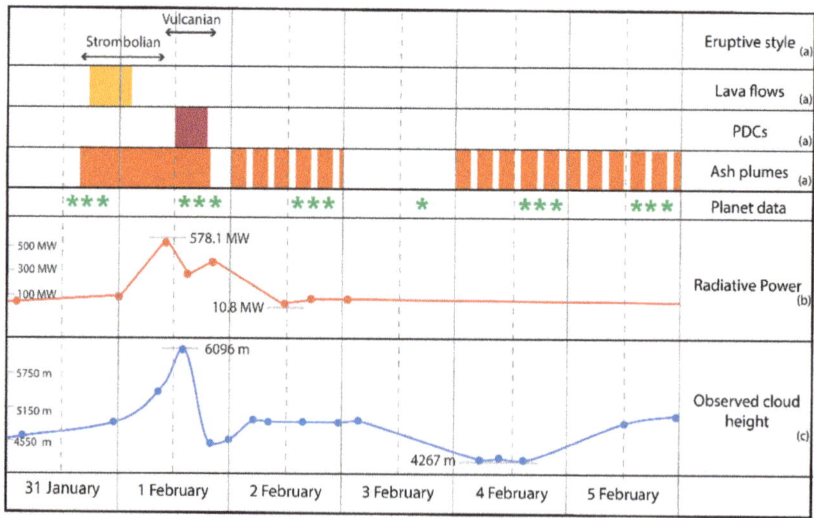

Figure 2. Timeline of the February 2018 eruption of Fuego and its main physical and volcanological features. Data sources: (**a**) Smithsonian Institution Global Volcanism Program (GVP) [24]; Planet Labs data coverage [26]; (**b**) MODVOLC website [36]; (**c**) VAAC database [37]. Planet labs imagery timeline is in Table S1.

3. Data and Methods

3.1. The Planet Labs CubeSat Constellation

The Planet Labs (PL) CubeSat constellation comprises (as of early 2019) over 150 CubeSats or 'Doves', each roughly 10 × 10 × 30 cm in size (i.e., three-unit or 3U CubeSats), orbiting in two near-polar, Sun-synchronous (SS) orbits with ~8° and ~98° inclination at an altitude of ~475 km. Each Dove carries a telescope and a 6600 × 4400 pixel CCD array, which acquires both visible (red–green–blue or RGB) and near-infrared (NIR) PS data with 12-bit radiometric resolution. PS scenes are frame images (i.e., an entire scene is imaged by the CCD at an instant in time, unlike pushbroom sensors) with approximate dimensions of 25–30 × 8–10 km, and a native spatial resolution of 3.7 m, resampled to 3 m for delivery. The Doves have been deployed since 2014 in 'flocks' from various launch vehicles. Early CubeSats (operational in 2014–2017) were released from the International Space Station (ISS) into the ISS orbit with 52° inclination at ~375 km altitude, whereas more recent PS data (including the data used in this study) are acquired by the near-polar SS-orbiting flocks (Table 2). The Dove constellation images the entire Earth landmass on a daily basis, with overpass times in the local morning hours. To achieve daily contiguous global coverage, the swaths of each consecutive Dove overlap in the across-track direction, providing the opportunity for multiple scene acquisitions within a few minutes in the overlap region [35]. Planet Labs also operates a fleet of five RapidEye satellites and 14 SkySats, which provide even higher spatial resolution.

Table 2. Specifications of Planet Labs deployments in International Space Station (ISS) and Sun synchronous orbits (SSO).

Criteria	PlanetScope (PS2)	
Deployment	ISS orbit	SSO orbit
Orbit altitude	400 km	475 km
Max/min latitude coverage	52°	81.5°
HFOV	21.8 km	24.6 km
VFOV	14.5 km	16.4 km
Area	316 km^2	405 km^2
Max image strip per orbit	8100 km^2	20,000 km^2
Analytic resolution	16 bit	16 bit

Several PS data products with different processing levels are available; here we use the 'Analytic_SR' data products which are 16-bit calibrated orthorectified surface reflectance data with a positional accuracy of better than 10 m.

3.2. Data Processing

This study of the February 2018 activity at Fuego is based on change detection techniques-a comparison of images pre, syn and post-eruption of the same area at different times. The activity at Fuego permits the evaluation of PS data for a variety of eruption products including lava flows, trephra fall deposits and PDCs.

PS imagery was browsed, and products downloaded using the Planet Explorer interface [26]. Images from before and after the 31 January–2 February eruption were identified, inspected for cloud cover and clarity, and downloaded. Visual inspection of the images was performed by selecting suitable RGB colour stretching values to highlight volcanic deposits and structures. Images were merged and cropped to the extent of the area of interest. Visual comparison of pre- and post-eruption images was performed to identify and map changes, including new deposits produced by tephra fall, lava flows and PDCs. Visual comparisons primarily used the visible bands, but the NIR band was also used in some cases (substituting for the green band in the RGB composite images), particularly when changes in vegetation cover due to the volcanic activity were involved. Where new eruptive material was deposited over areas previously covered by vegetation, a simple change detection strategy based on the

NDVI difference was applied. NDVI values were calculated for each image from the surface reflectance values, corresponding to the red and NIR bands [38,39]. We then subtracted the NDVI values in the pre-eruption images from the post-eruption NDVIs, for spatially collocated pixels, yielding a raster dataset of pre/post-eruption NDVI differences. A negative NDVI difference thus indicated areas where vegetation has been damaged and/or covered by volcanic deposits; the NDVI difference can have any value between −1 and 1, from completely unvegetated (or ash-covered) to completely vegetated (or ash-free). To simplify the interpretation a threshold was chosen, usually −0.1 or −0.125, to represent the change that was considered significant and interpreted as being produced unequivocally by the deposition of volcanic material (e.g., tephra fall). Cloud cover limits the use of reflectance based remote sensing data in the visible and NIR bands. Clouds and cloud shadows were visually identified and masked in all the images; such masks were then applied to exclude cloudy areas and any associated noise and biases from the data analysis. Finally, the raster maps obtained from the NDVI differences were simplified, by applying a low-pass filter (e.g., 15 × 15 neighbour pixel average operation) to the threshold maps and converting them to polygons that mapped the tephra-covered areas.

The NDVI difference approach in general is not suitable for detecting changes in areas that were originally non-vegetated, like the upper flanks of the volcano or the active barrancas. In such cases other band combinations can be tested to see if a suitable band or combination of bands could be used to detect and identify the new deposits. In particular, the visible and NIR bands of a pre-eruption image were subtracted from the corresponding band of the post-eruption image. A trial-and-error approach was adopted using single bands (e.g., pre-eruption Band 1 minus pre-eruption Band 1, etc.) but results were not satisfactory. Also, the same band on pre and post-eruption dates was analyzed with the same approach as NDVI differences for a time-dependent analysis.

Processing of the visible, NDVI and other band combinations yielded a series of maps of changes between the pre- and post-eruption images, which were then interpreted in terms of new eruptive deposits and other volcanic processes. Additionally, volcanic structures in the vent region and upper flanks of Fuego were also mapped visually using RGB images. PDCs can be highly erosive; they engrave channels and incorporate pre-existing material as they move downhill. This produces characteristic features along the PDC path that can be detected by comparing pre- and post-eruption images. Comparing images from a longer time-series with those captured during short-livedevents allow us to distinguish changes in crater morphology associated with background and paroxysmal activity.

4. Results

Eruptive deposits and volcanic structures mapped with the methods previously described were compiled into a GIS platform (ArcGis,® Environmental Systems Research Institute, Redlands, California, United States) for interpretations and analysis. Here we describe the different types of deposits and structures separately as interpreted from our analysis.

4.1. Lava Flows

Deposits optically identified and interpreted as two lava flows erupted during the February 2018 paroxysm are shown in Figure 3. These lava flows moved down the NW slope of Fuego. The longest of the two reaches a length of ~2000 m and they respectively cover areas of 0.12 and 0.18 km^2. They are representative of the beginning of the activity; a Strombolian style eruption characterized by sustained lava fountains and bombs. At the beginning of the activity the upper flank is characterized by spatter deposits and lava flows (Figure 3a). Distinguishing these deposits is difficult because of the lack of contrast between them. In this case, the morphology plays a crucial role for the distinction between the smoother lava flows and the rugged surface of the spatter deposit. Automated detection of different deposits on the highest part of the edifice is more difficult. In fact, neither the NDVI difference nor any other simpler band operation gives good results for this kind of deposits. It is still possible to map the darker material using single bands and appropriate thresholds. Figure 3b shows band-2 pixel values with digital numbers (DN) <3000 (in a 16-bit image) in the vent region; similar results were obtained for the other three bands. The result is closer to the optical map of the deposits shown in Figure 3a, but with some additional signal (likely due to tephra fall) on the upper west flank.

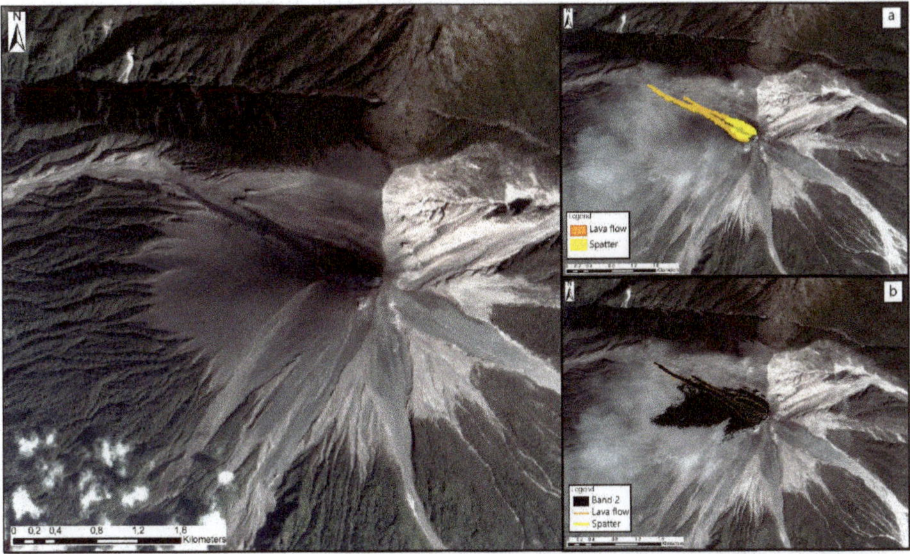

Figure 3. PlanetScope image of Fuego on February 2, 2018; 15:57 UTC. (**a**) Lava flows and spatter deposits optically mapped are shown in orange and yellow respectively. (**b**) Result for single Band 2, using a threshold DN < 3000, for comparison.

4.2. Tephra fall

During the February 2018 paroxysm the eruption column reached over 6000 m height (Figure 2) and was mostly dispersed to the west of the volcano. Figure 4 shows the tephra deposit extent obtained from the NDVI difference method. One other image source (Sentinel 2) was also processed in a similar way for comparison. For the PS images, the NDVI difference is calculated using an image collected in January 2018 and another right after the eruption. In order to have a meaningful comparison between PL and Sentinel 2, the post-eruption images are both from 4 February (Figure 5), with pre-eruption images from 9 January (PS) and 10 January (Sentinel 2). Slightly different thresholds were used to delineate the tephra fall covered areas for each image source; for the PS imagery a threshold of −0.125 was used for the NDVI difference values (i.e., pixels with an NDVI difference <−0.125 were considered to be covered with ash) which gives an area of 41.5 km^2 for the ash covered vegetation. For Sentinel 2 data, a NDVI difference threshold of −0.1 was used, which results in an ash covered area of 35.6 km^2. Deposits derived from the two data sources are similar to each other in shape and orientation (Figure 4), and both show limitations in detecting the very proximal tephra fall deposits (i.e., in non-vegetated areas).

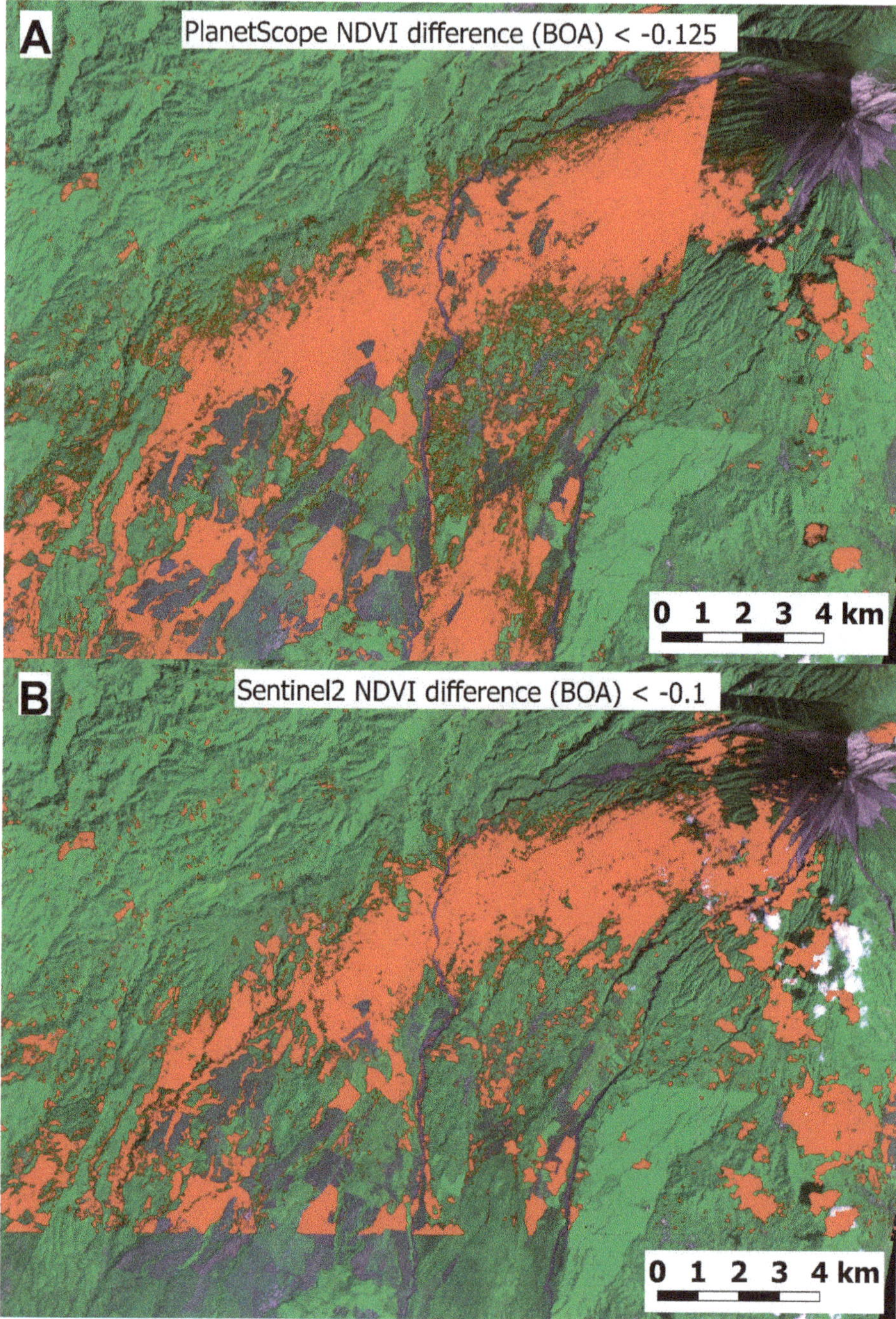

Figure 4. *Cont.*

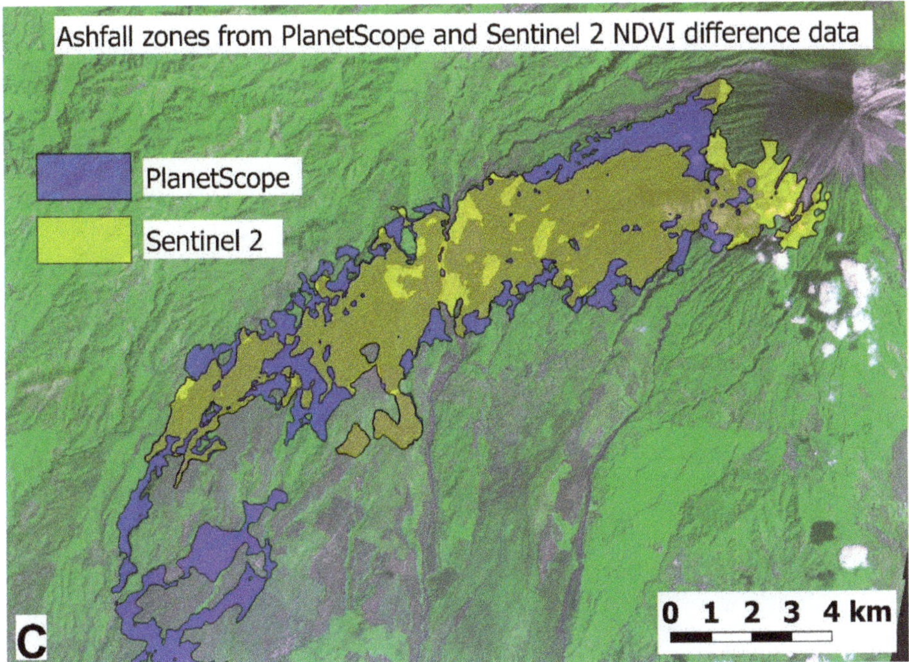

Figure 4. Tephra fall deposit mapped using NDVI difference analysis of (**A**) PlanetScope and (**B**) Sentinel 2 images from 9th 10th January and 4th February 2018. Threshold values of –0.1 for Sentinel 2, and–0.125 for PlanetScope were used to identify the ash covered areas shown in red in (**A**) and (**B**). (**C**) Shows the areas that can be identified as ash covered by the NDVI differences in each set of satellite images after applying a low-pass filter and converting them to polygons (only taking the connecting polygons along the expected dispersal axis). BOA = Bottom of Atmosphere. Background image is a PlanetScope image from February 4, 2018 at 15:58 UTC.

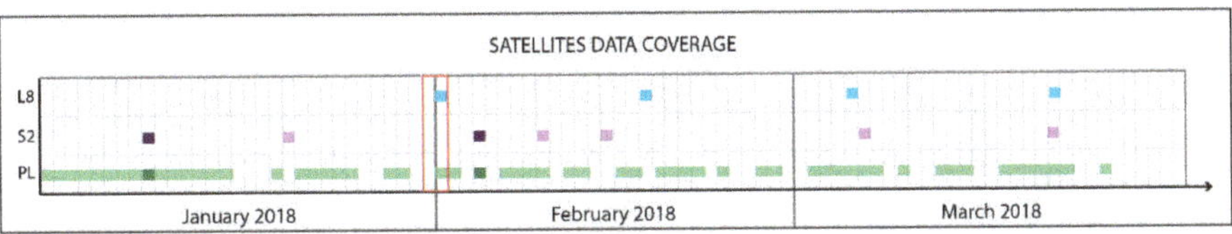

Figure 5. Schematic three-month timeline for satellite data coverage of Landsat 8 (L8), Sentinel 2 (S2) and Planet Labs (PL). Only usable data, with no cloud coverage over the study area are considered. Images selected for NDVI analysis are marked with darker colours. The red box shows the February 2018 event period.

4.3. Pyroclastic Density Currents

The PDC deposits from the eruption were mapped visually and are shown in Figure 6. The PDC material in Barranca Las Lajas and El Jute covers an area of 1.24 km^2. The second PDC deposited in Barranca Honda covered an area of 0.43 km^2. The use of automated methods (e.g., NDVI or other indices or single band differences) to delineate the deposits was not successful for the whole deposit in any of the barrancas, due to the lack of pre-existing vegetation in these zones. Figure 7 demonstrates the value of high spatial resolution PS imagery for identifying morphological changes associated with eruptive activity. A scar located SE of the Fuego vent region in the upper reaches of Barranca Honda indicates erosion and/or collapse of the steep upper flanks. The orientation of the scar suggests that the derived material must have contributed to the PDC emplaced in barranca Honda.

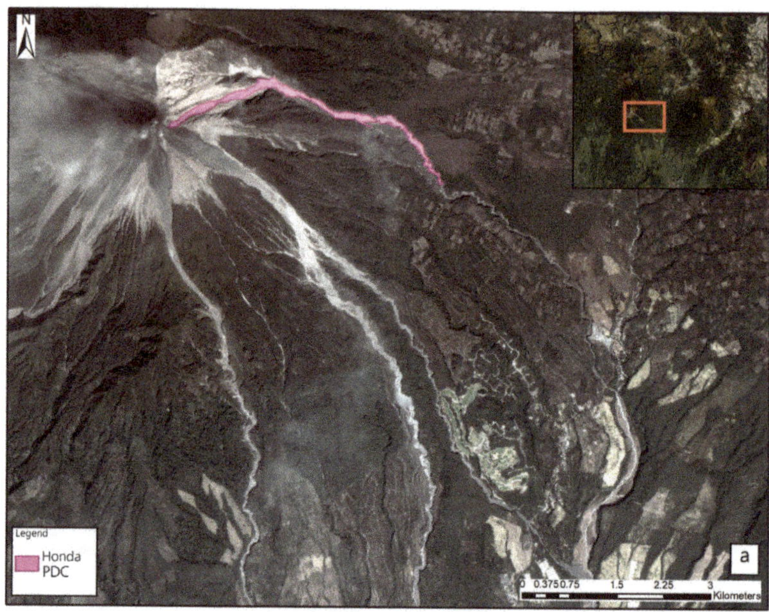

Figure 6. PDC deposits in (**a**) Barranca Honda and (**b**) Barranca Las Lajas and El Jute visually mapped using PlanetScope imagery. The inset shows the location of the displayed area. Background image is from February 2, 2018 at 15:57 UTC.

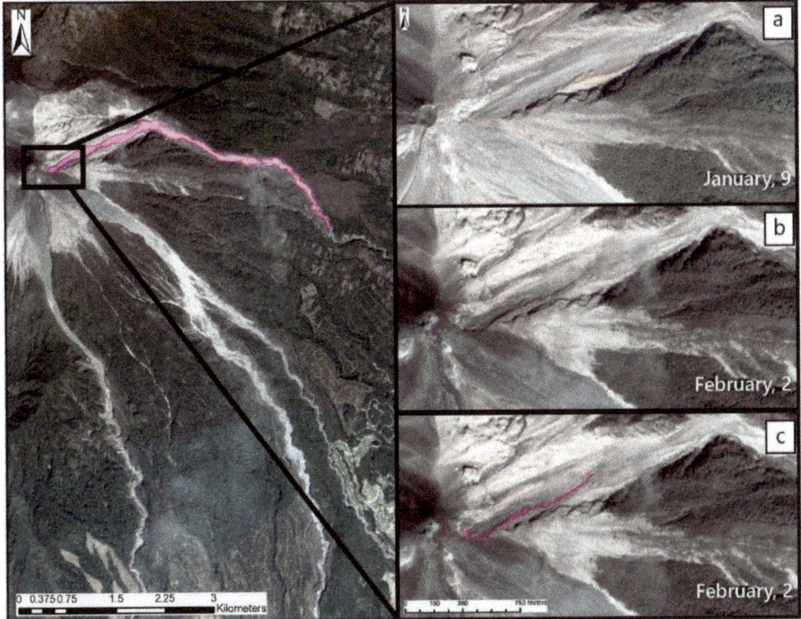

Figure 7. PlanetScope images of Fuego collected on January 9, 2018 at 16:57 UTC and February 2, 2018 at 15:57 UTC, showing morphological changes pre- (**a**) and post-eruption (**b**). The February 2 PlanetScope image (**c**) shows the presence of a scar formed during the February 2018 paroxysm, in the upper reaches of Barranca Honda.

4.4. Structures Near the Vent and the Upper Cone Area

The persistent activity of Fuego volcano, including background activity, causes frequent and often continuous variations in the morphology of the summit. Tracking such changes, which may provide important indications of future activity, is very challenging from the ground and typically requires overflights by manned or unmanned aircraft, which may be infrequent. Figure 8 shows some of these changes identified using PS imagery collected over a period of ~ 4 months, from December 2017 to March 2018. A roughly circular vent in December 2017 evolved into a well-defined circle with a clear summit crater in January 2018. During the paroxysmal activity at the end of January and beginning of February 2018 the crater morphology changes substantially and material appears to overspill down the upper west flank. In subsequent images, cycles of crater excavation and infilling can be discerned (Figure 8). The high temporal and spatial resolution of PS imagery thus provides a novel ability to track morphological variations in the inaccessible regions of active volcanoes, such as Fuego, before, during and after eruptions.

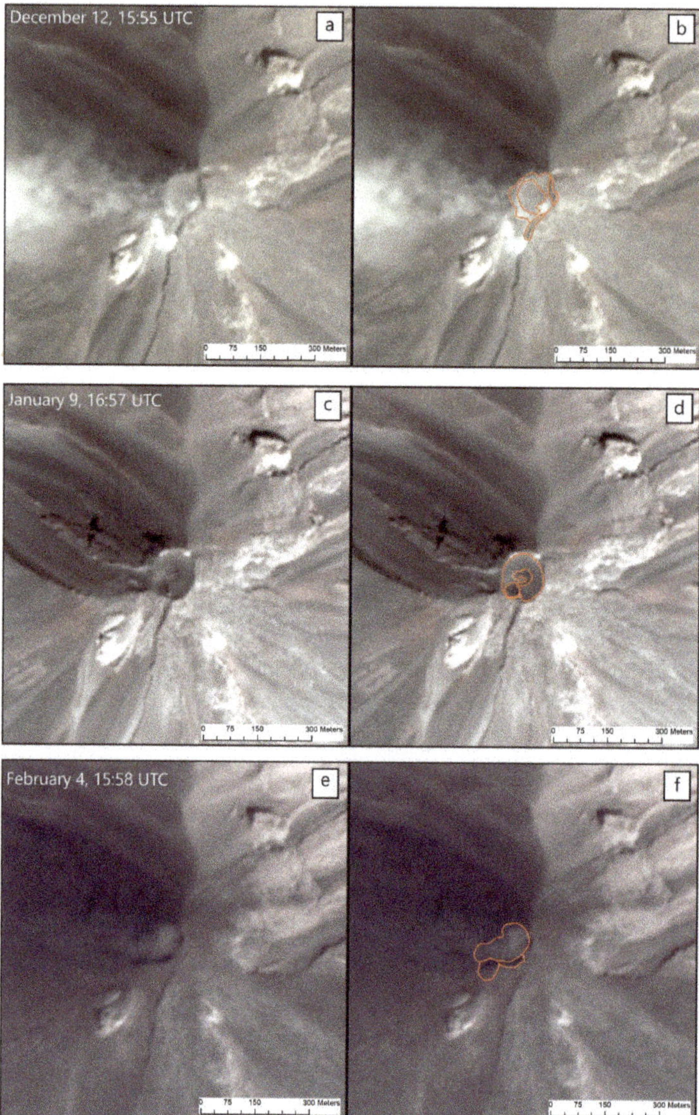

Figure 8. *Cont.*

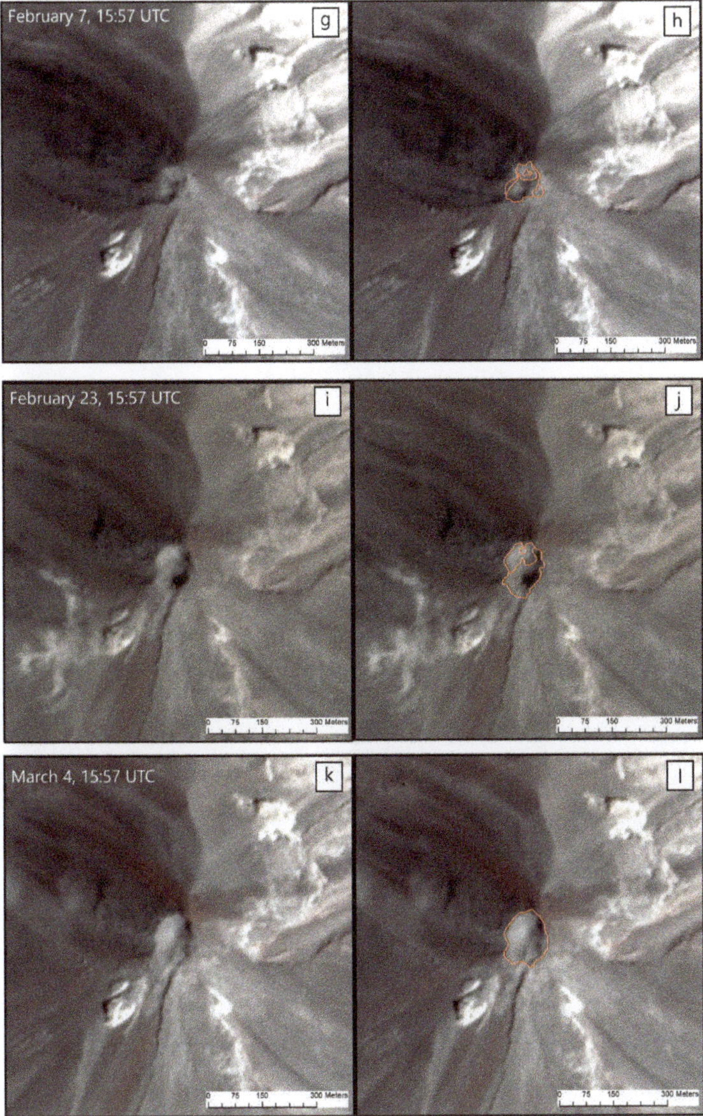

Figure 8. PlanetScope images of Fuego collected between December 2017 and March 2018, showing the evolution of the vent region of Fuego volcano. Left panels (**a**, **c**, **e**, **g**, **i**, and **k**) show the images without annotations; right panels (**b**, **d**, **f**, **h**, **j**, and **l**) show the mapped morphology and significant changes (orange lines). Figures a-d display the pre-eruption phase, whereas figures e–l represent the morphology right after the event to one month from it.

5. Discussion

The February 2018 eruption of Fuego lasted for approximately 20 h [24], and a summary of the activity is shown in Figure 2. The activity was captured by a monitoring webcam on the SE-flank of the volcano, located ~6 km from the vent. A video of the webcam images that tracks the full eruption can be accessed at https://youtu.be/JHwKFPTGAQk and https://youtu.be/4jdfmf3j4ww. Two types of activity can be distinguished: Strombolian to violent-Strombolian activity, characterized by lava fountains, lava flows and a small eruptive column developing over the active vent, and a second type of activity generating PDCs. The variety of eruptive processes and products generated during this short eruption presents a unique opportunity to test the capability of PL satellite data to identify and map volcanic deposits quickly after they have been emplaced.

In the case of the lava flows, the visual mapping using the visible bands gives good results. Attempts to use band differences or index-based methods (e.g., NDVI) did not work well. Although fresh (still hot) lava flows can be mapped using thermal (usually TIR and SWIR) bands [15], our results show that mapping lava flows with high resolution (~3 m pixel size) visible images can be a good alternative provided the new lava flows show enough contrast with the background. The Smithsonian Institution Global Volcanism Program (GVP) reported a maximum length of 800 m for the lava flows, but from the mapping, it is clear that the length reaches ~2000 m for the longest branch and ~1500 m for the shorter lobe (Figure 3). This shows that the detailed mapping possible with PS images can significantly improve the estimation of such parameters. The two branches in the lower part of the lava flow are easily identified from the visible bands but closer to the vent region it is necessary to do a more detailed analysis. The use of a threshold on a single band highlights a larger area which comprises spatter, lava flows and tephra. Visual mapping of the February 2018 lava flows allows us defining some important volcanological features and better understand the importance of this eruption in the history of the Fuego volcano. According to [33] the mean thickness of recent lava flows, which are mainly basaltic in composition, is 2 to 4 m, therefore we estimate an approximate total volume of 900,000 m^3 of lava was emitted during this eruption (the total area of lava flows is 0.30 km^2, with an average thickness of 3 m). Considering the eruption duration (~20 h), we derive an approximate effusion rate of 12 m^3/s. The total volume of lava flows and PDCs (the latter being more difficult to assess, as discussed below) confirms a VEI of 2 [24] for the studied eruption. According to [33], the lava flow length, volume and effusion rate of the February 2018 activity are values typical of paroxysmal eruptions of Fuego. It is worth noting that this paroxysmal eruption occurred only five months before a VEI 3 eruption (June 2018).

Similar considerations apply to the mapping of PDC deposits. Our attempt to map the PDCs using single bands or NDVI differences did not give good results, similar to the lava flows. For smaller eruptions of Fuego, such as in February 2018, the material is confined to the channels on the flanks of the edifice, which, for the most active barrancas, are usually non-vegetated, making the use of change detection techniques based on NDVI difference methods more difficult or impossible. A larger event, such as the June 2018 eruption (VEI 3), produces a larger volume of deposits that may overspill the barrancas or cover less frequently impacted areas downstream, so that the detection can be done not only visually but also based on change detection methods. In the February 2018 case, visual mapping based on visible band combinations worked well.

For the tephra fall mapping analysis the NDVI difference technique gave better results, and we consider this a much better alternative to visual mapping of the deposit, given the gradational boundaries of the tephra fall deposits, i.e., the transition from areas with heavy tephra fall to areas with no tephra fall is smooth and may be difficult to define visually. We highlight that this technique can be used with other sensors that include the necessary bands (visible and NIR). Figure 5 shows a comparison between the results obtained from the analysis performed on PL and Sentinel 2 images, and the results show a good agreement, however the practical implications of higher spatial resolution (~3 m PS pixel compared with the 10 m Sentinel-2 pixel) and temporal cadence (up to daily for PS compared with ca. five days for Sentinel 2) can be significant.

Deposition of volcanic material can change the reflectance of the land surface, depending on the type of surface cover present before the volcanic material is deposited, and on the type of volcanic deposit. If the new volcanic material has very different reflectance properties from the pre-existing surface material, the deposition of the new material will result in a strong reflectance change, and possibly also a strong reflectance contrast with the surrounding areas. For volcanoes in highly vegetated regions (a majority of the volcanoes in tropical and temperate regions), like Fuego volcano, the new volcanic material will often deposit on top of previously vegetated terrain which will result in a strong reflectance change, as volcanic material and vegetation have very different reflectance spectra. We exploit these characteristics by using a NDVI difference technique to detect changes in vegetated areas caused by new deposited material. Although the technique does not work well for non-vegetated areas, the detection of new deposits in such areas is often very difficult, in general, because the new volcanic deposits can have virtually the same reflectance as the pre-existing material and, therefore, a general spectral technique with a limited number of bands may not be able to detect those changes. In such cases a visual inspection may be a quick, simple and effective alternative. Our goal, then, is to focus on the deposition of material in vegetated areas, where we know the NDVI difference method has a high chance of success, as is the case for the tephra fall deposit, described in this paper. For other types of deposits in other areas we use a visual inspection approach, to illustrate that the images can still be useful when analyzed in that way.

It is quite apparent that the NDVI difference results have a sharp boundary near the vent area when showing the mapped tephra fall region (Figure 4). This is most likely due to the presence of ash in the air for the pre-eruption image (from 9 January 2018), as Fuego is a very active volcano. This illustrates that care must always be taken to interpret the data and, in this case, recognize such artefacts.

Finally, the higher spatial resolution of the PS images permits detailed mapping of structural and morphological changes associated with the volcanic activity. The scar at the head of Barranca Honda (Figure 7) is an example of the detail that PS data provides. This structure is interpreted as either a collapse or erosion feature associated with the generation or transit of PDCs through that area, and the identification of this structure in the image acquired immediately after the eruption, puts its possible origin (and the corresponding interpretation) in the necessary volcanological context.

The comparison between PS and other satellite data sources underlines the strength of the PL products. In addition to high spatial resolution, the high temporal cadence permits detailed monitoring and offers a higher probability of obtaining useful data (e.g., Figure 5). The ability to monitor the morphological evolution of otherwise inaccessible volcanic vents is a novel feature of PS images (e.g., Figure 8). A similar analysis to that shown here for Fuego could be applied to volcanoes where lava dome growth or the accumulation of potentially unstable material on the upper flanks constitutes a potential hazard (e.g., through the generation of PDCs) or, in general, where deposited material could be later mobilized, producing lahars. Such analysis could reduce the potential exposure of field-based scientists to volcanic hazards by highlighting potential threats and providing an alternative to field-based deposit mapping. Access and processing of PS data can be performed relatively quickly; locating, downloading, preprocessing, visual inspection and quantitative analysis (e.g., NDVI differences) can take < 1 h to a few hours for a trained analyst. Rapid assessment of volcanic hazards is particularly important in highly populated areas [40], where timely monitoring of volcanic activity (e.g., deposition of volcanic products) and frequently updated products are necessary for hazard mitigation.

6. Conclusions

Persistently active volcanoes such as Fuego provide an excellent test of PS images as a resource for mapping relatively small and ephemeral, yet significant, volcanic deposits. The paroxysmal activity of February 2018 produced a range of deposits with variable spatial and spectral characteristics which could be mapped using visual analysis and change detection techniques based on single bands or the NDVI. We have shown how such deposit mapping yields estimates of eruption volume and magnitude, with important implications for subsequent activity. The high spatial resolution and temporal cadence of PS imagery also permits the identification of new deposits and morphological changes on scales of a few meters, which are a poorly understood aspect of frequently active volcanoes. These characteristics also proved to be a valuable tool for monitoring pre-eruptive and eruptive phases, which is often impossible from the ground. Similar methods could be applied to other active volcanoes during volcanic crises. We conclude that PS images represent a valuable new resource for volcano monitoring that promises to increase our understanding of, and ability to observe, volcanic processes.

Supplementary Materials: The following are available online at http://www.mdpi.com/2072-4292/11/18/2151/s1, Table S1: Planet Labs imagery timeline data entry for Figure 2.

Author Contributions: Conceptualization: A.A., S.C. and R.E.-W.; formal analysis: A.A. and R.E.-W.; funding acquisition: S.C. and G.G.; investigation: A.A.; methodology: S.C. and R.E.-W.; resources: S.C.; supervision: S.C., R.E.-W. and G.G.; validation: R.E.-W.; visualization: A.A.; writing—original draft, A.A.; writing—review and editing: A.A., S.C., R.E.-W. and G.G.

Funding: This research received no external funding.

Acknowledgments: We thank Planet Labs Inc. for providing an exciting new source of Earth observation data. We also acknowledge the International Geological Masters in Volcanology and Geotechniques (INVOGE) program, which promoted international cooperation between the University of Milano Bicocca and Michigan Technological University. The comments of three anonymous reviewers significantly improved the paper.

Conflicts of Interest: The authors declare no conflict of interest.

References

1. Dean, K.; Dehn, J.; McNutt, S.; Neal, C.; Moore, R.; Schneider, D. Satellite imagery proves essential for monitoring erupting Aleutian Volcano. *Eos. Trans. Am. Geophys. Union.* **2002**, *83*, 241–247. [CrossRef]
2. Pieri, D.; Abrams, M. ASTER watches the world's volcanoes: A new paradigm for volcanological observations from orbit. *J. Volcanol. Geotherm. Res.* **2004**, *135*, 13–28. [CrossRef]
3. Ramsey, M.; Dehn, J. Spaceborne observations of the 2000 Bezymianny, Kamchatka eruption: the integration of high-resolution ASTER data into near real-time monitoring using AVHRR. *J. Volcanol. Geotherm. Res.* **2004**, *135*, 127–146. [CrossRef]
4. Thomas, H.E.; Watson, I.M. Observations of volcanic emissions from space: Current and future perspectives. *Nat. Hazards.* **2010**, *54*, 323–354. [CrossRef]
5. Herold, M.; Scepan, J.; Clark, K.C. The use of remote sensing and landscape metrics to describe structures and changes in urban land uses. *Environ. Plan. Econ. Space.* **2002**, *34*, 1443–1458. [CrossRef]
6. Carn, S.A.; Krueger, A.J.; Krotkov, N.A.; Yang, K.; Evans, K. Tracking volcanic sulfur dioxide clouds for aviation hazard mitigation. *Nat. Hazards.* **2008**, *51*, 325–343. [CrossRef]
7. Campion, R.; Martinez-Cruz, M.; Lecocq, T.; Caudron, C.; Pacheco, J.; Pinardi, G.; Hermans, C.; Carn, S.; Bernard, A. Space-and ground-based measurements of sulphur dioxide emissions from Turrialba Volcano (Costa Rica). *Bull. Volcanol.* **2012**, *74*, 1757–1770. [CrossRef]
8. Carn, S.A.; Clarisse, L.; Prata, A.J. Multi-decadal satellite measurements of global volcanic degassing. *J. Volcanol. Geotherm. Res.* **2016**, *311*, 99–134. [CrossRef]
9. Carter, A.J.; Ramsey, M.S.; Belousov, A.B. Detection of a new summit crater on Bezymianny Volcano lava dome: Satellite and field-based thermal data. *Bull. Volcanol.* **2007**, *69*, 811–815. [CrossRef]
10. Vaughan, R.G.; Keszthelyi, L.P.; Lowenstern, J.B.; Jaworowski, C.; Heasler, H. Use of ASTER and MODIS thermal infrared data to quantify heat flow and hydrothermal change at Yellowstone National Park. *J. Volcanol. Geotherm. Res.* **2012**, *233–234*, 72–89. [CrossRef]
11. Wessels, R.L.; Vaughan, R.G.; Patrick, M.R.; Coombs, M.L. High-resolution satellite and airborne thermal infrared imaging of precursory unrest and 2009 eruption at Redoubt Volcano, Alaska. *J. Volcanol. Geotherm. Res.* **2013**, *259*, 248–269. [CrossRef]
12. Wright, R.; Blackett, M.; Hill-Butler, C. Some observations regarding the thermal flux from Earth's erupting volcanoes for the period of 2000 to 2014. *Geophys. Res. Lett.* **2015**, *42*, 282–289. [CrossRef]
13. Norini, G.; Groppelli, G.; Sulpizio, R.; Carrasco-Núñez, G.; Dávila-Harris, P.; Pellicioli, C.; Zucca, F.; De Franco, R. Structural analysis and thermal remote sensing of the Los Humeros Volcanic complex: Implications for volcano structure and geothermal exploration. *J. Volcanol. Geotherm. Res.* **2015**, *301*, 221–237. [CrossRef]

14. Flynn, L.P.; Harris, A.J.L.; Wright, R. Improved identification of volcanic features using Landsat 7 ETM+. *Remote Sens. Environ.* **2001**, *78*, 180–193. [CrossRef]
15. Lu, D.; Li, G.; Valladares, G.S.; Batistella, M. Mapping soil erosion risk in Rondônia, Brazilian Amazonia: Using RUSLE, remote sensing and GIS. *Land Degrad. Dev.* **2004**, *15*, 499–512. [CrossRef]
16. Carter, A.J.; Ramsey, M.S. ASTER- and field-based observations at Bezymianny Volcano: Focus on the 11 May 2007 pyroclastic flow deposit. *Remote Sens. Environ.* **2009**, *113*, 2142–2151. [CrossRef]
17. Lanari, R.; Lundgren, P.; Sansosti, E. Dynamic deformation of Etna Volcano observed by satellite radar interferometry. *Geophys. Res. Lett.* **1998**, *25*, 1541–1544. [CrossRef]
18. Hooper, A.; Zebker, H.; Segall, P.; Kampes, B. A new method for measuring deformation on volcanoes and other natural terrains using InSAR persistent scatterers. *Geophys. Res. Lett.* **2004**, *31*, 23. [CrossRef]
19. Ramsey, M.S.; Fink, J.H. Estimating silicic lava vesicularity with thermal remote sensing: A new technique for volcanic mapping and monitoring. *Bull. Volcanol.* **1999**, *61*, 32–39. [CrossRef]
20. Kervyn, M.; Ernst, G.G.J.; Goossens, R.; Jacobs, P. Mapping volcano topography with remote sensing: ASTER vs. SRTM. *Int. J. Remote Sens.* **2008**, *29*, 6515–6538. [CrossRef]
21. Joyce, K.E.; Belliss, S.E.; Samsonov, S.V.; McNeill, S.J.; Glassey, P.J. A review of the status of satellite remote sensing and image processing techniques for mapping natural hazards and disasters. *Prog. Phys. Geogr. Earth Environ.* **2009**, *33*, 183–207. [CrossRef]
22. Arnous, M.O.; Green, D.R. GIS and remote sensing as tools for conducting geo-hazards risk assessment along Gulf of Aqaba coastal zone, Egypt. *J. Coast. Conserv.* **2011**, *15*, 457–475. [CrossRef]
23. Pallister, J.S.; Schneider, D.J.; Griswold, J.P.; Keeler, R.H.; Burton, W.C.; Noyles, C.; Newhall, C.G.; Ratdomopurbo, A. Merapi 2010 eruption—Chronology and extrusion rates monitored with satellite radar and used in eruption forecasting. *J. Volcanol. Geotherm. Res.* **2013**, *261*, 144–152. [CrossRef]
24. Smithsonian Institution, Recent bulletin report. Bulletin of the Global Volcanism Program. Available online: https://volcano.si.edu/volcano.cfm?vn=342090 (accessed on 9 September 2019).
25. Zhu, Z. Change detection using landsat time series: A review of frequencies, preprocessing, algorithms, and applications. *ISPRS J. Photogramm. Remote Sens.* **2017**, *130*, 370–384. [CrossRef]
26. Planet Team, Planet Application Program Interface: In Space for Life on Earth. San Francisco, CA. 2017. Available online: https://api.planet.com. (accessed on 9 September 2019).
27. Tucker, C.J.; Pinzon, J.E.; Brown, M.E.; Slayback, D.A.; Pak, E.W.; Mahoney, R.; Vermote, E.F.; El Saleous, N. An extended AVHRR 8-km NDVI dataset compatible with MODIS and SPOT vegetation NDVI data. *Int. J. Remote Sens.* **2005**, *26*, 4485–4498. [CrossRef]
28. Grainger, R.G.; Peters, D.M.; Thomas, G.E.; Smith, A.J.A.; Siddans, R.; Carboni, E.; Dudhia, A. Measuring volcanic plume and ash properties from space. *Geol. Soc. Lond. Spec. Publ.* **2013**, *380*, 293–320. [CrossRef]
29. Chesner, C.A.; Rose, W.I., Jr. Geochemistry and evolution of the fuego volcanic complex, Guatemala. *J. Volcanol. Geotherm. Res.* **1984**, *21*, 25–44. [CrossRef]
30. Manea, V.C.; Manea, M.; Ferrari, L. A geodynamical perspective on the subduction of Cocos and Rivera plates beneath Mexico and Central America. *Tectonophysics* **2013**, *609*, 56–81. [CrossRef]
31. Martin, D.P.; Rose, W.I., Jr. Behavioral patterns at Fuego volcano, Guatemala. *J. Volcanol. Geotherm. Res.* **1981**, *10*, 67–81. [CrossRef]
32. Davies, D.K.; Quearry, M.W.; Bonis, S.B. Glowing avalanches from the 1974 eruption of the volcano Fuego, Guatemala. *Geol. Soc. Am. Bull.* **1978**, *89*, 369–384. [CrossRef]
33. Lyons, J.J.; Waite, G.P.; Rose, W.I.; Chigna, G. Patterns in open vent, Strombolian behavior at Fuego volcano, Guatemala, 2005–2007. *Bull. Volcanol.* **2009**, *72*, 1–15. [CrossRef]
34. Waite, G.P.; Nadeau, P.A.; Lyons, J.J. Variability in eruption style and associated very long period events at Fuego volcano, Guatemala. *J. Geophys. Res. Solid Earth.* **2013**, *118*, 1526–1533. [CrossRef]
35. Kuenzi, W.D.; Horst, O.H.; McGehee, R.V. Effect of volcanic activity on fluvial-deltaic sedimentation in a modern arc-trench gap, southwestern Guatemala. *Geol. Soc. Am. Bull.* **1979**, *90*, 827–838. [CrossRef]
36. MODVOLC, near-real-time satellite monitoring of global volcanism using MODIS. Available online: http://modis.higp.hawaii.edu/ (accessed on 9 September 2019).
37. NOAA Satellites and Information Service, Current Volcanic Ash Advisories. Available online: https://www.ssd.noaa.gov/VAAC/messages.html (accessed on 9 September 2019).

38. Richardson, A.J.; Wiegand, C.L. Distinguishing vegetation from soil background information. *Photogramm. Eng. Remote Sens.* **1977**, *43*, 1541–1552.
39. Bannari, A.; Morin, D.; Bonn, F.; Heute, A.R. A review of vegetation indices. *Remote Sens. Rev.* **1995**, *13*, 95–120. [CrossRef]
40. Loughlin, S.C.; Sparks, R.S.J.; Sparks, S.; Brown, S.K.; Jenkins, S.F.; Vye-Brown, C. *Global Volcanic Hazard and Risk*; Cambridge University Press: Cambridge, UK, 2015; pp. 1–80.

MDPI
St. Alban-Anlage 66
4052 Basel
Switzerland
Tel. +41 61 683 77 34
Fax +41 61 302 89 18
www.mdpi.com

Remote Sensing Editorial Office
E-mail: remotesensing@mdpi.com
www.mdpi.com/journal/remotesensing

www.ingramcontent.com/pod-product-compliance
Lightning Source LLC
LaVergne TN
LVHW071617170726
843515LV00009B/2415